The Eupithecia of China

The Eupithecia of China

A Revision

By

Vladimir Mironov
Anthony Galsworthy

BRILL

LEIDEN · BOSTON
2014

Cover illustration: *Eupithecia succenturiata* Linnaeus; photograph by Alexey Matov (Zoological Institute RAS, Saint-Petersburg).

Library of Congress Cataloging-in-Publication Data

Mironov, Vladimir, 1957-
The Eupithecia of China : a revision / by Vladimir Mironov, Anthony Galsworthy.
pages cm
Includes bibliographical references and index.
ISBN 978-90-04-25451-0 (hardback : alk. paper) -- ISBN 978-90-04-25453-4 (e-book) 1. Eupithecia--China. 2. Geometridae--China. 3. Moths--China. I. Galsworthy, Anthony, 1944- II. Title.

QL561.G6M57 2014
595.780951--dc23

2013037840

This publication has been typeset in the multilingual "Brill" typeface. With over 5,100 characters covering Latin, IPA, Greek, and Cyrillic, this typeface is especially suitable for use in the humanities. For more information, please see www.brill.com/brill-typeface.

ISBN 978-90-04-25451-0 (hardback)
ISBN 978-90-04-25453-4 (e-book)

This book is printed on acid-free paper.

CONTENTS

ACKNOWLEDGEMENTS

We are extremely grateful to the many people around the world who have helped to make this book possible. In particular we wish to acknowledge the generous assistance given by colleagues in the Institute of Zoology of the Chinese Academy of Sciences in Beijing, including in particular Professor Xue Dayong, Dr. Han Hongxiang, and Dr. Jiang Nan. The Institute have made available on loan the whole collection of Chinese *Eupithecia* belonging to the Institute for a number of years. They have provided comprehensive facilities for the second author in the course of visits to Beijing, and invaluable assistance in interpreting label data and correcting Chinese geographical names in the text.

We are also grateful for the considerable help and encouragement we have received from colleagues in the Natural History Museum, London, in particular, Dr. Malcolm Scoble, Dr. Jeremy Holloway, Martin Honey, and John Chainey, and of course for continual access to the excellent collections maintained there.

We also wish to thank those responsible in the many other institutions outside mainland China who have generously allowed us access to material in the collections under their care, in many cases provided loans of material extending over many years, and frequently helped with taxonomic discussion and provision of information:

Institute of Agriculture and Environment, Estonian Agricultural University, Tartu, Estonia: Dr. Jaan Viidalepp.

Museum für Naturkunde, Zentralinstitut der Humboldt-Universitat zu Berlin, Germany: Dr. Wolfram Mey, Dr. Hans-Joahim Hanneman.

National Science Museum of Tokyo, Japan: Dr. Mamoru Owada.

National Science Museum of Taichung, Taiwan: Dr. Cheng-Shing Lin, Miss Mei-Ling Chan.

Staatliches Museum für Naturkunde Karlsruhe, Germany: Dr. Robert Trusch, Mr. Michael Falkenberg.

Termeszettudomanyi Muzeum Allattara (Hungarian Natural History Museum), Budapest, Hungary: Dr. László Ronkay.

United States National Museum of Natural History, Washington, USA: Dr. Scott Miller.

Zoologisches Forschungsinstitut und Museum Alexander Koenig, Bonn, Germany: Dr. Dieter Stüning, Mrs. Christina Blúme, Mrs. Ulrike Kleikamp.

Zoological Museum of the University, Helsinki, Finland: Dr. Lauri Kaila.

Zoologische Staatssammlung München, Germany: Dr. Axel Hausmann, Dr. Ulf Buchsbaum.

We also acknowledge with gratitude the considerable help we have received from the following individuals:

Dr. Evgeny Beljaev, Vladivostok (RU): Kind support of the first author during institutional visits. Loan of material from Asian Russia (Primorje) and Korea. Provision of literature. Discussion of taxonomic problems.

Dr. Alexey Bidzilya, Kiev (UA): Loan and gift of material from China (Inner Mongolia).

Mr. Konrad Ebert, Berlin (D): Various valuable and important help in the Museum für Naturkunde, Humboldt-Universität zu Berlin.

† Mr. Michael Fibiger, Sorø (DK): Loan of material from Nepal. Kind support on private visits. Provision of literature. Discussion of taxonomic problems.

Mr. C.M. Fu, Taichung (TW): Loan of material from Taiwan.

Dr. Reinhard Gaedike, Eberswalde (D): Loan of material from Taiwan (types of Bastelberger).

† Prof. Dr. Hiroshi Inoue, Iruma City (JP): Loan of material from Japan. Provision of literature. Discussion of taxonomic problems.

Mr. Igor Kostjuk, Kiev (UA): Kind support of the first author during private and Kiev Museum visits. Loan of material from Tajikistan, SE Kazakhstan and Transbaikalia. Provision of literature, and other very valuable help. Discussion of taxonomic problems.

Dr. Mikhail Kozlov, Turku (FI): Provision of literature. Various valuable and important help.

Dr. Victor Krivokhatsky, Saint-Petersburg (RU): Loan of material from China (Beijing).

Mr. Gyula M. László, Budapest (HU): Loan of extensive material from Taiwan.

Mr. Oleg Pekarsky, Budapest (HU): Loan of material from China and Altai Mts. Various valuable and important help.

Dr. Linda Pitkin, London (UK): Provision of important bibliographical and other information.

Mr. Łukasz Przybyłowicz, Kraków (PL): Loan of material from North Korea and Japan.

Dr. Ulrich Ratzel, Karlsruhe (D): Loan of material from Kashmir and Nepal. Provision of photos of types of Staudinger and Dietze from Berlin Museum and types of Vojnits from Karlsruhe Museum; other very valuable help. Kind support of the first author during private and Karlsruhe Museum visits. Discussion of taxonomic problems.

Dr. Sergey Sinev, Saint-Petersburg (RU): Loan of material from Asian Russia (Primorje). Provision of literature, and other very valuable help. Discussion of taxonomic problems.

Mr. Peder Skou, (DK): Kind support on private visits. Provision of literature. Various valuable and important help.

Dr. Petr Ustjuzhanin, Novosibirsk (RU): Loan of material from Altai Mts. (Russia) and Mongolia.

Dr. Sergey Vasilenko, Novosibirsk (RU): Loan of material from Kazakhstan, Kirghizstan, Asian Russia (Siberia and Far East) and Mongolia. Provision of literature. Discussion of taxonomic problems.

Mr. Roman Yakovlev, Barnaul (RU): Loan of material from Altai Mts. and Mongolia.

Dr. Shen-horn Yen, Taichung (TW): Loan of material from Taiwan.

Dr. Vadim Zolotuhin, Uljanovsk (RU): Loan of material from Uzbekistan and Vietnam.

Thanks are also due to the Deutsche Forschung Gemeinschaft (DFG; grants nos 436 RUS 17/99/02, STU 498/2-1 and TR 1086/1-1) which enabled Dr. V. Mironov to do extensive studies on Asian Eupitheciini in ZFMK (Bonn), SMNK (Karlsruhe) and MNHU (Berlin) in 2003, 2009 and 2011, and to the Royal Society, London, who provided a grant to enable Dr. V. Mironov to work in BMNH (London) in 2006.

Finally the authors express their gratitude to their wives and families for the patience which they have shown in putting up with the moths over the many years which this work has taken.

ABBREVIATIONS

AMNH	American Museum of Natural History, New York, USA.
BMNH	The Natural History Museum, London, United Kingdom.
DEI	Deutsches Entomologisches Institut, Eberswalde, Germany.
DEUU	Department of Entomology, University of Uppsala, Sweden.
EIHU	Entomological Institute, Hokkaido University, Sapporo, Japan.
EMS	E. von Mentzer collection, Ornstigen, Sweden.
IAET	Institute of Agriculture and Environment, Estonian Agricultural University, Tartu, Estonia.
ISEN	Institute of Animal Systematics and Ecology, Novosibirsk, Russia.
IZAR	Institute of Zoology, Academy of Sciences of Armenia, Erivan, Armenia.
IZCAS	Institute of Zoology, Chinese Academy of Sciences, Beijing, China.
LSL	Linnean Society of London, United Kingdom.
MCZ	Museum of Comparative Zoology, Harvard University, Cambridge, Massachusetts, USA.
MNHN	Museum National d'Histoire Naturelle, Paris, France.
MNHU	Museum für Naturkunde, Zentralinstitut der Humboldt-Universitat zu Berlin, Germany.
MRAC	Musée Royal de l'Afrique Centrale, Tervuren, Belgium.
MRSN	Museo Regionale di Scienze Naturali, Torino, Italy.
MZLU	University Museum of Zoology, Lund, Sweden.
NHMW	Naturhistorisches Museum Wien, Austria.
NHRS	Naturhistoriska Riksmuseet, Stockholm, Sweden.
NSMTO	National Science Museum of Tokyo, Japan.
NSMTA	National Science Museum of Taichung, Taiwan.
NSMW	Naturwissenschaftliche Sammlung, Museum Weisbaden, Germany.
SMNK	Staatliches Museum für Naturkunde Karlsruhe, Germany.
TTM	Termeszettudomanyi Muzeum Allattara (Hungarian Natural History Museum), Budapest, Hungary.
USNM	United States National Museum of Natural History, Washington, USA.
ZFMK	Zoologisches Forschungsinstitut und Museum Alexander Koenig, Bonn, Germany.
ZISP	Zoological Institute, Russian Academy of Sciences, Saint-Petersburg, Russia.
ZMH	Zoological Museum of the University, Helsinki, Finland.
ZMKU	Zoological Museum, National Shevchenko University, Kiev, Ukraine.
ZMMU	Zoological Museum, Moscow State University, Russia.
ZMTU	Zoological Museum of Tartu University, Estonia.
ZMUC	Zoological Museum of the University, Copenhagen, Denmark.
ZMUH	Zoologische Museum Universität, Hamburg, Germany.
ZSM	Zoologische Staatssammlung München, Germany.

TAXONOMIC CHANGES INTRODUCED IN THIS BOOK

New taxon

Eupithecia likiangi pedisequa **ssp. n.**

Changes of status

Eupithecia batangi Vojnits, 1984 from *Eupithecia placida batangi* Vojnits, 1984.
Eupithecia robiginascens flexicornuta Inoue, 1988 from *Eupithecia flexicornuta* Inoue, 1988.
Eupithecia paupera Dietze, 1910 from *Eupithecia innotata* f. *paupera* Dietze, 1910.
Eupithecia pallescens Dietze, 1910 from *Eupithecia sinuosaria* f. *pallescens* Dietze, 1910.

Replacement names

Eupithecia dolichoptera Galsworthy & Mironov for *Eupithecia longipennata* Inoue, 1988 (junior secondary homonym of *Eupithecia longipennata* Warren, 1904).
Eupithecia danielena Mironov & Galsworthy for *Eupithecia ronkayi* Mironov & Galsworthy, 2009 (junior primary homonym of *Eupithecia ronkayi* Vojnits, 1994).

New synonymy

Eupithecia pekingiana Vojnits, 1973 of *Eupithecia cichisa* Prout, 1939.
Eupithecia hoenei Vojnits, 1976 of *Eupithecia bohatschi* Staudinger, 1897.
Eupithecia yasudai Inoue, 1987 of *Eupithecia lineisdistincta* Vojnits, 1981.
Eupithecia acerba Vojnits, 1984 of *Eupithecia rubridorsata* Hampson, 1895.
Eupithecia konradi Vojnits, 1976 of *Eupithecia ustata* Moore, 1888.
Eupithecia eszterkae Vojnits, 1976 of *Eupithecia sinicaria* Leech, 1897.
Eupithecia yunnani Vojnits, 1976 of *Eupithecia sinicaria* Leech, 1897.
Eupithecia violacea Vojnits, 1981 of *Eupithecia szelenyica* Vojnits & de Laever, 1974.

Eupithecia larentimima Vojnits, 1974 of *Eupithecia rufescens* Butler, 1878.
Eupithecia stramineata Inoue, 2000 of *Eupithecia irambata* Warren, 1893.
Eupithecia noxia Vojnits, 1979 of *Eupithecia granata* Vojnits, 1979.
Eupithecia abrepta Vojnits, 1979 of *Eupithecia vana* Vojnits, 1979.
Eupithecia perfuscata Vojnits, 1975 of *Eupithecia dissertata* Püngeler, 1905.
Eupithecia infecta Vojnits, 1981 of *Eupithecia lariciata* Freyer, 1842.
Eupithecia studiosa Vojnits, 1979 of *Eupithecia emanata* Dietze, 1906.
Eupithecia epileptica Vojnits, 1984 of *Eupithecia pediba* Vojnits, 1981.
Eupithecia acuminata Mironov & Galsworthy, 2004 of *Eupithecia pediba* Vojnits, 1981.
Eupithecia incohata Vojnits, 1979 of *Eupithecia tripunctaria* Herrich-Schäffer, 1852.
Eupithecia exacerbata Vojnits, 1984 of *Eupithecia laudabilis* Vojnits, 1984.
Eupithecia subplacida Vojnits, 1984 of *Eupithecia refertissima* Vojnits, 1984.
Eupithecia testacea Mironov & Galsworthy, 2004 of *Eupithecia avara* Vojnits, 1984.
Eupithecia recens Dietze, 1904 of *Eupithecia uliata* Staudinger, 1897.
Eupithecia recens var. *creta* Dietze, 1908 of *Eupithecia uliata* Staudinger, 1897.
Eupithecia extrinseca Vojnits & de Laever, 1978 of *Eupithecia uliata* Staudinger, 1897.
Eupithecia infortunata Vojnits, 1984 of *Eupithecia eximia* Vojnits & de Laever, 1978.
Eupithecia albipunctata f. *privata* Dietze, 1913 of *Eupithecia virgaureata* Doubleday, 1861.
Eupithecia hundamoi Vojnits & de Laever, 1978 of *Eupithecia kobayashii* Inoue, 1958.
Eupithecia importuna Vojnits, 1981 of *Eupithecia kobayashii* Inoue, 1958.
Eupithecia nonferenda Vojnits 1981 of *Eupithecia kobayashii* Inoue, 1958.
Eupithecia pimpinellata f. *limbosignata* Dietze, 1910 of *Eupithecia denotata* Hübner [1813].
Eupithecia denotata f. *difficilis* Dietze, 1910 of *Eupithecia nigrilinea* Warren, 1896.
Eupithecia mortua Vojnits, 1984 of *Eupithecia lobblicherata* Schütze, 1961.
Eupithecia orba Mironov & Galsworthy, 2004 of *Eupithecia particeps* Vojnits, 1988.
Eupithecia nanata kozhantschikovi Wehrli, 1929 of *Eupithecia innotata* Hufnagel, 1767.
Eupithecia inexplicabilis Vojnits, 1982 of *Eupithecia innotata* Hufnagel, 1767.

Eupithecia mystica Vojnits, 1988 of *Eupithecia innotata* Hufnagel, 1767.
Eupithecia innotata contracta Vojnits, 1982 of *Eupithecia parallelaria* Bohatsch, 1893.
Eupithecia emittens Inoue, 1993 of *Eupithecia parallelaria* Bohatsch, 1893.
Eupithecia fulgurata Vojnits, 1982 of *Eupithecia corroborata* Dietze, 1908.
Eupithecia hysterica Vojnits, 1988 of *Eupithecia corroborata* Dietze, 1908.
Eupitheica innotata f. *omniparens* Dietze, 1908 of *Eupithecia praesignata* Bohatsch, 1893.
Eupithecia unedonata f. *famelica* Dietze, 1908 of *Eupithecia praesignata* Bohatsch, 1893.
Eupithecia omnigera Vojnits, 1982 of *Eupithecia praesignata* Bohatsch, 1893.
Eupithecia recentissima Vojnits, 1982 of *Eupithecia praesignata* Bohatsch, 1893.
Eupithecia paupera Vojnits, 1982 of *Eupithecia paupera* Dietze, 1910.
Eupithecia sclerata Vojnits, 1982 of *Eupithecia mitigata* Dietze, 1906.
Eupithecia sp. f. *privata* Dietze, 1910 of *Eupithecia vicariata* Dietze, 1904.
Eupithecia costisignata f. *lutulenta* Dietze, 1913 of *Eupithecia vicariata* Dietze, 1904.
Eupithecia proprivata Vojnits, 1982 of *Eupithecia vicariata* Dietze, 1904.
Eupithecia garrula Vojnits, 1984 of *Eupithecia perendina* Vojnits, 1980.
Eupithecia sinuosaria f. *modesta* Dietze, 1913 of *Eupithecia pallescens* Dietze, 1910.
Eupithecia morosa Vojnits, 1976 of *Eupithecia pallescens* Dietze, 1910.
Eupithecia finitima Vojnits, 1979 of *Eupithecia sacrimontis* Prout, 1938.
Eupithecia deformis Vojnits, 1979 of *Eupithecia sacrimontis* Prout, 1938.
Eupithecia aporia Vojnits, 1975 of *Eupithecia kuldschaensis* Staudinger, 1892.
Eupithecia inveterata Vojnits 1987 of *Eupithecia thalictrata* Püngeler, 1902.
Eupithecia selinata f. *tenebricosa* Dietze, 1910 of *Eupithecia addictata* Dietze, 1908.
Eupithecia subexiguata Vojnits, 1974 of *Eupithecia exiguata* Hübner, [1813].
Eupithecia manca Vojnits, 1979 of *Eupithecia hannemanni* Vojnits & de Laever, 1973.
Eupithecia pollens Vojnits, 1981 of *Eupithecia duplex* Sterneck, 1931.
Eupithecia helveticaria f./ab. *robusta* Dietze, 1910 of *Eupithecia mandschurica* Staudinger, 1897.
Eupithecia depressa Vojnits, 1979 of *Eupithecia mandschurica* Staudinger, 1897.

Eupithecia absinthiata f. *lunata* Dietze, 1910 of *Eupithecia intricata* Zetterstedt, 1839.
Eupithecia inculta Vojnits, 1975 of *Eupithecia satyrata* Hübner, [1813].
Eupithecia indecora Vojnits, 1981 of *Eupithecia impolita* Vojnits, 1980.
Eupithecia semicalva Vojnits, 1979 of *Eupithecia neosatyrata* Inoue, 1979.
Eupithecia subita Vojnits, 1979 of *Eupithecia indissolubilis* Vojnits, 1979.
Eupithecia assa Mironov, 1989 of *Eupithecia indissolubilis* Vojnits, 1979.
Eupithecia exactata f. *modesta* Dietze, 1910 of *Eupithecia exactata* Staudinger, 1882.
Eupithecia verecunda Vojnits, 1980 of *Eupithecia subfuscata* Haworth, 1809.
Eupithecia perpetua Vojnits, 1984 of *Eupithecia subfuscata* Haworth, 1809.
Eupithecia gibbosa Mironov & Galsworthy, 2004 of *Eupithecia costipicta* Warren, 1893.
Eupithecia iterata Vojnits, 1980 of *Eupithecia praecipitata* Vojnits, 1979.
Eupithecia tempestuosa Vojnits, 1984 of *Eupithecia praecipitata* Vojnits, 1979.
Eupithecia laudenda Vojnits, 1984 of *Eupithecia praecipitata* Vojnits, 1979.
Eupithecia fragmentaria Vojnits, 1984 of *Eupithecia lasciva* Vojnits, 1980.
Eupithecia depressa Vojnits, 1984 of *Eupithecia lasciva* Vojnits, 1980.
Eupithecia depressa disiuncta Vojnits, 1984 of *Eupithecia lasciva* Vojnits, 1980.
Eupithecia benigna Vojnits, 1984 of *Eupithecia lasciva* Vojnits, 1980.
Eupithecia seminuda Vojnits,1984 of *Eupithecia egena* Vojnits, 1984.
Eupithecia sublasciva Vojnits, 1984 of *Eupithecia egena* Vojnits, 1984.
Eupithecia ficta Vojnits, 1984 of *Eupithecia egena* Vojnits, 1984.
Eupithecia anteacta Vojnits, 1984 of *Eupithecia egena* Vojnits, 1984.
Eupithecia insana Vojnits, 1984 of *Eupithecia otiosa* Vojnits, 1981.
Eupitheica nodosa Vojnits, 1984 of *Eupithecia otiosa* Vojnits, 1981.

INTRODUCTION

Eupithecia is the largest genus of Geometrid moths, with well over 1400 current species, and more than 2200 available names. With few exceptions, moths in the genus are small and drably coloured, and usually difficult to identify with certainty without dissection of the male and female genitalia. Fortunately the latter are in most species distinctive in both males and females.

The genus is found almost worldwide, with the exception of Australia and New Zealand, where true *Eupithecia* is not known to occur. There are however abundant representatives of other genera within the Eupitheciini in these two countries, and the tribe thus has a worldwide distribution. This might perhaps suggest a Gondwanan origin for the tribe, but the absence of *Eupithecia* proper from Australasia is a little puzzling in this case, since it is well distributed in every other continent, including the southern part of the Neotropical region. The Neotropical and African *Eupithecia* species are not yet sufficiently well known to attempt a biogeographic analysis which might throw more light on this question.

Eupithecia in China

Although *Eupithecia* as a genus had been established by John Curtis in 1825, no species in the genus were described specifically from China until the last decade of the century. A limited number of species of wider distribution now known to occur in China had however been described earlier, initially from Europe, and later from areas contiguous with China, particularly the Himalayas, Japan and Siberia, by authors such as Achille Guénée, Eduard Friedrich von Eversmann and Hugo Christoph. In the last decade of the nineteenth century, however, a number of German collectors, in particular the brothers Fritz and Heinrich Dörries and the Rückbeil family, began to operate within or close to the western and northern borders of modern China, and brought back material of a number of species of *Eupithecia*. The Rückbeils penetrated as far as northern Tibet and Qinghai provinces. Their material found its way back to Otto Staudinger, and a little later, Karl Dietze, the latter of whom specialised in the genus *Eupithecia*. The first species in the genus to be described from China proper appears to have been *E. kuldschaensis*, described by Staudinger in

1892 from Kuldja in the modern Xinjiang autonomous region. In 1897 Staudinger, in his work on the fauna of the Amur region, described a number of further species which also occur in northern and western China, and in the same year Leech in the UK described two species from central western China which had found their way into the BMNH collection (*E. sinicaria* and *E. consortaria*).

It was however Karl Dietze who first took a systematic interest in the genus, and amassed a large specialized collection of *Eupithecia*. This collection, which is now in the Humboldt Museum, Berlin, together with the collections of Otto Staudinger himself, and Rudolf Püngeler, contain a wealth of material from central Asia and the Russian Far East, most of which is still in excellent condition, and which constitutes by far the best material from western China available anywhere even today. Between 1902 and 1910, Dietze, in his monumental work on the 'Biology of Eupithecia' described and illustrated what we now estimate to be somewhat over 20 good species in the Chinese fauna, drawing extensively on material collected in areas now forming parts of Xinjiang, Tibet and Qinghai. However, partly because of the intrinsic taxonomic difficulty of the material, partly because of the fashion prevailing at the time of assigning a latin name to every individual variation of pattern, and most importantly of all, because male and female genitalia were not examined at the time, the resulting picture was far from clear, and has remained so till the present day.

Louis Beethoven Prout surveyed the existing knowledge of the genus in Asia in two volumes of Seitz's Macrolepidoptera of the World (Prout 1915 and 1938), but for the most part did not greatly change the taxonomy. In his 1938 work he did however introduce a few new taxa, including *E. sacrimontis* from China.

In the two decades after the first world war, Hermann Höne, a German businessman resident in China, accumulated a very large collection of Chinese Lepidoptera, which is now housed in the Koenigs Museum in Bonn. This is by far the best collection of Chinese Lepidoptera available anywhere outside China, and contains extensive material of *Eupithecia*. The *Eupithecia* were not however examined critically until the advent of the Hungarian entomologist Dr András Vojnits, who worked extensively on this collection and those in Berlin in the 1970s and 80s, concentrating exclusively on *Eupithecia*. Vojnits described in all over 200 species of *Eupithecia,* more than half of which were from China.

Unfortunately Dr Vojnits' work was flawed in a number of respects. Although he examined older type material in Berlin, he did not until very late in the day examine such material from the Natural History Museum in London. This resulted in the redescription of a number of species which had earlier been described, mostly from India or Burma. Moreover, although he did examine genitalia, the dissection techniques used were not good, and in particular the specimens were not stained, so that important parts of them became invisible once mounted in medium on the slide. This applies particularly to the important eighth sternite in the male, and the bursa copulatrix in the female. In addition the specimens were often broken. Worse, many dissections were evidently carried out at the same time, and in a number of cases confusion resulted, so that genitalia on the slides do not always correspond to the specimen on the pin with which they are linked. The papers in which the new species were published contained no illustrations of the adults, rudimentary descriptions, and for the most part very poor drawings of the genitalia. Males and females were often incorrectly associated.

After Dr Vojnits left the Hungarian Natural History Museum at the end of the 1980s the material on which he had been working was left in Budapest in disorder. It included the type material of the vast majority of species described by him, much of which was on loan from other institutes and museums in Europe. It took a number of years for this material to be sorted and returned to the collections of the institutes and museums to which it belonged, and during this period taxonomic work on Chinese *Eupithecia* became virtually impossible. The large number of names introduced by Dr Vojnits could not be linked with other material because of the inadequate descriptions and inaccessibility of type material. Thus when Professor Xue Dayong produced volume 15 of the Insecta in the Fauna Sinica series in 1999, covering the Larentiinae, the genus *Eupithecia* had to be omitted altogether.

From the 1950s onwards, Professor Hiroshi Inoue in Japan took a special interest in the genus, which culminated in the publication of two very important papers describing the Japanese fauna (Inoue 1979 and 1980) and subsequently one on the Taiwanese fauna (Inoue 1988). All these papers include species found also on the mainland of China.

Over the last ten years the authors of the present volume have located and examined all extant type material of *Eupithecia* described from China, and have examined all available material in institutions outside China, together with the substantial collection of the Zoological Institute of the Chinese Academy of Sciences in Beijing. Ten species occurring in China were described in the eighteenth and early nineteenth centuries by Johann Siegfried Hufnagel (1767), Michael Denis & Ignaz Schiffermüller

(1775) and Jacob Hübner (1813, 1825): the original type material for these species appears to have been lost quite early on. In addition the type material of one further species, *E. subpulchrata* Alphéraky, 1883 has so far not been traced. Details of type material examined are given in the text, and a list of the main repositories is also given on p. xi.

The authors have been fortunate to have had access to all the material of *Eupithecia* collected by Chinese collectors for the Zoological Institute of the Chinese Academy of Sciences since 1949. In addition material collected more recently has been kindly provided by a number of international collectors, notably Mr Oleg Pekarsky, Budapest.

The authors have produced 7 joint papers describing 81 new species which have been discovered in the course of this work. In addition papers have been produced covering the genus in the Western Himalayas, SE Asia and in the Indian subcontinent, since many species occurring there have been found also in China, and the taxonomy very often needed to be updated.

The Chinese habitat

China (fig. 1) is a huge country and contains within its borders an enormous variety of climate zones and habitats. Most of south east China is tropical to subtropical and, relatively speaking, low lying. The south west rises to the highest mountains in the world, passing through a relatively moist, monsoon-affected zone in Yunnan and eastern Tibet to much more arid montane desert and grassland in the interior of Tibet and Qinghai, and from there northwards to the largely low-lying deserts of Xinjiang and Gansu, bordered in the west by the Karakorum and Pamir mountains, and in the north by the Tian Shan. Central China is for the most part subtropical in summer, but cool to relatively cold in winter, and in the interior holds a number of mountain ranges which provide a series of different altitudinal climatic and vegetation zones. Much of central northern China has a continental climate, with extremely hot, relatively dry summers, and cold very dry winters, while the north east is closer climatically to far eastern Russia and Korea, with short summers and rather moister, but very cold winters.

Not surprisingly, the Chinese *Eupithecia* fauna reflects these geographic and climatic differences. While a very small number of species, as for instance *E. subtacincta*, are found across much of the country, most species are readily assigned to specific zones. The fauna of the north east and

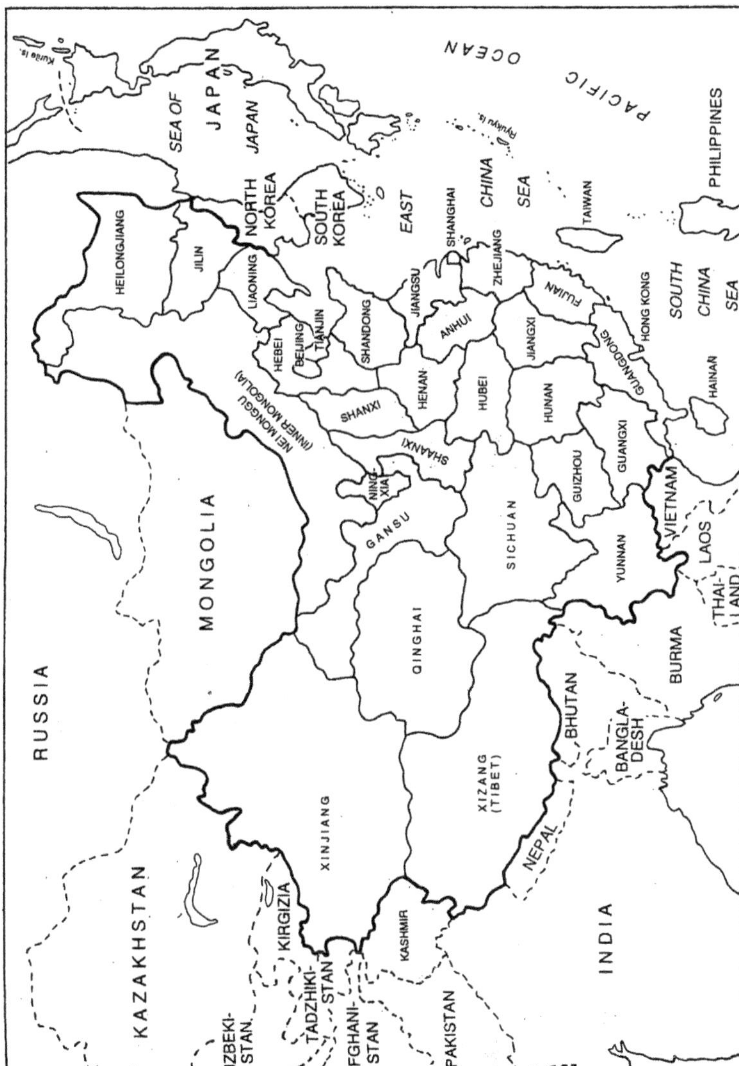

Fig 1. Map of China and adjacent countries showing the provinces of China.

central north has close affinities with that of the Primorye zone of far eastern Russia, Korea and northern Japan. Species characteristic of the Russian and Mongolian Altai extend into Inner Mongolia and northern Xinjiang, while most of the Xinjiang species are found widely across the central Asian republics, many extending into Pakistan. In Qinghai and Tibet, the fauna begins to have a more Himalayan look to it, and an increasing number of species found in northern Pakistan, Nepal and northern India crop up. Many of these extend into the Himalayan outliers in Yunnan, where species characteristic of the higher altitude areas of northern south east Asia also appear. Elements of this fauna are also present in Taiwan, which also has a number of endemic species.

The *Eupithecia* fauna of lowland southern and eastern China is poor, as it is also in south east Asia, and restricted to a few widespread species, such as *E. costalis* Walker. A number of species hitherto known only from Taiwan have been found in this area, generally in the relatively low mountain ranges from Fujian to Guangxi provinces. However a much richer fauna is present at altitude in the higher mountain ranges of central China, and much of this is endemic to China.

The taxonomy of *Eupithecia*

Precisely because *Eupithecia* is such a large genus, it is often suggested that it would be more convenient to find a way of splitting it into a number of smaller, more manageable, genera or subgenera. This, however, is far from straightforward. Despite the great diversity of *Eupithecia,* the ground plan of the species is remarkably uniform, so that most members of the genus can be recognised immediately by anyone familiar with it as belonging within *Eupithecia*. Within the overlying uniformity, a number of variations of pattern can however be recognised which suggest that those taxa which share them are more closely related to each other than to other members of the genus. For instance an important group of species has replaced the normal hook-shaped uncus in the male genitalia with a lamellar plate. Other similarities within this group in external appearance, in the genitalia, and where they are known, in the larval feeding habits, suggest that this represents a single divergence rather than parallel evolution. There are a number of other such cases, though the characterization of some groups is more robust than of others.

We have considered designating these groups as subgenera, but have decided against it for the following reason. While we can separate many species into such species groups, there are many *Eupithecia* species which cannot be assigned confidently to any group. It is thus not possible to construct a subgeneric system which satisfactorily covers the whole genus: one is inevitably left with a large rump group which would have to be designated 'subgenus *Eupithecia*'. No formal cladistic analysis has yet been done on the genus (a molecular study might be particularly informative). However we have little doubt that the groups which we can identify are derived from a more general *Eupithecia* pattern, and are likely in a cladistic analysis to nestle within the undefined species, rather than appear as sister groups to them. For instance, at least one species which is probably related to the 'lamellar uncus' group mentioned above is half way towards it, with a tiny uncus perched on the front of a lamellar plate. The rump subgenus comprising the rest of *Eupithecia* would thus inevitably be paraphyletic, and such a system would therefore not contribute helpfully to the taxonomic elucidation of the genus. We have dealt with this subject in a recent paper (Mironov & Galsworthy 2012).

As recorded there, we therefore decided to retain the species groups as such: many have already been defined elsewhere, but we define some new ones in this work. Our 2012 paper also separated from *Eupithecia* a few species which in our opinion did not belong there. They were omitted from the Fauna Sinica volume on the Larentiinae, and for completeness we therefore include them here under the genera *Mesoptila*, *Pareupithecia*, *Girida*, and *Eva*.

Arrangement of the book

We have grouped the species into species groups where we were able to do so, and these groups are dealt with first in the text. We are not yet able to assign any taxonomic order to these groups, and the arrangement therefore is not intended to have any taxonomic significance. The remaining species which we were unable to assign to groups follow in alphabetic order. All species are illustrated in colour, and drawings of male and female genitalia are provided where these are known. Each species is numbered and the numbers of the illustrations and genitalia drawings correspond to the species numbers. The species numbers are indexed at the end. To save space, species which are fully treated in the book are identified by generic and specific names only where they are mentioned elsewhere in the text, while other species are given author and year of description as well.

The entries under each species give a brief description of the adult, followed by descriptions of the male and female genitalia. The following three headings give details of the known ranges, and the phenology and altitude ranges where we have been able to make deductions from the specimens available. These are likely to give at best a partial view. In the few cases where anything is known of the biology, an entry is inserted after phenology. The next entry deals with diagnosis of the species from similar species: with a genus as notoriously confusing as *Eupithecia*, it is impracticable to list all the possible confusion species, but we have tried to list those where distinction is most difficult and to give pointers. In nearly all such cases, reference to the genitalia will be indispensable. We then list all type material which we have been able to examine, together with its locations, and all Chinese material in museums and collections known to us together with its label data. These lists may at times seem rather tedious, but they will be essential to future researchers working on the material, since they set out our conclusions on the available museum material, much of which can be extremely confusing. Where no country is specified in the label data, the specimens are from China. The numbers of genitalia slides generally follow those assigned in the repositories where they are kept, but in the case of the preparations at IZCAS, they are assigned a number in the 'ZL' series: this is a series of preparations made by the second author, in most cases deposited at IZCAS. Finally we add under 'Remarks' any additional points.

The genitalia diagrams are drawn by the first author, where possible, from the wet, so that it was possible to obtain views from various angles. In such cases the uncus of the male is shown in side profile as well as from the ventral perspective. The aedeagus and eighth sternite, and in a few cases, the eighth tergite, are shown separately. In most cases a magnified view of the papilla on the end of the anterior arm of the labides is also shown separately: in the text the word 'papilla' in descriptions of the male genitalia refers to this structure.

The female genitalia are drawn from the dorsal side, so that the shape of the eighth tergite can be fully shown. It should be noted that this is the opposite side to that shown in the normal genitalia preparation, where the ventral side is uppermost. In the text description, the terms 'left' and 'right' refer to the view from the ventral side, normally obtained when a preparation is viewed through the microscope.

Identifying *Eupithecia*

When in fresh condition, a few species of *Eupithecia* can reliably be identified by external appearance, and a rather larger number, with experience, tentatively identified at least to group. The shape of the wings is important, and relatively constant, as are the main pattern elements, despite some individual variation in the latter. However in our experience the vast majority of specimens in museums are relatively or very worn. Such specimens can usually only be identified by dissection of the genitalia, which is in any case advisable for certainty with the vast majority of species.

In most species groups both the male and female genitalia offer good distinctions: exceptions are the *innotata* and to some extent the *satyrata* groups, the latter particularly in the females.

In the male the shape of the valve is helpful in some cases, but many species have a very uniform valve: where the valve is not modified in some way, great caution should be exercised in judging its overall shape, since the membranous valve is very easily distorted in the mounting medium. The shape of the uncus can be very helpful, and is constant within species. The shape of the papillae on the ends of the anterior arms of the labides is also constant within species, and can help with identification in difficult cases: the length of the setae on these papillae is also useful, and is accurately shown in the drawings. The armature of the aedeagus vesica is probably the single most important feature in identifying species: for the most part the aedeagus is shown with vesica uneverted, which allows direct comparison with type material, the majority of which is mounted in this position (in making preparations, we recommend that the genital capsule and the eighth sternite are stained, but that the aedeagus is left unstained, since staining of the vesica can make the cornuti hard to distinguish). Finally the shape of the eighth sternite is often very helpful, though there are a number of patterns which recur regularly over a large number of species.

The female genitalia are more diverse, and features such as the sclerotization or striation of the bursa wall, as well as the overall shape of the bursa and the pattern of spining are very helpful. The position of attachment of the ductus seminalis to the bursa is often important. Other features which are often important are the shapes of the colliculum, antrum and eighth tergite, and the length and shape of the apophyses. In a few species, the shape of the papillae anales is diagnostic.

Envoi

Finally, although we believe that we have made very considerable progress in clarifying and describing the Chinese Eupithecia fauna, we do not for a moment suppose that ours will be the last word. We fully expect that some of our judgements, particularly those based on limited material, will need reexamination when better material is available. There will also undoubtedly be errors for future researchers to correct. Many areas of China remain very under-collected, and many species will be found to be more widespread than the ranges recorded here. We have no doubt that there are also more new species to be discovered.

SYSTEMATIC ACCOUNT

EUPITHECIA CURTIS, 1825

One of the largest geometrid genera, occurring almost worldwide. Adults of *Eupithecia* are small in size compared with other Geometridae, usually without sexual dimorphism in wing size, shape and pattern. The wingspan of the majority of species ranges from 12 to 35 mm. Most species are cryptically coloured in greyish or brownish hues, but some species have more striking coloration ranging from snowy white to ochreous and even green. Most *Eupithecia* are nocturnal. The biology is unknown for the majority of species, but where it is known, mainly for western Palaearctic, Japanese and Nearctic faunas, the caterpillars usually feed on the generative organs of plants: flowers and seeds in the fruits and seed-capsules.

The key male genitalic features which in our opinion define the genus are the combination of the shape of the juxta, the structure of the uncus, the setation of the valves, the structure of the labides with the papillae on the ends of the anterior arms, and the form of sclerotization of the eighth sternite. While none of these features on its own is completely unique within the Eupitheciini, the combination of them is not found in any other genus, and seems to be consistent throughout the genus. The female genitalia are less consistent, with the result that it is not always possible to distinguish an individual species as belonging to *Eupithecia* rather than another eupitheciine genus on the female genitalia alone.

The *haworthiata* species group

Vojnits (1972) first introduced a "*haworthiata*" group, but included within it a mixture of species externally similar to *E. haworthiata* Doubleday, without any clear definition of the group. In particular, he appears not to have recognised the most obvious apomorphy of the group, which is the reduction in the male genitalia of the uncus to a lamellar plate without the normal hook-like structure. Thus he included several species with a normal uncus in the group. He later (1976a) described a number of new species which are related to *haworthiata* under the name '*bohatschi*' species group, though he again included one species (*E. jermyi*) with a normal uncus. Inoue recognised this group, in particular in his publication on the *Eupithecia* of Taiwan (1988), and mentioned (p.338) the 'lack of spined process of uncus' as a characteristic. Mironov (1990) described this species group as subgenus *Lamelluncia*, including more than 30 species from Europe, Asia, Africa and North America. For reasons explained in the introduction, we are not here following the treatment of the group as a subgenus. Apart from the lamellar uncus, the group is characterized by a male eighth sternite with two narrow arms, and in the female genitalia by a rather small ovoid corpus bursae with the ductus seminalis attached close to the colliculum, and often elongate antrum and papillae, though none of these characters are unique to the group. The biology is known for only a few species, but there is some evidence of specialisation in the Ranunculaceae.

Though the group is very well characterized, its origin is uncertain. It is present both in the African and North American fauna (Mironov, 2003), and thus is probably quite ancient. We suspect that the lamellar uncus is derived from the expanded part of the tegumen which is normally present just below the hook-like uncus, and that the group is thus derived from an ancestor with an uncus more typical for *Eupithecia*. There is at least one species (*E. ochracea*) where the uncus is very reduced and appears as a small appendage on the expanded tegumen below it, which may represent an intermediate form. From similarities in the form of the female genitalia, *E. ochracea* is related to a further species, *E. idaeoides*, Mironov & Galsworthy, 2010, which has an entirely lamellar uncus. The group is represented in China by the following 20 species.

1. *Eupithecia subtacincta* Hampson, 1895

Eupithecia subtacincta Hampson, 1895, *Fauna Br. India* (Moths) 3: 399. Holotype ♀ (BMNH; examined), [India]: Dharmsála.

Eupithecia tabidaria Inoue, 1955, *Tinea* 2 (1/2): (82). Holotype ♂ (BMNH; examined), [Japan]: Tokyo, Takao-san (synonymised in Mironov et al., 2008a).

Eupithecia gozmanyi ussurii Vojnits, 1972, *Annls hist.-nat. Mus. natn. hung.* 64: 301. Holotype ♀ (ZFMK; examined), [Russia]: Ussuri, Wladiwostok occ., Sutschanski-Rudnik (synonymised in Mironov, 1990).

Eupithecia minibursae Vojnits, 1973, *Annls hist.-nat. Mus. natn. hung.* 65: 252, fig. 3a. Holotype ♀ (ZFMK; examined), [Russia]: Ussuri, Sutchan (synonymised in Mironov et al., 2008a).

Misidentifications: *Eupithecia meszarosi* Vojnits, 1973 (1♂ paratype in TTM); *Eupithecia pekingiana* Vojnits, 1973 (1♂ and 1♀ paratypes in ZFMK).

Unavailable names (incorrect subsequent spelling): *gozmanyi ussuriensis*: Vojnits (1973) [incorrect alteration of *gozmanyi ussurii* Vojnits, 1972, and homonym of *Eupithecia subfuscata ussuriensis* Dietze, 1910].

Description. A small broad winged species, blackish or dark brownish grey in colour, with some, usually inconspicuous, evenly curved transverse lines and small, blackish, ovoid or slightly elongate discal dots on all wings, often clearest on the hind wings. Abdomen, and often notum, mainly red or reddish ochreous.

Male genitalia. Uncus membranous, short, spade-shaped. Valve unmodified. Vinculum short and relatively broad, with deep medial hollow. Papillae short, rather narrow, broadened to apices. Aedeagus slim, a little shorter than valve length. Vesica covered with numerous spinules, and armed with an elongate, narrow, knife-shaped cornutus and a small, irregular cornutus near ductus ejaculatorius base. Sternite A8 two narrow rods, with expanded and blunt apices, connected by a narrow band at base.

Female genitalia. Bursa ovate, membranous, with two large, elongate patches of small spines on dorsal and ventral sides, and a patch of similar spines on posterior wall near colliculum. Ductus bursae not distinguishable. Ductus seminalis narrow, slightly broadened to base, attached to corpus bursae near base of colliculum on right. Colliculum short and narrow, displaced to the right side. Antrum membranous. Apophyses narrow, the anterior slightly expanded at apices. Tergite A8 square, with convex anterior margin. Papillae anales elongate and narrow, tapered. Both this and the following species have a narrow sclerotized strip placed longitudinally between the papillae anales.

Range. China (Yunnan, Sichuan, Gansu, Shaanxi, Hunan, Zhejiang, Guangdong, Taiwan, Hong Kong); Asian Russia (Priamurje, Primorje), Korea, Japan, Philippines (Luzon), Pakistan/India (Jammu & Kashmir), Sri Lanka.

Phenology. Univoltine in the north and in the highlands and hills: mid-June to September; possibly bivoltine in the south, especially in lowlands: from April to late November.

Biology. Reared from *Clematis* in Japan (Inoue et al., 1982). Larva found on *Cornus controversa* (Cornaceae) (Japan: Oh 1992): in our opinion this may not have been the food plant.

Altitude. From sea level to about 2200 m (Taiwan: Inoue 1988).

Similar species. Similar to *E. zibellinata* and *E. homogrammata*. Distinguished from both by the more distinct discal dots on all wings and the reddish abdomen. *E. zibellinata* is considerably darker and *E. homogrammata* usually paler. The combination of uncus, narrower apex to the valve, very distinct, deep medial hollow in the vinculum, longer cornuti on the vesica and longer apical arms to the eighth sternite in the male genitalia, and in the female the absence of a small patch of spines at the base of the ductus seminalis, shorter colliculum and apophyses, narrower, pointed papillae anales should distinguish this species from *E. zibellinata*. For distinction from *E. cichisa*, see under species below.

Material. ♀, *Eupithecia subtacincta* Hmpsn. type ♀, BM Geom. slide no 20937♀ (holotype, BMNH); ♂, [Japan], Takao-san, 30.vii.1949, H. Inoue (holotype of *E. tabidaria*, coll. H. Inoue in BMNH); ♀, [Russia], Sutschanski-Rudnik, Wladiwostock occ., Ussuri, August, 817., Vojnits slide no. 524♀ (holotype of *E. gozmanyi ussurii* [*E. gozmanyii* ssp. *ussuriensis* Vojnits on the label], ZFMK); 1♀, same data, 818, Vojnits slide no. 518♀ (paratype of *E. gozmanyi ussurii* [*E. gozmanyii* ssp. *ussuriensis* Vojnits on the label], ZFMK); 1♀, [Russia], Sutchan, Ussuri, Vojnits slide no. 496♀ (paratype of *E. gozmanyi ussurii* [*E. gozmanyii* ssp. *ussuriensis* Vojnits on the label], ZFMK); ♀, [Russia], Sutchan, Ussuri, *Eupithecia asperata* Brdt., Vojnits slide no. 521♀ (holotype of *E. minibursae*, ZFMK); 1♂, Szetschwan [Sichuan], Wassekou, Exp. Stötzner (paratype of *E. pekingiana*, ZFMK); 1♀, Szetschwan [Sichuan], Tschengtufu, Exp. Stötzner, Vojnits slide no. 500♀ (paratype of *E. pekingiana* without hind wings; ZFMK); 1♂, 4♀, West-China, Szetschwan [Sichuan], Exped. Stötzner, Coll. Dr. Wehrli, Basel, Vojnits slides nos. 1047♀, 1049♀, 1059♀ (ZFMK); 1♂, Pr. Kwangtung [Guangdong], Canton, 11.xi.[19]23, Höne (ZFMK); 12♂, 2♀, Prov. Chekiang [Zhejiang], West Tien-mu-shan, 2, 9, 16, 22, 23, 25, 29.iv., 8.v.1932, H. Höne, Vojnits slides nos. 1058♂, 1068♂, 11395♂, 1051♂, 11394♂, de Laever slides nos. 317♂, 319♂ (ZFMK); 1♂, same loc., 17.iv.1932, H. Höne, Vojnits slide no. 1043♂ (TTM); 1♀, China, Provinz Nord-Yuennan, Li-kiang [Lijiang], 6. ix.1935, H. Höne, Vojnits slide no. 1040♀ (paratype of *E. minibursae*, ZFMK); 13♂, 16♀, same loc., 1, 23, 29, 30.vi., 5, 7, 8, 9, 10, 11, 12, 13, 14, 18.vii., 5.viii.1934, 7.vi., 5.viii.1935, H. Höne, Vojnits slides nos. 1071♂, 1119♂, 1053♂, 1118♂, 11405♂, 11402♀, 1120♀, 11408♂, 1067♀, 1073♀, 11404♂, 1066♀, 1072♀, 1121♀, 1116♀, 1076♀, 11410♀, 1074♂, de Laever slide no. 278♂, Mironov slides nos. 755♀, 756♂ (ZFMK); 1♂, same loc., 6.viii.1934, H. Höne (SMNK); 1♂, 1♀, same loc., 1, 5.vii.1934, H. Höne, Vojnits slides nos. 1041♀, 1117♂ (TTM); 1♀, same loc., *ca.* 2000 m, 29.vi.1934, H. Höne (ZFMK); 1♂, China, Sued-Shensi [Shaanxi], Tapaishan [Taibaishan] im Tsinling [Qinling], 6.vii.1935, H. Höne (ZFMK); 1♂, same loc., *ca.* 1700 m, 20.v.1936, H. Höne, Vojnits slide no. 1081♂ (paratype of *E. meszarosi*, TTM); 2♂, 3♀, same loc. and alt., 17, 20, 22.v., 10.vi., 10.vii.1936, H. Höne, Vojnits slides nos. 1083♀, 1082♀, 11396♀, 1080♂ (ZFMK); 1♀, Taiwan, Prov. Ilan, Fu-Shan Bot. Garden, 24°45′13″ N, 121°35′39″ E, 700 m, 27.ix.1999, leg. A. Kun, L. Peregovits & L. Ronkay (ZSM); 1♀, same loc., 24°54′ N, 121°45′ E, 700 m, 24.viii.2000, leg. L. Ronkay (TTM); 1♀, Taiwan, Taipei County, Taipei, Nanshin Chiao, Han-Lo-Dae, 450 m, December 2001, leg. A. Kun & L. Ronkay (TTM); 1♂, Yunnan, Hekou, 200 m,

9.vi.1956, coll Huang Keren, slide no ZL58; 1♀, Sichuan, Dukou, Lanmuqiao, 2150 m, 31.viii.1980, coll Zhang Baolin, slide no. ZL430; 1♂, Gansu, Wenxian, 1450 m, 24.vii.1999, coll Zhu Chaodong, slide no. ZL557; 1♂, Gansu, Dangchang, 1800 m, 7.vii.1998, coll Yuan Decheng (IZCAS). Inoue collection material from Taiwan (BMNH): 6♂, 2♀, Hualien, Juisui, 5–6.vii.1973, M. Owada; 1♂, 5F, Lushan Spa, 1200 m, (Nantou Hsien), 9–11.iii.1985, 1♂, 2♀, same loc., 30.vii 1984, K. Yazaki; 1♀, Chiai [Chiayi] Hsien, Shihtyulu, 21–23. vii.1984, R. Sato; 2♀, Mt Alishan, 2200 m (Chiayi Hsien), 28.vii.1981, R. Sato; 1♂, 5♀, Nantou, Lushan, 28.vii.1981, 1♀, same loc. 22.vii.1981, R. Sato; 1♀, Hualien, Tayulin, 24.vii.1981, R. Sato; 1♀, Nantou hsien, Lushan Wenchuan, 26–28.vii.1981, R. Sato.

2. *Eupithecia cichisa* Prout, 1939

E[upithecia] cichisa Prout, 1939, *in* Seitz, *The Macrolepidoptera of the World* 4 (Suppl.): 253, pl. 18: c. Holotype ♂ (BMNH; examined), China (west): Mt. Omei, Gipfel.

Eupithecia pekingiana Vojnits, 1973, *Annls hist.-nat. Mus. natn. hung.* **65**: 250, fig. 1b. Holotype ♀ (ZFMK; examined), [China]: Peking [Beijing], Westberge. **New junior synonym.**

Misidentification: *Eupithecia gozmanyi ussuriensis* Vojnits, 1972 ([incorrect change of name] 1♀ paratype in ZFMK).

Description. A small unicolourous dark grey or grey species with distinct small discal dots on the wings (rounded on the fore wings and elongate on the hind wings), usually with dark angulate postmedial lines on the hind wings. Notum and abdomen grey or dark grey.

Male genitalia. Uncus membranous, spade-shaped, with shallow medial hollow. Valve unmodified. Vinculum short, with deep medial hollow. Papillae very short and wide. Aedeagus stout, shorter than valve length, with two broad lateral lobes containing numerous very thin spines. Vesica granulate; armed with a small, elongate, U-shaped cornutus and a smaller hooked cornutus near ductus ejaculatorius base. Sternite A8 two narrow, blunt and sclerotized rods connected by a very narrow curved basal band.

Female genitalia. Bursa large, oval, membranous, with two elongate lateral patches of small spines, connected in posterior part of bursa near colliculum. Ductus bursae not distinguishable. Ductus seminalis slightly broadened at base, attached to posterior part of bursa near colliculum on right. Colliculum short and narrow, displaced to right. Antrum membranous, narrow. Apophyses broad, slightly expanded at apices. Tergite A8 quadrate, with slightly convex anterior margin and rounded posterior corners. Papillae anales small, elongate and tapered, with a narrow sclerotized strip between them.

Range. China (Yunnan, Sichuan, Shaanxi, Beijing, Jiangsu, Hunan, Zhejiang, Jiangxi, Taiwan).
Phenology. Polyvoltine (?): late April to mid-July; late November.
Altitude. From sea level to about 1700 m in Taibaishan (China: coll ZFMK).
Similar species. Very similar to *E. subtacincta*, but distinguished by the angulate postmedial line on the hind wing (evenly curved in *subtacincta*) and the grey notum and abdomen (both reddish in *subtacincta*). The male genitalia have much shorter papillae on the anterior arms of the labides. The long cornutus present in the aedeagus vesica in *subtacincta* is also lacking. The female genitalia are very alike.
Material. ♂, Mt. Omei, 3500 m, 16.viii.[19]31, Franck, L.B. Prout Coll B.M. 1939–643, *Eupithecia cichisa* Prout type ♂, BM Geom. slide no. 20934♂ (holotype, BMNH); ♂ [*nec* ♀], Peking [Beijing], Westberge, Exp. Stötzner, *Eup. isogramaria*, 2.vi., (holotype of *E. pekingiana*, ZFMK; abdomen lost); 1♂, Peking [Beijing], 6.vii., *Eup. isogramaria* (paratype of *E. pekingiana*, ZFMK); 1♀, China, Shanghai, Vojnits slide no. 495♀ (paratype of *E. gozmanyi* ssp. *ussuriensis*, ZFMK); 1♀, Prov. Kiangsu [Jiangsu], Shanghai, 22.vi.[19]32, Höne (ZFMK); 3♂, 4♀, same loc., 24, 25.v., 2, 16, 22.vi., 13.vii., 15.viii.[19]33, H. Höne, Vojnits slides nos. 1050♀, 1064♀, 1065♂, de Laever slides nos. 166♂, 318♀, 320♂, 321♀ (paratypes of *E. pekingiana*, TTM); 1♂, 1♀, Prov. Kiangsu [Jiangsu], Lungtan b. Nanking [Nanjing], 18.vi., 25.vii.[19]33, H. Höne (paratypes of *E. pekingiana*, ZFMK); 1♂, 3♀, same loc., 22.vi.[19]32, 26.v., 18, 19.vi.[19]33, H. Höne, Mironov slide no. 718♂ (ZFMK); 1♂, same loc., 29.v.[19]33, H. Höne, Mironov slide no. 798♂ (SMNK); 1♂, Prov. Hunan, Hoengshan, 15.iv.1933, H. Höne, Vojnits slide no. 1077♂ (ZFMK); 1♀, China, [Jiangxi] Kuling, 2.v.1934, H. Höne, de Laever slide no. 21♀ (ZFMK); 1♀, same loc., 15.vii.1934, H. Höne, de Laever slide no. 282♀ (ZFMK); 1♀, Tibet [Sichuan], Batang, Im Tal des Yangtze (*ca.* 2800 m), 6. vii.1936, H. Höne, de Laever slide no. 23♂ (ZFMK); 2♂, 1♀, China, Sued-Shensi [Shaanxi], Tapaishan [Taibaishan] im Tsinling [Qinling], *ca.* 1700 m, 8, 9.vii.1936, H. Höne, de Laever slide no. 168♂ (ZFMK); 1♀, same loc. and alt., 8.vii.1936, H. Höne, Vojnits slide no. 12959♀ (TTM); 1♂, Prov. Chekiang [Zhejiang], West Tien-mu-shan, 25.iv.1932, H. Höne (ZFMK); 1♀, same loc., 17.vi.1932, H. Höne, Vojnits slide no. 1042♀ (TTM); 4♂2♀, Kwanhsien [Guanxian], Szechwan [Sichuan], W. China, 21.vii.1924, 12. viii.1925, 14.viii.1925, 21.vii.1928, 29.vii.1928, 3.viii.1928, G.M. Franck, BM Geometrid slides nos.. 8879, 8880(♂), 8881(♀) (BMNH); 1♀, Taiwan, Prov. Kao-Hsiung, 15 km NE Taoyuan, 1850 m, 7.vii.1995, leg. G. Csorba & L. Németh, coll. Mus. Witt (TTM); 1♂, Taiwan, Prov. Hualien, Taroko N.P., Hsipan, 270 m, 5.iv.1997, leg. Csorba & Ronkay (ZSM); 2♀, Taiwan,

Prov. Ilan, Fu-Shan Bot. Garden, 700 m, 24°45′13″ N, 121°35′39″ E, 27.xi.1999, leg. A. Kun, L. Peregovits & L. Ronkay (ZSM).
Remarks. The holotype of *E. pekingiana* is a male, not a female as stated by Vojnits: he did not examine its genitalia because the specimen had lost its abdomen. Two paratypes of *E. pekingiana* (male and female) belong to the previous species, *E. subtacincta*.

3. *Eupithecia aenigma* Mironov & Galsworthy, 2004

Eupithecia aenigma Mironov & Galsworthy, 2004, *Trans. lepid. Soc. Japan* 55 (3): 225, figs 1, 13. Holotype ♂ (ZFMK; examined), China: North Yunnan, Li-kiang [Lijiang].

Misidentification: *Eupithecia minibursae* Vojnits, 1973 (1♀ paratype in TTM).

Description. A small species. Fore wing ground colour pale warm brown; transverse lines only faintly marked in slightly warmer brown; discal dot faint, a short brown streak; antemedial, double medial, and postmedial present, represented by dark dots on the costa, continuing as four parallel lines, angled in the cell; a wavy white subterminal line; terminal line dark brown, broken on the veins. Hind wing very rounded, concolorous with fore wing, and with minute discal spot and three to four very faint transverse brown lines; wavy white subterminal line visible only at tornus.
Male genitalia. Uncus membranous, short and broad with small pointed tip. Valve unmodified. Vinculum short, without medial hollow. Papillae elongate and narrow. Aedeagus short and narrow, vesica armed with numerous minute spinules and an elongate, thin cornutus at ductus ejaculatorius base. Sternite A8 small, with two long, narrow, tapered rods.
Female genitalia. Bursa globular, covered with dense small spines except on left and right sides. Ductus bursae not distinguishable. Ductus seminalis narrow, attached to corpus bursae near colliculum on left. Colliculum short and narrow. Antrum membranous, narrow. Tergite A8 almost square. Apophyses narrow, slightly expanded and flattened at apices. Papillae anales narrow and elongate.
Range. China (Sichuan, Yunnan, Jiangsu).
Phenology. Univoltine: mid-June to mid-July.
Similar species. Similar to *E. subtacincta*, but distinguished by the lack of clear discal spots and the narrower and more pointed apex of the fore wing. The male genitalia are distinguished by the presence of a small pointed tip to the uncus, and by the lack of a deep medial hollow in the vinculum.
Material. ♂, China, Provinz Nord-Yuennan, Li-kiang [Lijiang], 20.vi.1934, H. Höne, Vojnits slide no. 11407♂ (holotype, ZFMK); 1♀, same loc., 30.

vi.1934, H. Höne, Vojnits slide no. 1054♀ (paratype of *E. minibursae*, TTM); 1♀, Prov. Kiangsu [Jiangsu], Lungtan b. Nanking [Nanjing], 3.vii.[19]33, H. Höne, Vojnits slide no. 1063♀ (paratype of *E. minibursae*, ZFMK); 1♂, China, Sichuan Prov., Xiangcheng, 2800 m, 13.vii.2001, Li Houhun leg. (coll. P. Skou).

4. *Eupithecia zibellinata* Christoph, 1881

Eupithecia zibellinata Christoph, 1881, *Bull. Soc. Imp. Nat. Moscou* 55 (3): 117. Lectotype ♂ (nec ♀) (TTM, examined and designated here to stabilize nomenclature), [Russia]: Wladiwostock.

Description. A small broad-winged, rather uniformly dark species, with wings, head and body varying from blackish brown or dark chocolate to almost black, with a series of usually indistinct, evenly curved lighter transverse lines and inconspicuous discal dots on all wings. Some specimens are completely dark and unicolourous, without pale transverse lines on the wings.

Male genitalia. Uncus membranous, spade-shaped. Valve unmodified. Vinculum short and narrow, with very shallow medial hollow. Papillae elongate and narrow. Aedeagus narrow, slightly shorter than valve length. Vesica armed with a small, elongate flat cornutus and a smaller irregular cornutus near ductus ejaculatorius base; surface of vesica covered with numerous spinules. Sternite A8 broadened at base, with two narrow rods, expanded to more sclerotized blunt tips at apex.

Female genitalia. Bursa ovoid, membranous, with two large, elongate patches of spines, one on dorsal part and one in posterior part, and one smaller patch near ductus seminalis base. Ductus seminalis attached to bursa near colliculum on right. Colliculum narrow, slightly elongate, displaced to right side. Antrum membranous. Apophyses narrow. Tergite A8 almost square with a small indentation in centre of dorsal margin. Papillae anales rather small, rounded, covered by sparse short setae.

Range. China (Heilongjiang); Asian Russia (Kamchatka peninsula, Sakhalin, southern Priamurje, Primorje), Japan (Hokkaido).

Phenology. Univoltine: early June to late July.

Similar species. This species is most easily confused with three other Asian *Eupithecia* species, *E. subtacincta*, *E. homogrammata* and *E. caliginea* Butler, 1878. *E. subtacincta* is paler and has a reddish abdomen; *E. homogrammata* has paler wings; lastly, *E. caliginea* has more elongate and pointed fore wings, usually with narrower light transverse lines. In the male genitalia the wide valve with broad apex, the shallow medial hollow in the vinculum, the shorter cornuti on the vesica and the shape of sternite A8 with shorter apical arms in the male are distinctive; in the female, the disposition of spines in the corpus bursae, the longer colliculum and apophyses and broader, rounded papillae anales without the narrow sclerotized strip between them are distinct from those of *E. subtacincta*. The genitalia in both sexes of *E. homogrammata* and *E. caliginea* are quite different from those of *E. zibellinata*.

Material. ♂, [Russia], Sutschanski-Rudnik, Wladiwostock occ., Ussuri, 815, August, *Zibellinata Chr.* (Christoph's hand-writing), Vojnits slide no. 1089♂ (lectotype, TTM); 1♀, [Russia], Raddeffskaja Chr., *Zibellinata* Chr., Origin., Zool. Mus. Berlin, Vojnits slide no. 15298♀, *Eupithecia* photo 84 A 10 det. A. Vojnits (paralectotype, coll. O. Staudinger in MNHU); 1♀, [Russia], Askold Dörr., Zool. Mus. Berlin, Vojnits slide no. 15300♀, *Eupithecia* photo 84 A 12 det. A. Vojnits (paralectotype, coll. O. Staudinger in MNHU); 1♀, [Heilongjiang], Charbin [Harbin], Mandschurei, Juli 150 m, Vojnits slide no. 544♀ (TTM); 1♂, Heilongjiang, Dailing, 390 m, 26.vi.1957, coll He Zhong, slide no. ZL213 (♂); 1♂, 1♀, Heilongjiang, 3.vii.1962, coll Bai Jiuwei, slides nos. ZL197♂, ZL214♀ (IZCAS).

Remarks. Not previously recorded from China.

5. *Eupithecia olgae* Mironov, 1986

Eupithecia olgae Mironov, 1986, *Trudy VEO* **67**: 96, figs 2a, b. Holotype ♀ (ZISP; examined), Kirghizskaya SSR [Kirghizstan]: Inner Tian-Shan, Naryn, 2030 m.

Description. A smallish species with narrow, elongate and rather pointed fore wings. Ground colour of fore wings ash grey or brownish grey (in older specimens), with numerous pale, very indistinct transverse lines; discal dots not present. Hind wing slightly paler, similarly marked. Fringes short, unicolourous, ash grey or brownish grey. Abdomen ash grey or pale brownish grey with narrow pale dorsal stripe.

Male genitalia. Uncus membranous, more or less short, triangular. Valve unmodified. Vinculum very short, relatively broad, with shallow medial hollow. Papillae small and narrow. Aedeagus slim, almost equal to valve length. Vesica armed with two small, elongate flat cornuti; surface of vesica covered with numerous spinules. Sternite A8 broadened at base, with two narrow elongate apical rods.

Female genitalia. Bursa ovoid, membranous, with two large, elongate lateral patches of spines, an additional patch of widely spaced similar spines in posterior part, and a smaller patch of close spines between ductus seminalis base and colliculum. Ductus seminalis attached to bursa near colliculum on right. Colliculum short and narrow, displaced to the right and slightly inclined to left. Antrum short, membranous. Apophyses narrow, relatively short. Tergite A8 small, rectangular, width longer than length. Papillae anales small, rather short and broad, rounded.

Range. China (Sichuan, Qinghai, Gansu, Shanxi); SE Kazakhstan, Uzbekistan, Kirghizstan, Tajikistan, Afghanistan (NE), Pakistan/India (Jammu & Kashmir), Mongolia.

Phenology. Univoltine: early June to early August. Larva: July and August. Overwintering as pupa in detached cocoon in the ground.

Biology. Larva found in the flower buds of *Clematis orientalis* and *C. glauca* (Ranunculaceae) (Kirghizstan: original data of V. Mironov).

Altitude. From about 1100 m in Ghissar Mts. (Tajikistan: coll. ZMKU) up to about 3800 m in Tibet (China: coll. ZFMK, IZCAS).

Similar species. Externally very similar to *E. illaborata*, which is usually of comparable size and coloration. *E. olgae* usually has broader wings, darker ground colour, less distinct and less oblique transverse lines on the fore wings, and a narrow whitish dorsal band on the abdomen. The male and female genitalia of *E. olgae* and *E. illaborata* are very similar. In the male genitalia of *E. olgae* the uncus is broader than in *E. illaborata*, and the base of sternite A8 is differently shaped, as illustrated. The female has larger spiniferous areas, and an accordingly smaller membranous zone in the base of the corpus bursae. The posterior apophyses are also shorter than in *E. illaborata*.

Material. ♀, Kirghizstan, Inner Tien-Shan Mts., Naryn, L, 2030 m, 25. vii.1983, leg. V. Mironov (holotype, ZISP); 1♂, Tibet [Sichuan], Batang, Untere Urwaldzone (*ca.* 3800 m), 21.vi.1936, H. Höne, Vojnits slide no. 12447♂ (ZFMK); 1♀, Prov. Shansi [Shanxi], Mien-shan, Obere Höhe *ca.* 2000 m, 29.vii.1937, H. Höne, de Laever slide no. 301♀ (ZFMK); 3♂, 1♀, China, Gansu, Xiahe, 2600–3000 m, 26.-30.vi.1986, Coll. Nr. 246, leg. Görgner, Mironov slide no. 701♂ (ZFMK); 1♂, China, Süd-Gansu, Linxia/ Xiahe, 2700–3100 m, 28.vi.-10.vii.1985, Coll. Nr. 221, leg. Görgner (ZFMK); 7♂, 11♀, Qinghai, Menyuan, Xianmi, 2800 m, 23.vii.1992, slides nos. ZL156, 316 ♂, 158, 270 ♀; 7♂, 7♀, Qinghai, Tongren, Maixiu, 2950 m, 30.vii.1992, slide no. 364♂; 1♂,1♀, Gansu, Yongdeng, Tulugou, 25–29.vii.1991, slide no. ZL318♀, all coll Xue Dayong; 1♀, Shanxi, Pangquangou, 7.vii.1988, coll Cao Tianwen, slide no. ZL318♀ (IZCAS).

Remarks. Not previously recorded from China. The majority of Chinese specimens from Qinghai are larger than "typical" specimens of *E. olgae* from the western Tien Shan.

6. *Eupithecia illaborata* Dietze, 1904

Eup[ithecia] illaborata Dietze, 1904, *Dt. ent. Z.. Iris* [1903] **16** (2): 357, pl. 3, fig. 34. Syntypes 4♂, 4♀ (coll. K. Dietze and R. Püngeler in MNHU; examined), [China]: Aksu, Chinese. West Turkestan; Korla.

Description. A small pale ash-grey species. Fore wing narrow, elongate with rather pointed apex; transverse lines oblique, angled near costa; terminal area with a more or less distinct, narrow, whitish, wavy subterminal line; discal dot small and pale, often invisible. Hind wing similarly coloured or slightly paler with indistinct transverse lines and a minute discal dot, sometimes lacking.

Male genitalia. Uncus membranous, elongate, spade-shaped. Valve unmodified. Vinculum short and narrow, with very shallow medial hollow. Papillae small, rather elongate and narrow. Aedeagus short and narrow, slightly shorter than valve length. Vesica armed with an elongate patch of numerous spinules, a small, elongate flat cornutus and a smaller irregular cornutus near ductus ejaculatorius base. Sternite A8 small, broadened at base with well marked exterior angles, and with two narrow parallel apical rods.

Female genitalia. Bursa bean-shaped, membranous, with two rather narrow, elongate lateral patches of sparse spines, a further similar patch in posterior part, and a small patch of smaller denser spines between ductus seminalis base and base of colliculum. Ductus seminalis attached to bursa near colliculum on right. Colliculum short and narrow, displaced to the right. Antrum short, membranous. Apophyses narrow. Tergite A8 small, almost square. Papillae anales small, and notably rounded.

Range. China (Xinjiang, Inner Mongolia); SE Kazakhstan, Kirghizstan (Tien-Shan Mts.).

Phenology. Bivoltine: early June to August.

Altitude. Found from about 1700 m (SE Kazakhstan: coll. ZMKU) up to about 2200–2300 m (Kirghizstan: Viidalepp 1988; original data).

Similar species. See under previous species.

Material. 1♂, [Xinjiang], mit. Mai, *illaborata* n. sp., Original, Mitte Mai, Aksu, Bohatsch gesehen, Type., K. Dietze (syntype, coll. K. Dietze in MNHU); 1♂, [Xinjiang], *illaborata* Korla, *illaborata* n. sp., Type., Korla, ♂, Rückbeil, durch Tancré, von R. Püngeler, K. Dietze, Gen.-Unters. Nr. [Schmidt slide no.] Eu 97a, Zool. Mus. Berlin (syntype, coll. K. Dietze in MNHU); 1♀, [Xinjiang], *illaborata* n. sp., Type, ♀, Maklan Wüste, Aksu, Mitte Mai, Rückbeil, durch Tancré, K. Dietze, Gen.-Unters. Nr. [Schmidt slide no.] Eu 97 (syntype, coll. K. Dietze in MNHU); 1♂, [Xinjiang], *illaborata*, Original, Aksu, K. Dietze, Mitte Mai (syntype, coll. K. Dietze in

MNHU); 2♀, [Xinjiang], Aksu, Ost-Turkestan, *illaborata*, Aksu 03, K. Dietze (syntypes, coll. K. Dietze in MNHU); 1♂, 1♀, China, Inner Mongolia, 80 km SW of Hohhot, Nantianmen, 13.viii.2007, [leg.] A. Bidzilya (ZISP).

Remarks. De Laever (1960) was the first to describe and illustrate the male and female genitalia of *E. illaborata*. However, he incorrectly referred to type specimens of this species from North Korea (Saichin). In fact, *E. illaborata* was originally described from Korla in NW China.

7. *Eupithecia rubellata* Dietze, 1904

Eup[ithecia] rubellata Dietze, 1904, *Dt. ent. Z.. Iris* [1903] **16** (2): 337, pl. 3, figs 3–5. Syntypes 5♂, 8♀ (coll. K. Dietze and R. Püngeler in MNHU; examined). [China]: Korla, Sinin [Xining], Aksu, Chines. West Turkestan.

Eup[ithecia] rubellata f. *deserticola* Dietze, 1910, *Biologie der Eupithecien* 1: pl. 73, fig. 372. Syntype ♂ (coll. K. Dietze in MNHU; examined). [China]: Altyn-tag.

Eup[ithecia] rubellata f. *mediopunctata* Dietze, 1910, *Biologie der Eupithecien* 1: pl. 73, fig. 373. Syntypes 2♀ (coll. K. Dietze in MNHU; examined). [China]: Aksu, Makan-Wüste.

Eup[ithecia] rubellata f. *scotaeata* Dietze, 1910, *Biologie der Eupithecien* 1: pl. 73, figs 374, 375. Syntypes 3♂, 2♀ (coll. K. Dietze in MNHU; examined). [China]: Ili, [Xinjiang], Juldus [Yining] region.

Eupithecia invenusta Vojnits, 1977, *Acta zool. Acad. Sci. hung.* **23** (3–4): 472, figs 9, 11. Holotype ♂ (TTM; examined), Mongolia: Gobi Altaj aimak, Chasat chajrchan ul Mts., 20 km S of Somon Zargalant (synonymised in Mironov, 1990).

Description. A relatively large species with narrow elongate wings. Fore wing ground colour rusty ochreous or clear reddish brown; costal margin usually darker, brownish grey, especially in basal half; basal and antemedial oblique, sharply angled onto costa; postmedial evenly curved, perpendicular to costa, sometimes forming a longitudinal, narrow and pointed zigzag on the stalk of vein R; terminal area with a fine, indented, white subterminal line; discal dot very small, rounded or ovoid, black with white area around it. Hind wing white or off-white, with a distinct dark postmedial, darker terminal area and small discal dot.

Male genitalia. Uncus elongate, membranous with a more strongly sclerotized, thickened, horseshoe-shaped base. Valve large, unmodified. Vinculum trapezoid, short and rather broad. Papillae short and broad. Aedeagus narrow, shorter than valve length. Vesica covered with numerous spinules, armed with an elongate, flat, U-shaped cornutus. Sternite A8 consisting of two long and narrow rods with tapered pincer-like tips, connected by a narrow band at base.

Female genitalia. Bursa ovate-oblong, membranous, with small ovate lateral patches of spines and a large patch of sparse spines from middle of bursa to colliculum. Ductus seminalis attached to bursa near colliculum. Colliculum very short and narrow. Antrum membranous, narrow and very long, almost equal to bursa in length. Apophyses broad and long, slightly expanded apically. Tergite A8 very large, rectangular. Papillae anales large, bean-shaped, sclerotized.

Range. China (Xinjiang, Qinghai, Tibet); SE Kazakhstan, Kirghizstan (Tien-Shan Mts.), Pakistan/India (Jammu and Kashmir), Mongolia.

Phenology. Univoltine (?): mid-June to early August.

Altitude. From about 2400 m in Tien-Shan (Kirghizstan: coll. ZMKU) up to about 3700 m in Jammu and Kashmir (Pakistan/India: coll. ZFMK).

Similar species. An easily recognized species, the combination of the strongly reddish fore wing ground colour, and the clear white hindwing ground colour being unique in the genus. Specimens from Xinjiang are sandier in colour (form *deserticola* Dietze, 1910). Though the adults are very different, the male and female genitalia of *E. rubellata* are very similar to those of *E. bohatschi*. We have found no significant structural differences between the male and female genitalia of these two species, except that the membranous uncus is rather longer in *E. rubellata*.

Material. 1♀, [Qinghai], *rubellata* n. sp., ♀ Original, Sinin, Nord Tibet, 1900., von Herrn Tancré (K. Dietze) (syntype, coll. K. Dietze in MNHU); 2♀, [Qinghai], n. sp. Stgr., ♀, Sinin, Tibet, *rubellata* n. sp., Original ♀ (K. Dietze), *Eupithecia rubellata* Dietze ♀, Sinin Tibet, prep. № 1653.4, E. de Laever det. 1957 (syntypes, coll. K. Dietze in MNHU); 1♂, [Xinjiang], *rubellata* var. Original, Aksu, Maklan., v. Tancré, K. Dietze, Gen.-Unters. Nr. [Schmidt slide no.] Eu 91a, Zool. Mus. Berlin (syntype, coll. K. Dietze in MNHU); 1♀, [Xinjiang], *rubellata*, Aksu ♀ n. sp., Gen.-Unters. Nr. [Schmidt slide no.] Eu 91, Zool. Mus. Berlin (syntype, coll. K. Dietze in MNHU); 1♀, [Xinjiang], Aksu, Ost-Turkestan, *rubellata* n. sp. ♀, Aksu, K. Dietze (syntype, coll. K. Dietze in MNHU); 2♂, [Xinjiang], *rubellata* Aksu ♂ (syntypes, coll. K. Dietze in MNHU); 1♂, [Xinjiang], Korla, Sai-Chin, 370, *rubellata* n. sp., ♂, K. Dietze (syntype, coll. K. Dietze in MNHU); 1♂, [Xinjiang], Korla, 97., Hbhr., n. sp. bei *Variostrigata* Alph. (Boh.), *Eup. rubellata* Original, Korla (K. Dietze), Vojnits slide no. 19029♂ (syntype, coll. O. Staudinger, MNHU); 1♂, [Xinjiang], Ost-Turkestan (Aksu), Rückbeil 1900, *rubellata* Dietze ♂, Original, 3/01 v. R. Tancré, *rubellata* n. sp. Original. ♂, (K. Dietze) (coll. R. Püngeler in MNHU); 1♂, 1♀, [Xinjiang], Asia centr., Altyn-Tagh, Rückbeil 1903, *rubellata* Dietze, cotype, 5/05 v. R. Tancré (coll. R. Püngeler in MNHU); 1♂, [Xinjiang], *rubellata* v. *scotaeata*, Juldus Ili [Yining], *Eupithecia rubellata* v. *scotaeata* ♂ Dietze, Ili [Yining], cotype, prep. № 1653.3 E. de Laever det. 1957 (coll. R. Püngeler in MNHU); 1♂, [Xinjiang], Aksu, *rubellata n. sp.* (hand writing of K. Dietze, ZFMK); 1♀, *rubellata n. sp.* (hand writing of K. Dietze, possibly from Aksu also, ZFMK); ♂, [Xinjiang], *rubellata* forma: *deserticola*, Typ., Altyn-tag [= Altun Shan], (372), Vojnits slide no. 10016♂ (type of *E. rubellata* f. *deserticola*, coll. K. Dietze in MNHU); 1♀, [Xinjiang], *rubellata* forma: *mediopunctata* Typ., Aksu (373) (syntype of *E. rubellata* f. *mediopunctata*, coll. K. Dietze in MNHU); 1♀, [Xinjiang], *rubellata* forma: *scotaeata* Typ. Kuldja (374), Vojnits slide no. 10017♀ (syntype of *E. rubellata* f. *scotaeata*, coll. K. Dietze in MNHU); 3♂, 1♀, [Xinjiang], Juldus, Ili [Yining] (syntypes of *E. rubellata* f. *scotaeata*, coll. K. Dietze in MNHU); 1♀, [Qinghai], *rubellata* n. sp., form. *fasciata*, Sinin, Tibet, 1900 (syntype, coll. K. Dietze in MNHU); ♂, Mongolia: Gobi Altaj aimak, Chasat chajrchan ul Gebirge, *cca.* 20 km S von Somon Zargalant, 2400 m, Exp. Dr. Z. Kaszab, 1966, Nr. 695, 15.vii.1966, Vojnits slide no. 10552♂ (holotype of *E. invenusta*, TTM); 14♀, Qinghai, Menyuan, Xianmi, 2800 m, 3.vii.1992; 9♂, 25♀, Qinghai, Tongren, Maixiu, 2950 m, 30.vii.1992, Xue Dayong, slides nos. ZL21, 590♂, 148♀; 1♀, same loc., 2800 m, 37°16′ N, 102°00′ E, 6.viii.2005, Xue Dayong, IOZ(E)1847007; 1♀, Tibet, Zham, Quxiang, 3300 m, 6.vii.1975, Wang Ziqing; 2♂, 3♀, Tibet, Zogang, 3800 m, 29.vi-2.vii.1976, Zhang Xuezhong, slides nos. ZL425♂, 593♀; 7♂, 17♀, Tibet, Jomda, 3400 m, 24–29.vii.1976, Han Yinheng, slides no. ZL323, 348♂, 261♀; 6♂, 4♀, Tibet, Riwoqê, 3750 m, 17–25.viii.1976, Zhang Xuezhong; 1♂, 23.viii.1976, 1♀, 5.vii.1976, Tibet, Zhag'yab, Gyitang, 3600 m, Han Yinheng, slide no. ZL155♀; 1♀, Tibet, Qamdo, 3900 m, 5.vii.1976, Zhang Xuezhong; 1♀, Qinghai, Qilian, 29.vii.1977, Zhang Baolin; 2♂, 1♀, 1.viii.1983, slide no. ZL471(♀); 1♂, 5.viii.1983, Renbu; 2♂, Tibet, Namling, Wuyu, 4–5.viii.1983, Yan Yaoguang et al, slide no. 427(♂); 1♂, 5♀, Xinjiang, Tomürt mountain, 20.vi-21.vii.1977, slide no. ZL584 (♂), 592 (♀); 2♀, Xinjiang, Yecheng Qipanxiang, 3000 m, 16.vii.1989, Zhang Xuezhong; 2♀, Xinjiang, Akto, 2650 m, 6.viii.1989; 1♀, Xinjiang, Taxkorgan, 4500 m, 13.viii.1989.

8. *Eupithecia bohatschi* Staudinger, 1897

Eup[*ithecia*] *bohatschi* Staudinger, 1897, *Dt. ent. Z.*, *Iris* **10** (1): 111, pl. 3, fig. 73. Lectotype ♀ (design. by Mironov, coll. O. Staudinger in MNHU; examined), [Russia]: Sutschan.

Eupithecia (*Pena*) *kawakamiana* Matsumura, 1925, *J. Coll. Agric. Hokkaido Imp. Univ.* **15** (3): 172, pl. 11, fig. 19. Syntypes 2♀ (EIHU), [Japan/Russia]: Saghalien (synonymised in Prout, 1939).

Eupithecia hoenei Vojnits, 1976, *Acta zool. Acad. Sci. hung.* **22** (1–2): 208, figs 1c, 2f. Holotype ♂ (TTM; examined), [China]: Prov. Shensi [Shanxi], Mien-shan, 2000 m. **New junior synonym.**

Eupithecia tuvinica Viidalepp, 1976, *Nasekomye Mongolii* **4**: 392, figs 18, 41–47. Holotype ♂ (IAET; examined), USSR [Russia]: Tuvinskaya ASSR [Tuva], near Kyzyl town (synonymised in Mironov, 1990).

Eupithecia sachalini Vojnits, 1981, *Annls hist.-nat. Mus. natn. hung.* **73**: 234, figs 11(*nec* 19), 17. Holotype ♂ (TTM; examined), [Russia]: Sahhalin [= Sakhalin], Tomovsk [= Tymovskoje] (synonymised in Mironov, 1990).

Description. A small species. Fore wing basal area along costa blackish brown, the rest of basal area and frontal half of medial area white; posterior half of medial area and terminal area chocolate brown; basal transverse line usually evenly curved, often inconspicuous; medial white, usually well visible from anal margin to discal dot; postmedial double, sinuate, evenly curved, white; terminal area with fine, narrow, indented white subterminal line, usually broadened near tornus and forming a small white tornal spot; discal dot small, black. Hind wing paler, whitish, with brownish transverse lines, including an especially distinct postmedial line, and with darker brownish grey terminal area; terminal area with narrow indented white subterminal line; discal dot small. Notum snowy white. Abdomen covered with white and brown scales with small dorsal tufts of white scales on each segment and clear white last segment.
Male genitalia. Uncus membranous, triangular with rounded tip. Valve unmodified. Vinculum short and rather wide, with very shallow medial hollow. Papillae short and broad. Aedeagus long and slim, shorter than valve length. Vesica armed with a broad flat cornutus and a smaller irregular cornutus near ductus ejaculatorius base; surface of vesica covered with numerous minute spinules. Sternite A8 with two long parallel apical rods, connected by a short and narrow band at base.
Female genitalia. Bursa ovate-oblong, membranous, with two small oval lateral patches of spines and a separate patch near colliculum. Ductus seminalis attached to bursa near colliculum. Colliculum short and narrow. Antrum membranous, narrow and very long, almost equal to bursa in length. Apophyses broad and long, slightly expanded apically. Tergite A8 transverse. Papillae anales bean-shaped, heavily sclerotized, covered with short, stout, pointed setae resembling spines, orientated towards eighth tergite.
Range. China (Beijing, Hebei, Heilongjiang, Inner Mongolia, Shanxi, Gansu, Sichuan, Qinghai,Yunnan, Tibet); Asian Russia (southern Siberia: Tuva, Transbaikalia, Dahuria, Priamurje, Primorje; Sakhalin), Mongolia, Korea.
Phenology. Univoltine: late June to early October.
Altitude. From sea level in Primorje (Russia: coll. ZISP) to about 4000 m in Yunnan (China: Vojnits 1976).
Similar species. Variable in depth of coloration and distinctness of markings, especially in the distribution of white and brown on all wings. In spite of this, usually an easily recognized species, but could be confused with *E. jezonica*, *E. ustata*, and *E. likiangi*. See under these species for differences. The male genitalia can be distinguished by the combination of the

large valve with broadly rounded apex, the shape of vinculum with very shallow medial hollow, the short and broad papillae on the anterior arms of the labides and the shape of the eighth sternite. The female genitalia can readily be distinguished by their extreme length, especially of the antrum. The male and female genitalia closely resemble those of *E. rubellata*, but the adults are readily distinguishable from one another.

Material. ♀, [Russia], Sutschan 90 Dörr., 73., Origin. (lectotype, design. by Mironov, 2009, coll. O. Staudinger in MNHU); 1♀, [Russia], Suifun 85 D., *bohatschi* ♀, Origin. (syntype, coll. O. Staudinger in MNHU); 1♀, [Russia], *bohatschi* Stgr., Suifun, Amur mer., Type Stgr., Coll. Bohatsch (syntype, coll. K. Dietze in MNHU); 1♀, [Russia], Saghalien Matsumura, Kawakami, July 30 '24, *Eupithecia* (*Pena*) *kawakamiana* n., Type Matsumura, R. Sato, 1995 ♀ Genitalia on slide no. Gmt-103 (syntype of *E. kawakamiana*, EIHU); 1♀, [Russia], Saghalien Matsumura, Ichinos.awa July 25 '24, Co-type Matsumura (syntype of *E. kawakamiana*, EIHU); ♂, Prov. Shansi [Shanxi], Mien-shan, Obere Höhe *ca.* 2000 m, 7.viii.1937, H. Höne, Schmidt slide no. Eu 182b (holotype of *E. hoenei*, TTM); 2♂, same loc. and alt., 1, 2.viii.1937, H. Höne, Vojnits slide no. 10426♂, Schmidt slide no. Eu 182a (paratypes of *E. hoenei*, TTM); 3♂, same loc. and alt., 7, 8.viii.1937, H. Höne (ZFMK); ♂, U.S.S.R. [Russia], Tuvinskaya ASSR [Tuva], environs of Kyzyl, at light, 2.viii.1972, leg. T. Ruben & J. Viidalepp (holotype of *E. tuvinica*, IZBE); 1♂, Mandschuria, Maoerschan (= Mauschan), Juli (ZFMK); 2♀, same data, Juli, Vojnits slides nos. 832♀, 833♀ (TTM); 1♀, Maoershan, Mandshuria or, Schmidt slide no. Eu 179 (TTM); 2♀, Charbin [Harbin], Mandschurei, 20. vii.1937, 30.vii.1939 (ZFMK); 1♀, same loc., Juli 150 m (ZFMK); 4♂, 2♀, Tibet [Sichuan], Batang, Im Tal des Yangtze (*ca.* 2800 m), 15.vii., 20, 27, 29, 31. viii., 3.ix.1936, H. Höne (ZFMK); 1♂, 2♀, same loc. and alt., 13.viii., 6, 24. ix.1936, H. Höne, de Laever slide no. 420♀ (TTM); 1♂, same loc., Untere Urwaldzone (*ca.* 3800 m), 3.x.1936, H. Höne (ZFMK); 1♂, same loc. and alt., 1.viii.1936, H. Höne, de Laever slide no. 87♂ (TTM); 4♂, Nord Yünnan, A-tun-tse [Dêqên], Mittlere Höhe (*ca.* 4000 m), 1, 18, 30.viii.1936, H. Höne (ZFMK); 2♂, same loc. and alt., 30.vii., 17.viii.1936, H. Höne, de Laever slide no. 427♂, Schmidt slide no. Eu 179b (TTM); 1♂, same loc., Talsohle (*ca.* 3500 m), 5.ix.1936, H. Höne (ZFMK); 1♂, 1♀, China, Inner Mongolia, 100 km SW of Hohhot, Mo-Tien-Ling, 12.viii.2007, [leg.] A. Bidzilya (ZISP); 20♂, 3♀, Tibet, Jomda military station, 3400 m, 23–31.vii.1976, Han Yinheng and Zhang Xuezhong, slides nos. ZL20, 43, 45, 492, 495, 597 ♂, 42, 597♀; 2♂, Tibet, Zhag'yab, Gyitang, 3600 m, 5 and 9.vii.1976, Han Yinheng; 2♀, same loc, 19.vii.1976 and 27.viii.1976, respectively Han Yinheng and Zhang Xuezhong; 3♂, 5♀, Qinghai, Menyuan, Xianmi, 2800 m, 23.vii.1992, Xue

Dayong, slide no. ZL39♀; 2♀, Qinghai, Tongren, Maixiu, 2950 m, 30.vii.1992, coll Xue Dayong, slide no. ZL37♀; 2♀ Beijing, Yanqing, 29.viii.1964, slides nos. ZL44, 596♀; 1♂, 1♀, Beijing, Sanpu, 19.viii.1964, coll Liao Subai, slides nos. ZL595♂, ZL493♀; 1♀, Beijing, Mentougou, Liyuanling, 1100 m, 20–21. viii.2002, 1♀, Beijing, Mentougou, Xiaolongmen, 1100 m, 22.viii.2002, Han Hongxiang; 1♂, same data but 24.ix.1998; 1♀, 2♂, Inner Mongolia, Yierto district, 26 and 28.vii and 5.viii.1983 respectively, slides nos. ZL40, 64♀; 1♂, Inner Mongolia, Yuancheng banner, 10.v.1986, slide no. ZL448♂; 1♂, 11. viii.1987; 3♀, Heilongjiang, Dailing, 390 m, 9.vii and 2.viii. 1962, Bai Jiuwei and Zhou Shixiu, slides nos. ZL41, 46♀; 1♀, Gansu, Yongdeng, Tulugou, 2280 m, 25–29.vii.1991, Xue Dayong (IZCAS).

9. *Eupithecia boneta* Mironov & Galsworthy, 2011

Eupithecia boneta Mironov & Galsworthy, 2011, *Lep. Science* **62** (1): 16, figs 2, 22. Holotype ♀ (ZFMK; examined), China, Yunnan, Li-kiang [Lijiang].

Description. A smallish species with a predominantly chocolate-brown fore wing with clear white transverse lines. Hind wing rather dark, chocolate brown in terminal areas, becoming abruptly paler at line of postmedial, and shading to off white towards base. First three abdominal segments white with scattered brown scales, all other segments rusty-brown.

Female genitalia. Bursa relatively large, globular, completely covered with small spines, with an oblique, narrow, heavily sclerotized appendage, slightly bisected at the apex, and with a membranous tip, on the left side. Ductus bursae short, sclerotized. Ductus seminalis broad and sharply curved at base, attached to ductus bursae near base of colliculum on left. Colliculum rather short and narrow. Antrum membranous. Apophyses elongate, narrow, expanded at apices. Tergite A8 almost quadrate. Papillae anales narrow and rounded.

Male unknown.

Range. China (Yunnan).

Similar species. Similar in size to *E. bohatschi*, but distinguishable by the full chocolate-brown fore wing with fine, whitish, double transverse lines between basal, medial and terminal areas, and the darker hind wing with straighter white transverse band between medial and terminal areas. The female genitalia of *E. boneta* are distinguished from those of all other Chinese *Eupithecia* species by the sclerotized appendage on the left side of the bursa.

Material. ♀, China, Provinz Nord-Yuennan, Li-kiang [Lijiang], 7.ix.1934, H. Höne, Mironov slide no. 697♀ (holotype, ZFMK).

Remarks. Known only from the single holotype with a missing right fore wing.

10. *Eupithecia jezonica* Matsumura, 1927

Eupithecia jezonica Matsumura, 1927, *Insecta matsumur.* 1 (4): 184. Lectotype ♀ (EIHU; examined by Sato, 1995), [Japan]: [Hokkaido], Sapporo.

Eupithecia viidaleppi Vojnits, 1981, *Annls hist.-nat. Mus. natn. hung.* 73: 232, figs 16, 18. Holotype ♂ (TTM; examined), [Russia]: S Ussuri, Lazo rj., Benevskoje (synonymised in Mironov et al., 2004d).

Eupithecia catosophia Inoue, 1988, *Bull. Fac. domest. Sci. Otsuma Wom. Univ.* 24: 336, figs 9a-f. Holotype ♂ (BMNH; examined), [Taiwan]: Nantou Hsien, Yushih, near Wushe (synonymised in Mironov & Galsworthy, 2007).

Misidentification: *Eupithecia hoenei* Vojnits, 1976 (1♂ paratype in ZFMK).

Description. Fore wing usually predominantly rusty brown, with white areas only around base and discal dot, costal margin of basal and medial areas rusty brown, transverse lines white, evenly curved, but medial line obtusely angled onto costa; terminal area with a narrow dentate white subterminal line; discal dot very small, often not visible. Hind wing paler, white with pale but distinct brownish grey, evenly curved transverse lines; terminal area brownish grey with broad, wavy white subterminal line and light, whitish medial blotch; discal dot very small, sometimes invisible. Abdomen bright white when fresh.

Male genitalia. Uncus membranous, broad, spade-shaped with obtuse apex. Valve unmodified, with concave dorsal margin. Vinculum short and broad, with deep medial hollow. Papillae short and stout. Aedeagus short and narrow, broadened anteriorly, shorter than valve length. Vesica armed with a small, flat cornutus and an irregular cornutus near anterior end of aedeagus; surface of vesica multiply granulate. Sternite A8 broad, with two narrow apical rods connected by a narrow basal band.

Female genitalia. Bursa copulatrix ovate-oblong, membranous, with two lateral patches of spines and a small patch of sparse spines near colliculum and ductus seminalis base. Ductus seminalis broadened basally, attached to corpus bursae near colliculum. Colliculum elongate and narrow. Antrum membranous, narrow. Apophyses rather short and narrow. Tergite A8 almost square. Papillae anales elongate and tapered.

Range. China (Jilin, Shanxi, Taiwan); Asian Russia (Altai Mts., Transbaikalia, southern Priamurje, Primorje), E. Kazakhstan, Pakistan/India (Jammu & Kashmir), Nepal, Korea, Japan (Hokkaido, Honshu, Sado).

Phenology. Polyvoltine (?): mid-May to late October. Larva: early August to mid-September. Overwintering as pupa in flimsy cocoon in the ground.

Biology. Larva found in the seed-balls of *Clematis fusca* (Ranunculaceae) where it feeds on the unripe seeds (Russia, Primorje: original data of Mironov).
Altitude. From sea level to about 2400 m in the Himalayas (Nepal: coll C. Herbulot in ZSM).
Similar species. Similar to *E. bohatschi* in size but distinguishable by the rusty brown colour, which is usually more dominant than white, the smaller, often invisible discal dot on the fore wing, the more distinct transverse lines, the presence of a white blotch in the middle of the terminal area on the hind wing, and the white abdomen. In the male genitalia, in comparison with *E. bohatschi*, the uncus is broader, the valve more tapered, the papillae on the anterior arms of the labides longer and narrower, the medial hollow in the vinculum deeper, the aedeagus shorter and narrower, and the eighth sternite considerably broader. In the female, the antrum is much shorter than in *E. bohatschi*, and the colliculum about twice the length.
Material. ♀, [Japan], Sapporo Matsumura, 13/viii[19]'18, Type Matsumura, *Eupithecia jezonica* n., det. Matsumura, R. Sato, 1994 ♀ Genitalia on slide No. Gmt-77 (lectotype; EIHU); ♂, Taiwan, Yushin, Nantou Hsien, 3–4. viii.1987, A. Kawabe, BM Geom. slide no. 16542♂ (holotype of *E. catosophia*, coll. H. Inoue in BMNH); 1♂, Prov. Shansi [Shanxi], Mien-shan, Obere Höhe *ca.* 2000 m, 14.viii.1937, H. Höne, Schmidt slide no. Eu 182♂ (paratype of *E. hoenei*, ZFMK); 1♂, 1♀, same loc. and alt., 9.vii., 3.viii.1937, H. Höne (ZFMK); 1♂, Taiwan, Taichung Co., Lishan, alt. 2020 m, 23/30.ix.1992, F. Aulombard et J. Plante (ZFMK); 2♂, 1♀, Taiwan, Prov. Nantou, Meifong, 2,250 m, 24°05′ N, 121°10′ E, 25.ix.1999, leg. G. Csorba and B. Herczig (ZSM); 1♂, Central Taiwan, Nantou Co., Wu-Shi, *ca.* 850 m NN, 24°00′35″ N / 121°06′13″ E, LF, 1.ix.2002, leg. M.-Y. Chen & U. Buchsbaum (ZSM); 8♂, 2♀, Central Taiwan, Nantou Co., Meifeng, *ca.* 2100 m NN, 24°05′55″ N / 121°10′48″ E, LF, 03, 04, 05, 09, 10. September 2002, leg. U. Buchsbaum (ZSM, ZISP); 1♂, same loc. and alt., 12. July 2003, leg. Mei-Yu Chen (ZSM); 2♂, same loc. and alt., 24°05′00″ N / 121°10′45″ E, LF, 15, 20. Sept. 2002, leg. M.-Y. Chen & U. Buchsbaum (ZSM); 1♂, Central Taiwan, Nantou Co., 1300 m NN, Chun-Yang, near Lushan Hot Spring, 24°01′25″ N / 121°09′54″ E, LF, 13. Sept. 2002, leg. M.-Y. Chen & U. Buchsbaum (ZISP); 1♂, Central Taiwan, near Puli, Nantou Co., Meifeng, *ca.* 2200 m NN, 24°05′11″ N / 121°10′21″ E, LF, 20. Sept. 2004, leg. Mei-Yu Chen (ZSM); 4♂, same loc. and alt., 24°05′25″ N / 121°10′21″ E, LF, 15. October 2006, leg. Mei-Yu Chen & U. Buchsbaum (ZSM); 1♀, Jilin, Changbaishan, 3.vii.1974, coll Yang Zhuming, slide no. ZL47 (♀) (IZCAS).

11. *Eupithecia sophia* Butler, 1878

Eupithecia sophia Butler, 1878, *Ann. Mag. nat. Hist.* (5) 1 (6): 444. Holotype ♂ (BMNH; examined), [Japan]: Yokohama.

Misidentification: *Eupithecia bohatschi* Staudinger, 1897 (1♀ syntype in MNHU); *Eupithecia hoenei* Vojnits, 1976 (1♂ paratype in ZFMK).

Description. A small broad-winged species. Fore wing almost completely rusty brown with small white areas near base and in the middle around discal dot; usually costal margin and sometimes terminal area darker, dark brown, with two very small white tornal and central subterminal spots; discal dot black, small, slightly elongate. Hind wing off-white, usually with distinct light brown, evenly curved transverse lines; subterminal line distinct, wavy and white; discal dot small but conspicuous, black, ovoid or comma-like. Head, thorax and abdomen snowy white.

Male genitalia. Uncus membranous, relatively large and broad, spade-shaped. Valve unmodified. Vinculum rather short and narrow, trapeziform, without medial hollow. Papillae short and broad. Aedeagus very narrow, much shorter than valve length. Vesica armed with one extremely short stick-like cornutus near ductus ejaculatorius base. Sternite A8 U-shaped, with two narrow, tapered and sclerotized apical rods connected by a narrow basal band.

Female genitalia. Bursa ovate, membranous, almost completely covered with numerous microscopic spinules except for small area near ductus seminalis base. Ductus bursae very short and narrow. Ductus seminalis short and narrow, slightly broadened basally, attached to bursa near ductus bursae base on left. Colliculum lightly sclerotized, very short and narrow. Antrum membranous. Apophyses narrow, relatively long. Tergite A8 elongate. Papillae anales rather small, elongate and tapered.

Range. China (Shanxi); Far East Russia (S Priamurje, Primorje, Kuril Islands: Kunashiri), Japan (Hokkaido, Honshu, Sado, Shikoku, Kyushu, Tsushima).

Phenology. Polyvoltine: April to October.

Altitude. From sea level in Russia to at least 1300 m in Japan (Inoue, 1979). The only Chinese specimen is from 2000 m.

Similar species. Most closely resembles *E. jezonica*, but distinguished by the darker hind wings, which in *E. jezonica* are white, with shorter, straighter postmedial transverse lines. The male genitalia of *E. sophia* are distinguished from those of *E. jezonica* by the absence of a medial hollow in the vinculum, the narrower and longer aedeagus and the shape the of the eighth sternite, with narrower lateral arms and no basal lobes.

The most distinctive difference in the female genitalia is the much shorter and narrower colliculum.
Material. ♂, Japan 77–9, *E. sophia* Butler Type, BM Geom. slide no. 3520♂ (holotype, BMNH); 1♀, [Russia], Wladiwstk Chr., Origin. (syntype of *E. bohatschi*, coll. O. Staudinger in MNHU); 1♂, Prov. Shansi [Shanxi], Mien-shan, Obere Höhe *ca.* 2000 m, 9.viii.1937, H. Höne, Vojnits slide no. 10425♂ (paratype of *E. hoenei*, ZFMK).
Remarks. Not previously recorded from China.

12. *Eupithecia lineisdistincta* Vojnits, 1981

Eupithecia lineisdistincta Vojnits, 1981, *Acta zool. Acad. Sci. hung.* 27 (1–2): 217, figs 1, 2. Holotype ♂ (ZSM; examined), Nepal: Thodung, 3100 m.
Eupithecia yasudai Inoue, 1987, *Bull. Fac. domest. Sci. Otsuma Wom. Univ.* **23**: 241, figs 63: G, 66: A, B. Holotype ♂ (BMNH; examined), [Nepal], Taplejung, Walungchung Gola, 3310 m. **New junior synonym.**

Description. An average-sized species, with rather blunt, wide fore wings of grey or dark grey ground colour with numerous, very distinct, wavy, dark transverse lines and small, narrow, oblique discal dots. Medial area of fore wing broad, bordered by very distinct, wavy ante- and postmedial lines, forming dark brownish grey costal blotches. Hind wing paler, off-white, but darker grey along anal margin and near tornus, with conspicuous, fine, wavy transverse lines.
Male genitalia. Uncus membranous, relatively large, broad and elongate, with slightly sclerotized lateral margins. Valve unmodified. Vinculum short and rather narrow. Posterior arms of juxta narrow, bifurcate at apices. Papillae short and broad, tapered. Aedeagus short and narrow, shorter than length of valve, with two latero-apical setaceous lobes. Vesica armed with two very small, weakly sclerotized, plate-like cornuti, one short and triangular, the other elongate and narrow, near ductus ejaculatorius base. Sternite A8 lyre-shaped, with two narrow, posteriorly sclerotized rods connected by a narrow band at base.
Female genitalia. Bursa small, globular, almost completely covered with small spines except for a short, membranous ductus bursae. Ductus seminalis short and narrow, slightly broadened near base, attached to ductus bursae near colliculum on right. Colliculum short and narrow. Antrum elongate and narrow, membranous, with a patch of dense spinules behind eighth tergite. Tergite A8 elongate, slightly broadened posteriorly. Apophyses long and narrow, expanded and flattened at apices. Papillae anales relatively large, elongate, tapered.
Range. China (Yunnan); Nepal.

Phenology. Bivoltine (May and July).
Altitude. Recorded from about 3100 m to 3500 m in the Himalayas (Nepal: Vojnits 1981, coll. TTM, ZISP).
Similar species. Unlikely to be confused with any other Asian *Eupithecia* species, and can usually be easily distinguished by the shape, coloration and maculation of the wings (but see note below on *E. matura*). The male genitalia are similar to those of *E. sinicaria*, *E. fletcheri* Prout, 1926 and *E. lactibasis* Inoue, 2000, especially in the shapes of the vinculum and eighth sternite. However, *E. sinicaria* has longer and narrower papillae on the anterior arms of the labides; *E. fletcheri* has shorter posterior arms to the juxta and a smaller aedeagus; lastly *E. lactibasis* has a narrower valve with evenly rounded apex and a different shape to the papillae on the anterior arms of the labides. The female genitalia are similar to those of *E. ustata*, *E. sinicaria* and *E. albigutta*, but distinguished from all these by the globular shape of the bursa copulatrix and the larger and broader papillae anales.
Material. ♂, Nepal, Thodung, 3100 m, 22.v.1962, leg. G. Ebert u. H. Falkner, Vojnits slide no. 11771♂ [= ZSM Genitalprp. no. G 10620] (holotype, ZSM); 1♀, China, Yunnan Prov., Zhongdian, 3100 m, 14.vii.2001, Li Houhun leg. (coll. P. Skou).
Remarks. This species is very similar to the Himalayan *E. matura* Vojnits, 1981, which is known only from the single holotype male. No significant structural differences can be found in the male genitalia of the two taxa. It seems possible that the holotype specimen of *E. matura* is a variety or aberration of *E. lineisdistincta*. However since at present only the male holotype is available, we think it better to defer consideration of this until more material is available from the Himalayas.

13. *Eupithecia albigutta* Prout, 1958

Eupithecia albigutta Prout, 1958, *Bull. Br. Mus. nat. Hist.* (Ent.) **6** (12): 393. Holotype ♀ (BMNH; examined), India: Simla, 7000 ft.
Eupithecia pulla Vojnits, 1988, *Acta zool. hung.* **34** (1): 38, pl. 2, figs G-H; figs. 1–3, 10. Holotype ♂ (NSMTO), Nepal (east): Sagarmatha Solukhumbu Thame Og, 3800 m (synonymised in Inoue, 2000).

Description. An average sized species showing little variation in size. Fore wing reddish brown with a white or yellowish white basal spot, with small white traces of transverse lines on the costa and anal margin, a narrow broken white subterminal line, and a distinct oblique ovoid or subulate black discal dot with a white spot on the outside. Hind wing white with a distinct brown basal area along anal margin, brown tornal blotch, brownish apex and with a small, oblique comma-like discal dot. Head, notum and at least rear abdominal segment white.

Male genitalia. Uncus broad, spade-shaped, membranous. Valve unmodified. Vinculum short and narrow, rounded with a shallow medial hollow. Papillae relatively small, narrow, slightly tapered. Aedeagus long and narrow, but shorter than length of valve. Vesica covered with minute dentate spicules, and armed with one short and broad, apically tapered, plate-like, U-shaped cornutus with short hooked basal arms, placed near ductus ejaculatorius base. Sternite A8 broad, elongate, lyre-shaped, consisting of two long and narrow, curved arms, tapered and connected by a narrow band at base.

Female genitalia. Bursa small, ovate, completely covered with dense small spines. Ductus bursae membranous, short and tapered. Ductus seminalis narrow, attached to posterior part of bursa near base of the membranous ductus bursae on right. Colliculum short and narrow. Antrum membranous, long and narrow; area between posterior margin of eighth tergite and papillae anales densely covered by small setae. Tergite A8 long and narrow. Apophyses long and narrow, slightly expanded and flattened at apices. Papillae anales relatively small, elongate, very narrow, and tapered.

Range. China (Tibet, Sichuan, Yunnan, Guangdong, Taiwan); Nepal, India (Punjab).

Phenology. Bivoltine: mid-March to late May; mid-October.

Altitude. From about 1600 to 3800 m in the Himalayas (China: Vojnits 1988; Nepal: Inoue 2000).

Similar species. *E. albigutta* is an easily recognized species, but it could occasionally be confused with *E. rubridorsata*. See differential diagnosis under this species.

Material. ♀, [India], Simla, 7000 ft., Oct. '97, Pilcher, *Eupithecia albigutta* Prout ♀ type, 99–206, BM Geom. slide no. 20139 (holotype, BMNH); 1♂, E. Nepal, Sagarmatha Solukhumbu, Thame Og, 3800 m, 1–2.x.1979, leg. M. Owada, Vojnits slide no. 14656♂ (paratype of *E. pulla*, TTM); 1♀, China, [Guangdong], Canton (ZFMK); 1♂, China, Provinz Nord-Yuennan, Li-kiang [Lijiang], 16.iv.1935, H. Höne, de Laever slide no. 257♂ (ZFMK); 1♂, 5♀, China, Sichuan, Daxue Shan, Gongga Shan, NW Moxi, 29°41′ N, 101°58′ E, 2600–3200 m, 23.iv.-15.v.2001, leg. V. Siniaev & A. Plutenko, *ex* coll. Dr. A. Schintlmeister (ZSM); 1♀, Taiwan, Nantou, Meifeng, 9–10.xi.1998, leg. C.S. Lin & W.T. Yang, NMNS ENT 3042–440, slide no. ACG TW17; 1♂, same loc., 21–22.x.1991, leg. C.S. Lin, NMNS ENT 1202–122, slide no. ACG TW26 (NSMTA); 7♂, 2♀, China, W. Sichuan, road Ya'an / Kangding, Erlang

Shan Mt., 2100 m, 29°51′ N / 102°18′ E, 11.x.2009, Floriani & Saldaitis leg. (coll. O. Pekarsky); 2♂, China, W. Sichuan, S. from Kangding, near Menghugang, 15.x.2009, 2800 m, Floriani & Saldaitis leg. (coll. O. Pekarsky); 1♀, Tibet, Zhag'yab, Gyitang, 3600 m, 5.viii.1976, coll Han Yinheng, slide no ZL589♀.

Remarks. Not previously recorded from China. We think it likely that *albigutta* is synonymous with *Eupithecia dolia* West, 1929, described from the Philippines, and if so *dolia* would be the senior name, but since no male material is yet available from the type locality of the latter for comparison, we have not formally synonymised the two names.

14. *Eupithecia rubridorsata* Hampson, 1895

Eupithecia rubridorsata Hampson, 1895, *Fauna Br. India* (Moths) **3**: 403. Holotype ♀ (BMNH; examined), [India]: Sikhim [= Sikkim].

Eupithecia circumscriptrix Vojnits, 1983, *Acta zool. Acad. Sci. hung.* **29** (1–3): 279, fig. 22. Holotype ♂ (ZSM; examined), Nepal: Khumbu, Khumdzung, 3900 m (synonymised in Inoue, 2000).

Eupithecia acerba Vojnits, 1984, *Acta zool. hung.* **30** (1–2): 228, fig. 20. Holotype ♂ (TTM; examined), China: Sued-Shensi, Tapaishan [Taibaishan] im Tsinling. **New junior synonym.**

Description. Rather a small species with characteristic ochreous head, notum, abdomen and similarly coloured basal spots on the fore wings. Fore wing ground colour of basal and medial areas blackish brown; medial area irrorated by numerous white scales; terminal area reddish brown with small white subterminal dots; transverse lines inconspicuous, except evenly curved postmedial; discal dot small but distinct, circular, black. Hind wing paler than fore wing, off-white; basal area blackish along anal margin; transverse lines wavy, well marked along anal margin only; terminal area narrow, slightly darker, especially near tornus, with wavy inner border; discal dot minute, sometimes absent.

Male genitalia Uncus membranous, relatively broad, spade-shaped. Valve unmodified. Vinculum small, rounded, with shallow medial hollow. Papillae relatively short and stout. Aedeagus narrow and elongate, shorter than valve length. Vesica covered with longitudinal wrinkles, armed with two small irregular, plate-like cornuti near ductus ejaculatorius base. Sternite A8 two narrow, elongate, sinuate arms with pointed, pincer-like apices, connected by a narrow basal band.

Female genitalia. Bursa small, globular, completely covered with spines. Ductus bursae short and narrow, membranous, spineless. Ductus seminalis slightly broadened at base, attached to ductus bursae on left side. Colliculum very short and narrow. Antrum membranous, long and narrow.

Tergite A8 rather rectangular and elongate, with rounded posterior corners. Apophyses long and narrow, tapered. Papillae anales narrow and elongate, tapered.
Range. China (Yunnan, Sichuan, Shaanxi); Nepal, India (Sikkim).
Phenology. Bivoltine (?): mid-April, mid-July, early August and early October.
Altitude. Recorded from about 700 to 3900 m in the Himalayas (Nepal: Vojnits 1983).
Similar species. Superficially similar to *E. albigutta*, but easily distinguishable by the clear ochreous basal blotches on the fore wings and similarly coloured head, notum and abdomen, the rounded discal dots usually without adjacent white blotches, and the off-white hind wing with less distinct blackish brown tornal blotch. The male genitalia of *E. rubridorsata* are smaller than those of *E. albigutta*, but otherwise similar. The female genitalia are more distinct, especially in the spherical corpus bursae with larger spines, and the longer antrum, tergite A8 and apophyses.
Material. ♀, Sikkim, 1.iii.1889, Knyvett, Collectio H.J. Elwes, *Eupithecia rubridorsata* Hmpsn. Type ♀, BM Geom. slide no. 20126 (holotype, BMNH); ♂, Nepal, Khumbu, Khumdzung, 3900 m, 18.vi.1962, leg. G. Ebert u. H. Falkner, Vojnits slide no. 10653♂ [= ZSM slide no. G 10625] (holotype of *E. circumscriptrix*, ZSM); 1♂, Nepal, Thodung, 3100 m, 22.v.1962, leg. G. Ebert u. H. Falkner, Vojnits slide no. 10651♂ (paratype of *E. circumscriptrix*, ZSM); ♂, China, Sued-Shensi [Shaanxi], Tapaishan [Taibaishan] im Tsinling [Qinling], 2.vii.1935, H. Höne, Vojnits slide no. 13270♂ (holotype of *E. acerba*, TTM); 1♂, 4♀, China, Sichuan, Daxue Shan Mts., Gongga Shan, W Moxi, 29°34′ N, 101°59′ E, 3200 m, 20.-21.vii.1999, leg. V. Siniaev & A. Plutenko (ZFMK); 1♂, Yunnan, Yulong mountain, 27.vi.1962, Song Shimei, slide no. ZL38♂; 1♀, Sichuan, Kangding, 6.vii.1979, Chen Tailu (IZCAS).
Remarks. Not previously recorded from China.

15. *Eupithecia ustata* Moore, 1888

Eupithecia ustata Moore, 1888, *Descr. new Indian lepid. Insects Colln late Mr W.S. Atkinson* (3): 268. Lectotype ♀ (coll. O. Staudinger in MNHU; examined), [India]: Darjiling [= Darjeeling].
Eupithecia konradi Vojnits, 1976, *Acta zool. Acad. Sci. hung.* 22 (1–2): 199, figs 1f, 2b, 3d. Holotype ♀ (TTM; examined), [China]: North Yunnan, A-tun-tse [Dêqên], 4000 m. **New junior synonym.**

Description. A smallish species with rusty brown fore wings and white hind wings. Fore wing basal area mostly white, but brown along costa;

basal and medial areas separated by a double, indented white band; medial area rusty brown with white blotches near and around discal dot; medial white, distinct from discal dot to costa only; postmedial present only as a white costal spot; terminal area rusty brown, with narrow, indented, whitish subterminal line, broadened near apex, at the middle and near tornus, with a black shade along inner side; discal dot small, black, linear. Hind wing with brown transverse lines distinct, wavy; basal and postmedial broadened near anal margin; terminal area darker, brownish white, especially near tornus, with wavy inner border and broad and wavy white subterminal line; discal dot pale, small.

Male genitalia. Uncus membranous, short and broad, spade-shaped. Valve unmodified. Vinculum short, trapezoid, with shallow medial hollow. Papillae slightly tapered to apices. Aedeagus short and narrow, slightly broadened at anterior end, shorter than length of valve. Vesica armed with two very small and narrow cornuti near ductus ejaculatorius base. Sternite A8 two narrow, sinuate, pointed rods, heavily sclerotized at apices, connected by a narrow basal band.

Female genitalia. Bursa small, ovoid, membranous, almost completely covered with small spines except for a narrow longitudinal area along right side. Ductus bursae short, membranous. Ductus seminalis broadened basally, attached to ductus bursae near colliculum on right. Colliculum short and narrow. Antrum long and narrow, membranous, with two lateral patches of scales behind posterior margin of eighth tergite. Tergite A8 elongate and narrow. Apophyses long and narrow, the posterior slightly expanded and flattened at apices. Papillae anales elongate, very narrow.

Range. China (Sichuan, Yunnan, Tibet); Pakistan/India (Jammu & Kashmir), India (Sikkim).

Phenology. Bivoltine: late April to mid-June; early July to late August.

Altitude. From about 2600 to 4000 m (China: coll. ZSM; Vojnits 1976 [as *E. konradi*]).

Similar species. Similar to *E. likiangi*, but distinguished by the absence of blackish brown blotches in the medial and terminal areas on the fore wings and the less angulate hind wings with less oblique dark transverse lines. The male and female genitalia of *E. ustata* are similar to those of *E. sinicaria*. However, the uncus is narrower in its basal part, the papillae on the anterior arms of labides are shorter in the male, and in the female the apophyses are longer than in allied species.

Material. ♀, [India], Darjeeling 7000' Coll. Atkinson, Origin., 3792., Coll. Staudinger, K. 326 (lectotype, design. by Mironov, coll. O. Staudinger in

MNHU); ♀, Nord Yünnan, A-tun-tse [Dêqên], Mittlere Höhe (*ca.* 4000 m), 24.vii.1936, H. Höne, Schmidt slide no. Eu 180c (holotype of *E. konradi*, TTM; its slide in ZFMK); 2♀, same loc. and alt., 24.vii., 6.viii.1936, H. Höne (paratypes of *E. konradi*, both without abdomens, ZFMK); 1♂, 2♀, same loc. and alt., 5, 23.vii., 5.viii.1936, H. Höne, de Laever slide no. 69♂, Vojnits slides nos. 10410♀, 10420♀ (paratypes of *E. konradi*, TTM); 2♂, same loc., Obere Höhe (*ca.* 4500 m), 9.vii.1936, 17.vii.1937, H. Höne, Vojnits slide no. 10409♂, Schmidt slide no. Eu 180a (paratypes of *E. konradi*, TTM; Schmidt slide no. Eu 180a in ZFMK); 1♂, 2♀, same loc., Talsohle *ca.* 3000 m, 18.v., 2.vii.1937, H. Höne, Schmidt slide no. Eu 180 (paratypes of *E. konradi*, ZFMK); 1♂, 2♀, same loc. and alt., 4, 9.vii.1937, H. Höne, de Laever slides nos. 68♀, 423♂ (paratypes of *E. konradi*, TTM); 1♀, China, Provinz Nord-Yuennan, Li-kiang [Lijiang], 26.vii.1935, H. Höne (SMNK); 1♂, Tibet [Sichuan], Batang, Im Tal des Yangtze (*ca.* 2800 m), 18.v.1936, H. Höne; 1♂, same loc., Untere Urwaldzone (*ca.* 3800 m), 1.x.1936, H. Höne (without abdomen, ZFMK); 1♂, China, Szetschwan [Sichuan], Wassekou, Exped. Stötzner (ZFMK); 2♀, China, Sichuan, Daxue Shan, Gongga Shan, NW Moxi, 29°41′ N, 101°58′ E, 2600–3200 m, 23.iv.-15.v.2001, leg. V. Siniaev & A. Plutenko, *ex* coll. Dr. A. Schintlmeister (ZSM); 3♂, 2♀, Tibet, Jomda, 3400 m, 26–29.vii.1976, Zhang Xuezhong and Han Yinheng, slides nos. ZL84,ZL600 ♂, ZL85, 599♀ (IZCAS); 7♂, China, N. Yunnan, Diqing Tibetan Aut. Pref., 5 km SE of Deqin, elev. 3356 m, 16–17.vi.2009, leg. B. Benedek, Pekarsky slide no. OP 0495♂ (coll. O. Pekarsky); 1♀, China, W. Sichuan, road Yajiang/Litang, elev. 3600 m, 29°59′ N / 100°52′ E, 15.vii.2009, leg. I. & A. Floriani, Saldaitis, Pekarsky slide no. OP 0481♀ (coll. O. Pekarsky); 1♂, 1♀, China, W. Sichuan, road Xinngcheng/Zhongdian, 3300 m, 17.vii.2009, 28°46′ N; 99°51′ E, leg. I. & A. Floriani, Saldaitis, Pekarsky slide no. OP 0480♂ (coll. O. Pekarsky); 2♂, China, N. Yunnan, road Xinngcheng/ Zhongdian, 3500 m, 28°32′ N / 99°49′ E, 18.vii.2009, I. & A. Floriani, Saldaitis (coll. O. Pekarsky).

Remarks. Mironov et al. (2008) recorded that specimens standing under this name in the BMNH collection were identical to specimens of *E. delaeveri* Vojnits, but that the holotypes of neither species had at that time been found. Both now have. Examination of these has shown that the BMNH series is not conspecific with the lectotype of *E. ustata* Moore (which we believe was a unique specimen), and are referable to *E. delaeveri*. The lectotype of *E. ustata* lacks left fore and hind wings. De Laever (1960) described and illustrated the purported male genitalia of *E. ustata* on the basis of the wrongly identified specimens from the Khasia Hills in BMNH.

16. *Eupithecia delaeveri* Vojnits, 1976

Eupithecia delaeveri Vojnits, 1976, *Acta zool. Acad. Sci. hung.* 22 (1–2): 201, figs 1a, 2c, 3c. Holotype ♂ (TTM; examined), [China]: North Yunnan, A-tun-tse [Dêqên], Talsohle, 3000 m.

Eupithecia supersophia Inoue, 1987, *Bull. Fac. domest. Sci. Otsuma Wom. Univ.* 23: 243, figs 63h, 66c & d. Holotype ♀ (BMNH; examined), [Nepal (north-east)]: Taplejung, Walungchung Gola, 3350 m (synonymised in Inoue, 2000).

Description. A handsome species with a sharply clear fore wing pattern. Fore wing basal and medial areas white with large rusty brown costal spots; transverse lines usually distinct, indented or wavy, narrow, brown; postmedial twice right angled onto costa, broadly shaded from within by brown or rusty brown; terminal area rusty brown with a large, white costal spot near apex, and a dentate, narrow, white subterminal line, usually forming a white spot near apex; discal dot absent. Hind wing white, with pale, brownish wavy transverse lines; basal area brown along anal margin; discal dot usually absent. Head, thorax and patagia snowy white. Abdomen white with two broad, rusty brown rings and a rusty brown tip to last segment.
Male genitalia. Uncus membranous, spade-shaped, relatively short and broad. Valve unmodified. Vinculum short, rounded, with shallow medial hollow. Papillae large, cone-shaped. Aedeagus narrow, shorter than length of valve, with two latero-apical lobes, covered internally by dense setae. Vesica armed with two small, hooked cornuti near ductus ejaculatorius base. Sternite A8 elongate, slightly broadened at base, with two parallel arms with slightly expanded and rounded apices.
Female genitalia. Bursa relatively small, ovate, completely covered with dense spines. Ductus bursae membranous, short, tapered. Ductus seminalis broadly attached to ductus bursae on right. Colliculum short and narrow, offset to left. Antrum membranous, very long and narrow. Tergite A8 large, almost quadrate with convex lateral margins. Apophyses very long and stout, slightly expanded at apices. Papillae anales heavily sclerotized, elongate, narrow and tapered.
Range. China (Tibet, Yunnan, Sichuan); Pakistan/India (Jammu & Kashmir), Nepal.
Phenology. Bivoltine or polyvoltine (?): mid-May to early October.
Altitude. From about 2600 to 4000 m in the Himalayas (China: Vojnits 1976; Nepal: Inoue 2000).
Similar species. Very similar externally to *E. likiangi*, sharing with it a bold pattern, in particular with two prominent dark brown blotches in the medial area of the costa. *E. likiangi* is usually smaller, and the terminal area of the fore wing a more even brown, whereas some paler markings

are usually present in *E. delaeveri*, particularly a doubly angled postmedial. It is also broadly similar to *E. ustata* and *E. jezonica*. It can be distinguished from both these species by the larger size, the brighter colour of the fore wings generally, with more angulate transverse lines, and the much smaller areas of matt brown free of white scales. In the male genitalia *E. delaeveri* has a broader valve than any of the other species, and a characteristic shape to the 8th sternite, with rather broad arms bowed inwards in the middle. The cone-shaped papillae on the anterior arms of the labides are also distinct. The female genitalia are closest to those of *E. likiangi*, but can be distinguished by the different disposition of spines in the corpus bursae.

Material. 1♂, Nord Yünnan, A-tun-tse [Dêqên], Talsohle *ca.* 3000 m 26.v.1937, H. Höne (holotype, TTM); 44♂, 4♀, same loc. and alt., 22, 24, 25, 26, 29, 30, 31.v., 1, 2, 3, 4, 5, 13, 15.vi., 2, 3.vii.1937, H. Höne, de Laever slides nos. 66♂, 67♀ (paratypes, ZFMK; 3♂ and 1♀ without abdomens); 8♂, 1♀, same loc. and alt., 18, 19, 22, 29, 31.v., 1, 4, 16.vi.1937, H. Höne, de Laever slide no. 429♂, Vojnits slide no. 10414 (paratypes, TTM); 2♂, same loc., Mittlere Höhe (*ca.* 4000 m), 12, 18.viii.1936, H. Höne (paratypes, ZFMK); 2♀, same loc. and alt., 18.viii.1936, H. Höne, Vojnits slide no. 10413♀ (paratypes, TTM); 1♂, 1♀, same loc., Obere Höhe (*ca.* 4500 m), 14, 28.vii.1937, H. Höne (paratypes, ZFMK); 1♀, same loc. and alt., 3.viii.1936, H. Höne (paratype, TTM); 6♂, 11♀, Tibet [Sichuan], Batang, Untere Urwaldzone (*ca.* 3800 m), 1, 2, 3, 4, 5, 6, 7, 9, 11, 13.viii., 12.ix.1936, H. Höne (paratypes, ZFMK); 2♂, 7♀, same loc. and alt., 1, 6, 7, 9, 10, 13, 16.viii.1936, H. Höne, De Laever slides nos. 53♂, 54♀, Vojnits slide no. 10411 (paratypes, TTM); 1♂, 3♀, same loc. and alt., 2, 5, 12, 14.viii.1936, H. Höne, de Laever slide no. 201♀ (ZFMK); ♀, N.E. Nepal, Taplejung: Walungchung Gola, alt. 3350 m, 2.vii.1962, T. Yasuda, BM Geom. slide no. 16525 (holotype of *E. supersophia*, coll. H. Inoue in BMNH); 1♂, China, N. Yunnan, Diqing Tibetan Aut. Pref., 8 km NNE of Shangri La at Nairi village, elev. 3300 m, 14.vi.2009 leg. B. Benedek (coll. O. Pekarsky); 2♂, 2♀, China, North Yunnan, Diqing Tibetan Aut. Pref., 5 km SE of Deqin, 3356 m, 19.vi.2009, leg. B. Benedek (coll. O. Pekarsky); 1♀, China, W. Sichuan, Kangding, 3000 m, 29°53′ N / 101°55′ E, 13.vii.2009, leg. I. & A. Floriani, Saldaitis, Pekarsky slide no. OP 0483♀ (coll. O. Pekarsky); 1♂, China, W. Sichuan, Xindugiao, evn 3500 m, 30°04′ N / 101°26′ E, 14.vii.2009, leg. I. & A. Floriani, Saldaitis (coll. O. Pekarsky); 1♂, China, W. Sichuan, Litang, 4000 m, 29°50′ N / 100°21′ E, 16.vii.2009, leg. I. & A. Floriani, Saldaitis, Pekarsky slide no. OP 0482♂ (coll. O. Pekarsky). 31♂, 18♀, Tibet, Jomda, 3400 m, 9.vii. and 22–31.vii.1976, Han Yinheng and Zhang Xuezhong, slides nos. ZL22, 63, 338, 345♂, 594♀;

1♂, 9.vii.1976, 3♀, 16–27.viii.1976, Tibet, Zhag'yab Gyitang, 3600 m, Han Yingheng, slide no. 48♀; 1♀, Tibet, Qamdo, 23.v.1976, Zhang Xuezhong; 1♀, Tibet, Markam, Rongxi, 4000 m, 21.vii.1976, 1♀, Tibet, Riwoqê, Sangdo, 29.viii.1976, Zhang Xuezhong; 2♀, Sichuan, Yajiang, 3300 m, 24.viii.1982, Wang Shuyong and Zhang Xuezhong, slide no. ZL475♀; 1♀, Sichuan, Batang, Yidun, 3370 m, 17.viii.1982, Wang Shuyong (IZCAS).
Remarks. The holotype specimen of *E. delaeveri* remains undissected.

17. *Eupithecia likiangi likiangi* Vojnits, 1976

Eupithecia likiangi Vojnits, 1976, *Acta zool. Acad. Sci. hung.* **22** (1–2): 203, figs 1e, 2d, 3b. Holotype ♂ (ZFMK; examined), China: North Yunnan, Li-kiang.

Description. Fore wing with four broad blackish brown costal blotches divided by narrow white bands; basal area brown, with white blotch; antemedial broken three times, obtuse angled onto costa; medial and postmedial sinuate, postmedial twice angled near costa; medial area with white, brown and rusty brown blotches; terminal area broad, rusty brown, but dark brown in apical half and near tornus; white, dentate subterminal line very distinct, broadened in apical half; discal dot relatively small, rounded or comma-like, blackish brown. Hind wing ground colour white; transverse lines dark brown, oblique; basal and postmedial broad, two other lines narrow; terminal area rusty brown, with distinct wavy inner border and broad wavy white subterminal line; discal dot small, indistinct, rounded.
Male genitalia. Uncus membranous, tapered. Valve unmodified. Vinculum short, with shallow medial hollow. Papillae short. Aedeagus narrow, slightly shorter than valve length. Vesica armed with one V-shaped, plate-like cornutus near ductus ejaculatorius base. Sternite A8 two narrow, sclerotized rods connected by a short, rather broad basal band.
Female genitalia. Bursa ovoid, membranous, covered with slim spines forming two large lateral patches connected by a transverse patch of shorter spines in the posterior half. Ductus bursae short, membranous, spineless. Ductus seminalis broadened basally, attached to ductus bursae near base of colliculum on right. Colliculum short and narrow. Antrum long, narrow, membranous, with large patch of spinules behind tergite A8. Tergite A8 large, almost quadrate with rounded posterior corners. Apophyses very long, relatively broad, expanded and flattened at apices. Papillae anales elongate and narrow, tapered.
Range. China (Tibet, Yunnan); India (north), Nepal.
Phenology. Univoltine (?): mid-July to early September.

Altitude. Found from about 2200 to 3000 m above sea-level in Tibet (China: colls ZFMK, IZCAS).
Similar species. Externally closely similar to *E. delaeveri*, but usually has a terminal area of a more solid brown, without clear transverse pale lines. In this it resembles *E. ustata*, and *E. jezonica*, but can be distinguished from both by the sharper pattern generally with clear dark blotches on the costa, and by the more distinct dark transverse lines on the hind wing, which make the hind wing generally darker. The male genitalia have a narrower apex to the valve than the other species, and a subtly different shape to the eighth sternite, as illustrated. The female genitalia are of similar dimensions to those of *E. delaeveri*, but have unspined areas at the anterior end of the more ovate corpus bursae. In this they resemble those of *E. bohatschi*, but the lateral spiniferous areas in the corpus bursae are larger than in the latter species.
Material. ♂, China, Provinz Nord-Yuennan, Li-kiang [Lijiang], 7.ix.1935, H. Höne (holotype, ZFMK); 1♂, 2♀, same loc., 2, 3.ix.1934, 9.ix.1935, H. Höne (paratypes, ZFMK); 3♂, same loc., 28.viii., 2, 7.ix.1935, H. Höne (all marked as allotypes, TTM); 1♀, same loc., 11.x.1934, H. Höne, de Laever slide no. 339♀ (ZFMK); 1♀, same loc., 9.ix.1935, H. Höne (SMNK); 1♂, same loc., *ca.* 3000 m, 30.viii.1934, H. Höne, Schmidt slide no. Eu 185♂ (marked as allotype, specimen in TTM; gen. slide in ZFMK); 1♂, Tibet [Sichuan], Batang, Im Tal des Yangtze (*ca.* 2800 m), 26.viii.1936, H. Höne (ZFMK); 1♂, Tibet, Nyingchi, Bayi, 2999 m, 29°39′ N, 94°21′ E, 2.viii.2006, Lang Songyun, IOZ(E)1847009.
Remarks. Schmidt slides nos. Eu 185a♂ and Eu 185b♀ in ZFMK also belong to *E. likiangi*, but no specimens with these genitalia slide numbers could be found in the collection of the Koenig Museum.

17a *Eupithecia likiangi pedisequa* Mironov & Galsworthy ssp. n.

Description. Specimens from Taiwan are very similar to nominate *likiangi*, but have a larger white medial area on the fore wing with no central red-brown blotch.
Male genitalia. Similar to those of nominate *likiangi*, but valve slightly longer, and much broader at apex. Aedeagus with two cornuti instead of one. Sternite A8 broader at base and with converging arms.
Female genitalia. Very similar to those of nominate *likiangi*, but apophyses shorter, and lateral patches of spines in bursa not quite meeting spine field in posterior part of bursa.
Range. China (Taiwan).

Phenology. Univoltine (?): late October to early November.
Altitude. All specimens found at altitude about 2300 m above sea-level (Taiwan: coll. MNHU).
Similar species. Not likely to be confused with any other species known from Taiwan.
Material. ♀, Taiwan, Nantou, Meifeng, *ca.* 2300 m, 29.x.-3.xi.1998, leg. Mey & Speidel, Zool. Mus. Berlin (holotype, MNHU); 1♂, 1♀, same data, Mironov slides nos. 815♂, 812♀ (paratypes, MNHU).
Remarks. The nearest known population of *E. likiangi* is in the mountains of Yunnan in south west China. The Taiwan population has evidently been isolated for a very long time, but even so the morphological changes do not in our opinion yet merit specific separation.

18. *Eupithecia sinicaria* Leech, 1897

Eupithecia sinicaria Leech, 1897, *Ann. Mag. nat. Hist.* (6) **20**: 68. Holotype ♂ (BMNH; examined), China (western): Che-tou.
Eupithecia fulcrata Prout, 1958, *Bull. Br. Mus. nat. Hist.* (Ent.) **6** (12): 393. Holotype ♂ (BMNH; examined), Burma (north-western): Htawgaw, 6000 ft.
Eupithecia eszterkae Vojnits, 1976, *Acta zool. Acad. Sci. hung.* **22** (1–2): 198, figs 1d, 2a. Holotype ♂ (TTM; examined), China: Sued-Shensi, Tapaishan in Tsinling. **New junior synonym.**
Eupithecia yunnani Vojnits, 1976, *Acta zool. Acad. Sci. hung.* **22** (1–2): 204, figs 1b, 2e, 3a. Holotype ♂ (TTM; examined), [China]: North Yunnan, A-tun-tse [Dêqên], 4000 m. **New junior synonym**

Misidentification: *Eupithecia subplacida* Vojnits, 1984 (1♂ paratype in ZFMK).

Description. A comparatively large, narrow winged species. Costal margin brown interrupted by white; white basal and leading part of medial areas with a series of narrow brown transverse lines; a broad terminal area and the rearmost part of medial area light reddish brown; a small but distinct ovoid black discal dot on the fore wing. Hind wing paler, light brownish white with some wavy brownish transverse lines.
Male genitalia. Uncus membranous, broadly spade-shaped. Valve unmodified. Vinculum short and narrow, trapezoid, with shallow medial hollow. Papillae elongate and narrow. Aedeagus narrow, shorter than length of valve. Vesica armed with one small and a second even smaller, irregular, weakly sclerotized, plate-like cornutus near ductus ejaculatorius base. Sternite A8 consisting of two narrow sinuate arms, more strongly sclerotized and pointed at apices, connected by a narrow basal band.
Female genitalia. Bursa small, globular or egg-shaped, completely covered with small spines except small area at base. Ductus bursae short, membranous, tapered to colliculum. Ductus seminalis short and narrow,

slightly broadened at base, attached to ductus bursae on right. Colliculum narrow and slightly elongate. Antrum membranous, elongate and narrow. Tergite A8 elongate, rectangular or spade-shaped. Apophyses long and stout, slightly expanded at apices. Papillae anales narrow and elongate, with rounded apices.

Range. China (Tibet, Qinghai, Gansu, Shaanxi, Sichuan, Yunnan); Northern Myanmar.

Phenology. Bivoltine (?): April to late August.

Altitude. From about 2800 to 4500 m above sea-level in Tibet and Yunnan (China: coll. ZFMK).

Similar species. Similar in general maculation to *E. ustata*, but larger with white colour dominating in basal area and an elongate discal dot on the fore wing; also with a darker hind wing with more evenly curved wavy transverse lines, which are lighter in their basal parts than in *E. ustata*. The male genitalia are distinguished by the absence of basal lobes in the eighth sternite and the much broader uncus than in *E. ustata*. The female genitalia of both species are similar, but in *E. sinicaria* tergite A8 is longer, and the apophyses longer and stouter than in *E. ustata*.

Material. ♂, Che-tou, 11.070ft., Native coll., July & Aug. 1890, *E. sinicaria* sp. n. Type ♂, Leech Coll. 1900–64, BM Geom. slide no. 3523 (holotype, BMNH); ♂, Htawgaw, N.E. Burma, 6000ft., A.E. Swann, iv-v.(19)23, *Eupithecia fulcrata* Prout, ♂ type, abdomen guaranteed, L.B.P., L.B. Prout Coll., B.M. 1939–643, Vojnits slide no. 19314♂, BM Geom. slide no. 20138 (holotype of *E. fulcrata*, BMNH); 1♂, Sued-Shensi [Shaanxi], Tapaishan [Taibaishan] im Tsinling [Qinling], L, 29.vii.1935, H. Höne, Vojnits slide no. 10408♂ (paratype of *E. eszterkae*, TTM); 1♂, Nord Yünnan, A-tun-tse [Dêqên], Obere Höhe (*ca.* 4500 m), 14.viii.1936, H. Höne, de Laever slide no. 421♂ in ZFMK (holotype of *E. yunnani*, specimen in TTM); 1♂, same loc. and alt., 15.viii.1936, H. Höne (paratype of *E. yunnani*, ZFMK); 3♂, same loc. and alt., 29.vii., 9.viii.1936, 13.vii.1937, H. Höne, Vojnits slide no. 10419♂, other specimens without abdomens (paratypes of *E. yunnani*, TTM); 3♂, same loc. and alt., 8.vii., 23.viii.1936, 29.vii.1937, H. Höne, de Laever slides nos. 76♂, 83♂ (ZFMK); 1♂, same loc., Mittlere Höhe (*ca.* 4000 m), 17.vii.1936, H. Höne, Schmidt slide no. Eu 183c♂ in ZFMK (paratype of *E. yunnani* in TTM); 1♂, same loc., Talsohle (*ca.* 3500 m), 25.viii.1936, H. Höne (paratype of *E. yunnani*, ZFMK); 1♂, same loc., Talsohle *ca.* 3000 m, 4.vi.1937, H. Höne, Vojnits slide no. 13309♂ (paratype of *E. subplacida*, wrongly labelled gen. slide of *E. refertissima* Vojnits, ZFMK); 1♂, China, Provinz Nord-Yuennan, Li-kiang [Lijiang], 22.vii.1935, H. Höne (paratype of *E. yunnani*, ZFMK); 2♂, 2♀, same loc., 22, 31.vii., 1, 25.viii.1935, H. Höne, Vojnits slides nos. 10418♀,

10423♂, 10424♀ in TTM, Schmidt slide no. 183a♂ in ZFMK (paratypes of *E. yunnani*, specimens in TTM); 1♂, same loc., 1.vii.1935, H. Höne, de Laever slide no. 376♂ (ZFMK); 1♀, Tibet, Ta-Ho, Chasseurs indigènes, Printemps 1895, Vojnits slide no. 839♀ (TTM); 1♂, Tibet [Sichuan], Batang, Im Tal des Yangtze (*ca.* 2800 m), 10.v.1936, H. Höne, Vojnits slide no. 12421♂ (ZFMK); 1♂, [Sichuan] Chasseurs Thibétains de Ta-tsien-loû, Eté 1896, reçu du R.P. Déjean, Vojnits slide no. 837♂ (TTM); 1♂, [China, Sichuan], Ta-tsien-Loû, 1897, ex. R. P. Déjean, Vojnits slide no. 838♂ (TTM); 1♀, China, Tibet, Jomda military post, 3400 m, 26.vii.1976, leg. Han Yienheng, Galsworthy slide no. ZL237♀ (IZCAS); 4♂, 13♀, China, North Yunnan, Diqing Tibetan Aut. Pref., 5 km SE of Deqin, 3356 m, 16–17.vi.2009, leg. Balázs Benedek, Pekarsky slides nos. OP 0465♀, OP 0469♀, OP 0470♂, OP 0471♂, OP 0472♀, OP 0539♂, OP 0846♂, OP 0848♀ (coll. O. Pekarsky); 3♀, same loc. and alt., 19.vi.2009, leg. B. Benedek (coll. O. Pekarsky); 2♂, 2♀, China, W. Sichuan, road Yajiang/Litang, 3600 m, 15.vii.2009, 29°59′ N / 100°52′ E, I. & A. Floriani, Saldaitis leg., Pekarsky slides nos. OP 0539♂, OP 0696♂ (coll. O. Pekarsky); 1♀, China, N. Yunnan, road Xinngcheng/Zhongdian, 3500 m, 18.vii.2009, 28°32′ N / 99°49′ E, I. & A. Floriani, Saldaitis leg., Pekarsky slide no. OP 0541♀ (coll. O. Pekarsky). 2♀, Tibet, Markam, Rongxi, 4000 m, 23.vii.1976, slide no. ZL526♀, and Jomda, 3400 m, 21.vii.1976, Zhang Xuezhong slide no. ZL604♀; 1♀, Jomda, 3400 m, 26.vii.1976, Han Yinheng, slide no. ZL237♀; 1♂, Tibet, Nyingqi, Bayi, 3040 m, 2–3.viii.2002, Han Hongxiang, slide no. ZL640♂; 1♂, Sichuan, Yajiang, 3300 m, 24.viii.1982, Wang Shuyong, slide no. ZL36♂; 1♂, Qinghai, Huzhu north mountain, 2300 m, 30.vii.1991, slide no. ZL491♂; 1♀, same loc., 2600 m, 36°55′ N, 102°23′ E, 2.viii.2006, 1OZ(E)1847077, both Xue Dayong; 1♀, Gansu, Yongdeng, Tulugou, 2280 m, 25–29.vii.1991; 1♀, same loc.and alt., 36°41′ N, 102°42′ E, 8.viii.2005, IOZ(E)1847037, both Xue Dayong (IZCAS).
Remarks. The photos of the aedeagi of *E. sinicaria* and *E. consortaria* were reversed in de Laever's publication (1960, figs. 44 and 57).

19. *Eupithecia ruficorpus* Warren, 1897

Tephroclystia ruficorpus Warren, 1897, *Novit. zool.* 4: 230. Holotype ♀ (BMNH; examined), [India]: Khasi Hills.

Description. A small species. Fore wing ground colour pale grey, sometimes with reddish tinge, especially in posterior half, and with a basal, two medial and a larger apical dark costal blotch; terminal area with two similar medial and tornal dark blotches or sometimes entirely dark; discal dot small, rounded. Hind wing slightly paler, pale grey, with broad light transverse band between medial and terminal areas, darker terminal area and very small, rounded or slightly elongate discal dot. First two and eighth abdominal segments pale grey or pale brownish grey; all other segments (A3-A7) deep pink with distinct blackish brown dorsal spots on each segment.

Male genitalia. Uncus membranous, trowel-shaped, elongate and narrow, slightly tapered, with thickened lateral margins. Valve rather short and narrow, unmodified. Vinculum relatively short and narrow with deep medial hollow. Papillae elongate, narrow, curved. Aedeagus narrow, slightly shorter than length of valve, with one apical lobe covered by long setae. Vesica covered with numerous denticules, and armed with one elongate, folded, weakly sclerotized, V-shaped cornutus. Sternite A8 with two narrow arms, pointed at apices and broadened at bases, connected by a short and very narrow basal band.

Female genitalia. Bursa large, ovoid, membranous, with two ovoid lateral patches of sparse spines and a row of slim spines between base of ductus seminalis and colliculum. Ductus bursae not distinguishable. Ductus seminalis slightly broadened at base, curved, attached to posterior part of corpus bursae near base of colliculum on right. Colliculum short and narrow, slightly inclined to left. Antrum membranous, relatively short and narrow. Tergite A8 small, almost quadrate, slightly tapered posteriorly, with convex anterior margin and rounded posterior corners. Apophyses relatively short and narrow, not expanded at apices. Papillae anales elongate and narrow, tapered.

Range. China (Tibet); Nepal, India (Assam).

Phenology. Very little information. October in northern India, July in Tibet.

Altitude. No information other than Tibetan records from 2400 to 2800 m.

Similar species. *E. ruficorpus* is a distinctive species: it has some external similarity to the Nepalese species *E. seditiosa* Vojnits, 1981, but has deep pink abdominal segments (A3-A7). The male and female genitalia of the two species are entirely different.

Material. ♀, [India], Khasis, Oct. 1896, Nat. Coll., *Tephroclystia ruficorpus* Warr., Type ♀(holotype, BMNH); 2♂, 1♀, Tibet, Gyirong, 2400 m, 22.vii.1975, Huang Fusheng, slides nos. ZL218, 607♂; 1♀, Gyirong, 2800 m, 27.vii.1975, slide no. ZL605♀, Wang Ziqing (IZCAS).

20. *Eupithecia szelenyica* Vojnits & de Laever, 1974

Eupithecia szelenyica Vojnits & de Laever, 1974, *Folia ent. hung.* 27 (1): 229, fig. 1. Holotype ♀ (ZFMK; examined), China: North Yunnan, Li-kiang.

Eupithecia violacea Vojnits, 1981, *Acta zool. Acad. Sci. hung.* 27 (1–2): 220, figs 3, 4. Holotype ♂ (ZSM; examined), Nepal: Prov. Nr. 3 East Khumjung, 3800 m. **New junior synonym.**

Description. A small elegant species with pale ash-grey ground colour. Fore wing with slightly arched costal margin, evenly curved, convex terminal margin, rounded tornus and narrow apex; costal margin darker, fuscous grey; transverse lines inconspicuous; two orange blotches, each bordered by dark grey, are placed over inconspicuous postmedial line, one between R_5 and M_1, and the other between M_3 and CuA_2; terminal area slightly darker in the middle and near tornus; discal dot minute, pale, rounded. Hind wing slightly paler with indistinct transverse lines and minute discal dot.
Male genitalia. Uncus membranous, short and broad, tapered to bifid apex. Valve rather broad, strongly broadened medially, unmodified. Vinculum short and broad, without central indentation. Papillae short and broad. Aedeagus very narrow, broadened at anterior end, with two lateral hirsute lobes at apex, length shorter than valve. Vesica armed with one long and narrow plate-like cornutus near ductus ejaculatorius base and a second smaller, hook-like cornutus near anterior end of aedeagus. Sternite A8 short and broad, broadened at base, with two very narrow arms connected by a very narrow band at base.
Female genitalia. Bursa relatively small, ovate or ovate-oblong, membranous, almost completely covered with spines except for base and left side beneath base of ductus seminalis. Ductus bursae short, tapered to colliculum. Ductus seminalis slightly broadened basally, attached to posterior part of corpus bursae. Colliculum short and narrow. Antrum elongate, narrow, membranous. Tergite A8 small, more or less quadrate, with sinuate and heavily sclerotized anterior margin, sclerotized lateral margins and slightly convex posterior margin. Apophyses narrow, the anterior slightly expanded and flattened at apices. Papillae anales fusiform, elongate and narrow.
Range. China (Tibet, Shaanxi, Sichuan, Yunnan); Nepal.
Phenology. Univoltine: early July to mid-August.
Altitude. From about 1700 to 3900 m in Qinling Mts. and Himalayas (China: Vojnits 1981a; coll. ZFMK, SMNK).
Similar species. Unlikely to be confused with any other Chinese *Eupithecia* species, the two orange blotches on the postmedial of each fore wing being unique. In the male, the double-pointed uncus, and broad vinculum without central indentation are useful characters. The female genitalia have no very clear distinguishing features, but the relatively short apophyses and

antrum are useful pointers, together with the disposition of spines in the corpus bursae, which is very uniform between specimens.

Material. ♀, China, Provinz Nord-Yuennan, Li-kiang [Lijiang], O, 1.vii.1935, H. Höne (holotype, ZFMK); 1♀, same loc., 11.viii.1935, H. Höne, Vojnits slide no. 139♀ (paratype, TTM); 1♂, same loc., 5.vii.1934, H. Höne, de Laever slide no. 233♂ (ZFMK); ♂, Nepal, Prov. Nr. 3, East Khumjung, 3800 m, 12.vii.1964, leg. W. Dierl, Vojnits slide no. 11774♂ (holotype of *E. violacea*, ZSM); 1♀, Nepal, Dudh Kosi Tal, 3500 m, 22.-23.vii.1962, leg. G. Ebert u. H. Falkner, Staatsslg. München, Vojnits slide no. 11772♀ (paratype of *E. violacea*, TTM); 1♀, Nepal, Khumbu, Khumdzung, 3900 m, 11.vii.1962, leg. G. Ebert u. H. Falkner, Staatsslg. München, Vojnits slide no. 10652♀ (paratype of *E. violacea*, TTM); 1♂, 1♀, Sued-Shensi [Shaanxi], Tapaishan [Taibaishan] im Tsinling [Qinling], *ca.* 1700 m, 8, 9.viii.1936, H. Höne, Vojnits slide no. 12276♂ (gen. slide labeled by Vojnits as *E. invenusta* Vojnits), Mironov slide no. 645♀ (ZFMK); 1♂, same loc. and alt., 19.viii.1936, H. Höne, Mironov slide no. 796♂ (SMNK); 1♂, 1♀, China, Sichuan, Daxue Shan Mts., Gongga Shan, W Moxi, 29°34′ N, 101°59′ E, ~3200 m, 20.-21.vii.1999, leg. V. Siniaev & A. Plutenko (ZFMK); 1♀, China, Yunnan Prov., Lijiang, 2650 m, 18.vii.2001, Li Houhun leg. (coll. P. Skou); 1♀, China, N. Yunnan, road Zhongdian/Lijiang, 3000 m, 19.vii.2009, 27°28′ N / 99°53′ E, I. & A. Floriani, Saldaitis leg. (coll. O. Pekarsky); 1♂, 3♀, Tibet, Gyirong, 3300 m, 2–4.viii.1975, coll Wang Ziqing and Huang Fusheng, slides nos. ZL229♂, 228, 428, 523♀ (IZCAS).

The *kamburonga* species group

Mironov & Galsworthy (2009a, p.108) introduced this species group to cover a number of species from the Far East and south east Asia. The group is characterized by the shape of the uncus in the male genitalia, which is divided into two lobes in a dorso-ventral plane; this feature is not known to be shared with any other member of the genus. The relationship between the members of the group is also supported by the similarities in shape of the eighth sternite, peg-like with two apical horns, and the presence of two strong cornuti of varying shapes in the aedeagus vesica, as well as by the general pattern of the female genitalia. The latter have a strongly reduced corpus bursae, a diverticulum of various shapes on the ductus bursae, and a broad colliculum. Externally members of the group have a tendency for the most part to strong contrast between dark and pale areas of the wings.

Apart from the four species included below, one species from Thailand *E. schnitzleri* Mironov & Galsworthy, 2009 and one from Borneo *E. kamburonga*

Holloway, 1976 belong to this species group. The wider relationships of this Asian group of *Eupithecia* are unclear.

21. *Eupithecia liberata* Inoue, 2000

Eupithecia liberata Inoue, 2000, *Tinea* **16** (Suppl. 1): 35, pl. 165, fig. 24; fig. 1301 [*nec* 1342]. Holotype ♀ (NSMTO; examined), [Nepal]: [Kathmandu], Godavari, 1600 m.

Misidentification: *Eupithecia perendina* Vojnits, 1980 (1♂ paratype in ZFMK).

Description. A fairly small moth with a striking brown and white pattern, the white occupying almost all the hind wing and the medial part of the fore wing, but only extending narrowly to the costa. Discal dot large, relatively broad and elongate, intensely black. Fringes chequered off-white and light brown. Abdomen whitish brown with black lateral stripes; second segment black.

Male genitalia. Uncus as described for species group. Valve elongate and relatively narrow, with a small prominence near the apex. Vinculum short and centrally pointed. Papillae very short and broad. Aedeagus short and narrow, much shorter than valve. Vesica armed with one narrow, twisted horn-like cornutus almost equal to aedeagus length, one small elongate plate and a smaller irregular plate near ductus ejaculatorius base. Sternite A8 elongate, peg-like with two short blunt apical rods.

Female genitalia. Bursa very small, membranous, ovoid, covered with dense spinules in the right half and a row of the same spines along left side from base to colliculum; two small membranous, thick-walled diverticula on bursa, one basal and tapered and the other broader and blunt on left. Ductus bursae short, with longitudinal striations. Ductus seminalis long, medially broad; attached to ductus bursae near colliculum on right. Colliculum elongate and relatively broad. Antrum short and broad, membranous. Tergite A8 large, transverse. Apophyses long and broad, slightly expanded at apices, the anterior most markedly. Papillae anales large, elongate and pointed.

Range. China (Sichuan, Yunnan); Nepal.

Phenology. Bivoltine (?): late March to early October.

Altitude. From about 1600 to 3600 m (Nepal: Inoue 2000; China: coll. O. Pekarsky).

Similar species. Externally similar to the Taiwanese *E. yoshimotoi* Inoue, 1988, but usually smaller; fore wing a little more slender (Inoue 2000) with less dark basal and terminal areas without reddish tinge, and also with darker medial area. The genitalia are distinguished from those of *E. yoshimotoi* by the shape of the valve with a straight anterior margin, the pointed tip of the vinculum, the much shorter papillae on the anterior arms of the labides, the different cornuti on the vesica and the slightly longer apical arms of the eighth sternite in the male, and by the broader and shorter eighth tergite in the female.

Material. ♀, Nepal, Kathmandu, Godavari, 1600 m, 29.iii.1992, M.S. Limbu leg., BM Geom. slide no. 21970♀ (holotype, coll. H. Inoue in BMNH); 1♀, Nepal, 20 km SSE Katmandu, Route du Phulchoki, 2400 m, 3.x.1983, C. Herbulot (ZSM); 1♂, China, Provinz Nord-Yuennan, Li-kiang [Lijiang], 26.iv.1935, H. Höne, Vojnits slide no. 12239♂ (ZFMK); 2♂, Nord-Yünnan, A-tun-tse [Dêqên], Talsohle *c.* 3000 m, 25.v.1937, H. Höne, de Laever slide no. 88♂, Vojnits slide no. 12502♂ (paratype of *E. perendina*, ZFMK); 1♂, same loc. and alt., 4.vi.1937, H. Höne, Vojnits slide no. 12995♂ (TTM); 8♀, China, Sichuan, Daxue Shan, Gongga Shan, NW Moxi, 29°41′ E, 101°58′ N, 2600–3200 m, 23.iv.-15.v.2001, leg. V. Siniaev & A. Plutenko, *ex* coll. Dr. A. Schintlmeister (ZSM); 1♀, China, Yunnan Prov., Lijiang, 2650 m, 17.vii.2001, Li Houhun leg. (coll. P. Skou); 2♀, China, W. Sichuan, road Yajiang/Litang, 3600 m, 15.vii.2009, 29°59′ N / 100°52′ E, I. & A. Floriani, Saldaitis leg. (coll. O. Pekarsky), 2♀, China, W. Sichuan, road Xinngcheng/Zhongdian, 3300 m, 17.vii.2009, 28°46′ N / 99°51′ E, I. & A. Floriani, Saldaitis leg., Pekarsky slides nos. OP 0496♀, OP 0501♀ (coll. O. Pekarsky).

Remarks. Mironov & Galsworthy (2010a) illustrated the true genitalia of *E. liberata*, based on the holotype, and recorded that the female genitalia published by Inoue (2000; p. 59, fig. 1342) came from one of the paratypes, which was not conspecific with the holotype female, which Inoue had not dissected. Not previously recorded from China.

22. *Eupithecia yoshimotoi* Inoue, 1988

Eupithecia yoshimotoi Inoue, 1988, *Bull. Fac. domest. Sci. Otsuma Wom. Univ.* **24**: 340, figs 12a-b. Holotype ♀ (BMNH; examined), [Taiwan]: Hualien Hsien, Tayuling, 2600 m.

Misidentification: *Eupithecia costimacularia* Chang, 1990 (*nec* Leech, 1897).

Description. A handsome species with broad brown or reddish brown basal and terminal areas and a strongly contrasting white medial area on the fore wing. Fore wing slightly elongate; medial white fascia sinuate, touching discal dot and sharply angled onto costa; fainter white postmedial twice angled onto costa; terminal area with fine, indented, white subterminal line and medial and tornal white spots; discal dot black. Hind wing almost entirely white, with three fine brown wavy transverse lines; discal dot small but distinct.

Male genitalia. Uncus as described for species group. Valve unmodified. Vinculum short and broad. Papillae long and very narrow. Aedeagus narrow, considerably shorter than length of valve. Vesica armed with two horn-like cornuti curved basally towards opposite sides, one blunter, shorter and more weakly sclerotized, and the other long, sinuous, and broadened at the middle and base; also an elongate, longitudinally folded plate and a smaller, irregular plate near ductus ejaculatorius base. Sternite A8 peg-like, narrow, broadened at base, with two small, more strongly sclerotized, diverging apical horns.

Female genitalia. Bursa very small, sclerotized, covered with very small, dense spines in the right half and with a row of the same spines along left side; a cone-shaped, lightly sclerotized, more or less elongate diverticulum at base. Ductus bursae short, sclerotized, with longitudinal striations, a short row of small spines on the left side and a short, broad and weakly sclerotized diverticulum on the left side. Ductus seminalis long and broad, curved, attached to ductus bursae on right. Colliculum elongate and relatively broad. Antrum short and broad, membranous. Tergite A8 large, almost square. Apophyses relatively long and broad, slightly expanded at apices. Papillae anales heavily sclerotized, elongate and pointed at apices.

Range. China (Taiwan), N Thailand.

Phenology. Univoltine (?): late March.

Altitude. From about 1550 to 3200 m in Daxue Shan and Taiwan (Inoue 1988; coll. ZSM).

Similar species. Generally similar to preceding species, but easily distinguished by the darker basal and terminal areas and whitish medial area on the fore wing, as well as by the lighter, whitish hind wings with distinct, narrow transverse lines. Male genitalia distinguished by the longer valve with slightly upcurved apex, the rounded vinculum, much longer papillae on the anterior arms of the labides, and the presence of two horn-like cornuti of different shape on the vesica (one in *E. liberata*). The female genitalia of both species are very similar, but tergite A8 is longer and narrower in *E. yoshimotoi*.

Material. ♀, Taiwan, Ta-yu-ling, Hualien Hsien, 2600 m, 28–31.iii.1981, leg. H. Yoshimoto, BM Geom. slide no. 16544 (holotype, coll. H. Inoue in BMNH); 3♀, data as holotype, Inoue slide no. 12160; 1♀, Taiwan, Alishan, 2200 m, Chia Yi hsien, 15–16.iii.1981, leg. H. Yoshimoto; 1♀, Rantaizan, Formosa, May 1909, BM Geometrid slide no. 20954; 1♂, 2♀, Taiwan, Suyuanyako, 2800 m, Ilan Hsien, 31.viii.1999, H. Y. Wang, Inoue slide no. 16935(♀) (all BMNH); 1♀, Taiwan, Prov. Ilan, Suyuan near Pinan, at the Road 7/1, 1,550 m, 29.iii.1997, leg. Csorba & Ronkay (TTM); 2♂, 4♀, Taiwan,

Prov. Nantou, Hohuachi, 1950 m, 24°13′ N, 121°16′ E, 31.iii.2000, leg. A. Kun & L. Peregovits (ZSM).

23. *Eupithecia buysseata* Mironov & Galsworthy, 2011

Eupithecia buysseata Mironov & Galsworthy, 2011, *Lep. Science* **62** (1): 22, figs 8, 31. Holotype ♂ (ZFMK; examined), Nord-Yünnan, A-tun-tse [Dêqên], 3000 m.

Description. Compared with other species in the group, a very uniformly brown moth, with paler patches only outside the forewing discal dot, which is black and elongate.

Male genitalia. Uncus as described for species group, but with pointed horns rather than lobes. Valve unmodified. Vinculum short and narrow, slightly pointed at apex. Papillae elongate and narrow. Aedeagus relatively short, shorter than length of valve. Vesica armed with two scalpel-shaped cornuti, one close to apex short, curved at base with longitudinal striations and the other long and shaped like a cutlass, and two twisted, plate-like cornuti near ductus ejaculatorius base. Sternite A8 peg-like with heavily sclerotized lateral margins in apical half, and two short, asymmetrical apical horns.

Female unknown.

Range. China (Yunnan).

Altitude. Unique specimen from 3000 m above sea-level in North Yunnan (China: coll. ZFMK).

Similar species. *E. buysseata* is easily distinguished externally from *E. yoshimotoi* and *E. liberata* by the unicolorous pale brown wings with inconspicuous transverse lines and pale blotches only near discal dots on the fore wings. The male genitalia are closest to those of *E. yoshimotoi*, but the valve is shorter, and the apex of the uncus and the larger cutlass-shaped cornutus in the vesica are differently shaped.

Material. ♂, Nord-Yünnan, A-tun-tse [Dêqên], Talsohle *ca.* 3000 m, 30.v.1937, H. Höne, Vojnits slide no. 13141♂ (holotype, ZFMK).

Remarks. The single holotype male is extremely worn, and it is possible that a fresh specimen would show stronger patterning.

24. *Eupithecia kuroshio* Inoue, 1980

Eupithecia kuroshio Inoue, 1980, *Bull. Fac. domest. Sci. Otsuma Wom. Univ.* **16**: 178, fig. 46A; figs 47R, S; fig. 48A; fig. 49F; fig. 53E; fig. 59B. Holotype ♂ (BMNH; examined), [Japan]: Kagawa Pref(ecture), Mt. Zôzu.

Description. A rather brightly coloured species with roundish wings and a strong contast between predominantly ash to dark grey ground colour and numerous pale transverse lines, especially the white medial band on the fore wing, which is sharply angled onto the costa behind the small black rounded discal dot. The hind wing is similarly coloured with some distinct whitish, wavy transverse lines and a small rounded discal dot.

Male genitalia. Uncus as in species group description. Valve large, unmodified. Vinculum short, rounded, with slight emargination at centre. Papillae long, narrow, slightly tapered. Aedeagus narrow, considerably shorter than valve length. Vesica armed with two slim horn-like cornuti, one long, slightly sinuate, curved near apex, the other short, sharply curved, and a smaller narrow, plate-like cornutus near ductus ejaculatorius base. Sternite A8 peg-like, elongate, broadened near base, with two very small close-set horns at apex.

Female genitalia. Bursa copulatrix very small, oval, completely covered with small spines. Ductus bursae elongate and broad with a large patch of small spines connecting with bursa and a prominent membranous diverticulum on left. Ductus seminalis slightly broadened basally, attached to base of ductus bursae on right. Colliculum not present. Antrum slightly sclerotized and covered with numerous pores. Tergite A8 small, almost square. Apophyses rather narrow and short. Papillae anales small, rounded.

Range. China (Taiwan); Japan (Shikoku, Kyushu, Izu Isls., Yakushima, Amami-ôshima).

Phenology. Univoltine: March to May.

Altitude. From about 200 to 1200 m above sea-level (Taiwan: Inoue 1988; coll. TTM).

Similar species. Rather similar externally to *E. lariciata* (Freyer, 1842). The narrow, oblique whitish transverse band on the forewing, sharply angled behind the discal dot, is a useful distinguishing feature. The male genitalia are similar to those of *E. yoshimotoi* but easily distinguished by the shape of the cornuti in the vesica. The female genitalia are similar in general pattern to other species in this group, but the pattern of spining and the absence of a colliculum are distinctive.

Material. ♂, [Japan], Kagawa, Mt. Zôzu, 28.iii.1967, M. Toshima, BM Geom. slide no. 16377♂ (holotype, coll. H. Inoue in BMNH); 1♀, Taiwan, Prov. Ilan, Ming Chyr Forest Recreation Area, 1200 m, 30–31.iii.1997, leg. Csorba & Ronkay (TTM); 1♀, Taiwan, Huali Hsien, Hungyeh Spa, 200 m, 29–30.iii.1984, A. Kawabe; 1♀, Taiwan, Kukuan, Taichung Hsien, 25.iii.1982, H. Yoshimoto (BMNH).

The *brevifasciaria* species group

This small East Asian species group was introduced in Mironov & Galsworthy (2012). Members of the group are broad-winged, often with patterns more resembling representatives of the genus *Horisme* Hübner, [1825] than typical *Eupithecia*. The male genitalia have a stout, uniapical uncus with a broad, flattened basal part, and a transparent bubble-like 'window' free of setae at the base of the valve. They also lack elongate papillae on the anterior arms of the labides, having instead a setose membrane on the lower face of the arms, a characteristic which they share with some, but not all, members of the following group. The calcar of the juxta is usually broad and rounded. In species other than *E. chingana*, sternite A8 is poorly sclerotized, broad, and horseshoe-shaped. The female genitalia are diverse, but usually have a spineless, membranous or thick-walled anterior part, and a sclerotized posterior part with stout, large spines, short apophyses and short and broad papillae anales. Three species occur in China; the Himalayan species *E. subrubescens* Warren, 1888 also belongs to the group, and may well occur in Tibet.

25. *Eupithecia chingana* Wehrli, 1926

Chloroclystis chingana Wehrli, 1926, *in* O. Bang-Haas, *Horae macrolepidopt. Reg. palaearct.* 1: 95, pl. 11, fig. 27. Holotype ♂ (coll. O. Bang-Haas in MNHU; examined), Mongolia: Chingan Mts, Inn Shan, 2000 m.

Horisme sternecki Prout, 1938, *in* Seitz, *The Macrolepidoptera of the World* 4 (Suppl.): 213, pl. 18, fig. g. Holotype ♂ (BMNH; examined), [China]: Pekin, Western Hills (synonymised in Mironov & Galsworthy, 2012).

Unavailable names (incorrect subsequent spelling): *chinganga*: Wehrli (1927).

Description. Fore wing broad, ground colour mid-brown, area between ante- and postmedial, and terminal area darker brown; basal and antemedial lines faint; medial sharply angled close to or touching discal dot; postmedial with several zigzags outside discal dot, and with a prominent pale band outside it in costal half; a pale dentate subterminal line; discal dot prominent, elongate. Hind wing concolorous, with a series of darker, curved transverse lines, and a more prominent double pale postmedial.

Male genitalia. Uncus digitate, uniapical. Valve unusually long, unmodified, basal 'window' large. Vinculum very broad, square. Papillae as in species group description, with a long row of setae on the arm of the labides along their edge. Calcar of juxta broad. Aedeagus about half valve length, vesica covered with numerous minute teeth. Sternite A8 broad, tapering posteriorly, with two rounded apical lobes.

Female genitalia. Bursa large, oval, covered with numerous short spines in posterior half, with a spine-free window edged with stronger spines linked at their bases beneath the ductus bursae. Latter short, tapering, with spines in lower half, and attached at upper side of bursa. Ductus seminalis slightly broadened at base, attached to central part of bursa. Antrum tapered, lightly sclerotized, with a weak colliculum at junction with ductus bursae. Tergite A8 rectangular. Apophyses relatively broad. Papillae anales broad and rounded.
Range. China (Inner Mongolia, Hebei, Beijing).
Phenology. ? Univoltine. May to August.
Altitude. From about 1300 m in Beijing to 2000 m in Chingan Range (Wehrli 1926).
Similar species. Distinguished from most other *Eupithecia* by the complete transverse lines on the hind wings, a feature shared with *E. eurytera*, which often flies with it. The latter has a much more clear cut medial fascia on the fore wing. The transparent 'window' at the base of the valve in the male genitalia is diagnostic for this group, and the shape of the eighth sternite is unique within the group. The female genitalia are distinctive.
Material. ♂, [China: Inner Mongolia], Inn Shan, Chingan mont., Mongolei or., 2000 m Juli, Horae Macrolep. Vol. I. abgebildet, t. 11, f. 27, beschrieb. p. 95, *Chloroclystis chingana* Wrli sp. n., Type, O. Bang-Haas (holotype, coll. O. Bang-Haas in MNHU); 1♂, Western Hills, Pekin [Beijing]. 7.Aug.1911. F.S. Hughes, 1913–174, *Horisme sternecki* Prout, ♂ type, Type, Figured in Seitz. Vol. 4 (holotype of *H. sternecki*, BMNH); 1♀, same data but 18.viii.1911 (allotype, BMNH); 1♂, same data, but 14.viii.1911 (paratype, BMNH); 1♂, 1♀, [China]: Beijing, Sanpu, 18.viii.1964, Liao Subai (BMNH); 12♂, 22♀, China, Dong Ling Shan, Beijing, 1300 m, 22.vi.2001, A.C. Galsworthy (BMNH); 1♂1♀, Beijing, Baihuashan, 31.v.73, Han Yinheng, slide no. L-2138♂; 1♀, same loc., 12.vii.73, slide no. L-2140♀, 1♂, Beijing, Fang Shan, 23.v.72; 1♀, Beijing, Jinshan, 21.vi.90, Zhao Jie; 1♂, 1♀, Beijing, Baihuashan, 14.vii.73 and 29.v.73, Han Yinheng, slides nos. L-2137♂, L-2139♀ (IZCAS).

26. *Eupithecia brevifasciaria* Leech, 1897

Eupithecia brevifasciaria Leech, 1897, *Ann. Mag. nat. Hist.* (6) **20**: 70. Holotype ♂ (BMNH; examined), China (western): Chow-pin-sa.

Description. Fore wing broad, ground colour pale brown, costal area between ante- and postmedial, apical area, and three blotches in terminal area dark brown; antemedial sharply angled before discal dot; medial faint, angled at discal dot; postmedial angled sharply outwards at one third, and sinuous to costa; a wavy white subterminal line; discal dot elongate. Hind wing concolorous with multiple wavy brown transverse lines, and a wider pale fascia between medial and postmedial; discal dot small.

Male genitalia. Uncus slender, uniapical. Valve elongate, unmodified, basal 'window' rather small. Vinculum broad, rounded. Papillae as in species group description, with a row of elongate setae. Juxta with broad, semicircular calcar. Aedeagus shorter than length of valve. Vesica armed with a lightly sclerotized plate-like cornutus near ductus ejaculatorius base and numerous minute denticules. Sternite A8 with two broad lateral arms, connected by a broad anteriorly emarginate transverse band.

Female genitalia. Bursa oval-oblong, with a transverse belt of stout spines; anterior membranous, chaotically covered with poorly sclerotized spines; posterior sclerotized, with longitudinal striations and a large patch of spines on one side. Ductus bursae very short. Colliculum broad. Antrum membranous. Tergite A8 small. Apophyses relatively short and narrow. Papillae anales small, slightly elongate and tapered.

Range. China (Sichuan).

Phenology. ?Bivoltine. April – early June and July – September.

Altitude. Most known specimens are from 1000 to 2000 m in Sichuan (coll. IZCAS).

Similar species. Similar in pattern to *E. chingana*, but ground colour of both wings much paler, giving more contrast with costal and terminal blotches. The male genitalia are similar to those of *E. chingana*, but have a differently shaped eighth sternite. The female genitalia are distinguished from those of all other allied species by the shape of the bursa, its sclerotization, the disposition of spines and the presence of a collar-like colliculum.

Material. ♂, [China: Sichuan], Chow-pin-sa, *Coremia brevifasciaria* sp. n. type ♂, Type, Leech Coll. 1900–64, Geometridae genitalia slide no. 3514♂ (holotype, BMNH); 1♂, 1♀, [China]; Szetschwan [Sichuan], Tatsienlu, Exp. Stötzner, BM 1934–120 (BMNH); 1♀, same data but Kwanhsien [Guanxian] (BMNH); 5♂, 3♀, [China: Sichuan], Tse Kou, R.P. Dubernard, 1895, ex Oberthür coll. 1927–3, BM Geom. slide no. 1932/28 (BMNH); 3♂, Sichuan, Emeishan, Qinyinge 800–1000 m, 23–30.iv.1957, slides nos. ZL233, 627♂; 1♂, same loc., 27.v.1957; 2♂, same loc., 15–16.vii.1957; 3♂, same loc., 15, 16, 20, and 23.vii.1957, Zhu Fuxing and Huang Keren, slide no. L-2141♂; 1♂, Sichuan, Luding Xinxing, 1900 m, 14.vi.1983, Wang Shuyong, slide no. ZL467♂; 1♀, Sichuan, Emeishan, 15.viii.1977, slide no. ZL152♀; 1♂, same loc, 1000 m, 4.vi.1979, Shang Jinwen; 1♂, Sichuan, Qingchengshan, 1000 m, 4.vi.1979, Shang Jinwen (IZCAS).

27. *Eupithecia eurytera* Prout, 1938

Horisme eurytera Prout, 1938, *in* Seitz, *The Macrolepidoptera of the World* 4 (Suppl.): 212, pl. 17, fig. e. Holotype ♂ (BMNH; examined), China (west): Pehlinting, 6000 ft., 50 miles NNW of Cheng-tu.

Description. Fore wing narrower than in *E. chingana*, ground colour reddish brown; transverse lines black; basal evenly curved, medial zigzag, postmedial wavy to level of cell, then angled and straight to costa; area between medial and postmedial dark brown; subterminal line pale, wavy; discal dot linear, oblique. Hind wing ash grey, crossed by a series of close-set, zigzag transverse lines.

Male genitalia. Uncus short and stout, uniapical. Valve short with narrow apex, and large basal 'window'. Vinculum short, broad. Papillae as in species group description. Aedeagus stout, with ductus arising centrally. Vesica armed with two very small plates at base and five narrow apical cornuti. Sternite A8 short, with two pointed apical arms connected basally by a broad, transverse, scarcely emarginate band.

Female genitalia. Bursa very long, heavily sclerotized, partially broadened on right side, with many longitudinal rows of stout spines, and with a large membranous diverticulum at anterior end. Ductus bursae very short, forming a trapezoid plate connecting with antrum. Ductus seminalis narrow, arising from medial projection of bursa on right. Antrum short, broad, lightly sclerotized. Tergite A8 transverse. Apophyses very short. Papillae anales broad, rounded.

Range. China (Beijing, Sichuan, Tibet).

Phenology. Possibly univoltine: June in Beijing and July to August in Sichuan and Tibet.

Altitude. All records from between 1000 and 2500 m (coll. BMNH, IZCAS).

Similar species. Easily distinguished by the clear characteristically shaped fore wing median fascia and the smoky grey ground colour of the hind wings. The male genitalia are distinctive in the short uncus and the (usually) five slim spines on the aedeagus vesica. The female genitalia are unmistakable.

Material. ♂, [China: Sichuan], Pehlintin 6000ft., 50 miles N.N.W. of Chen-tu, vii-viii, G.M. Franck, *Horisme eurytera* Prout ♂ type, Type, L.B. Prout Coll. BM 1939–643, Geometridae genitalia slide no. 20924♂ (holotype, BMNH); 1♀, same data, (allotype, BMNH); 1♀, [China: Sichuan], Mt. Omei, 7000ft, 17.vii.1931, G.M. Franck (BMNH); 1♂, [China]: Szechwan [Sichuan], Mt Omei, 4000–5000ft, 9.viii.1929, G.M. Franck, BM Geom. slide no. 1936–29 (BMNH); 1♀, [China]: Szechwan [Sichuan], Kwanhsien [Guanxian], W. China, 17.vii.1928, G.M. Franck, Geometridae genitalia slide no. 23858♀ (BMNH); 1♂, China, Tibet, Bomi, Yi'ong, 2300 m, 15.viii.1983, coll Han Yinheng, slide no. ZL404♂ (IZCAS); 3♂, 4♀, China, Dong Ling Shan, Beijing, 1300 m, 22.vi.2001, A.C. Galsworthy (BMNH); 3♂, Sichuan, Emeishan, Jiulaodong, 1800–1900 m, 6.vii.57, Zhu Fuxing, slides nos. ZL626, L-2142♂; 1♂, Sichuan, Emeishan, Qinyingge, 800–1600 m, Huang Keren; 2♂, Beijing, Sanpu, 20.viii.64, Zhou Qin; 2♀, same data but 18.viii.64, Liao Subai; 1♂, Beijing, Baihuashan, 7.vii.73, Han Yinheng (IZCAS).

Remarks. Prout's original type series was from Sichuan. The species has since been found to be fairly common in the hills around Beijing, but no material is available from areas between. The specimens from Sichuan and Tibet are darker, and have the outer line of the fore wing postmedial fascia less prominent than specimens from Beijing, though there is some variation in both. The male and female genitalia from both areas appear almost identical; there are however two male preparations, one from Beijing, illustrated in Xue & Zhu (1999, fig. 1167) as *Horisme brevifasciaria*, and one from Tibet (ZL404, fig. 27a) in which only the three basal cornuti are present: it is possible that the more apical cornuti are deciduous. More information on distribution is needed, but if the two populations prove to be widely separated, it is possible that the northern one might best be described as a separate subspecies.

The *abietaria* species group

This Holarctic species group was introduced by Inoue (1979) as group 'A', with *E. abietaria* as the first species included within it. Later Bolte (1990) described a similar "*mutata*" group on the basis of five North American species: *E. albicapitata* Packard, 1876, *E. helena* Taylor, 1906, *E. cocoata* Pearsall, 1908, *E. mutata* Pearsall, 1908 and *E. spermaphaga* Dyar, 1917. The group is mainly characterized in the male genitalia by the long uniapical uncus, the elongate strap-like valve, and the single long cornutus in the aedeagus vesica. Some, but not all, species lack the normal digitate papillae on the anterior arms of the labides. The female genitalia share a large corpus bursae with several discrete patches of spines, and a row of spine-like setae on the posterior margin of tergum A8. Where the biology is known, the larvae are specialist feeders on the cones of evergreen trees (Pinaceae and Taxodiaceae). The group is represented in China by the following six species.

28. *Eupithecia rufescens* Butler, 1878

Eupithecia rufescens Butler, 1878, *Ann. Mag. nat. Hist.* (5) 1 (6): 445. Holotype ♀ (BMNH; examined), Japan: Yokohama.

Eupithecia larentimima Vojnits, 1974, *Acta Zool. Acad. Sci. Hung.* **20** (1–2): 221, figs 2a-e. Holotype ♀ (ZFMK; examined), [China]: Prov. Chekiang [Zhejiang], West-Tien-Mu-Shan, 1600 m. **New junior synonym.**

Description. An easily recognized species, with broad triangular fore wings. On the fore wing the basal and terminal areas are purplish or rusty brown, and the medial area light brownish grey to brown, bordered by pale double transverse lines; the discal dot is small, ovoid to elongate, and brown. The hind wing usually has distinct rather close-set transverse lines, with a pale double postmedial. The first abdominal segment is purplish or rusty brown.

Male genitalia. Uncus narrow, uniapical. Valve long and narrow with parallel and evenly curved dorsal and ventral margins. Vinculum short and broad, rounded or slightly tapered. Papillae broad and very short. Aedeagus elongate, slightly curved. Vesica armed with one tapered, sabre-shaped semi-cylindrical cornutus about 4/5 length of aedeagus and with numerous tiny sclerotized irregular shapes. Sternite A8 long and broad with almost parallel lateral margins and with two elongate and heavily sclerotized, sometimes asymmetrical, apical horns terminating in narrow prongs.

Female genitalia. Bursa copulatrix bean-shaped, membranous, with four patches of large spines, one near base, two medial on dorsal and ventral sides, and one narrow and elongate on left side, and with one patch of usually seven small spines with rounded flat bases near base of colliculum. Ductus bursae not distinguishable. Ductus seminalis sharply recurved, narrow, arising from broad prominence on left side of posterior half of bursa. Colliculum broadened basally and tapered. Antrum broad and elongate, campaniform, lightly sclerotized. Tergite A8 with lightly sclerotized anterior margin and with convex posterior margin, densely covered with numerous long setae. Apophyses of average length and thickness. Papillae anales short and broad, tapered.

Range. China (Zhejiang); Japan (Honshu, Sado, Shikoku, Kyushu, Tsushima, Yakushima, Okinawa).

Phenology. Probably bivoltine: March to mid-July. Overwintering as pupa.

Biology. Larva found in the cones of *Cryptomeria japonica* (Taxodiaceae) (Japan: Inoue 1979).

Altitude. Found at about 1600 m in Zhejiang (China: Vojnits 1974; coll. ZFMK).

Similar species. Among Asian species, *E. rufescens* is only likely to be confused with the following species, but it can usually be distinguished by its smaller size, the clear purplish or rusty brown basal and terminal areas, the narrower, less distinct ante- and postmedial transverse lines and the smaller discal dot on the fore wing. The genitalia are similar to those of *E. abietaria*, but in *E. rufescens* the valve is shorter (less than 2 mm) in the male and in the female genitalia there is a short transverse row of short stout spines in the corpus bursae near the base of the colliculum.

Material. ♀, Japan 77–9, *E. rufescens* Butler Type (holotype, BMNH); ♀, Pz. Chekiang [Zhejiang], West-Tien-Mu-Shan (1600 m), 22.iv.1932, H. Höne (holotype of *E. larentimima*, ZFMK); 1♂, 4♀, same loc. and alt., 22, 24.iv., 5.v., 1.vi., 10.vii.1932, H. Höne, Schmidt slides nos. Eu 145a♂, Eu 145b♀ (paratypes of *E. larentimima*, ZFMK); 2♂, 1♀, same loc. and alt., 24, 28.iv., 10.v.1932, H. Höne, Vojnits slides nos. 10034♀, 10035♂ (paratypes of *E. larentimima*, TTM); 3♂, same loc. and alt., 1, 25.v.1932, H. Höne, Mironov slide no. 650♂ (ZFMK); 5♂, Prov. Chekiang [Zhejiang], West-Tien-mu-shan, 26.iv., 3, 18, 27.v., 1.vi.1932, H. Höne, Schmidt slide no. Eu 145♂ (paratypes of *E. larentimima*, ZFMK); 3♂, 3♀, same loc., 28.iv., 3, 8, 16, 22, 29.v.1932, H. Höne, de Laever slides nos. 325♂, 326♀, Mironov slide no. 649♀ (ZFMK); 1♂, [China, Zhejiang], Ost Tien Mu Shan, April – 1931, 1500 m, H. Höne (paratype of *E. larentimima*, ZFMK).

29. *Eupithecia leamariae* Ratzel, 2011

Eupithecia leamariae Ratzel, 2011, *Ent. Z., Stuttgart* **121** (4): 174, fig. 5. Holotype ♀ (coll. U. Ratzel, SMNK; examined), Bhutan (west): Prov. Paro, Road to Jomolhari, near Lemda, 2680 m.

Description. Fore wing broad, dark grey, sometimes with purplish tinge; terminal area purplish with grey subterminal line; antemedial faint, touching discal dot and obtusely angled onto costa; postmedial double, evenly curved; discal dot large, oblique. Hind wing paler, grey, with numerous concentric curved transverse lines; postmedial a pale double band bordered internally with dark dots; discal dot small, linear. Abdomen ash-grey, second segment black.

Male genitalia. Uncus elongate, uniapical. Valve elongate with parallel margins. Vinculum broad, rounded. Papillae limited to a small crescent on underside of labides. Juxta with very characteristic, almost rhomboid caulis. Aedeagus long and narrow. Vesica armed with an elongate cornutus connected at base with a rounded cornutus near ductus ejaculatorius base. Sternite A8 broad, rectangular, terminating apically in two narrow, pointed rods.

Female genitalia. Bursa copulatrix ovoid, broadened near ductus seminalis base, with a medial transverse patch of elongate spines and a smaller patch of similar spines near base of colliculum. Ductus bursae not distinguishable. Ductus seminalis narrow, slightly broadened basally, attached to prominence on left side of bursa. Colliculum narrow and elongate. Antrum elongate, membranous. Tergite A8 rectangular, elongate, with convex anterior margin and lateral patches of setae at posterior margin. Apophyses very long, narrow, broadened and flattened at apices. Papillae anales very narrow, elongate.

Range. China (Yunnan); Bhutan.

Phenology. Probably univoltine: mid-May.

Altitude. Found from about 2700 to 3200 m (Bhutan: Ratzel 2011; coll. SMNK, Yunnan: BMNH).

Similar species. Closest to the Palaearctic species *E. analoga* Djakonov, 1926, from which distinguished by shorter labial palpi, dark ash-grey colour of fore wings, ovoid discal dot touching antemedial line on the fore wing and smaller discal dots on the hind wings. The male and female genitalia are similar to those of other representatives of the *abietaria* group, especially *E. analoga*. The male has a different shape to the base of the uncus, a characteristic shape of caulis in the juxta, and almost no basal lobes to the eighth sternite. The pattern of spines in the female genitalia are distinctive, as are the unusually long apophyses.

Material. ♀, Bhutan (West), Prov. Paro, Weg z. Jomolhari, bei Lemda, (Abzw. n. Lhading Gönpa), LF, N 27°33′50″, E 89°18′27″, 18.v.2010, 2680 m, leg. Ratzel, Hauenstein, Falkenberg & R.T., /w *Eupithecia leamariae*, Holotypus, GU20311/4♀, det. U. Ratzel (holotype, coll. U. Ratzel); 1♂, West-Bhutan, Provinz Paro, Jomolhari-Trek, N Shinkarap, Abzweig zum Tremo La, N 27°39′08″, E 89°15′31″, 16.v.2010, 3120 m NN, LF, leg. Hauenstein, Falkenberg, Ratzel, Trusch; SMNK E-Lep 268, Mironov slide no. 870♂ (SMNK); 2♂, 1♀, Yunnan, KIB/RBGE Research station, Lijiang, 3200 m, 15–20.vi.2009, A.C. Galsworthy (BMNH).

30. *Eupithecia abietaria* Goeze, 1781

Phalaena (Geometra) abietaria Goeze, 1781, *Ent. Beytr.* 3 (3): 439. Syntype(s), [Europe].

Phalaena (Tinea) pini Retzius, 1783, *Genera et Species Insectorum*: 50 [Europe] *nec* Linnaeus, 1758 (identity: *Dendrolimus pini*, Lasiocampidae). Junior primary homonym (synonymised in Prout, 1915).

Phalaena (Geometra) strobilata Borkhausen, 1794, *Natur. Eur. Schmett.* 5: 352. Syntype(s). [Europe] (synonymised in Borkhausen, 1794).

Geometra togata Hübner, [1817], *Samml. Eur. Schmett.* 5 (Geometrae 2), pl. 90: 464. Syntype(s) lost. [Europe] (synonymised in Prout, 1907).

Acidalia bilunulata Zetterstedt, 1839, *Insecta Lapponica* 4 (6): 960. Syntype(s) ♀ (MZLU), [Sweden ?]: Lappland (synonymised as *E. bilunata* [sic!] Zetterstedt, 1839 to *E. togata* Hübner, 1817 in Staudinger & Rebel, 1901).

Eup[ithecia] togata var. *debrunneata* Staudinger, 1897, *Dt. ent. Z., Iris* **10** (1): 109. Syntypes 2♀ (coll. O. Staudinger in MNHU; examined), [Russia]: Ussuri (treated as form of *E. abietaria* in Dietze, 1908).

Misidentification: *Eupithecia rufescens* Matsumura, 1925 (*nec* Butler, 1878).
Unavailable names (incorrect subsequent spellings): *togaria*: Boisduval (1840); *strobilaria*: Boisduval (1840); *abjetaria*: Rebel (1910); *strobinata*: de Laever (1956).

Description. Fore wing with oblique transverse vein in areole; light liver-brown or light greyish brown; transverse lines slightly sinuate, dark brown; basal line angled near costa, with broad purplish brown outer stripe; antemedial with dark brown costal spot; postmedial broadened at costa, usually weakly angled, and with a series of short, triangular dark brown spots on the veins; terminal area purplish brown, usually ash-grey near apex or on outer side and with a slightly paler blotch between M_3 and CuA_1; pale wavy line conspicuous, dentate and unbroken; discal spot large, dark brown or black. Hind wing slightly paler; transverse lines darkish, clouded, well marked only near dorsum; postmedial well defined along its entire length, twice curved, broad, clouded, usually with a series of short, triangular spots; terminal area darker, light purplish brown; pale wavy line more or less distinct; discal spot large, ovate-acuminate or ovate. Abdomen light greyish brown, usually with ochreous tinge, and with small, dark brown dorsal and lateral spots on each segment.
Male genitalia. Uncus elongate, uniapical. Valve long and narrow, slightly curved, with parallel dorsal and ventral margins. Vinculum short and broad, rounded. Papillae very short and broad. Aedeagus elongate and rather narrow, shorter than length of valve. Vesica armed with one long narrow, twisted, sabre-shaped cornutus. Sternite A8 long and broad, broadened at base, with two narrow and blunt apical horns.
Female genitalia. Bursa copulatrix large, ovate-oblong, with two patches of large spines, one at base and the other smaller, obliquely longitudinal in its apical part. Ductus bursae not distinguishable. Ductus seminalis broadened basally, attached to posterior part of corpus bursae close to colliculum on right. Colliculum small, slightly elongate and narrow. Antrum broad, campaniform, sclerotized. Tergite A8 broad, trapezoid, with broadened posterior margin covered by setae. Apophyses relatively long and broad, with slightly expanded and rounded apices. Papillae anales large, tapered.
Range. China (Xinjiang, Qinghai, Sichuan, Yunnan); Europe (except Iceland, Iberian peninsula, Albania, Greece and all Mediterranean islands), Caucasus, western Transcaucasus (Georgia), Asian Russia (Urals, southern Siberia, Yakutia, Priamurje, Primorje, Kamchatka peninsula, Sakhalin

Isl., Southern Kuril Islands), Japan (Hokkaido, Honshu). Likely also to occur in north eastern China.
Phenology. Univoltine: late June to early August.
Biology. Larva found in the cones of *Picea jezoensis* (Pinaceae) (Japan: Inoue 1979) and *Pinus pumila* (Pinaceae) (Russia, Kamchatka peninsula: Kurentzov & Ivliev, 1960). It has been found feeding in the ripening cones of *Picea abies* (= *excelsa*), *Abies alba*, *Pinus cembra* and *P. sylvestris* (Pinaceae) in Europe (Dietze 1913; Mironov 2003), often on diseased or damaged trees, eating unripe seeds.
Altitude. From sea level to about 3200 m in Daxue Shan (China: colls. ZFMK and IZCAS).
Similar species. Easily confused with another palaearctic species *E. analoga* Djakonov, 1926 (which may well occur in northern China, since it is known from the Russian Altai mountains) but is usually distinguished from it by the larger size, the shorter labial palpi, the less rounded discal spot on the fore wing and by the presence of dark triangular spots on the veins along the postmedial line on the hind wing, as well as the less dark second abdominal tergite. The male, and especially the female genitalia reliably separate *E. abietaria* from *E. analoga*, but they are rather similar to those of the following two, much larger, species.
Material. ♀, *debrunneata* Stgr., Origin., Lectotype, Sieh Lempke, Tÿdschr. Entom. vol. 112, 1969 (lectotype of *E. togata* var. *debrunneata*, design. by Lempke, coll. O. Staudinger in MNHU); 1♀, [Russia], Ussuri Dörr. 82, Origin., 146, (paralectotype of *E. togata* var. *debrunneata*, design. by Mironov, 2009, coll. O. Staudinger in MNHU); 1♂, China, Sichuan, Daxue Shan Mts., Gongga Shan, W Moxi, 29°34′ N, 101°59′ E, ~3200 m, 20.-21. vii.1999, leg. V. Siniaev & A. Plutenko, Mironov slide no. 726♂ (ZFMK); 1♂, 1♀, Qinghai, Tongren, Maixiu, 2950 m, 30.vii.1992, coll Xue Dayong, slide no. ZL283 ♂; 1♂, Xinjiang, Tomürt Peak, 2600 m, 21.vii.1977, coll Li Changjiang, slide no. ZL230 ♂ (IZCAS); 1♀, China, Yunnan Prov., Zhongdian, 3100 m, 14.vii.2001, Li Houhun leg. (coll. P. Skou).
Remarks. Not previously recorded from China.

31. *Eupithecia gigantea* Staudinger, 1897

Eup[ithecia] gigantea Staudinger, 1897, *Dt. ent. Z,. Iris* **10** (1): 109, pl. 3, fig. 70. Holotype ♀ (coll. O. Staudinger in MNHU; examined), [Russia]: Ussuri.
Cidaria (*Euphyia*) *karafutonis* Matsumura, 1925, *J. Coll. Agr. Hokkaido Imp. Univ.* **15** (3): 167, pl. 10, fig. 13. Syntypes 2♂, 5♀ (EIHU), [Japan/ Russia]: South-Saghalien, (Ichinos.awa) (synonymised in Inoue, 1956).

Description. A large broad winged species. Fore wing grey; transverse lines distinct, black; basal line oblique, angled onto costa; broad transverse band between basal and antemedial purple to blackish brown; antemedial wavy, evenly curved, forming a black costal spot; postmedial more distinct, black, broken, once or twice angled, broadened near costa; terminal area broad, divided by pale grey wavy subterminal line; inner part of terminal area purple brown, its outer part between subterminal and terminal lines ash-grey; discal dot large, intensely black. Hind wing unicolourous grey; dark transverse lines black, distinct, especially basal and postmedial; latter broken or sinuate, sometimes forming a series of black dashes on the veins; terminal area broad, unicolourous; discal dot comma-like, relatively large.

Male genitalia. Uncus elongate, uniapical. Valve long and narrow, tapered distally. Vinculum broad, rounded. Papillae short and broad. Aedeagus stout and long, but shorter than length of valve. Vesica armed with one long and narrow, twisted, sabre-shaped cornutus. Sternite A8 long and broad, with two elongate apical horns with rounded apices.

Female genitalia. Bursa very large, oval, with five separate patches of large spines, one at base, two in the central area and two in the posterior part. Ductus bursae not distinguishable. Ductus seminalis broadened basally, attached to posterior part of bursa near colliculum on right. Colliculum relatively short and narrow. Antrum membranous. Tergite A8 broad, trapezoid, with broadened posterior margin covered by long setae. Apophyses long and stout, with slightly expanded apices. Papillae anales large, tapered.

Range. China (Heilongjiang, Shaanxi, Gansu); Asian Russia (Priamurje, Primorje, Sakhalin, Kuril Islands: Kunashiri), Korea, Japan (Hokkaido, Honshu, Shikoku, Kyushu, Tsushima, Yakushima).

Phenology. Univoltine: late June to late August. Possibly bivoltine in Yakushima Island (Japan).

Biology. Larva found in the cones of *Pinus strobus* and *Abies sachalinensis* (Pinaceae) (Japan: Inoue 1979).

Similar species. Could only be confused with the two other large allied Asian species *E. interrubrescens*, and *E. dalhousiensis* Mironov & Galsworthy, 2008 (the latter not known from China), but distinguished from them by the evenly curved antemedial line, which is sharply angled onto the costa in the other two species, and the broader, ovoid discal dot on the fore wing, as well as the lack of a dentate subterminal line on the hind wing. In the male genitalia, compared with *E. interrubescens*, the present species has a broader base to the uncus, larger and broader papillae on the ends of anterior arms of the labides and longer apical horns to sternite A8. In the female the trapezoid tergite A8 of *E. gigantea* is distinct from the more rectangular shape in both *E. interrubrescens* and

E. dalhousiensis: in both the latter species the colliculum is also much longer than in *E. gigantea*.

Material. ♀, [Russia], Ussuri Dörr., *gigantea* ♀, Origin., 70., bon. spec. bei *Togata* (Boh.), Typus, Gen. vérifiés 1456[2] *Eupithecia gigantea* ♀, Ussuri, dét. E. de Laever 1954 (holotype, coll. O. Staudinger in MNHU); 3♂, 2♀, Heilongjiang, Dailing, 390 m, 19.ii.1964, coll Bai Jiuwei, slides nos. ZL254 ♂, 343♀ (IZCAS).

32. *Eupithecia interrubrescens* Hampson, 1902

Phibalapteryx interrubrescens Hampson, 1902, *J. Bombay nat. Hist. Soc.* 14 (3): 513. Syntypes ♂, ♀ (BMNH; examined), [India]: Dalhousie; [India/China]: [Sikkim/Tibet border], Yatong.

Description. A very large species with greyish brown basal and medial areas, a broad purple-brown transverse band between them and a purple-brown terminal area to the fore wing; ante- and postmedial transverse lines forming dark, blackish brown costal spots; discal dot usually large, oblique, ovoid or narrow and elongate. Hind wing paler, brownish grey with broad dark, twice angled postmedial transverse line, with paler, usually comma-like discal dot and slightly darker terminal area.

Male genitalia. Uncus lightly sclerotized, uniapical. Valve straplike, slightly curved and tapered, with almost parallel dorsal and ventral margins. Vinculum short and broad, rounded, slightly tapered. Juxta with two broad, slightly bifurcate processes. Papillae very short and broad. Aedeagus large, elongate but shorter than length of valve, with one membranous apical lobe, covered with dense soft setae. Vesica covered with numerous denticules, and armed with a long, narrow, sinuate, sabre-shaped cornutus. Sternite A8 wide strongly broadened at base, with two strongly sclerotized, blunt apical horns.

Female genitalia. Bursa ovoid, membranous, with three patches of large spines, one rounded at base, a second longitudinal along the right side and a third shorter and oblique near base of ductus seminalis, and a small row of spines with flattened bases close to colliculum. Ductus bursae not distinguishable. Ductus seminalis broadened near base, recurved, attached to posterior part of bursa near colliculum on left. Colliculum elongate, broadened medially. Antrum large, funnel-shaped, heavily sclerotized. Tergite A8 large, almost square. Apophyses long and broad, the posterior slightly expanded and flattened at apices; lobes of the branches of posterior apophyses very large and broad. Papillae anales broad, rounded.

Range. China (Tibet); Pakistan/India (Kashmir), India (Sikkim).

Phenology. Probably univoltine: mid-June to early July.
Altitude. From about 2600 to 3200 m in the Himalayas (Kashmir: coll. ZFMK).
Similar species. See under preceding species.
Material. ♀, Tibet, Yatung, A.E. Hobson, 98–201, *Phibalapteryx interubrescens* Hmpsn. type ♀, BM Geom. slide no. 2941 (lectotype, BMNH); 1♂, same data, allotype, BM Geom. slide no. 2942♂ (paralectotype, BMNH).

33. *Eupithecia costalis* Walker, 1863

Pena costalis Walker, 1863, *List Specimens lepid. Insects Colln Br. Mus.* 27: 130. Holotype ♀ (BMNH; examined), Borneo: Sarawak.
Leiocera ferrisparsata Hampson, 1893, *Illust. typical Specimens Lepid. Heterocera Colln Br. Mus.* 9: 41, 152, pl. 171, fig. 8. Holotype ♂ [*nec* ♀] (BMNH; examined), [Ceylon] [= Sri Lanka]: Pundaloya (synonymised in Hampson, 1895).

Description. An average-sized species with unusual maculation and coloration. Fore wing ground colour delicate, pale ash-grey; costal margin darker, purple brown; transverse lines very distinct, fine, narrow, blackish; basal, postbasal and antemedial obliquely evenly curved; postmedial very characteristic, oblique, forming a strong zigzag between CuA_2 and A veins, and two more on leading half of wing; terminal area with strong reddish streaks and black scales on veins; a distinct wavy, whitish subterminal line; discal dot invisible. Hind wing pale ash-grey with very distinct close-set transverse lines, the double postmedial particularly marked.
Male genitalia. Uncus small and narrow, uniapical. Valve narrow, with almost parallel dorsal and ventral margins. Vinculum short and relatively narrow, tapered. Papillae short and broad, rounded. Aedeagus narrow, slightly shorter than length of valve. Vesica covered with numerous spicules, and armed with an elongate, medially broadened cornutus with pointed apex. Sternite A8 relatively large, broadened medially, with two diverging apical arms and a large, heart-shaped, membranous patch in posterior half.
Female genitalia. Bursa relatively small, spherical, sclerotized, with two patches of large, pointed spines. Ductus bursae sclerotized, short and narrow. Ductus seminalis broadened and sclerotized at base, sharply curved, attached to posterior part of bursa on right. Colliculum short and narrow. Antrum membranous, elongate. Tergite A8 small, almost quadrate, with sinuate anterior margin, its posterior margin and broadened basal lobes of branches densely covered with long setae. Branches of anterior apophyses connected with each other by a narrow, heavily sclerotized band, which encircles the antrum (as in *Perizomini* species). Apophyses broad, slightly

expanded and flattened at apices. Papillae anales narrow and slightly elongate, tapered.

Range. China (Tibet, Hong Kong, Taiwan, and probably much of south eastern China); Sri Lanka, North India, Malaysia Peninsula, Borneo, Sumatra.

Phenology. Polyvoltine: mid-April to late September.

Altitude. From about 1000 to 2200 m above sea-level in the Pacific Islands (Sumatra: coll. ZSM; Taiwan: coll. ZSM, TTM).

Similar species. Unmistakable, and more similar externally to some tropical species of *Gymnoscelis* than to *Eupithecia*. We place *E. costalis* in this group mainly on genitalia characteristics: the elongate, uniapical uncus, the elongate and narrow valve, the broad and very short papillae on the ends of the anterior arms of the labides, the short and broad vinculum, the single pointed cornutus on the vesica and the shape of the eighth sternite in the male, and the disposition of large, stout spines in the bursa copulatrix, and the presence of rows of long setae on the posterior margin of tergite A8 in the female genitalia. The close set transverse lines are also a little reminiscent of *E. rufescens.*

Material. ♀, Type, SAR., Sarawak, Saunders' Coll. 94–68, *Pena costalis* (holotype, BMNH); ♂, Ceylon, Green Coll. 92–26, Pundaloya Ceylon, *Eupithecia ferrisparsata* Hmpsn. type ♂ (holotype of *L. ferrisparsata* [*nec* ♀], BMNH); 1♀, Taiwan, Prov. Ilan, Suyuan, Yakou, 2100 m, 9.v.1997, leg. Gy.M. László & G. László (TTM); 1♀, Central Taiwan, Nantou Co., Meifeng, *ca.* 2100 m NN, 24°05′55″ N / 121°10′48″ E, LF, 23. April 2003, leg. Mei-Yu Chen (ZISP); 1♂, 1♀, Central Taiwan, near Puli, Nantou Co., Meifeng, *ca.* 2200 m NN, 24°05′25″ N / 121°10′21″ E, LF, 18. April 2004, 19. May 2004, leg. Mei-Yu Chen (ZSM; ZISP); 1♀, Tibet, Mêdog, 1091 m, 29°19′ N/ 95° 19′ E, 22.viii.2006, coll. Lang Songyun (IZCAS); 1♀, Taiwan, Lushan Spa, Nantou, 19–21.iii.1972, K. Naketami; 1♀, Rantaizan, Formosa, 7500 ft, 10.v.1909, E. Wileman (BMNH).

The *proterva* species group

This is one of the most species-rich species groups, and is principally East Asian. It was introduced by Inoue (1979) as group 'B', including eleven Japanese species. He pointed to the pair of 'slender whip-like cornuti' in the male aedeagus vesica as the main distinguishing feature. Mironov (1990) introduced a similar group under the name *proterva*, but without any diagnosis. The morphological similarities between the genitalia and external appearance of many of the species included leaves no doubt that many of them are closely related to one another, but at the same time

there is no clear apomorphy to distinguish them, and the boundaries of the group must be regarded as uncertain. The group includes mainly univoltine spring or late autumn species and we currently include in it the 22 following Chinese species. The group may possibly be related to the predominantly western palaearctic *graphata* group.
Male genitalia. Uncus from short to long, biapical or uniapical. Valve variable in size and shape, often with a broad, sometimes obtusely angled, ventral process near base. Vinculum very variable. Papillae on the anterior arms of the labides usually small. Aedeagus usually stout, large, and longer than length of valve. Vesica covered by numerous very small spinules and usually armed with one long, sometimes curved, sabre-shaped or V-shaped cornutus and with a small, irregular, folded cornutus at the anterior part of aedeagus. Sternite A8 usually small, short, peg-like, from narrow to broad; basal hollow shallow; apical hollow very shallow or absent.
Female genitalia. Bursa copulatrix usually large, always partly spined, with a tendency for the spines to occupy one side only. Ductus bursae not clearly distinguishable. Ductus seminalis attached to medial or basal part of corpus bursae, in many species the base of the bursa itself. Colliculum large and broad in the majority of species. Antrum sclerotized, rarely membranous. Tergite A8 of variable shape, usually large and broad. Apophyses narrow, generally rather short, in some species slightly expanded and flattened at apices. Papillae anales large, rounded in the majority of species.

34. *Eupithecia jermyi* Vojnits, 1976

Eupithecia jermyi Vojnits, 1976, *Acta zool. Acad. Sci. hung.* **22** (1–2): 206, figs 1H, 2H. Holotype ♂ (ZFMK; examined), [China]: Kuatun [Guadun], 2300 m.

Description. An ornate and strikingly coloured species. Fore wings with base, terminal and most of costal areas deep reddish brown, and central area ash grey, with numerous fine oblique dark lines; antemedial line double, white, sharply angled onto costa; postmedial similar, bounded internally by crenulate dark border; discal dot small and black. Hind wing paler, whitish grey with numerous elegant, distinct dark wavy transverse lines and a small rounded discal dot.
Male genitalia. Uncus short and pointed, uniapical. Valve broadened basally, with very narrow apex. Vinculum short and narrow, tapered. Papillae stout, rather clavate and slightly curved. Aedeagus stout, slightly curved, with thickened anterior end, longer than length of valve. Vesica armed with one long, slightly curved, longitudinally forked and apically

blunt cornutus, a small medial patch of short, weakly sclerotized spinules and a small, folded, irregular cornutus near anterior end of aedeagus. Sternite A8 short, almost triangular, lightly sclerotized, with two short and blunt apical arms.

Female genitalia. Bursa small, pear-shaped, about ¾ covered with slim spines on right side with an oblique, more strongly sclerotized area from base of ductus seminalis to colliculum on left side. Ductus bursae not clearly distinguishable. Ductus seminalis attached to base of corpus bursae on left side. Colliculum relatively large, elongate and broad, slightly tapered. Antrum membranous, short and broad. Tergite A8 rectangular, narrow and elongate. Apophyses short and narrow, branches arising from anterior apophyses near apices. Papillae anales narrow, slightly elongate, with rounded apices, covered with short setae.

Range. China (Fujian); Vietnam (Tonkin).

Phenology. Univoltine: early February to mid-March.

Altitude. Found at about 2100–2650 m (North Vietnam: coll. ZFMK, TTM, Mironov & Galsworthy 2009b).

Similar species. A handsome species, generally easy to recognise by the distinctive ornamentation and delicate colours of the wings. The general pattern resembles the next species, and two other south east Asian species. For distinctions see under the next species. The long aedeagus with blunt cornutus and the squat sternite A8 are the most distinctive features in the male genitalia. Though the general shape of the female genitalia is unremarkable, the strongly sclerotized area on one side of the corpus bursae is distinctive.

Material. ♂, [Fujian], Kuatun [Guadun], 2300 m, 27, 40 n. Br., 117, 40 ö. L., J. Klapperich, 4.iii.1938 (Fukien), de Laever slide no. 426♂ (holotype, ZFMK); 1♀, same loc. and alt., 11.iii.1946 (Fukien) (ZFMK; abdomen missing).

Remarks. This species was erroneously included by Vojnits (1976) in the "*bohatschi*" group (our *haworthiata* group). In de Laever's slide no. 426♂, belonging to the holotype of *E. jermyi*, the uncus was bent forward under the tegumen, giving the impression that it was absent or membranous.

35. *Eupithecia lini* Mironov & Galsworthy, 2007

Eupithecia ***lini*** Mironov & Galsworthy, 2007, *Trans. lepid. Soc. Japan* 58 (3): 344, figs 3, 15. Holotype ♂ (NSMTA; examined), Taiwan: Nantou, Hohuanshan.

Description. Fore wing ground colour pale ash grey; basal area reddish brown with vein R blackened; basal line double, sharply angled onto costa;

medial area with two narrow reddish brown costal spots; antemedial inconspicuous; medial triple, slightly sinuate, oblique, sharply angled onto costa behind discal dot; postmedial inconspicuous, forming a series of dark dashes on the veins, sharply angled onto costa; terminal area with strongly crenulate inner margin, bordered distally with reddish brown, and with three larger apical, subapical and tornal reddish brown blotches; discal dot small, ovoid. Hind wing ground colour white; transverse lines brown, well marked only along anal margin, postmedial and subterminal lines markedly crenulate; discal dot small and round.

Male genitalia. Uncus short, uniapical. Valve relatively elongate and narrow, with narrowly rounded apex. Vinculum short and narrow, triangular. Papillae rather stout. Aedeagus long and narrow, almost equal to length of valve. Vesica armed with a long and narrow, slightly sclerotized cornutus, with two curved arms at base, and a folded, irregular cornutus near base of ductus ejaculatorius. Sternite A8 lightly sclerotized, with two parallel, elongate and pointed apical rods; basal lobes pointed laterally.

Female unknown.

Range. China (Taiwan).

Phenology. Univoltine: early November to early January.

Altitude. Found at about 2200–2300 m above sea-level (China: Vojnits 1976; Vietnam: coll. ZFMK).

Similar species. This species has similar external patterning to the preceding species, and also to *E. costinotata* Inoue, 2000, and a Vietnamese species *E. peregovitsi* Mironov & Galsworthy, 2009. It is considerably larger than *E. jermyi*, with a much more pointed fore wing with an oblique rather than rounded terminal margin. The dark marginal areas are less extensive, but the clear fascia between antemedial and medial lines is much broader. The male genitalia are similar to those of *E. jermyi*, but distinguished by the broader apex of the valve, the shorter, stouter aedeagus with shorter large cornutus on the vesica, and the different shape of the eighth sternite, which is longer, with longer and pointed apical rods and laterally pointed basal lobes.

Material. ♂, Taiwan, Nantou, Hohuanshan, 8–9.xi.1989, leg. C.S. Lin, NMNS ENT 493–23 (holotype, NSMTA); 1♂, Taiwan, Nantou, Meifeng, 1–3.xii.1988, leg. C.S. Lin, NMNS ENT 1116–478, Galsworthy slide no. ACG TW4 (paratype, NSMTA); 1♂, Taiwan, Hualien Hsien, Hungyeh spa, 200 m, 3.i.[19]89, leg. A. Kawabe (paratype, BMNH).

36. *Eupithecia lucigera* Butler, 1889

Eupithecia lucigera Butler, 1889, *Illust. typical Specimens Lepid. Heterocera Colln Br. Mus.* 7: 24, 115, pl. 137, fig. 11. Holotype ♂ (BMNH; examined), [India]: [Kangra district], Dharmsala.

Description. A relatively small dark species with broad triangular fore wings, distinguished by the shiny brownish black ground colour of basal, medial and terminal areas relieved by two broad white bands in the costal half of the medial area, by the presence of a small whitish tornal spot and a very distinct, intensely black discal dot on the fore wing. Some specimens have a rusty brown tornal area. The hind wing is usually dark, but paler than the fore wing, shiny brownish black, darker along anal and terminal margins with an inconspicuous transverse postmedial and a small discal dot. The notum is usually rusty brown.
Male genitalia. Uncus stout, uniapical, with hooked tip. Valve long and narrow, evenly tapered to the narrowly rounded apex. Vinculum short and narrow, with shallow medial hollow. Papillae of average size. Aedeagus stout and short. Vesica armed with a flat, slightly curved and folded, sabre-shaped cornutus, bipartite at the anterior end. Sternite A8 peg-like, broadened at base, with obtuse apex.
Female genitalia. Bursa small, ovate, with dorsal and a ventral longitudinal chain of large, stout spines. Ductus bursae not distinguishable. Ductus seminalis broadened at base, attached to left of bursa near base. Colliculum broad, slightly tapered. Antrum membranous. Tergite A8 transverse. Apophyses short and broad, the posterior with short spurs near bases. Papillae anales short and broad, rounded.
Range. China (Yunnan, Sichuan); Nepal, India (Punjab, Sikkim).
Phenology. Bivoltine (?): May to mid-August; late September to early October.
Altitude. From about 1750 to 2500 m in the Himalayas; one specimen from Sichuan about 1000 m (Inoue 2000; coll. C. Herbulot in ZSM; IZCAS).
Similar species. Usually easily recognized, but could be confused with another dark coloured species, *E. atrisignis*; the two broad whitish costal patches in the medial area of the fore wing, and the absence of a reddish blotch under the discal dot, which usually characterizes *E. atrisignis*, should be decisive. In the male genitalia, the shape of the uniapical uncus, the short aedeagus with broad, sinuous cornutus, and the shape of the eighth sternite are characteristic. In the female, the disposition of spines in the corpus bursae, and the very short apophyses readily distinguish the species.
Material. ♂, [India], Dharmsala 87–59, *Eupithecia lucigera* Butler type, BM Geom. slide no. 3518 (holotype, BMNH); 6♂, China, Provinz Nord-Yuennan, Li-kiang [Lijiang], 10.vii, 5, 10.viii, 6.ix.1934, 23.v, 24.vi.1935, H. Höne, de Laever slides nos.140♂, 142♂, 292♂, Vojnits slides nos. 122240♂, 122241♂, 122242♂ (ZFMK); 1♀, Sichuan, Emeishan, Qingyinge, 800–1000 m, 6.v.1957, coll Wang Sungyuan, slide no. ZL56♀ (IZCAS).
Remarks. Not previously recorded from China.

37. *Eupithecia missionerata* Mironov & Galsworthy, 2011

Eupithecia ***missionerata*** Mironov & Galsworthy, 2011, *Lep. Science* **62** (1): 17, figs 3, 18. Holotype ♂ (ZFMK; examined), Thibet, Tà-tsień-loû.

Description. A very distinctive species, with basal area of fore wing as far as medial line brown, and the rest mainly pale yellowish, with terminal area darker, especially three blotches, costal, tornal and medial, which are the same shade as the basal area. Discal dot distinct, but small. Hind wing colour much as paler part of fore wing, darker in terminal area.
Male genitalia. Uncus short and narrow, biapical. Valve unmodified, sacculus slightly broadened and gently angulate. Vinculum short, with very shallow medial hollow. Papillae relatively narrow. Aedeagus slightly shorter than length of valve. Vesica granulate and armed with an elongate plate-like cornutus, tapered at anterior end, and a small, pointed plate-like cornutus near ductus ejaculatorius base. Sternite A8 peg-like, slightly broadened at apex, with sclerotized lateral margins and two short blunt apical arms.
Female unknown.
Range. China (Sichuan).
Phenology. Univoltine (?): May-June.
Similar species. A very distinctive species, most reminiscent of *E. atuni*, but the latter has the basal area of the fore wing pale, and the hind wing much darker. The male valve shape is similar to that of *E. subbreviata*, but in *E. missionerata* the uncus is biapical. The aedeagus and its cornutus are closer to those of *E. lucigera*, but the cornutus is straighter, and the eighth sternite longer, with two apical horns.
Material. ♂, Thibet [Sichuan], Tà-tsień-loû, Mai, Juin 1896, Chasséurs Thibétains, Mironov slide no. 636♂ (holotype, ZFMK).

38. *Eupithecia proterva* Butler, 1878

Eupithecia proterva Butler, 1878, *Ann. Mag. nat. Hist.* (5) 1 (6): 445. Holotype ♂ (BMNH; examined), Japan: Yokohama.
Eup[ithecia] ? *detritata* f. *inconstans* Dietze, 1910, *Biologie der Eupithecien* 1: pl. 78, figs 787–790. Holotype ♂ (design. by Fletcher, coll. K. Dietze in MNHU; examined), [Japan]: Yokohama (synonymised in Prout, 1939).

Misidentification: *Eupithecia subbreviata* Inoue, 1959 (*nec* Staudinger, 1897).

Description. An average sized species with triangular fore wings and angulate hind wings. Fore wing basal, medial and terminal areas grey divided by pale, ash-grey transverse bands; basal, postbasal and antemedial lines sharply angled onto costa; postmedial obtuse or only slightly angled onto costa; medial area often with veins covered by black scales; terminal area narrow, usually with weakly visible wavy whitish subterminal line and more or less well marked small whitish tornal spot; discal dot rounded to ovoid, black. Hind wing pale ash-grey, with dark transverse lines well marked along anal margin; terminal area very narrow, slightly darker with wavy inner border; discal dot minute.

Male genitalia. Uncus short and narrow, uniapical. Valve broadened at base with short and blunt ventral process near base, tapered to narrow apex. Vinculum long and narrow, tapered to apex. Papillae narrow. Aedeagus narrow, slightly curved and broadened to anterior end, about equal to length of valve. Vesica minutely granulate, armed with a long, broad and curved U-shaped cornutus with thickened longitudinal margins and a small, folded, irregular plate near anterior end of aedeagus. Sternite A8 almost egg-shaped, emarginate at base and with a shallow U-shaped notch at apex.

Female genitalia. Bursa globular or ovate, membranous, densely covered with slim spines for $^2/_3$ of total area over base and left side, and with a long row of slim, longer spines in the membranous area along right side from base of ductus seminalis almost to base of colliculum. Ductus bursae not distinguishable. Ductus seminalis narrow, slightly broadened at base, strongly recurved, attached to central part of bursa from right side. Colliculum short and narrow. Antrum lightly sclerotized. Tergite A8 transverse, with shallow medial hollow in posterior margin. Apophyses narrow. Papillae anales slightly elongate, tapered.

Range. China (Shaanxi, Shanghai, Taiwan); Asian Russia (southern Primorje), Korea, Japan (Hokkaido, Honshu, Sado, Shikoku, Kyushu, Yakushima).

Phenology. Univoltine: early February to early May (late May according to Oh, 1992). Possibly overwintering as pupa.

Biology. Larva polyphagous. Found on *Acer mono* (Aceraceae), *Quercus mongolica* var. *grosseserrata* (Fagaceae) and *Viburnum plicatum* (Caprifoliaceae) (Japan: Inoue 1979).

Altitude. From close to sea level in Japan and eastern China to about 2600 m in Taiwan (Inoue 1988).

Similar species. Very similar to *E. subbreviata*, but usually smaller, with less elongate fore wings, and with darker medial areas and smaller discal dots on the fore wings. Determination by male and female genitalia is safest: the expanded sacculus and elongate vinculum on the male are distinctive, as is the shorter row of large spines between the bases of ductus seminalis and colliculum in the female.

Material. ♂, Japan 77–9, *E. proterva* Butler Type, BM Geom. slide no. 1950–590 (holotype, BMNH); ♂, [Japan], ♀ ? *subbreviata* v. *accessata*, Yokohama, Typus, Geom. slide no. 1950–589A (holotype [♂] of *E. inconstans* Dietze, coll. K. Dietze in MNHU); 1♂, [Japan], *accessata* ? n. sp., Yokohama, Geom. slide no. 1950–590A (paratype of *E. inconstans*, coll. K. Dietze in MNHU); 1♂, China, Sued-Shensi [Shaanxi], Tapaishan [Taibaishan] im Tsinling [Qinling], L, 2.vii.1935, H. Höne, Vojnits slide no. 10909♂ (TTM); 1♀, Taiwan, Nantou, Jenai Chunyang, 9–11.ii.1999, leg. C.S. Lin and H.T. Shih, NMNS ENT 3124–348, Galsworthy slide no. ACG TW20 (NSMTA); 1♀, Shanghai, Heude museum, 4–6.iv.1934, coll A. Savio, slide no ZL444♀ (IZCAS).
Remarks. Mironov & Galsworthy (2007) dealt with the identification of type material of *E. proterva*.

39. *Eupithecia subbreviata* Staudinger, 1897

Eupithecia subbreviata Staudinger, 1897, *Dt. ent. Z., Iris* **10** (1): 120, pl. 3, fig. 78. Syntypes 2♂ (coll. O. Staudinger in MNHU; examined), [Russia]: Askold.

Misidentifications: *Eupithecia proterva* Prout, 1915 (part., *nec* Butler, 1878); *Eupithecia detritata* Staudinger, 1897 (1♀ paralectotype in coll. O. Staudinger in MNHU).

Description. A narrow-winged, ash-grey species. Fore wing basal, medial and terminal areas dark ash-grey; transverse lines straight; basal, postbasal and antemedial sharply angled onto costa; postmedial right or obtuse angled onto costa; terminal area narrow, usually with well marked, wavy or dentate, pale subterminal line; discal dot small. Hind wing pale ash-grey, paler than fore wing; basal and postmedial transverse lines usually well marked near anal margin only; discal dot minute, rounded, smaller and paler than on the fore wing.
Male genitalia. Uncus short, uniapical. Valve evenly tapered to narrow apex, sacculus lightly sclerotized and slightly broadened at base. Vinculum relatively short, tapered. Papillae rather narrow. Aedeagus narrow, broadened anteriorly, almost equal to length of valve. Vesica armed with an elongate, V-shaped cornutus with apically broadened arms, and a small, irregular plate near anterior end of aedeagus. Sternite A8 small, almost triangular: two variations are illustrated.
Female genitalia. Bursa ovate-oblong, membranous, densely covered with spines over base and left side, and with a long separate row of spines from base of ductus seminalis almost to colliculum, the main spine field extended almost to meet this at its apex. Ductus bursae not distinguishable. Ductus seminalis narrow, attached to corpus bursae close to base on right. Colliculum short and narrow. Antrum lightly sclerotized. Tergite A8

almost square with shallow, membranous medial hollow on posterior margin. Apophyses relatively short and narrow. Papillae anales slightly elongate, rounded apically.

Range. China (Shaanxi); Asian Russia (Priamurje, S. Primorje), Korea, Japan (Hokkaido, Honshu, Sado, Shikoku, Kyushu, Tsushima, Yakushima).

Phenology. Univoltine: late March to late May.

Altitude. From sea level to about 1600 m in Qinling Mts. (China: coll. ZFMK).

Similar species. See differential diagnosis under the previous species.

Material. 1♂, [Russia], Askold Dörr., *Subbreviata* Stgr., Origin., bon. spec. bei *abbreviata* (Boh.), Vojnits slide no. 19020♂ (syntype, MNHU); 1♂, [Russia], Askold, 78., *Subbreviata* Stgr., Origin. (syntype, MNHU); 1♀, [Russia], Wladiwstk Chr., 11/5, 24., *detritata* ♂, Origin., Gen. vérifiés 1449[1] *Eupithecia detritata* Stgr., ♀, dét. E. de Laever 1954 (paralectotype of *E. detritata*, design. by Mironov, 2009, MNHU); 2♀, China, Shaanxi, Taibai Shan, 33°53′ N, 107°49′ E, 1500 m, April 2000, leg. Siniaev & C° (ZSM); 2♂, 3♀, C. China, prov. Shaanxi, Mt. Tai bei shan, Hao Djenzy village, 33°48′ N, 107°43′ E, 1400 m, 1.-5.iv.1999, leg. V. Siniaev & A. Plutenko (ZFMK); 4♂, 31♀, C. China, prov. Shaanxi, Foping Natural Reserve, Panda area, 33°45′ N, 107°48′ E, 1600 m, 6.-11.iv.1999, leg. V. Siniaev & A. Plutenko (ZFMK); 3♂, 7♀, same loc. and alt., 15.-20.iv.1999, leg. V. Siniaev & A. Plutenko (ZFMK).

Remarks. Not previously recorded from China.

40. *Eupithecia megaproterva* Inoue, 1988

Eupithecia megaproterva Inoue, 1988, *Bull. Fac. domest. Sci. Otsuma Wom. Univ.* 24: 326, figs 2A-F. Holotype ♂ (BMNH; examined), [Taiwan]: Hualien Hsien, Tayuling, 2600 m.

Description. Fore wing elongate with pointed apex, ground colour soft grey with two pale orange transverse bands between basal, medial and terminal areas; dark transverse lines indistinct; terminal area with narrow, wavy whitish subterminal line and white tornal spot; discal dot small, intensely black. Hind wing with shallow medial hollow at the termen; ground colour paler, whitish grey; transverse lines inconspicuous; terminal area slightly darker with indistinct whitish subterminal line forming two white spots near tornus; discal dot smaller and paler. First abdominal segment pinkish orange.

Male genitalia. Uncus narrow, biapical. Valve relatively small, with well-sclerotized sacculus strongly broadened at base. Vinculum elongate, tapered. Papillae unusually short and stout. Aedeagus broad, considerably longer than valve length. Vesica covered with numerous minute, oblique striations, and armed with a broad, U-shaped cornutus with anteriorly broadened arms and a small, bowed, irregular plate near anterior end of aedeagus. Sternite A8 short and broad, pear-shaped with a broad obtuse apex.

Female genitalia. Bursa ovate, almost completely covered with spines except for a small medial area and patches near ductus seminalis base and colliculum on dorsal and ventral sides. Ductus bursae not distinguishable. Ductus seminalis broadened basally, directed forward, attached to middle of corpus bursae from right side. Colliculum short and narrow. Antrum lightly sclerotized, covered by numerous minute pores. Tergite A8 trapezoid, broadened medially, with rounded posterior corners and sclerotized, band-like, concave anterior margin. Apophyses relatively short. Papillae anales broad and rounded.

Range. China (Yunnan, Taiwan).

Phenology. Univoltine: mid-February to early April.

Altitude. From about 1700 to 2800 m (Taiwan: coll. TTM; ZSM).

Similar species. This species is superficially similar to *E. costiconvexa*, but can be distinguished from it by the more extensive orange colouring on the postmedial and the presence of two distinct white marginal spots on the fore wing and one on the hind wing. The latter species also generally has a much paler hind wing. The male genitalia have a less prominent ventral process, a longer vinculum, shorter papillae on the ends of anterior arms of the labides, a longer aedeagus and a considerably broader eighth sternite than in *E. costiconvexa*. The female genitalia lack the separate chain of spines in the corpus bursae near the base of the ductus seminalis, which is present in *E. costiconvexa*.

Material. ♂, Taiwan, Tayuling, 2600 m, Hualien Hsien, 12–14.iii.1985, Coll. K. Yazaki, BM Geom. slide no. 16658 (holotype, BMNH); 1♂, China, Provinz Nord-Yuennan, Li-kiang [Lijiang], 9.iii.1935, H. Höne, de Laever slide no. 211♂ (ZFMK); 1♂, Taiwan, Prov. Tai-Chung, 2 km S of Shengkuang, 1960 m, 26.ii.1996, leg. Gy. Fábián & L. Németh (ZSM); 1♂, 1♀, Taiwan, Prov. Nan-Tou, Tayuling, 2550 m, 27.ii.1996, leg. Gy. Fábián & L. Németh (ZSM); 1♂, Taiwan, Prov. Tai-Chung, An-ma-Shan, Hooping, 2000 m, 2.iii.1996, leg. Gy. Fábián & L. Németh (TTM); 1♀, Taiwan, Prov. Hualien, Taroko N.P., Kuanyuan, 2400 m, 4.iv.1997, leg. Csorba & Ronkay (TTM); 2♂, Taiwan, Prov. Nantou, Road no. 7, 1700 m, 24°24′ N, 121°21′ E, 14.ii.1999, leg. L. Peregovits & G. Ronkay (TTM); 1♂, Taiwan, Prov. Nantou, Yushan N.P., Tatachia, 2610 m, 23°29′ N, 120°54′ E, 21.ii.1999, leg. L. Peregovits & G. Ronkay (TTM); 1♂, Taiwan, Prov. Nantou, Yuanfeng, SW of Hohuan Pass, 2760 m, 24°07′ N, 121°15′ E, 1.iv.2000, leg. A. Kun & L. Peregovits

(ZSM); 6♂, 1♀, Central Taiwan, near Puli, Nantou Co., Meifeng, *ca.* 2200 m NN, 24°05'25" N / 121°10'21" E, LF, 09, 10. January, 16, 23. February 2004, leg. Mei-Yu Chen (ZSM; 1♂, 1♀ in ZISP).

41. *Eupithecia verprota* Mironov & Galsworthy, 2006

Eupithecia ***verprota*** Mironov & Galsworthy, 2006, *Trans. lepid. Soc. Japan* 57 (4): 338, figs 2, 14, 18. Holotype ♂ (ZFMK; examined), C. China: prov. Shaanxi, Foping Natural Reserve, Panda area, 33°45' N 107°48' E, 1600 m.

Description. Fore wing ground colour pale grey; basal area brownish; medial lines inconspicuous, forming two broad, brown costal spots; terminal area darker, brownish with pale, narrow tornal spot; discal dot moderately large, circular, black. Hind wing ground colour pale, whitish grey; transverse lines clear, wavy, brownish; basal area brown near anal margin; discal dot circular, small.

Male genitalia. Uncus short, relatively slender, biapical. Valve narrow, broadened basally with prominent rather square ventral process. Vinculum large, wide, rounded. Papillae short and broad. Aedeagus stout, slightly curved, equal to length of valve. Vesica covered with numerous minute dentate spikes, armed with a long curved plate and a small irregular plate near ductus ejaculatorius base. Sternite A8 very wide, almost triangular, with prominent and wavy lateral margins, more sclerotized medially.

Female genitalia. Bursa ovate, almost completely covered with dense spines except near base of ductus seminalis, colliculum, and a curved area between base of ductus seminalis and colliculum on the ventral side; medial spines larger than basal and apical. Ductus bursae not clearly distinguishable. Ductus seminalis broadened near base, recurved, attached to bursa near its base on the left. Colliculum elongate and relatively narrow. Antrum slightly sclerotized, covered by numerous pores. Tergite A8 rather elongate, slightly broadened posteriorly, with a shallow medial hollow on the posterior margin, rounded posterior angles and a narrow more sclerotized medial area. Apophyses relatively long and narrow, slightly expanded at tips. Papillae anales rather narrow and tapered.

Range. China (Shaanxi).

Phenology. Univoltine: April.

Altitude. All known specimens come from about 1600 m in the Qinling Mts. (China: coll. ZFMK).

Similar species. Distinguished from other species of this group by the paler grey colour of the wings, the indistinct medial lines and the presence of dark costal spots on the fore wings. It most resembles large pale specimens of *E. subbreviata*, which is easily distinguishable in the male genitalia, since the valves lack the developed saccular extension. In the male, the shape of the valves, with a more prominent, blunter ventral process, the very broad eighth sternite, and in the female the much longer colliculum distinguish this species from two other closely related species *E. proterva* and *E. niphonaria* Leech, 1897.

Material. ♂, C. China, prov. Shaanxi, Foping Natural Reserve, Panda area, 1600 m, 6.-11.iv.1999, 33°45′ N, 107°48′ E, leg. V. Siniaev & A. Plutenko (holotype, ZFMK); 2♀, same data (paratypes, ZFMK); 1♂, 1♀, same data (paratypes, ZISP); 3♂, same loc. and alt., but date 15.-20.iv.1999 (paratypes, ZFMK).

42. *Eupithecia clavifera* Inoue, 1955

Eupithecia clavifera Inoue, 1955, *Tinea* 2 (1/2): (82), pl. 6, fig. 12; pl. 7, fig. 16. Holotype ♂ (BMNH; examined), [Japan]: Kanagawa Pref(ecture), Tanzawa Mts., Fudakake.

Eupithecia breviclava Inoue, 1988, *Bull. Fac. domest. Sci. Otsuma Wom. Univ.* 24: 328, figs 3A-F. Holotype ♂ (BMNH; examined), [Taiwan]: Hualien Hsien, Tayuling, 2600 m (synonymised in Mironov & Galsworthy, 2007).

Misidentification: *Eupithecia detritata* Staudinger, 1897 (1♀ paralectotype in coll. O. Staudinger in MNHU).

Unavailable names (incorrect subsequent spelling): *brevivlava*: Chang (1989–1991), Heppner & Inoue (1992).

Description. Fore wing strikingly narrow and elongate, and rather pointed at apex; ground colour almost unicolourous dark grey; transverse lines indistinct; terminal area with weak trace of a whitish, wavy subterminal line and small but distinct white tornal spot; discal dot small, black. Hind wing paler than fore wing, ground colour grey; transverse lines inconspicuous except postmedial, which well marked along anal margin; whitish indented subterminal line well marked near tornus; discal dot small, rounded or invisible.

Male genitalia. Uncus narrow, biapical. Valve rather short and narrow. Vinculum relatively small, with a slight indentation at apex. Aedeagus very long, about twice as long as valve, narrow and curved, broadened at base. Vesica densely granulate, with a very long narrow, curved cornutus, longer than the aedeagus tube, with a spear-shaped tip, and a small, curved, irregular plate near anterior end of aedeagus. Sternite A8 small, peg-like, broadened near base, with truncate basal lobes and two very small, rounded apical teeth.

Female genitalia. Bursa very long, spindle-shaped, membranous, with two narrow elongate patches of spines on the dorsal and ventral sides, connected by a large spiniferous area with smaller spines extending almost to colliculum. Ductus bursae short, with longitudinal striations.

Ductus seminalis long, broadened at base, attached to base of corpus bursae. Colliculum moderately elongate and narrow. Antrum membranous. Tergite A8 relatively small, square. Apophyses short and narrow. Papillae anales short and rounded.

Range. China (Gansu, Shaanxi, Taiwan); Asian Russia (southern Primorje), Korea, Japan (Hokkaido, Honshu, Sado, Shikoku, Kyushu, Tsushima).

Phenology. Univoltine: late March to late June.

Biology. Larva found on *Cornus controversa* (Cornaceae) and *Quercus glauca* (Fagaceae) (Japan: Inoue 1979; Funakoshi 2008).

Altitude. From sea level to about 2600 m (Taiwan: Inoue 1988).

Similar species. Externally similar to *E. proterva*, but rather unicolourous, darker, with less distinctly marked medial areas on the fore wings. In the male genitalia, the extreme length of the curved aedeagus (longer than in any other known Asian *Eupithecia*) is diagnostic. The female genitalia are equally distinctive, with the elongate half-spined corpus bursae narrowing directly into the ductus seminalis at its anterior end. They can be distinguished from those of *E. signigera*, which have a similar general shape, by the anteriorly divided spine field.

Material. ♂, Japan, Mt. Tanzawa Sagami, 12.iv.1939, Coll. H. Inoue (holotype, coll. H. Inoue in BMNH); 1♀, 29/5, Origin., Vojnits slide no. 19025♀ (paralectotype of *E. detritata*, design. by Mironov, 2009, coll. O. Staudinger in MNHU); ♂, Taiwan, Tayuling, 2600 m, Hualien Hsien, 23–24.iii.1982, H. Yoshimoto leg., BM Geom. slide no. 16538♂ (holotype of *E. breviclava*, coll. H. Inoue in BMNH); 2♂, Taiwan, Prov. Ilan, Suyuan near Pinan, at the Road 7/1, 1550 m, 29.iii.1997, leg. Csorba & Ronkay (TTM); 1♀, Taiwan, Prov. Hualien, Taroko N.P., Kuanyuan, 2400 m, 4.iv.1997, leg. Csorba & Ronkay (TTM); 3♂, 2♀, C. China, prov. Shaanxi, Mt. Tai bei shan, Hao Djenzy village, 33°48′ N, 107°43′ E, 1400 m, 1.-5.iv.1999, leg. V. Siniaev & A. Plutenko (ZFMK); 24♂, 43♀, C. China, prov. Shaanxi, Foping Natural Reserve, Panda area, 33°45′ N, 107°48′ E, 1600 m, 6.-11.iv.1999, leg. V. Siniaev & A. Plutenko (ZFMK); 7♂, 6♀, same loc. and alt., 15.-20.iv.1999, leg. V. Siniaev & A. Plutenko (ZFMK); 1♂, 2♀, same loc. and alt., 20.iv.-11.v.1999, leg. V. Siniaev & A. Plutenko (ZFMK); 1♀, China, Shaanxi, Tsinling Mts. [Qinling], Foping Nature Reserve, 33°51′ N, 107°57′ E, 1600 m, 15.-20.iv.1999, leg. V. Siniaev & A. Plutenko, *ex* coll. Dr. A. Schintlmeister (ZFMK); 1♂, 1♀, N. Taiwan, Taoyuan Co., Mengchi Recreation Area, *ca.* 1100 m, 4.iii.2000, lux., N.Y. Wang leg., Mironov slides nos. 721♂, 723♀ (ZFMK); 1♀, Taiwan, Prov. Nantou, Hohuachi, 1950 m, 24°13′ N, 121°16′ E, 31.iii.2000, leg. A. Kun & L. Peregovits (ZSM); 1♂, China, Shaanxi, Taibai Shan, 33°53′ N, 107°49′ E, 1500 m, April 2000, leg. Siniaev & C° (ZSM); 2♀, Gansu, Tianshui Dongcha forestry reserve, Taohuaping, 990 m, 15–16.iv.2005, Wang Xuejian, IOZ(E)1847082 and 3, slide no. ZL694 (IZCAS).

43. *Eupithecia wangi* Mironov & Galsworthy, 2011

Eupithecia ***wangi*** Mironov & Galsworthy, 2011, *Lep. Science* **62** (1): 18, figs 4, 19, 21. Holotype ♂ (NSMTO; examined), China, Guangxi, Guilin, Mao'ershan, 500–1400 m.

Description. Fore wing ground colour greyish brown; transverse lines almost invisible, except postmedial; latter a pale diffuse double stripe bordered internally by a series of short black lines on veins and twice angled close to costa, where it forms a distinct pale spot; terminal area with indistinct pale dentate subterminal line; discal dot small, rounded or ovoid, black. Hind wing ground colour paler, whitish grey; tornal area darker; dark transverse lines wavy, well marked in anal half only; posterior half of wing with a row of subterminal dark arrow-shaped marks; discal dot small, pale, ovoid or linear.

Male genitalia. Uncus short and narrow, uniapical. Valve slightly elongate and rather narrow. Vinculum elongate, tapered. Papillae elongate and stout. Aedeagus very large and stout, considerably longer than length of valve. Vesica covered with numerous minute spicules, armed with one very long, longitudinally folded cornutus, the apex curved and truncate, and with a smaller, folded, irregular cornutus near anterior end of aedeagus. Sternite A8 peg-like, broadened basally with obtuse apex.

Female genitalia. Bursa large, ovoid-oblong, densely covered with small spines in basal half from base of ductus seminalis diagonally to base of ductus bursae on right, and with a longitudinal row of spines along left side and a diagonal band of spines in the middle of the bursa. Ductus bursae short and relatively narrow. Ductus seminalis long and narrow, broadened at base, attached to anterior part of bursa near base on left. Colliculum not present. Antrum broad with marginal rows of slim, pointed spinules. Tergite A8 quadrate. Apophyses tapered; the posterior narrower than the anterior. Papillae anales elongate and narrow, with rounded apices.

Range. China (Guangxi).

Phenology. Univoltine: late March.

Altitude. Found from about 500–1400 m above sea-level in Mao'ershan Mts. (China: coll. NSMTO).

Similar species. Externally superficially similar to *E. flavoapicaria* (Japan, Taiwan), but easily distinguished by the much shorter pale spot on the fore wing costa, as well as the paler hind wings with distinctly bordered terminal areas. Both male and female genitalia are quite distinct from those of *E. flavoapicaria*. The male genitalia most resemble those of *E. jermyi* and *E. lini*, but the adult is externally dissimilar to both. The female genitalia are closest to those of *E. costiconvexa*, but the disposition of spines in the corpus bursae is different.

Material. ♂, China, Guangxi, Guilin, Mao'ershan, 500–1400 m, 26–29.iii.2005, M. Wang et al. leg., Galsworthy slide nos. NSMT-1 (holotype, SCAU). 2♂, 2♀, same locality, date and collector, paratypes, Galsworthy slide nos. NSMT-2 (NSMT, 1♂, 1♀, BMNH).

44. *Eupithecia signigera* Butler, 1879

Eupithecia signigera Butler, 1879, *Ann. Mag. nat. Hist.* (5) 4 (24): 442. Lectotype ♀ (BMNH; examined), [Japan]: [Yokohama].

Eupithecia carearia Leech, 1897, *Ann. Mag. nat. Hist.* (6) **20**: 69. Lectotype ♂ (BMNH; examined), Japan (synonymised in Inoue, 1977).

Unavailable names (incorrect subsequent spelling): *signifera*: Matsumura (1905).

Description. A smallish species with narrow elongate fore wings. Fore wing ground colour pale brownish ochreous, basal and medial areas brown or brownish grey; paler antemedial straight, oblique, sharply angled onto costa; postmedial pale, rather broad, angled onto costa behind discal dot; stalk of R vein covered by black scales in basal area; terminal area brown, with wavy whitish subterminal line forming a small white tornal spot with dark brown inner shade; discal dot small and black. Hind wing ground colour paler, whitish grey with ochreous tinge; anal margin darker, brown from base to postmedial line with traces of transverse lines; discal dot small.

Male genitalia. Uncus biapical, narrow. Valve relatively short and narrow, with a broad and blunt ventral process at base. Vinculum narrow and elongate, tapered to apex, sometimes with a shallow medial hollow. Papillae elongate and narrow. Aedeagus broadened at anterior end, considerably longer than length of valve. Vesica granulate, armed with a very long, narrow cornutus connected at base with a small, folded, irregular plate near anterior end of aedeagus. Sternite A8 small, peg-like, broadened basally, apex narrow.

Female genitalia. Bursa spindle-shaped, membranous, densely covered with small spines along left side from ductus bursae to base of ductus seminalis. Ductus bursae narrow, membranous, thick-walled with numerous wrinkles. Ductus seminalis long, evenly broadened to base, attached to base of corpus bursae. Colliculum quite elongate, narrowed posteriorly. Antrum sclerotized, covered by numerous minute pores. Tergite A8 short

and broad with a more sclerotized, band-like anterior margin. Apophyses relatively short. Papillae anales rounded.
Range. China (Shaanxi); Korea, Japan (Honshu, Sado, Shikoku, Kyushu, Tsushima, Yakushima).
Phenology. Univoltine: February to late May.
Biology. Larva polyphagous. Found on *Abelia spathulata* (Caprifoliaceae), *Symplocos chinensis* (Symplocaceae), *Pieris japonica*, *Rhododendron kaempferi* (Ericaceae) and *Euonymus alata* (Celastraceae) (Japan: Inoue 1979).
Altitude. From sea level to about 1600 m in Qinling Mts. (China: coll. ZFMK).
Similar species. Variable in size and in the depth of colour and distinctness of markings. Can be distinguished from other allied representatives of the *proterva* group by the distinctive pale brownish ochreous ground colour and the presence of a dark, blackish brown tornal spot on the fore wing. The male genitalia are broadly similar to those of *E. megaproterva* but can be easily distinguished by the the longer, narrower aedeagus with a much narrower cornutus on the vesica and the shape of the eighth sternite which is much narrower than in *E. megaproterva*. The female genitalia resemble those of *E. clavifera,* but have an anteriorly undivided spine field in the corpus bursae.
Material. ♀, Japan 79–48, 574, *E. signigera* Butler Type, BM Geom. slide no. 3904 (lectotype, BMNH); ♂, H. Preyer Coll., Japan, Leech Coll. 1900–64, *E. carearia* sp. n., *Eupithecia carearia* Leech. Lectotype ♂ [*nec* ♀] designated by H. Inoue, 1979, *Bull. Fac. domestic Sci., Otsuma Woman's Univ.* 15: 180, BM Geom. slide no. 3903 ♂ (lectotype of *E. carearia*, BMNH); 1♂, C. China, prov. Shaanxi, Mt. Tai bei shan, Hao Djenzy village, 33°48′ N, 107°43′ E, 1400 m, 1–5.iv.1999, leg. V. Siniaev & A. Plutenko (ZFMK); 5♀, C. China, prov. Shaanxi, Foping Natural Reserve, Panda area, 1600 m, 6.-11. iv.1999, 33°45′ N, 107°48′ E, leg. V. Siniaev & A. Plutenko (ZFMK); 1♀, China, Shaanxi, Taibai Shan, 33°53′ N, 107°49′ E, 1500 m, April 2000, leg. Siniaev & C° (ZSM).
Remarks. Not previously recorded from China.

45. *Eupithecia russula* Mironov & Galsworthy, 2004

Eupithecia russula Mironov & Galsworthy, 2004, *Trans. lepid. Soc. Japan* 55 (1): 52, figs 8, 24 (*nec* 25). Holotype ♀ (ZFMK; examined), [China]: Tibet [Sichuan], Batang, Yangtse Valley, 2800 m.

Description. Fore wing ground colour pale reddish brown; basal and medial areas of wing suffused with dark brown scales; antemedial and medial obscure or suppressed; discal dot dark brown, slightly elongate; a triple postmedial line, angled twice close to costa, the inner two faint and strongly angled close to the discal dot, the outer stronger and thicker; subterminal area almost unmarked, but with slight brownish terminal suffusion at apical one third and at tornus. Hind wing ground colour white, with a very small discal dot; anal margin to cell strongly suffused with brown; terminal line brown; fringes as fore wing. Abdomen covered with greyish brown scales with rusty tinge dorsally and pale grey scales ventrally.

Female genitalia. Bursa globular, sclerotized, completely covered with small spines on the ventral side and on the dorsal side other than the spineless medial part. Ductus bursae slightly broadened near colliculum and membranous on left side. Ductus seminalis incurved, attached to upper part of corpus bursae. Colliculum short and rather narrow. Antrum slightly sclerotized, with numerous pores. Tergite A8 relatively large and broad. Apophyses short. Papillae anales short and wide.

Male unknown.

Range. China (Sichuan).

Phenology. Univoltine (?): mid-May.

Altitude. The two specimens were collected at about 2800 m in Tibet (China: coll. ZFMK, TTM).

Similar species. Rather similar externally to the European *E. cocciferata* Millière, 1864 and *E. abbreviata* Stephens, 1831. The female genitalia can be distinguished by the shape of the bursa copulatrix, the disposition of the spiniferous area, and the shape of tergite A8.

Material. ♀, Tibet [Sichuan], Batang, Yangtse Valley, *ca.* 2800 m, 13.v.1936, H. Höne, Mironov slide no. 507♀ (holotype, ZFMK); 1♀, same loc. and alt., 16.v.1936, H. Höne, Vojnits slide no. 1160♀ (TTM).

Remarks: the attribution of this species to the *proterva* species group is rather speculative.

46. *Eupithecia flavoapicaria* Inoue, 1979

Eupithecia flavoapicaria Inoue, 1979, *Bull. Fac. domest. Sci. Otsuma Wom. Univ.* 15: 182, figs 4M, 4N, 5K, 11D, 13A. Holotype ♂ (BMNH; examined), [Japan]: Fukuoka Pref[ecture], Omuta City.

Eupithecia mantissa Inoue, 1988, *Bull. Fac. domest. Sci. Otsuma Wom. Univ.* 24: 330, figs 5A-E. Holotype ♂ (BMNH; examined), [Taiwan]: Alishan, 2200 m (synonymised in Mironov & Galsworthy, 2007).

Description. Fore wing light brownish ochreous; whole basal part of stalk R and bases of veins in medial and terminal areas, especially R_5, picked out with black scales; a brownish medial area along costa and a lighter, rusty ochreous blotch between postmedial line and apex; a distinct small white tornal spot shaded internally by blackish brown; the largish black discal dot rounded to ovoid. Hind wing paler, with transverse lines faintly marked in darker brown along anal margin and tornus, with a small white tornal dot shaded by brown. First abdominal segment forms a dark brown ring on abdomen.

Male genitalia. Uncus short, sharply pointed, with rounded basal part. Valve elongate, narrow, with narrow apex and a short, almost square, prominently protruding sacculus. Vinculum tapered. Papillae short and broad. Aedeagus curved, shorter than length of valve. Vesica covered with numerous denticules and armed with one elongate, V-shaped cornutus with two apically broadened arms, and a shorter, bowed, irregular cornutus near ductus ejaculatorius base. Sternite A8 peg-like, sharply broadened at base, narrow with parallel lateral margins in posterior half and two short apical horns.

Female genitalia. Bursa large, ovoid, densely covered with spines over whole area apart from right side, which is membranous from colliculum to close to base. Ductus bursae not distinguishable. Ductus seminalis slightly broadened at base, attached to central part of corpus bursae on right. Colliculum elongate and narrow. Antrum membranous, long and narrow. T ergite A8 small, elongate. Apophyses narrow. Papillae anales small, relatively short and narrow, tapered.

Range. China (Taiwan); Japan (Kyushu).

Phenology. Univoltine: early December to late January.

Altitude. Found at about 2200 m above sea-level (Taiwan: Inoue 1988).

Similar species. *E. flavoapicaria* has a distinctive maculation which is unlike that of any other *Eupithecia* species. The male genitalia broadly resemble those of *E. proterva*, but are easily distinguished by the shorter sacculus and the shape of the eighth sternite. The female bursa lacks the separate row of spines between base of ductus seminalis and colliculum, which is present in *E. proterva*.

Material. ♂, [Japan], Fukuoka Pref., Ōmuta City, 14.xii.1972, T. Sata, BM Geom. slide no. 16360 (holotype, BMNH); ♂, Taiwan, Alishan, Naw-two hsien, 1–3.i.1986, Coll. H. Nakajima, BM Geom. slide no. 16540 (holotype of *E. mantissa*, BMNH); 1♀, details as holotype, Inoue slide no. 13110 (paratype of *E. mantissa*, BMNH).

47. *Eupithecia blenna* Mironov & Galsworthy, 2006

Eupithecia ***blenna*** Mironov & Galsworthy, 2006, *Trans. lepid. Soc. Japan* 57 (4): 335, figs 1, 13, 16. Holotype ♀ (ZFMK; examined), C. China: prov. Shaanxi, Foping Natural Reserve, Panda area, 33°45′ N 107°48′ E, 1600 m.

Description. Fore wing ground colour dark brownish grey; transverse lines inconspicuous except for oblique reddish postmedial, which is right angled near costa behind discal dot; terminal area slightly darker with narrow, crooked whitish subterminal line and small tornal spot; discal dot small and black. Hind wing off-white along costa and in the middle and brownish grey along terminal and anal margins; postmedial line well marked near anal margin; discal dot minute, rounded, dark brown.

Male genitalia. Uncus short and narrow, biapical. Valve narrow, slightly curved, sharply tapered at narrow apex; broad ventral processes close to bases of valves asymmetrical, narrower and more rounded on the right valve, broader and obtuse on the left valve. Vinculum rather short and wide. Papillae elongate and narrow. Aedeagus equal to length of valve. Vesica granulate, with numerous oblique wrinkles, armed with a long plate with thickened longitudinal margins and a small irregular plate near ductus ejaculatorius base. Sternite A8 short, peg-like, with relatively broad and rounded, narrowly sclerotized apex and a less sclerotized medial area close to base.

Female genitalia. Bursa large, ovate, covered with dense spines of varying sizes for three quarters of the right side. Ductus bursae not distinguishable. Ductus seminalis uncinate, attached to corpus bursae near its base on the left side. Colliculum short and relatively broad. Antrum membranous. Tergite A8 almost square, slightly broadened towards the posterior edge. Apophyses relatively short and narrow. Papillae anales broad and rounded.

Range. China (Sichuan, Shaanxi).

Phenology. Univoltine: early April to mid-May.

Altitude. From about 1600 m in Qinling Mts. to 3200 m in Daxue Shan (China: coll. ZSM).

Similar species. Externally this species is difficult to distinguish from *E. clavifera* and *E. megaproterva*, though it has a slightly larger wingspan. The male and female genitalia are distinct. The male genitalia are more or less similar to those of other species with a saccular protrusion, but the asymmetrical valves are unique in this group. The female genitalia are rather similar to those of *E. subfumosa* Inoue, 1965: however, the latter has almost uniform spines in the bursa copulatrix, a narrower colliculum and longer apophyses.

Material. ♀, C. China, prov. Shaanxi, Foping Natural Reserve, Panda area, 1600 m, 6.-11.iv.1999, 33°45′ N, 107°48′ E, leg. V. Siniaev & A. Plutenko, Mironov slide no. 534♀ (holotype, ZFMK); 1♂, same loc. and alt., 15.-20. iv.1999, 33°45′ N, 107°48′ E, leg. V. Siniaev & A. Plutenko, Mironov slide

no. 654♂ (paratype, ZFMK); 1♂, China, Sichuan, Daxue Shan, Gongga Shan, NW Moxi, 29°41′ E, 101°58′ N, 23.iv.-15.v.2001, 2600–3200 m, leg. V. Siniaev & A. Plutenko, *ex* coll. Dr. A. Schintlmeister (paratype, ZSM).
Remarks. The two paratype males are in worn condition; their eighth sternites differ slightly from one another in the shapes of the basal lobes.

48. *Eupithecia convexa* Inoue, 1988

Eupithecia convexa Inoue, 1988, *Bull. Fac. domest. Sci. Otsuma Wom. Univ.* **24**: 328, figs 4A-F. Holotype ♂ (BMNH; examined), [Taiwan]: Hualien Hsien, Tayuling, 2600 m.

Description. A typical representative of the *proterva* species group with narrow, pointed wings. Fore wing brown; transverse lines indistinct, oblique; postmedial line twice sharply angled near costa; basal, medial and terminal areas divided by two lighter, brownish red, oblique transverse bands; terminal area with a whitish wavy subterminal line, very dentate near apex and forming a small white tornal spot; discal dot intensely black, small, linear or ovoid. Hind wing ground colour white; transverse lines rather inconspicuous, visible close to anal margin only; discal dot very small, circular.
Male genitalia. Uncus narrow, biapical. Valve short and broad, ventral margin with two broad blunt processes with a deep indentation between them. Vinculum broad and elongate. Papillae short and stout. Aedeagus slightly sinuate, broadened at anterior end, longer than valve length. Vesica minutely granulate with oblique striations, armed with one long, broad, slightly curved, V-shaped cornutus with broadened apical arms, a small, spiniferous, apical plate and a small, folded, irregular plate near anterior end of aedeagus. Sternite A8 peg-like, broadened basally, with two short blunt apical rods.
Female genitalia. Bursa globular, almost completely covered with dense spines except for posterior area near base of colliculum and ductus seminalis. Ductus bursae short and narrow, membranous, thick-walled, covered with numerous wrinkles, with an obliquely elongate patch of microspines. Ductus seminalis evenly broadened to base, attached to posterior part of corpus bursae on left. Colliculum short and narrow, tapering. Antrum sclerotized, covered by numerous minute pores. Tergite A8 almost square, slightly broadened medially. Apophyses rather short. Papillae anales relatively small, slightly elongate, tapered.
Range. China (Taiwan).
Phenology. Univoltine: mid-February to late March.

Altitude. From about 450 to 2600 m above sea-level (Taiwan: Inoue 1988; coll. BMNH, TTM, ZSM).
Similar species. See comments under the very similar Taiwanese *E. concava*.
Material. ♂, Taiwan, Tayuling, 2600 m, Hualien Hsien, 23–24.iii.1982, H. Yoshimoto leg., BM Geom. slide no. 16539 (holotype, coll. H. Inoue in BMNH); 1♀, Taiwan, Prov. Kao-Hsiung, 15 km NE of Taoyuan, 1850 m, 12.iii.1996, Gy. Fábián & L. Németh (TTM); 2♀, Taiwan, Prov. Taipei, 10 km SE of Pinglin, Pihou, 450 m, 26.iii.1997, leg. Csorba & Ronkay (TTM); 1♂, 1♀, Taiwan, Prov. Ilan, Ming Chyr Forest Recreation Area, 1200 m, 30–31.iii.1997, leg. Csorba & Ronkay (TTM); 1♀, Taiwan, Prov. Nantou, Road no. 7, 1700 m, 24°24′ N, 121°21′ E, 14.ii.1999, leg. L. Peregovits & G. Ronkay (TTM); 1♂, Taiwan, Prov. Taitung, Hsiangyang, 2300 m, 23°14′ N, 120°58′ E, 17.ii.1999, leg. L. Peregovits & G. Ronkay (TTM); 1♀, Taiwan, Prov. Nantou, Yushan N.P., Tatachia, 2610 m, 23°29′ N, 120°54′ E, 21.ii.1999, leg. L. Peregovits & G. Ronkay (TTM); 1♀, Taiwan, Prov. Nantou, Hohuachi, 1950 m, 24°13′ N, 121°16′ E, 31.iii.2000, leg. A. Kun & L. Peregovits (ZSM).

49. *Eupithecia opicata* Mironov & Galsworthy, 2004

Eupithecia ***opicata*** Mironov & Galsworthy, 2004, *Trans. lepid. Soc. Japan* 55 (4): 294, figs 7, 20. Holotype ♂ (IZCAS; examined), China: Gansu, Yongdeng, Tulugou, 2280 m.

Description. Fore wing very elongate, apex pointed; ground colour mid brown, blotched darker, especially along costa and in terminal area; transverse lines not visible; discal dot dark brown, narrow, elongate. A strong, pale oblique postmedian fascia, twice sharply angled close to costa; terminal area crossed by a pale wavy line; terminal line dark brown, broken at veins. Hind wing ground colour off-white, contrasting with fore wing; discal dot minute; transverse lines marked at anal margin only; terminal area slightly darker.
Male genitalia. Uncus long and narrow, biapical. Valve with broad ventral process, strongly sclerotized. Vinculum pointed. Papillae tapering. Aedeagus very stout, longer than valve length. Vesica covered with numerous small sclerotized wrinkles, and armed with a long, laterally bowed cornutus and a small V-shaped plate near anterior end of aedeagus. Sternite A8 elongate, peg-like, with heavily sclerotized apical cap, uneven lateral margins and short basal lobes.
Female genitalia. Bursa ovate-oblong, densely covered with spines except for a narrow area running obliquely from below colliculum to the ductus seminalis. Ductus bursae broad and membranous, heavily wrinkled.

Ductus seminalis very broadly attached to base of corpus bursae on right, narrowing rapidly. Colliculum large, very broad. Antrum partly sclerotized with numerous pores. Tergite A8 square. Apophyses short, broad, spatulate at tips. Papillae anales broad and rounded.
Range. China (Gansu, Shaanxi).
Similar species. Resembles *E. megaproterva* from Taiwan, but much smaller, with a pale median fascia on the fore wing. The male genitalia are closest to those of *E. costiconvexa*, but are distinguished from it by the longer sacculus (the distance between base of valve and tip of ventral process), the broader aedeagus with broader large cornutus on the vesica, and the narrower and slightly longer eighth sternite, with very shallow basal hollow. The female genitalia are distinct from those of other members of the group.
Material. ♂, China, Gansu, Yongdeng, Tulugou, 2280 m, 1–16.viii.1991, coll. Meng Feng; Galsworthy slide no. ZL518♂ (holotype IZCAS); 1♂, same data (paratype, IZCAS); 2♂, 2♀, C. China, prov. Shaanxi, Mt. Tai bei shan, Hao Djenzy village, 33°48′ N, 107°43′ E, 1400 m, 1.-5.iv.1999, leg. V. Siniaev & A. Plutenko, Mironov slides nos. 739♀, 741♀, 744♂ (ZFMK); 1♂, 1♀, same data (ZISP); 5♂, C. China, prov. Shaanxi, Foping Natural Reserve, Panda area, 33°45′ N, 107°48′ E, 1600 m, 6.-11.iv.1999, leg. V. Siniaev & A. Plutenko, Mironov slides nos. 740♂, 742♂, 743♂, 745♂ (ZFMK); 2♂, 1♀, same data, (ZISP).

50. *Eupithecia costiconvexa* Inoue, 1979

Eupithecia costiconvexa Inoue, 1979, *Bull. Fac. domest. Sci. Otsuma Wom. Univ.* 15: 178, figs 4J, 4K, 5I, 11C, 12B. Holotype ♂ (BMNH; examined), [Japan]: Kochi Pref[ecture], Muroto City, Sakinohama.

Description. Fore wing ground colour rather uniform brown, with transverse lines obscure; postmedial marked by a broad slightly paler fascia, twice angled near costa, and becoming a rhomboid yellow spot on the costa; terminal area slightly darker with an inconspicuous light, wavy subterminal line, and a large dark brown blotch near tornus; discal dot smallish, black. Hind wing slightly paler brown, transverse lines represented by two rows of small, dark brown dots on the veins along terminal margin; discal dot small, brown.
Male genitalia. Uncus very short, broad, biapical. Valve relatively short, sharply narrowed at apex, and with sclerotized sacculus forming a broad, blunt protruding asymmetrical flange, slightly longer on the right valve than on the left. Vinculum rounded, slightly tapered. Papillae elongate and narrow. Aedeagus longer than valve, broad and slightly curved,

broadened at anterior end, with a single long needle shaped cornutus, and a sclerotized knot at base. Sternite A8 peg-like, evenly tapered to apex, with two small blunt processes at apex, and a very small central hollow at base.

Female genitalia. Bursa large, ovoid with a rather flattened base, covered with small spines in anterior part and up right side, with two separate vertically orientated rows of spines. Ductus bursae short, membranous. Ductus seminalis broadened at base, sharply recurved, attached to anterior part of bursa on left. Colliculum not present. Antrum short, broad, slightly sclerotized, with numerous pores and a transverse row of numerous small spines. Tergite A8 trapezoid, broadened posteriorly. Apophyses relatively short, slightly expanded and flattened at apices. Papillae anales slightly elongate, rounded apically.

Range. China (Taiwan); Japan (SW Honshu, Shikoku, Kyushu, Tsushima, Yakushima), N. Thailand (Chiang Mai).

Phenology. Bivoltine: late January to first half of April and mid-August.

Altitude. Found at about 700 m in Taiwan (coll. TTM).

Similar species. Superficially similar to Taiwanese *E. flavimacula*, but distinguished from it by the lack of yellowish markings in the basal and terminal areas. *E. hashimotoi* is also similar, but is greyer, and has a much paler hind wing. The male genitalia are closest to those of *E. opicata*: see under that species for distinction. The female genitalia are similar to those of the mainland Chinese species *E. wangi*, but differ in the larger bursa, the larger spines of the separate row along the left side, the trapezoid tergite A8 and the slightly expanded and flattened apices of the apophyses.

Material. ♂, [Japan], Sakinohama, Muroto City, Kōchi Pref., 21.iii.1977, T. Masui, Inoue coll. B.M. 1992–71, type status verified K. Buckmaster 1992 (holotype, coll. H. Inoue in BMNH); 2♂, 1♀, Taiwan, Ilan County, Fu-Shan, Botanical Garden, 700 m, 24°54′02″ N, 121°45′27″ E, 21.i.2002, leg. L. Ronkay (TTM).

51. *Eupithecia concava* Mironov & Galsworthy, 2007

Eupithecia concava Mironov & Galsworthy, 2007, *Trans. lepid. Soc. Japan* 58 (3): 342, figs 1, 13, 20. Holotype ♂ (TTM; examined), Taiwan, Prov. Taoyuan, Ming Chyr Forest Recreation Area, 1160 m.

Description. Fore wing ground colour reddish brown, costal margin blackish brown, especially near apex; basal, medial and terminal areas divided by broad, lighter yellowish brown stripes; transverse lines inconspicuous; postmedial oblique, twice sharply angled near costa, situated in broadest of lighter brown transverse stripes, forming a conspicuous fascia; terminal area slightly darker, brown with white subterminal line much indented near apex and forming one small white medial spot and a larger, bisected white tornal spot; discal dot large and rounded, intensely black. Hind wing ground colour lighter, off-white with slightly darker terminal and anal areas, and a small, white tornal spot.

Male genitalia. Uncus very short and narrow, biapical. Valve broadened basally, sharply tapered, with two broad, blunt ventral processes. Vinculum broad and rounded. Papillae short and stout. Aedeagus stout, slightly curved and broadened at anterior end, almost equal to length of valve, with two broad lateral lobes at posterior end, covered with numerous spinules. Vesica with numerous, minute oblique striations, and armed with a long, narrow, slightly curved, V-shaped cornutus and a small, folded, irregular cornutus near anterior end of aedeagus. Sternite A8 short and ovate, broadened near base, its apical margin sclerotized and square at the corners.

Female genitalia. Bursa ovate-orbicular, broadened in posterior part, almost completely covered with dense small spines on dorsal side and with similar small spines on ventral side to about half way from base; posterior part on ventral side with a long, transverse band-like patch of small spines almost in the middle and another shorter chain of longer spines near base of ductus seminalis on left. Ductus bursae short, sharply narrowed near colliculum. Ductus seminalis evenly broadened to base, recurved, attached to posterior part of bursa on left. Colliculum short and narrow. Antrum sclerotized, covered with numerous pores. Tergite A8 small, almost quadrate. Apophyses relatively short and narrow, tapered. Papillae anales small, narrow with rounded apices.

Range. China (Taiwan).

Phenology. Univoltine: late February to late April.

Altitude. From about 1100 to 2150 m above sea-level (Taiwan: coll. ZFMK and MNHU).

Similar species. Very easily confused with the allied Taiwanese *E. convexa*, but distinguishable by the larger, rounded discal dots on the fore wings. The male genitalia are easily distinguishable from those of *E. convexa* by the less prominent and broader ventral processes on the valve, and the different shape of the eighth sternite. The female genitalia are distinguished by the presence of a transverse chain of small spines on the ventral side of the bursa and the separate short chain of longer spines near the ductus seminalis base, as well as by the eighth tergite being narrower than in *E. convexa*.

Material. ♂, Taiwan, Prov. Taoyuan, Ming Chyr Forest Recreation Area, 1160 m, 29.ii.1996, leg. Gy. Fábián & L. Németh (holotype, TTM); 1♂, N. Taiwan, Taoyuan Co., Mengchi Recreation Area, *ca.* 1100 m, 4.iii.2000, lux., H. Y. Wang leg., Mironov slide no. 651♂ (paratype, ZFMK); 1♀, Taiwan, Prov. Hualien, Taroko NP, 1400 m, 2.iv.1997, leg. Csorba & Ronkay, Galsworthy slide no. ACG GL6 (paratype, coll. GL); 1♀, Taiwan, Nantou, Piluhsi Station, 29.iv.2004, leg. W.T. Yang, NSMT ENT 4409–401, Galsworthy slide no. ACG TW18 (paratype, NSMTA); 1♀, Taiwan, Nantou, Meifeng, *ca.* 2150 m, 27.-30.iv.1999, leg. Mey & Ebert, Zool. Mus. Berlin, Mironov slide no. 813♀ (MNHU).

52. *Eupithecia aritai* Inoue, 1977

Eupithecia aritai Inoue, 1977, *Bull. Fac. domest. Sci. Otsuma Wom. Univ.* **13**: 321, fig. 44. Holotype ♂ (BMNH; examined), [Japan]: Aichi Pref(ecture), Nagoya, Yagoto.

Description. Fore wing ground colour pale brownish grey; transverse lines very distinct and dark brown near costa; basal line twice right angled near costa; antemedial slightly wavy, touching discal dot and sharply angled from discal dot to costa; postmedial obtusely curved onto costa; terminal area darker, with darker blotch near tornus and a wavy, whitish subterminal line; discal dot relatively large. Hind wing ground colour paler, transverse lines distinct, especially along anal margin, forming a series of dark dashes on veins; terminal area darker near tornus with distinct, light, dentate subterminal line; discal dot small. Abdomen pale brownish grey with black lateral stripes.

Male genitalia. Uncus short, pointed, uniapical. Valve regularly tapered to narrowly rounded apex, but with a small rounded protuberance at base of sacculus. Vinculum narrow, sharply tapered. Papillae short, slightly tapered. Aedeagus lightly curved, broadened at anterior end, narrow and longer than length of valve. Vesica armed with two narrow, longitudinally bifurcate, ribbon-like cornuti and a small, folded, irregular cornutus near ductus ejaculatorius base. Sternite A8 short, slightly sclerotized, with two narrow pincer-like apical arms.

Female genitalia. Bursa ovoid, thick-walled, with a single large patch of spines in the anterior part between left side and base of ductus seminalis. Ductus bursae not distinguishable. Ductus seminalis broadened and sharply curved, attached on right side of base. Colliculum short and narrow. Antrum short and wide, membranous with numerous pores. Tergite A8 elongate, with a medial hollow in the anterior margin, and a small, pointed medial projection on the posterior margin. Anterior apophyses short, about twice as broad as posterior. Papillae anales very rounded.

Range. China (Taiwan); Japan (SW Honshu, Shikoku, Kyushu, Yakushima, Iriomotejima).
Phenology. Univoltine: February to mid-May.
Altitude. Found at an altitude of about 2200 m (Taiwan: coll. ZSM).
Similar species. The rather broad wings and distinctive maculation immediately separate *E. aritai* from other members of this species group. The male genitalia are broadly similar to those of other members of the group which lack a more developed sacculus, but the shape of the eighth sternite is distinctive. In the female, the shape of the spiniferous area in the bursa, and the presence of a medial projection on the eighth tergite are diagnostic.
Material. ♂, [Japan], Yagoto, Nagoya Aichi Pref., 11. Apr. 1970, Y. Arita leg., Inoue Coll. B. M. 1992–71, Type status verified K. Buckmaster 1992 (holotype, coll. H. Inoue in BMNH); 1♀, Central Taiwan, near Puli, Nantou Co., Meifeng, *ca.* 2200 m, 24°05′11″ N/121°10′21″ E, LF, 11.v.2005, leg. Mei-Yu-Chan & U. Buchsbaum, Mironov slide no. 632♂ (ZSM).
Remarks. Not previously recorded from Taiwan.

53. *Eupithecia admiranda* Mironov & Galsworthy, 2011

Eupithecia ***admiranda*** Mironov & Galsworthy, 2011, *Lep. Science* **62** (1): 18, figs 5, 24, 26. Holotype ♂ (ZFMK; examined), C. China: prov. Shaanxi, Foping Natural Reserve, Panda area, 33°45′ N, 107°48′ E, 1600 m.

Description. Fore wing ground colour reddish ochreous; two conspicuous dark brown markings on costa at base, wedge-shaped, and at one half, triangular, and a further less well defined triangular mark at centre of dorsum; centre of medial area paler, reddish ochreous; discal dot conspicuous at apex of costal triangle, brownish black, ovoid; terminal area reddish ochreous with soft brownish tinge and two medial and tornal brownish blotches; whitish subterminal line relatively broad. Hind wing paler, whitish with ochreous tinge, with quadrate brown patches on anal margin representing basal and antemedial, and a more diffuse reddish brown mark representing postmedial; terminal area bordered by a wavy line; discal dot rounded, blackish.
Male genitalia. Uncus narrow, biapical. Valve rather narrow, with narrowly rounded apex. Vinculum elongate and relatively narrow, tapered. Papillae of average size, untapered. Aedeagus narrow, broadened anteriorly, longer than length of valve. Vesica covered with oblique striations and armed with one long, narrow, plate-like cornutus, bifurcate at base. Sternite A8 peg-like, with two short and narrow, lightly sclerotized lateral thickenings near apex.

Female genitalia. Bursa ovate, membranous, covered on basal four fifths of left side only with dense spines. Ductus bursae not distinguishable, area around colliculum spineless. Ductus seminalis narrow, broadened at base, attached at anterior end of bursa. Colliculum narrow and elongate. Antrum narrow, lightly sclerotized. Tergite A8 transverse, with a shallow medial hollow in posterior margin. Apophyses short and narrow, the posterior slightly expanded and flattened at apices. Papillae anales relatively small, rounded.

Range. China (Shaanxi, Tibet).

Phenology. Univoltine: second half of April.

Altitude. Found at about 1600 m above sea-level (China: Mironov & Galsworthy 2011).

Similar species. A very distinctly marked species, easily separable from most *Eupithecia*. The dark central fascia gives it a superficial resemblence to *E. fletcheri* Prout, 1926, which is smaller and more colourful, and has entirely different genitalia. The male genitalia resemble those of *E. subbreviata* and *E. hashimotoi*, but the former has a uniapical uncus and lacks horns on the eighth sternite, and the latter has a much larger aedeagus, considerably longer than the length of the valve. The female genitalia are closer to those of *E. hashimotoi*, but are easily distinguished by the much longer colliculum.

Material. ♂, C. China, prov. Shaanxi, Foping Natural Reserve, Panda area, 1600 m, 15.-20.iv.1999, 33°45′ N, 107°48′ E, leg. V. Siniaev & A. Plutenko, Mironov slide no. 630♂ (holotype, ZFMK); 2♀, same data, Mironov slide no. 631♀ (paratypes, ZFMK); 2♀, same data (paratypes, ZISP); 2♀, same loc. and alt., 20.iv.-11.v.1999, 33°45′ N, 107°48′ E, leg. V. Siniaev & A. Plutenko (paratypes, ZFMK); 1♀, same data (paratype, ZISP); 1♀, Tibet, Tschang-Tang Dsagar-Berge, 4500 m, Juli (paratype, ZFMK).

54. *Eupithecia hashimotoi* Inoue, 1988

Eupithecia hashimotoi Inoue, 1988, *Bull. Fac. domest. Sci. Otsuma Wom. Univ.* 24: 330, figs 6A-E. Holotype ♂ (BMNH; examined), [Taiwan]: Hualien Hsien, Tayuling, 2600 m.

Description. Fore wing ground colour grey usually with olive tinge; oblique transverse lines pale and straight, sharply angled onto costa; postmedial wavy, twice sharply angled onto costa; costal margin with prominent square, rich yellow spot at end of postmedial; terminal area with distinct wavy whitish subterminal line; discal dot rounded or ovoid, intensely black. Hind wing ground colour paler, white to ash-grey, terminal and anal margins darker; transverse lines grey, well marked; discal dot small and pale; fringes distinctly spotted light and dark grey on all wings.

Male genitalia. Uncus short and narrow, biapical. Valve slightly sinuate and tapered to narrow rounded apex. Vinculum conical. Papillae narrow and parallel-sided. Aedeagus stout and considerably longer than length of valve. Vesica with numerous minute, oblique wrinkles, and armed with a long, broad, twisted, V-shaped cornutus with heavily sclerotized lateral margins, bifid for anterior half, and a small, irregular plate near anterior end of aedeagus. Sternite A8 peg-like, broadened basally with wavy lateral margins and two short, obtuse, almost membranous apical rods.

Female genitalia. Bursa large, ovate, evenly covered with slim spines on right side from base to colliculum. Ductus bursae not distinct. Ductus seminalis slightly broadened basally, attached to base of bursa on left. Colliculum short, almost quadrate. Antrum funnel-shaped, slightly sclerotized. Tergite A8 broadened medially, with lateral margins narrowly sclerotized in anterior half and emarginate posterior corners. Apophyses short and narrow. Papillae anales rounded.

Range. China (Taiwan).

Phenology. Univoltine: late February to early March.

Altitude. From about 1500 to 2600 m above sea-level (Taiwan: Inoue 1988; coll. TTM, ZSM).

Similar species. Generally an easily recognized species. The conspicuous yellow costal spot distinguishes it from all congeners in Taiwan except *E. costiconvexa* and *E. flavimacula*. The former is a much darker species, and the latter has extensive yellow maculations. The male genitalia resemble those of *E. subbreviata* and *E. admiranda*, neither of which is externally similar; the combination of the large aedeagus and the shape of the eighth sternite are decisive. The female genitalia resemble those of *E. admiranda*, but have a much shorter and broader colliculum.

Material. ♂, Taiwan, Hualien county, Tayuling, 7.iii.1982, S. Hashimoto, BM Geom. slide no. 16541 (holotype, coll. H. Inoue in BMNH); 2♀, Taiwan, Prov. Nan-Tou, Tayuling, 2550 m, 27.ii. and 16.iii.1996, leg. Gy. Fábián & L. Németh (ZSM; TTM); 1♂, Taiwan, Prov. Ilan, Suyuan near Pinan, at the Road 7/1, 1550 m, 29.iii.1997, leg. Csorba & Ronkay (TTM); 2♀, Taiwan, Prov. Taitung, Hsiangyang, 2300 m, 23°14′ N, 120°58′ E, 17.ii.1999, leg. L. Peregovits & G. Ronkay (TTM); 2♀, Taiwan, Prov. Tai-Chung, Anma-Shan, Hooping, 2000 m, 2.iii.1996, leg. Gy. Fábián & L. Németh (TTM).

55. *Eupithecia flavimacula* Mironov & Galsworthy, 2007

Eupithecia flavimacula Mironov & Galsworthy, 2007, *Trans. lepid. Soc. Japan* 58 (3): 343, figs 2, 14. Holotype ♂ (ZFMK; examined), N. Taiwan: Taoyuan Co., Mengchi Recreation Area, 1100 m.

Description. Fore wing ground colour grey-brown with soft yellowish tinge, basal area with two yellow blotches at costal and anal margins; antemedial evenly curved; postmedial wavy, sharply angled near costa; medial area relatively broad, delimited from terminal area by a yellowish band with large yellow costal and smaller anal spots; terminal area broad, dark grey, with narrow, wavy subterminal line and three yellow spots, costal, medial and tornal; discal dot small, rounded, black; fringes long, chequered grey and pale yellowish grey. Hind wing ground colour, terminal line and fringes as fore wing; transverse lines inconspicuous, except for crenulate postmedial; terminal area with two small medial and tornal yellow spots; discal dot smaller and paler than on fore wing, rounded.
Male genitalia. Uncus short and narrow, biapical. Valve short, sharply narrowed from an obtuse angle at apex of sacculus; latter fairly heavily sclerotized, forming a round protrusion with a fold at centre. Vinculum long and broad, with a shallow medial hollow. Papillae broad, parallel-sided. Aedeagus stout, slightly curved, broadened anteriorly, longer than length of valve. Vesica covered with oblique striations and armed with one long curved plate with thickened longitudinal margins and one small irregular plate near ductus ejaculatorius base. Sternite A8 lightly and evenly sclerotized, short and broad with two parallel, broad, pointed apical rods shaped like pincers.
Female unknown.
Range. China (Taiwan).
Phenology. Univoltine: early March.
Altitude. Found at about 1000 m above sea-level (Taiwan: coll. ZFMK).
Similar species. The well developed yellow costal spot gives this species a slight resemblance to *E. costiconvexa* and *E. hashimotoi* but it is distinguished from both by the presence of more extensive yellow maculations on all wings. The male genitalia are similar to those of an allied Japanese species *E. antivulgaria* Inoue, 1965, but the shape of the valve is somewhat different, the cornutus more slender and the shape of the eighth sternite quite different. *E. antivulgaria* is a late autumn species whilst *E. flavimacula* is a spring species.
Material. ♂, N. Taiwan, Taoyuan Co., Mengchi Recreation Area, *ca.* 1100 m, 4.iii.2000, lux., H.Y. Wang leg. (holotype, ZFMK).

The *irambata* species group

This species group is introduced here for the first time. It consists of four rather large broad-winged species, all from the Himalayan-Tibetan area.

The close relationship of these externally rather similar species is supported by features of the genitalia which are common to them. In the male, the valves are rather elongate, and unmodified, and the eighth sternite consists of two sclerotized arms, joined anteriorly only by a narrow sclerotized strip. The female genitalia in the three species where they are known are very similar, with a small globular, partly spined corpus bursae, a narrow elongate colliculum, a rather transverse eighth tergite, with spreading anterior apophyses, and rather squat papillae anales. Three of the four species occur in China; the fourth, *E. irambata* Warren, 1893 is known from the Indian Himalayas only.
Male genitalia. Uncus relatively short and narrow, uni- or biapical. Valve usually elongate, without modifications; sacculus thickened but lightly sclerotized. Vinculum semicircular or slightly tapered to apex. Papillae on the anterior arms of the labides short. Aedeagus rather small and narrow, considerably shorter than length of valve. Vesica with a variety of ornamentation, ranging from a bank of spines to a small U-shaped cornutus. Sternite A8 two parallel, elongate and narrow rods, connected with each other by a short and narrow basal band.
Female genitalia. Bursa copulatrix small, membranous, egg-shaped or ovoid, about 2/3 covered with relatively small spines on the right side. Ductus bursae not distinguishable. Ductus seminalis narrow, attached to posterior part of corpus bursae near base of colliculum on left side. Colliculum narrow and unusually elongate. Antrum short, membranous. Tergite A8 usually large, broad. Apophyses short and stout, pointed at apices. Papillae anales very large, broad and usually obtuse.
Remarks. This species group may well be related to the *graphata* species group on the basis of male and female genitalic characteristics. However, the form of sternite A8 with two long and narrow rods narrowly joined at the base in the male, and the stout apophyses and very large and broad papillae anales in the female constitute clear and consistent differences from that group.

56. *Eupithecia amplificata* Inoue, 2000

Eupithecia amplificata Inoue, 2000, *Tinea* **16** (Suppl. 1): 28, pl. 165, fig. 4; figs 1284, 1326. Holotype ♂ (coll. H. Nakajima), [Nepal]: [Bagmati], Kyangjin, Langtang Himal, 3900–3910 m.

Description. A large handsome species with broad light ochreous fore wings with distinct, oblique, brownish transverse lines; basal, antemedial and medial sharply angled onto costa, postmedial internally with a series of darker, brown dashes on the veins, and acutely curved onto costa;

terminal area slightly darker, brownish with a distinct, fine, very dentate, white subterminal line and a distinct white tornal spot; black discal dot small and elongate. Hind wing paler, with distinct brownish transverse lines, the postmedial especially distinct, and a darker brownish ochreous terminal margin; discal dot small and pale. Fringes of all wings distinctly spotted with off-white and brown. Abdomen clear brownish ochreous.
Male genitalia. Uncus narrow and pointed, uniapical. Valve long and relatively narrow. Vinculum short and broad, rounded. Papillae short and narrow. Aedeagus narrow, considerably shorter than length of valve. Vesica wrinkled, armed with an elongate, longitudinally folded, U-shaped cornutus and a smaller, irregular cornutus near ductus ejaculatorius base. Sternite A8 two narrow, parallel arms, inwardly curved at apices and with blunt tips, connected by a short, narrow basal band.
Female genitalia. Bursa membranous, ovate-oblong, covered with small spines from base to colliculum and base of ductus seminalis on right half. Ductus bursae not present. Ductus seminalis narrow, sharply curved near base, attached to posterior part of bursa near colliculum on left. Colliculum long and narrow. Antrum membranous, narrow. Tergite A8 broad, rather trapezoid, with a longer and more strongly sclerotized anterior margin, and a slightly convex posterior margin. Apophyses short and stout, the posterior expanded close to their point of origin. Papillae anales very broad and flat.
Range. China (Yunnan); Nepal.
Phenology. All records from August or September.
Altitude. The only data is from 3900 m in Nepal.
Similar species. Similar to *E. irambata*, Warren, 1893, but distinguishable by its darker, brownish ochreous colour with distinct transverse lines on all wings and the larger discal dots on the fore wings. The narrower plate-like base of the uncus and the narrower papillae on the anterior arms of the labides in the male genitalia distinguish *E. amplificata* from *E. irambata*. There are no obvious differences between the female genitalia of the two species.
Material. ♂, Nepal, Langtang Himal, Bagmti Zone, Kyangjin, 3900 m, 11.viii.1993, Nakajima Hideo leg. (coll. H. Nakajima, Tokyo); 1♂, 7♀, China, Provinz Nord-Yuennan, Li-kiang [Lijiang], 29, 31.viii, 1, 4, 6, 7, 20, 21.ix.1935, H. Höne, de Laever slides nos. 109♂, 110♀, Mironov slide no. 461♀ (ZFMK).
Remarks. Not previously recorded from China.

Inoue (2000) illustrated the female genitalia of *E. amplificata* (fig. 1326) and *E. irambata* (fig. 1355), the latter of which he described as a new species, *E. stramineata*. Having examined the holotypes of both, we conclude that *E. irambata* and *E. stramineata* are conspecific, and consequently sink *stramineata* (**new junior synonym**). From the illustrations it would appear that the papillae anales of *amplificata* and *irambata* are of an entirely different shape. This is however an artefact of the preparation of *E. amplificata* shown in fig. 1326, where the papillae anales are squashed and folded in two. In fact they are indistinguishable between the two taxa.

57. *Eupithecia asempiterna* Inoue, 2000

Eupithecia asempiterna Inoue, 2000, *Tinea* **16** (Suppl. 1): 29, pl. 165, fig. 7; figs 1286, 1328. Holotype ♂ (NSMTO; examined), [Nepal]: Sagarmatha, Thame Og, 3800 m. [Replacement name for *Eupithecia sempiterna* Vojnits, 1988, praeocc., *nec* Vojnits, 1984.]

Eupithecia sempiterna Vojnits, 1988, *Acta zool. hung.* **34** (1): 48, figs 20–22, 25; pl. 2, figs A-C. Holotype ♀ (NSMTO; examined), Nepal (east): Sagarmatha Solukhumbu, Thame Og, 3800 m. [Junior primary homonym of *Eupithecia sempiterna* Vojnits, 1984].

Description. A large species with quite a variable pattern. Fore wing ground colour brownish grey or brownish ochreous; basal and antemedial transverse lines almost straight, sharply angled onto costa; postmedial wavy, twice angled onto costa, usually with a broad, buff fascia outside it; terminal area usually with two large apical and medial greyish ochreous blotches; discal dot distinct, relatively large; fringe chequered pale ochreous and brownish grey. Hind wing much paler, off-white, usually with distinct brownish grey transverse lines; inner margin of terminal area indented; discal dot small, pale, rounded.

Male genitalia. Uncus rather elongate, biapical. Valve long, relatively narrow, evenly tapered to narrow rounded apex. Vinculum short and broad, rounded. Papillae relatively short. Aedeagus stout, slightly curved, shorter than length of valve. Vesica armed with an elongate patch of short dentate spines in the middle and a small, hook-like, flat cornutus near ductus ejaculatorius base. Sternite A8 two narrow, elongate, apically blunt lateral rods, tapered and sclerotized to apices and connected by a narrow basal band.

Female genitalia. Bursa ovoid, about three quarters of anterior part covered with small spines, and with a sclerotized, transverse fold in the membranous area. Ductus bursae not distinguishable. Ductus seminalis broadened at base, attached to bursa near base of colliculum on left. Colliculum broad and elongate, slightly tapered posteriorly. Antrum sclerotized. Apophyses elongate and narrow, the anterior slightly expanded and flattened at apices. Tergite A8 almost quadrate. Papillae anales broad and tapered.

Range. China (Yunnan, Sichuan, ?Tibet); Nepal, India (Himachal Pradesh).

Phenology. Univoltine: early to late October.
Altitude. From about 2350 to 3850 m above sea-level in the Himalayas (India: coll. ZFMK; Nepal: coll. Gy. M. László; coll. C. Herbulot in ZSM; Vojnits, 1988 [as *E. sempiterna*]; Inoue, 2000).
Similar species. Externally similar to Himalayan *E. noncoacta* Vojnits, 1988, but the latter has a very elongate discal dot and a dark streak at the apex of the fore wing. The male and female genitalia are similar to those of *E. irambata* (Warren, 1893) and *E. amplificata*. However, in the male the presence in the aedeagus vesica of a large patch of dentate spines is diagnostic. The best characters in the female are the much broader colliculum and the presence of a strong fold in the corpus bursae.
Material. ♂, [Nepal]: Sagarmatha, Thame Og, 3800 m, 1–2.x.1979, [M. Owada leg.] (holotype, NSMTO); ♀, E Nepal: Sagarmatha Solukhumbu, Thame Og, 3800 m, 1–2.x.1979, M. Owada [leg.], Vojnits slide no. 14660♀, photo 85 D 16 (holotype of *E. sempiterna* Vojnits, 1988; nec *E. sempiterna* Vojnits, 1984, NSMTO); 1♀, China, North Yunnan, Mekong River Valley, Yubeng village, 3200 m, 14–16.x.2009, leg. Balázs Benedek (coll. U. Ratzel); 1♀, China, W. Sichuan, S. from Kangding, near Menghugang, 15.x.2009, 2800 m, Floriani & Saldaitis leg., Pekarsky slide no. OP 0549♀ (coll. O. Pekarsky); 1♂, [Sichuan] Chasseurs Thibétains de Ta-tsien-lou, Eté 1894, reçu du R.P.Dejean (BMNH).
Remarks. Not previously recorded from China.

58. *Eupithecia wardi* Prout, 1958

Eupithecia wardi Prout, 1958, *Bull. Br. Mus. nat. Hist.* (Ent.) **6** (12): 394. Holotype ♂ [*nec* ♀] (BMNH; examined), [China]: Tibet (south-east), Tsangpo Valley, Tya La, 14000ft.

Description. A large species with broad fore wings of almost unicolourous brownish grey. Fore wing with inconspicuous transverse lines, the medial and postmedial lines oblique, straight, twice angled onto costa and forming two slightly darker costal spots; veins covered with dark scales, the stalk of R especially distinct; terminal area broad with a weak whitish wavy subterminal line; discal dot small, comma-like, black. Hind wing paler, greyish white, slightly darker along anal and terminal margins; discal dot pale, comma-like.
Male genitalia. Uncus elongate, biapical. Valve narrow, angulate at mid point of ventral margin, sharply tapered from middle to narrowly rounded apex. Vinculum short and narrow, slightly tapered to apex. Papillae relatively small. Aedeagus very short and narrow, slightly curved. Vesica covered with numerous wrinkles, and armed with an elongate, sinuate,

plate-like cornutus and a very small irregular cornutus near ductus ejaculatorius base. Sternite A8 short and broad, with two almost parallel arms, pointed at apices, connected by a narrow band at base.
Female unknown.
Range. China (Tibet).
Altitude. The holotype was collected at about 4200 m above sea-level (China: Prout 1958).
Similar species. Externally this species does not resemble any other Chinese species, though it is rather like the Spanish *E. intricata* f. *longipterata* Mironov, 2003. The very small male genitalia, and particularly aedeagus, are very distinctive.
Material. ♂, S.E. Tibet, Tsangpo Valley, Tya La, 14.000', 20.ix.1924, F. Kingdon Ward, Brit. Mus. 1925–189, *Eupithecia wardi* Prout ♂ type, BM Geom. slide no. 20309♂ (holotype, BMNH).
Remarks. The holotype remains the only known specimen to date.

The *ultimaria* species group

A small species group first introduced by Mironov (1990) and subsequently elaborated in Mironov (2003). The group is reasonably well characterized: in the male, the eighth sternite consists of two elongate rods, the apical part more heavily sclerotized, joined at the base by a very narrow strip of sclerotization. The uncus is biapical, with the ventral arm enlarged and rounded. The female genitalia are uniform in general pattern, with an elongate oval corpus bursae covered by spines on one side only, and with the ductus seminalis attached at mid point on the right side. The group is represented in China definitely by two, and more tentatively by a third species.

59. *Eupithecia minusculata* Alphéraky, 1883

Eup[ithecia] minusculata Alphéraky, 1883, *Horae Soc. ent. Ross.* [1882] **17** (3/4): 225, pl. 9, fig. 91. Holotype ♀ (ZISP; examined), [China]: Kuldja [= Yining (Gulja)].
Eupithecia inversaria Turati, 1930, *Atti Soc. ital. Sci. nat.* **69** (1): 63, pl. 2: 1–2. Syntype(s) (MRSN?), Libya: Cyrenaica, Zuara (synonymised in Scoble, 1999).
Eupithecia penultimaria Wiltshire, 1985, *The Entomologist's Gaz.* **36** (4): 265, pl. 9: 2, 3, 5a-c, 6. Holotype ♂ (BMNH; examined), Iraq: Baghdad, on a Tigris Isl. (synonymised in Scoble, 1999).
Eupithecia herrenschmidti Mentzer & Moberg, 1992, *Nota lepid.* **14** (4): 313, figs 2, 4, 6. Holotype ♀ (EMS), Spain: Andalusia, Sierra Nevada, between Motril and Granada, Orjiva (synonymised in Dantart, 2000).

Description. A small species. Fore wing grey or brownish grey, mottled and irrorate with pale grey scales, especially near apex and along termen; transverse lines fine, slightly sinuate, dark grey, broadened at costa; two medial lines sharply curved near costa beyond discal dot; first medial line usually touching discal dot; postmedial and basiterminal lines sharply curved near costa with a broad paler stripe between them; terminal area narrow, sharply broadened to costa; pale wavy subterminal line usually weakly defined or obsolete; discal dot narrow, elongate, black. Hind wing the same colour; transverse lines weakly defined except paler medial stripe, sharply angled at centre; terminal area slightly darker, narrow, broadened at apex; pale wavy subterminal line often ill-defined or obsolete; discal dot linear, blackish grey. Abdomen brownish grey.

Male genitalia. Uncus biapical, the ventral arm strongly broadened. Valve regularly tapered to narrow apex. Vinculum short. Papillae long and broad. Aedeagus narrow, slightly shorter than length of valve. Vesica armed with three sclerotized strips and one elongate, V-shaped cornutus at ductus ejaculatorius base. Sternite A8 two narrow sinuate rods, connected at base by a narrow sclerotized strip.

Female genitalia. Bursa membranous, pear-shaped, with one large patch of small spines covering two thirds of left side and a smaller patch of spinules above it close to colliculum. Ductus seminalis with base broadened, attached to central part of bursa from right side. Colliculum very short and narrow. Antrum strongly sclerotized. Tergite A8 transverse, spade-like, with a prominent medial hollow in posterior margin. Apophyses short and narrow, not expanded at apices. Papillae anales short and rounded.

Range. China (Xinjiang, Hebei); Southern Europe (Spain; Ukraine: Crimea; Russia: Lower Volga, Stavropol kraj, Daghestan), North Africa (Algeria; Tunisia; Libya), Asia (north-eastern Turkey; Syria; Israel; Jordan; Georgia; Armenia; Azerbaijan; Iraq; Turkmenistan; Uzbekistan; Tajikistan; Kazakhstan; Mongolia).

Phenology. Bi- or probably polyvoltine: late March to late October. Larva: April to October. Overwintering as pupa.

Altitude. From sea level up to about 1200–1300 m (Uzbekistan: Weisert 1998).

Similar species. See under following species.

Material. ♀, [Xinjiang], Guldscha, 1.5.1879, Alph., *minusculata*, Origin. (holotype, ZISP); ♂, Iraq, Baghdad, 10.v.1937, E.P.Wiltshire, *Eupithecia penultimaria* ♂ holotype, E.P.Wiltshire prep. no. 2386 (holotype, *E. penultimaria*); 1♂, Tientsin, N. China, F.N. Thomson, July 1921 (BMNH).

60. *Eupithecia opistographata* Dietze, 1906

Eupithecia ultimaria f. *opistographata* Dietze, 1906, *Dt. ent. Z. Iris* **19** (2): 66 (non binominal). Lectotype ♂ [*nec* holotype] (coll. K. Dietze in MNHU; examined), [China]: Aksu.
Tephroclystia arenicola Rothschild, 1913, *Novit. zool.* **20**: 133. Syntypes 1♂, 3♀ (BMNH; examined), [Algeria]: halfway between Ouargla and El-Golea; South Oued Mya (synonymised in Scoble, 1999).

Description. A small species. Fore wing grey; transverse lines narrow, sinuate; basal, postbasal and antemedial obtusely angled onto costa; medial line oblique, right angled onto costa behind discal dot; postmedial line broken, two or three times angled; terminal area narrow sometimes with weak trace of pale subterminal line; discal dot linear, blackish. Hind wing similarly coloured to fore wing; transverse lines oblique; postmedial right angled; terminal area narrow, slightly darker, with angled inner border; discal dot similar to fore wing. Underside of fore and hind wings with dark, black or blackish, angulate terminal areas and with rows of small white spots on the veins instead of subterminal lines, or with veins entirely covered by white scales.
Male genitalia. Uncus narrow, biapical with ventral tip rounded and longer than dorsal. Valve small, evenly tapered to narrowly rounded apex. Vinculum semicircular. Papillae narrow and clavate. Aedeagus shorter than length of valve. Vesica densely covered with minute spinules; armed with three short, pointed cornuti and a small U-shaped plate near ductus ejaculatorius base. Sternite A8 two narrow and sinuate rods, inwardly bowed and strongly sclerotized near apex, connected by a very narrow band at base.
Female genitalia. Bursa pear-shaped, the left half densely covered with spines. Ductus bursae not distinguishable. Ductus seminalis narrow, recurved, attached to central part of corpus bursae on right. Colliculum short and narrow. Antrum membranous. Tergite A8 rectangular with concave anterior and convex posterior margins. Apophyses short and narrow. Papillae anales small, relatively broad, rounded.
Range. China (Xinjiang, Gansu); Algeria, Saudi Arabia, Iraq, Iran, Turkmenistan, Uzbekistan, Tajikistan, SE Kazakhstan, Mongolia.
Phenology. Bivoltine or polyvoltine: late April to mid-September.
Biology. Larvae found on the flowers of *Tamarix leptostachys* and *T. ramosissima* (Tamaricaceae) in the second half of August in Turkmenistan (Mironov: original data).
Similar species. This species is difficult to separate from *E. minusculata*. There are no significant differences between the two in shape of wings, colour or maculation on the upper side of wings. However, the underside

in *E. opistographata* has a more distinct black or blackish pattern, and the black terminal area on the fore wing has a row of distinct white spots on the veins; the same black terminal area on the hind wing has veins covered with white scales, and all discal dots larger than those on the underside of *E. minusculata*. In the male genitalia the differing shapes of the uncus, of the papillae on the anterior arms of the labides, and especially the cornuti on the aedeagus vesica are reliable features to distinguish the two; in the female the differing patterns of spines and the shape of the eighth tergite are definitive.

Material. ♂, [Xinjiang], *ultimaria* Forma: *opistographata* Original, Aksu, K. Dietze, von Tancré, Zool. Mus. Berlin, Holotypus *opistographata* Dietze ♂, E. Schütze, Kassel, *opistographata* präp. Nr. A 838♂ det. E. Schütze, Kassel, *Eupithecia arenicola* (Rothschild, 1913) ♂, J. Dantart det. 92 (lectotype [*nec* holotype] of *E. ultimaria* f. *opistographata*, coll. K. Dietze in MNHU); 1♀, [Xinjiang], *ultimaria* Boisd. form: *opistographata* Typ. Aksu. (675), Zool. Mus. Berlin, Allotypus *Eupith. opistographata* Dietze ♀, E. Schütze, Kassel, *opistographata* präp. Nr. A 837♀ det. E. Schütze, Kassel, *Eupithecia arenicola* (Rothschild, 1913) ♀, J. Dantart det. 92 (paralectotype [*nec* paratype] of *E. ultimaria* f. *opistographata*, coll. K. Dietze in MNHU); 1♀, [Xinjiang], *ultimaria* Forma: *opistographata* Aksu, Original, K. Dietze (paralectotype of *E. ultimaria* f. *opistographata*, coll. K. Dietze in MNHU); 1♂ Lanzhou, Gansu, PRC, 16.vi.[19]98, A.C. Galsworthy, slide no. G309♂ (BMNH).

Remarks. In accordance with article 74.1 of the International Code of Zoological Nomenclature, the male specimen listed above, which was designated by E. Schütze as holotype, should be regarded as lectotype of *E. opistographata*. Two other syntypes are paralectotypes of this species.

61. *Eupithecia viata* Mironov & Galsworthy, 2004

Eupithecia viata Mironov & Galsworthy, 2004, *Trans. lepid. Soc. Japan* 55 (4): 290, figs 5, 18. Holotype ♂ (IZCAS; examined), China: Gansu, Zhoni, 2500 m.

Description. Fore wing ground colour pale buff, transverse lines very pale brown; basal line straight, basal area darker than ground colour; antemedial double, sinuous; a faint medial running straight from discal dot to inner margin; discal dot small, dark, rounded; double postmedial curved towards costa at level of discal dot; terminal area slightly darker. Hind wing ground colour white, discal dot faint; a double, gently curved postmedial, strongest near anal margin.

Male genitalia. Uncus biapical. Valve rather narrow, with narrowly rounded apex. Vinculum trapezoid. Papillae short and tapered. Aedeagus narrow, considerably shorter than length of valve. Vesica armed with an elongate patch of small scale-shaped cornuti, a narrow U-shaped cornutus and a smaller, irregular cornutus near ductus ejaculatorius base. Sternite A8 consisting of two long and narrow tapered rods connected by a narrow band at base.

Female unknown.

Range. China (Gansu).

Similar species. The unique holotype is too worn to allow meaningful comparison with other species. For the moment identification is necessarily by the male genitalia, which are distinctive.

Material. ♂, China, Gansu, Zhoni, 2500 m, 4.ix.1990, coll. Xue Dayong, slide no. ZL391♂ (holotype, IZCAS).

Remarks. Characters of the male genitalia, particularly the short biapical uncus, the shapes of the valves and eighth sternite, and the cornuti on the vesica, lead us tentatively to place this species in the *ultimaria* group.

The *leucostaxis* species group

This is a small Central and South Asian group introduced here for the first time. It is characterized primarily by the shape of the cornuti in the aedeagus vesica and of the eighth sternite in the male, and in the female by the pear-shaped bursa with the basal half covered by spines, and in addition a separate vertical row or patch of spines running towards the colliculum. It is represented in China by the following two species. Two other Asian mountain species also belong to the group, *E. thomasi* Mironov & Galsworthy, 2008 from the Indian Himalayas and *E. solianikovi* Viidalepp, 1988 from the mountains of Central Asia (Ghissar, Darvaz, Alai and western Pamir Mts.).

Male genitalia. Uncus short, biapical. Valve short, broadened medially, without ventral process; sacculus lightly sclerotized. Vinculum short and relatively narrow. Papillae on the anterior arms of the labides relatively narrow, and curved. Aedeagus shorter than length of valve. Vesica armed with one or two apical, usually short, horn-like cornuti, a characteristic medial elongate, twisted, V-shaped cornutus and a smaller, folded, irregular cornutus near ductus ejaculatorius base. Sternite A8 small, slightly elongate, broadened to base, with two narrow, parallel apical rods about one half of sternite length.

Female genitalia. Bursa copulatrix rather small, elongate, pear-shaped, about 1/2–2/3 covered with small spines at the base and also with a medial patch or row of larger spines, placed on a sclerotized elongate blotch on the dorsal side. Ductus seminalis sharply curved backward, attached to ductus bursae from right side. Colliculum narrow, of varying lengths. Antrum membranous. Tergite A8 quadrate or rectangular. All apophyses narrow, elongate. Papillae anales usually elongate, narrow, rounded or tapered to apices.

Remarks. The relationship of this species group is not clear, but the shape of the valve and eighth sternite and also the cornuti on the vesica in the male, might indicate a relationship with the *undata* species group. However the particular pattern of spining in the female bursa is distinct from that group.

62. *Eupithecia mustangata* Schütze, 1961

Eupithecia mustangata Schütze, 1961, *Veröff. zool. StSamml. Münch.* **6**: 181, pl. 29, figs 4, 9; pl. 31, fig. 4; pl. 32, fig. 1. Holotype ♂ (ZSM; examined), [Nepal]: Mustangbhot, 29°11′ N, 83°58′ E, Muktinath, 3500 m.

Eupithecia emikoae Inoue, 1996, *Trans. lepid. Soc. Japan* 47 (4): 237, figs 1, 2, 8, 11. Holotype ♂ (BMNH; examined), [India/Pakistan]: Kashmir, Pahalgam-Kolohoi, Lidderwat (synonymised in Mironov et al., 2008a).

Description. Fore wing ground colour pale ash-grey to greyish brown, costal margin usually darker; medial area paler, especially behind discal dot; transverse lines more or less distinct, oblique and sharply angled onto costa; ante- and postmedial usually more distinct than the others; postmedial twice angled near costa; terminal area narrow with a distinct, narrow, wavy whitish subterminal line; discal dot small, black. Hind wing paler, especially in anterior half; transverse lines well marked along dorsum; postmedial evenly curved and forming a series of dark dashes on the veins; terminal area narrow, darker, with traces of a light, indented subterminal line; discal dot pale, sometimes invisible.

Male genitalia. Uncus stout, pointed, biapical. Valve with broadly rounded apex. Vinculum short, with shallow medial hollow. Papillae elongate and narrow. Aedeagus narrow, shorter than length of valve. Vesica armed with a small apical cornutus, a narrow V-shaped cornutus at centre and an irregular cornutus near ductus ejaculatorius base. Sternite A8 elongate and narrow, with two apically tapered, parallel arms.

Female genitalia. Bursa very small, pear-shaped, covered with relatively large sparse spines at the base and with a sclerotized longitudinal band armed with long stout spines. Ductus bursae sclerotized with numerous longitudinal striations and a small pouch-like membranous diverticulum near the base of the colliculum on the right. Ductus seminalis slightly broadened near base, sharply recurved, attached to ductus bursae on left

side. Colliculum slightly elongate, narrow. Antrum elongate, membranous. Tergite A8 elongate and narrow. Apophyses stout, elongate and apically rounded. Papillae anales rather elongate and narrow.
Range. China (Tibet); Pakistan/India (Jammu & Kashmir), Nepal.
Phenology. Univoltine (?): late June to mid-August.
Altitude. From about 2400 to 4300 m in the Himalayas (coll. ZFMK, ZSM).
Similar species. Very variable in size, coloration and distinctness of markings. Best recognised by the paler medial area, especially near the discal dot on the fore wing. The genitalia resemble those of the immediately following species, and *E. solianikovi* Viidalepp, 1988. The pattern of cornuti on the vesica of the aedeagus and the different shape of the eighth sternite in the male, and in the female the larger corpus bursae lacking the heavily sclerotized medial spiniferous band will readily separate *E. leucostaxis*, which is externally also quite distinct; *E. solianikovi* has a shorter eighth sternite in the male, and in the female a larger area of spining in the corpus bursae, a sclerotized longitudinal band similar to that of the present species, but with very short spines, and shorter and narrower apophyses.
Material. ♂, Nepal, Mustangbhot, Muktinath, 3500 m, 29°11′ n.Br., 83°58′ ö.L., 5.viii.[19]55, leg. F. Lobbichler, Präp. Nr. A 707♂, det. E. Schütze, Kassel [= ZSM slide no. G 13367] (holotype, ZSM); ♂, Lidderwat (3000 m), Pahalgam-Kolohoi, Kashmir, NW India, Aug. 8–9.1982, leg. E. Yoshimoto (holotype of *E. emikoae*, coll. H. Inoue in BMNH); 1♂, Tibet: Gautsa, Chumbi Valley, 11500ft., 21.vii.1924, Maj. R.W.G. Hingston, Everest Exp. Brit. Mus. 1924–386, BM Geom. slide no. 21475♂ (BMNH); 1♀, Tibet, Zham, Quxiang, 3300 m, 1.viii.1975, coll Wang Ziqing, slide no. ZL105♀ (IZCAS).

63. *Eupithecia leucostaxis* Prout, 1926

Eupithecia leucostaxis Prout, 1926, *J. Bombay nat. Hist. Soc.* **31** (2): 318. Holotype ♀ (BMNH; examined), [Burma (upper)]: Hpimaw Fort.
Eupithecia deprima Vojnits, 1974, *Acta zool. Acad. Sci. hung.* **20** (1–2): 223, figs 3a-e. Holotype ♀ (ZFMK; examined), China: North Yunnan, Li-kiang (synonymised in Inoue, 2000).

Description. Ground colour of all wings, head and body a soft, pale yellow, with numerous, delicate, darker ochreous transverse lines and lighter, yellowish white postmedial lines; terminal areas slightly darker, ochreous, with a series of distinct white spots along terminal margins of all wings; discal dots minute or absent.
Male genitalia. Uncus broad, pointed, biapical. Valve short and broad, tapering sharply to rounded apex. Vinculum short and flat. Papillae clavate, curved. Aedeagus stout, shorter than length of valve. Vesica armed with two horn-like cornuti, one longer, slightly sinuate and the

other shorter, with a pectinate base on one side, an elongate, V-shaped cornutus and a small, hooked cornutus near ductus ejaculatorius base. Sternite A8 peg-like, strongly broadened at base, with two rather close apical horns.

Female genitalia. Bursa pear-shaped, the basal half completely covered with dense small spines, and with a small discrete patch of sparse stout spines placed on a sclerotized area on the dorsal side. Ductus bursae elongate, tapered, spineless, with numerous longitudinal striations and a short, broad diverticulum on the left. Ductus seminalis broad, strongly recurved, attached to ductus bursae on right side. Colliculum elongate and narrow. Antrum membranous. Tergite A8 almost square with convex anterior margin. Apophyses narrow, slightly expanded at apices. Papillae anales rather small and rounded.

Range. China (Sichuan, Yunnan, Tibet); India, Myanmar.

Phenology. Univoltine: early July to mid-August.

Altitude. Found at about 3500 m in the Himalayas (Nepal: coll. M. Fibiger).

Similar species. Only likely to be confused with *E. ochracea.* Externally it is distinguished by the paler yellow ground colour and the presence of a series of distinct white marginal spots on all wings. The male and female genitalia are similar to those of other members of the group, but the presence of two apical cornuti in the aedeagus and the shape of sternite A8 distinguish *E. leucostaxis.* The female genitalia are similar to those of *E. thomasi* Mironov & Galsworthy, 2008, but in the latter the large spines are distributed at the posterior end of the bursa.

Material. ♀, Upper Burma: Hpimaw Fort, Nr. Myitkyina, 8000ft., 9–13.viii.1923, Capt. A.E. Swann, L.B. Prout Coll. B.M. 1939–643, *Eupithecia leucostaxis* Prout ♀ type, BM Geom. slide no. 20090♀ (holotype, BMNH); ♀, China, Provinz Nord-Yuennan, Li-kiang [Lijiang], 21.vii.1935, H. Höne (holotype of *E. deprima*, ZFMK); 3♂, 12♀, same loc., 7.vii., 28.viii.1934, 21, 23, 24, 28, 31.vii., 6.viii.1935, H. Höne, de Laever slide no. 220♀, Schmidt slides nos. Eu 144♀, Eu 144a♂, Eu 144b♂, Eu 160♀, Eu 160a♀ (paratypes of *E. deprima*, ZFMK); 2♂, 2♀, same loc., 21, 23, 27, 31.vii.1935, H. Höne, Vojnits slides nos. 10036♂, 10037♀ (paratypes of *E. deprima*, TTM); 1♀, Likiang, Provinz Nord-Yuennan, 24.vii.1935, H. Höne, slide ZL76 (paratype of *E. deprima*, IZCAS); 1♀, same loc., 18.viii.1934, H. Höne, Mironov slide no. 790♀ (SMNK); 5♂, 6♀, same loc., 27.viii.1934, 22, 23, 26.vii., 3, 7, 16.viii.1935, H. Höne, de Laever slide no. 205♂ (ZFMK); 1♂, [Sichuan], Chasseurs Indigènes des Missionnaires de Ta-tsien-Loû, 1906 (ZFMK); 1♂, Tsekou, Th. Monbeig, 1909 (ZFMK); 1♀, Tibet, Gyirong, 3300 m, 2.viii.1978, coll. Wang Ziqing (IZCAS).

The *undata* species group

This species group was originally proposed by Mironov (1990), initially without a definition. It was subsequently elaborated in Mironov (2003). It is fairly well characterized: the male valve is generally broad and elongate; the aedeagus contains one or more long heavily sclerotized cornuti; the eighth sternite consists of two long arms joined at the base by a band of sclerotization of variable width. In the female the corpus bursae is usually irregularly shaped, with a heavily spined posterior portion, often with distinct rings of large spines, and a membranous area at base. There is often a distinct diverticulum at about mid point, with the ductus seminalis situated at the end of it.

This is a very large group, and is particularly abundant in northern and high mountain Palaearctic and Nearctic areas: it is also very species-rich in the mountains of Asia. At present we place the following twenty six species within the group. Several other species from the group, for instance *E. pygmaeata*, Hübner, 1799, are found in southern Siberia and Primorje near the Chinese border, and are likely to be found in China in the future.

64. *Eupithecia ochracea* Warren, 1888

Asthena ochracea Warren, 1888, *Proc. zool. Soc. Lond.* **1888** (3): 321. Lectotype ♀ (BMNH; examined), Western India [Pakistan]: Thundiani [incorrect]; [India]: Khasi Hills, Shillong.

Eupithecia nova Vojnits, 1974, *Acta zool. Acad. Sci. hung.* **20** (1–2): 219, figs 1a-e. Holotype ♂ (ZFMK; examined), China: North Yunnan, Li-kiang (synonymised in Mironov et al., 2008a).

Description. A small species with relatively broad and rounded ochreous wings crossed by some whitish transverse lines, and lacking discal dots; terminal areas darker, with a well marked whitish subterminal line, white tornal spot, and apical and tornal brownish patches. Hind wing similarly coloured, paler, with distinct transverse lines, darker terminal area, especially near tornus, and a whitish tornal spot shaded by brown.

Male genitalia. Uncus broad with a very small pointed uniapical tip. Valve short, broadened medially, angulate, and sharply tapered to narrow apex. Vinculum semicircular. Papillae parallel-sided. Aedeagus broadened anteriorly, almost equal to length of valve. Vesica with strong wrinkles, and armed with two stout horn-like cornuti and a lightly sclerotized, folded, U-shaped cornutus. Sternite A8 elongate, with long diverging apical arms, and rather shorter basal arms; apical arms pincer-like at apices.

Female genitalia. Bursa elongate oval, with a membranous spineless base, and a transverse ring of spines between base and base of ductus seminalis;

centre of right side with a patch of slim spines, and a patch of smaller spines in posterior part near colliculum on left. Ductus bursae not distinguishable. Ductus seminalis broadened at base, attached to central part of bursa on dorsal side. Colliculum short and relatively broad. Antrum membranous, short and broad. Tergite A8 small, square, with rounded posterior corners. Apophyses relatively short and narrow. Papillae anales very small and rounded.

Range. China (Yunnan, Guizhou, Hubei); India (Assam).

Phenology. Univoltine: mid-August to late September.

Altitude. Captured at altitudes ranging from 920 m in Hubei to about 2100 m on Mount Fanjingshan (China: coll. P. Skou).

Similar species. The ochreous colour and unique maculation of the wings distinguish this species from all other Chinese *Eupithecia*, except the preceding species, which is a paler yellow, and has a series of distinct white marginal spots on all wings. The male genitalia are unique among Chinese species in having a tiny uncus; the eighth sternite is also distinctive. The pattern of spining in the female bursa is also very distinctive, resembling only that of the externally very different *E. idaeoides* Mironov & Galsworthy 2010.

Material. ♀, Thundiani, 17/8/86, Thundiani 86–128, Type, BM Geom. slide no. 20091♀ (lectotype ♀, designated in Mironov et al., 2008a); ♂, China, Provinz Nord-Yuennan, Li-kiang [Lijiang], 7.ix.1934, H. Höne (holotype of *E. nova*, ZFMK); 13♂, 5♀, same loc., 27.viii., 1, 2, 5, 10, 15, 17, 23.ix.1934, 2, 6, 8, 9, 19, 21.ix.1935, H. Höne, Schmidt slides nos. Eu 143♂, Eu 143a♂, Eu 143b♂, Eu 143c♂, Eu 143d♀, de Laever slides nos. 215♂, 216♀, 217♀ (paratypes of *E. nova*, ZFMK; one paratype male lacking abdomen; Schmidt slides nos. Eu 143♂, Eu 143a♂ marked as unpublished *Eupithecia hara* Vojnits); 3♂, 1♀, same loc., 22.ix.1934, 30.viii., 3, 14.ix.1935, H. Höne, Vojnits slides nos. 10032♂, 10033♀ (paratypes of *E. nova*, TTM); 1♂, 1♀, Provinz Nord-Yuennan, Li-kiang, 19.viii.1934 and 7.ix.1935 respectively, slides nos. ZL74♂ and ZL75♀ (paratypes of *E. nova*, IZCAS); 16♂, 3♀, same loc., 9.vii., 31.viii., 1, 5, 6, 7, 10, 12, 15.ix.1934, 26.vii., 6, 8, 9, 10, 14, 24.ix.1935, H. Höne (ZFMK; 1♂ without abdomen); 1♂, China, Guizhou Prov., Mount Fanjingshan, 2100 m, 30. vii.2001, Li Houhun leg. (coll. P. Skou); 1♀, Hubei, Shennongjia, Songluo, 920 m, 31.viii.1981, coll. Han Yinheng, slide no. ZL52♀ (IZCAS).

65. *Eupithecia homogrammata* Dietze, 1906

Eup[ithecia] homogrammata Dietze, 1906, *Dt. ent. Z., Iris* **19** (2): pl. 2, figs 1, 2; *ibidem* (1908), **21** (3): 191, pl. 2, figs 3, 4. Syntypes 32, at least 7♂, 19♀ (coll. K. Dietze in MNHU; examined), [Russia]: Radde; Amur; Kasakewitsch; Ussuri.

Eupithecia meszarosi Vojnits, 1973, *Annls hist.-nat. Mus. natn. hung.* 65: 251, fig. 2b. Holotype ♀ (ZFMK; examined), [Russia]: Ussuri, Sutschanski Rudnik (synonymised in Mironov, 1990).
Eupithecia bicornuta Oh, 1992, *Korean J. Appl. Entomol.* 31 (4): 466, figs 10, 21, 36. Holotype ♂ (coll. Oh ?), [Korea]: Gyeonggi Prov., Mt. Suri-san (synonymised in Oh, 1993).

Misidentification: *Eupithecia sidemii* Vojnits, 1973 (1♂ paratype in ZFMK; 2♂ paratypes in TTM).

Description. A small dark, brownish grey species with broad blunt fore wings; transverse lines usually inconspicuous except evenly curved postmedial; terminal area unicolourous with very weak subterminal line, often invisible; discal dot very small, again often invisible. Hind wing rounded, same colour as fore wing; transverse lines indistinct; discal dot minute or absent. Body dark brownish grey.
Male genitalia. Uncus biapical. Valve small, parallel-sided with relatively broad round apex. Vinculum short and broad. Papillae short and narrow. Aedeagus short, vesica armed with two curved cornuti of almost equal size, and a very small, bowed, irregular plate near ductus ejaculatorius base. Sternite A8 narrow, with two long blunt apical rods.
Female genitalia. Bursa a long oval, bulging in the centre, with a membranous basal part and a spiniferous, sclerotized posterior part. Ductus bursae sclerotized, with longitudinal striations and a transverse patch of small spines. Ductus seminalis sharply curved, with sclerotized base, attached to ductus bursae near colliculum on ventral side. Colliculum broad but very short. Antrum membranous. Tergite A8 small, square, with rounded posterior corners. Apophyses short and narrow. Papillae anales small and rounded.
Range. China (Heilongjiang, Inner Mongolia, Shaanxi, Yunnan); Asian Russia (Tuva, Transbaikal region, Kamchatka peninsula, Kuril Islands, Sakhalin Isl., Priamurje, Primorje), Korea, Japan (Hokkaido, Honshu, Shikoku, Kyushu).
Phenology. Bivoltine (?): mid-June to early September.
Altitude. From sea level to about 3000 m in Yunnan (China: coll. ZFMK, SMNK, TTM).
Similar species. Externally very similar to *E. subtacincta* Hampson, 1895 but with broader wings, less distinct transverse lines and a greyish brown abdomen (reddish in *E. subtacincta*). The male and female genitalia easily distinguish these two species.
Material. 1♀, [Russia], Radde, m. Amur, 25 Juli 03, Korb, *homogrammata* n. sp., Type ♀, Amur, K. Dietze, Original, Zool. Mus. Berlin, Vojnits slide no. 14630♀ (syntype, coll. K. Dietze in MNHU); 1♂, [Russia], same loc. and

date, *homogrammata* n. sp. Type ♂., Amur., K. Dietze, Original (syntype, coll. K. Dietze in MNHU); ♀, [Russia], Sutschanski-Rudnik, Ussuri, 185, Vojnits slide no. 497♀ (holotype of *E. meszarosi*, ZFMK); 1♀, [Russia], Sutchan, Ussuri, Vojnits slide no. 519♀ (paratype of *E. meszarosi*, ZFMK); 1♀, [Heilongjiang], Mandschuria, Maoreschan (= Mauschan), Juli, Vojnits slide no. 1048♀ (paratype of *E. meszarosi*, TTM); 1♂, [Russia], Sidemi (Mantschourie), M. Jankowski 1882, Vojnits slide no. 1045♂ (paratype of *E. sidemii*, ZFMK); 1♀, Sued-Shensi [Shaanxi], Tapaishan [Taibaishan] im Tsinling [Qinling], *ca.* 1700 m, 7.viii.1936, H. Höne, Vojnits slide no. 1078♀ (paratype of *E. meszarosi*, TTM); 2♂, same loc. and alt., 24.vi.1935, 6. viii.1936, H. Höne, Vojnits slides nos. 1056♂, 1070♂ (paratypes of *E. sidemii*, TTM); 2♂, 1♀, same loc. and alt., 14.vii., 7, 9.viii.1936, H. Höne, de Laever slide no. 148♂, Mironov slide no. 719♂ (ZFMK); 1♀, same loc. and alt., 14.vii.1936, H. Höne (MNHU); 1♀, same loc. and alt., 6.viii.1936, H. Höne (TTM); 1♂, same loc. (without alt.), 1.vii.1935, H. Höne, Vojnits slide no. 1079♂ (ZFMK); 2♂, same loc., F, 21.vi., 6.vii.1935, H. Höne, Mironov slide no. 794♂ (SMNK); 1♀, Nord-Yuennan, A-tun-tse [Dêqên], Talsohle *ca.* 3000 m, 10.vii.1937, H. Höne, Vojnits slide no. 12961♀ (TTM); 4♀, China, Shaanxi, San Tai Shan, 100 km SE Xian, 1600 m, 33°42′ N, 109°37′ E, 15–18. vii.2008, leg. Viktor Sinyaev (coll. O. Pekarsky); 1♀, Inner Mongolia, Yakeshi, Oragxan, slide no. ZL209♀; 1♀, Shaanxi, Foping, 950 m, 23.7.1998, coll. Yao Jian, slide no. ZL550♀ (IZCAS).

66. *Eupithecia plumbeolata* Haworth, 1809

Phalaena plumbeolata Haworth, 1809, *Lepid. Br.* (2): 360. Syntype(s), [United Kingdom].
Larentia isogrammata Treitschke, 1828, *Die Schmett. Eur.* 6 (2): 100. Holotype ♀ (TTM; examined), Hungary: Ofen (synonymised in Walker, 1862).
Eupithecia begrandaria Boisduval, 1840, *Genera Index meth. eur. Lepid.*: 211. Syntype(s), [France]: Paris (synonymised in Herrich-Schäffer, 1848).
[*Eupithecia*] *singularia* Herrich-Schäffer, 1848, *Syst. Bearb. Schmett. Eur.* 3 (32): 121, 132, *ibidem* (1846), pl. 23: 141, 142 (non binominal). Syntypes lost. [Germany] (synonymised in Staudinger & Wocke, 1861).
Eupithecia immundata var. *reikjavikaria* Staudinger & Wocke, 1861, *Cat. Lepid. Eur. Faunengeb.* 1: 197. Syntypes 2♂, 2♀ (coll. O. Staudinger in MNHU; examined), Iceland: Reikjavik (synonymised in Lindroth, 1931).
Eupithecia plumbeolata var. *lutosaria* Bohatsch, 1893, *Dt. ent. Z., Iris* 6: 29. Syntypes 1♂, 1♀ (coll. O. Staudinger in MNHU; examined), [Georgia]: Achalzich.
Eup[*ithecia*] *plumbeolata* f. *enucleata* Dietze, 1910, *Biologie der Eupithecien* 1, pl. 72: 248–250; *ibidem* (1913), 2: 30. Syntypes 3♂, 2♀ (coll. K. Dietze in MNHU; examined), [Germany]: Bav(aria), Allgäu, Oberdorf.
Eupithecia dshungarica Viidalepp, 1988, *Fauna pjadenits gor Srednej Azii*: 91, pl. 2, fig. 5; text-pl. 14: 5, 6. Holotype ♂ (IAET; examined), Kazakhstan: Dzhungarskij Alatau Mts., Tekeli (synonymised in Mironov, 1990).

Misidentifications: *Eupithecia valerianata* Staudinger, 1857 (*nec* Hübner, 1813); *Eupithecia* ? *immundata* Wolff, 1929 (*nec* Lienig & Zeller, 1846).

Unavailable names (infrasubspecific): *obscura*: Lempke (1969). – (incorrect subsequent spellings): *isogrammaria*: Boisduval (1840); *plumbeolaria*: Doubleday (1849); *begrandata*: Bruand (1850).

Description. A small species. Fore wing pale brownish grey; transverse lines darker brown, fine, slightly sinuous, usually evenly curved and slightly more emphasised at costa, forming a pattern of close curved lines; discal dot absent. Hind wing unicolourous brownish grey with inconspicuous transverse lines. Fringes unicolourous brownish grey or slightly spotted with brownish. Abdomen unicolourous grey.

Male genitalia. Uncus short, biapical. Valve slightly broadened medially, with bowed dorsal margin and broadly rounded apex. Vinculum short, semicircular. Papillae rather short. Aedeagus stout, shorter than valve length. Vesica armed with two cornuti near apex, one narrow, one broad, and a small plate at ductus ejaculatorius base. Sternite A8 lightly sclerotized, elongate and narrow, with two parallel, apical rods.

Female genitalia. Bursa elongate oval, sclerotized, all but the membranous basal part covered with small spines. Ductus bursae very short with longitudinal striations. Ductus seminalis stout, heavily sclerotized at base, sharply curved around bursa, and attached to posterior part of bursa near colliculum on ventral side. Colliculum very short, but relatively broad. Antrum membranous. Tergite A8 square, with rounded posterior corners. Apophyses short and narrow. Papillae anales small, rounded.

Range. China (Xinjiang); SE Kazakhstan (Dzhungarsky Alatau Mts.), Kirghizstan (Tien-Shan Mts.), Europe except Portugal, Albania and Mediterranean islands, NE Asian Turkey, Georgia, Russia (European part, Caucasus, Urals, W Siberia, Altai and Sajan Mts., Transbaikalia).

Phenology. Uni- or bivoltine (?): late May (late April) to mid-July (late August). Larva: mid-July (early July) to early September (late September). Overwintering as pupa.

Biology. Larva found in the seed capsules of *Rhinanthus songaricus* (Scrophulariaceae) (Kirghizstan: Mironov 2003); also found on the flowers and in the seed capsules of *Melampyrum pratense*, *M. sylvaticum*, *M. arvense*, *M. nemorosum Rhinanthus alectorolophus*, *R. minor* (Scrophulariaceae), probably occasionally on *Clematis vitalba* (Ranunculaceae), *Calluna vulgaris* (Ericaceae) and *Galeopsis* (Lamiaceae) (Europe: Mironov 2003).

Altitude. Found at about 2200–2300 m in Tien-Shan Mts. (Kirghizstan: original data of Mironov).

Similar species. Similar to *E. homogrammata* but larger and paler, with characteristic close striation on the fore wings. The male and female

genitalia of this species are also very similar to those of *E. homogrammata*. In the male, the difference in size of the two apical cornuti in the aedeagus vesica is the best distinguishing feature. In the female the characteristic bulge at the centre of the bursa in *E. homogrammata* separates the two.
Material. ♂, Kazakhskaya SSR [Kazakhstan], Dzhungarsky Alatau Mts., Tekeli, 11.vi.1971, (leg.) Remm, Viidalepp slide no. 1393 (holotype of *E. dshungarica*, IAET); 1♂, [Xinjiang], *plumbeolata* ?, Juldus, Ili, K. Dietze v. Staudinger 1908 (coll. K. Dietze in MNHU).
Remarks. Not previously recorded from China.

67. *Eupithecia severa* Mironov & Galsworthy, 2006

Eupithecia severa Mironov & Galsworthy, 2006, *Trans. lepid. Soc. Japan* 57 (4): 350, figs.10, 32. Holotype ♀ (ZSM; examined), China: Shaanxi, Tsinling [Qinling] Mts., South Taibai Shan, Mt. Houzhenzi, 33°53′ N 107°49′ E, 1500 m.

Description. Fore wing ground colour brownish grey with yellowish tinge, very uniform; transverse lines inconspicuous, mainly evident as darker brown costal blotches; medial and postmedial lines faintly visible, bluntly angled near costa; terminal line brown, continuous and narrow; discal dot rather vaguely marked in darker brown Hind wing concolourous with fore wing; dark transverse lines faintly marked along anal margin only; discal dot inconspicuous.
Male genitalia (not previously described). Uncus elongate, biapical. Valve relatively short and broad, tapered to narrow, almost pointed apex; sacculus thickened but lightly sclerotized. Vinculum relatively broad, rounded. Papillae elongate and very narrow, slightly clavate. Aedeagus short and stout, shorter than length of valve. Vesica armed with three stout, horn-like cornuti with broadened bases, an elongate, U-shaped, plate-like cornutus and a large, irregular cornutus near ductus ejaculatorius base. Sternite A8 with two elongate basal lobes and two narrower, slightly diverging, sclerotized and apically obtuse apical rods.
Female genitalia. Bursa small, membranous, with sclerotized spineless base, armed with one patch of large spines near base and a separate smaller patch of small short spines near base of ductus seminalis. Ductus bursae not distinguishable. Ductus seminalis curved, basally broadened, attached to middle of corpus bursae on right. Colliculum short and broad, lightly sclerotized. Antrum membranous. Tergite A8 broad, trapezoid, with narrowly sclerotized anterior margin. Apophyses narrow, tapered, relatively long. Papillae anales short and narrow, rounded.
Range. China (Shaanxi).
Phenology. Univoltine: late May – early June.

Altitude. From about 1000 to 1600 m in Qinling Mts. (China: coll. O. Pekarsky).
Similar species. This species is distinguished by its uniform coloration with only faint transverse lines. It is similar to *E. thalictrata* in size and colour, but easily distinguished from it by the structure of both male and female genitalia. It can also be distinguished externally from *E. thalictrata* by the slightly narrower wings with more pointed apices and the more obtusely angled medial line near the costa of the fore wing. The female genitalia are similar to those of *E. chesiata* but the disposition of spines differs, and the eighth tergite lacks a narrow and deep medial gap on its posterior margin.
Material. ♀, China, Shaanxi, South Taibai Shan, Tsinling Mts., Houzhenzi, 33°53′ N, 107°49′ E, 16.-21.v.2000, 1500 m, leg. Siniaev & Plutenko (holotype, ZSM); 1♀, China, Shaanxi prov., Qinling Mts., N from Taibai, h – 1000 m, N 34°14.124′ / E 107°18.482′, 30.v.2009, leg. Floriani & Saldaitis, Pekarsky slide no. OP 0676♀ (coll. O. Pekarsky); 1♂, China, Shaanxi prov., Qinling Mts., Hauzherze vill., h – 1600 m, N 33°52.423′ / E 107°48.779′, 2–3.vi.2009, leg. Floriani & Saldaitis, Pekarsky slide no. OP 0678♂ (coll. O. Pekarsky).

68. *Eupithecia spissata* Mironov & Galsworthy, 2006

Eupithecia spissata Mironov & Galsworthy, 2006, *Trans. lepid. Soc. Japan.* 57 (4): 339, figs 3, 15, 17. Holotype ♀ (ZFMK; examined), China: Sichuan, Daxue Shan Mts., Gongga Shan, W Moxi, 29°34′ N 101°59′ E, 3200 m.

Description. Fore wing ground colour brownish grey; transverse lines indistinct, except oblique postmedial which forms a series of black dashes on the veins and is obtusely angled onto the costa; terminal area slightly darker with obscure pale wavy subterminal line and a very clear rounded, whitish tornal spot; discal dot moderately large, ovoid, black. Hind wing ground colour pale whitish grey along costa and in the centre, and darker grey along terminal and anal margins; terminal area with clear, narrow, wavy, whitish subterminal line and a small, white tornal spot; dark discal dot small, circular or ovoid.
Male genitalia. Uncus pointed, biapical, with ventral apex longer than dorsal. Valve evenly tapered to narrowly rounded apex. Vinculum rather flat. Papillae rather broad and slightly curved. Aedeagus short and stout, shorter than valve length. Vesica armed with a broad fusiform cornutus, a smaller pointed, obliquely lying cornutus and a very small, irregular plate near ductus ejaculatorius base. Sternite A8 short, with two narrow, tapered apical rods.

Female genitalia. Bursa membranous, broadened basally, and heavily sclerotized at base, especially on the right side, with a short thick-walled diverticulum near its base; an S-shaped patch of relatively small dense spines covers one side of the bursa, while a separate patch of dense stout spines is placed at its ventral side near the base. Ductus bursae not clearly separate and colliculum absent. Ductus seminalis long and narrow; attached to bursa towards its base on left. Antrum broad, membranous. Tergite A8 small, trapezoid with rounded posterior angles. Apophyses elongate and tapered. Papillae anales rounded.
Range. China (Sichuan).
Phenology. Univoltine (?): second half of July.
Altitude: Captured at an altitude of about 3200 m above sea-level in Daxue Shan Mts. (China: coll. ZFMK).
Similar species. *E. spissata* is distinctive in this group by virtue of the large white tornal dot, but it could easily be confused with other species outside the group which share this feature, for instance *E. cotidiana*. Identification is therefore most reliable on the basis of examination of the genitalia. The male genitalia most resemble those of *E. plumbeolata*, which has however a narrower, more tapered valve and is readily distinguishable externally. The female are similar to those of *E. fervida*, but can be distinguished by the more elongate eighth tergite, which is transverse in the latter species.
Material. ♀, China, Sichuan, Daxue Shan Mts., Gongga Shan, W Moxi, 29°34′ N, 101°59′ E, 3200 m, 20–21.vii.1999, V. Siniaev & A. Plutenko (holotype, ZFMK); 1♂, same data, Mironov slide no. 653♂ (paratype, ZFMK).

69. *Eupithecia cordata* Mironov & Galsworthy, 2004

Eupithecia cordata Mironov & Galsworthy, 2004, *Trans. lepid. Soc. Japan* 55 (3): 227, figs 2, 14, 17. Holotype ♀ (ZFMK; examined), [China]: Yunnan, A-tun-tse [Dêqên], 4000 m.

Description. Fore wing ground colour pale buff-brown; antemedial and medial double; postmedial also double, twice angled close to costa, and sinuous to inner margin; discal dot small, dark brown; postmedial bounded on exterior by a broader pale fascia; terminal area darker brown, crossed by a faint pale line parallel to postmedial, and a sinuous pale subterminal line. Hind wing ground colour rather uniform grey brown; discal dot small; anal margin with wedge of darker brown extending from base to one half; postmedial brown, slightly zig-zag, almost complete; terminal area darker, crossed by wavy pale line.
Male genitalia. Uncus rather short, biapical. Valve fairly broad and short, tapering to a narrow apex. Vinculum very short with shallow medial

hollow. Papillae parallel sided. Aedeagus shorter than valve, vesica armed with a long horn-like cornutus, an elongate, slightly curved plate and a small irregular knot near ductus ejaculatorius base. Sternite A8 with two narrow rods, apically slightly sclerotized and tapered, connected by a narrow band at base.

Female genitalia. Bursa heart-shaped, heavily sclerotized; irregularly covered with spines except areas at left side and near base of ductus seminalis, with a medial longitudinal row of larger spines set on a sclerotized base, and a small membranous diverticulum at base. Ductus bursae very short, with longitudinal striations. Ductus seminalis long and narrow, attached to posterior part of bursa from right side. Colliculum short and rather wide. Antrum membranous. Tergite A8 almost square, with small membranous medial hollow on the posterior margin. Apophyses short and tapered. Papillae anales broad.

Range. China (Yunnan, Sichuan).

Phenology. Univoltine (?): early June to mid-August.

Altitude. Found from about 3800–4000 m above sea-level in Sichuan and Yunnan (China: coll. ZFMK).

Similar species. Best identified externally by the contrast between the highly patterned fore wings and rather plain hind wings, a little like some specimens of the European *E. undata*, Freyer, 1842. The male genitalia resemble those of *E. plumbeolata*, but the valve is narrower and the cornuti larger. The female genitalia are very similar to those of *E. undata* and *E. fennoscandica* Knaben, 1949, but are distinguishable by the presence of a colliculum and the transverse shape of the eighth tergite.

Material. ♀, Nord Yünnan, A-tun-tse [Dêqên], Mittlere Höhe (*ca.* 4000 m), 31.vii.1936, H. Höne, de Laever slide no. 147♀ (holotype, ZFMK); 1♂, 5♀, same loc. and alt., 19, 24, 29, 30 and 31.vii.1936 and 2.viii.1936, H. Höne, Vojnits slides nos. 12479♂, 12480♀, 12484♀, 12485♀, 12486♀ and 13240♀ (paratypes, ZFMK); 2♂, 2♀, Tibet [Sichuan], Batang, Untere Urwaldzone (*ca.* 3800 m), 10, 25, 26.vi, 12.viii.1936, H. Höne, Vojnits slides nos. 12402♂, 12432♀, 13216♂, 13823♀ (paratypes, ZFMK); 1♂, same loc. and alt., 11.vi.1936, H. Höne, Mironov slide no. 716♂ (ZFMK).

Remarks. Slides nos. 12479♂, 12480♀, 12484♀, 12485♀ and 12486♀ are marked as "paratypus of *Eupithecia contagiosa* Vojnits". No species under this name was described by Vojnits, nor do the specimens themselves bear this label. The membranous diverticulum of the bursa has been removed in Vojnits slides nos. 12485♀ and 13823♀, and is damaged in slide no. 12480♀ and de Laever slide no. 147♀.

70. *Eupithecia turpis* Vojnits, 1981

Eupithecia turpis Vojnits, 1981, *Acta zool. Acad. Sci. hung.* 27 (3–4): 422, figs 39–41. Holotype ♂ (ZFMK; examined), [China]: North Yunnan, A-tun-tse [Dêqên], 4000 m.

Description. A readily recognized species with characteristic elongate, narrow fore wings. Fore wing ground colour a combination of dark brown and rusty ochreous; costal margin rusty ochreous at middle; basal area rusty ochreous in posterior half and along stalk of Cu; a broad oblique diagonal rusty ochreous streak narrowing from anal margin to apex; transverse lines oblique, sharply angled at level of discal dot; terminal line black, interrupted at vein ends; discal dot relatively large, almost rounded, intensely black. Hind wing paler, brownish grey; dark transverse lines well marked along anal margin and near tornus only; terminal area unicolourous, slightly darker; discal dot small, pale, rounded, almost invisible.
Male genitalia. Uncus short, biapical. Valve broad, with medially convex dorsal margin and blunt apex. Vinculum broad, with shallow medial hollow. Papillae slightly clavate, with thickened apices. Aedeagus stout, broadened anteriorly, longer than valve length. Vesica armed with one long, slightly sinuate cornutus, and five smaller ones, lying parallel in the apical part of the aedeagus, as well as a large irregular cornutus at the anterior end of the aedeagus. Sternite A8 peg-like, broadened basally, with two narrow apical rods.
Female unknown.
Range. China (Yunnan).
Phenology. Univoltine (?): early July to mid-August.
Altitude. From about 3000 to 4000 m above sea-level in Yunnan (China: Vojnits 1981b).
Similar species. Could only be confused with the next species, but can be distinguished from it by the less rich rusty ochreous colour of the fore wing, the less curved postmedial transverse line between CuA_2 and A veins, the smaller discal dot and the absence of a blackish tornal blotch on the fore wing, as well as the darker hind wing, with smaller, almost invisible discal dot. The male genitalia, especially the narrower valve, the larger, longer horn-like cornuti on the vesica and the shape of the eighth sternite readily distinguish the two species.
Material. ♂, Nord Yünnan, A-tun-tse [Dêqên], Mittlere Höhe (*ca.* 4000 m), 13.viii.1936, H. Höne, Vojnits slide no. 13128♂ (holotype, ZFMK); 1♂, same loc. and alt., 26.vii.1936, H. Höne, Vojnits slide no. 13126♂ (paratype, ZFMK); 1♂, same loc., Talsohle *ca.* 3000 m, 2.vii.1937, H. Höne, Vojnits slide no. 13129♂ (paratype, TTM).

71. *Eupithecia peguensis* Prout, 1958

Eupithecia peguensis Prout, 1958, *Bull. Br. Nat. Hist.* (*Ent*) **6** (12): 392. Holotype ♀ (BMNH; examined), Burma: E. Pegu.
Eupithecia torva Vojnits, 1983, *Acta zool. Hung.* **29** (1–3): 271, figs 14, 17. Holotype ♂ (ZSM; examined), Nepal: Kathmandu (synonymised in Mironov et al., 2009b).

Description. Fore wing very elongate, ground colour reddish brown; transverse lines very lightly marked or invisible, apart from prominent double pale postmedial, strongly angled near dorsum, then broadly rounded to costa; terminal area slightly darker, with dark streaks on veins and broader dashes at apex and tornus, and crossed by a rather straight pale submarginal; discal dot large, very dark. Hind wing paler, off-white with darker anal and terminal margins; transverse lines very faintly visible, rounded; discal dot almost invisible.
Male genitalia. Uncus short, biapical. Valve unusually broad, especially at middle, tapering abruptly to a narrowly rounded apex. Vinculum broad, rather square. Aedeagus stout, longer than valve length, strongly broadened at anterior end. Vesica very wrinkled, armed with a chain of short broad spines at anterior end, and a twisted knot at base. Sternite A8 two narrow, sinuate, pincer-like arms, broadened at base, and joined by a very narrow band.
Female genitalia. Bursa shaped like a retort, the ductus bursae with a row of long spines which expand to cover all the body except for one side where the ductus seminalis is attached; on the opposite side from the ductus a small rounded thick-walled, spineless diverticulum. Ductus bursae very elongate and furrowed. Colliculum absent. Antrum very short, membranous. Apophyses narrow, pointed. Papillae anales elongate, rather rounded.
Range. China (Tibet); Nepal, Myanmar, Thailand, Vietnam (Upper Tonkin).
Phenology. Appears to be a univoltine spring species. Records from January and February in Thailand, March and April in Myanmar and Nepal, and May in Tibet.
Altitude. Ranges from about 1000 m in Thailand to 2250 m in Tibet.
Similar species. See differential diagnosis under preceding species.
Material. ♀, Type, E. Pegu, 4–5000ft., March-April 1890, W. Doherty, *Eupithecia peguensis* ♀ Prout type, BM Geom. slide no. 11446 (holotype, BMNH); ♂, Nepal, Kathmandu, 1400 m, 15.iv.1962, leg. G. Ebert u. H. Falkner, Vojnits slide no. 10667♂ (= ZSM slide no. G10630) (holotype of *E. torva*, ZSM); 1♀, Tibet, Nyalam, Zham, 2250 m, 12.5.1974, coll Zhang Xuezhong, slide no. ZL421♀ (IZCAS).

72. *Eupithecia apta* Mironov & Galsworthy, 2004

Eupithecia apta Mironov & Galsworthy, 2004, *Trans. lepid. Soc. Japan* 55(1): 54, figs, 11, 30. Holotype ♀ (ZFMK; examined), [China]: Yunnan, Li-kiang [Lijiang].

Misidentification: *Eupithecia inepta* Vojnits, 1979 (1♀ paratype in ZFMK) [Junior primary homonym of *Eupithecia inepta* Prout, 1922].

Description. Fore wing ground colour pale yellow-brown; discal dot dark brown, prominent, slightly elongate; traces of an antemedial visible on holotype; postmedial well distal of discal dot, curving from costa parallel to the termen, then straightening to meet inner margin at right angle; terminal area slightly darker. Hind wing uniform pale yellow brown; discal dot darker brown, elongate; traces of antemedial, medial and postmedial lines running from anal margin to posterior margin of cell; subterminal area slightly darker.

Male genitalia. Uncus elongate, biapical. Valve broad, with a wide, blunt apex. Vinculum broad with a deep medial hollow. Papillae tapered. Aedeagus stout, elongate, slightly curved, almost equal to length of valve. Vesica armed with one stout, curved horn-like cornutus, about equal to length of aedeagus, an elongate, U-shaped cornutus with thickened and folded lateral sides, and a small, irregular cornutus, both the latter placed near ductus ejaculatorius base. Sternite A8 long and narrow, consisting of two narrow arms with pointed tips, slightly broadened and diverging near apices, connected by a short, narrow basal band.

Female genitalia. Bursa broad at anterior end, tapered towards ductus, sclerotized, with a large prominent sclerotized diverticulum on the right side and a membranous diverticulum at base; almost completely covered with spines apart from small areas along left and right sides. Ductus bursae short and broad, completely covered with spines. Ductus seminalis narrow, attached to centre of corpus bursae. Colliculum short and broad. Tergite A8 trapezoid, with rounded posterior corners. Apophyses narrow, the anterior relatively long, the posterior rather short. Papillae anales short and rounded.

Range. China (Yunnan).

Phenology. Univoltine (?): mid-July to late August.

Altitude. From about 2800 to 4500 m in mountains of Sichuan and Yunnan (China: coll. ZFMK).

Similar species. Externally similar to *E. consortaria*, but easily distinguished by the structure of the female genitalia. The female genitalia of *E. consortaria* are larger and have very large spines in the corpus bursae, and a finer structure of the diverticulum at the base of the bursa copulatrix.

The female genitalia of this species are similar to those of *E. undata* Freyer, 1942 and *E. fennoscandica* Knaben, 1949. They are distinguished from the genitalia of these two species by the shape of the corpus bursae, the presence of a large prominent sclerotized diverticulum on the right side of the bursa copulatrix, and by shorter posterior apophyses.
Material. ♀, China, Provinz Nord-Yuennan, Li-kiang [Lijiang], 28.viii.1934, H. Höne, Vojnits slide no. 11467♀ (holotype [= paratype of *E. inepta* Vojnits, 1979], ZFMK); 1♀, same loc., 30.viii.1934, H. Höne, Vojnits slide no. 13115♀ (TTM); 1♀, Nord Yünnan, A-tun-tse [Dêqên], Obere Höhe (*ca.* 4500 m), 6.viii.1936, H. Höne, Vojnits slide no. 12478♀ (paratype, ZFMK); 1♂, Sichuan, Daxue Shan Mts, Gongga Shan, NW Moxi, 14–19.vii.1999, 29°41′ N 101°58′ E, leg. V. Sinaev & A. Plutenko (ZFMK).
Remarks. The holotype was originally published in Vojnits (1979a) as a female paratype of *E. inepta* Vojnits, 1979. It is not, however, conspecific with the male holotype of that species.

73. *Eupithecia burselongata* Mironov & Galsworthy, 2004

Eupithecia burselongata Mironov & Galsworthy, 2004, *Trans. lepid. Soc. Japan* 55 (2): 124, figs 6, 18, 21. Holotype ♀ (ZFMK; examined), [China]: North Yunnan, A-tun tse [Deqing], upper heights, 4500 m.

Description. Fore wing ground colour very pale buff; basal area brown, a brown suffusion continuing on the costa to about half; faint undulating pale brown antemedial and medial lines below discal dot, the latter dark brown and slightly elongate; postmedial two narrow pale brown lines following contour of termen, curving gradually to meet costa at right angles; an irregular darker brown suffusion present on both sides of postmedial, most marked near costa; terminal area with darker brown blotches apically, medially and tornally. Hind wing short, ground colour dingy white; discal dot absent to small; anal margin darkened to mid brown as far as lower margin of cell; a faint pale brown postmedial.
Male genitalia. Uncus elongate, biapical. Valve narrow, tapering in distal half. Vinculum short, rounded. Papillae narrow and elongate. Aedeagus slim, shorter than valve length. Vesica armed with one slightly sinuate horn-like cornutus and one small, curved plate near base of ductus ejaculatorius. Sternite A8 with two narrow, parallel apical rods with blunt tips.
Female genitalia. Bursa fusiform, sclerotized, covered with small spines on right side; a small membranous diverticulum near base. Ductus bursae short with separate small patch of minute spines. Ductus seminalis broadened at base, attached to ductus bursae near colliculum. Colliculum

relatively large, elongate. Antrum membranous. Tergite A8 nearly quadrate. Apophyses short and narrow. Papillae anales short and narrow.
Range. China (Yunnan).
Phenology. Bivoltine (?): mid-June to mid-August.
Altitude. Found between about 3000 m and 4500 m in North Yunnan (China: coll. O. Pekarsky; coll. ZFMK).
Similar species. Externally rather similar to *E. consortaria* Leech, 1897. In the male, the relatively elongate and narrow valve, and the single horn-like cornutus on the vesica, approximately half the length of the aedeagus, separate the species well. In the female, the fusiform bursa copulatrix with a solid diverticulum at its base, the disposition of small spines in the corpus bursae, and the elongate and relatively broad colliculum are distinctive.
Material. ♀, Nord Yünnan, A-tun-tse [Dêqên], Obere Höhe (*ca.* 4500 m), 31.vii.1936, H. Höne, de Laever slide no. 86♀ (holotype, ZFMK); 1♂, same loc. and alt., 11.viii.1936, H. Höne, Vojnits slide no. 12491♂ (paratype, ZFMK); 1♀, same loc. and alt., 23.vii.1936, H. Höne, Vojnits slide no. 12256♀ (TTM); 1♀, same loc., Mittlere Höhe (*ca.* 4000 m), 18.viii.1936, H. Höne, Vojnits slide no. 12497♀ (paratype, ZFMK); 2♀, China, N. Yunnan, Diqing Tibetan Aut. Pref., 8 km NNE of Shangri La at Nairi village, elev. 3300 m, 14.vi.2009, leg. B. Benedek, Pekarsky slide no. OP 0466♀, OP 0467♀ (coll. O. Pekarsky); 3♀, KIB/RBGE research station, Lijiang, Yunnan, 3200 m, 15–20.vi.2009, A.C. Galsworthy, BM Geom. slide no. 22688♀ (BMNH); 1♀, same data (IZCAS).

74. *Eupithecia minutula* Mironov & Galsworthy, 2004

Eupithecia minutula Mironov & Galsworthy, 2004, *Trans. lepid. Soc. Japan* 55 (2): 126, figs 7, 22. Holotype ♀ (ZFMK; examined), [China]: North Yunnan, Li-kiang [Lijiang].

Description. Very small. Fore wing ground colour dark brown, speckled liberally with pale scales; transverse lines, where visible, paler brown; pale, sinuous medial most distinct, curling round outside discal dot; latter darker than ground colour, elongate. Hind wing evenly brown, slightly darker than fore wing, unmarked except for faint discal dot.
Female genitalia. Bursa fusiform, sclerotized, covered with an irregular pattern of small spines, with a short row of larger spines and a small membranous diverticulum near the base. Ductus bursae short, with a separate patch of minute spines and a small sclerotized bulla. Ductus seminalis broadened at base, attached to ductus bursae near colliculum. Colliculum wide and quite elongate. Antrum membranous. Tergite A8 small, nearly

quadrate. Apophyses short and narrow. Papillae anales relatively large, short but broad.
Male unknown.
Range. China (Yunnan, Tibet, Sichuan).
Phenology. Univoltine: mid-June to early July.
Altitude. Found from about 3000 m to 5000 m in Tibet and Yunnan (China: coll. ZFMK, TTM).
Similar species. The very small size, and uniformly dark hind wing of this species, darker than the fore wing, are very distinctive, and unlike any other Chinese species. The female genitalia are similar to those of *E. burselongata* Mironov & Galsworthy, 2004 but distinguished by the presence of a short row of larger spines near the base of the bursa.
Material. ♀, China, Provinz Nord-Yuennan, Li-kiang [Lijiang], 25.vi.1935, H. Höne, de Laever slide no. 153♀ (holotype, ZFMK); 1♀, Tibet, Nyingchi, Bayi, 3040 m, 2–3.vii.2002, leg. Han Hongxiang, Galsworthy slide no. ZL652♀ (paratype, IZCAS); 1 ex., Tibet [Sichuan], Batang, Alpine Zone (*ca.* 5000 m), 18.vi.1939, H. Höne, Vojnits slide no. 12252[2] (TTM; gen. slide absent in TTM and ZFMK); 1♀, KIB/RBGE research station, Lijiang, Yunnan, 3200 m, 15–20.vi.2009, A.C. Galsworthy (BMNH).

75. *Eupithecia impavida* Vojnits, 1979

Eupithecia impavida Vojnits, 1979, *Acta zool. Acad. Sci. hung.* 25 (1–2): 206, fig. 18. Holotype ♂ (ZFMK; examined), China: North Yunnan, Li-kiang.
Eupithecia secura Vojnits, 1979, *Acta zool. Acad. Sci. hung.* 25 (3–4): 428, fig. 4, 5. Holotype ♀ (ZFMK; examined), [China]: Sued-Shensi, Tapaishan [Taibaishan] im Tsinling, 1700 m (synonymised in Mironov et al., 2008b).
Eupithecia commiserenda Vojnits, 1983, *Acta zool. Acad. Sci. hung.* **29** (1–3): 270, fig. 13. Holotype ♂ (ZSM; examined), Nepal: Prov. Nr. 3, East Junbesi, 2750 m (synonymised in Mironov et al., 2008b).

Misidentifications: *Eupithecia pacifica* Inoue, 1980 (paratypes).

Description. Fore wing unicolorous greyish brown; transverse lines narrow, wavy, conspicuous only near costa, where they form darker spots; two medials touching discal dot; postmedial almost straight, marked with a row of dark dashes on the veins, and bounded distally by a faint paler fascia, angled onto costa; subterminal whitish wavy line with a small whitish tornal spot; discal dot distinct, black, rounded or slightly elongate. Hind wing concolorous with rather inconspicuous transverse lines, and a relatively broad transverse lighter band between medial and broad terminal areas; discal dot small and pale.
Male genitalia. Uncus elongate, biapical. Valve very broad, tapering sharply in distal third to narrowly rounded apex. Vinculum with a distinct

medial hollow. Papillae narrow and strongly curved. Aedeagus stout and longer than valve. Vesica armed with a stout, slightly sinuate cornutus as long as the aedeagus, minutely dentate close to apex, a smaller and much shorter cornutus with blunt apex, and an elongate, V-shaped cornutus with more strongly sclerotized margins. Sternite A8 very long, with two parallel arms, each with a cap-shaped tip, connected at base by a narrow band.

Female genitalia. Bursa elongate, with a heavily sclerotized left side, a medially prominent, membranous right side, and a broad membranous diverticulum at base; corpus bursae covered with numerous spines, especially densely in its posterior part and along lateral sides, the larger spines often forming an oval wheel shape. Ductus bursae not distinguishable. Ductus seminalis narrow, attached to membranous prominence of the medial part of bursa from ventral side. Colliculum very lightly sclerotized, covered with numerous pores. Antrum membranous. Tergite A8 broadened at base, with convex and narrowly sclerotized anterior margin and rounded posterior margin. Posterior apophyses slightly expanded and flattened near their apices and much narrower than anterior apophyses. Papillae anales rather small and slightly elongate.

Range. China (Tibet, Yunnan, Sichuan, Gansu, Shaanxi, Henan, Hubei, Hunan); Pakistan, Nepal, Japan.

Phenology. Univoltine: late June to mid-August.

Altitude. Found from about 1500 m in Gansu to about 3000 m in Sichuan (China: coll. O. Pekarsky).

Similar species. This widespread species has few external distinguishing characteristics: it is rather uniformly dark, and lightly marked, and resembles a number of other species similarly lacking in distinctive details. It is best distinguished by examination of the genitalia. The sinuate heavy cornutus, dentate along one side near the apex, in the male aedeagus vesica, and in the female the broad, heavily sclerotized bursa, usually with an oval of large spines in the centre are the best distinguishing features. See also below under *E. consortaria* and under next species.

Material. ♂, China, Prov. Nord-Yuennan, Li-kiang [Lijiang], X, 14.viii.1935, H. Höne, Vojnits slide no. 11470♂ (holotype, ZFMK); 1♂, same loc., 30. viii.1934, H. Höne, Vojnits slide no. 12535♂ (paratype of unpubl. *E. idonea* Vojnits, [nec *E. idonea* Vojnits, 1982 from Armenia], ZFMK); 1♂, 1♀, same loc., 15, 17.viii.1935, H. Höne, Vojnits slide no. 13174♂, Mironov slide no. 818♀ (TTM); 2♂, 2♀, same loc. and alt., 5, 9, 11.viii.1936, H. Höne, de Laever slide no. 308♀, Vojnits slide no. 12310♀, Mironov slides nos. 667♂,

668♂ (ZFMK); 1♀, same loc., *ca.* 3000 m, 13.vii.1936, H. Höne, Vojnits slide no. 12209♀ (paratype, ZFMK); 3♂, same loc. and alt., 10, 11.viii.1936, H. Höne, Vojnits slides nos. 12312♂, 11491+, 12309♂ (ZFMK); 1♀, same loc. (without alt.), 1.vii.1935, H. Höne, Vojnits slide no. 12239♀ (ZFMK); ♀, Sued-Shensi [Shaanxi], Tapaishan [Taibaishan] im Tsinling [Qinling], *ca.* 1700 m, 24.vii.1936, H. Höne, Vojnits slide no. 12215♀ (holotype of *E. secura*, ZFMK); 1♀, same loc. and alt., 12.vii.1936, H. Höne, Vojnits slide no. 12218♀ (paratype of *E. secura*, TTM); 1♂, same loc. and alt., 8.viii.1936, H. Höne, Vojnits slide no. 11425♂ (paratype of *E. impavida*, TTM); ♂, Nepal, Prov. Nr. 3, East Junbesi, 2750 m, 25–31.vii.1964, leg. W. Dierl, Vojnits slide no. 11874♂ [= ZSM slide no. G 10613♂] (holotype of *E. commiserenda*, ZSM); 2♂, China, Sichuan, Lao Lin Kou, 1900 m, 28°21′ N, 103°26′ E, 26.vi.-12.vii.2008, leg. Viktor Sinyaev, Pekarsky slides nos. OP 0245♂, OP 0246♂ (coll. O. Pekarsky); 1♂, China, W. Sichuan, Kangding, 3000 m, 29°53′ N / 101°55′ E, 13.vii.2009, I. & A. Floriani, Saldaitis leg., Pekarsky slide no. OP 0574♂ (coll. O. Pekarsky); 2♂, Yunnan, Lijiang, Yulongshan, 22.vii.1962, slides nos. ZL171♂, ZL611♂; 1♂, Yunnan, Weixi Pantiange, 2500 m, 25.vii.1981, Wang Shuyong; 1♀, Yunnan, Yongsheng Liude, 2250 m, 10.vii.1984, Liu Dajun, slide no. ZL412♀; 2♂, Yunnan, Tengchong, Tengchong Dahaoping, 2020 m, 98°43′ N 24°58′ E, 5–7.viii.2007, Xue Dayong/Wu Chunguang; 1♂, Sichuan, Emeishan, 1800–2000 m, 12.vii.1957, Wang Songyuan, slide no. ZL54♂; 1♂, Tibet, Gyirong, 2800 m, 30.vii.1975, Wang Ziqing, slide no. ZL606♂; 3♂, 1♀, Hubei, Shennongjia, Dajiuhu, 1800 m, 4–5.viii.1981, Han Yinheng, slides nos. ZL154, 198, 227♂, ZL151♀; 1♂, Hunan, Tianpingshan, 25.vii.1981; 4♂, 5♀, Gansu, Wenxian, Qiujiaba, 2350 m, 21–22.vii.1999, Yao Jian, slides nos. ZL559♂, ZL552, 556, 567, 570♀; 1♀, same loc. and collector, but Wangshan temple, 1500 m, 28.vii.1999, slide no. ZL577♀; 1♀, Henan, Funiushan, 7–9.viii.1993, Wang Zhiguo, slide no. ZL676♀ (IZCAS).

Remarks. The synonymy of this species, and Professor Inoue's subsequent description of it as *E. pacifica*, of which the undissected holotype was in fact a specimen of *E. consortaria*, were dealt with in Mironov et al. (2008b). This and the next two species have proved to be very difficult to separate and to assign males confidently to females. We believe that we have arrived at the right conclusions, but it is possible that fresher material will shed further light on the complex.

76. *Eupithecia fervida* Vojnits, 1984

Eupithecia fervida Vojnits, 1984, *Acta zool. Acad. Sci. hung.* **30** (1–2): 216, figs 4, 7. Holotype ♂ (TTM; examined), [China]: North Yunnan, A-tun-tse [Dêqên], 4000 m.

Description. Fore wing ground colour grey with more or less distinct transverse lines; basal and antemedial oblique, angled or curved near costa; medial very wavy, sharply angled onto costa behind discal dot; postmedial oblique, partly formed by a row of small dark dashes on the veins, twice angled onto costa; terminal area slightly darker, with narrow, wavy, whitish subterminal line and small whitish tornal spot; discal dot relatively small, slightly elongate, black. Hind wing paler than fore wing, whitish grey, darker along anal and terminal margins; transverse lines marked only along anal margin, with postmedial rather prominent; tornus with small whitish spot; discal dot small and pale.
Male genitalia. Uncus slightly elongate, biapical. Valve rather broad, and medially broadened, with broadly rounded apex. Vinculum broad, with shallow medial hollow. Papillae elongate, slightly curved, and tapered. Aedeagus equal to length of valve. Vesica armed with a stout, slightly curved horn-like cornutus about equal to length of aedeagus, a small weakly sclerotized, folded, V-shaped cornutus and a smaller, curved irregular cornutus near ductus ejaculatorius base. Sternite A8 with two very long and narrow rods connected by a narrow basal band.
Female unknown.
Range. China (Sichuan, Yunnan).
Phenology. Univoltine: early July to mid-August.
Altitude. Found at about 4000 m above sea-level in North Yunnan (China: Vojnits 1979).
Similar species. Not easy to distinguish from other allied *Eupithecia* species, in particular *E. impavida* and *E. inexhausta*. *E. impavida* has darker hind wings, less well marked transverse lines on the fore wings, and larger discal dots. *E. inexhausta* has a strong contrast between the pale upper part of the hind wing and the extensively darkened anal margin. The male genitalia are very similar to those of *E. inexhausta* but can be distinguished by the broader apex to the valve and the stouter horn-like cornutus on the vesica.
Material. ♂, Nord Yünnan, A-tun-tse [Dêqên], Mittlere Höhe (*ca.* 4000 m), 3.vii.1936, H. Höne, Vojnits slide no. 14097♂ (holotype, TTM); 1♂, same loc. and alt., 21.vii.1936, H. Höne, Vojnits slide no. 14067♂ (paratype, ZFMK); 1♂, same loc. and alt., 17.vii.1936, H. Höne, Vojnits slide no. 14071♂ (paratype, TTM); 1♂, Szetschwan [Sichuan], Kunkalaschan, *Eupithecia acyrtoterma* Prt., de Laever slide no. 431♂ (ZFMK); 2♂, [Sichuan], Ta-tsien-lu, Szetschwan-China, Exped. Stötzner, Vojnits slides nos. 12528♂, 12536♂ (paratypes of unpubl. *E. idonea* Vojnits, [nec *E. idonea* Vojnits, 1982 from Armenia], ZFMK); 1♂, Yunnan, Lijiang, Gan He Ba, Lijiang, 23.vi.2009, A.C. Galsworthy, BMNH slide no. 22691♂ (BMNH).

Remarks. The female described by Vojnits (1984) as a paratype of *E. fervida* (Vojnits slide no. 14053♀) belongs to the next species *E. inexhausta.*

77. *Eupithecia inexhausta* Vojnits, 1984

Eupithecia inexhausta Vojnits, 1984, *Acta zool. Acad. Sci. hung.* **30** (1–2): 218, fig. 6. Holotype ♂ (ZFMK; examined), [China]: North Yunnan, A-tun-tse [Dêqên], Talsohle, 3500 m.

Description. Fore wing ground colour light grey-brown, transverse lines slightly darker, and lightly marked; medial straight, oblique, sharply angled onto costa behind discal dot; postmedial evenly curved, twice angled near costa; discal dot small, black, rounded. Hind wing clear white in upper two thirds, close set transverse lines dark, well marked along anal margin only, and forming a contrasting dark border; terminal area slightly darker than rest of wing; discal dot smaller and paler than on fore wing.
Male genitalia. Uncus biapical. Valve broad, sharply tapering in distal third to narrowly rounded apex. Vinculum short, usually with very shallow medial hollow. Papillae elongate and tapering. Aedeagus stout and rather narrow, almost equal to length of valve. Vesica armed with one long and relatively narrow, slightly curved horn-like cornutus, and a narrow elongate, folded, H-shaped cornutus near ductus ejaculatorius base. Sternite A8 two long and narrow arms, diverging posteriorly, with pointed apices, connected by a narrow basal band.
Female genitalia. Bursa elongate, strongly broadened in basal part; left side heavily sclerotized, right side membranous, with a broad membranous diverticulum at base; corpus bursae covered with numerous spines, especially densely in its posterior part and along sides. Ductus bursae not distinguishable. Ductus seminalis narrow, broadened at base, attached to membranous prominence of the medial part of the corpus bursae from the ventral side. Colliculum short, lightly sclerotized, covered with numerous pores. Antrum membranous. Tergite A8 almost rectangular. Papillae anales rather short and rounded.
Range. China (Sichuan, Yunnan, Gansu).
Phenology. Univoltine: mid-July to early September.
Altitude. Found from about 3500 to 4500 m above sea-level in Yunnan (China: Vojnits 1984).
Similar species. Externally very similar to *E. impavida* and *E. fervida*: for distinction see under preceding species. All known material is very worn: fresh material might be more easily recognizable. Male genitalia very similar to those of *E. fervida*: see under that species. The female genitalia are distinct and easily separable from other related species.

Material. ♂, Nord Yünnan, A-tun-tse [Dêqên], Talsohle (*ca.* 3500 m), 5.ix.1936, H. Höne, Vojnits slide no. 14115♂ (holotype, ZFMK); 2♂, A-tun-tse [Dêqên] (Nord Yunnan) Mittlere Höhe, *ca.* 4000 m, 7.ix.1936, and same loc., *ca.* 4500 m, 9.vii.1936, H. Höne, respectively Vojnits slides nos. 12461♂ (labeled as holotype of unpubl. *E. fraudulenta*), 12462♂ (labeled as paratype of unpubl. *E. fraudulenta*) (ZFMK); 1♂, same loc., Mittlere Höhe (*ca.* 4000 m), 15.viii.1936, H. Höne, Vojnits slide no. 14118♂ (paratype, TTM); 1♀, same loc., Obere Höhe (*ca.* 4500 m), 21.vii.1936, H. Höne, Vojnits slide no. 14053♀ (paratype of *E. fervida*, ZFMK); 1♀, same loc. and alt., 26.vii.1936, H. Höne, Mironov slide no. 833♀ (TTM); 2♀, same loc. and alt., 26.vii., 9.viii.1936, H. Höne, Vojnits slides nos. 12452♀, 14087♀ (ZFMK); 2♂, China, Prov. Nord-Yuennan, Li-kiang [Lijiang], 8.viii.1934, 26.vii.1935, H. Höne, Vojnits slides nos. 13041♂, 13184♂ (TTM); 1♀, [Sichuan], Ta-tsien-Loû, Chasseurs du P. Déjean 1904, Vojnits slide no. 855♀ (TTM); 1♀, China, Sichuan, Yajiang, 3300 m, 24.viii.1982, coll Zhang Xuezhong, slide no. ZL478♀; 1♀, Gansu, Wenxian, Qiujiaba, 2350 m, 21.vii.1999, coll Zhu Chaodong, slide no. ZL613♀ (IZCAS); 2♀, China, W. Sichuan, road Yajiang / Litang, 3600 m, 15.vii.2009, 29°59′ N / 100°52′ E, I. & A. Floriani, Saldaitis leg., Pekarsky slides nos. OP 0593♀, OP 0655♀ (coll. O. Pekarsky).

78. *Eupithecia hongxiangae* Mironov & Galsworthy, 2011

Eupithecia hongxiangae Mironov & Galsworthy, 2011, *Lep. Science* **62** (1): 21, figs 6, 23, 27. Holotype ♀ (IZCAS; examined), China: Gansu, Wenxian, Qiujiaba, 2350 m.

Description. Fore wing ground colour grey; transverse lines inconspicuous except on costa, where they form a series of dark spots; postmedial consisting of a series of dark dashes on veins, with a narrow pale line outside them, sharply angled near costa; terminal area slightly darker with fine, wavy, whitish subterminal line; discal dot distinct, rounded or ovoid, black. Hind wing paler, whitish grey but shaded darker in anal and terminal areas; terminal area with a wavy whitish subterminal line and small whitish tornal dot; discal dot smaller and paler, rounded or comma-like.
Male genitalia. Uncus elongate, biapical. Valve broad, with bowed dorsal margin near base, slightly sinuate ventral margin and broadly rounded apex. Vinculum broad with distinct medial hollow. Papillae long and narrow, narrowed medially. Aedeagus very stout, equal to length of valve. Vesica armed with a long, stout horn-like cornutus as long as aedeagus, a short, stout cornutus, and a poorly sclerotized narrow, H-shaped cornutus, both the latter near apex. Sternite A8 consisting of two almost parallel, long and relatively narrow rods, with blunt and curved, more

sclerotized, cap-shaped tips, the rods connected by a short narrow band at base.

Female genitalia. Bursa elongate and narrow, the posterior two thirds completely covered with large spines forming longitudinal rows; at base a narrow, rounded unspined diverticulum, and on the right side a much broader square-ended diverticulum, almost completely covered with chaotic, mainly small spines. Ductus bursae and colliculum not present. Ductus seminalis slightly broadened at base, attached to lateral diverticulum. Antrum membranous, short and broad. Tergite A8 relatively small, spade-shaped with rounded posterior margin. Apophyses short, narrow, and pointed. Papillae anales short and rounded.

Range. China (Gansu, Qinghai, Henan).

Phenology. Univoltine (?): all known specimens from July and August.

Altitude. Found between 2000 and 3000 m in Gansu and Qinghai, (China: coll. IZCAS).

Similar species. Easily confused with *E. impavida* and *E. inexhausta*. The hind wings are paler than those of *E. impavida*, with a pale area contrasting with the darker anal margin, more like those of *E. inexhausta*. It is smaller than the latter, but examination of the genitalia is likely to be necessary for safe determination. The long cornutus in the male vesica is stouter and straighter than in *E. impavida*, and lacks the indented margin near the apex; the second apical cornutus is characteristic, and easily distinguished from its equivalent in either of the other two species. In the female the square-ended lateral diverticulum separates the species readily: the bursa resembles that of *E. alishana*, but both sets of apophyses are strikingly shorter and narrower in the present species.

Material. ♀, China, Gansu, Wenxian, Qiujiaba, 2350 m, 21–22.vii.1999, Yao Jian, slide no. ZL570♀ (holotype, IZCAS); 1♂, China, Gansu, same data, slide no. ZL559♂; 1♂, China, Henan, Funiushan, 7–9.viii.1993, Wang Zhiguo, slide no. ZL678♂ (paratypes, IZCAS); 1♀, Qinghai, Tongren, Maixiu, 2950 m, 30.vii.1992, coll Xue Dayong, slide no. ZL392♀ (BMNH).

79. *Eupithecia alishana* Inoue, 1970

Eupithecia alishana Inoue, 1970, *Bull. Fac. domest. Sci. Otsuma Wom. Univ.* 6: 3, pl. 1, figs 5, 6; pl. 2, fig. 13; pl. 3, figs 16–18. Holotype ♂ (BMNH; examined), [Taiwan]: Chiayi, Alishan, 2200 m.

Description. Fore wing ground colour pale ash grey. Fore wing relatively broad with weakly visible transverse lines, which form a series of dark spots on the costa; discal dot black, large and rounded; postmedial line formed by a series of short dark dashes on the veins; terminal area usually darker, grey, with fine whitish subterminal line and small whitish tornal spot. Hind wing similarly coloured, with darker anal and terminal margins; discal dot smaller and paler.

Male genitalia. Uncus narrow, biapical. Valve broad, with narrowly rounded apex. Vinculum short and relatively broad, with shallow medial hollow. Papillae long and medially narrowed. Aedeagus rather stout, but shorter than length of valve. Vesica armed with a stout, slightly curved horn-like cornutus as long as aedeagus, a narrow H-shaped cornutus and a smaller, elongate plate lying alongside the large cornutus. Sternite A8 two long parallel rods with blunt tips, connected by a narrow band at base.

Female genitalia. Bursa elongate and narrow, heavily sclerotized, mostly covered with dense spines, the largest of which form a long oval shape; at base, a membranous, spineless area; on left side, a large square-ended diverticulum mostly covered with spines. Ductus bursae not distinguishable. Ductus seminalis broadened at base, broadly attached to diverticulum. Colliculum absent. Antrum broad, membranous. Tergite A8 rectangular and elongate. Apophyses elongate and stout, the posterior with small spurs. Papillae anales elongate and narrow.

Range. China (Hubei, Taiwan).

Phenology. Univoltine: late May to mid-July.

Altitude. Found in Taiwan from about 2200–2400 m (Taiwan: Inoue 1970; coll. ZSM, TTM). Hubei record is from 1660 m.

Similar species. Externally much greyer than other members of this group, and rather similar to *E. quadripunctata,* but has shorter broader wings and a grey abdomen, whereas the abdomen of *E. quadripunctata* is covered with ochreous or reddish ochreous scales. The long, stout and almost straight horn-like cornutus without denticules on its apex in the male genitalia is distinctive. The female genitalia most closely resemble those of the preceding species, but have much longer and stouter apophyses.

Material. ♂, Formosa [Taiwan], Alishan, 2200 m, Chiayi, 9–11.vii.1964, H. Inoue, BM Geom. slide no. 16313 (holotype, coll. H. Inoue in BMNH); 1♂, Taiwan, Prov. Hualien, Taroko N.P., Kuanyuan, 2400 m, 9.vi.1997, leg. B. Herczig & L. Ronkay (ZSM); 1♂, Taiwan, Prov. Taitung, Hsiangyang, upper forest zone, 2200 m, 13–14.vi.1997, leg. B. Herczig & L. Ronkay (TTM); 1♀, Central Taiwan, near Puli, Nantou Co., Meifeng, *ca.* 2200 m NN, 24°05′25″ N / 121°10′21″ E, LF, 20. May 2006, leg. Mei-Yu Chen & U. Buchsbaum (ZSM); 1♀, Hubei, Shennongjia, 1660 m, 16.vii.1981, coll Han Yinheng, slide no. ZL392♀ (IZCAS).

Remarks. The specimen from Hubei is the only mainland record: both externally and in the genitalia it is a very good match for Taiwan material, but further material would be useful to confirm the presence of this species on the mainland.

80. *Eupithecia antiqua* Mironov & Galsworthy, 2004

Eupithecia antiqua Mironov & Galsworthy, 2004, *Trans. lepid. Soc. Japan.* 55 (1): 46, figs 5, 20. Holotype ♂ (ZFMK; examined), [China]: Yunnan, Li-kiang [Lijiang].

Description. Fore wing ground colour grey, all markings dark brown; basal, postbasal, medial and postmedial well marked, all angled close to costa; postmedial with a series of short dark dashes on veins; discal dot round, placed close to costal angle of medial line; terminal area shaded dark brown, crossed by a pale zigzag submarginal line. Hind wing ground colour white, contrasting with fore wing; discal dot faint and small; brownish medial and postmedial present from anal margin to cell; terminal area brown from anal margin to vein M_1, with zigzag white submarginal line.
Male genitalia. Uncus long, possibly biapical. Valve broad, saccular area slightly protruding. Vinculum short and broad. Papillae tapered. Aedeagus stout, shorter than length of valve. Vesica armed with two large horn-like cornuti, broadening at their midpoints, a slim and narrow U-shaped plate, and a large bowed and irregular plate near ductus ejaculatorius base. Sternite A8 two elongate and relatively broad arms with narrow spine-like tips. Female unknown.
Range. China (Yunnan).
Phenology. The only known specimen was caught early in May.
Similar species. The greyish tone to the wings is similar to that of *E. alishana* or *E. mustangata*, but the maculated medial area of the fore wing distinguishes this species from both. The male genitalia are distinctive, especially the pattern of cornuti in the vesica and the shape of the eighth sternite. In this species group only *E. homogrammata* has two similarly sized cornuti, but those are strongly curved.
Material. ♂, China, Provinz Nord-Yuennan, Li-kiang [Lijiang], 3.v.1935, H. Höne, de Laever slide no. 263♂ (holotype, ZFMK).

81. *Eupithecia assulata* Bastelberger, 1911

Eupithecia assulata Bastelberger, 1911, *Int. ent. Z.* (Guben) 4 (45): 241. Lectotype ♀, herewith designated to stabilize nomenclature (DEI; examined), Formosa [Taiwan]: [Arizan].

Description. Fore wing ground colour pale ash-grey; transverse lines mostly absent, leaving a clear grey central area to the wing; postmedial formed by short dark dashes on the veins; costal margin with four darker spots; terminal margin slightly darker with fine, wavy white subterminal line, broadened near tornus; discal dot ovoid, black. Hind wing slightly paler, whitish grey with darker, soft ash-grey terminal margin, a distinct, wavy, white subterminal line and a small conspicuous white tornal spot; discal dot very small, pale, rounded. Abdomen light grey with a series of small black lateral dots, and in male with last two segments white.

Male genitalia. Uncus elongate, biapical. Valve unmodified. Vinculum short and broad. Papillae elongate, narrowed medially. Aedeagus equal to valve length. Vesica armed with two apical and one medial cornuti, two with hooked tips, and a small, irregular cornutus near ductus ejaculatorius base. Sternite A8 lightly sclerotized, with two parallel rods.

Female genitalia. Bursa small, elongate, lightly sclerotized, spineless, with a very large membranous diverticulum on right side, which has two small patches of spinules centrally. Ductus bursae elongate, more sclerotized than corpus bursae, densely covered with spinules. Ductus seminalis narrow, attached to corpus bursae under base of diverticulum. Colliculum inconspicuous, short and relatively broad. Antrum membranous. Tergite A8 rectangular. Apophyses short and narrow. Papillae anales small, with rounded apices.

Range. China (Taiwan).

Phenology. Univoltine (?): early August to late October.

Altitude. From about 2200 to 2600 m above sea-level in Taiwan (Inoue 1988; coll. ZSM).

Similar species. Rather distinctive due to the clear grey, relatively unmarked fore wings, but similar externally to *E. vojnitsi* Inoue, 2000 (Nepal), which has entirely different genitalia.

Material. ♀, Formosa [Taiwan], Arrizan (Kagi) (= Alishan) (lectotype, DEI); 1♂, Taiwan, Prov. Nantou, 2250 m, Meifong, 24°05′ N, 121°10′ E, 25.ix.1999, leg. G. Csorba and B. Herczig (ZSM).

Remarks. The illustration of the male of this species in Inoue, 1988, p. 350, fig. 17A is of the holotype of *E. masculina*, evidently included here in error, since the same illustration is shown in fig 24A on p. 363 as *E. masculina*.

82. *Eupithecia adoranda* Vojnits, 1984

Eupithecia adoranda Vojnits, 1984, *Acta zool. Acad. Sci. hung.* **30** (1–2): 231, fig. 23. Holotype ♂ (TTM; examined), [China]: Prov. Shansi [Shanxi], Mien-shan, 2000 m.

Description. Fore wings broad and rather blunt, unicolourous, dark brownish grey; transverse lines inconspicuous, but present as darker brown lines, broadly angled near costa, and forming dark costal spots; terminal area darker, with wavy pale subterminal line and very small light tornal spot; discal dot moderately large, rounded, black. Hind wing slightly paler, unicolourous without transverse lines; terminal area darker, with small, whitish V-shaped tornal spot; discal dot minute.

Male genitalia. Uncus rather stout, biapical. Valve broad, with bowed dorsal margin, and broadly rounded apex. Vinculum short and broad, with shallow medial hollow. Papillae narrow and slightly curved. Aedeagus stout, slightly shorter than length of valve. Vesica armed with two parallel, elongate, basally united cornuti, and a smaller, irregular cornutus near ductus ejaculatorius base. Sternite A8 broadened basally, with two relatively broad apical rods.

Female genitalia (Here described for the first time). Bursa copulatrix entirely membranous, without spines, consisting of a very large oval corpus bursae, and a globular, membranous diverticulum near colliculum on the right side. Ductus bursae not distinguishable. Ductus seminalis long and narrow, broadened near base, attached to base of corpus bursae. Colliculum short and broad, tapered posteriorly, displaced to the right side. Antrum short, membranous. Tergite A8 slightly elongate with rounded posterior angles. Apophyses narrow, relatively short. Papillae anales short and rounded.

Range. China (Shanxi, Shaanxi, Henan).

Phenology. Univoltine: late July to first half of August.

Altitude. Found at about 1900–2000 m in Qinling Mts. and Mien-shan (China: Vojnits 1984a; coll. ZFMK).

Similar species. Externally very similar to other all-dark, weakly marked species, such as *E. impavida*, but distinguishable with difficulty by its short, rather rounded wing shape, and the absence even of rudimentary transverse lines on the hind wing. The male genitalia have a distinct combination of cornuti in the aedeagus vesica; the female genitalia are unique and could not be confused with those of any other species.

Material. ♂, Prov. Shansi [Shanxi], Mien-shan, Obere Höhe *ca.* 2000 m, 30.vii.1937, H. Höne, Vojnits slide no. 13741♂ (holotype, TTM); 1♂, same loc. and alt., 3.viii.1937, H. Höne, Vojnits slide no. 11498♂ (TTM); 1♂, Sued-Shensi [Shaanxi], Tapaishan [Taibaishan] im Tsinling [Qinling], *ca.* 1700 m, 5.viii.1936, H. Höne, Vojnits slide no. 12962♂ (TTM); 1♀, same loc. and alt., 7.viii.1936, H. Höne, Vojnits slide no. 12362♀ (ZFMK); 2♀, China, Shaanxi, South Taibashan, 1900 m, Tsinling Mts. [Qinling], near Houzhenzi, 33°53′ N, 107°49′ E, 1.-12.viii.1999, leg. Siniaev & Plutenko, Mironov slides nos. 549♀, 658♀ (ZFMK); 1♂, 6♀, China, Henan, Funiushan, 7–9.viii.1993, Wang Zhiguo, slides nos. ZL675♂, 670, 672, 673, 674, 675, 677♀ (IZCAS).

Remarks. The valves are damaged and the eighth sternite is crumpled in Vojnits slide no. 13741♂ belonging to the male holotype.

83. *Eupithecia granata* Vojnits, 1979

Eupithecia granata Vojnits, 1979, *Acta zool. Acad. Sci. hung.* 25 (1–2): 193, figs 1, 2. Holotype ♂ (ZFMK; examined), China: North Yunnan, Li-kiang.

Eupithecia nonpurgata Vojnits, 1979, *Acta zool. Acad. Sci. hung.* 25 (1–2): 195, fig. 3. Holotype ♀ [*nec* ♂] (ZFMK; examined), China: North Yunnan, Li-kiang (synonymised in Mironov et al., 2004d).

Eupithecia noxia Vojnits, 1979, *Acta zool. Acad. Sci. hung.* 25 (1–2): 196, fig. 5. Holotype ♀ [*nec* ♂] (ZFMK; examined), China: North Yunnan, Li-kiang. **New junior synonym.**

Description. Fore wing ground colour brownish grey; costal margin with three blackish spots; basal, antemedial and medial transverse lines narrowly marked in darker brown with blackish dots on the Cu and A veins; double medial sharply angled onto costa behind discal dot; postmedial oblique, angled onto costa, with a series of black dashes on the veins; terminal area darker with a fine, narrow, wavy white subterminal line and whitish tornal dash with a dark inner shade; discal dot relatively large, ovoid, intensely black. Hind wing paler, transverse lines marked along anal margin; postmedial more distinct, regularly curved; terminal area darker with distinct whitish tornal spot; discal dot distinct.

Male genitalia. Uncus elongate, biapical. Valve almost parallel sided with broadly rounded apex. Vinculum short, with very shallow medial hollow. Papillae very long and narrow. Aedeagus strongly curved, almost equal to length of valve. Vesica armed with a long, curved horn-like cornutus, and a small, folded H-shaped cornutus near ductus ejaculatorius base. Sternite A8 elongate and narrow, with two broad parallel arms with spine-shaped apices.

Female genitalia. Bursa oval-oblong, almost completely covered with dense small spines except for a broad, membranous but thick-walled base; the spiniferous area on the ventral side of corpus bursae with a longitudinal spineless gap along left side. Ductus bursae short and broad, sclerotized, almost completely covered with small spines. Ductus seminalis broadened at base; attached to posterior part of corpus bursae on right side. Colliculum not present. Antrum membranous, broad and short. Tergite A8 quadrate, with shallow medial hollow in anterior margin. Apophyses short and narrow. Papillae anales short and broad.

Range. China (Yunnan, Sichuan).

Phenology. Univoltine: early August to mid-September.

Similar species. Externally very similar to other species in the group, such as *E. impavida*, *E. fervida* and *E. amicula*. In good condition separable with care by the hind wings being paler than the fore wings (almost concolorous in *E. impavida*), but less pale than those of *E. fervida*, and by the ground colour being brownish rather than greyish as in *E. amicula*. Identification by the genitalia is however advisable. In the male, the strongly curved aedeagus with one stout cornutus is distinctive. In the female the pattern of spining and the shape of the bursa are unlike those of any other member of the group.

Material. ♂, China, Provinz Nord-Yuennan, Li-kiang [Lijiang], 26.viii.1934, H. Höne, Vojnits slide no. 11446♂ (holotype, ZFMK); 9♂, 4♀, same loc., 7, 28.viii., 1, 2, 3, 5, 6, 7, 10.ix.1934, 3, 17.viii., 15.ix.1935, H. Höne, Vojnits slides nos. 12060♂, 11414♂, 11419♀, 11442♂, 12056♂, 12058♀, 12054♂, 12055♀, 12059♂, 11448♂, 11440♂, 11449♂, 12057♀ (paratypes, ZFMK); 1♂, 4♀, same loc., 1, 4, 5, 11.ix.1934, H. Höne, Vojnits slides nos. 11415♀, 11416♀, 11417♂, 11418♀, 11441♀ (paratypes, TTM); ♀ [*nec* ♂], same loc., 6.ix.1934, H. Höne, Vojnits slide no. 11569♂ (holotype of *E. nonpurgata*, but referenced in error to gen. slide of *E. egena*, ZFMK); 1♀, same loc., 1.ix.1934, H. Höne, Vojnits slide no. 11568♀ (paratype of *E. nonpurgata*, but referenced in error to gen. slide of *E. particeps*); ♀ [*nec* ♂], same loc., 1.ix.1934, H. Höne, Vojnits slide no. 11567♂ (holotype of *E. noxia*, referenced in error to gen. slide of *E. lobbichlerata*, ZFMK); 1♀, same loc., 13.ix.1934, H. Höne, referenced in error to Vojnits slide no. 11561♀, which contains the genitalia of *E. otiosa*; 6♂, 13♀, same loc., 22.v., 9.vii., 13, 18, 26, 30, 31.viii., 1, 3, 6, 10.ix.1934, 3, 21.ix.1935, H. Höne, Vojnits slides nos. 11409♀, 11401♀, 11403♂, 13249♀, de Laever slides nos. 231♂, 234♀, 322♀, Mironov slides nos. 714♀, 715♂ (ZFMK); 3♂, same loc., 29.viii.1934, 18.viii., 3.ix.1935, H. Höne, Vojnits slides nos. 13255♂, 13261♂, 13262♂ (TTM); 2♂, same loc., X, 3.viii., 3.ix.1935, H. Höne, Mironov slide no. 795♂ (SMNK); 2♂, same loc., *ca.* 2000 m, 14.viii., 3.ix.1934, H. Höne, Vojnits slide no. 13188♂, de Laever slide no. 315♂ (a further labelling error: the slide is of a *Pasiphila* sp., ZFMK); 3♂, China, Sichuan, Daxue Shan Mts., 80 km W Mianning, 28°34′ N, 102°00′ E, ~ 2750 m, 7.-8.vii.1999, leg. Siniaev & Plutenko (ZFMK); 1♂, China, Sichuan, Daxue Shan Mts., Gongga Shan, W Moxi, 29°34′ N, 101°59′ E, ~ 3200 m, 20.-21.vii.1999, leg. V. Siniaev & A. Plutenko (ZFMK).

Remarks. We dealt earlier with the synonymy of *E. nonpurgata* (Mironov et al., 2004d), and the associated problem of a mislabelled genitalia slide. As will be apparent from the list of material above, we have now found multiple errors in labelling genitalia slides associated with specimens in this series. Evidently something had gone very wrong with the preparation process at the time.

84. *Eupithecia amicula* Mironov & Galsworthy, 2004

Eupithecia amicula Mironov & Galsworthy, 2004, *Trans. lepid. Soc. Japan* 55 (1): 40, figs 1, 13, 16. Holotype ♂ (BMNH; examined), [China]: [Sichuan], Pu-tsu-fang, 9820ft.
Misidentification: *Eupithecia* ***infensa*** Vojnits, 1979 (1♀ paratype in ZFMK).

Description. Fore wing ground colour pale grey-brown, with most veins, especially lower border of cell, picked out in darker brown; an undulating brown basal line; postbasal and antemedial pale, with an angle close to the costa; medial area clear grey-brown, crossed by an undulating double medial; discal dot prominent, bordered by a semicircular brown marking on the costa; a pale postmedial fascia, bordered proximally by dark markings on the veins, curving to an abrupt zigzag close to the costa; terminal area slightly darker, crossed by an undulating pale subterminal line. Hind wing ground colour grey-white, with a small discal dot; anal margin as far as discal area tinged with brown mottling; terminal area pale brown, with two pale wavy subterminal lines.
Male genitalia. Uncus long, biapical. Valve large and broad. Vinculum short, rather broad with shallow medial hollow. Papillae elongate and narrow. Aedeagus shorter than length of valve. Vesica armed with two horn-like cornuti, one long and sinuate, the other smaller and bowed, and a thin elongate plate. Sternite A8 two long, very narrow and apically strongly diverging rods.
Female genitalia. Bursa bipartite, mostly membranous, but its central part heavily sclerotized and irregularly covered with numerous spines ranging from small to large. Antrum membranous. Ductus seminalis missing in slide. Colliculum absent. Tergite A8 elongate, slightly tapered to apex. Apophyses long and narrow. Papillae anales relatively short and narrow.
Range. China (Sichuan, Yunnan, Shaanxi).
Phenology. Univoltine: mid-June to early August.
Altitude. From about 1700 to 3200 m above sea-level in Daxue Shan (China: coll. ZFMK).
Similar species. Generally recognisable from other species of this group, except possibly *E. alishana*, by the paler, greyer cast to the wings, with the centre of the wing rather uniform, but crossed by narrow but distinct transverse lines. Very similar however to the next species, from which it needs to be distinguished by the genitalia. In *E. hoenehermanni* the smaller cornutus is much longer, and not curved as in *E. amicula*, and there is a small blunt ventral process on the valve. The female genitalia are very distinctive and could not be confused with any other species.
Material. ♂, [Sichuan], Pu-tsu-fang, 9,820ft., native collector, vi and vii.1890, Leech coll. 1900–64, BM Geom. slide no. 3517 (holotype, BMNH);

1♂, same data (paratype, BMNH); 1♂, Yunnan, Yulongshan, 10.vii.1962, coll. Song Shimei, Galsworthy slide no. ZL185♂ (paratype, IZCAS); 1♂, same data but 27.vii.1962 (IZCAS); 1♂, 1♀, Sued-Shensi [Shaanxi], Tapaishan [Taibaishan] im Tsinling [Qinling], *ca.* 1700 m, 25.vii., 4.viii.1936, H. Höne, Vojnits slides nos. 12393♂, 11483♀ (paratypes, ZFMK; gen. slide no. 11483♀ belongs to paratype of *E. infensa* Vojnits, 1979 [= *korbi* Dietze, 1908]); 1♂, China, Provinz Nord-Yuennan, Li-kiang [Lijiang], 12.vi.1935, H. Höne, Vojnits slide no. 12178♂ (TTM); 4♀, China, Sichuan, Daxue Shan Mts., Gongga Shan, NW Moxi, 29°41′ N, 101°58′ E, 2850 m, 14.–19.vii.1999, leg. V. Siniaev & A. Plutenko, Mironov slides nos. 702♀, 703♀, 704♀, 705♀ (ZFMK); 1♂, 1♀, same data (ZISP); 1♀, same loc., 29°34′ N, 101°59′ E, 3200 m, 20.-21.vii.1999, leg. V. Siniaev & A. Plutenko, Mironov slide no. 706♀ (ZFMK); 2♂, China, N. Yunnan, road Zhongdian / Lijiang, 3000 m, 19. vii.2009, 27°28′ N / 99°53′ E, I. & A. Floriani, Saldaitis leg., Pekarsky slides nos. OP 0585♂, OP 0586♂ (coll. O. Pekarsky); 3♀, China, W. Sichuan, road Ya'an / Kangding, Erlang Shan Mt., 2100 m, 29°51′ N / 102°18′ E, 12.vii.2009, I. & A. Floriani, A. Saldaitis leg., Pekarsky slides nos. OP 0645♀, OP 0646♀, OP 0647♀ (coll. O. Pekarsky).

Remarks. The female from Taibaishan detailed above was published by Vojnits (1979a) as the female of *E. infensa*, but is not conspecific with the male holotype of this taxon, which belongs to the *satyrata* species group.

85. *Eupithecia hoenehermanni* Mironov & Galsworthy, 2011

Eupithecia hoenehermanni Mironov & Galsworthy, 2011, *Lep. Science* **62** (1): 21, figs 7, 25, 28. Holotype ♀ (ZFMK; examined), [China]: North Yunnan, A-tun tse [Deqing], 4000 m.

Description. Fore wing ground colour light ochreous brown; transverse lines distinct, oblique, brown; basal line right-angled onto costa; medial sharply angled onto costa near discal dot; postmedial twice angled onto costa with broad brown inner shade; terminal area darker, brown with light ochreous brown apical streak; a similarly coloured large medial blotch between M_3 and CuA_1 and a light elongate tornal spot; discal dot relatively large, intensely black. Hind wing off-white with weak brownish tinge; dark brown transverse lines distinct in anal third; discal dot small.

Male genitalia. Uncus relatively elongate, biapical. Valve large, with very short and blunt medial ventral process, apex narrowly rounded. Vinculum short and broad. Papillae relatively short, slightly tapered. Aedeagus tapered at anterior end, stout, shorter than valve length. Vesica armed with two parallel, sabre-shaped, slightly sinuate cornuti,one longer than the other, an elongate, twisted plate-like cornutus and a smaller, crumpled cornutus at ductus ejaculatorius base. Sternite A8 elongate, consisting of two narrow, lyre-shaped rods, pointed and sclerotized at apices, connected by a narrow basal band.

Female genitalia. Bursa ovoid, sclerotized, almost completely covered with small spines except on the right side which carries a sclerotized patch with a short, blunt, membranous diverticulum in the middle. Ductus bursae not clearly differentiated. Ductus seminalis broadened, with an elongate area of small spines at base, attached to central part of bursa on left. Colliculum short and broad, slightly inclined to the left. Antrum short and broad. Tergite A8 slightly elongate. Anterior apophyses narrow, short, apically rounded. Posterior apophyses slightly broadened at bases and pointed at apices. Papillae anales short and rounded.

Range. China (Yunnan).

Phenology. Univoltine: late May to early July.

Altitude. Found from about 3000 to 4000 m in Yunnan (coll. ZFMK; O. Pekarsky).

Similar species. Very similar to the preceding species, from which it should be distinguished by examination of the genitalia. For distinction of the male genitalia, see under preceding species. The female genitalia are distinctive, and do not closely match those of any other known species.

Material. ♀, Nord Yünnan, A-tun-tse [Dêqên], Mittlere Höhe (*ca.* 4000 m), 2.vii.1936, H. Höne, Vojnits slide no. 11539♀ (holotype, ZFMK); 1♂, same loc., Talsohle *ca.* 3000 m, 29.v.1937, H. Höne, Vojnits slide no. 13133♂ (paratype, ZFMK); 1♂, 3♀, China, North Yunnan, Diqing Tibetan Aut. Pref., 5 km SE of Deqin, 3356 m, 19.vi.2009, leg. B. Benedek, Pekarsky slides nos. OP 0850♀, OP 0851♀, 0853♀, OP 0855♂ (paratypes, coll. O. Pekarsky).

86. *Eupithecia extinctata* Dietze, 1904

Eup[ithecia] extinctata Dietze, 1904, *Dt. ent. Z. Iris* [1903] **16** (2): 356, pl. 3, fig. 33. Syntypes 5♂ (coll. K. Dietze, O. Staudinger and R. Püngeler in MNHU; examined), [China]: Northern Tibet, Kuku-Noor [= Qinghai Hu].

Description. Fore wing ground colour pale ash-grey, irrorated with brownish scales; transverse lines brownish, broadened at costa; antemedial straight, evenly curved onto costa; double medial sinuate, touching discal dot and sharply angled onto costa; postmedial sinuate, evenly curved to costa behind discal dot; terminal area broad, darker, with distinct wavy whitish subterminal line broadened near tornus; discal dot relatively small, linear. Hind wing paler, off-white; transverse lines inconspicuous except evenly curved postmedial; terminal area darker, greyish brown with distinct, indented white subterminal line; discal dot smaller and paler, almost invisible.

Male genitalia. Uncus relatively short, biapical. Valve rather small and curved, with parallel sides and narrowly rounded apex. Vinculum narrow, slightly tapered. Papillae very short. Aedeagus unusually long and narrow, curved, with ductus ejaculatorius attached to central part of aedeagus. Vesica armed with two pointed cornuti, one apical, short, pen-shaped and the other narrow and curved, almost equal to length of aedeagus. Sternite A8 small, narrowed to apex, with two narrow sclerotized apical rods.

Female unknown.

Range. China (Qinghai).

Similar species. Relatively easily recognised by the rather cloudy pale brown maculation of the fore wing, contrasting strongly with the pale hind wing. In the male genitalia, the narrow elongate aedeagus is unique.

Material. 1♂, [Qinghai], Tibet, Kuku-Noor, *Eupithecia extinctata* Stgr. ♂, Tancré 22/4 95, *extinctata* Original ♂, Kuku-Noor, Tibet, Geschenk d. Herrn Püngeler (Rückbeil 94) durch Tancré, K. Dietze, Typus, Zool. Mus. Berlin, Gen.-Unters. Nr. [Schmidt slide no.] Eu 116, Zool. Mus. Berlin (syntype, coll. K. Dietze in MNHU); 1♂, [Qinghai], Tibet, Kuku-Noor, *Eupithecia extinctata* Stgr. ♂, Tancré 22/4 95, *extinctata* n. sp. Original. (K. Dietze), Typus, Zool. Mus. Berlin, Gen.-Unters. Nr. [Schmidt slide no.] Eu 116a, Zool. Mus. Berlin (syntype, coll. K. Dietze in MNHU); 1♂, [Qinghai], Kuku-Noor 94. Rückb., *Extinctaria* Stgr., *Eup. extinctata* Stgr., Original. Kuku-Noor, *extinctata* ♂ Original (K. Dietze), Gen. vérifiés 1458,[1] *Eupithecia extinctata* Stgr. ♂, Kuku-Nor, dét. E. de Laever 1954 (syntype, coll. O. Staudinger in MNHU); 1♂, [Qinghai], Tibet (Kuku-Noor) Rückbeil 1894, *extinctata* Dietze Type ♂ (v. R. Tancré 4.95), *extinctata* ♂, Original (K. Dietze), Cotype *extinctata* Dietze ♂ (syntype, coll. R. Püngeler in MNHU); 1♂, [Qinghai], Tibet (Kuku-Noor), *extinctata* Dietze ♂, v. R. Tancré 5.05 (syntype, coll. R. Püngeler in MNHU); 1♂, [Qinghai], *Extinctata* Dietze, Kuku-Noor (coll. R. Püngeler in MNHU); 1♂, [Qinghai], *Extinctata* Stgr. ♂ *Altaicata* var. Kuku Noor, Vojnits slide no. 1193♂ (TTM); 1♂, [Qinghai], *Extinctata* Stgr. ♂, Sining Tibet, Vojnits slide no. 1192♂ (TTM).

Remarks. No modern material of this species is known. The male genitalia were first described and illustrated by de Laever (1956).

87. *Eupithecia consortaria* Leech, 1897

Eupithecia consortaria Leech, 1897, *Ann. Mag. nat. Hist.* (6) **20**: 66. Holotype ♂ (BMNH; examined), China (western): Moupin.

Eupithecia pacifica Inoue, 1980, *Bull. Fac. domest. Sci. Otsuma Wom. Univ.* **16**: 163, fig. 39T; fig. 40A; fig. 42I; fig. 53A; fig. 58A. Holotype ♂ (BMNH; examined), [Japan]: Nagano Pref(ecture), Dakesawa (synonymised in Mironov et al., 2008b).

Misidentification: *Eupithecia selinata* Inoue, 1958, male (*nec* Herrich-Schäffer, 1861).

Description. Fore wing ground colour grey or brownish grey. Fore wing with slightly darker costal margin, usually with two dark spots subapically; transverse lines inconspicuous, the postmedial usually more distinct, oblique, right angled onto costa; terminal area darker with weak trace of light subterminal line; discal dot relatively small, distinct, ovoid. Hind wing grey with more distinct, angulate postmedial line, slightly darker terminal area, small whitish tornal spot and very small, pale, rounded discal dot.
Male genitalia. Uncus narrow, biapical. Valve broad, further broadened medially, and with broadly rounded apex. Vinculum short and broad. Papillae long and narrow. Aedeagus long and stout. Vesica armed with a very long and sinuate, pointed horn-like cornutus, broadened near apex, and a weakly sclerotized, band-like cornutus near ductus ejaculatorius base. Sternite A8 with two parallel apical arms connected by a narrow band at base.
Female genitalia. Bursa spindle-shaped, heavily sclerotized, almost completely covered with dense spines except area near base of ductus seminalis, with a V-shaped chain of large stout spines in the middle, and a long, anteriorly broadened, membranous diverticulum at base. Ductus bursae broad, heavily sclerotized, completely covered with dense large spines. Ductus seminalis narrow, broadened at base, recurved, attached to medial part of corpus bursae from left side. Colliculum not present. Antrum membranous. Tergite A8 small, square. Apophyses short and narrow. Papillae anales short and rounded.
Range. China (Qinghai, Sichuan); Asian Russia (S. Priamurje, Primorje), Korea, Japan (Honshu, Shikoku).
Phenology. Univoltine: early July to early September.
Altitude. Up to about 1600 m in Primorje, and 3000 m in Qinghai (coll. ZISP, IZCAS).
Similar species. Very similar to other species, such as *E. impavida*, with darker hind wings and faint or absent transverse lines, and best separated by examination of the genitalia. In the male the best distinction is the long and sinuate shape of the cornutus in the aedeagus vesica, which is, importantly, strongly broadened close to the apex. In *E. impavida* the cornutus is broad throughout, while in *E. inexhausta*, where the cornutus can appear more like that of the present species, the shape of the eighth sternite is

different. The female genitalia are similar to those of *E. impavida*, but the shape of the basal diverticulum is different, and the apophyses much shorter and narrower in the present species.

Material. ♂, [Sichuan], Moupin, Kricheldorff coll., July 1890, Leech Coll. 1900–64, *E. consortaria* sp. n. Type ♂, BM Geom. slide no. 3516 (holotype, BMNH); ♂, [Japan], Dakesawa, Nagano Pref., 1.viii.1958, Y. Kobayashi (holotype of *E. pacifica*, BMNH); 2♂, Qinghai, Huzhu beishan, 2300 m, 30.vii.1991; Tongren, Maixiu, 2950 m, 30.vii.1992, coll Xue Dayong, slides nos. ZL380, 387♂ (IZCAS).

Remarks. Photos of the aedeagi of *E. consortaria* and *E. sinicaria* were transposed in de Laever (1960, figs. 44 and 57). Mironov et al. (2008b) dealt with the synonymy with *E. pacifica*, explaining that the undissected holotype of the latter species was referable to the present species, while Inoue's genitalia illustrations under this name were all referable to *E. impavida*.

88. *Eupithecia vana* Vojnits, 1979

Eupithecia vana Vojnits, 1979, *Acta zool. Acad. Sci. hung.* 25 (1–2): 200, figs 8, 9. Holotype ♂ (ZFMK; examined), China: Sued-Shensi, Tapaishan in Tsinling, 1700 m.

Eupithecia abrepta Vojnits, 1979, *Acta zool. Acad. Sci. hung.* 25 (3–4): 431, fig. 7. Holotype ♀ (ZFMK; examined), [China]: Sued-Shensi, Tapaishan in Tsinling, 1700 m. **New junior synonym.**

Description. Fore wing ground colour even fuscous grey, with three to four small dark spots on costa; transverse lines inconspicuous, ante- and postmedial marked as blackish costal spots only; terminal area slightly darker with narrow interrupted whitish subterminal line and distinct white tornal spot; discal dot large, intensely black. Hind wing almost concolorous, but slightly paler in anterior part, with darker anal margin; transverse lines invisible; tornus with small whitish spot; discal dot small and pale.

Male genitalia. Uncus narrow, biapical. Valve broad with broadly sclerotized dorsal margin and a short spine on the strongly broadened base of the dorsal margin near apex of valve. Vinculum tapered anteriorly, with shallow medial hollow. Papillae long and narrow. Aedeagus stout, longer than length of valve. Vesica armed with one narrow horn-like cornutus equal to length of aedeagus, and strongly broadened near apex; two parallel, weakly sclerotized styloid cornuti and a twisted, U-shaped cornutus with broadened arms near ductus ejaculatorius base. Sternite A8 peg-like, broadened basally, with two narrow, parallel apical rods.

Female genitalia. Bursa elongate, heart-shaped, heavily sclerotized, covered with mainly small spines in medial part and a V-shaped chain of large spines in the anterior half; at base of bursa a long and narrow thick-walled

membranous basal diverticulum and a second large, thin-walled, membranous diverticulum, narrowly attached to corpus bursae near base from ventral side. Ductus bursae not distinguishable. Ductus seminalis narrow, sharply broadened at base, attached to central part of bursa from dorsal side. Colliculum plate-like, long and broad, with rounded apical area covered by numerous pores. Antrum membranous. Tergite A8 transverse, with broadly sclerotized anterior margin and rounded posterior corners. Apophyses very short and stout. Papillae anales short and broad.
Range. China (Shaanxi, Henan).
Phenology. Bivoltine (?): mid-April to mid-May and early to late August.
Altitude. Found at about 1600–1700 m in Qinling Mts. (China: Vojnits 1979a, 1979b; coll. ZFMK).
Similar species. Similar in shape to European *E. assimilata* Doubleday, 1856 and hence with broader rounder wings than other similarly plain Chinese species. Also with less prominent fore wing maculation. However, best separated by examination of the genitalia. The spine on the dorsum of the male valve is shared only with *E. nirvana*. Similarly, the very distinctive female bursa could only be confused with that of *E. nirvana*. See under following species for distinction between them.
Material. ♂, Sued-Shensi [Shaanxi], Tapaishan [Taibaishan] im Tsinling [Qinling], *ca.* 1700 m, 24.viii.1936, H. Höne, Vojnits slide no. 11434♂ (holotype, ZFMK); ♀, same loc. and alt., 12.v.1936, H. Höne, Vojnits slide no. 12205♀ (holotype of *E. abrepta*, ZFMK); 1♀, same loc. and alt., 7.viii.1936, H. Höne, Vojnits slide no. 12285♀ (paratype of *E. abrepta*, TTM); 2♂, same loc. and alt., 10.viii.1936, H. Höne, Vojnits slides nos. 12214♂, 12968♂ (ZFMK); 2♂, same loc. and alt., 12, 17.v.1936, H. Höne, Vojnits slides nos. 1199♂, 12213♂ (TTM); 1♂, same loc., *ca.* 3000 m, 10,viii.1936, H. Höne, Vojnits slide no. 12330♂ (ZFMK); 1♂, same loc., L, 31.viii.1935, H. Höne, Vojnits slide no. 12314♂ (ZFMK); 3♂, 2♀, C. China, prov. Shaanxi, Foping Natural Reserve, Panda area, 33°45′ N, 107°48′ E, ~1600 m, 20.iv.-11.v.1999, leg. V. Siniaev & A. Plutenko (ZFMK; 1♂ and 1♀ in ZISP); 1♂, Henan, Funiushan, 7–9.viii.1993, Wang Zhiguo, slide no. ZL671♂ (IZCAS).
Remarks. The holotype of this species is extremely worn. However, further specimens among modern material from China in ZFMK (Bonn), have enabled us confidently to associate the well-preserved holotype female of *E. abrepta* with *E. vana*.

89. *Eupithecia nirvana* Mironov & Galsworthy, 2006

Eupithecia nirvana Mironov & Galsworthy, 2006, *Trans. lepid. Soc. Japan* 57 (4): 341, figs 4, 19, 22. Holotype ♂ (ZFMK; examined), China, Sichuan, Daxue Shan Mts., Gongga Shan, NW Moxi, 29°41′ N 101°58′ E, 2850 m.

Description. Fore wing rather elongate, ground colour brownish grey; transverse lines inconspicuous, forming three dark spots on costal margin; white tornal spot large and clear, with dark inner shade; discal dot large, ovate or semicircular, black. Hind wing ground colour as fore wing but lighter near costal margin; transverse lines inconspicuous; clear whitish tornal spot and dark discal dot, but smaller than those on the fore wing.
Male genitalia. Uncus short, biapical. Valve broad with widely sclerotized costal margin and a single spine on the costa near the apex; ventral margin with a shallow medial hollow. Vinculum short, tapered anteriorly. Papillae elongate. Aedeagus stout, slightly shorter than length of valve. Vesica armed with one long horn-like cornutus, equal to length of aedeagus, with a broad and curved base and slightly broadened near apex, a parallel styloid cornutus and a twisted plate with a hooked ending near ductus ejaculatorius base. Sternite A8 broadened basally, with two narrow, parallel apical rods with weakly broadened tips.
Female genitalia. Bursa heavily sclerotized, funnel-shaped, expanded on both sides close to colliculum and with a spineless thick-walled sack near base, covered with mainly small spines in the medial part and with a V-shaped chain of large spines on the middle of the ventral side of the corpus bursae; a large, spherical, membranous diverticulum with a narrow tubulate base attached to central part of bursa from ventral side. Ductus bursae not well developed. Ductus seminalis short, broadened basally, attached to medial part of bursa from dorsal side. Colliculum almost plate-like, broad, heavily sclerotized as bursa. Antrum broad, poriferous, heavily sclerotized. Tergite A8 transverse. Apophyses very short and stout. Papillae anales relatively large, broad and rounded.
Range. China (Sichuan).
Phenology. A series of six specimens was taken in the middle of July.
Altitude. Found at about 2850–3200 m in Daxue Shan (China: coll. ZFMK, ZISP).
Similar species. A rather nondescript species in which only the large tornal dot stands out. It has longer more pointed fore wings than the preceding species, as well as a larger fore wing discal dot. The male genitalia could only be confused with those of *E. vana*: in *E. nirvana* the dorsal margin of the valve is much less broadened under the spine, and the spine is longer and narrower. In addition the vinculum lacks a medial hollow, the cornutus in the aedeagus vesica is shorter and less flared at the apex, and only one styloid cornutus is present. The female genitalia of *E. nirvana* are distinguishable from those of *E. vana* by the funnel-shaped bursa copulatrix with a different disposition of spines, the shorter and broader diverticulum at the base of the corpus bursae and the shorter anterior apophyses.

Material. ♂, China, Sichuan, Daxue Shan Mts., Gongga Shan, NW Moxi, 29°41′ N, 101°58′ E, 14.-19.vii.1999, ~2850 m, leg. V. Siniaev & A. Plutenko (holotype, ZFMK); 1♀, same data (paratype, ZFMK); 2♀, same data (paratypes, ZISP); 2♀, same loc., 29°34′ N, 101°59′ E, 3200 m, 20.-21.vii.1999, leg. V. Siniaev & A. Plutenko (paratypes, ZFMK).

The *venosata* species group

This group was introduced by Dietze (1913), and subsequently described by Schütze (1960) as subgenus *Bohatschia*. Mironov (1990) relegated it to a species group.

The species are mainly characterized externally by a more or less well marked cellular pattern consisting of prominent dark transverse lines and dark lines on some of the veins. In the male genitalia the valves are unmodified, and the aedeagus vesica has one to three cornuti. The eighth sternite always has two apical arms, usually curved into a pincer shape, and often with an elongate base, itself sometimes with shorter basal arms. The female genitalia are diverse in the shape of the bursa, but in many cases have a small to large diverticulum at the posterior end, and always have the ductus seminalis placed close to the colliculum. This is one of the most species rich Mediterranean and Central Asian *Eupithecia* species group, but it is comparatively species poor in East Asia, and is represented in China by the following six species.

90. *Eupithecia hilariata* Dietze, 1908

Eupithecia hilariata Dietze, 1908, *Dt. ent. Z., Iris* **21** (3): 180. Syntypes 2♂, 1♀ [*nec* 1♂, 2♀] (coll. K. Dietze and R. Püngeler in MNHU; examined), [Tajikistan]: [Ispajran]; [China]: [Juldus-Gebirge], [Ili Gebiet].

Description. Ground colour pale grey usually with weak ochreous tinge. Fore wing dark transverse lines narrow, sinuate; antemedial three or four times sharply angled; double medial touching or passing through discal dot; postmedial sinuate; terminal area slightly darker without pale subterminal line; discal dot blackish brown, elongate. Hind wing slightly paler, with transverse lines well marked along dorsum, sometimes with distinct postmedial; terminal area darker without light subterminal line; discal dot considerably smaller and paler. Third abdominal segment with black dorsal half ring.

Male genitalia. Uncus elongate, uniapical. Valve elongate and narrow, with narrowly rounded apex. Vinculum slightly elongate, tapered. Papillae very short. Aedeagus narrow, shorter than length of valve. Vesica armed with two long and narrow horn-like cornuti and a small, irregular cornutus at anterior end of aedeagus. Sternite A8 two narrow parallel arms, connected by a narrow transverse band at base.

Female genitalia. Bursa small, ovoid, almost completely covered with small spines, and with two membranous diverticula, one large and broad on the left side and the other small, at the middle of the right side. Ductus bursae not distinguishable. Ductus seminalis narrow; attached to posterior part of bursa near colliculum on left side. Colliculum very small. Antrum membranous. Tergite A8 transverse with slightly convex anterior margin. Apophyses notably short and narrow. Papillae anales broad and rounded.

Range. China (Xinjiang); Turkmenistan (Kugitang Mts.), Uzbekistan (Karzhan Tau Mts., Chimgan Mts.), Kirghizstan (Tien-Shan Mts.), Tajikistan (Pamirs Mts.), S and SE Kazakhstan.

Phenology. Bivoltine (?): early May to mid-August.

Altitude. Found from about 980 m in Syugeti Mt. (S Kazakhstan: coll. O. Pekarsky) to about 2800 m in Kugitang Mts. (Turkmenistan: coll. ISEN).

Similar species. The external features, colour and maculation of the wings and the black half ring on the third abdominal segment of *E. hilariata* distinguish it from other Chinese *Eupithecia* species, the closest resemblance being to *E. carpophilata*. In the male genitalia, the uniapical uncus, the short papillae on the anterior arms of the labides, the cornuti on the vesica and the shape of the eighth sternite separate the species from other members of the group. In the female, the two diverticula, and the very short anterior and posterior apophyses, are distinctive.

Material. 1♂, [Tajikistan], Ispajran [= Isfara], Alai sept., 3400 m, August, 124., *hilariata* Dietze, Jspajran [= Isfara] Alai sept., Orig., ♂, Typus, Vojnits slide no. 1178♂ (syntype, coll. K. Dietze in MNHU); 1♂, [Xinjiang], *hilariata* ♂, Juldus, Ili [Yining], Zool. Mus. Berlin, *hilariata* Dtze., präp. Nr. A 835♂, det. E. Schütze, Kassel (syntype, coll. K. Dietze in MNHU); 1♀, [Xinjiang], *hilariata* ♀, Palpen lang, Juldus, Ili [Yining], Zool. Mus. Berlin, *Eupithecia hilariata* ♀, Juldus Ili [Yining], prep. № 1657.2 E. de Laever det. 1957 (syntype, coll. K. Dietze in MNHU); 1♂, [Kirghizstan], Asia centr., Togus-torau, Rückbeil 1901, ? *Schiefereri* var. ♂ (*hilariata*), v. R. Tancré, 2.02, *hilariata*, Sehr nache *schiefereri*; palpen spitzer u. langer (Dtze.), *Eupithecia hilariata* ♂, Type, Prep. № 1657.1 E. de Laever det. 1957 (coll. R. Püngeler in MNHU).

Remarks. Mr. de Laever in 1957 erroneously designated a specimen of this species from Togus-torau (Kirghizstan) in the coll. R. Püngeler in MNHU as a type of *E. hilariata*. This specimen was not part of the original type series (Dietze, 1913, page 59).

91. *Eupithecia hesperina* Mironov & Galsworthy, 2004

Eupithecia hesperina Mironov & Galsworthy, 2004, *Trans. lepid. Soc. Japan* 55 (3): 228, figs 3, 15, 18. Holotype ♂ (ZFMK; examined), North Yunnan, Li-kiang [Lijiang].

Description. Fore wing ground colour buff, well-scattered with mid-brown scales, giving uniform pale brown appearance; basal and antemedial lines marked by indistinct brown spots on costa; discal dot dark brown, elongate; medial oblique, with a darker dot on costa above discal dot; postmedial double, with a dark dot on the costa, angled outwards close to costa; terminal area darker, crossed by a paler slightly wavy line. Hind wing slightly paler; discal dot fairly large; anal margin with denser brown scales, and indistinct transverse lines; postmedial double, almost complete; terminal area darker.

Male genitalia. Uncus long, biapical. Valve with parallel dorsal and ventral margins, apex rounded. Vinculum short with shallow medial hollow. Papillae elongate and narrow. Aedeagus much shorter than valve length. Vesica armed with a small apical horn-like cornutus, a bowed H-shaped plate and a smaller irregular plate near ductus ejaculatorius base. Sternite A8 with two narrow, lyre-shaped apical rods, broadly linked at base.

Female genitalia. Bursa globular, completely covered with small spines. Ductus bursae membranous, with broad membranous diverticulum. Ductus seminalis attached to ductus bursae from right side. Colliculum short and broad. Antrum lightly sclerotized. Tergite A8 rectangular, slightly elongate. Apophyses narrow. Papillae anales small, rounded.

Range. China (Sichuan, Yunnan).

Phenology. Univoltine: mid-July to mid-September.

Altitude. The only data is from 2800 m in Sichuan.

Similar species. More lightly marked than most species in this group, but still with the fore wing pattern broken up by cellular markings. Both male and female genitalia are distinct from other members of the group, but a little similar to those of *E. contraria*. The male is distinguished by the shorter, less hooked cornutus on the vesica, and the shape of the eighth sternite, and the female genitalia of the present species has more large spines in the bursa and a broader colliculum than in *E. contraria*.

Material. ♂, China, Nord-Yuennan, Li-kiang [Lijiang], 30.vii.1934, H. Höne, Vojnits slide no. 13233♂ (holotype, ZFMK); 2♂, 1♀, same loc., respectively 23.vii., 12.viii. and 13.ix.1934, H. Höne, Vojnits slides nos. 13043♂, 13170♂, 13251♀ (paratypes, ZFMK); 1♂, China, Sichuan Prov., Xiangcheng, 2800 m, 13.vii.2001, Li Houhun leg. (coll. P. Skou).

92. *Eupithecia dissertata* Püngeler, 1905

Tephroclystia dissertata Püngeler, 1905, *Societas ent.* **20** (18): 138. Syntypes 2♂, (coll. R. Püngeler and K. Dietze in MNHU; examined), [Switzerland]: Wallis, Zermatt.
Eupithecia kaszabi Vojnits, 1974, *Acta zool. Acad. Sci. hung.* **20** (3–4): 445, figs 1a, 2a. Holotype ♂ (TTM; examined), Mongolia: Central aimak, SE of Somon Bajanzogt, 1600 m (synonymised in Mironov, 2003).
Eupithecia perfuscata Vojnits, 1975, *Acta zool. Acad. Sci. hung.* **21** (3–4): 451, fig. 3. Holotype ♀ (TTM; examined), Mongolia: Cojbalsan aimak, Chamardavaa ul., 80 km SE of Somon Chalchingol, 600 m. **New junior synonym.**
Eupithecia certa Vojnits, 1981, *Acta zool. Acad. Sci. hung.* **27** (3–4): 420, fig. 31. Holotype ♀ (ZFMK; examined), [China]: Prov. Shansi [Shanxi], Mien-shan, 2000 m (synonymised in Mironov, 2003).

Misidentification: *Eupithecia carpophilata* Vojnits, 1975, 1977 (*nec* Staudinger, 1897).

Description. Fore wing ash-grey or reddish brown usually with darker, greyish brown or reddish brown terminal area; transverse lines fine, broadened at costa; basal line angled near costa; postbasal and antemedial twice angled; two parallel medials sinuate, crossing, and sharply angled near discal dot; postmedial sinuate, curved near costa; whitish subterminal line from inconspicuous to very distinct and from wavy to very indented, usually forming a small white tornal spot; discal dot small. Hind wing pale ash-grey to dark brownish grey with darker terminal area; three fuscous transverse lines well marked along dorsum only; discal spot minute.
Male genitalia. Uncus biapical. Valve large, broadened at the middle, and apex narrowly rounded. Vinculum semicircular. Papillae relatively short. Aedeagus shorter than length of valve. Vesica armed with two horn-like cornuti, one large, elongate, slightly curved, the other smaller and hook-like near apex, and a small, lightly sclerotized cornutus in medial part of aedeagus. Sternite A8 with two pincer-like apical horns about half of sternite length.
Female genitalia. Bursa fusiform, almost half covered with small spines, and with a large, membranous diverticulum at base. Ductus bursae not present. Ductus seminalis narrow, attached to bursa near colliculum. Colliculum short and broad. Antrum membranous. Tergite A8 quadrate. Apophyses narrow. Papillae anales short and broad, rounded.
Range. China (Qinghai, Gansu, Shanxi); northern Mongolia, Asian Russia (Siberia: Altai, Sajan Mts., Tuva, Transbaikalia, Dahuria, Magadan region), SE Kazakhstan (Saur Mts.), Central Europe (Alps and Carpathians).
Phenology. Univoltine: late June to early August.
Altitude. From about 500–550 m in Magadan region (Russia: coll. ZMH) up to about 2300 m in Altai Mts. (Russia: coll. ZISP), and to nearly 3000 m in Qinghai (China: coll. IZCAS).

Similar species. Variable in depth of colour from light grey to brown or reddish brown. This species is usually larger and more lightly marked than *E. carpophilata*, the most likely Chinese confusion species, and has more elongate and pointed fore wings. The cornuti in the male vesica, particularly the short hooked one at apex, are distinctive. The fusiform shape of the female bursa with the small round diverticulum at base is also characteristic.

Material. ♂, [Switzerland], Zermatt, an Thymus-Blütten, 5 Uhr N.M. 27.6.01, R. Püng., *dissertata* Püng. ♀, Original (holotype, coll. R. Püngeler in MNHU); ♂, Mongolia, Central aimak, SO von Somon Bajanzogt, 1600 m, 27.vii.1966, Exp. I. Kaszab, 1966, Nr. 751, Vojnits slide no. 10181♂ (holotype of *E. kaszabi*, TTM); ♀, Mongolia, Cojbalsan aimak, Chamardavaa ul., 80 km SO von Somon Chalchingol, 600 m, 12.viii.1965, Exp. Dr. Z. Kaszab, 1965, Nr. 399, Vojnits slide no. 10294♀ (holotype of *E. perfuscata*, TTM); ♀, Prov. Shansi [Shanxi], Mien-shan, Obere Höhe *ca.* 2000 m, 10.viii.1937, H. Höne, Vojnits slide no. 13121♀ (holotype of *E. certa*, ZFMK); 2♀, same loc. and alt., 29.vii.1937, H. Höne, de Laever slides nos. 94♀, 281♀ (ZFMK); 1♀, same loc. and alt., 2.viii.1937, H. Höne, Vojnits slide no. 11495+ (slide absent in TTM and ZFMK); 1♀, Qinghai, Tongren, Maixiu, 2950 m, 30.vii.1992, Xue Dayong, slide no. ZL271♀; 1♀, Qinghai, Huzhu Beishan Linchang, Langshidang, 2600 m, 36°55′ N, 102°23′ E, 7.viii.2005, Xue Dayong, slide no. ZL700♀; 2♀, Gansu, Yongdeng Tulugou, 2280 m, 20–26.vii.1991, Meng Feng, slide no. ZL313♀ (IZCAS).

93. *Eupithecia carpophilata* Staudinger, 1897

Eup[*ithecia*] *carpophilata* Staudinger, 1897, *Dt. ent. Z., Iris* **10** (1): 108, pl. 3, fig. 69. Lectotype ♂ (design. by Mironov, coll. O. Staudinger in MNHU; examined), [Russia]: Uss(uri).

[*Eupithecia*] *carpophilata* var. *collega* Dietze, 1908, *Dt. ent. Z., Iris* **21** (3): 199, (non binominal). Syntypes 8♂, 3♀ (coll. K. Dietze and O. Staudinger in MNHU; examined), [China]: Aksu (synonymised in Mironov, 1989a).

Description. Fore wing ground colour mid brown to grey-brown, very uniform over wing. Fore wing with rounded apex, transverse lines distinct, narrow, blackish; basal line evenly curved, antemedial usually angulate; medial double, sinuate, crossing elongate black discal dot; postmedial sinuate, evenly curved. Hind wing same colour or slightly paler with weak traces of dark, narrow transverse lines, often only along anal margin; terminal area sometimes slightly darker with trace of pale subterminal line; discal dot small.

Male genitalia. Uncus broad, biapical. Valve broadened in middle, with squarish apex. Vinculum short and narrow. Papillae short. Aedeagus shorter than length of valve. Vesica armed with two long sinuate horn-like cornuti. Sternite A8 broad with two long, pincer-like apical rods and two short anterior arms.

Female genitalia. Bursa small, more or less globular, completely covered with small spines. Ductus bursae very long with longitudinal striations and rows of larger spines and a patch of small spines near colliculum. Ductus seminalis narrow, attached to ductus bursae near colliculum on right side. Colliculum wide, plate-like. Antrum membranous. Tergite A8 transverse. Apophyses short. Papillae anales short and broad, rounded.

Range. China (Xinjiang, Qinghai, Shanxi, Shaanxi); Central and SE Kazakhstan, Kirghizstan (Tien-Shan Mts.), Asian Russia (S. Ural, Altai Mts., Sajan Mts., Transbaikalia, Dahuria, Priamurje, Primorje, Sakhalin Isl.), Mongolia, Korea.

Phenology. Bivoltine (?): early June to early September.

Altitude. From sea level to about 2400 m in Kyrghyzsky Mts. (Kirghizstan: coll. ZMKU), and 2800 m in Qinghai (China: coll. IZCAS).

Similar species. Similar to brownish specimens of the preceding species, but distinguishable by the smaller size, and the broader and more obtuse fore wings, usually with evenly curved postmedial lines. In the male genitalia, the shape of the eighth sternite is very distinctive. The female genitalia, with a small round bursa situated at the end of the elongate ductus, are also immediately recognizable.

Material. ♂, [Russia], Ussuri Dörr. 82, *Carpophilata* Stgr., Origin., 69. (syntype, coll. O. Staudinger in MNHU); 1♂, [Xinjiang], *carpophilata* form. *collega* ♂, Typ. Aksu, *carpophilata* Stgr. präp. Nr. A 831♂ det. E. Schütze, Kassel (syntype of *E. carpophilata* var. *collega*, coll. K. Dietze in MNHU); 2♂, [Xinjiang], v. *collega* ♂, Type, Aksu (syntypes of *E. carpophilata* var. *collega*, coll. K. Dietze in MNHU); 1♀, [Xinjiang], v. *collega* ♀ Aksu, Type, Anfg. Juni, K. Dietze (syntype of *E. carpophilata* var. *collega*, coll. K. Dietze in MNHU); 5♂, 1♀, [Xinjiang], form or v. *collega*, Aksu 03♂ (syntypes of *E. carpophilata* var. *collega*, coll. K. Dietze in MNHU); 1♂, [Xinjiang], Korla, Sai-chin, 373, forma: *collega* Korla (coll. K. Dietze in MNHU); 1♀, [Xinjiang], Aksu, Ost-Turkestan, *collega*, Aksu, 382, *carpophilata* Stgr., Präp. Nr. A 825♀, det. E. Schütze, Kassel (syntype of *E. carpophilata* var. *collega*, coll. O. Staudinger in MNHU); 1♂, [Xinjiang], Asia centr., Saichin-Kette, nördl. von Korla, Rückbeil 1902, *alliaria* var. *carpophilata* Stgr. ♂, v. R. Tancré 11.02, sehr ähnlich *carpophilata* Stgr. = ? *alliaria* var. (coll. R. Püngeler in MNHU); 1♂, [Xinjiang], Aksu, Chin Turk., 194., 10/1914 von B.-Haas, Püng., *carpophilata*; forma *collega* (coll. R. Püngeler in MNHU); 1♀, [Xinjiang], Mus-tag-ata Yarkend, ♀ *collega* (coll. K. Dietze in MNHU); 1♀, [Xinjiang], *carpophilata* form. *inclinans* Typ., Juldus [Yining] Gebirge (coll. K. Dietze in MNHU); 1♂, 2♀, [Xinjiang], Asia centr., Thian-Schan, Juldus [Yining] Geb., Coll. Wagner, v. *inclinans* (coll. K. Dietze in MNHU); 1♀, [Xinjiang], Juldus [Yining], *inclinans* Juldus, Ili [Yining] (coll. K. Dietze in MNHU); 5♂, 1♀, China, Sued-Shensi [Shaanxi], Tapaishan [Taibaishan] im Tsinling [Qinling], 23, 24, 26, 27, 29.vi., 1.vii.1935, H. Höne, de Laever slides nos. 59♂, 60♂, 62♂, 63♀ (ZFMK); 6♂, same loc., 21, 22, 24, 26.vi., 2.vii.1935, H. Höne, Vojnits slides nos. 1109♂, 11478♂ (TTM); 1♀, same loc., *ca.* 1700 m, 15.vi.1936, H. Höne (TTM); 1♀, Prov. Shansi [Shanxi], Mien-shan, Obere Höhe *ca.* 2000 m, 30.vi.1937, H. Höne, de Laever slide no. 64♀ (ZFMK); 2♀, same loc. and alt., 30.vi., 5.viii.1937, H. Höne (TTM); 9♂, 23♀, Qinghai, Menyuan, Xianmi, 2800 m, 23.vii.92, Xue Dayong, slide no. ZL66, 302♂, 67, 294♀; 1♀, Qinghai, Menyuan, Xianmi, 2800 m, 37°16′ N, 102°00′ E, 6.viii.2005, Xue Dayong, slide no. ZL712♀; 1♀, Xinjiang,Yecheng, Qipanxiang, 16.vii.1989, coll Zhang Xuezhong (IZCAS).

Remarks. The shape of the eighth sternite of the male is very variable (Vojnits 1975).

94. *Eupithecia albispumata* Warren, 1893

Eupithecia albispumata Warren, 1893, *Proc. zool. Soc. Lond.* **1893** (2): 384, pl. 30, fig. 23. Lectotype ♀ (BMNH; examined), [India]: Khasi Hills.

Eupithecia pengata Schütze, 1961, *Veröff. zool. StSamml. Münch.* **6**: 180, pl. 29, fig. 2; pl. 30, figs 3, 3a, 3b. Holotype ♂ (ZSM; examined), Nepal: Mustangbhot, 29°11′ N, 83°58′ E, Penga, 3800 m (synonymised in Mironov et al., 2008b).

Eupithecia acseszteri Vojnits, 1988, *Acta zool. hung.* **34** (1): 50, figs 30–32, pl. 1: C. Holotype ♂ (NSMTO; examined), Nepal (east): Solukhumbu Sagarmatha, Kharikhola, 1980 m (synonymised in Mironov et al., 2008b).

Misidentifications: *Eupithecia lobbichlerata* Schütze, 1961 (1♀ paratype in ZSM); *Eupithecia anteacta* Vojnits, 1984 (1♀ paratype in TTM).

Description. Fore wings rusty brown, often with darker costal and terminal margins. Fore wing with fine oblique transverse lines, angled onto costa; area near discal dot often paler; terminal area with indistinct wavy whitish subterminal line, broadened near tornus and usually forming a small whitish tornal spot; discal dot relatively small, ovoid or slightly elongate, black. Hind wing paler, off-white, with darker anal margin showing indistinct transverse lines and very small discal dot.

Male genitalia. Uncus broad, biapical. Valve relatively short, evenly tapered to apex. Vinculum trapezoid. Papillae short, broadened basally. Aedeagus narrow, but broadened at level of ductus ejaculatorius. Vesica armed with one long horn-like cornutus, a chain of minute teeth and two smaller cornuti near ductus ejaculatorius base. Sternite A8 with two broad, lyre-shaped horns, broadened at apex.

Female genitalia. Bursa small, globular, covered with small spines mainly on dorsal side. Ductus bursae elongate, with longitudinal striations, two small patches of spines near apex, and a narrow, thick-walled diverticulum attached at middle on right side. Ductus seminalis long, sharply recurved, attached to posterior part of ductus bursae near colliculum. Colliculum elongate and narrow. Antrum membranous. Tergite A8 trapezoid. Apophyses stout, broadened at apices. Papillae anales elongate and pointed, spinuliferous.

Range. China (Sichuan, Yunnan, Qinghai, Gansu); Kashmir, India (Kumaon, Assam), Nepal.

Phenology. Bivoltine (?): mid-July to mid-October.

Altitude. From about 1980 to 4000 m in the Himalayas (Nepal: Vojnits 1988; China [Sichuan]: coll. O. Pekarsky).

Similar species. In good condition, only likely to be confused with *E. russeola* Prout, 1926 from Myanmar, which has similar maculation, but a more reddish subterminal fascia on the fore wing. Male and female genitalia are both very distinctive.

Material. ♀, [India], Khasia, 6000ft, Sept. 1886, H.J. Elwes, Collectio H.J. Elwes, *Eupithecia albispumata* Warr. Type ♀, BM Geom. slide no. 20133 (lectotype, BMNH); ♂, Nepal, Mustangbhot, Penga, 29°11′ n, Br., 83°58′ ö.L., 3800 m, 8.viii.[19]55, leg. F. Lobbichler, *pengata* Schütze, Holotypus, Präp. Nr A746♂, det. E. Schütze, Kassel (holotype of *E. pengata*, ZSM); 1♀, Nepal, Mustangbhot, Kehami, 3,700 m, 29°11′ n.Br., 83°58′ ö.L., 20.viii. [19]55, leg. F. Lobbichler, *lobbichlerata* Schü., Allotypus, Präp. Nr. A 712♀, det. E. Schütze, Kassel [= ZSM slide no. G 13370] (allotype of *E. lobbichlerata*, ZSM); ♂, E. Nepal, Kharikhola, 1980 m, Solukhumbu Sagarmatha, 7.x.1979, leg. M. Owada, Vojnits slide no. 15706♂ (holotype of *E. acseszteri*, NSMTO); 1♀, China, Provinz Nord-Yuennan, Li-kiang [Lijiang], 9.ix.1934, H. Höne, Vojnits slide no. 13768♀ (paratype of *E. anteacta*, TTM); 11♂, same loc., 20, 22, 23, 26, 27.vii., 4.viii., 8, 9.ix.1935, H. Höne, Vojnits slides nos. 14199♂, 14203♂, 14175♂, 14204♂, 14195♂, 14202♂, 14206♂, 11048♂, 13253♂, 14193♂, de Laever slide no. 383♂ (ZFMK); 1♀, same loc., 3.x.1934, H. Höne, Mironov slide no. 834♀ (TTM); 2♂, Nord Yünnan, A-tun-tse [Dêqên], Obere Höhe (*ca.* 4500 m), 10.vii., 6.viii.1936, H. Höne, Vojnits slides nos. 11548♂, 11543♂ (holotype and paratype of unpubl. *E. liquida* Vojnits, ZFMK); 4♂, 1♀, same loc. and alt., 25, 30.vii., 3, 11, 15.viii.1936, H. Höne, Vojnits slides nos. 13948♂, 14112♂, 14065♂, 13747♂, 12183♀ (wrongly labelled gen. slide of *E. refertissima*, ZFMK); 1♂, same loc., Mittlere Höhe (*ca.* 4000 m), 11.viii.1936, H. Höne, Vojnits slide no. 14070♂ (ZFMK); 1♂, same loc., Talsohle *ca.* 3000 m, 2.vii.1937, H. Höne, Mironov slide no. 835♂ (TTM); 1♂, Tibet [Sichuan], Batang, Untere Urwaldzone (*ca.* 3800 m), 30.vii.1936,

H. Höne, Vojnits slide no. 12419♂ (paratype of unpubl. *E. liquida* Vojnits, ZFMK); 5♂, same loc. and alt., 13.vi., 3, 7.viii.1936, H. Höne, de Laever slides nos. 467♂, 471♂, 483♂, Vojnits slides nos. 13775♂, 13839♂ (ZFMK); 2♂, 5♀, same loc. and alt., 2, 3, 4, 7, 8.viii.1936, H. Höne, Vojnits slide no. 13856♂ (TTM); 3♂, same loc., Im Tal des Yangtze (*ca.* 2800 m), 11, 13, 30.viii.1936, H. Höne, Vojnits slides nos. 13265♂, 13777♂, de Laever slide no. 473♂ (ZFMK); 1♂, [Sichuan], Ta-tsien-Loû, Chasseurs du P. Déjean 1904, Vojnits slide no. 870♂ (TTM); 1♂, [Sichuan], Sung-pan-ting, Szetschwan-China, Exped. Stötzner, Vojnits slide no. 12533♂ (paratype of unpubl. *E. liquida* Vojnits, ZFMK); 1♂, [Sichuan], Wassekou, Szetschwan-China, Exped. Stötzner, Vojnits slide no. 12534♂ (holotype of unpubl. *E. idonea* Vojnits [homonym of *Eupithecia idonea* Vojnits, 1982 from Armenia]; slide marked as "Paratypus *Eupithecia liquida* Vojnits", ZFMK); 1♂, 1♀, China, Sichuan, Daxue Shan Mts., Gongga Shan, NW Moxi, 29°41′ N, 101°58′ E, 2850 m, 14.-19.vii.1999, leg. V. Siniaev & A. Plutenko, Mironov slides nos. 729♀, 731♂ (ZFMK); 1♂, China, W. Sichuan, Kangding, 3000 m, 29°53′ N / 101°55′ E, 13.vii.2009, I. & A. Floriani, Saldaitis leg., Pekarsky slide no. OP 0684♂ (coll. O. Pekarsky); 2♀, China, W. Sichuan, Litang evn 4000 m, 29°50′ N / 100°21′ E, 16.vii.2009, I. & A. Floriani, Saldaitis leg., Pekarsky slides nos. OP 0709♀, OP 0712♀ (coll. O. Pekarsky); 2♂, China, W. Sichuan, road Xinngcheng / Zhongdian, 3300 m, 17.vii.2009, 28°46′ N / 99°51′ E, I. & A. Floriani, Saldaitis leg., Pekarsky slides nos. OP 0803♂, OP 0808♂ (coll. O. Pekarsky); 2♀, Qinghai, Menyuan, Xianmi, 2800 m, 23.vii.1992; 12♂, 18♀, Qinghai, Tongren, Maixiu, 2950 m, 30.vii.1992, Xue Dayong, slides no. ZL275, 359, 361♂, 161, 269♀; 2♂, 3♀, Gansu, Yongdeng, Tulugou, 2280 m, 25.vii.1991, Xue Dayong, slide no. ZL514♀; 1♂, 2♀, same loc, 20.vii and 11.viii. 1991, Meng Feng, slides no. 163, 272♀; 2♂, Gansu, Zhouqu Shatan forestry reserve, 2400 m, 14–16.vii.1999, Yao Jian; 2♀, Gansu, Yongdeng Liancheng Linchang, Tulugou, 2600 m, 36°41′ N, 102° 42′ E, 7–8.viii.2005, Han Hongxiang, IOZ(E)1847036, and 1847058, slide no. ZL692♀; 1♂, Sichuan, Xiangcheng, 3800 m, 1.vii.1982, slide no. ZL122♂; 1♂, Sichuan, Luding, Moxi, 1550 m, 18.ix.1982, Wang Shuyong, slide no. ZL165♂; 1♀, Sichuan, Pingwu, Sanlang, 2830 m, 2.viii.1999, Zhou Xin, ZL666♀; 1♂, Yunnan, Lijiang, Yulongshan, 21.vii.1962, Song Shimei, slide no. ZL53♂(IZCAS).

Remarks. The synonymy of this species was extensively discussed in Mironov et al., 2008b.

95. *Eupithecia astricta* Inoue, 1988

Eupithecia astricta Inoue, 1988, *Bull. Fac. domest. Sci. Otsuma Wom. Univ.* **24**: 370, figs 30a-e. Holotype ♂ (BMNH; examined), [Taiwan]: Hualien Hsien, Tayuling, 2600 m.

Description. A relatively small species with triangular pointed fore wings; the fore wing has a narrow postmedial transverse line crossing the narrow elongate black discal dot, and two weak darker costal spots medially and apically.
Male genitalia. Uncus narrow, biapical. Valve broadened medially, with a narrow pointed apex. Vinculum tapered. Papillae elongate. Aedeagus narrow, shorter than length of valve. Vesica armed with two short horn-like cornuti at apex, and a folded U-shaped cornutus with two curved and apically broadened arms. Sternite A8 with two lyre-shaped, apically sclerotized and bifid horns.
Female unknown.
Range. China (Taiwan).
Phenology. Apparently a spring species, both known specimens being taken in March.
Altitude. Found at about 2600 m above sea-level (Taiwan: Inoue 1988; coll. NSMTA).
Similar species. Similar in shape to Taiwanese *E. dolichoptera*, but much smaller. Given the present lack of information on appearance when fresh, determination should be by examination of the genitalia, which are distinct from any other Taiwanese species.
Material. ♂, Taiwan, Tayuling, Hualien county, Tayuling, 5.iii.1982, leg. S. Hashimoto, BM Geom. slide no. 16557♂ (holotype, coll. H. Inoue in BMNH); 1♂, Taiwan, Taichung Co., Anmashan, 2600 m, 16.iii.2001, C.M. Fu, Galsworthy slide no. TW10 (NSMTA).
Remarks. Only two very worn specimens of this species with almost transparent wings are so far known, so that the description is necessarily very sketchy.

The *interruptofasciata* species group

This species group was first introduced and described in detail by Bolte (1990) after examining two North American species. It was further developed in Mironov (2003). It is a large Holarctic group, especially numerous in the Mediterranean area. It is represented in China only by the following four species. The group is fairly diverse in external appearance, but with a tendency to have a prominent medial transverse line on the fore wing. In the male genitalia, the valves of many species are either angled or produced at the tip of the sacculus, and the eighth sternite is always peg-like in shape, usually with two very short sclerotized horns at apex. The female genitalia are diverse, but always have a substantial area at the posterior end of the bursa which is thickly membranous and well wrinkled. The ductus seminalis is attached high on the right side of the bursa, and the colliculum is usually weakly marked and broad.

96. *Eupithecia abbreviata* Stephens, 1831

Eupithecia abbreviata Stephens, 1831, *Illust. Br. Ent.* (Haustellata) 3: 283. Syntypes (BMNH), [United Kingdom]: [England], Kent, Darenth; York.

Phalaena nebulata Haworth, 1809 (*nec* Scopoli, 1763), *Lepid. Br.* (2): 360. Syntype(s) (BMNH), [United Kingdom].

Eupithecia subfasciata Stephens, 1831, *Illust. Br. Ent.* (Haustellata) 3: 284. Holotype ♀ (BMNH), [United Kingdom]: [England], Kent, Darenth Wood.

Misidentification: *Eupithecia guinardaria*: Duponchel, 1842 (*nec* Boisduval, 1840).

Unavailable names (infrasubspecific): *hirschkei*: Bastelberger (1908); *striata*: Lempke (1951); *nigra*: Cockayne (1953); *pallifasciata*: Lempke (1969); *impuncta*: Lempke (1969); *obsoleta*: Lempke (1969). – (incorrect subsequent spellings): *abbreviaria*: Doubleday (1849); *abreviata*: Millière (1870).

Description. Fore wing ground colour light yellowish brown, transverse lines more or less distinct; basal, postbasal and antemedial lines oblique, sharply angled near costa; medial strongly sinuous; medial area between antemedial and medial darker than rest of wing; postmedial irregularly curved and twice right angled near costa, with a series of black dashes on the veins on its inner side; pale wavy subterminal line weakly visible, forming a small pale tornal spot; discal dot narrow, elongate. Hind wing distinctly paler than fore wing, off-white with ochreous tinge; transverse lines inconspicuous; postmedial represented by series of dark dashes on the veins; discal dot minute.

Male genitalia. Uncus biapical. Valve basally very broad, with a heavily sclerotized pointed ventral process, and tapering to a narrowly rounded apex. Vinculum elongate, with shallow medial hollow. Papillae of average size. Aedeagus stout, longer than length of valve. Vesica armed with one long and one short horn-like cornutus and two flat, curved cornuti near ductus ejaculatorius base. Sternite A8 peg-like, basally broadened, with two small, tapered apical lobes and two pointed basal lobes.

Female genitalia. Bursa thick walled, almost completely covered with spinules, and with larger, stout marginal spines. Ductus bursae long and broad, strongly twisted, sclerotized, with an elongate patch of spinules and longitudinal striations. Ductus seminalis attached to base of ductus bursae on right side. Colliculum very broad and undefined. Antrum membranous, short and broad. Tergite A8 quadrate. Apophyses short and narrow. Papillae anales broad and rounded.

Phenology. In Europe a spring species. There is no date on the single Chinese specimen.
Range. China (Sichuan); almost all Europe (except northern part), Tunisia, Georgia.
Similar species. The single known Chinese specimen is distinguished externally from European material only by the paler pattern, yellowish tinge and less clear pattern details. The male genitalia differ in the following respects: the vinculum is longer, the ventral process of the valve also longer, the aedeagus slightly shorter, the long horn-like cornutus more pointed, the broad horn-like cornutus longer, and sternite A8 has pointed rather than rounded basal lobes.
Material. ♂, [Sichuan], Chasseurs Indigènes des Missionnaires de Ta-tsien-Loû 1906, Vojnits slide no. 665♂ (TTM).
Remarks. This is a rather surprising find in China. With this in mind, we have carefully examined this specimen, including its label and entomological pin. They appear to be authentic, and the pin is certainly similar to other pins which were used for Chinese material by the missionaries in Ta-tsien-Loû in 1906. If more material is eventually obtained, in particular females, it is possible that it will warrant description at least as a distinct subspecies. The specimen in question is shown in fig. 96b, beneath a typical European specimen (fig. 96a).

97. *Eupithecia pusillata* [Denis & Schiffermüller], 1775

Geometra pusillata [Denis & Schiffermüller], 1775, *Syst. Verz. Schmett. Wien.*: 110. Syntype(s) lost, [Austria]: Vienna district.
Phalaena laevigata Haworth, 1809, *Lepid. Br.* (2): 362. Syntype(s) (BMNH), [United Kingdom]. [Junior primary homonym of *Phalaena laevigata* Scopoli, 1763] (synonymised to *E. sobrinata* Hübner, 1817 in Walker, 1862).
Geometra sobrinata Hübner, [1817], *Samml. Eur. Schmett.* 5 Geometrae (2), pl. 90: 465. Syntype(s) lost, [Europe] (synonymised in Prout, 1915).
Eupithecia expressaria Herrich-Schäffer, 1848, Syst. Bearb. Schmett. Eur. 3 (32): 121; pl. 46: 284, 285. Syntypes. [Europe] (treated as var. of *E. sobrinata* Hübner, 1817 in Staudinger, 1861).
Eupithecia scoriata Staudinger, 1857, *Stett. Ent. Z.* 18 (7–9): 265. Syntypes, at least 2♂, 6♀ (coll. O. Staudinger in MNHU; examined), [Iceland] (treated as var. of *E. sobrinata* Hübner, 1817 in Bohatsch, 1893).
Eupithecia anglicata Herrich-Schäffer, 1863, *KorrespBl. zool.-min. Ver. Regensburg* 17: 23. Syntype(s). [United Kingdom]: [Kent], Dover (synonymised to *E. sobrinata* Hübner, 1817 in Prout, 1907).
Eupithecia latoniata Millière, 1882, *Lépidoptérologie* 8: 13, pl. 2: 10. Syntypes ♂, ♀ (?), [France]: near Saint-Martin-Lantosque (synonymised to *E. sobrinata* var. *graeseriata* Rätzer, 1882 in Bohatsch, 1893).
Eupithecia sobrinata var. *graeseriata* Rätzer, 1882, in Frey, *Mitt. Schweiz. ent. Ges.* **6** (7): 361. Syntype(s), [Switzerland]: Gadmen.

Eupithecia stevensata Webb, 1896, *The Entomologist's Rec. J. Var.* **8**: 10. Syntypes, at least 1♀ (coll. K. Dietze in MNHU; examined), [United Kingdom]: [Kent], Dover (synonymised to *E. sobrinata* Hübner, 1817 in Staudinger & Rebel, 1901).

Eupithecia lavicaria Fuchs, 1902, *Stett. Ent. Z.* **63** (1): 318. Syntypes 2, not traced (Aarvik pers. comm.), Norway: Lavik (synonymised in Mironov, 2003).

Eup[ithecia] sobrinata f. *luneburgensis* Dietze, 1910, *Biologie der Eupithecien* **1**, pl. 77, figs 679, 680; *ibidem* (1913), **2**: 147. Syntypes, at least 6♀ (coll. K. Dietze in MNHU; examined), Germany: Lüneburg.

Eup[ithecia] sobrinata f. *scotica* Dietze, 1910, *Biologie der Eupithecien* **1**, pl. 77, figs 681–683; *ibidem* (1913), **2**: 147. Syntypes, at least 1♂, 3♀ (coll. K. Dietze in MNHU; examined), [United Kingdom], Scotland: Archimore; England.

Eup[ithecia] sobrinata f. *castiliana* Dietze, 1910, *Biologie der Eupithecien* **1**, pl. 77, figs 684, 685; *ibidem* (1913), **2**: 147. Syntypes, at least 1♂, 3♀ (coll. K. Dietze in MNHU; examined), (Spain: Castilia; Andalusia, Sierra Nevada).

[Eupithecia] sobrinata f. *confluens* Dietze, 1910, *Biologie der Eupithecien* **1**, pl. 77, figs 689, 690; *ibidem* (1913), **2**: 146. Syntypes, at least 2♂, 1♀ (coll. K. Dietze in MNHU; examined), Austria: Stainz.

Eupithecia masuii Inoue, 1980, *Bull. Fac. domest. Sci. Otsuma Wom. Univ.* **16**: 158, fig. 35D; fig. 38B; figs 39J-K; fig. 42E. Holotype ♂ (coll. H. Inoue in BMNH; examined), [Japan]: Kagawa Pref(ecture), Takamatsu, Mt. Iwaseo (synonymised in Mironov et al., 2010b).

Misidentification: *Eupithecia satyrata* Fristrup, 1943 (*nec* Hübner, 1813) (identity: *Eupithecia satyrata*).

Unavailable names (infrasubspecific): *conjuncta*: Dietze (1913); *nigrofasciata* Dietze (1910); *impuncta*: Lempke (1951). – (incorrect subsequent spellings): *pusillaria*: Hübner, [1825]; *sobrinaria*: Boisduval (1840); *variata* Pagenstecher (1902; misspelling of *scoriata* Staudinger, 1857); *graessariata*: Dietze (1910); *pussillata*: Dantart (1989).

Description. Fore wing ground colour varying from off-white to fuscous grey; antemedial, medial and upper part of postmedial usually strongly marked, with area between the former two darker; postmedial bordered internally with a series of dark dashes on the veins; whitish wavy subterminal line variable but usually present, with a whitish tornal spot with dark shade on inner side; discal spot small. Hind wing paler, greyish brown, transverse lines weakly marked; discal spot minute or absent. Abdomen pale greyish brown, usually with narrow black lateral stripes, second segment often darker, brown or blackish brown.

Male genitalia. Uncus relatively narrow, biapical. Valve triangular, sharply tapered to apex, with a sclerotized pointed ventral process. Vinculum short and broad. Papillae stout and elongate. Aedeagus broadened at anterior end, equal to valve length. Vesica armed with two horn-like cornuti, a broad flat cornutus and a smaller, irregular cornutus at ductus ejaculatorius base. Sternite A8 elongate, broadened at base, with two narrow, sometimes diverging apical rods.

Female genitalia. Bursa rounded, completely covered with small spines and larger marginal spines in basal half. Ductus bursae elongate and broad, sclerotized, covered with longitudinal striations, and with a short

row of large spines in basal half on left side. Ductus seminalis sharply recurved, attached to posterior part of bursa on right. Colliculum short and broad. Antrum membranous. Tergite A8 almost quadrate. Apophyses narrow. Papillae anales broad, with rounded apices.
Range. China (Gansu, Sichuan); almost all Europe (westward to Canary Islands), Asian Russia (Urals, Siberia, Transbaikalia, Priamurje, Sakhalin, Kamchatka), Japan (Honshu, Shikoku), Pakistan/India (Jammu & Kashmir, subsp. *kashmirica*).
Phenology. Univoltine: early July to late September. Larva: late March to mid-July. Overwintering as egg.
Biology. Larva found on the needles and flowers of *Juniperus communis, J. communis* ssp. *alpina* and *J. oxycedrus* (Europe: Mironov 2003).
Altitude. From sea level to about 2500 m in mountains of Europe (Sierra Nevada: coll. ZMUC and Alps: Weigt 1993), and to about 3350 m in China (coll. IZCAS).
Similar species. Despite its considerable variability, *E. pusillata* is one of the most easily recognized species amongst Asian representatives of the genus *Eupithecia*, usually recognizable by the darker medial area or transverse lines on the fore wing. The shape of the male valve, coupled with the disposition of cornuti in the aedeagus vesica and shape of the eighth sternite are distinctive. In the female, the broad ductus bursae with a single row of large spines is characteristic.
Material. ♂, [Japan], Mt. Iwaseo, Takamatsu, Kagawa Pref., 19.x.1973, T. Masui [leg.], *masuii*, Inoue slide no. 6501♂, BM Geom. gen. slide no. 16374♂, Inoue Coll. B.M. 1992–71, Type status verified K. Buckmaster 1992 (holotype of *E. masuii*, coll. H. Inoue in BMNH); 2♂, 1♀, Gansu, Yongdeng Liancheng Linchang, Tulugou, 2600 m, 36°41′ N, 102° 42′ E, 8. viii.2005, Han Hongxiang and Li Jing, IOZ(E)1847019, 1847032, and 1847063, slides nos. ZL702, 710♂, 711♀; 1♀, Sichuan, Batang Yidun, 3350 m, 16.viii.1982, Zhang Xuezhong, slide no. ZL172♀ (IZCAS).

98. *Eupithecia kozlovi* Viidalepp, 1973

Eupithecia kozlovi Viidalepp, 1973, *Ent. Obozr.* 52 (2): 398, fig. 2. Holotype ♂, (ZISP; examined), [Mongolia]: South Gobi aimak, 60 km E of the Talyn-Bil'gekh-Bulak spring.

Description. A small species with narrow elongate pale brownish grey fore wings, and dark, very distinct, oblique basal, antemedial, medial and postmedial transverse lines, only slightly curved close to costa; discal dot small, usually elongate; terminal area slightly darker with conspicuous, multiple wavy pale subterminal lines. Hind wing paler with indistinct transverse lines, well marked along anal margin only and with a small pale rounded discal dot.

Male genitalia. Uncus stout, biapical. Valve slightly broadened medially, apex narrow and rounded. Vinculum semicircular. Papillae short and tapered. Aedeagus equal to valve length, narrow, markedly curved, with ductus ejaculatorius attached to central part of aedeagus. Vesica armed with one long slim, slightly curved, horn-like cornutus. Sternite A8 with two narrow, parallel apical arms with rounded tips.

Female genitalia. Bursa oval, with a small membranous diverticulum on right side, covered with small spines except at base, and with some large spines near base of ductus bursae. Ductus bursae elongate with longitudinal striations. Ductus seminalis narrow, attached to ductus bursae near colliculum on left side. Colliculum short and narrow. Tergite A8 quadrate. Apophyses short and narrow. Papillae anales small.

Range. China (Xinjiang, Inner Mongolia, Qinghai and Gansu); Tajikistan, Kirghizstan, SE Kazakhstan, Mongolia.

Phenology. Univoltine: early July to mid-August.

Altitude. From about 1100 m in Ghissar Mts. (Tajikistan: Viidalepp, 1988) up to about 3000 m in West Tien-Shan and mountains of Qinghai (Kirghizstan: coll. M. Sommerer; China: coll. IZCAS).

Similar species. Closely related to *E. pusillata*, but easily distinguished by smaller size, paler coloration and by the maculation on the fore wings. In the male, the curved aedeagus with a single long cornutus, and in the female the rather pointed diverticulum at the base of the bursa make this a straightforward species to recognise.

Material. ♂, [Mongolia], South Gobi aimak, 60 km E of the Talyn-Bil'gekh-Bulak spring, L, 17–19.viii.[1]969, leg. M. Kozlov, ZISP slide no. 11892♂ (holotype, ZISP); 3♂, 5♀, Qinghai, Menyuan, Xianmi, 2800 m, 23–24.vii.1992, slides nos. ZL305, 507♂, 303, 503♀; 1♀, same loc, 2800 m, 37°16′ N, 102°00′ E, 6.viii.2005, slide no. ZL713♀1♂, Qinghai, Tongren, Maixiu, 2950 m, 30.vii-1.viii.1992, slide no. ZL504♂; 1♀, Gansu, Yongdeng Tulugou, 25–29.vii.1991, slide no. ZL506♀, Xue Dayong (IZCAS); 1♀, China, Inner Mongolia, 100 km SW of Hohhot, Mo-Tien-Ling, 12.viii.2007, [leg.] A. Bidzilya (ZISP).

Remarks. Not previously recorded from China. Chinese specimens from Qinghai are paler and usually smaller than specimens from western Tien-Shan.

99. *Eupithecia exrubicunda* Inoue, 1988

Eupithecia exrubicunda Inoue, 1988, *Bull. Fac. domest. Sci. Otsuma Wom. Univ.* **24**: 358, figs 21a-b. Holotype ♀ (BMNH; examined), [Taiwan]: Hualien Hsien, Tayuling, 2600 m.

Description. Fore wing with brown basal and medial areas divided from each other by a rusty transverse band; basal and antemedial lines oblique, the latter right angled onto costa; medial area broad with two pale sinuate transverse lines and a small, intensely black discal dot; postmedial perpendicular to anal and costal margins, slightly curved near costa; terminal area reddish, with a whitish, wavy subterminal line. Hind wing paler, darker near anal margin and tornus, with a distinct postmedial and a small, rounded discal dot.

Male genitalia. Uncus narrow, biapical. Valve small, with a relatively narrow rounded apex. Vinculum narrow, tapered. Papillae curved and clavate. Aedeagus longer than length of valve. Vesica armed with one long, narrow horn-like cornutus, a broad U-shaped cornutus with one basal arm longer than the other and a smaller, irregular cornutus near ductus ejaculatorius base. Sternite A8 peg-like, with a broad base.

Female genitalia. Bursa almost globular, completely covered with small spines. Ductus bursae long, twisted near colliculum, with numerous longitudinal striations, a longitudinal row of stout spines at base and a patch of small spines in posterior half. Ductus seminalis long and relatively broad, attached to central part of ductus bursae. Colliculum short and narrow, slightly inclined. Antrum membranous. Tergite A8 transverse. Apophyses narrow, pointed. Papillae anales short and rounded.

Range. China (Taiwan).

Phenology. Univoltine (?): early May.

Altitude. The only altitude data is for one specimen collected at 2600 m (Taiwan: Inoue 1988).

Similar species. A very distinctive species owing to the contrast between the reddish terminal area and the basal brown ground colour. The male genitalia are very small, and relatively easily recognized from the aedeagus vesica cornuti. The female genitalia resemble those of *E. pusillata*, but can be separated by the ductus seminalis arising from the central part of the ductus bursae, and not from the corpus bursae itself, as in all other allied species of this group.

Material. ♀, [Taiwan], Tayuling, 2600 m, Hualien Hsien, 2–4.v.1984, leg. H. Yoshimoto, BM Geom. slide no. 16547 (holotype, coll. H. Inoue in BMNH); 1♂, Taiwan, prov. Nantou, 6 km SW Hohuan Pass, Yuanfeng, 2760 m, 31.v.1997, leg. Gy.M. László & G. László, Galsworthy slide no. ACG GL70 (coll. GL); 1♂, Taiwan, Nantou, Jenai, Yuanfeng, 19–20.v.1998, leg. C.S. Lin & W.T. Yang, Galsworthy slide no. ACG TW15 (NSMT); 1♂, Taiwan, prov. Nantou, Hohuachi, 1950 m, 24°13′ N, 121°16′ E, 15–19.ii.1999, leg. L. Peregovits & G. Ronkay (TTM).

The *lanceata* species group

This species group was introduced by Mironov (2003). Here, we unite the *lanceata* species group with the *tantillaria* group, which was introduced by Mironov (2003), on the basis of two European species, *E. tantillaria* Boisduval, 1840, and *E. liguriata* Millière, 1884, and one East Asian species *E. tantilloides* Inoue, 1958. On examination of a wider range of species associated with these groups, we find that there are no significant structural differences in the male and female genitalia between the two. The male genitalia are characterized by a highly sclerotized and modified sacculus, usually with one, or more often two ventral processes. The aedeagus is usually stout, with one or more large cornuti; sternite A8 is most often peg-like, usually divided at apex, but in some species divided to the base. In the female genitalia, the bursa copulatrix is quite varied, but usually with one or more rows of stout spines, in addition to a field of smaller spines; in many species there is a spineless diverticulum at base. The group is holarctic, but most numerous in East Asia, especially in China, where it is represented by the following fifteen species.

100. *Eupithecia lanceata* Hübner, [1825]

Dyscymatoge lanceata Hübner, [1825], *Verz. bekannt. Schmett.* (21): 324. Syntype(s) lost, [Europe].
Larentia hospitata Treitschke, 1828, *Die Schmett. Eur.* 6 (2): 120. Syntype(s) (TTM, not traced), [Austria; Hungary] (synonymised in Lederer, 1853).

Misidentification: *Geometra succenturiata* Hübner, 1799 (*nec* Linnaeus, 1758).
Unavailable names (incorrect subsequent spellings): *hospitaria*: Boisduval (1840); *hospitarla*: Herrich-Schäffer [1848].

Description. A small species with narrow, pointed triangular fore wings. Ground colour ochreous brown, with slightly darker basal and narrow medial areas; ante- and postmedial distinct, blackish; antemedial usually oblique and sharply angled onto costa; postmedial slightly sinuate, almost straight near costa, usually touching discal dot on outer side; discal dot small, elongate; outer half of terminal area darker brown. Hind wing ovoid, paler, usually with distinct basal and postmedial and darker, brownish terminal area.

Male genitalia. Uncus elongate, biapical. Valve rather broad, sacculus heavily sclerotized with one pointed and one rounded apical ventral process. Vinculum rectangular with a shallow medial hollow. Papillae clavate, curved. Aedeagus equal to length of valve. Vesica armed with a large, slightly curved horn-like cornutus with a plate-like cornutus at its base, two or three small spine-like cornuti, a V-shaped cornutus and a smaller, irregular cornutus at ductus ejaculatorius base. Sternite A8 peg-like, relatively narrow and elongate, narrowing to apex.

Female genitalia. Bursa elongate, broadened in medial part, cylindrical at base, with numerous spines at base and with rows of large spines and a small patch of spinules in apical part. Ductus bursae not distinct. Ductus seminalis short, broadly attached to corpus bursae near base, forming a large membranous diverticulum. Colliculum elongate, plate-like. Antrum short and broad. Tergite A8 almost square. Apophyses short and very narrow. Papillae anales short and rounded.

Range. China (Gansu); Asian Russia (Southern Siberia: from Krasnojarsky kraj and Altai Mts. to Transbaikalia and Priamurje), Northern and Central Europe.

Phenology. Univoltine: early March to mid-June in Europe. The Chinese specimens were collected in late April.

Altitude. From sea level in Europe to about 2300 m in Gansu province (China: coll. IZCAS).

Similar species. *E. lanceata* can be distinguished easily from other *Eupithecia* species by the bipectinate antennae in the male; the fore wing pattern is also distinct, especially the narrow darker medial area between the antemedial line, which is almost straight, sharply angled near the costa, and the slightly sinuate postmedial line. The male and female genitalia are both distinctive.

Material. 3♀, China, Gansu, Yongdeng Tulugou, 2280 m, 24–29. iv.1992, Meng Feng, slide no. ZL528♀ (IZCAS).

Remarks. Not previously recorded from China. The three Chinese specimens are smaller and slightly paler than European specimens.

101. *Eupithecia kudoi* Inoue, 1983

Eupithecia kudoi Inoue, 1983, *Tinea* 11 (16): 141, figs 5, 6a-e. Holotype ♂ (BMNH; examined), [Taiwan]: Hualien Hsien, Tayuling, 2565 m.

Description. Fore wing pale brown with darker brown medial fascia with characteristic shape between inwardly curved antemedial and oblique undulating postmedial, which often touches discal dot; basal transverse line distinct, curved; terminal area strongly broadened in apical part, usually with medial and tornal blackish blotches and a pale, wavy subterminal line; discal dot oblique, linear. Hind wing much paler, white with yellowish tinge; transverse lines inconspicuous except oblique postmedial from termen to middle of costa; discal dot small.

Male genitalia. Uncus elongate, biapical. Valve long, sacculus with two prominent ventral processes, slightly shorter and narrower on the left valve than on the right. Vinculum broad, elongate with shallow medial hollow. Papillae long, narrow, slightly broadened to apex. Aedeagus vesica armed with a hooked horn-like basally broadened cornutus and a large, elongate U-shaped plate. Sternite A8 short, peg-like with two very short, blunt apical arms.
Female genitalia. Bursa elongate, the base almost completely covered with large spines, and with a broad blunt diverticulum on the left side. Ductus bursae broad and sclerotized, with numerous longitudinal striations and a transverse patch of large, stout spines at the level of the ductus seminalis. Ductus seminalis broad, attached to central part of ductus bursae from right side. Colliculum very broad. Antrum broad, membranous. Tergite A8 spade-shaped. Sternite A8 heavily sclerotized, horseshoe-shaped, with thickened posterior margin, completely covered by numerous pores. Apophyses narrow. Papillae anales rather short and rounded.
Range. China (Taiwan).
Phenology. Univoltine: late March to early May.
Altitude. From about 2500 to 2800 m above sea-level (Taiwan: Inoue 1988; coll. ZSM).
Similar species. This and the following species are distinctive because of the dark fore wing central fascia. For differences between them, see under the following species.
Material. ♂, Taiwan, Tayuling, 2565 m, Hualien Hsien, 27–29.iii.1981, Coll. K. Kudo, BM Geom. slide no. 16449 (holotype, coll. H. Inoue in BMNH); 1♂, Taiwan, Prov. Nan-Tou, Tayuling, 2550 m, 16.iii.1996, leg. Gy. Fábián & L. Németh (TTM); 3♂, Taiwan, Prov. Hualien, Taroko N.P., Hohuan Pass, 3000 m, 3.iv.1997, leg. Csorba & Ronkay (TTM); 1♂, 17♀, Taiwan, Prov. Nantou, 6 km SW Hohuan Pass, Yuanfeng, 2760 m, 10.v.1997, leg. Gy.M. László & G. László (TTM; ZSM); 5♂, 15♀, Taiwan, Prov. Nantou, Yuanfeng, SW to Hohuan Pass, 2760 m, 24°07′ N, 121°15′ [E], 1.iv.2000, leg. A. Kun & L. Peregovits (ZSM; ZISP); 1♀, Taiwan, Prov. Nantou, Hohuachi, 1950 m, 24°13′ N, 121°16′ E, 31.iii.2000, leg. A. Kun & L. Peregovits (ZSM).

102. *Eupithecia daemionata* Dietze, 1904

Eup[ithecia] daemionata Dietze, 1904, *Dt. ent. Z.,. Iris* [1903] 16 (2): 339, pl. 3, fig. 9. Holotype ♂ [*nec* ♀] (coll. R. Püngeler in MNHU; examined), Japan: Nikko.

Description. Fore wing ground colour ash-grey or brownish; antemedial and postmedial distinct, black, the former sharply bent inwards to the costa, and the latter undulating, the area between forming a darker brown fascia; terminal area also dark, broadened in apical part, with a more or less distinct whitish medial blotch and a distinct, zigzag, whitish subterminal line; discal dot black, elongate, sometimes touching postmedial. Hind wing off-white, ante- and postmedial narrow, distinct, postmedial usually evenly curved; discal dot small and pale. Abdomen rusty brown, with black dorsal spots and black lateral bands on the A4-A7 segments; A3 segment blackish brown; A8 segment paler than others.

Male genitalia. Uncus elongate, biapical. Valve heavily sclerotized, with two ventral processes, the basal one small and pointed and the other broad and blunt. Vinculum elongate and broad. Papillae relatively long and broadened at apices. Aedeagus shorter than valve length. Vesica multiply granulate, armed with a slim, horn-like cornutus and a narrow, longitudinally folded, U-shaped plate near ductus ejaculatorius base. Sternite A8 small, peg-like, sometimes with short blunt apical rods.

Female genitalia. Bursa almost completely covered with large spines from base to base of ductus seminalis, with a short, blunt membranous diverticulum on the left side. Ductus bursae broad, sclerotized, with numerous longitudinal striations and an oblique patch of large, stout spines. Ductus seminalis broad, evenly curved around bursa, attached to central part of ductus bursae from right side. Colliculum short and broad, with a broad sclerotized lobe on ventral side of antrum. Antrum membranous. Tergite A8 almost square with a small, pointed medial tip on posterior margin. Sternite A8 broad, strongly curved and heavily sclerotized, completely covered by numerous pores. Apophyses narrow. Papillae anales rounded.

Range. China (Shaanxi, Sichuan, Jilin, Taiwan); Asian Russia (Priamurje, Primorje, Sakhalin Isl.), Japan (Hokkaido, Honshu, Shikoku, Kyushu, Tsushima).

Phenology. Univoltine: mid-March to mid-June.

Altitude. From about 1500 to 3200 m above sea-level in Taibai Shan and Daxue Shan (China: coll. ZSM).

Similar species. Very similar to preceding species. Distinguished from it by smaller size, the greyish tinge of all wings, usually the smaller ovoid (not linear) discal dot, and the presence of a whitish medial blotch in the terminal margin on the fore wing, as well, usually, as the more curved postmedial transverse line on the hind wing. In the male genitalia, *E. kudoi* is distinguished by the longer ventral processes on the valve, and the two horn-like cornuti on the vesica, of which one is curved near the apex, and in the female by the triangular tergite A8.

Material. ♂, Japonia (Nikko), W. Blüthgen, *daemionata* Dietze ♀ ♂, Type, abgebildet Iris 1902 v. Gv. 414, 5/4 95, *daemionata* Original (Dietze) (holotype, coll. R. Püngeler in MNHU); 1♂, [Russia], Ussuri, Kasakewitsch 1907 Korb, *daemionata ? detritata* Stg. Ussuri, Gen. slide Geom. 1951–7A (holotype ♂ of *E. daemionata accessata* Dietze, coll. K. Dietze in MNHU); 1♂, [Russia], Ussuri, Kasakewitsch 1907 Korb, *daemionata* n. sp. Ussuri (paratype ♂ of *E. daemionata accessata* Dietze, coll. K. Dietze in MNHU); 1♂, China, Shaanxi, Taibai Shan, 33°53′ N, 107°49′ E, 1500 m, April 2000, leg. Siniaev & C° (ZSM); 2♂, 4♀, China, Sichuan, Daxue Shan, Gongga Shan, NW Moxi, 29°41′ N, 101°58′ E, 2600–3200 m, 23.iv.-15.v.2001, leg. V. Siniaev & A. Plutenko, *ex* coll. Dr. A. Schintlmeister (ZSM); 1♂, Jilin, Changbaishan, 8.vi.1974, coll Yang Liming, slide no. ZL447♂ (IZCAS).

103. *Eupithecia habermani* Viidalepp & Mironov, 1988

Eupithecia habermani Viidalepp & Mironov, 1988, *Eesti NSV Tead. Akad. Toim.* (Biol.) 37 (3): 209, figs 3a-d. Holotype ♀ (IAET; examined), [Russia]: Sakhalin, Novo-Aleksandrovskoye.5

Description. Fore wing rather elongate. Ground colour soft ochreous brown; transverse lines fine, blackish; basal, postbasal and antemedial lines obliquely straight, sharply angled onto costa; postmedial almost straight, double, touching posterior end of discal dot; terminal area broad, slightly darker near apex and tornus, with a pale dentate subterminal line; discal dot oblique, elongate, black. Hind wing paler with distinct, oblique postmedial; terminal area darker near tornus; discal dot distinct, small, rounded.

Male genitalia. Uncus long and narrow, biapical. Valve long, with two prominent ventral processes, the proximal one a semicircular flap, and the distal one broad and rounded. Vinculum elongate and tapered. Papillae very short, cone-shaped. Aedeagus shorter than length of valve, vesica armed with a small apical patch of spinules, two heavily sclerotized, dentate cornuti near apex, a small stout cornutus, an elongate U-shaped cornutus, and a smaller, sharply recurved, V-shaped cornutus near ductus ejaculatorius base. Sternite A8 short, peg-like, with two narrow apical arms.

Female genitalia. Bursa small, rectangular, heavily sclerotized, with transverse wrinkles in posterior part and one or two longitudinal rows of large, stout spines at the middle; base of bursa narrow and membranous with a large oval thin-walled diverticulum attached close to base. Ductus bursae and colliculum not distinguishable. Ductus seminalis narrow, attached to bursa near base on left side. Antrum membranous, short and broad.

Tergite A8 large, almost square. Apophyses very narrow. Papillae anales large, broad and rounded.
Range. China (Qinghai, Gansu); Asian Russia (southern Priamurje, Primorje, Sakhalin Isl.).
Phenology. Univoltine: August to mid-September.
Similar species. Broadly similar to a group of related species which have the median fascia between the antemedial and postmedial sharply broadening, often with the anterior parts of the lines darkened to give the effect almost of a dark crescent on the costa, including *E. dolichoptera, E. phantastica, E. fortis,* and the Japanese *E. takao* Inoue, 1955. Best distinguished by the genitalia. In the male genitalia, the shape of the ventral processses of the valve, and the ornamentation of the aedeagus vesica are distinctive; in the female, the presence of a large membranous diverticulum near the base of the bursa copulatrix and extending well beyond it is unique in this group.
Material. ♀, [Russia]: Sakhalin, Novo-Alexandrovskoye, 27.viii.1970, leg. H. Remm, Viidalepp slide no. 209/1970 (holotype, IZBE); 1♂, [Russia], Valley of Khor river, Tundimajani Mt., 30.viii.[19]37, leg. Kurentzov, ZISP slide no. 17103♂ (paratype, ZISP); 1♂, 1♀, Qinghai, Huzhu Beishan Linchang, Langshidang, 2600 m, 36°55′ N, 102°23′ E, 7.viii.2005, Han Hongxiang, slide no. ZL718♂; 1♂, Gansu, Yongdeng, Tulugou, 11–16.viii.1991, coll Meng Feng, slide no. ZL292♂; 1♀, Gansu, Yuzhong, Xinglong shan, 31.viii.1990, coll Xue Dayong, slide no. ZL140♀; 1♀, Gansu, Yongdeng, Tulugou, no date (IZCAS).
Remarks. Not previously recorded from China.

104. *Eupithecia emendata* Vojnits, 1983

Eupithecia emendata Vojnits, 1983, *Acta zool. Acad. Sci. hung.* **29** (1–3): 268, figs 11, 16. Holotype ♂ (ZSM; examined), Nepal: Prov. Nr. 3, East Junbesi, 2750 m.
Eupithecia mediargentata Inoue, 1987, *Bull. Fac. domest. Sci. Otsuma Wom. Univ.* **23**: 243, figs 63i, 67a. Holotype ♀ (BMNH; examined), [Nepal (north-east)]: between Yangma and Nup, 3310 m (synonymised in Inoue, 2000).

Description. An easily recognized species with elongate fore wings of a dominantly white ground colour with a broad, oblique basal brown band, two distinct brown medial costal spots and three large brown apical, medial and tornal terminal blotches; discal dots round and black. Hind wing also white with brownish transverse lines well marked along anal margin only and with a very small but distinct discal dot.
Male genitalia. Uncus stout, biapical. Valve broad, with a heavily sclerotized hook-like medial process on ventral margin and relatively broad apex. Vinculum elongate, tapered. Papillae elongate and narrow. Aedeagus

broadened medially, shorter than length of valve. Vesica armed with a large apical patch of stout teeth, a large curved horn-like cornutus with a narrow U-shaped cornutus at its base, and a large, irregular cornutus near ductus ejaculatorius base. Sternite A8 two long and narrow, almost parallel rods, connected by a broad basal band.

Female genitalia. Bursa long and narrow, well sclerotized, with a dense patch of spines in posterior half, and a membranous ovoid, basal diverticulum covered with sparse spinules. Ductus bursae not clearly distinguishable. Ductus seminalis very strongly broadened in basal part, attached to central part of bursa on left side. Colliculum relatively broad, almost equal to diameter of posterior part of bursa. Antrum short and broad. Tergite A8 quadrate, with convex anterior margin. Apophyses short and stout, the posterior slightly expanded at apices. Papillae anales narrow and elongate.

Range. China (Tibet); Nepal.

Phenology. Univoltine (?): end June to late July.

Altitude. Found from about 2200 to 3300 m in the Himalayas (Nepal: Vojnits 1983; Inoue 1987).

Similar species. Externally similar to *E. succenturiata* f. *exalbidata* Staudinger, 1901, but easily distinguished from it by the lack of a brown border to the hind wing. Could also be confused with *E. liberata*, but brown markings much less extensive. In the male genitalia, the hook-like process on the valve is unique. The female genitalia could be confused with species in the *undata* group, such as *E. hongxiangae*, but the disposition of spines and shape of the papillae anales are distinct.

Material. ♂, Nepal, Prov. Nr. 3, East Junbesi, 2750 m, 25–31.vii.1964, leg. W. Dierl, Vojnits slide no. 11749♂ [= ZSM slide no. G 10607♂] (holotype, ZSM); ♀, N.E. Nepal, Tamur Valley, Between Yangma & Nup, 3310 m, 25.vii.1963, T. Haruta et al., BM Geom. slide no. 16526 (holotype of *E. mediargentata*, coll. H. Inoue in BMNH); 1♀, China, Tibet, Zham, 2200 m, 30.vi.1975, Wang Ziqing, slide no. ZL483♀ (IZCAS).

Remarks. Not previously recorded from China.

105. *Eupithecia dolichoptera* Galsworthy & Mironov, 2013 nom. nov.

Eupithecia longipennata Inoue, 1988, *Bull. Fac. domest. Sci. Otsuma Wom. Univ.* **24**: 354, figs 19a-d. Holotype ♂ (BMNH; examined), [Taiwan]: Hualien Hsien, Tayuling, 2600 m (junior secondary homonym of *Eucymatoge longipennata* Warren, 1904).

Description. A large species. Fore wing basal and postmedial areas pale brownish grey; medial area ochreous brown; basal line and antemedial near discal dot sharply angled onto costa; postmedial almost straight; ante- and postmedial with dark brown shades near costa; terminal area darker at apex and near tornus, with a broad, wavy, white subterminal line forming a white tornal spot; discal dot oblique, intensely black. Hind wing paler, whitish grey with a soft brownish tinge, darker near tornus; brownish transverse lines well marked along anal margin and tornus; discal dot small, circular.

Male genitalia. Uncus triangular, biapical. Valve with two heavily sclerotized ventral processes, one digitate, the other rounded. Vinculum elongate, tapered. Papillae long and clavate. Aedeagus stout, shorter than length of valve. Vesica strongly wrinkled, armed with a long horn-like cornutus, an elongate U-shaped plate and a small, irregular plate near ductus ejaculatorius base. Sternite A8 peg-like, broadened basally, with two short pincer-like apical rods.

Female genitalia. Bursa narrow and elongate, covered with small spines on whole ventral side and basal half of dorsal side, with a characteristic U-shaped gap in between, and with a long, curved and apically broadened membranous diverticulum on the right side; base of diverticulum sclerotized, with a short row of spines on inner side. Ductus bursae long and broad, with numerous longitudinal striations on dorsal side, a short row of small spines on ventral side near colliculum, and a slightly prominent diverticulum near base of colliculum on left side. Ductus seminalis narrow, attached to posterior margin of diverticulum on ventral side. Colliculum very broad. Antrum short. Tergite A8 slightly narrowed posteriorly, with a sinuate anterior margin and a small membranous hollow on posterior margin. Apophyses short and narrow. Papillae anales short and broad, rounded.

Range. China (Taiwan).

Phenology. Univoltine: early February to late March.

Altitude. From about 500 to 2550 m above sea-level (Inoue 1988; coll. ZSM).

Similar species. Similar in wing pattern to *E. takao* Inoue, 1955, but paler. The male genitalia of *E. dolichoptera* also resemble those of *E. takao*, but can be distinguished by having only one long cornutus on the vesica and by the shape of the eighth sternite. The female genitalia are entirely different.

Material. ♂, Taiwan, Hualien County, Tayuling, 6.iii.1982, leg. S. Hashimoto, BM Geom. slide no. 16546 (holotype, coll. H. Inoue in BMNH); 1♀, Taiwan, Tienchyr, 2300 m, Kaohsiung Hsien, 23.iii.1991, leg. H.Y. Wang, 1282–13281, Galsworthy slide no. ACG TW1 (NSMTA); 2♂, 2♀, Taiwan, Prov. Nan-Tou,

Tayuling, 2550 m, 27.ii.1996, leg. Gy. Fábian & L. Németh (ZSM); 1♀, Taiwan, Prov. Nan-Tou, 15 km N of Puli, 500 m, 15.iii.1996, leg. Gy. Fábián & L. Németh (TTM); 1♀, [Taiwan], Taichung County, Anmashan, 2350 m, 9.iii.[19]96, C.M. Fu (coll. CMF); 1♀, same loc., 2100 m, 14.ii.[19]97, C.M. Fu (coll. CMF); 1♀, same loc. and alt., 10.ii.[19]98, leg. C.M. Fu (coll. CMF).
Remarks. *Eupithecia longipennata* Warren was described (Warren, 1904) from South America. This name was omitted from Scoble et al. (1999), presumably in error, since a card exists for the species in the BMNH index. The new name is derived from the equivalent Greek words meaning *longipennata*, i.e. long-winged.

106. *Eupithecia phantastica* Mironov & Galsworthy, 2006

Eupithecia phantastica Mironov & Galsworthy, 2006, *Trans. lepid. Soc. Japan* 57 (4): 347, figs 8, 23, 28. Holotype ♂ (ZFMK; examined), [China]: Pz. Chekiang [Zhejiang], West Tien-Mu-Shan, 1600 m.

Description. Fore wing broad, ground colour pale brown; basal and antemedial oblique, sharply angled near costa; postmedial oblique; medial fascia sharply narrowed near discal dot and darker near costa, forming a brown triangle on costa, its apex crossed by the elongate black discal dot; terminal area darker, its inner border crenulate, marked with dark spots on veins, and crossed by a wavy whitish subterminal line. Hind wing ground colour whitish grey; transverse lines marked by dark dashes on veins; discal dot narrow and elongate.
Male genitalia. Uncus biapical. Valve elongate with two blunt ventral processes. Vinculum narrow, tapered anteriorly. Papillae sclerotized (unusually), elongate, S-shaped. Aedeagus shorter than valve. Vesica armed with two slim, horn-like cornuti, one elongate, clavate apical cornutus with minute teeth at apex and a weakly twisted elongate plate near ductus ejaculatorius base. Sternite A8 peg-like, with two sclerotized lateral blotches near apex.
Female genitalia. Bursa small, bifid, left part heavily sclerotized with an apical globular, thick-walled diverticulum and a small patch of spines posteriorly, right part membranous with longitudinal striations and an elongate diverticulum at apex. Ductus bursae not distinguishable. Ductus seminalis very narrow, attached to central part of membranous diverticulum. Colliculum broadly trapezoid. Antrum plate-like, slightly sclerotized. Tergite A8 elongate and narrow. Apophyses narrow, the anterior slightly expanded at apices. Papillae anales rounded.
Range. China (Zhejiang, Taiwan).

Phenology. All known specimens were collected in November.
Altitude. Found from about 1600 m in Zhejiang to almost 2000 m in Taiwan (China: Mironov & al., 2006).
Similar species. See under *E. habermani*. *E. phantastica* is smaller, has a lighter ground colour, and broader fore wings than similar species. The male and female genitalia are closest to those of *E. dolichoptera*, but in the male the ventral processes of the valve are differently shaped, and in the female the bursa is much smaller and differently ornamented.
Material. ♂, West-Tien-Mu-Shan, (1600 m), Pz. Chekiang [Zhejiang], 10. xi.1932, H. Höne, 35721 (holotype, ZFMK); 1♂, 1♀, Taiwan, Prov. Nantou, Hohuachi, 1950 m, 24°13′ N, 121°16′ E, 28.xi.1999, leg. A. Kun, L. Peregovits & L. Ronkay (paratypes, ZSM).
Remarks. The male genitalia of the Taiwanese specimens are larger than in the single known mainland specimen: the valve, vinculum and aedeagus are slightly longer, and the eighth sternite has two short and blunt apical rods.

107. *Eupithecia atuni* Mironov & Galsworthy, 2006

Eupithecia atuni Mironov & Galsworthy, 2006, *Trans. lepid. Soc. Japan* 57 (4): 346, figs 7, 27, 30. Holotype ♂ (ZFMK; examined), [China]: North Yunnan, A-tun-tse [Dêqên], 3000 m.

Description. Fore wing ground colour pale buff; antemedial area mainly pale; transverse lines brown, oblique, sharply angled near costa; medial area brown, forming a transverse band across wing; postmedial slightly curved inward and obtusely angled onto costa; terminal area pale buff in middle and brown near apex, tornus and along terminal margin; whitish subterminal line narrow; discal dot circular, brownish black. Hind wing pale brown; basal, ante- and postmedial lines and terminal area darker; a whitish wavy subterminal line; discal dot small.
Male genitalia. Uncus broad, biapical. Valve elongate with digitate ventral process. Vinculum trapezoid. Papillae broad. Aedeagus broadened anteriorly, longer than valve. Vesica armed with a long, twisted plate, an irregular plate near ductus ejaculatorius base and a broad apical plate, with blunt tip. Sternite A8 peg-like, with two short apical rods.
Female genitalia. Bursa thin-walled, large and almost oblong, with a characteristic small protrusion near colliculum, almost completely covered with small spines except short ductus bursae and close to attachment of ductus seminalis. Latter broadened basally, curved and adpressed to corpus bursae, attached to bursa near base on left side. Colliculum narrow, slightly elongate. Antrum membranous. Tergite A8 trapezoid, with a small

medial hollow on posterior margin. Apophyses short and narrow. Papillae anales rounded and broad.

Range. China (Sichuan, Yunnan).

Phenology. Univoltine: mid-May to late July.

Altitude. Found from about 2500 to 4500 m above sea-level (China: coll. P. Skou and ZFMK).

Similar species. Unmistakable. Similar to *E. daemionata* and *E. kudoi* in the dark median fascia, but has darker hind wings. The male genitalia are distinctive, in particular the digitate ventral process on the valve, and the aedeagus vesica ornamentation. The female genitalia resemble those of the following species, which is externally quite different.

Material. ♂, Nord Yünnan, A-tun-tse [Dêqên], Talsohle *ca.* 3000 m, 30.v.1937, H. Höne (holotype, marked as holotype of unpubl. *E. atuni* Vojnits, ZFMK); 3♂, 13♀, same loc. and alt., 16, 18, 24, 26, 30, 31.v., 1, 6, 10, 30.vi., 4.vii.1937, H. Höne, Schmidt slides nos. Eu 146♀, Eu 146a♂, Mironov slides nos. 665♂, 664♀ (paratypes, marked as paratypes of unpubl. *E. atuni* Vojnits, ZFMK); 6♂, 6♀, same loc. and alt., 24, 28, 30, 31.v., 1, 3, 13.vi.1937, H. Höne, de Laever slides nos. 51♂, 52♀, 122♀, Vojnits slides nos. 10427♀, 10428♂, 10429♂, 10430♀ (all marked as paratypes of unpubl. *E. atuni* Vojnits, TTM); 2♀, same loc., Mittlere Höhe (*ca.* 4000 m), 18, 28.vii.1936, H. Höne (marked as paratypes of unpubl. *E. atuni* Vojnits, ZFMK); 2♀, same loc. and alt., 12.viii., 16.ix.1936, H. Höne, Vojnits slides nos. 10431♀, 11476♀ (marked as paratypes of unpubl. *E. atuni* Vojnits, TTM); 1♀, same loc., Obere Höhe (*ca.* 4500 m), 13.vii.1936, H. Höne (paratype, marked as paratype of unpubl. *E. atuni* Vojnits, ZFMK); 1♀, China, Sichuan, Prov., Batang, 2500 m, 10.vii.2001, Li Houhun leg. (coll. P. Skou); 2♂, China, North Yunnan, Diqing Tibetan Aut. Pref., 5 km SE of Deqin, 3356 m, 16–17.vi.2009, leg. Balázs Benedek (coll. O. Pekarsky); 2♀, same loc. and alt., 19.vi.2009, leg. B. Benedek (coll. O. Pekarsky).

Remarks. The name "*atuni*" was an unpublished MS name of Dr. A. Vojnits, and refers to the type locality of A-tun-tse [Dêqên].

108. *Eupithecia funerea* Inoue, 1988

Eupithecia funerea Inoue, 1988, *Bull. Fac. domest. Sci. Otsuma Wom. Univ.* **24**: 369, figs 29a-b. Holotype ♀ (BMNH; examined), [Taiwan]: Nantou Hsien, Lushan Spa, 1200 m.

Description. A relatively small dark species. Fore wing ground colour dark brown, with ochreous tinge medially; antemedial almost perpendicular; double medial sinuate, evenly curved onto costa behind discal dot; postmedial oblique, broad, blackish brown, evenly curved; terminal area

slightly paler medially, with fine whitish subterminal line and distinct, small, white tornal spot; discal dot relatively large, intensely black. Hind wing paler, brownish, darker along anal margin, with dark postmedial and small discal dot.
Male genitalia. Uncus narrow, biapical. Valve with a short, acutely angled ventral process. Vinculum short and broad, medially hollowed. Papillae short, very broad. Aedeagus slightly longer than valve, vesica armed with three horn-like cornuti, a long, narrow, marginally folded cornutus and a small, irregular cornutus near ductus ejaculatorius base. Sternite A8 small, broadened basally, with two short apical arms.
Female genitalia. Bursa elongate, completely covered with spines. Ductus bursae short, spineless, with heavily sclerotized prominence near colliculum on left side. Ductus seminalis slightly broadened at base, attached to central part of corpus bursae on left side. Colliculum short and broad. Antrum membranous. Tergite A8 quadrate. Apophyses short and narrow, possibly asymmetric in length. Papillae anales short and rounded.
Range. China (Taiwan).
Phenology. Univoltine: mid-March to late March.
Altitude. From about 1200 m to 1850 m above sea-level (Taiwan: Inoue 1988; coll. TTM).
Similar species. *E. funerea* is superficially similar to *E. atrisignis* but has smaller discal dots on the fore wings. The male genitalia are distinct in the ventral ornamentation of the valve, and the multiple cornuti in the aedeagus vesica. The female genitalia are most similar to those of the preceding species.
Material. ♀, Formosa [Taiwan], Lushan Spa, Nan Tou, 1200 m, 19–21. iii.1972, leg. K. Nakatomi, BM Geom. slide no. 16556♀ (holotype, coll. H. Inoue in BMNH); 1♂, Taiwan, Prov. Kao-Hsiung, 15 km NE of Taoyuan, 1850 m, 12.iii.1996, leg. Gy. Fábián & L. Németh (TTM).
Remarks. Regrettably, the aedeagus of what is so far the only known male was lost after the drawing had been made by the first author. For the moment, the drawing is therefore the only record.

109. *Eupithecia fulvidorsata* Mironov & Galsworthy, 2006

Eupithecia fulvidorsata Mironov & Galsworthy, 2006, *Trans. lepid. Soc. Japan* 57 (4): 343, figs 6, 21. Holotype ♂ (ZFMK; examined), C. China: prov. Shaanxi, Foping Nature Reserve, Panda area, 33°45′ N 107°48′ E, 1600 m.

Description. A small species. Fore wing ground colour grey; transverse lines invisible; costal margin almost black, in outer half; terminal area dark grey with pale wavy subterminal line, and a series of obscure ochreous spots between veins; discal dot relatively large, ovate, black. Hind wing ground colour paler; transverse lines inconspicuous; discal dot small and pale. Head, notum and abdomen covered with rusty ochreous scales.

Male genitalia. Uncus biapical. Valve with two prominent, blunt ventral processes. Vinculum relatively long, trapezoid. Papillae parallel sided. Aedeagus about length of valve. Vesica armed with three short horn-like cornuti, and a twisted V-shaped plate with two pointed ends, connecting with a small irregular plate near ductus ejaculatorius base. Sternite A8 heavily sclerotized, short, triangular, basally broadened, with two short diverging apical rods.

Female genitalia. Bursa globular, completely covered with spines. Ductus bursae heavily sclerotized, sinuate, broad at base but tapered to colliculum, with sclerotized spinose prominence at base on right side. Ductus seminalis broadened at base, attached to base of ductus bursae on left side. Colliculum short and broad. Antrum short and broad, membranous. Tergite A8 almost quadrate. Apophyses narrow and tapered. Papillae anales elongate and narrow, tapered.

Range. China (Shaanxi); Nepal.

Phenology. Univoltine: mid-April to late May.

Altitude. From about 1600 m in the Qinling Mts. (China: coll. ZFMK) up to about 2700 m in the Himalayas (Nepal: coll. TTM).

Similar species. Darker and more uniform than most other species, and distinguished by the rusty ochreous head, notum and abdomen. The ventral processes of the male valve are unique. The shape of the eighth sternite is rather similar to that of the Himalayan *E. concinna* Vojnits, 1983, but this species is quite different both externally and in the rest of the male genitalia structure.

Material. ♂, C. China, prov. Shaanxi, Foping Natural Reserve, Panda area, 33°45′ N, 107°48′ E, 1600 m, 20.iv-11.v.1999, leg. V. Siniaev & A. Plutenko (holotype, ZFMK); 1♂, same data, Mironov slide no. 652♂ (paratype, ZFMK); 2♀, Nepal, Ganesh Himal, 2700 m, Nesukharka, 12 km S Somdang, 85°11′ E, 28°08′ N, 20–21.v.1995, leg. G.Y.Fábián & L.Ronkay, Galsworthy slides nos. GL25 and 57♀ (TTM).

Remarks. The Nepalese females are referred to this species somewhat tentatively, in view of the geographical separation from the males, but they are an excellent match, including the unusual series of ochreous spots in the terminal area.

110. *Eupithecia okadai* Inoue, 1958

Eupithecia okadai Inoue, 1958, *Tinea*. 4 (2): 251, pl. 34: 21, fig. 8. Holotype ♂ (BMNH; examined), [Japan]: Ehime Pref., Mt. Takanawa.

Description. Ground colour of all wings pale ash grey; fore wing with five blackish brown costal spots; discal dot very large, rounded; terminal margin darker. Hind wing with prominent, but smaller discal dot and slightly darkened anal margin.
Male genitalia. Uncus stout, biapical. Valve with a heavily sclerotized, blunt ventral process near base of left valve and two smaller processes on right valve. Vinculum broad, trapezoid, with shallow medial hollow. Papillae parallel sided. Aedeagus shorter than valve length. Vesica granulate, with an elongate apical plate, a small hook-like medial cornutus, and a small flat H-shaped plate connecting with a smaller irregular plate near ductus ejaculatorius base. Sternite A8 almost X-shaped, with apical lobes longest.
Female genitalia. Bursa globular, heavily spinose except in posterior part. Ductus bursae not distinguishable. Ductus seminalis slightly broadened at base, attached near base of colliculum. Colliculum short and narrow. Antrum membranous. Tergite A8 transverse. Apophyses narrow. Papillae anales short and rounded.
Range. China (Shaanxi); Korea, Japan (Honshu, Sado, Shikoku, Kyushu).
Phenology. Univoltine: late April and May.
Biology. Larva found on *Symplocos chinensis* var. *leucocarpa* (Symplocaceae) (Japan: Inoue 1979).
Altitude. From about 750 to 1600 m above sea-level (Japan: Inoue 1979; China: coll. ZFMK).
Similar species. Externally closest to *E. conterminata*, but larger, with more distinct costal spots, larger, rounded discal dots and paler terminal areas. In the male genitalia, the heavily modified and asymmetric ventral area of the valves is unique. The female genitalia lack unique characters, but the pattern of spines is unlike those of any species of similar external appearance.
Material. ♂, [Japan], Shikoku, Mt. Takanawa, 10.v.1953, M. Okada (holotype, coll. H. Inoue in BMNH); 1♂, C. China, prov. Shaanxi, Foping Natural Reserve, Panda area, 1600 m, 15.-20.iv.1999, 33°45′ N, 107°48′ E, leg. V. Siniaev & A. Plutenko, Mironov slide no. 659♂ (ZFMK).
Remarks. Not previously recorded from China.

111. *Eupithecia delicata* Mironov & Galsworthy, 2004

Eupithecia delicata Mironov & Galsworthy, 2004, *Trans. lepid. Soc. Japan* 55 (3): 239, figs 10, 27, 30. Holotype ♀ (ZSM; examined), China: Sichuan, Daxue Shan, Gongga Shan, NW Moxi, 29°41′ E, 101°58′ N, 2600–3200 m.

Description. Fore wing ground colour light greyish brown; common stalk of Cu, M_2 and CuA_2 covered with black scales; transverse lines narrow, brown; basal line angled; basal area pale brown; postbasal and antemedial lines oblique, straight, sharply angled in cell, meeting costa at oblique angle; postmedial sinuate; medial area soft light brown; terminal area wide, darker, rusty brown; whitish subterminal wavy line forming a tornal spot; discal dot prominent, black. Hind wing paler, brownish grey; transverse lines marked along dorsum especially near tornus; discal spot small. Abdomen light greyish brown with reddish tinge, with two dark lateral bands and a row of small dark brown dorsal dots.
Male genitalia. Uncus very narrow, biapical. Valve with a blunt ventral process. Vinculum elongate, with shallow medial hollow. Papillae long, clavate. Aedeagus almost equal to length of valve. Vesica with two slim horn-like cornuti and an elongate, H-shaped cornutus at ductus ejaculatorius base. Sternite A8 peg-like, very small and weakly sclerotized. Tergite A8 elongate, bell-shaped, with a rigid, hook-like apical projection.
Female genitalia. Bursa ovate, with a partial covering of sparse spines. Ductus bursae large, membranous, with numerous striations and two oblique, heavily sclerotized wrinkled plates. Ductus seminalis very broad, attached to ductus bursae from right side. Colliculum plate-like. Antrum sclerotized, covered by numerous pores. Tergite A8 square. Apophyses relatively short and narrow. Papillae anales small, rounded.
Range. China (Sichuan).
Phenology. Univoltine: late April to mid-May.
Altitude. Found at about 2600–3200 m above sea-level in Daxue Shan (China: coll. ZSM).
Similar species. This and the next two species form a closely related subgroup. The present species is well separated from allied Asian species by the distinctive pattern. In the male genitalia, the shape of the ventral processes on the valve is distinctive, with a slight resemblance only to *E. funerea*, and the modified eighth tergite separates it from all but the two following species. The female genitalia are unmistakable.
Material. ♀, China, Sichuan, Daxue Shan, Gongga Shan, NW Moxi, 29°41′ E, 101°58′ N, 2600–3200 m, 23.iv-15.v.2001, leg. V. Siniaev & A. Plutenko, *ex.* coll. Dr. A. Schintlmeister (holotype, ZSM); 3♂, 7♀, same data, Mironov slides nos. 513♂, 514♀ (paratypes, ZSM); 2♀, same data (paratypes, ZISP).

112. *Eupithecia camilla* Mironov & Galsworthy, 2004

Eupithecia camilla Mironov & Galsworthy, 2004, *Trans. lepid. Soc. Japan* 55 (3): 240, figs 11, 28, 31. Holotype ♂ (ZFMK; examined), China: Shaanxi, Foping Natural Reserve, Panda area, 33°45′ N, 107°48′ E, 1600 m.

Description. Fore wing ground colour brownish; basal line sharply angled, dark brown; basal area brown; postbasal and antemedial lines oblique, sharply angled in cell to meet costa obliquely; postmedial slightly sinuate; medial area narrow, darker brown, broadened at costa and dorsum; terminal area broad, brown, with a paler ochreous blotch between M_3 and CuA_2, and two elongate black dashes on CuA_2 and between CuA_2 and A; whitish wavy line distinct, forming a crescent-shaped whitish tornal spot; discal dot black, ovate-acuminate. Hind wing pale grey, transverse lines dark brown, conspicuous along dorsum only; discal spot small. Abdomen brownish grey with black lateral stripes; third segment dark brown; last segment pale grey.

Male genitalia. Uncus biapical. Valve with two ventral processes, the proximal one pointed. Vinculum elongate. Papillae long, slightly clavate. Aedeagus stout, equal to length of valve. Vesica armed with two large horn-like cornuti and a wide, slightly folded, V-shaped cornutus at ductus ejaculatorius base. Sternite A8 peg-like, very small but heavily sclerotized. Tergite A8 elongate, with almost parallel lateral margins, and a heavily sclerotized, hook-like apical projection.

Female genitalia. Bursa oblong, heavily sclerotized, with small membranous diverticulum at base, covered with spines at base, the marginal spines larger than the others. Ductus bursae heavily sclerotized, with a V-shaped chain of small, short spines and with two heavily sclerotized oblique processes directed towards the right side, the posterior one with a row of spines. Ductus seminalis very broad, attached to ductus bursae from right side. Colliculum plate-like. Antrum sclerotized with numerous pores. Tergite A8 short and broad. Apophyses relatively short and narrow, tapered. Papillae anales rounded.

Range. China (Shaanxi).

Phenology. Bivoltine: April and September.

Altitude. Found from about 1400–1600 m in Shaanxi (China: coll. ZFMK).

Similar species. Distinguished from other species with dark central fascia (*E. lanceata*, *E. daemionata* and *E. takao* Inoue, 1955) by the narrower central fascia and presence of two black dashes near fore wing tornus. In the male genitalia the ventral process is longer than in *E. delicata*, and slightly resembles that in *E. dolichoptera*, from which it is distinguished by the modified eighth tergite. The female genitalia are very distinct, with a

general resemblance to the other two species in the subgroup, but much difference in detailed ornamentation.
Material. ♂, C. China, Shaanxi, Foping Natural Reserve, Panda area, 1600 m, 6.-11.iv.1999, 33°45′ N, 107°48′ E, leg. V. Siniaev & A. Plutenko, Mironov slide no. 515♂ (holotype, ZFMK); 1♂, C. China, Shaanxi, Mt. Tai bei shan, 1400 m, Hao Djenzy village, 33°48′ N, 107°43′ E, ix.1998, leg. loc. coll., Mironov slide no. 516♂ (paratype, ZFMK); 1♀, China, Shaanxi, Taibai Shan, 33°53′ N, 107°49′ E, 1500 m, iv.2000, leg. V. Siniaev & A. Plutenko, Mironov slide no. 517♀ (paratype, ZSM).

113. *Eupithecia julia* Mironov & Galsworthy, 2004

Eupithecia julia Mironov & Galsworthy, 2004, *Trans. lepid. Soc. Japan* 55 (3): 241, figs 12, 29, 32. Holotype ♀ (ZSM; examined), China: Sichuan, Daxue Shan, Gongga Shan, NW Moxi, 29°41′ E, 101°58′ N, 2600–3200 m.

Description. Fore wing ground colour whitish grey, all transverse lines inconspicuous; basal area brown with rusty brown blotch near dorsal margin; antemedial oblique, sharply angled in cell; postmedial slightly curved behind discal spot; medial area pale grey, but brown along costa; terminal area darker, rusty brown with a large pale grey medial blotch, a semicircular whitish tornal spot, and an inconspicuous whitish wavy line; discal spot small, but distinct. Hind wing paler, transverse lines marked along dorsum only; discal spot small.
Male genitalia. Uncus biapical. Valve asymmetrical, ventral processes on left valve parallel, but diverging widely on right valve. Vinculum elongate. Papillae elongate and narrow, broadened apically. Aedeagus almost equal to length of valve. Vesica armed with three slim horn-like cornuti and a smaller irregular plate near ductus ejaculatorius base. Sternite A8 very short, peg-like with broad apex. Tergite A8 heavily sclerotized and poriferous, with sinuously tapering lateral margins, and a rigid, hook-like and ridged apical projection.
Female genitalia. Bursa oblong, heavily sclerotized except basal $^1/_3$, covered centrally and along right side with spines and with a patch of smaller spines at base. Ductus bursae heavily sclerotized, with oblique, heavily sclerotized process towards right side and two patches of spines. Ductus seminalis very broad, attached to ductus bursae from right side. Colliculum plate-like, elongate and broad. Antrum membranous. Tergite A8 transverse. Apophyses short and narrow. Papillae anales small, rounded.
Range. China (Sichuan).
Phenology. Univoltine: late April to mid-May.

Altitude. Found from about 2600–3200 m in Daxue Shan (China: coll. ZSM).
Similar species. Externally a little similar to *E. delicata*, but with paler ground colour, and a considerably larger whitish tornal spot on the fore wing. The male and female genitalia are most similar to those of *E. camilla*. In the male, the valve is asymmetrical with much longer ventral processes, and the cornuti on the vesica are slimmer. In the female, the shape of the bursa copulatrix is different, and there is a membranous diverticulum at its base.
Material. ♀, China, Sichuan, Daxue Shan, Gongga Shan, NW Moxi, 29°41′ E, 101°58′ N, 2600–3200 m, 23.iv-15.v.2001, leg. V. Siniaev & A. Plutenko, *ex.* coll. Dr. A. Schintlmeister (holotype, ZSM); 1♂, 3♀, same data, Mironov slides nos. 518♂, 519♂, 520♀ (paratypes, ZSM); 1♂, 1♀, same data (paratypes, ZISP).

114. *Eupithecia fortis* Mironov & Galsworthy, 2004

Eupithecia fortis Mironov & Galsworthy, 2004, *Trans. lepid. Soc. Japan* 55 (1): 55, figs 12, 31. Holotype ♀ (ZFMK; examined), [China]: Yunnan, Li-kiang [Lijiang].

Description. Fore wing ground colour pale yellowish brown; antemedial brown, acutely angled close to large, intense black, discal dot; postmedial touching edge of discal dot, and almost straight to inner margin; central costal area darker brown, forming a triangular mark on costa; terminal area with dark brown apical, subapical and tornal patches, and all veins picked out in dark brown. Hind wing ground colour grey-white, discal dot small; anal margin suffused with dark brown, terminal area with paler brown; a pale postmedial fascia, and a pale sinuous subterminal line.
Female genitalia. Bursa pyriform, heavily sclerotized, covered with long spines at anterior end, where there is a membranous diverticulum. Ductus bursae broad and elongate, with longitudinal striations. Ductus seminalis narrow, attached to central part of corpus bursae from left. Colliculum broad. Antrum membranous. Tergite A8 large. Apophyses narrow, but fairly long. Papillae anales broad, rounded.
Male unknown.
Range. China (Yunnan).
Similar species. Externally similar to the Japanese *E. takao* Inoue, 1955. The female genitalia are very distinctive. The membranous diverticulum at base of the bursa is found in very few species, none of which resemble the present one.
Material. ♀, China, Provinz Nord-Yuennan, Li-kiang [Lijiang], 17.ix.1935, H. Höne, de Laever slide no. 77♀ (holotype, ZFMK).

Remarks. This species is still known only from the holotype female. External appearance suggests that it should belong to this species group, but in the absence of the male, this should be regarded as tentative.

The *lariciata* species group

This species group was introduced by Bolte (1990). While there is considerably variation in external appearance, the genitalia within the group are very uniform. In the male the uncus is normally swollen at base, and often bell-shaped; the valve has a single small ventral process, and the eighth sternite is entire, usually with convex sides (flask-shaped) and two small apical arms. The aedeagus vesica frequently has a group of spines at the apex. In the female, the bursa copulatrix is globular and spined throughout. The ductus bursae and antrum have various modifications, which are important in specific diagnosis. The ductus seminalis is attached to the ductus bursae. The group is species-rich in East Asia, and is represented in China by the following fourteen species.

115. *Eupithecia lariciata* Freyer, 1842

Larentia lariciata Freyer, 1842, *Neuere Beitr. Schmettkde* **4** (61): 135, pl. 366. Syntype(s). [Germany]: Wellenburg; Leitershausen.

Geometra residuata Hübner, [1817], *Samml. Eur. Schmett.* **5** Geometrae (1), pl. 91, fig. 467 ([Europe]). Syntype(s) lost (synonymised in Staudinger & Rebel, 1901).

Eupithecia luteata Packard, 1867, *Proc. Boston Soc. Nat. Hist.* **11**: 45. Lectotype ♂ (MCZ), [Canada]: [Labrador], Cariboo Island, Straits of Belle Isle (synonymised in Bolte, 1990).

Tephrocystis [sic] *luteata bifasciata* Dyar, 1904, *Proc. U.S. Natn. Mus.* **27** (1376): 891. Holotype ♀ (USNM), [Canada]: British Columbia, Kaslo (synonymised in Bolte, 1990).

Eupithecia catskillata Pearsall, 1908, *Ent. News* **19** (5): 192. Lectotype ♂ (AMNH), [U.S.A.]: [New York], Catskill Mts., Big Indian Valley (synonymised to *E. luteata* Packard, 1867 in McDunnough, 1929).

Eupithecia fasciata Taylor, 1910, *Can. Ent.* **42** (3): 79. Holotype ♀ (USNM), [Canada]: Ontario, Ottawa (synonymised to *E. luteata* Packard, 1867 in McDunnough, 1929).

[*Eupithecia*] *lariciata* f. *uniformis* Dietze, 1913, *Biologie der Eupithecien* **2**: 155. Syntypes, at least 1♀ (coll. K. Dietze in MNHU; examined), [Turkey]: Taunus (type locality not published).

[*Eupithecia*] *lariciata* f. *luxuriosa* Dietze, 1913, *Biologie der Eupithecien* **2**: 155. Syntype(s) (coll. K. Dietze in MNHU, not traced), [Europe ?: type locality not stated).

[*Eupithecia*] *lariciata* f. *strigata* Dietze, 1913, *Biologie der Eupithecien* **2**: 155; *ibidem* (1910), 1, pl. 78, fig. 734. Syntypes, at least 1♂ (coll. K. Dietze in MNHU; examined), [Germany]: Frankfurt a.M.

[*Eupithecia*] *lariciata* f. *basifasciata* Dietze, 1913, *Biologie der Eupithecien* **2**: 155; *ibidem* (1910), 1, pl. 78, fig. 736. Syntypes, at least 1♂ (coll. K. Dietze in MNHU; examined), [Germany]: Sachsen.

[*Eupithecia*] *lariciata* f. *mediopallens* Dietze, 1913, *Biologie der Eupithecien* **2**: 155; *ibidem* (1910), 1, pl. 78, fig. 733. Syntypes, at least 1♂ (coll. K. Dietze in MNHU; examined), [Switzerland]: Wallis; [Germany]: Sachsen.

Eupithecia laboriosa Vojnits, 1977, *Acta zool. Acad. Sci. hung.* **23** (3–4): 479, fig. 15. Holotype ♀ (TTM; see remarks below), Mongolia: Central Aimak, SE of Somon Bajanzogt, 1600 m (synonymised in Mironov, 2003).

Eupithecia infecta Vojnits, 1981, *Acta zool. Acad. Sci. hung.* 27 (3–4): 417, figs 27–29. Holotype ♂ (ZFMK; examined), [China]: Sued-Shensi, Tapaishan in Tsinling, 3000 m. **New junior synonym.**

Misidentification: *Eupithecia verecunda* Vojnits, 1980 (1♀ paratype in ZFMK); *Eva petulans* Vojnits, 1981 (1♀ paratype in ZFMK).
Unavailable names (incorrect subsequent spellings): ?*residuaria*: Boisduval (1840); *lariciaria*: Herrich-Schäffer (1848).

Description. Fore wing light grey; transverse lines distinct, dark grey, slightly sinuate and sharply angled near costa; medial sharply angled beyond discal dot or touching it, with delicate paler double stripe beyond medial and with pale ochreous cloud under discal dot in some specimens; postmedial slightly curved and right-angled near costa, usually with dark dashes on the veins; pale wavy subterminal line sometimes present; discal dot relatively small, intensely black. Hind wing ground colour slightly paler, transverse lines weakly marked; terminal area usually darker; discal dot faint. Abdomen light grey, with black lateral stripes or spots, usually beginning with fourth segment.
Male genitalia. Uncus biapical, very broad in lateral view. Valve long and broad with a small, blunt, sclerotized, ventral process. Vinculum rather short, trapezoid, with shallow medial hollow. Papillae elongate and narrow, slightly tapered. Aedeagus slim, considerably shorter than valve length. Vesica armed with a patch of spinules near apex, two horn-like cornuti, one hooked at base, an elongate, U-shaped cornutus and a small, irregular cornutus at ductus ejaculatorius base. Sternite A8 large, peg-like with bowed lateral margins and short apical horns.
Female genitalia. Bursa globular, completely covered with dense, uniform spines. Ductus bursae long and narrow, spineless, with longitudinal striations. Ductus seminalis broad at base; attached to middle of ductus bursae from right side. Colliculum funnel-shaped, present at most as a band of weak sclerotization. Antrum membranous. Tergite A8 trapezoid with rounded posterior corners. Apophyses short and narrow. Papillae anales rounded.
Range. China (Inner Mongolia, Shanxi, Shaanxi, Hebei); Europe, Urals, Asian Russia (W Siberia, Altai Mts., Tuva, Yakutia, Burjatia, Transbaikalia, Dahuria, Priamurje, Primorje, Sakhalin, Kamchatka peninsula), Mongolia, Korea, Japan (Hokkaido, Honshu); N. America (Canada and USA).
Phenology. Bivoltine (?): early May to mid-September.
Biology. Larva found on the needles of *Larix decidua* (= *europaea*), *L. sibirica*, *L. kaempferi* (= *leptolepis*) and occasionally on *Picea abies* (Pinaceae) and *Juniperus* (?) (Cupressaceae) in Europe. In North America

on *Picea glauca*, *P. mariana*, *Abies balsamea* and *Larix laricina*, occasionally on *Picea rubens*, *Tsuga canadensis*, *Pinus strobus*, *Pseudotsuga menziesii* (Pinaceae) and *Thuja occidentalis* (Cupressaceae) (Canada: McGuffin 1958; Bolte 1990).
Altitude. From sea level to about 3000 m in Tapaishan (China: Vojnits 1981b).
Similar species. Easily confused with *E. subfuscata* but usually paler with more distinct pale wavy lines on all wings. The presence of a white thoracic spot in *E. lariciata* is diagnostic. The male and female genitalia of the two species are quite distinct. Best distinguished from similarly patterned Asian species on the genitalia: in the male aedeagus vesica, the strong basally hooked cornutus and small patch of spines at the apex of the aedeagus are diagnostic. In the female, the colliculum is almost lacking: this characteristic is shared with *E. pediba* and *E. primitiva*: the former has a much longer ductus, and the latter a sclerotized tapering antrum.
Material. ♀, Mongolia, Central aimak, SO von Somon Bajanzogt, 1600 m, Exp. Dr. Z. Kaszab, 1964, Nr. 267, 4.vii.1964, Vojnits slide no. 10559♀ (holotype of *E. laboriosa*, TTM); ♂, China, Sued-Shensi [Shaanxi], Tapaishan [Taibaishan] im Tsinling [Qinling], L, 20.vi.1935, H. Höne, Vojnits slide no. 12949♂ (holotype of *E. infecta*, ZFMK); 1♀, same data, Vojnits slide no. 12384♀ (ZFMK); 1♀, same loc., 27.vi.1935, H. Höne, Vojnits slide no. 11457♀ (paratype of *E. verecunda*, ZFMK); 1♀, same loc., 2.vii.1935, H. Höne, Vojnits slide no. 13159♀ (paratype of *Eva petulans*, ZFMK; gen. slide wrongly labelled, belonging to *Eva flexa* Vojnits, 1981); 1♀, Tientsin, N. China, July 1922, F.M. Thomson (BMNH); 4♀, Inner Mongolia, Oragxan, 11.viii.1983, slides nos. ZL100, 138, 142♀; 1♀, same loc., 31.v.1984, 2♂, 1♀, same loc., 28.vi.1985; 2♀, Manguei, 680 m, 29.vi.1985; 1♂, 2♀, Inner Mongolia, Jiwen, 4.vii.1985, slide nos. ZL110♂, ZL144♀, coll Xue Dayong; 1♂, 2♀, Shanxi, Pangquangou, 26.vi.1988, slide no. ZL295♀ (IZCAS).
Remarks. Vojnits slide no. 12949♂, which belongs to the holotype of *E. infecta*, was incorrectly marked by him as a paratype. We have examined the female type specimen of *E. laboriosa* Vojnits, 1977, and believe that it is a specimen of *E. lariciata*. However the genitalia slide linked to it (Vojnits slide no. 10559♀, TTM) contains female genitalia which appear to belong to a species of *Pasiphila*, presumably as a result of an error in the dissection process.

116. *Eupithecia emanata* Dietze, 1906

Eup[*ithecia*]? *lariciata* f. *emanata* Dietze, 1906, *Dt. ent. Z., Iris*, **19** (2): pl. 2, fig. 11; *ibidem* (1908) **21** (3): 192. Syntypes 1♂, 1♀, (coll. K. Dietze in MNHU; examined), [Russia]: Radde, Amur; Kasakewitsch, Ussuri.

Eupithecia sordidata Wileman, 1911, *Trans. ent. Soc. Lond.* **1911** (2): 329. Lectotype ♂ (BMNH; examined), Japan: Oshima, Tōbetsu (synonymised in Prout, 1915).
Eupithecia subicterata Prout, 1930, *Novit. zool.* **35** (4): 306. Holotype ♀ (BMNH; examined), [Japan]: Takao-San (synonymised in Inoue, 1977).
Eupithecia studiosa Vojnits, 1979, *Acta zool. Acad. Sci. hung.* **25** (3–4): 434, figs 11, 12. Holotype ♂ (ZFMK; examined), [China]: Sued-Shensi, Tapaishan [Taibaishan] im Tsinling, 3000 m. **New junior synonym.**

Misidentifications: *Eupithecia diffisata* Dietze: Vojnits, 1978; *Eupithecia daemionata* Dietze: Inoue, 1980, fig. 56A.

Description. Fore wing ground colour brownish grey; medial part of medial and terminal areas usually paler, ochreous brown; transverse lines distinct, blackish; basal, postbasal and antemedial lines oblique, twice sharply angled; double medial less distinct, sinuate, sharply angled onto costa behind discal dot; postmedial almost straight, then right angled onto costa; terminal area broad, unicolorous or darker near apex and tornus; light subterminal line usually distinct, wavy, indented near apex; discal dot relatively large. Hind wing paler, anal margin shaded with brown; terminal area usually darker; discal dot small but distinct.
Male genitalia. Uncus stout, tapered, biapical. Valve broad, with triangular ventral process. Vinculum short. Papillae long and narrow. Aedeagus slim, shorter than valve length. Vesica with a dense apical patch of stout spines, a narrow, Y-shaped cornutus, and an elongate, U-shaped plate close to ductus ejaculatorius base. Sternite A8 flask-shaped, with two small apical horns.
Female genitalia. Bursa globular, completely covered with dense spines. Ductus bursae short and narrow with a small patch of spines. Ductus seminalis evenly broadened to base, attached to ductus bursae from left side. Colliculum short and narrow. Antrum triangular and sclerotized, with edges more heavily so. Tergite A8 almost square with convex posterior margin. Apophyses short and narrow. Papillae anales rounded.
Range. China (Tibet, Zhejiang, Guangdong, Shanxi, Shaanxi, Beijing); Asian Russia (southern Priamurje, Primorje), Korea, Japan (Hokkaido, Honshu, Sado, Shikoku, Kyushu).
Phenology. Bivoltine: late March to early June; late July to mid-September.
Altitude. From sea level up to about 3000 m in Tapai-shan (China: Vojnits 1979b).
Similar species. Very similar externally to Himalayan *E. conjunctiva* but in this latter the postmedial usually forms a distinct, pointed angle near costa on the fore wing. The presence of a substantial patch of relatively large spine-like cornuti at the vesica apex in the male genitalia distinguishes

E. emanata from other representatives of this species group. The female genitalia are similar to others in the group, but the sclerotized antrum, with the edges even more heavily sclerotized, is distinctive.

Material. 1♀, [Russia], Radde, m. Amur, 15.juli.03 Korb, ? *lariciata, emanata* ♀ Type., Radde, Amur, 15.7.03 (von Korb) K. Dietze, Zool. Mus. Berlin, Geom. slide 1950–585A, Typus (syntype, coll. K. Dietze in MNHU); 1♂, [Russia], Ussuri, Kasakewitsch 1907 Korb., *emanata* ♂, Ussuri, K. Dietze, Original Type, Fühler scheinbar lockerer bewimpert cilien, länger. Type. *emanata* ♂, Zool. Mus. Berlin, Geom. slide 1950–584A, Typus (syntype, coll. K. Dietze in MNHU); ♂, [Japan], Tobetsu prov., Ishikari, Hokkaido plains, 6.vi.1902, A.E. Wileman, 695, Wileman Coll. B.M. 1929–261, Tobetsu Toshima, vi/6/902♂, *Eupithecia sordidata* sp. n. Type, BM Geom. slide no. 1950–587 (lectotype of *E. sordidata*, BMNH); ♀, [Japan], Takao-San, W. of Tokio, 2 May, 26 (M. Aigner), *Eupithecia subicterata* Prout ♀ type, BM Geom. slide no. 6114 (holotype of *E. subicterata*, BMNH); ♀, [Korea], Prov. Ryang-gang, plateau Chann-Pay, Sam-zi-yan, 1600 m, 28.viii.1971, No. 219, leg. S. Horvatovich et J. Papp, Vojnits slide no. 10827♀ (as *E. diffisata*, TTM); ♂, Sued-Shensi [Shaanxi], Tapaishan [Taibaishan] im Tsinling [Qinling], *ca.* 3000 m, 25.viii.1936, H. Höne, Vojnits slide no. 12269♂ (holotype of *E. studiosa*, ZFMK); 2♂, same loc. and alt., 21, 25. viii.1936, H. Höne, Vojnits slides nos. 12281♂, 12272♂, 12275♂ (paratypes of *E. studiosa*, ZFMK); 2♀, same loc. and alt., 22, 25.viii.1936, H. Höne, Vojnits slides nos. 12268♀, 12277♀ (paratypes of *E. studiosa*, TTM); 1♀, same loc. and alt., 7.ix.1936, H. Höne, Vojnits slide no. 12975♀ (TTM); 1♂, 1♀, same loc., *ca.* 1700 m, 24, 25.viii.1936, H. Höne, Vojnits slides nos. 12282♂, 12280♀ (paratypes of *E. studiosa*, slide no. 12282♂ absent in ZFMK); 1♂, same loc., 19.viii.[19]36, (H. Höne), Vojnits slide no. 12278♂ (paratype of *E. studiosa*, TTM); 1♀, same loc., 30.viii.1935, H. Höne, Vojnits slide no. 12279♀ (paratype of *E. studiosa*, ZFMK); 1♀, same loc. and alt., 20.viii.1936, H. Höne, Vojnits slide no. 11570♀ (gen. slide of *E. lasciva* erroneously linked to specimen, TTM); 1♀, same loc., 28.viii.1935, H. Höne, Vojnits slide no. 12274♀ (paratype of *E. studiosa*, TTM); 1♀, same loc., 31.viii.1935, H. Höne, Vojnits slide no. 11571♀ (gen. slide of *E. egena* erroneously linked to specimen, TTM); 1♂, same loc., 31.viii.1935, H. Höne, Vojnits slide no. 12356♂ (ZFMK); 1♂, Pz. Chekiang [Zhejiang], West-Tien-Mu-Shan (1600 m), 17.iv.1932, H. Höne, de Laever slide no. 82♂ (ZFMK); 1♂, same loc., (400 m), Mitte April 1936, H. Höne, de Laever slide no. 385♂ (ZFMK); 1♀, China, [Shaanxi] Shangxi prov., Qiling Mts., S. from Baoji, 1600 m, N 34°13,932′, E 106° 58,080′, 31.v.-01.vi.2009, leg. Floriani & Saldaitis, Pekarsky slide no. OP 0489♀ (coll. O. Pekarsky); 1♀, Shanxi, Wutaishan, 1992, slide no. ZL354♀;

1♀, Tibet, Bomi, 2300 m, 22.viii.1983, Han Yinheng, slide no. ZL488♀; 1♀, Beijing, Mentougou, Xiaolongmen, 1100 m, 22.viii.2002, Han Hongxiang, slide no. ZL664;1♀, Beijing, Mentougou, Liyuanling 1100 m, 20–21.viii.2002, Han Hongxiang, IOZ(E)1847042, slide no. ZL696♀ (IZCAS); 1♂, Guangdong, Shaoguan, Nanling, 800–1400 m, 5–11.ix.2005, M. Owada, slide no. ACG NSMT19 (NSMTO).
Remarks. Not previously recorded from China.

117. *Eupithecia honesta* Mironov & Galsworthy, 2004

Eupithecia honesta Mironov & Galsworthy, 2004, *Trans. lepid. Soc. Japan* 55 (1): 41, figs 2, 14, 17. Holotype ♂ (BMNH; examined), [China]: eastern frontier of Tibet.

Misidentification: *Eupithecia captiosa* Vojnits, 1984 (1♀ paratype in ZFMK).

Description. Fore wing ground colour pale orange brown; basal area mid-brown; basal and antemedial when visible consisting of pale oblique double lines; medial dark, touching discal dot; lower edge of cell dark brown; discal dot square, with a square orange brown spot distally and below it; postmedial brown, sharply angled at vein R_5, bordered distally by a pale line; terminal area darker brown, crossed by a wavy white subterminal line. Hind wing ground colour white, discal dot minute; anal margin with brown mottling; terminal area brown, crossed by a pale zigzag line.
Male genitalia. Uncus very broad, biapical. Valve broad with a finger-like, sclerotized ventral process. Vinculum broad, with shallow medial hollow. Papillae elongate. Aedeagus vesica armed with a horn-like cornutus, an elongate, U-shaped plate, and a small, irregular plate near ductus ejaculatorius base. Sternite A8 with curved edges tapering to two short apical arms.
Female genitalia. Bursa globular, covered closely with spines. Ductus bursae elongate, with a short sclerotized diverticulum near base. Ductus seminalis broadened at base, attached to central part of ductus bursae. Colliculum short and narrow. Tergite A8 trapezoid. Apophyses narrow, tapered. Papillae anales short and broad.
Range. China (Sichuan,Yunnan).
Phenology. Univoltine (?): late June to early August.
Altitude. Found at about 3000 m above sea-level (China: coll. ZFMK, IZCAS).
Similar species. Similar to *E. lariciata*, but distinguished by larger discal dots, and orange spots on fore wings and whitish hind wings. The male genitalia differ from those of *E. lariciata* by a more elongate valve with

a larger ventral process and a larger horn-like cornutus on the vesica. The female genitalia of this species are easily distinguished from those of *E. lariciata* by the heavily sclerotized diverticulum on the ductus bursae and the narrow colliculum.

Material. ♂, [Sichuan], eastern frontier of Tibet, native collectors of P. Dejean, 1905, Oberthür coll. 1927–3, BM Geom. slide no. 1950–589 (holotype, BMNH); 1♂, China, Provinz Nord-Yuennan, Li-kiang [Lijiang], 1.vii.1935, H. Höne, de Laever slide no. 270♂ (paratype, ZFMK); 2♀, same loc., 20, 30.vii.1935, H. Höne, respectively de Laever slides nos. 245♀, 264♀ (paratypes, ZFMK); 1♀, Nord-Yünnan, A-tun-tse [Dêqên], Talsohle *ca.* 3000 m, 21.vi.1937, H. Höne, Vojnits slide no. 14189♀ (paratype of *E. captiosa*, ZFMK); 1♀, Yunnan, Zhongdian, 3150 m, 6.viii.1981, Liao Subai, Galsworthy slide no. ZL135♀ (paratype, IZCAS).

Remarks. The female paratype from Dêqên referred to above was published by Vojnits (1984a) as the female of *E. captiosa* Vojnits. It is not however the same species as the male holotype of *E. captiosa*, which is conspecific with *E. robiginascens* Prout, 1926.

118. *Eupithecia pediba* Vojnits, 1981

Eupithecia pediba Vojnits, 1981, *Acta zool. Acad. Sci. hung.* 27 (3–4): 416, fig. 19. Holotype ♀ (ZFMK; examined), [China]: Tibet [Sichuan], Batang, 5000 m.

Eupithecia epileptica Vojnits, 1984, *Acta zool. Acad. Sci. hung.* **30** (1–2): 221, figs 11, 14. Holotype ♀ (TTM; examined), [China]: North Yunnan, A-tun-tse [Dêqên], 4500 m. **New junior synonym.**

Eupithecia acuminata Mironov & Galsworthy, 2004, *Trans. lepid. Soc. Japan.* 55 (3): 233, figs 6, 20. Holotype ♀, (ZFMK; examined), Tibet [Sichuan], Batang, 5000 m. **New junior synonym.**

Description. Fore wing markedly elongate; ground colour pale brown, all markings a uniform dark brown; transverse lines mainly indistinct; a curved antemedial; discal dot elongate, lying on pale sinuate medial; postmedial brown, edged paler distally, marked proximally by dashes on veins; terminal area suffused with dark brown scales. Hind wing also elongate, triangular, termen slightly emarginate, paler than fore wing; discal dot well marked; area between cell and anal margin and termen irregularly suffused with abundant dark brown scales.

Male genitalia. Uncus broad, biapical. Valve with a short sclerotized process on ventral margin. Vinculum short, trapezoid. Papillae slightly clavate. Aedeagus shorter than valve, vesica armed with a broad, curved, irregular, plate-like cornutus and a smaller, folded cornutus at ductus ejaculatorius base. Sternite A8 broad, sharply narrowed at apex, with two small apical arms.

Female genitalia. Bursa small, globular, covered with dense spines. Ductus bursae long and narrow, with numerous longitudinal striations. Ductus seminalis slightly broadened basally, attached to central part of ductus bursae from left side. Colliculum a broad sclerotized thickening. Antrum slightly sclerotized, covered by numerous minute pores. Tergite A8 small, trapezoid, with narrowly sclerotized anterior margin and rounded posterior corners. Apophyses narrow. Papillae anales small, with rounded apices.
Range. China (Sichuan, Yunnan).
Phenology. Univoltine (?): mid-June to mid-July.
Altitude. Found at about 4000–5000 m in Tibet and Yunnan (China: Vojnits 1981b; 1984 [as *epileptica*]; coll. ZFMK).
Similar species. All material seen of this species is worn. Can be distinguished from other members of the present species group by its elongate wing shape. In shape and pattern similar to *E. molestissima*, but with leading half of hind wing much paler. Otherwise rather lacking in distinguishing characters, and best determined by examination of the genitalia. Latter are similar to those of *E. lariciata,* but the male aedeagus vesica has a broad plate instead of horn-like cornuti. The female has a longer ductus bursae and narrower ductus seminalis.
Material. ♀, Tibet [Sichuan], Batang, Alpine Zone (*ca.* 5000 m), 456, 26.vi.1938, H. Höne, Vojnits slide no. 13137♀ (holotype, ZFMK); 2♀, same loc. and alt., 16, 23.vi.1938, H. Höne, Vojnits slides nos. 13135♀, 13136♀ (paratypes, ZFMK); 1♀, same loc. and alt., 17.vi.1938, H. Höne, Vojnits slide no. 13138♀ (paratype, TTM); ♀, same loc., 18.vi.1936, de Laever slide no. 460♀ (holotype of *E. acuminata*, ZFMK); 1♂, 1♀, same loc. and alt., 15, 28.vi.1938, de Laever slides nos. 103♀, 95♂ (ZFMK); ♀, Nord-Yuennan, A-tun-tse [Dêqên], Obere Höhe (*ca.* 4500 m), 17.vii.1936, H. Höne, Vojnits slide no. 13935♀ (holotype of *E. epileptica*, TTM); 1♂, same loc. and alt., 26.vii.1936, H. Höne, de Laever slide no. 79♂ (ZFMK).
Remarks. According to the original label, the holotype specimen of this species was captured on 28.vi.1938, not 26.vi.1936, as in Vojnits, 1981.

119. *Eupithecia argentea* Mironov & Galsworthy, 2004

Eupithecia argentea Mironov & Galsworthy, 2004, *Trans. lepid. Soc. Japan* 55 (4): 286, figs 2, 14. Holotype ♀ (IZCAS; examined), China: Sichuan, Emeishan, Qingyinge, 800–1000 m.

Description. Fore wing ground colour off-white; basal area flecked with brown scales, extending outwards at the costa; costa darkened except for short area beyond postmedial; two prominent dark blotches at antemedial and postmedial, and a less prominent one at medial position; discal dot large, dark brown; antemedial, medial, and postmedial faint, narrow, slightly sinuate; a line of spots between veins outside postmedial; terminal area broadly dark brown, narrower in middle. Hind wing ground colour as fore wing; discal dot faint and small.

Female genitalia. Bursa copulatrix globular, very small, covered with spines and heavily sclerotized near colliculum. Latter very small. Antrum long and narrow, membranous, but broadened and heavily sclerotized at the end, with a few large spines. Tergite A8 transverse. Apophyses short and narrow. Papillae anales broad.

Male unknown.

Range. China (Sichuan).

Similar species. Known only from the holotype. Externally similar to *E. emendata* Vojnits 1983, but with very different genitalia. The heavily sclerotized apex of the antrum is unique.

Material. ♀, China, Sichuan, Emeishan, Qingyinge, 800–1000 m, 17.v.1957, coll. Wang Songyuan, slide no. ZL486♀ (holotype, IZCAS).

Remarks. The ductus seminalis, antrum and one of the posterior apophyses are damaged in the genitalia slide. The attribution of this species to the present species group is tentative.

120. *Eupithecia primitiva* Mironov & Galsworthy, 2004

Eupithecia primitiva Mironov & Galsworthy, 2004, *Trans. lepid. Soc. Japan* 55 (4): 294, figs 8, 21. Holotype ♂ (IZCAS; examined), China: Gansu, Wenxian, Qiujiaba, 2350 m.

Description. Fore wing ground colour pale brown; no transverse markings visible, but costa with a series of squarish dark brown blotches marking ends of lines; discal dot small, dark, round; terminal area slightly darker. Hind wing paler than fore wing, ground colour off-white; discal dot tiny; some dark shading on anal margin and in terminal area.

Male genitalia. Uncus stout, biapical. Valve wide, with small and blunt ventral process. Vinculum short, with shallow medial hollow. Papillae narrow and elongate. Aedeagus slightly shorter than valve length, vesica armed with an apical patch of very small spines, two slim, slightly curved horn-like cornuti, an elongate, folded U-shaped cornutus and a large, irregular cornutus near ductus ejaculatorius base. Sternite A8 flask-shaped, with two short and curved, asymmetrical apical rods.

Female genitalia. (see remarks below) Bursa small, globular, covered with dense spines. Ductus bursae short and narrow. Ductus seminalis broadened near base, attached to central part of ductus bursae on right side.

Colliculum absent. Antrum relatively broad, funnel-shaped, sclerotized. Tergite A8 small, with convex and sclerotized anterior margin. Apophyses narrow and short. Papillae anales broad and rounded.
Range. China (Gansu); ? Kashmir.
Phenology. Univoltine (?): late June to early July.
Altitude. Found at about 2300–2350 m above sea-level (China: coll. IZCAS; Kashmir: coll. ZFMK).
Similar species. The male genitalia are similar to those of the Nepalese *E. mira* Vojnits, 1988 (see Inoue 2000), but can be distinguished by the more pointed valve, almost equal horn-like cornuti, and longer, slimmer eighth sternite. The female genitalia are very similar to those of *E. lariciata* (Freyer, 1842) and distinguished only by the shorter ductus bursae and the larger sclerotized antrum.
Material. ♂, China, Gansu, Wenxian, Qiujiaba, 2350 m, 28.vi.1998, coll. Yao Jia, Galsworthy slide no. ZL580♂ (holotype, IZCAS); 2♀, Indien, J & K, Kashmir, Daksum, 2300 m, 6+7.vii.1987, leg. W. Thomas (ZFMK).
Remarks. The two worn females from Kashmir listed above in the collection of ZFMK are a good match for this species, and belong to the right species group, but in view of the geographical separation, we assign them only tentatively to this species.

121. *Eupithecia sacrivicae* Mironov & Galsworthy, 2004

Eupithecia sacrivicae Mironov & Galsworthy, 2004, *Trans. lepid. Soc. Japan* 55 (3): 232, figs 5, 19, 23. Holotype: ♂ (IZCAS; examined), China, Hubei, Shennongjia, Dajiu Lake, 1800 m.

Description. Fore wing ground colour pale buff, costal area darker brown, especially at beginnings of transverse lines; area along inner margin suffused with pale orange; transverse lines brown; basal and antemedial double; discal dot black, slightly elongate; medial curving outside discal dot, then sinuous to inner margin, area between this line and outer line of double antemedial suffused darker brown, forming a darker fascia; postmedial double, ill-defined; terminal area suffused with darker brown subapically, subtornally and at tornus. Hind wing ground colour white; discal dot small; area between cell and anal margin suffused with dark brown scales, strongest in area of antemedial and medial; a faint curving postmedial; terminal area suffused with dark brown.
Male genitalia. Uncus broad, biapical. Valve with blunt, sclerotized ventral process. Vinculum short and broad. Papillae very long and narrow. Aedeagus vesica with a patch of spinules near apex, one long horn-like cornutus, a folded H-shaped cornutus and an irregular folded cornutus at

ductus ejaculatorius base. Sternite A8 flask-shaped, with two small apical lobes.

Female genitalia. Bursa globular, completely covered with small spines. Ductus bursae long, relatively narrow. Ductus seminalis broadened near base, attached to middle of ductus bursae from ventral side. Colliculum rather open. Antrum broad, membranous. Tergite A8 almost square. Apophyses narrow, spurs on the anterior arising near base. Papillae anales rounded.

Range. China (Hubei, Shaanxi).

Phenology. Early August.

Altitude. Found at about 1800 m above sea-level in Hubei (China: coll. IZCAS).

Similar species. Externally the orange suffusion on the inner margin separates from all except *E. stomachosa*, from which it is separated by the lack of the double white median lines. The male genitalia are similar to those of *E. lariciata* but distinguished by the longer ventral process on the valve, and the considerably longer horn-like cornutus on the vesica. The female genitalia are very similar to those of *E. lariciata*, but may perhaps be distinguishable by having a longer ductus bursae.

Material. ♂, China, Hubei, Shennongjia, Dajiu Lake, 1800 m, 1.viii.1981, coll. Han Yinheng, slide no. ZL520♂ (holotype, IZCAS); 1♂, same data (paratype, IZCAS); 1♀, Shaanxi, Wutaishan, 1992, slide no. ZL459♀ (paratype, IZCAS).

122. *Eupithecia horrida* Vojnits, 1980

Eupithecia horrida Vojnits, 1980, *Acta zool. Acad. Sci. hung.* **26** (4): 443, figs 11, 12, 22. Holotype ♂ (ZFMK; examined), China: North Yunnan, Li-kiang.

Description. Fore wing with blackish brown basal half from base to first medial, and blackish brown apical area and tornus; outer part of medial and medial part of terminal area ochreous; transverse lines distinct; medial double, very sinuate, the inner line touching discal dot and sharply angled onto costa; postmedial sinuate, twice angled onto costa; terminal area with dentate, whitish subterminal line and elongate, whitish tornal spot; discal dot large, intensely black. Hind wing ground colour white; area along anal margin blackish brown, with traces of broad transverse lines; discal dot small but distinct. Notum with whitish posterior spot.

Male genitalia. Uncus very stout, biapical. Valve with a shallow wrinkled medial hollow in the ventral margin. Vinculum short, with shallow medial hollow. Papillae long and narrow. Aedeagus narrow, shorter than valve

length. Vesica armed with a small, heavily sclerotized apical patch of short, basally united, spiniform cornuti, an elongate, U-shaped cornutus and a smaller, irregular cornutus near ductus ejaculatorius base. Sternite A8 flask-shaped with two short apical horns.
Female genitalia. Bursa globular, covered with dense, small spines except small posterior area, which carries a discrete patch of spines. Ductus seminalis narrow, attached to posterior part of corpus bursae near colliculum on ventral side. Colliculum short and narrow. Antrum membranous with sclerotized broadening in its anterior part near colliculum. Tergite A8 trapezoid, with narrowly sclerotized anterior margin. Apophyses rather short, narrow. Papillae anales small, ovate-oblong.
Range. China (Yunnan).
Phenology. Bivoltine: early March to mid-April; late July to early August.
Similar species. Externally similar to *E. stomachosa*, but distinguished by the reddish ochreous fascia on the fore wing, and the paler hind wings. In the male genitalia, the shallow wrinkled ventral hollow on the valve and the ornamentation of the vesica are diagnostic. The female is distinguished by the lack of a ductus bursae and the presence of the sclerotized patch on the antrum.
Material. ♂, China, Provinz Nord-Yuennan, Li-kiang [Lijiang], 8.iii.1935, H. Höne, Vojnits slide no. 12230♂ (holotype, ZFMK); 1♂, 5♀, same loc., 17, 27.iii., 4, 15, 16.iv.1935, H. Höne, Vojnits slides nos. 12229♀, 12231♀, 12233♀, 12234♂, 12235♀, 12236♀ (paratypes, ZFMK); 1♂, 1♀, same loc., 19.iii., 22. iv.1935, H. Höne, Vojnits slides nos. 12232♂, 12237♀ (paratypes, TTM); 2♂, 4♀, same loc., 29.vii., 4, 6.viii.1934, 20, 22.iii., 17.iv.1935, H. Höne, de Laever slides nos. 155♂, 259♂, 260♀, Vojnits slides nos. 13009♀, 13179♀ (ZFMK).

123. *Eupithecia stomachosa* Vojnits, 1980

Eupithecia stomachosa Vojnits, 1980, *Acta zool. Acad. Sci. hung.* **26** (4): 435, figs 3, 4, 18. Holotype ♂ (ZFMK; examined), [China]: Sued-Shensi, Tapaishan in Tsinling, 1700 m.

Description. Fore wing ground colour dark brown with weak reddish tinge, area along anal margin reddish; double medial line very characteristic, white, sharply angled onto costa behind discal dot and again near inner margin; postmedial sinuate, evenly curved, with a series of black dashes on veins; terminal area with a series of short black marginal streaks between veins and a wavy white subterminal line; discal dot relatively large. Hind wing paler, brownish grey with reddish tinge, especially along terminal margin; transverse lines well marked along anal margin; terminal area with weak whitish subterminal line and white tornal spot; discal dot distinct.

Male genitalia. Uncus stout, biapical. Valve with a small blunt ventral process. Vinculum relatively narrow, with very shallow medial hollow. Papillae elongate, narrow. Aedeagus narrow, shorter than valve length, vesica with one very slim horn-like cornutus, a broad, plate-like cornutus and a smaller irregular cornutus near ductus ejaculatorius base. Sternite A8 peg-like, rather narrow, broadened basally, with two short apical horns.
Female genitalia. Bursa globular, covered with small spines. Ductus bursae elongate and narrow, with numerous longitudinal striations. Ductus seminalis broad, attached to central part of ductus bursae from right side. Colliculum elongate and narrow. Antrum lightly sclerotized. Tergite A8 elongate, with broadly sclerotized anterior margin. Apophyses rather short and narrow. Papillae anales rounded.
Range. China (Shaanxi).
Phenology. Univoltine (?): first half of August.
Altitude. Found from about 1700–3000 m above sea-level in Tapaishan (China: coll. ZFMK).
Similar species. Easily recognised by the prominent white medial lines. The male genitalia have only a short ventral process on the valve, and a narrower sternite A8 than most members of the group; the female genitalia are distinguished by the elongate colliculum.
Material. ♂, Sued-Shensi [Shaanxi], Tapaishan [Taibaishan] im Tsinling [Qinling], *ca.* 1700 m, 8.viii.1936, H. Höne, Vojnits slide no. 12262♂ (holotype, ZFMK); 1♂, same loc. and alt., 9.viii.1936, H. Höne, Vojnits slide no. 12261♂ (paratype, ZFMK); 1♂, same loc. and alt., 7.viii.1936, Vojnits slide no. 12260♂ (paratype, TTM); 1♂, same loc. and alt., 8.viii.1936, H. Höne, Vojnits slide no. 12380♀ (wrongly referenced gen. slide of ♀ *E. subfuscata*, ZFMK); 3♂, same loc. and alt., 9.viii.1936, H. Höne, Vojnits slides nos. 11455♂, 12967♂, 13018♂ (TTM); 1♀, same loc., *ca.* 3000 m, 11.viii.1936, H. Höne, Vojnits slide no. 12259♀ (paratype, ZFMK); 1♂, same loc. and alt., 8.viii.1936, H. Höne, Vojnits slide no. 12289♂ (ZFMK); 1♀, Hubei, Shennongjia Daiuhu, 1800 m, 4.viii.1981, coll Han Yinheng, slide no. ZL179♀ (IZCAS).

124. *Eupithecia sola* Vojnits, 1983

Eupithecia sola Vojnits, 1983, *Acta zool. Acad. Sci. hung.* 29 (1–3): 275, fig. 19. Holotype ♂ (ZSM; examined), Nepal: Dudh Kosi Valley, 3500 m.

Description. Fore wing ground colour rusty ochreous with darker brown costal margin, median and tornal blotches in terminal area, a small white tornal spot, and a large black discal dot. Hind wing white with rusty anal

margin and tornus, and an inconspicuous brownish postmedial, a small white tornal spot and a small, black discal dot.
Male genitalia. Uncus triangular, biapical. Valve with a large blunt ventral process. Vinculum narrow, trapezoid. Papillae rather long. Aedeagus narrow, shorter than valve length, vesica armed with numerous short apical spines, two narrow, parallel horn-like cornuti, a small plate-like cornutus at centre and a small irregular cornutus at anterior end. Sternite A8 flask-shaped with two short apical horns.
Female genitalia. Bursa ovoid, densely covered with small spines. Ductus bursae very long, with a hook-like membranous but thick-walled appendix at base on right side. Ductus seminalis broadened at base, attached to central part of bursa on left side. Colliculum very short. Antrum broad, funnel-shaped, membranous, with a broad transverse band of numerous pores. Tergite A8 almost quadrate. Apophyses narrow. Papillae anales broad and rounded.
Range. China (Tibet); Nepal, India (Uttar Pradesh).
Phenology. Bivoltine (?): early March and late July.
Altitude. From about 1200–1600 m to about 3500 m (India: coll. U. Ratzel; Nepal: Vojnits 1983).
Similar species. Similar to *E. marmorea*, but distinguished by the larger rusty ochreous area on the fore wing. The male genitalia are best distinguished by the ornamentation of the aedeagus. The female genitalia are more elongate than most members of the group, and distinguished by the very short colliculum and the appendix at base of the ductus bursae.
Material. ♂, Nepal, Dudh Kosi Tal, 3500 m, 22–23.vii.1962, leg. G. Ebert & H. Falkner, Vojnits slide no. 10670♂ (holotype, ZSM); 1♀, Tibet, Yatung, 10000ft., 23.vii.1924, Maj, R.W.G. Hingston, Everest Exp. Brit. Mus. 1924–386, BM Geom. slide no. 21534♀ (BMNH).
Remarks. Not previously recorded from China.

125. *Eupithecia marmorea* Vojnits, 1983

Eupithecia marmorea Vojnits, 1983, *Acta zool. Acad. Sci. hung.* **29** (1–3): 274, fig. 18. Holotype ♀ (ZSM; examined), Nepal: Thodung, 3100 m.

Description. Fore wing ground colour fawn brown, with orange wash along inner margin and a central patch from discal dot to terminal margin; transverse lines dark brown, antemedial and double medial sinuate, very close, forming a dark oblique band through discal dot; postmedial strongly marked, sharply angled to costa; discal dot large. Hind wing off-white, with distinct dark transverse lines along anal margin, and a rounded discal dot.

Male genitalia. Uncus stout, biapical. Valve with an obtuse ventral process. Vinculum narrow, with very shallow medial hollow. Papillae long and narrow. Aedeagus narrow, shorter than length of valve. Vesica armed with a plate-like cornutus with four or more short spines at apex and an irregular cornutus near base of ductus ejaculatorius. Sternite A8 flask-like, with two short apical horns.
Female genitalia. Bursa globular, covered with small spines. Ductus bursae membranous, slightly shorter than corpus bursae. Ductus seminalis broadened at base, attached to lower part of ductus bursae. Colliculum not clearly distinguishable. Antrum broad, rectangular, sclerotized. Tergite A8 slightly elongate. Apophyses short and narrow. Papillae anales elongate with rounded apices.
Range. China (Tibet); Nepal.
Phenology. Univoltine (?): late May.
Altitude. From about 3100 to 3400 m in the Himalayas (Nepal: Vojnits 1983; coll. ZSM).
Similar species. Most similar to *E. sola*, but with less extensive orange patches and more distinct sinuous transverse lines. The male genitalia are easily distinguished by the shape of the ventral process of the valve, and the ornamentation of the aedeagus vesica, and the female by the expanded and sclerotized antrum.
Material. ♀, Nepal, Thodung, 3100 m, 21.v.1962, leg. G. Ebert u. H. Falkner, Vojnits slide no. 11660♀ [= ZSM slide no. G 10617♀] (holotype, ZSM); 1♂, Nepal, Prov. Nr. 3, East Dudh Kosi Tal unter Thangpoche, 3400 m, 29.v.1964, leg. W. Dierl (ZSM); 2♀, Tibet, Yatung, A.E. Hobson, 98–201, BM Geom. slide no. 21478♀ (BMNH).
Remarks. Not previously recorded from China.

126. *Eupithecia albimedia* Mironov & Galsworthy, 2004

Eupithecia albimedia Mironov & Galsworthy, 2004, *Trans. lepid. Soc. Japan* 55 (1): 43, figs 3, 15, 18. Holotype ♂ (ZFMK; examined), [China]: North Yunnan, Li-kiang [Lijiang].

Description. Fore wing ground colour yellowish tan brown; central part of medial fascia white, bounded internally by a very sinuous medial line, and externally by evenly curved dark buff postmedial; discal dot prominent, black. Hind wing off-white; anal margin as far as cell mottled with pale brown, with incomplete double antemedial lines just visible.
Male genitalia. Uncus stout, biapical. Valve with a small blunt ventral process. Vinculum short and narrow, with very shallow medial hollow. Papillae elongate. Aedeagus slim, shorter than valve length. Vesica armed

with a narrow straight horn-like cornutus, a broad H-shaped plate and a large irregular plate near ductus ejaculatorius base. Sternite A8 peg-like, rather narrow.

Female genitalia. Bursa globular, completely covered with small spines. Ductus bursae membranous, about one half length of corpus bursae. Ductus seminalis broad, attached to ductus bursae near colliculum. Colliculum elongate and relatively narrow. Antrum lightly sclerotized, elongate. Tergite A8 trapezoid. Apophyses narrow, the posterior with very short spurs near middle. Papillae anales rounded.

Range. China (Shaanxi, Sichuan, Yunnan).

Phenology. Univoltine: early August to early September.

Altitude. From about 1700 to 2200 m above sea-level (China: coll. ZFMK, IZCAS).

Similar species. Easily distinguished by the extensive white in the centre of the fore wing. In the male genitalia, the shape of the valve ventral process, the ornamentation of the vesica and the narrower sternite A8 in the male distinguish the species from others in the group. The female genitalia are similar to those of *E. lariciata*, but are distinguished by the shorter ductus bursae, narrow colliculum and the shape of the antrum. Both are similar to those of *E. stomachosa*, but the two species are easily separable externally.

Material. ♂, China, Provinz Nord-Yuennan, Li-kiang [Lijiang], 2.ix.1935, H. Höne, de Laever slide no. 152♂ (holotype, ZFMK); 2♂, 1♀, same loc., *ca.* 2000 m, 13.viii.1935, 14, 29.viii.1934, H. Höne, respectively de Laever slides nos. 295♂, 143♂, Vojnits slide no. 12238♀ (paratypes, ZFMK); 1♀, Sichuan, Dukou, 2150 m, 22.viii.1980, coll. Zhang Baolin, slide no. ZL608♀ (paratype, IZCAS); 1♂, Sued-Shensi [Shaanxi], Tapaishan [Taibaishan] im Tsinling [Qinling], *ca.* 1700 m, 8.viii.1936, H. Höne, Mironov slide no. 720♂ (ZFMK).

127. *Eupithecia hebes* Mironov & Galsworthy, 2004

Eupithecia hebes Mironov & Galsworthy, 2004, *Trans. lepid. Soc. Japan* 55 (3): 238, figs 9, 26. Holotype ♂ (ZFMK; examined), Shansi [Shanxi], Mianshan, 2000 m.

Description. Fore wing ground colour pale grey-brown; basal area and three small patches on costa darker brown; antemedial acutely angled close to costa; discal dot dark brown, elongate; medial sinuous, curving round outside discal dot; postmedial a dark broken fascia curved close to costa, and bordered outside by a narrow brown line; terminal area blotched darker brown. Hind wing paler, discal dot small, well marked; anal margin suffused with buff and dark brown scales, broadening towards tornus.

Male genitalia. Uncus triangular, possibly biapical. Valve with a very small medial process on the ventral margin and a longitudinal wrinkle along ventral margin. Vinculum short, with shallow medial hollow. Papillae narrow and elongate. Aedeagus slim, slightly shorter than length of valve. Vesica armed with a slim horn-like cornutus, a larger, broad cornutus, and a small irregular plate near ductus ejaculatorius base. Sternite A8 peg-like. Female unknown.
Range. China (Shanxi).
Altitude. The only known specimen was collected at about 2000 m in Mianshan (China: coll. ZFMK).
Similar species. Similar in general appearance to *E. lariciata*, but with the postmedial curved rather than angled close to costa. The male genitalia are close to those of the preceding species, but the ventral process of the valve is much smaller and the uncus narrower. Externally it is quite distinct from *E. albimedia*.
Material. ♂, Prov. Shansi [Shanxi], Mien-Shan, Obere Höhe *ca.* 2000 m, 29.vii.1937, H. Höne, Vojnits slide no. 13190♂ (holotype, ZFMK).
Remarks. The eighth abdominal sternite is damaged in Vojnits' slide preparation.

128. *Eupithecia stataria* Inoue, 1988

Eupithecia stataria Inoue, 1988, *Bull. Fac. domest. Sci. Otsuma Wom. Univ.* **24**: 367, figs 27a-f. Holotype ♀ (BMNH; examined), [Taiwan]: Hualien Hsien, Tayuling, 2600 m.

Description. Fore wing ground colour pale ash grey, transverse lines oblique, inconspicuous; antemedial and medial sharply angled onto costa; postmedial usually with a series of short dark dashes on veins and right angled onto costa; discal dot distinct, subulate. Hind wing paler with indistinct transverse lines, slightly darker, greyish terminal margin and pale discal dot. Abdomen grey or whitish grey with black lateral stripes except last segment.
Male genitalia. Uncus biapical, broad and bell-shaped. Valve with a cone-shaped ventral process. Vinculum short, semicircular. Papillae elongate. Aedeagus shorter than length of valve. Vesica armed with numerous stout teeth, an elongate, folded, V-shaped plate and a smaller, curved irregular plate near ductus ejaculatorius base. Sternite A8 peg-like, almost triangular, with two tiny apical horns.
Female genitalia. Bursa globular, covered with small spines. Ductus bursae elongate and broad, sclerotized and with numerous longitudinal wrinkles basally. Ductus seminalis inflated at base, attached to base of

ductus bursae on right side. Colliculum short and broad. Antrum very broad, funnel-shaped, lightly sclerotized, with numerous longitudinal wrinkles in posterior part. Tergite A8 trapezoid. Apophyses narrow. Papillae anales relatively small, slightly elongate and tapered.
Range. China (Taiwan).
Phenology. Bivoltine: mid-March to early May; late July to late August.
Altitude. Found from about 1900–2600 m above sea-level (Taiwan: Inoue 1988; coll. TTM).
Similar species. A smallish, rather plain species, similar to *E. quadripunctata*, which however, has an ochreous or rusty brownish abdomen. The male and female genitalia unmistakeably belong to this species group, but the former are distinguished by the spiny plates at apex of the aedeagus, and the latter by the heavily sclerotized antrum.
Material. ♀, Taiwan, Tayuling, 2,600 m, Hualien Hsien, 28–29.viii.1983, H. Yoshimoto leg. (holotype, coll. H. Inoue in BMNH); 1♂, Taiwan, Prov. Nantou, Tayuling, 2550 m, 16.iii.1996, leg. Gy. Fábián & L. Németh (TTM); 2♂, Taiwan, Prov. Nantou, Hohuachi, 1950 m, 24°13′ N, 121°16′ E, 31.iii.2000, leg. A. Kun & L. Peregovits (ZSM).

The *tripunctaria* species group

This species group was introduced and described by Mironov (2003). The male genitalia are similar to those of the preceding species group in the ventral process of the valve, and the flask-shaped eighth sternite, but have a characteristic armature in the vesica with a slender apical spine-like cornutus. With the exception of *E. tripunctaria* itself, the female genitalia have a characteristic more or less globular corpus bursae set on the end of a very long recurved ductus. The group is represented in China by the following six species. We have concluded that *E. cotidiana*, which was mentioned in Mironov (2003) as belonging to this group, is more appropriately placed in the *fletcherata* species group.

129. *Eupithecia tripunctaria* Herrich-Schäffer, 1852

Eupithecia tripunctaria Herrich-Schäffer, 1852, *Syst. Bearb. Schmett. Eur.* 6 (55): 77; *ibidem* (1851), 3 (Suppl.), pl. 75, fig. 461 (non binominal). Syntype(s). [Germany]: Frankfurt a/M.
Phalaena albipunctata Haworth, 1809, *Lepid. Br.*: 360. Syntype(s) (BMNH), [United Kingdom]: London. [Junior primary homonym of *Phalaena albipunctata* Hufnagel, 1767] (synonymised in Herrich-Schäffer, 1861 and Staudinger, 1861).
Eupithecia albipunctata var. *angelicata* Barrett, 1877, *The Entomologist's Monthly Mag.* (1876–1877) 8: 278. Syntype(s), [United Kingdom]: York.
Eupithecia promulgata Pearsall, 1909, *Proc. Ent. Soc. Wash.*, 11: 121 (partim ♀; *nec* ♂). Syntype ♀ (AMNH), [U.S.A.]: [New York], Big Indian Valley, Catskill Mts. (synonymised to *E. albipunctata* Haworth, 1809 in McDunnough, 1936).

Eupithecia incohata Vojnits, 1979, *Acta zool. Acad. Sci. hung.* 25 (3–4): 435, figs 13, 14. Holotype ♂ (ZFMK; examined), [China]: Sued-Shensi, Tapaishan [Taibaishan] im Tsinling, 3000 m. **New junior synonym.**

Unavailable names (incorrect subsequent spellings): *tripunctata*: Crewe (1862); *inchoata*: Vojnits (1984a). – (infrasubspecific): *aestiva*: Dietze (1910); *intermedia*: Lempke (1951).

Description. Fore wing unicolourous dark greyish fuscous with weak dark costal spots; all transverse lines indistinct; veins black, but usually chequered; white subterminal line multiply indented, often broken, with two or three conspicuous white spots, of which the tornal is largest; discal dot rather small. Hind wing grey, veins chequered black and white, basal area darker along dorsum; two small white tornal spots often distinct; discal dot small. Abdomen fuscous grey, with small white dorsal crests on the posterior margin of each segment and with black lateral stripes.
Male genitalia. Uncus stout, biapical. Valve broad, with a small pointed ventral process. Vinculum short with a shallow medial hollow. Papillae rather stout, tapered to apices. Aedeagus stout, shorter than length of valve. Vesica with transverse striations, armed with a slim, apical horn-like cornutus, and a smaller, flat cornutus at ductus ejaculatorius base. Sternite A8 flask-shaped with two pointed apical horns.
Female genitalia. Bursa elongate, covered with numerous small, mainly dorsal, spines in central area, membranous basally. Ductus bursae thick-walled, sclerotized, especially at base, evenly tapered to colliculum, with numerous longitudinal striations. Ductus seminalis broadened medially, attached to ductus bursae near colliculum on right. Colliculum relatively small. Antrum membranous. Tergite A8 quadrate. Apophyses narrow. Papillae anales short and rounded.
Range. China (Qinghai, Shaanxi, Shanxi, Sichuan, Zhejiang); Europe, Caucasus, Transcaucasus, Urals, Asian Russia (western and southern Siberia, Priamurje, Primorje, Sakhalin, Kuril Islands: Kunashiri), Korea, Japan (Hokkaido, Honshu, Shikoku). Also widespread in North America from eastern Newfoundland to western British Columbia; north to Labrador, and south to San Bernadino Mts., California (Bolte 1990).
Phenology. Bivoltine: late April to late June; mid-July to mid-September, overwintering as pupa.
Biology. Larva usually found on the flowers and developing seeds of various Apiaceae and occasionally on the flowers and seeds of *Solidago virgaurea, Eupatorium cannabinum* (Asteraceae), *Knautia arvensis* (Dipsacaceae), *Filipendula ulmaria* (Rosaceae); the second generation has been found on *Viburnum, Sambucus nigra, S. ebulus* (Caprifoliaceae) in Europe. In North America found on flower-heads of *Viburnum* (Caprifoliaceae),

umbels of *Heracleum maximum* (= *lanatum*) (Apiaceae) and one stunted adult was reared on *Abies balsamea* (Pinaceae), probably in error.
Altitude. From sea level to about 3500 m in Tapai-shan and Daxue Shan (China: Vojnits 1979b; coll. ZSM; IZCAS).
Similar species. *E. tripunctaria* is best distinguished externally by the combination of grey ground colour and prominent white spots on all wings. In the male genitalia the armature of the vesica and the shape of the eighth sternite are reasonably distinctive. The female genitalia are well differentiated from those of other members of the group, and distinguished by the central band of spining and the strong striations of the ductus bursae.
Material. ♂, Sued-Shensi [Shaanxi], Tapaishan [Taibaishan] im Tsinling [Qinling], *ca.* 3000 m, 11.viii.1936, H. Höne, Vojnits slide no. 12288♂ (holotype of *E. incohata*, ZFMK); 1♂, same loc. and alt., 12.viii.1936, H. Höne, Vojnits slide no. 12284♂ (paratype of *E. incohata*, ZFMK); 2♀, same loc., *ca.* 1700 m, 16, 18.v.1936, H. Höne, Vojnits slides nos. 12286♀, 12291♀ (paratypes of *E. incohata*, ZFMK); 4♀, same loc. and alt., 26.iv., 16, 17.v.1936, H. Höne, Vojnits slides nos. 12283♀, 12292♀, 12294♀, 12296♀ (paratypes of *E. incohata*, TTM); 1♀, same loc. and alt., 9.viii.1936, H. Höne, Vojnits slide no. 12290♀ (ZFMK); 1♂, Prov. Chekiang [Zhejiang], West Tien-mu-shan, 23.iv.1932, H. Höne, Mironov slide no. 712♂ (ZFMK); 1♂, 1♀, Prov. Shansi [Shanxi], Mien-shan, Obere Höhe *ca.* 2000 m, 1, 14.viii.1937, H. Höne, Vojnits slide no. 12971♀ (TTM); 2♀, China, Sichuan, Daxue Shan, Gongga Shan, NW Moxi, 29°41′ E, 101°58′ N, 2600–3200 m, 23.iv.-15.v.2001, leg. V. Siniaev & A. Plutenko, *ex.* coll. Dr. A. Schintlmeister (ZSM); 1♀, Sichuan, Batang, Yidun, 3500 m, 17.viii.1982, coll. Chai Huaicheng, slide no. ZL150♀; 1♀, Qinghai, Tongren, Maixiu, 2950 m, 30.vii.1992, coll. Xue Dayong, slide no. 369♀; 3♀, Qinghai, Huzhu Beishan Linchang, Langshidang, 2600 m, 36°55′ N, 102°23′ E, 7.viii.2005, Xue Dayong and Han Hongxiang, slide no. ZL690♀ (IZCAS).

130. *Eupithecia barteli* Dietze, 1908

Eupithecia barteli Dietze, 1908, *Dt. ent. Z., Iris* 21 (2): 162, pl. 2, figs 13, 14. Syntypes 1♂, 2♀ (coll. K. Dietze and R. Püngeler in MNHU; examined), [Kazakhstan]: Uralsk.
Eupithecia artshae Viidalepp, 1988, *Fauna pyadenits gor Srednej Azii* [*Geometridae fauna of the Central Asian mountains*]: 124, pl. 2, fig. 21; text-pl. 28, figs 2, 3, 6–8. Holotype ♂ (ZISP; examined), Tadzh. SSR [Tajikistan]: SW Pamirs, Khorog, botanical garden, 2300 m (treated as subsp. of *E. barteli* Dietze, 1908 in Mironov, 1990).

Description. Fore wing ground colour greyish fuscous with a soft ochreous tinge; costal margin darker; transverse lines indistinct, cloudy; medial sinuate, sharply angled from discal dot towards costa; postmedial twice right-angled near costa; medial area with distinct, pale ochreous cloud under discal dot; a double pale ochreous stripe between medial and terminal areas usually distinct; whitish wavy subterminal line, often broken, forming a series of small triangular spots or sometimes obsolete; discal dot large, very distinct. Hind wing paler, transverse lines more or less distinct; terminal area darker; discal dot conspicuous. Abdomen grey, with narrow black lateral stripes on posterior segments.

Male genitalia. Uncus very broad near base, tapering towards apex, biapical. Valve broad, with a small blunt ventral process. Vinculum short, rounded. Papillae elongate. Aedeagus considerably shorter than valve length. Vesica with transverse striations, armed with a slim horn-like cornutus near apex, and two smaller, irregular cornuti. Sternite A8 flask-shaped, less bulging than in preceding species, with two short blunt apical rods.

Female genitalia. Bursa globular, recurved, covered with dense spines except small area near base. Ductus bursae long, cylindrical, sclerotized, covered with longitudinal striations, with a short row of small spines on inner side, strongly sinuate. Ductus seminalis long and broad, attached to ductus bursae near colliculum on right. Colliculum short and narrow. Antrum membranous. Tergite A8 almost square with convex anterior and posterior margins. Apophyses relatively short and narrow. Papillae anales slightly elongate, rounded apically.

Range. China (Shaanxi, Yunnan); European Russia (Lower Volga), Kazakhstan (north west and south east: Tarbagatai Mts., Saur Mts), Kirghizstan (Alai Mts., Tien-Shan Mts.), Tajikistan (Ghissar Mts., western Pamir), Afghanistan, Pakistan (Swat), Pakistan/India (Jammu and Kashmir), India (Himachal Pradesh).

Phenology. Univoltine: early July to mid-September (early October). Larva in August and September. Overwintering as pupa.

Biology. Larva found in the ripening seedballs of *Cuscuta monogyna* and *C. lupuliformis* (Cuscutaceae) (Kirghizstan and southern Kazakhstan: Mironov 2003); feeds at night.

Altitude. From sea level (Kazakhstan: coll. ZISP) to about 2700 m (Afghanistan: coll. TTM).

Similar species. Similar to *E. lariciata* and *E. subfuscata*, but distinguished by the darker colour, pale ochreous medial blotch and double stripe beyond the postmedial line, and by the less distinct transverse lines on the fore wing. The genitalia of *E. barteli* are very similar to those of the following species, which has a rather different external facies;

but *E. barteli* can usually be distinguished by the shape of sternite A8 in the male.

Material. 1♀, [Kazakhstan], Uralsk, 2.viii.07 ♀ Max Bartel, *Barteli* ♀ Type., Uralsk., Karl Dietze, Zool. Mus. Berlin, Mironov slide no. 480♀ (syntype, coll. K. Dietze in MNHU); 1♀, [Kazakhstan], Uralsk, 24.7.05, 4/06 von M. Bartel, n. sp. (Dietze), n. sp. (Bohatsch), *Barteli* ♀ Type., K. Dietze (syntype, coll. R. Püngeler in MNHU); ♂, [Tajikistan], Pamir, Chorog, Hortes Botan., 2300 m, 29.viii.[1]965, [leg.] M. Zaprjagaew (holotype of *E. artshae*, ZISP); 1♂, 1♀, China, Provinz Nord-Yuennan, Li-kiang [Lijiang], 3, 15.viii.1935, H. Höne, Vojnits slides nos. 13182♂, 13238♀ (ZFMK); 1♀, Sued-Shensi [Shaanxi], Tapaishan [Taibaishan] im Tsinling [Qinling], *ca.* 1700 m, 8.viii.1936, H. Höne, Vojnits slide no. 12390♀ (ZFMK).

Remarks. Not previously recorded from China.

131. *Eupithecia tenuisquama* Warren, 1896

Tephroclystia tenuisquama Warren, 1896, *Novit. zool.* 3: (317). Lectotype ♀ (BMNH; examined), [India]: Darjeeling.

Eupithecia toshimai Inoue, 1980, *Bull. Fac. domest. Sci. Otsuma Wom. Univ.* **16**: 177, figs 44C, 46B, 47O-Q, 49G, 53D. Holotype ♂ (BMNH; examined), [Japan]: Kagawa Pref(ecture), Mt. Zôzu (synonymised in Inoue, 2000).

Description. A relatively large species with brindled dark grey coloration and veins covered with black scales on all wings. Fore wing with dark transverse lines oblique, usually indistinct; medial double, the inner one touching discal dot, sharply angled onto costa; postmedial twice angled; terminal area broad with white, wavy subterminal line broadened near tornus into a white spot; discal dot distinct. Hind wing paler, dark transverse lines distinct in anal half; a white tornal dot; discal dot small. Notum with white posterior spot.

Male genitalia. Uncus very broad near base and tapering towards apex, biapical. Valve broad, with small blunt ventral process. Vinculum short and broad, with shallow medial hollow. Papillae large, elongate. Aedeagus stout, but shorter than valve. Vesica with transverse striations, armed with a curved horn-like cornutus near apex, and two smaller, irregular, flat cornuti. Sternite A8 flask-shaped with a long neck, and two short blunt apical rods.

Female genitalia. Bursa ovate, covered with small spines, recurved. Ductus bursae cylindrical, sinuate, with strong longitudinal striations and a longitudinal row of small stout spines. Ductus seminalis broad, attached to ductus bursae near colliculum on right. Colliculum short and narrow. Antrum short, membranous. Tergite A8 almost quadrate with anterior and posterior margins medially produced into a point. Apophyses very narrow. Papillae anales angulate, slightly tapered.

Range. China (Tibet, Hunan, Zhejiang, Fujian, Sichuan, Shaanxi, Hong Kong); Nepal, India (Sikkim), Thailand, Japan (Honshu, Shikoku, Kyushu, Yakushima).

Phenology. Polyvoltine: March to October-November.

Biology. Larvae found on flowers of *Camellia sasanqua* and *C. sinensis* (Theaceae) (Japan: Murase 2000; as *E. toshimai*).

Altitude. Found from about 900 m in Hunan (coll. TTM) up to at least 2800 m in Tibet (coll. ZFMK).

Similar species. Best distinguished from *E. tripunctaria* by the brindled coloration of the fore wings: *E. tripunctaria* tends to be more uniform grey. The male genitalia are similar to *E. tripunctaria*, but the shape of the uncus is different, and the eighth sternite is narrower apically, with shorter, obtuse apical horns. The female genitalia are close to those of other Chinese members of the group. They can be distinguished from *E. invicta* Vojnits, 1981 and *E. puella* by the presence of a row of small spines in the ductus bursae and from *E. barteli* and *E. blandula* by the complete spining of the corpus bursae.

Material. ♀, Sikkim, 1.iv.1889, J.G. Pilcher, *tenuisquama* Warr. Type ♀, BM Geom. slide no. 20124♀ (lectotype, BMNH); ♂, [Japan], Kagawa, Mt. Zôzu, 14.iv.1967, H. Toshima (holotype of *E. toshimai*, coll. H. Inoue in BMNH); 1♂, Pz. Chekiang [Zhejiang], West-Tien-Mu-Shan (1600 m), 18. iv.1932, H. Höne, de Laever slide no. 306♂ (ZFMK); 1♂, 4♀, China, Provinz Hunan, Hoeng-Shan (900 m), 18, 26, 27.iii., 2.iv.1933, H. Höne, de Laever slides nos. 254♀, 258♀, Vojnits slides nos. 12369♀, 12370♀, 12543♂ (ZFMK); 2♀, same loc. and alt., 22.iii, 13.iv.1933, H. Höne, Mironov slide no. 820♀ (TTM); 1♂, China, Sued-Shensi [Shaanxi], Tapaishan [Taibaishan] im Tsinling [Qinling], F, 24.vi.1935, H. Höne, Mironov slide no. 713♂ (ZFMK); 1♂, Tibet [Sichuan], Batang, Im Tal des Yangtze (*ca.* 2800 m), 15.v.1936, H. Höne, de Laever slide no. 465♂ (ZFMK); 1♀, [Fujian], Kuatun [Guadun] (2300 m), 27, 40 n. Br. 117, 40 ö. L., J. Klapperich, 8.iv.1938 (Fukien), Vojnits slide no. 12371♀ (ZFMK).

132. *Eupithecia amandae* Galsworthy & Mironov, 2011

Eupithecia amandae Galsworthy & Mironov, 2011, *Lep. Science* **62**: 24, figs 10, 29, 34. Holotype ♂ (IZCAS; examined), China: Yunnan.

Description. Ground colour of all wings uniform pale, slightly reddish, brown, transverse lines darker brown; fore wing antemedial double; medial straight, passing just behind discal dot; postmedial thicker, parallel to medial in posterior half, rounded basad at level of discal dot, and sharply angled onto costa; three patches of darker brown scales subdorsally on antemedian, postmedian, and subtornally; terminal area slightly darker, crossed by very weak wavy pale submarginal line; discal dot large, intense black and elongate Hind wing with medial and postmedial lines clear right across wing, others visible only at anal margin; discal dot clear, slightly elongate.

Male genitalia. Uncus broad and biapical. Valve with a small, blunt medial process on ventral margin. Vinculum short, with shallow medial hollow. Papillae relatively short and slightly curved. Aedeagus narrow, shorter than length of valve. Vesica with transverse striations, armed with a slim, horn-like apical cornutus and two plate-like, U-shaped cornuti. Sternite A8 peg-like, with two short blunt apical horns.

Female genitalia. Bursa very small, globular, covered with small spines, recurved. Ductus bursae long, sclerotized, slightly sinuate, unspined, with longitudinal striae near colliculum. Ductus seminalis very large and broad, membranous, with a small patch of spinules near base. Colliculum relatively large. Antrum short and broad, membranous. Tergite A8 transverse. Apophyses narrow, tapered. Papillae anales rounded.

Range. China (Yunnan).

Phenology/Altitude. Only known specimens taken at 3200 m in mid June.

Similar species. Reasonably distinctive externally thanks to the prominent, rather broad postmedial. The male genitalia are similar to those of *E. invicta* Vojnits, 1981 (southern Himalayas), but the vinculum is narrower, the horn-like apical cornutus on the vesica shorter and narrower, and the eighth sternite narrower in its basal part. The moth is considerably larger than *E. invicta*. The female genitalia are similar in general shape to those of *E. tenuisquama* and *E. invicta*, but distinguished by the much smaller corpus bursae, and the narrower and shorter ductus bursae without a row of spines (present in *E. tenuisquama*), and above all by the massive ductus seminalis with a patch of very small spines at its base.

Material. ♂, KIB/RBGE research stn., Lijiang, Yunnan prov., PR of China, 15–20 June 2009, 3200 m, A. C. Galsworthy, Galsworthy slide no. LJ5♂ (holotype, IZCAS); 1♀, same loc. and data, B.M. Geom. slide no. 22620♀ (paratype, BMNH).

133. *Eupithecia blandula* Mironov & Galsworthy, 2007

Eupithecia blandula Mironov & Galsworthy, 2007, *Trans. lepid. Soc. Japan* 58 (3): 347, figs 4, 18, 21. Holotype ♂ (TTM; examined), Taiwan: Prov. Taitung, Hsiangyang, 2200 m.

Misidentifications: *Eupithecia ryukyuensis* Inoue, 1988 (*nec* Inoue, 1971); *Eupithecia toshimai* Inoue, 1988 (*nec* Inoue, 1980).

Description. Fore wing ground colour pale whitish grey; stalk of Cu with a black streak; transverse lines inconspicuous, medial sharply angled behind discal dot onto costa, postmedial forming a series of black dashes on the veins; terminal area slightly darker with wavy whitish subterminal line forming medial and tornal white spots; discal dot small. Hind wing slightly paler; transverse lines, especially postmedial, well marked along anal margin; terminal area darker, with wavy white subterminal line forming two large white spots near tornal angle; discal dot small and pale. Abdomen whitish grey with black lateral stripes except last segment.

Male genitalia. Uncus biapical, very broad in lateral view. Valve with a pointed, thorn-like ventral process. Vinculum short and narrow. Papillae tapered. Aedeagus stout, shorter than length of valve. Vesica with transverse striations, armed with a curved horn-like apical cornutus, and two smaller, irregular, flat cornuti. Sternite A8 flask-shaped with a long neck, and two unequal, spiniform, diverging apical horns.

Female genitalia. Bursa oval, recurved, armed with short and broad spines on two sides only. Ductus bursae long, with strong longitudinal striations and a row of sparse, stout spines. Ductus seminalis broad, attached to ductus bursae near base of colliculum. Colliculum short and narrow. Antrum short, membranous. Tergite A8 small, square. Apophyses short and narrow. Papillae anales small, rounded.

Range. China (Taiwan).

Phenology. Univoltine: early April to mid-May.

Altitude. Found at about 2150–2760m above sea-level (Taiwan: coll. MNHU and TTM).

Similar species. Externally similar to *E. dealbata*, but smaller, and with postmedial less curved onto the costa of the fore wing, and transverse lines lacking a series of black dashes on the hind wing. Male genitalia similar to those of *E. tripunctaria*, but with a longer, narrower ventral process on the valve, and a differently shaped sternite 8 with longer apical horns. Female genitalia similar to those of *E. tenuisquama*, but distinguished by the less spiniferous corpus bursae, and the more obvious longitudinal striations on the ductus bursae.

Material. ♂, Taiwan, Prov. Taitung, Hsiangyang, 2200 m, 10.iv.1997, leg. Csorba & Ronkay (holotype, TTM); 1♀, same data (paratype, TTM); 1♀, Taiwan, Prov. Nantou, 6 km SW Hohuan Pass, Yuanfeng, 2760 m, 10.v.1997, leg. Gy.M. László & G. László (paratype, TTM); 1♂, 1♀, Tayuling (2600 m), Hualien Hsien, Taiwan, 2–4.v.1984, leg. H. Yoshimoto, Inoue slides nos. 11194♂ and 11195♀ (BMNH); 1♀, Taiwan, Nantou, Meifeng, *ca.* 2150 m, 27.-30. iv.1999, leg. Mey & Ebert, Zool. Mus. Berlin, Mironov slide no. 814♀ (MNHU).

Remarks. Inoue (1988) published the last mentioned two paratypes as, respectively, *E. ryukyuensis* Inoue, 1971 and *E. toshimai* (= *E. tenuisquama* Inoue, 1980).

134. *Eupithecia puella* Mironov & Galsworthy, 2004

Eupithecia puella Mironov & Galsworthy, 2004, *Trans. lepid. Soc. Japan* 55 (3): 235, figs 8, 22, 25. Holotype ♂ (IZCAS; examined), Yunnan, Tengchong, Danjia forestry farm, 2500 m.

Description. Fore wing ground colour grey, heavily irrorated with dark brown scales; transverse lines indistinct; antemedial acutely angled near costa; discal dot well marked; medial reduced to a dark mark on costa above discal dot; postmedial ill-defined, brown; medial area an ill-defined pale fascia; apical and terminal areas dark brown, with a pale wavy transverse line, joining three pale subapical, subtornal and medial dots. Hind wing ground colour whitish, discal dot small; anal margin irrorated with brown scales, organised into transverse lines; terminal area slightly darker.

Male genitalia. Uncus bell-shaped, biapical. Valve with a small blunt medial process on ventral margin. Vinculum short and broad, with shallow medial hollow. Papillae long and narrow. Aedeagus shorter than valve. Vesica with transverse striations, armed with one large and one smaller horn-like cornuti, and a small irregular plate near ductus ejaculatorius base. Sternite A8 elongate, peg-like, with two narrow apical arms.

Female genitalia. Bursa globular, recurved, covered with large sparse spines. Ductus bursae broad and unridged, with a small patch of minute spines near colliculum. Ductus seminalis very broad, attached to ductus bursae near colliculum on ventral side. Colliculum short and narrow. Antrum short, membranous. Tergite A8 trapezoid. Apophyses narrow and tapered. Papillae anales short and narrow.

Range. China (Yunnan).

Phenology. Both known specimens from early June.

Altitude. Found at about 2500 m above sea-level in Yunnan (China: coll. IZCAS).

Similar species. Closest externally to *E. tenuisquama*, though with a less brindled appearance. The male genitalia are distinguished from other related species by the shape of the uncus, the longer ventral process of the valve and apical cornutus in the aedeagus, and from all but *E. invicta* by the shape of sternite A8. In the female genitalia the long ductus bursae without longitudinal striation and the small patch of minute spines near the colliculum are distinctive.

Material. ♂, Yunnan, Tengchong, Danza, 2500 m, 2–4.vi.1992, coll. Xue Dayong, Galsworthy slide no. ZL484♂ (holotype, IZCAS); 1♀, same data, Galsworthy slide no. ZL487♀ (paratype, IZCAS).

The *propagata* species group

This species group was originally introduced by Vojnits (1984) as the *inepta-sacrosancta* species group, based on *E. inepta*, Vojnits 1979, and nine other Chinese species, some of which we place in other groups. However, the species name *inepta* was preoccupied by Prout (1922), and there were considerable problems, detailed below, with the type material of *E. sacrosancta*. In our recent publications, beginning with Mironov et al. (2008c), we have therefore renamed this species group the *propagata* group on the basis of the earliest species within it to be described. The majority of species in this group were described by Dr. Vojnits: the series on which he based his descriptions are very mixed, and scattered between Bonn and Budapest. There are also, apparently, a large number of cases where slides or specimens have been mislabelled, leaving incorrect associations. We have done our best to sort through this morass, as will be evident from the material listed as having been examined, and to produce a coherent result, but would not be surprised if there are further undetected errors.

The male genitalia in the group are characterized by a more or less sclerotized flattened flange on the basal ventral area of the valve. The aedeagus vesica has a complex series of cornuti, which are very important in specific identification. The eighth sternite consists of two elongate arms, joined at the base. The female bursa is generally elongate-oval, in most cases with a combination of small spines covering the basal part, and a group of very large spines close to the centre. The female genitalia are sometimes difficult to separate.

The group, so far as is known, is confined to south and east Asia. It is represented in China by the following 14 species. In addition, *E. persuastrix* Mironov, 1990 has been found in Far East Russia (Primorje) and Japan, and should occur in NE China. The Himalayan species *E. karnaliensis* Inoue, 2000 may well occur in Tibet. Both also belong to this group.

135. *Eupithecia werneri* Vojnits, 1985

Eupithecia inepta Vojnits, 1979, *Acta zool. Acad. Sci. hung.* 25 (1–2): 208, fig. 19, (nec 20). Holotype ♂ (ZFMK; examined), China: North Yunnan, Li-kiang. [Junior primary homonym of *Eupithecia inepta* Prout, 1922.]

Eupithecia werneri Vojnits, 1985, *Acta zool. hung.* 31 (1–3): 272. [Replacement name for *Eupithecia inepta* Vojnits, 1979.]

Eupithecia ronkayi Vojnits, 1994, *Entomofauna* 15 (32): 362. [Second replacement name for *Eupithecia inepta* Vojnits, 1979.]

Misidentifications: *Eva petulans* Vojnits, 1981 (1♀ paratype in ZFMK); *Eupithecia placida* Vojnits, 1984 (1♀ paratype in ZFMK).

Description. Fore wing ground colour rusty brown; transverse lines mostly rather inconspicuous; antemedial and medial visible obliquely between discal dot and costa; postmedial with inner blackish brown shade, sharply and sinuously curved onto costa; terminal area relatively broad, slightly darker than rest of wing, with trace of light subterminal line; discal dot relatively large. Hind wing paler with distinct postmedial transverse line; terminal area and anal margin darker, especially near tornus, with trace of light, wavy subterminal line; discal dot relatively large, pale.

Male genitalia. Uncus biapical with long upper arm. Valve with a short deep flange in ventral margin. Vinculum short, with very shallow medial hollow. Papillae elongate, narrow. Aedeagus stout, shorter than valve length. Vesica armed with two stout apical cornuti, one strongly curved, and three cornuti near ductus ejaculatorius base, one comparatively large and horn-like. Sternite A8 two long parallel, apically pointed arms, connected with each other by a relatively narrow basal band.

Female genitalia. Bursa globular, basal part covered with small spines, with some separate larger marginal spines near ductus seminalis base, a sclerotized blotch and patch of stout spines opposite ductus seminalis, and a patch of smaller spines near colliculum. Ductus seminalis attached to upper part of bursa on right side. Colliculum broad. Antrum short and broad, membranous. Tergite A8 rectangular. Apophyses narrow. Papillae anales short and rounded.

Range. China (Tibet, Yunnan, Sichuan).

Phenology. Univoltine: mid-July to early September.

Altitude. Found from about 2000 to 4000 m in Tibet, Yunnan and Sichuan (China: coll. ZFMK and O. Pekarsky).

Similar species. Similar to *E. refertissima*, but distinguished from it by the more uniform rusty brown ground colour, the larger discal dots on all wings, and the darker, rusty brown hind wings. The male genitalia are closer to those of *E. placida*, but have a deeper ventral flange on the valve, a basal cornutus in the aedeagus vesica, and a longer eighth sternite. The female genitalia of the two species are possibly indistinguishable.

Material. ♂, China, Provinz Nord-Yuennan, Li-kiang [Lijiang], 4.ix.1934, H. Höne, Vojnits slide no. 11468♂ (holotype of *E. inepta* Vojnits, 1979, ZFMK);

1♂, same loc., 28.viii.1935, H. Höne, Vojnits slide no. 11466♂ (paratype of *E. inepta* Vojnits, 1979, TTM); 1♀, same loc., 5.ix.1934, H. Höne, Vojnits slide no. 13058♀ (paratype of *E. placida*, ZFMK); 3♂, 1♀, same loc., 27.viii., 3, 7, 8.ix.1934, H. Höne, Vojnits slides nos. 13034♂, 13065♂, 13354♀ (TTM); 1♂, same loc., 6.ix.1935, H. Höne, Mironov slide no. 817♂ (TTM); 1♂, same loc., *ca.* 2000 m, 10.viii.1934, H. Höne, Vojnits slide no. 13161♀ (paratype of *Eva petulans* with incorrectly associated genitalia slide, ZFMK); 6♂, 9♀, China, W. Sichuan, Litang, evn 4000 m, 29°50′ N / 100°21′ E, 16.vii.2009, I. & A. Floriani, Saldaitis leg., Pekarsky slides nos. OP 0533♂, OP 0534♀, OP 0535♀, OP 0536♀, OP 0718♀, OP 0723♂, OP 0724♂, OP 0725♀, OP 0726♀, OP 0736♂, OP 0737♂, OP 0738♂, OP 0740♀, OP 0742♀, OP 0747♀ (coll. O. Pekarsky); 7♂, 7♀, Tibet, Zhag'yab, Gyitang, 3600 m, 19–27.vii.1976; 4♂, 3♀, same loc., 3600 m, 16–17.viii.1976, slides nos. ZL141, ZL255, ZL423, ZL654♂, ZL183, ZL424, ZL442, ZL466, ZL623♀ (all coll Han Yinheng); 1♂, Tibet, Jomda, 3400 m, 27.vii.1976, coll Zhang Xuezhong, slide no ZL462♂ (IZCAS).

Remarks. Dr. Vojnits inexpicably chose the most worn specimen in the Höne material as holotype of this species. He subsequently replaced the original name twice in 1985 as *E. werneri* and in 1994 as *E. ronkayi*.

136. *Eupithecia batangi* Vojnits, 1984, stat. n.

Eupithecia placida batangi Vojnits, 1984, *Acta zool. Acad. Sci. hung.* **30** (3–4): 526, figs 6, 14, 22. Holotype ♂ (TTM; examined), [China]: Tibet [Sichuan], Batang, Tal des Yangtze, 2800 m.

Misidentifications: *Eupithecia garrula* Vojnits, 1984 (1♀ paratype in ZFMK); *Eupithecia ficta* Vojnits, 1984 (3♀ paratypes in ZFMK); *Eupithecia studiosa* Vojnits, 1984 (2♂ paratypes in TTM).

Description. Fore wing ground colour brownish grey; transverse lines blackish, oblique; basal and antemedial sharply angled onto costa; three medial lines sharply angled onto costa beyond discal dot; postmedial oblique, straight or slightly sinuous, twice angled near costa; medial area usually with light, reddish blotch between discal dot and postmedial line; terminal area darker than rest of wing, with a wavy, whitish subterminal line and small white tornal spot; discal dot distinct, elongate. Hind wing slightly paler, except anal margin, with distinct, curved transverse lines and a double, evenly curved, pale transverse band between medial and terminal areas; terminal area with traces of white subterminal line and a small, white tornal spot; discal dot small, pale.

Male genitalia. Uncus slender. Valve with relatively broad and long ventral flange. Vinculum rounded. Papillae elongate, narrow. Aedeagus

shorter than length of valve. Vesica armed with two stout apical, horn-like cornuti, one strongly curved, an elongate, U-shaped cornutus and a smaller, irregular cornutus near ductus ejaculatorius base. Sternite A8 two narrow, parallel, apically pointed arms, connected by a narrow basal band.
Female genitalia. Bursa ovoid, basal half covered with small spines, bordered with much larger spines near base of ductus seminalis, and a separate patch of large spines on the heavily sclerotized wall opposite base of ductus seminalis. Ductus bursae short, broad, with longitudinal wrinkles and a patch of very small spines. Ductus seminalis broadened at base, attached to ductus bursae on right side. Colliculum short and broad. Antrum membranous. Tergite A8 small, almost square. Apophyses relatively broad. Papillae anales small and narrow.
Range. China (Sichuan, Yunnan).
Phenology. From mid-July to early October.
Altitude. Found from about 2800 to 4500 m in Tibet and Yunnan (China: Vojnits 1984b; coll. TTM).
Similar species. Externally similar to *E. placida*, but transverse lines more oblique, and fore wing basal area less mottled. *E. placida* also lacks white tornal spots on all wings. In the male genitalia, the uncus is narrower, the ventral flange of the valve much broader, and the eighth sternite much longer. The female genitalia are very similar to those of *E. werneri* and *E. placida*, but the spines in the patch opposite the ductus seminalis are shorter than in either.
Material. ♂, Tibet [Sichuan], Batang, Im Tal des Yangtze (*ca.* 2800 m), 27.viii.1936, H. Höne, Vojnits slide no. 13796♂ (holotype of *E. placida batangi*, TTM); 2♂, same loc. and alt., 9, 21.ix.1936, H. Höne, Vojnits slides nos. 13864♂, 13838♂ (paratypes of *E. placida batangi*, ZFMK); 1♂, same loc. and alt., 26.viii.1936, H. Höne, Vojnits slide no. 13879♂ (paratype of *E. placida batangi*, TTM); 1♀, same loc. and alt., 18.ix.1936, H. Höne, Vojnits slide no. 13882♀ (paratype of *E. garrula*, ZFMK); 1♀, same loc. and alt., 6. ix.1936, H. Höne, Vojnits slide no. 14046♀ (paratype of *E. ficta*, ZFMK); 2♂, same loc. and alt., 4, 7.ix.1936, H. Höne, Vojnits slides nos. 11558♂, 11559♂ (paratypes of *E. studiosa*, but the latter incorrectly associated with genitalia slide of *E. emanata*, TTM); 2♂, 6♀, same loc. and alt., 1, 4, 8, 16, 23, 26.ix., 2.x.1936, H. Höne, Vojnits slides nos. 12203♂, 13232♀, 13228♀, 13227♀, 13231♀, 13221♀, 13215♀, de Laever slide no. 459♂ (ZFMK); 2♂, same loc. and alt., 25.viii., 16.ix.1936, H. Höne, Vojnits slides nos. 11464♂, 11563♂(paratype of unpubl. *E. effecta*, but incorrectly associate with gen. slide of *E. coniurata*; all in TTM); 2♂, same loc., Untere Urwaldzone (*ca.* 3800 m), 1, 12.viii.1936, H. Höne, Vojnits slides nos. 14042♂, 13912♂ (paratypes of *E. placida batangi*,

ZFMK); 5♂, same loc. and alt., 14.viii., 2, 12.ix., 6.x.1936, H. Höne, Vojnits slides nos. 13807♂, 13819♂, 13850♂, 14040♂, 14047♂ (paratypes of *E. placida batangi*, TTM); 2♀, same loc. and alt., 5, 6.x.1936, H. Höne, Vojnits slides nos. 13875♀, 13873♀ (paratypes of *E. ficta*, ZFMK); 5♂, 5♀, same loc. and alt., 2, 4, 12.viii., 2, 17, 26, 29.ix.1936, H. Höne, de Laever slides nos. 445♂, 433♀, 72♂, 476♀, 478♂, Vojnits slides nos. 12200♀, 13224♂, 12417♂, 12430♀, 12445♀ (ZFMK); 1♂, same loc. and alt., 6.viii.1936, H. Höne, Vojnits slide no. 12197♂(TTM); 1♂, Nord Yünnan, A-tun-tse [Dêqên], Obere Höhe (*ca.* 4500 m), 18.vii.1936, H. Höne, Vojnits slide no. 12181♂ (TTM).
Remarks. As pointed out above, there are significant differences between this taxon and *E. placida*.We therefore raise it to full specific status. In the female genitalia the quantity of large spines in the separate patch in the corpus bursae is highly variable, for example three in Vojnits slide no. 13227♀ to thirteen in slide no. 12200♀ (ZFMK).

137. *Eupithecia sacrosancta* Vojnits, 1979

Eupithecia **sacrosancta** Vojnits, 1979, *Acta zool. Acad. Sci. Hung* 25(1–2), 206, fig 17. Holotype ♂, Vojnits slide no. 11424♂ (ZFMK; examined), China (provenance uncertain).

Description. Only known complete specimen very worn, and only minimal description possible. Fore wing ground colour greyish brown with yellowish tinge; ante- and postmedial transverse lines slightly oblique, right angled onto costa; discal dot more or less rounded. Hind wing concolorous with fore wing.
Male genitalia. Uncus tapered, biapical. Valve with a short but deep ventral flange. Vinculum broad, semicircular. Papillae long and narrow. Aedeagus stout, shorter than length of valve. Vesica armed with two long apical horn-like cornuti, a broad, U-shaped cornutus and a smaller, irregular cornutus near ductus ejaculatorius base. Sternite A8 elongate, with two convergent apical rods.
Female unknown.
Range. China (Shaanxi).
Altitude. Found at about 1300–1500 m on Mt. Tai Bei Shan (China: coll. ZFMK).
Similar species. The only known complete specimen of this species is very worn, and little can be said about external appearance. The male genitalia are generally similar to those of *E. placida*, *E. batangi* and *E. caduca*, but distinguished by the larger valve with narrowly rounded apex, and the longer apical cornuti on the vesica, though shorter than in *E. caduca*. The shape of the eighth sternite is also distinctive.

Material. ♂, Vojnits slide no 11424♂, provenance unknown, incorrectly associated with a specimen of *E. lasciva* (holotype, ZFMK); 1♂, Vojnits slide no. 12482♂, provenance unknown, incorrectly associated with a specimen of *E. fenita*; 1♂, Central China, prov. Shaanxi, Mt. Tai Bei Shan, 33°35′ N, 107°43′ E, 1300–1500 m, 10.viii.-4.ix.1998, leg. V. Murzin & V. Siniaev, Mironov slide no. 746♂ (ZFMK).

Remarks. Mironov & Galsworthy (2011) discussed the difficulties surrounding the type material of this species, which is based on a male genitalia slide wrongly associated with a specimen of a different species. They based the species on the genitalia slide.

138. *Eupithecia placida* Vojnits, 1984

Eupithecia placida Vojnits, 1984, *Acta zool. Acad. Sci. hung.* **30** (3–4): 525, figs 2, 5, 13, 21. Holotype ♂ (TTM; examined), [China]: North Yunnan, A-tun-tse [Dêqên], 4000 m.

Misidentifications: *Eupithecia subplacida* Vojnits, 1984 (1♂ paratype in ZFMK); *Eupithecia sublasciva* Vojnits, 1984 (1♂ paratype in TTM; 1♀ paratype in ZFMK); *Eupithecia depressa* Vojnits, 1984 [Junior primary homonym of *Eupithecia depressa* Vojnits, 1979] (1♀ paratype in ZFMK).

Description. Fore wings distinctively narrow and pointed and hind wings elongate and narrow. Fore wing ground colour dark greyish brown; dark transverse lines distinct, oblique, sharply angled onto costa; ante- and postmedial forming dark costal spots; double medial touching discal dot; postmedial twice angled near costa; medial area with distinct reddish blotch under discal dot; terminal area with fine, wavy, whitish subterminal line broadened near tornus and forming a small, elongate white tornal spot; discal dot elongate and oblique. Hind wing slightly paler than fore wing; transverse lines faint except along anal margin; terminal area slightly darker with trace of indented pale subterminal line; discal dot small, almost invisible.

Male genitalia. Uncus bell-shaped, biapical. Valve with long but very narrow ventral flange. Vinculum broad, rounded. Papillae rather narrow. Aedeagus stout, shorter than valve length. Vesica armed with two stout apical cornuti, very constant in shape, a broad U-shaped cornutus and a smaller, irregular cornutus near ductus ejaculatorius base. Sternite A8 broadened at base, with two narrow apical rods.

Female genitalia. Bursa globular, almost completely covered with spines, with some separate larger spines near ductus seminalis base. A heavily sclerotized blotch and patch of large, stout spines opposite ductus seminalis. Ductus bursae short and broad with a patch of very small spines near base of colliculum. Ductus seminalis broadened basally, attached to

ductus bursae on right side. Colliculum short and broad. Antrum membranous. Tergite A8 transverse. Apophyses rather short. Papillae anales relatively short, rounded.
Range. China (Yunnan).
Phenology. Univoltine (?): from late July to mid-August.
Altitude. Found from about 3500 to 4500 m in Yunnan (China: Vojnits 1984; coll. TTM).
Similar species. Similar to *E. batangi* but distinguished by the wing shape, generally less reddish tone, and also the much stronger transverse lines in the basal fore wing area. The male genitalia are close to those of *E. batangi*, but the ventral flange of the valve is much narrower, and the eighth sternite a different shape. The female genitalia resemble those of *E. refertissima*, *E. caduca* and *E. werneri*. However, the former two can be distinguished by the broadened and flattened apices of their posterior apophyses. *E. werneri* is difficult to separate, but has a larger corpus bursae with shorter stout spines in the patch opposite the ductus seminalis.
Material. ♂, Nord Yünnan, A-tun-tse [Dêqên], Mittlere Höhe (*ca.* 4000 m), 16.viii.1936, H. Höne, Vojnits slide no. 13788♂ (holotype, TTM); 2♂, same loc. and alt., 9, 18.viii.1936, H. Höne, Vojnits slides nos. 13277♂, 13273♂ (paratypes, ZFMK); 5♂, same loc. and alt., 4, 11, 16, 17, 18.viii.1936, H. Höne, Vojnits slides nos. 13275♂, 13283♂, 13297♂, 13316♂, 13789♂ (paratypes, TTM); 1♂, same loc. and alt., 18.viii.1936, H. Höne, Vojnits slide no. 13313♂ (paratype of *E. subplacida*, incorrectly associated with gen. slide of *E. refertissima*, ZFMK); 1♀, same loc. and alt., 1.ix.1936, H. Höne, Vojnits slide no. 13284♀ (paratype of *E. sublasciva*, ZFMK); 1♂, same loc. and alt., 18.viii.1936, H. Höne, Vojnits slide no. 13298♂ (paratype of *E. sublasciva*; incorrectly associated with gen. slide of *E. egena*, TTM); 6♂, 1♀, same loc. and alt., 12.viii.1936, H. Höne, Vojnits slide no. 13326♀ (paratype of *E. depressa* Vojnits, 1984, ZFMK); 1♀, same loc. and alt., 12, 14, 16.viii.1936, H. Höne, Vojnits slides nos. 11540♂, 11541♀, 12128♂, 12129♂, 12130♂, 12164♂, 12175♂ (TTM); 1♀, same loc., Talsohle (*ca.* 3500 m), 18.vi.1936, H. Höne, Vojnits slide no. 12174♀ (ZFMK); 1♀, same loc., Obere Höhe *ca.* 4500 m, 23.vii.1936, H. Höne, Vojnits slide no. 10888♀ (TTM).
Remarks. According to the original label, the holotype specimen of *E. placida* was collected on 16.viii.1936, not 1916 as in Vojnits 1984.

139. *Eupithecia laudabilis* Vojnits, 1984

Eupithecia laudabilis Vojnits, 1984, *Acta zool. Acad. Sci. hung.* **30** (3–4): 527, figs 8, 16, 24. Holotype ♂ (TTM; examined), [China]: North Yunnan, A-tun-tse [Dêqên], 4000 m.

Eupithecia exacerbata Vojnits, 1984, *Acta zool. Acad. Sci. hung.* **30** (3–4): 528, figs 11, 19, 27. Holotype ♂ (TTM; examined), [China]: North Yunnan, A-tun-tse [Dêqên], 4000 m. **New junior synonym.**

Misidentification: *Eupithecia depressa* Vojnits, 1984 [Junior primary homonym of *Eupithecia depressa* Vojnits, 1979] (1♀ paratype in ZFMK; 2♀ paratypes in TTM).

Description. Fore wing ground colour mid brown; transverse lines often obscure, but double medial sometimes visible touching discal dot; curved postmedial marked by black dashes on veins; discal dot large, round, with a large reddish blotch immediately below it; fainter reddish apical and medial streaks beyond postmedial;a faint wavy white subterminal line ending in a tornal dot. Hind wing off white, sharply contrasting with dark brown area along anal margin extending onto terminal margin; discal dot small but distinct.

Male genitalia. Uncus rather short, biapical. Valve with ventral flange very short and narrow. Vinculum broad, rounded. Papillae long. Aedeagus shorter than length of valve. Vesica armed with three slim horn-like cornuti, an elongate, U-shaped cornutus and an irregular cornutus, both placed near ductus ejaculatorius base. Sternite A8 broadened at base, with two short apical horns.

Female genitalia. Bursa rounded, completely covered with spines. Ductus bursae slightly sclerotized, equal to length of corpus bursae, with a patch of 4–6 large spines at the middle and numerous small spines near colliculum. Ductus seminalis broad, attached to ductus bursae on right side. Colliculum relatively short and narrow. Antrum membranous. Tergite A8 almost square. Apophyses narrow. Papillae anales rounded.

Range. China (Yunnan, Sichuan).

Phenology. Bivoltine (?): mid-June to early September.

Altitude. From about 3000 to 4500 m (China: coll. O. Pekarsky; Vojnits 1984).

Similar species. Distinctive within the group by the reddish fore wing patches, though these occur in species in different groups, eg. *E. honesta*, which is very similar. Best recognised by examination of genitalia. The male genitalia are distinguished from those of *E. placida* by the presence of three horn-like cornuti on the vesica, and the shorter narrow ventral flange on the valve. The female genitalia resemble those of *E. intolerabilis* and *E. mentita*, but differ in the more rounded corpus bursae and the smaller separate patch of spines in the ductus bursae. *E. intolerabilis* also has elongate and pointed papillae anales.

Material. ♂, Nord-Yünnan, A-tun-tse [Dêqên], Mittlere Höhe (*ca.* 4000 m), 9.viii.1936, H. Höne, Vojnits slide no. 14159♂ (holotype, TTM); ♂, same loc. and alt., 3.vii.1936, H. Höne, Vojnits slide no. 14058[2]♂ (holotype of

E. exacerbata, TTM); 1♀, same loc. and alt., 30.viii.1936, H. Höne, Vojnits slide no. 14176♀ (paratype of *E. depressa* Vojnits, 1984, ZFMK); 1♀, same loc. and alt., 6.ix.1936, H. Höne, Vojnits slide no. 13786♀ (paratype of *E depressa* Vojnits, 1984, TTM); 2♀, same loc. and alt., 31.vii., 12.viii.1936, H. Höne, Vojnits slides nos. 10989♀, 12165♀ (ZFMK); 1♂, same loc. and alt., 13.vi.1936, H. Höne, Vojnits slide no. 12132♂ (TTM); 1♂, same loc., Obere Höhe (*ca.* 4500 m), 18.vii.1936, H. Höne, Vojnits slide no. 14088♂ (paratype of *E. laudabilis*, TTM); 2♂, 4♀, same loc. and alt., 24.vii., 11, 13, 15, 18, 23.viii.1936, H. Höne, Vojnits slides nos. 12185♀, 12186♀, 12493♂, 13247♀, 13785♀, de Laever slide no. 81♂ (ZFMK); 3♂, 1♀, same loc. and alt., 8, 10.vii., 3, 12. viii.1936, H. Höne, Vojnits slides nos. 11542♀, 12171♂, 12187♂, 12188♂ (TTM); 1♀, same loc. and alt., 23.viii.1936, H. Höne, Vojnits slide no. 13922♀ (paratype of *E depressa* Vojnits, 1984, TTM); 2♂, China, Provinz Nord-Yuennan, Li-kiang [Lijiang], 5, 23, 25.vii.1935, H. Höne, Vojnits slide no. 12170♂, Mironov slide no. 825♂ (TTM); 2♂, China, W. Sichuan, road Yajiang / Litang, 3600 m, 15.vii.2009, 29°59′ N / 100°52′ E, I. & A. Floriani, Saldaitis leg., Pekarsky slides nos. OP 0633♂, OP 0634♂ (coll. O. Pekarsky); 1♀, Batang, Yidun, 3350 m, 16.viii.1982, Zhang Xuezhong and Wang Shuyong, slide no. ZL436♀ (IZCAS); 1♀, KIB/RBGE research stn., 1♂, Gan He Ba, both Lijiang, Yunnan prov., PRC, 15–20.vi.2009, A. C. Galsworthy, BM Geometrid slide no. 22687♂ (BMNH).

Remarks. The female genitalia of *E. laudabilis* were illustrated as *E. depressa* (Vojnits 1984b, fig. 32).

140. *Eupithecia manifesta* Mironov & Galsworthy, 2011

Eupithecia manifesta Mironov & Galsworthy, 2011, *Lep. Science* **62**: 25, figs 12, 33. Holotype ♀ (ZFMK; examined), China: Yunnan.

Description. Labial palpi, frons and vertex brown. Notum light grey-brown with dark brown transverse band in front. Fore wing ground colour brownish with rusty tinge; transverse lines inconspicuous, postmedial twice right angled onto costa; discal dot very distinct, ovoid, black; terminal area with traces of narrow, light subterminal line and a series of blackish dashes along termen between veins. Hind wing off-white in anterior half and brown along anal and terminal margins; transverse lines indistinct; terminal area with small white tornal spot; discal dot small and pale.

Female genitalia. Bursa small, oval, completely covered with small spines. Ductus bursae elongate, more strongly sclerotized along right side, with two membranous diverticula on opposite sides, a small patch of large spines on the right side, and a patch of minute spines near base of colliculum.

Ductus seminalis long, slightly broadened near base, attached to base of ductus bursae on left side. Colliculum short. Antrum membranous, broadening posteriorly. Tergite A8 almost square. Apophyses narrow. Papillae anales small, ovoid.
Male unknown.
Range. China (Yunnan).
Altitude. Found at about 3000 m in Yunnan (China: coll. ZFMK).
Similar species. Closest to *E. caduca* and distinguishable from it with difficulty by the larger discal dots on the fore wings. The female genitalia are distinctive, with only a remote resemblance to those of *E. laudabilis*, from which it is easily distinguished by the presence of the two membranous diverticula on the ductus bursae.
Material. ♀, China, Provinz Nord-Yuennan, Li-kiang [Lijiang], *ca.* 3000 m, 4.x.1934, H. Höne, Vojnits slide no. 13117♀ (holotype, ZFMK); 1♀, same data, Vojnits slide no. 11447♀ (paratype, ZFMK).
Remarks. This species, with the male as yet unknown, is tentatively assigned to the present species group.

141. *Eupithecia intolerabilis* Vojnits, 1981

Eupithecia intolerabilis Vojnits, 1981, *Acta zool. Acad. Sci. hung.* 27 (3–4): 423, fig. 45. Holotype ♀ (ZFMK; examined), China: North Yunnan, Li-kiang.

Misidentifications: *Eupithecia subplacida* Vojnits, 1984 (1♂ paratype in ZFMK); *Eupithecia ficta* Vojnits, 1984 (1♀ paratype in TTM).

Description. Fore wing ground colour greyish brown, transverse lines pale, oblique, sharply angled onto costa; medial slightly sinuate; postmedial doubly angled and with a series of dark dashes on the veins; basal half of medial area slightly darker; terminal area with distinct, wavy, whitish subterminal line; discal dot relatively small, linear. Hind wing greyish white, dark transverse lines well marked along anal margin only; terminal area slightly darker; discal dot small.
Male genitalia. Uncus tapered, biapical. Valve with very narrow and short ventral flange. Vinculum broad, semicircular. Papillae short, relatively broad. Aedeagus shorter than length of valve. Vesica armed with one stout and straight, and one narrow curved cornutus at apex, a broad, U-shaped plate and an irregular cornutus near ductus ejaculatorius base. Sternite A8 two lyre-shaped, lateral rods connected by a narrow basal band.
Female genitalia. Bursa rounded, almost completely covered with slim spines. Ductus bursae merging with corpus, tapered to colliculum, its left half membranous with numerous striations; the right half sclerotized,

with a basal row of large spines and a patch of numerous spinules from middle to colliculum. Ductus seminalis broadened at base, attached to ductus bursae from left side. Colliculum short and broad. Antrum lightly sclerotized, covered by numerous, minute pores. Tergite A8 almost quadrate, with medially convex and broadly sclerotized anterior margin. Posterior apophyses slightly expanded at apices. Papillae anales elongate and narrow, pointed.

Range. China (Sichuan, Yunnan).

Altitude. Found at about 3800–4000 m in Tibet and Yunnan (China: coll. ZFMK and TTM).

Similar species. Generally similar to other members of the group which lack a reddish patch on the fore wing, but distinguished from most by the more contrasting pale hind wings. Usually larger than *E. werneri*. Most similar to *E. refertissima* and best distinguished by examination of genitalia. In the male, the ventral flange of the valve is narrow, as in *E. placida* and *E. laudabilis*, but the apical cornutus is straight and the eighth sternite a different shape. The female genitalia are more elongate, similar to those of *E. laudabilis* and *E. mentita*, but differ in the shape of the large patch of spines and the pointed papillae anales.

Material. ♀, China, Provinz Nord-Yuennan, Li-kiang [Lijiang], 15.viii.1935, H. Höne, Vojnits slide no. 13176♀ (holotype, ZFMK); 1♂, Nord Yünnan, A-tun-tse [Dêqên], Mittlere Höhe (*ca.* 4000 m), 23.vii.1936, H. Höne, Vojnits slide no. 14155♂ (gen. slide of *E. refertissima* incorrectly associated, TTM; paratype of *E. subplacida*, specimen in ZFMK); 1♂, same loc. and alt., 4.ix.1936, H. Höne, Vojnits slide no. 11546♂ (TTM); 1♀, Tibet [Sichuan], Batang, Untere Urwaldzone (*ca.* 3800 m), 7.viii.1936, H. Höne, Vojnits slide no. 14048♀ (paratype of *E. ficta*, TTM).

142. *Eupithecia mentita* Vojnits, 1984

Eupithecia mentita Vojnits, 1984, *Acta zool. Acad. Sci. hung.* **30** (1–2): 235, figs 29, 32. Holotype ♂ (TTM; examined), [China]: Tibet [Sichuan], Batang, 3800 m.

Misidentification: *Eupithecia depressa disiuncta* Vojnits, 1984 (1♀ paratype in ZFMK).

Description. Rather a broad winged species. Fore wing ground colour greyish brown with reddish blotch below discal dot; transverse lines pale and distinct; basal sharply angled onto costa; antemedial right angled; double medial oblique, sharply angled onto costa and touching discal dot; postmedial slightly sinuate, twice right angled near costa; lighter transverse band between medial and terminal areas with reddish tinge; terminal area relatively broad with wavy, white subterminal line and oblique, reddish apical streak; discal dot distinct, rather small. Hind wing off-white; dark transverse lines distinct, especially evenly curved postmedial; anal margin darker from base to postmedial; terminal area darker with wavy, whitish subterminal line; light transverse band between medial and terminal areas distinct; discal dot relatively large.

Male genitalia. Uncus short, stout, biapical. Valve with a broad, angled, elongate flange. Vinculum broad, with medial hollow. Papillae elongate, narrow. Aedeagus relatively stout, shorter than length of valve. Vesica with elongate patch of spinules, and armed apically with a hooked plate-like cornutus, and a very small horn-like cornutus, medially with one long narrow and one short broad cornutus, and basally with one U-shaped cornutus and one smaller irregular cornutus. Sternite A8 elongate with two parallel rods broadened and rounded apically.

Female genitalia. Bursa narrow and ovoid, covered with small spines in basal half, with some marginal spines near ductus seminalis base slightly larger; a patch of very long spines opposite ductus seminalis, the latter strongly broadened basally, attached to central part of corpus bursae on right. Colliculum slightly oblique. Antrum short and broad, membranous. Tergite A8 almost square. Apophyses narrow, the posterior broadened and flattened at apices. Papillae anales rounded.

Range. China (Sichuan, Yunnan).

Phenology. August.

Altitude. Found at about 380 m above sea-level in Tibet (China: Vojnits 1984a).

Similar species. This species has a contrasting hind wing similar to *E. intolerabilis*, but broader wings, and extensive reddish markings. The male genitalia are easily distinguishable, with a very broad ventral flange to the valve, a large plate at apex of the vesica, and distinctive apical arms on the eighth sternite. The female genitalia are also distinct, with a shortened oval shape and an outsize bunch of spines opposite the ductus seminalis.

Material. ♂, Tibet [Sichuan], Batang, Untere Urwaldzone (*ca.* 3800 m), 6. viii.1936, H. Höne, Vojnits slide no. 13832♂ (holotype, TTM); 1♂, China, Provinz Nord-Yuennan, Li-kiang [Lijiang], 14.viii.1935, H. Höne, Vojnits slide no. 13352♂ (paratype, ZFMK); 1♀, same data, Vojnits slide no. 13346♀ (paratype of *E. depressa disiuncta*, ZFMK).

143. *Eupithecia fenita* Mironov & Galsworthy, 2004

Eupithecia fenita Mironov & Galsworthy, 2004, *Trans. lepid. Soc. Japan* 55 (2): 122, figs 5, 17, 20. Holotype ♂ (ZFMK; examined), [China]: North Yunnan, A-tun-tse [Dêqên], 4500 m.

Misidentification: *Eupithecia placida* Vojnits, 1984 (2♂, 2♀ paratypes in ZFMK; 1♂, 3♀ paratypes in TTM); *Eupithecia subplacida* Vojnits, 1984 (3♂ paratypes in ZFMK; 5♂ paratypes in TTM).

Description. Fore wing ground colour pale brown; most transverse lines indistinct, but basal half of wing irrorated with darker brown scales; postmedial dark brown, angled close to costa, paralleled distad by a pale buff line; discal dot large, dark brown, with a large orange marking below it; terminal area darker, with a pale apical streak, and crossed by a wavy white submarginal. Hind wing white, with well marked discal dot; anal margin with all transverse lines well marked, but terminating abruptly at disc, except for double postmedial, which continues faintly towards costa.

Male genitalia. Uncus stout, biapical. Valve with a heavily sclerotized sacculus, but no flange, and a small blunt process above it. Vinculum narrow, with shallow medial hollow. Papillae elongate and narrow. Aedeagus stout, vesica armed with two apical and one basal horn-like cornuti, a bowed U-shaped plate, an irregular plate at ductus ejaculatorius base, and an elongate patch of minute spines. Sternite A8 rather narrow, with two blunt apical rods.

Female genitalia. Bursa pear-shaped, with a restricted area of spines at base, and a row of large spines on the left. Ductus bursae with longitudinal striations and a large patch of minute spines. Ductus seminalis broadened at base, attached to ductus bursae on right side. Colliculum narrow. Antrum membranous. Tergite A8 transverse. Apophyses short and narrow, the posterior expanded at tips. Papillae anales short and narrow.

Range. China (Yunnan, Sichuan).

Phenology. Univoltine: early July to early September.

Altitude. From about 2800 to 4500 m above sea-level in North Yunnan (China: coll. ZFMK).

Similar species. The orange fore wing mark is distinctive, and similar in this group only to *E. laudabilis*, from which it can be distinguished by the more distinct transverse lines in the anal area of the hind wing. See under latter for other confusion species. The male genitalia have a thickened sclerotized sacculus in place of the usual flange, but otherwise resemble those of *E. werneri*. In the female the bursa is slightly elongate, and most resembles that of *E. mentita* in the more restricted area of spining, but the spines opposite the ductus seminalis are smaller.

Material. ♂, Nord Yünnan, A-tun-tse [Dêqên], Obere Höhe (*ca.* 4500 m), 14.vii.1937, H. Höne, Vojnits slide no. 14066♂ (holotype, [labelled as paratype of *E. subplacida*], ZFMK); 2♂, 2♀, same loc. and alt., 21, 22.viii.1936,

H. Höne, Vojnits slides nos. 14092♂, 13957♀, 14111♂, 14098♀ (paratypes [all labelled as paratypes of *E. placida*], ZFMK); 1♂, same loc. and alt., 10.viii.1936, H. Höne, Vojnits slide no. 13748♂ (paratype [labelled as paratype of *E. subplacida*], ZFMK); 2♂, same loc. and alt., 17.vii., 12.viii.1936, H. Höne, Vojnits slide no. 12482♂ (incorrectly associated with gen. slide of *E. sacrosancta*), de Laever slide no. 78♂ (paratypes, ZFMK); 1♂, 2♀, same loc., 7.vii., 9, 14.viii.1936, H. Höne, Vojnits slides nos. 13318♀, 14058♀, 14093♂ (paratypes of *E. placida*, TTM); 1♀, same loc., Talsohle, *ca.* 3500 m, 11. vii.1936, H. Höne, Vojnits slide no. 12172♀ (paratype, ZFMK); 1♂, same loc., Mittlere Höhe (*ca.* 4000 m), 11.viii.1936, H. Höne, Vojnits slide no. 13291♀ (incorrectly associated with gen. slide of *E. subfuscata*; paratype, ZFMK); 2♂, China, Provinz Nord-Yuennan, Li-kiang [Lijiang], X, 19.viii.1934, 15.viii.1935, H. Höne, Vojnits slides nos. 13345♂, 13353♂ (paratypes [labelled as paratypes of *E. subplacida*], ZFMK); 1♀, same loc., 9.ix.1935, H. Höne, Vojnits slide no. 12532♀ (paratype [labelled as paratype of unpubl. *E. effecta*, ZFMK); 5♂, same loc., 17.viii.1934, 4, 10, 14, 28.viii.1935, H. Höne, de Laever slides nos. 135♂, 286♂, 314♂ (incorrectly associated with gen. slide of *E. granata*), 137♂, Mironov slide no. 510♂ (paratypes, ZFMK); 1♀, same loc., 27.viii.1934, H. Höne, Vojnits slide no. 12985♀ (paratype of *E. placida*, TTM); 5♂, same loc., 1, 6, 28.viii., 6.ix.1935, H. Höne, Vojnits slides nos. 13073♂, 13078♂, 13250♂, 13254♂, 13363♂ (paratypes of *E. subplacida*, TTM); 9♂, same loc., 19, 28.viii.1934, 3, 14, 18, 24.viii.1935, H. Höne, Vojnits slides nos. 12131♂, 12136♂, 12138♂, 12140♂, 12159♂, 12162♂, 13150♂ (last slide an incorrectly associated slide of *E. coniurata*; TTM); 1♂, same loc., *ca.* 2000 m, 8.viii.1934, H. Höne, Vojnits slide no. 12139♂ (TTM); 1♀, China, W. Sichuan, Litang evn 4000 m, 29°50′ N / 100°21′ E, 16.vii.2009, I. & A. Floriani, Saldaitis leg., Pekarsky slide no. OP 0729♀ (coll. O. Pekarsky).

Remarks. Some of the males of this species were listed by Vojnits (1984) as paratypes of *E. placida*, and some as paratypes of *E. subplacida*. The females were all listed as paratypes of *E. placida*. One worn specimen was included in the type series of *E. fenita* in error (Mironov et al. 2004b). It belongs to an unidentified different species but has an incorrectly associated slide of *E. fenita* (Vojnits slide no. 12490♀, ZFMK).

144. *Eupithecia refertissima* Vojnits, 1984

Eupithecia refertissima Vojnits, 1984, *Acta zool. Acad. Sci. hung.* **30** (1–2): 223, figs 16, 25. Holotype ♂ (TTM; examined), [China]: Tibet [Sichuan], Batang, 3800 m.

Eupithecia subplacida Vojnits, 1984, *Acta zool. Acad. Sci. hung.* **30** (3–4): 526, figs 3, 7, 15, 23. Holotype ♂ (TTM; examined), [China]: North Yunnan, A-tun-tse [Dêqên], 4000 m. **New junior synonym.**

Misidentifications: *Eupithecia placida* Vojnits, 1984 (3♂ paratypes in ZFMK); *Eupithecia placida batangi* Vojnits, 1984 (2♂ paratypes in ZFMK); *Eupithecia subplacida* Vojnits, 1984 (20♂, 5♀ paratypes in ZFMK; 14♂, 6♀ paratypes in TTM); *Eupithecia sublasciva* Vojnits, 1984 (1♀ paratype in ZFMK); *Eupithecia ficta* Vojnits, 1984 (16♀ paratypes in ZFMK; 14♀ paratypes in TTM).

Description. Fore wing brownish with ochreous tinge, with numerous, fine, oblique transverse lines, all sharply angled near costa; postmedial double, twice angled onto costa; a pale wavy subterminal line; discal dot black, rounded, normally with a streak of orange scales on vein below it. Hind wing white, with smudgy brown transverse lines along anal margin only; discal dot very small.

Male genitalia. Uncus stout, biapical. Valve with a fairly broad ventral flange. Vinculum short and broad. Papillae elongate and narrow. Aedeagus stout, shorter than valve length. Vesica armed with three apical and one central horn-like cornutus, a narrow patch of minute spinules, a U-shaped cornutus and a large irregular cornutus at ductus ejaculatorius base. Sternite A8 elongate and narrow, slightly broadened at base, with two parallel tapered arms.

Female genitalia. Bursa pear-shaped, basal 2/3 covered with spines, with a marginal patch of larger spines on left side, and a separate patch of large spines on right side. A transverse patch of short spinules runs from the base of the colliculum to the base of the ductus seminalis. Ductus bursae not separately distinguishable. Ductus seminalis very broad, attached to posterior part of corpus bursae near base of colliculum on left side. Colliculum short and broad. Antrum not expanded. Tergite A8 transverse, broadened to posterior margin. Apophyses elongate and stout, the posterior expanded and flattened at apices. Papillae anales sclerotized, elongate and narrow.

Range. China (Sichuan, Yunnan, Tibet).

Phenology. Univoltine: early August to late September.

Altitude. Found between about 3000 and 4000 m above sea-level in Tibet (China: Vojnits 1984).

Similar species. Within the group, rather similar to *E. mentita*, but with less prominent ochreous markings. Outside the group, almost indistinguishable from *E. egena*, except for, usually, a more ochreous tone. Identification is best made by examination of the genitalia. *E. egena* belongs to the *subfuscata* species group and has entirely different genitalia. The male genitalia of *E. refertissima* resemble those of *E. werneri* but are distinguished by the presence of a third, rather small, apical cornutus on the vesica. The female genitalia are similar to those of *E. werneri*, but distinguishable by the larger spiniferous area below the colliculum, and the broader apophyses.

Material. ♂, Tibet [Sichuan], Batang, Untere Urwaldzone (*ca.* 3800 m), 22.ix.1936, H. Höne, Vojnits slide no. 13883♂ (holotype, TTM); 16♀, same loc. and alt., 7.viii., 2, 10, 12, 17, 22, 29.ix., 5.x.1936, H. Höne, Vojnits slides nos. 14039♀, 13814♀, 13816♀, 13818♀, 13887♀, 13842♀, 13897♀, 13869♀, 13878♀, 13892♀, 13817♀, 13857♀, 13810♀, 13908♀, 13848♀, 13874♀ (paratypes of *E. ficta*, ZFMK); 2♂, same loc. and alt., 4.viii., 29.ix.1936, H. Höne, Vojnits slides nos. 13805♂, 13806♂ (paratypes of *E. placida batangi*, ZFMK); 3♂, 14♀, same loc. and alt., 12, 13.vi., 4, 7, 12.viii., 10, 12, 15, 17, 28, 29.ix., 5.x.1936, H. Höne, Vojnits slides nos. 12143♀, 12148♀, 13218♂, 12441♀, 13223♂, 12202♀, 12437♀, 12448♀, 12411♀, 12416♀, 12410♀, 12438♀, 12425♀, de Laever slides nos. 299♀, 454♂, 480♀, 73♀ (ZFMK); ♂, Nord Yünnan, A-tun-tse [Dêqên], Mittlere Höhe (*ca.* 4000 m), 17.viii.1936, H. Höne, Vojnits slide no. 13330♂ (holotype of *E. subplacida*, TTM); 19♂, 3♀, same loc. and alt., 23, 26, 27, 28, 31.vii., 1, 2, 4, 11, 12, 17, 18.viii.1936, H. Höne, Vojnits slides nos. 13939♂, 14072♂, 14051♂, 14060♂, 13945♀, 14057♂, 13340♂, 13951♂, 13954♂, 13959♂, 14063♂, 14158♂, 13339♂, 14064♂, 14157♀, 14091♂, 13311♀, 13332♂, 13337♂, 13341♂, 13342♂, 14177♂ (paratypes of *E. subplacida*, ZFMK); 14♂, 5♀, same loc. and alt., 23, 26, 27, 28, 30.vii., 2, 3, 4, 10, 11, 17, 18.viii.1936, H. Höne, Vojnits slides nos. 13272♂, 13274♂, 13276♂, 13278♂, 13279♂, 13281♂, 13282♂, 13290♂, 13310♂, 13314♂, 13315♂, 13328♀, 13336♂, 13934♀, 13953♀, 14052♀, 14055♂, 14059♀, 14069♂ (paratypes of *E. subplacida*, TTM); 1♀, same loc. and alt., 16.viii.1936, H. Höne, Vojnits slide no. 13299♀ (paratype of *E. sublasciva*, ZFMK); 3♂, same loc. and alt., 16, 18, 30.viii.1936, H. Höne, Vojnits slides nos. 13296♂, 13312♂, 13335♂ (all three incorrectly associated with gen. slides of *E. placida*) (paratypes of *E. placida*, ZFMK); 1♀, same loc. and alt., 1.viii.1936, H. Höne, Vojnits slide no. 12473♂ (incorrectly associated with gen. slide of ♂ *Eva flexa*; paratype of unpubl. *E. suavissima* Vojnits, ZFMK); 3♂, 4♀, same loc. and alt., 1, 4, 5, 11, 12, 18.viii., 6.ix.1936, H. Höne, Vojnits slides nos. 12514♀, 12474♂, 12520♂, 13242♂, 13244♀, 13241♀, 13288♀ (incorrectly assoicated with gen. slide of *E. subfuscata*) (ZFMK); 2♂, same loc. and alt., 26.vii., 7.ix.1936, H. Höne, Vojnits slides nos. 12158♂, 13070♂ (TTM); 1♀, same loc. and alt., 9.viii.1936, H. Höne, Vojnits slide no. 12454♀ (TTM); 2♀, same loc., Obere Höhe (*ca.* 4500 m), 11, 16.viii.1936, H. Höne, Vojnits slides nos. 13303♀, 13325♀ (paratypes of *E. subplacida*, ZFMK); 1♀, same loc. and alt., 14.vii.1937, H. Höne, Vojnits slide no. 13947♀ (paratype of *E. subplacida*, TTM); 1♂, 1♀, same loc. and alt., 28.vii., 11.viii.1936, H. Höne, Vojnits slides nos. 12156♀, 12467♂ (ZFMK); 1♀, same loc. and alt., 9.viii.1936, H. Höne, Vojnits slide no. 12454♀ (TTM); 1♂, China,

Provinz Nord-Yuennan, Li-kiang [Lijiang], 23.vii.1935, H. Höne, Vojnits slide no. 13944♂ (paratype of *E. subplacida*, labelled as paratype by Mironov in 2009, ZFMK); 1♂, same loc., 11.viii.1934, H. Höne, Vojnits slide no. 11530♂ (incorrectly associated with gen. slide of *E. lobbichlerata*, ZFMK); 14♀, Tibet [Sichuan], Batang, Untere Urwaldzone (*ca.* 3800 m), 3, 3, 6, 11.viii., 2, 12, 22, 26, 29.ix., 1, 6.x.1936, H. Höne, Vojnits slides nos. 13790♀, 13815♀, 13846♀, 13849♀, 13855♀, 13872♀, 13877♀, 13890♀, 13893♀, 13894♀, 13888♀, 13900♀, 13902♀, 14043♀ (paratypes of *E. ficta*, TTM); 1♂, same loc. and alt., 15.ix.1936, H. Höne, de Laever slide no. 454♂ (ZFMK); 9♂, 2♀, Tibet, Jomda, 3400 m, 22–31.vii.76, slides nos. ZL33, 108♂, ZL622♀; 2♂, 4♀, Tibet, Zhag'yab, Gyitang, 3600 m, 18–28.viii.1976, all Zhang Xuezhong and Han Yinheng, slides nos. 35, 446, 468, 182♀; 2 ♂, 2♀, Yunnan, Lijiang, Yulongshan, 11–21.vii.62, coll Song Shimei, slides nos. ZL106, 624♀; 1♀, Sichuan, Dukou, Lanmuqiao, 2150 m, 21.viii.80, coll Han Yinheng. slide no. ZL190♀ (IZCAS).

Remarks. Two of the apical cornuti are broken in Vojnits slide no. 13883♂ belonging to the holotype specimen.

145. *Eupithecia avara* Vojnits, 1984

Eupithecia avara Vojnits, 1984, *Acta zool. Acad. Sci. hung.* **30** (3–4): 530, figs 4, 10, 18, 26. Holotype ♂ (TTM; examined), China: North Yunnan, Li-kiang [Lijiang].

Eupithecia ***testacea*** Mironov & Galsworthy, 2004, *Trans. lepid. Soc. Japan* **55** (2): 120, figs 4, 16. Holotype ♂ (ZFMK; examined), [China]: North Yunnan. **New junior synonym.**

Misidentification: *Eupithecia mentita* Vojnits, 1984 (1♀ paratype in TTM).

Description. Fore wing ground colour pale brown, tinged orange distally along inner margin, and below discal spot; postmedial line doubly angled close to costa, bordered externally by a narrow pale fascia; discal dot dark brown, elongate; terminal area slightly darker, crossed by a wavy pale subterminal line ending in a white tornal spot. Hind wing ground colour off-white, suffused with darker brown in anal area; discal dot small and faint.

Male genitalia. Uncus tapered, biapical. Valve with ventral flange moderately broad, and with a strong prominence beyond it. Vinculum broad, trapezoid. Papillae elongate, clavate. Aedeagus shorter than length of valve, vesica with two apical cornuti, one strongly hooked, a U-shaped plate-like cornutus and a smaller, irregular, cornutus near ductus ejaculatorius base. Sternite A8 elongate, with two parallel, blunt-tipped apical rods.

Female genitalia. Bursa relatively small, globular, mostly covered with small spines. Ductus bursae broad, anteriorly heavily sclerotized, with numerous wrinkles, an elongate patch of spinules and posteriorly a broad

membranous diverticulum on the right side. Ductus seminalis attached to central part of ductus bursae on left under membranous diverticulum. Colliculum not present. Antrum relatively short and broad, membranous, with numerous pores in posterior part. Tergite A8 small, almost square. Apophyses very narrow, elongate. Papillae anales small, narrow, slightly elongate and apically rounded.
Range. China (Yunnan).
Phenology. Univoltine: August.
Similar species. Similar to *E. caduca* but distinguished by the larger size, more elongate wings and the more distinct white tornal spots on the fore wings. The short, strongly curved apical cornutus in the aedeagus vesica is distinctive. The female genitalia lack a separate group of large spines in the bursa, but have a very distinctive sclerotized central area.
Material. ♂, China, Provinz Nord-Yuennan, Li-kiang [Lijiang], 29.viii.1934, H. Höne, Vojnits slide no. 13059♂ (holotype, TTM); ♂, same loc., 19.viii.1934, H. Höne, de Laever slide no. 131♂ (holotype of *E. testacea*, ZFMK); 1♂, same loc., 19.viii.1934, H. Höne, de Laever slide no. 354♂ (paratype of *E. testacea*, ZFMK); 1♀, same loc., 21.viii.1935, H. Höne, Vojnits slide no. 14194♀ (paratype of *E. mentita*, TTM); 1♂, same loc., 25.viii.1935, H. Höne, Vojnits slide no. 13351♂ (*E. sacrosancta* Vojnits det., TTM); 2♂, same loc., 16.viii.1934, 12.viii.1935, H. Höne, Vojnits slides nos. 13021♂, 13177♂ (TTM).
Remarks. The red labels marking the holotype male and paratype female of *E. avara* were exchanged on the two specimens by the original author or his assistants. The two specimens belong to different species. The description by us of a supposed new species, *E. testacea*, (Mironov & Galsworthy, 2004) was a result of this confusion: we here synonymise it with *E. avara*. In our opinion, the paratype female of *E. avara* belongs to *E. vasta*, which belongs to a different species group.

146. *Eupithecia caduca* Vojnits, 1984

Eupithecia caduca Vojnits, 1984, *Acta zool. Acad. Sci. hung.* **30** (3–4): 529, figs 9, 17, 25. Holotype ♂ (TTM; examined), China: North Yunnan, Li-kiang.

Description. Fore wing ground colour ochreous brown; dark transverse lines inconspicuous, narrow, wavy and rather oblique, medial touching discal dot and sharply angled onto costa, postmedial twice angled onto costa; terminal area relatively broad, slightly darker, with wavy whitish subterminal line; discal dot relatively large, almost circular. Hind wing off-white, shaded with brown along anal and lower part of terminal margins, with a small white spot at tornus; discal dot pale brown, substantial.

Male genitalia. Uncus narrow, biapical. Valve with a relatively long and deep ventral flange. Vinculum broad, with shallow medial hollow. Papillae elongate and narrow. Aedeagus stout, shorter than valve length, vesica armed with two apical cornuti, one unusually long and hooked, and a broad, V-shaped cornutus connected with a smaller, sharply curved, irregular cornutus near ductus ejaculatorius base. Sternite A8 elongate, with two narrow parallel tapered arms.
Female genitalia. Bursa ovoid, basal half covered with small spines, and with two compact patches of larger spines on the edge of the spiniferous area. Ductus bursae short and broad, longitudinally striate and with a large patch of spinules. Ductus seminalis broadened basally, attached to corpus bursae near short and broad colliculum. Antrum broad, membranous. Tergite A8 transverse with a shallow medial hollow in posterior margin. Apophyses stout, the anterior slightly and the posterior strongly expanded and flattened at apices. Papillae anales narrow and tapered.
Range. China (Yunnan).
Phenology. Univoltine (?): early and mid-September.
Similar species. Externally very similar to a number of relatively lightly marked species, such as *E. avara* in this group, and *E. sacrimontis* and *E. coniurata* outside it, and cannot be reliably determined without examination of the genitalia. In the male genitalia the long apical cornutus with a hook-like apex is distinctive. The female genitalia are separable from other species in the group with difficulty by the shorter spines in the separate patch of large spines, but are quite fortunately quite distinct from those of the most similar species, *E. avara*.
Material. ♂, China, Provinz Nord-Yuennan, Li-kiang [Lijiang], 8.ix.1934, H. Höne, Vojnits slide no. 13045♂ (holotype, TTM); 1♂, 1♀, same loc., 11, 16.ix.1934, H. Höne, Vojnits slides nos. 13210♂, 12481♀ (ZFMK).

147. *Eupithecia propagata* Prout, 1926

Eupithecia propagata Prout, 1926, *Mem. Dep. Agric. India* (Ent.) **9** (8): 252. Holotype ♂ (BMNH; examined), [India]: Kumaon [= Kumaun], Muktesar [= Mukteswar], 7,500ft.
Eupithecia iracunda Vojnits, 1988, *Acta zool. hung.* **34** (1): 44, fig. 13; pl. 3, fig. B. Holotype ♀ (NSMTO; examined), Nepal (east): Solukhumbu Sagarmatha, Junbesi, 2760 m (synonymised in Mironov et al., 2008c).
Eupithecia beneficiaria Vojnits, 1988, *Acta zool. hung.* **34** (1): 46, figs 17–19; pl. 3, fig. E. Holotype ♂ (NSMTO; examined), Nepal (east): Solukhumbu Sagarmatha, Kharikhola, 1980 m (synonymised in Mironov et al., 2008c).

Unavailable names (incorrect subsequent spelling): *beneficaria*: Vojnits (1988).

Description. Fore wing ground colour rusty brown; basal, postbasal, antemedial and medial transverse lines oblique, sharply angled onto costa, medial sometimes and postmedial always twice angled; terminal part of medial area lighter, with faint ochreous blotch under discal dot; terminal area darker, with dark subapical blotch, black dashes between veins, and a wavy, whitish subterminal line, forming a small white tornal spot; discal dot black, elongate. Hind wing ground colour off-white, darker along anal margin and tornus; discal dot small but distinct.

Male genitalia. Uncus tapered, biapical. Valve with rounded flange, almost ear-shaped. Vinculum relatively short and broad. Papillae elongate. Aedeagus stout, shorter than length of valve. Vesica covered with numerous denticules, armed with a short, hook-shaped apical cornutus, a broad central cornutus, a curved U-shaped cornutus and a small irregular cornutus near ductus ejaculatorius base. Sternite A8 elongate, with two narrow parallel arms.

Female genitalia. Bursa pear-shaped, anterior part covered with small spines, and with one or two large spines on the wall opposite the ductus seminalis and a semicircular, heavily sclerotized area near base of ductus seminalis. Ductus bursae tapered, with a large patch of small spines. Ductus seminalis broadened at base, attached to centre of corpus bursae. Colliculum short and broad, tapered. Antrum short, membranous. Tergite A8 quadrate, with sclerotized anterior margin. Apophyses relatively broad, expanded and flattened at apices. Papillae anales short and narrow.

Range. China (Yunnan); India (Kumaon, Uttar Pradesh, West Bengal, Sikkim), Nepal, Bhutan.

Phenology. Univoltine (?): late July to late October.

Altitude. Found from about 1200 to 4000 m in the Himalayas (N. India: coll. U. Ratzel; China: coll. ZFMK).

Similar species. Very similar externally to *E. refertissima*, but rather smaller and darker. In the male genitalia the small hook-like apical cornutus in the aedeagus is shared only with *E. balintzsolti*, but the rounded ventral flange on the valve in the present species is distinctive. The female genitalia are very similar to a number of other species in the group, but have far fewer spines in the patch opposite the ductus seminalis.

Material. ♂, [India], Kumaon, 7500ft., Muktesar, Sept. 1922, Fletcher coll., *Eupithecia propagata* Prout type ♂, Brit. Mus. 1927–124, BM Geom. slide no. 16418 (holotype, BMNH); ♀, E Nepal, Junbesi 2760 m, Solukhumbu Sagarmatha, 10–11.x.1979, M. Owada, Vojnits slide no. 14688♀, photo 85 D 26 (holotype of *E. iracunda*, NSMTO); ♂, E Nepal, Kharikhola 1980 m, Solukhumbu Sagarmatha, 7.x.1979, M. Owada [leg.], Vojnits slide no. 14683♂, photo 85 D 21 (holotype of *E. beneficiaria*, NSMTO); 1♀, Nord

Yünnan, A-tun-tse [Dêqên], Mittlere Höhe (*ca.* 4000 m), 26.vii.1936, H. Höne, Vojnits slide no. 13933♀ (paratype of *E. placida*, ZFMK).
Remarks. Not previously recorded from China.

148. *Eupithecia balintzsolti* Vojnits, 1988

Eupithecia ***balintzsolti*** Vojnits, 1988, *Acta zool. hung.* 34 (1): 49, figs 27–29; pl. 1, fig. H. Holotype ♂ (NSMTO; examined), Nepal (east): Bagmati Zone Sindhu Palati, 1200 m.

Unavailable names (incorrect subsequent spelling): *belintzsolti*: Inoue (2002).

Description. Ground colour of all wings rusty brown. Fore wing markings reduced to a series of indistinct brown spots along costa, a narrow white wavy subterminal line, and large rounded discal dots. Hind wing only slightly paler, with area along anal margin heavily marked with dark brown scales, and substantial discal dots.
Male genitalia. Uncus elongate, biapical. Valve with ventral flange very broad, extending to base of sacculus. Vinculum broad and rounded. Papillae elongate, clavate. Aedeagus narrow, slightly shorter than length of valve. Vesica armed with a hook-shaped apical cornutus, a plate-like U-shaped cornutus and a small, irregular cornutus near ductus ejaculatorius base. Sternite A8 narrow, with two narrow apical horns.
Female genitalia. Bursa globular, completely covered with small spines. Ductus bursae elongate and broad, with a patch of small spines in posterior part near colliculum. Ductus seminalis membranous, bubble-shaped basally with a small patch of tiny spines near ductus bursae. Colliculum short and broad. Antrum long, lightly sclerotized with numerous pores. Tergite A8 almost quadrate with rounded posterior corners. Apophyses stout. Papillae anales slightly elongate.
Range. China (Yunnan); Nepal, North Thailand, Laos, North Vietnam.
Phenology. From mid-September to late November.
Altitude. From about 1100 m in Nepal to about 2050 m in Northern Thailand (Inoue 2000; coll. ZFMK).
Similar species. Easily recognized by its plain facies with large discal dots on all fore wings, slightly reminiscent only of *E. raniata* and *E. conterminata*, which are both smaller. In the male vesica, the short hook-shaped apical cornutus is shared only with the preceding species, which has the ventral flange of the valve differently shaped. The female genitalia are easily distinguished by the large globular base of the ductus seminalis, and the lack of a patch of larger spines.
Material. ♂, E. Nepal, Bagmati Zone, Sindhu, Palati 1200 m, 14.xi.1979, M. Owada, Vojnits slide no. 14692♂ (holotype, NSMTO); 1♀, China, Provinz

Nord-Yuennan, Li-kiang [Lijiang], 13.ix.1934, H. Höne, Mironov slide no. 632♀ (TTM).
Remarks. Not previously recorded from China.

The *fletcherata* species group

The *fletcherata* group was first established by Bolte (1990) to contain three North American species of *Eupithecia*. Mironov (2003) placed a single European species, *E. virgaureata* in this group, but also brought a large number of other species into a further group, which he termed the *egenaria* group. Following study of many more related Asian species, we have come to the conclusion that these two groups should be united into a single species group, which, in deference to Bolte, who first identified it, we term the *fletcherata* group.

The group is large and diverse, but its members share a number of genitalic features. In the male, the eighth sternite is nearly always quadrate, with two apical horns (occasionally one). The sacculus of the valve is frequently modified, with projections varying from a single spike to a large projection or flange. The aedeagus vesica always contains largish, usually very large, cornuti, ranging in number from one to five or many (as in *E. vivida* and *E. acutangula* Hampson, 1895). The female genitalia usually have an irregularly shaped bursa, with multiple patches of large spines, often formed into one or more bands, and in many cases large patches of strong sclerotization.

This group is very species-rich, and is represented in China by the following 36 species.

149. *Eupithecia hainanensis* Mironov & Galsworthy, 2004

Eupithecia ***hainanensis*** Mironov & Galsworthy, 2004, *Trans. lepid. Soc. Japan* 55 (2): 127, figs 8, 23, 26. Holotype ♂ (IZCAS; examined), [China]: Hainan, Jianfengling.

Description. Fore wing ground colour brown; antemedial faint, oblique; postmedial broad, dark, evenly curved outside large, slightly elongate discal dot; wavy white submarginal line faintly visible, ending in a well-marked white tornal spot. Hind wing concolorous with a well marked discal dot; anal margin suffused with darker scales, forming short transverse lines; a faint white wavy submarginal line, and a large white tornal spot.
Male genitalia. Uncus long and stout, biapical. Valve broad, sacculus heavily sclerotized with an asymmetrical ventral process, pointed on the right valve and blunt on the left. Juxta and vinculum very short and broad. Papillae long and narrow. Aedeagus about valve length, vesica armed with

one narrow, horn-like cornutus and a small irregular plate near ductus ejaculatorius base. Sternite A8 long and narrow, broadened at both ends, with two diverging apical lobes.

Female genitalia. Bursa spherical, completely covered with small spines. Ductus bursae very long, membranous with ductus seminalis attached centrally. A short diverticulum between ductus bursae and antrum. Antrum very long and wide, funnel-shaped, heavily sclerotized and wrinkled posteriorly. Tergite A8 spade-shaped. Apophyses narrow. Papillae anales rather short.

Range. China (Hainan); Thailand.

Phenology. A winter species: November and December.

Similar species. The dark concolorous wings are distinctive. The shape of sternite A8 of the male is similar only to that of *E. groenblomi*, which has an unmodified sacculus on the valve. The female genitalia are unmistakable.

Material. ♂, Hainan, Jianfengling, 21.xii.1983, coll. Liu Yuanfu, slide no. ZL60♂ (holotype, IZCAS); 2♀, same data, Galsworthy slide no. ZL61♀ (paratypes, IZCAS).

150. *Eupithecia groenblomi* Urbahn, 1969

Eupithecia groenblomi Urbahn, 1969, *Ann. Ent. Fenn.* 35 (4): 186, figs 1a, 1b, 2, 3. Holotype ♂ (ZMUH; examined), Suomi [Finland]: EH, Pälkäne.

Eupithecia tsushimensis Inoue, 1980, *Bull. Fac. domest. Sci. Otsuma Wom. Univ.* **16**: 167, fig. 40C; (*nec* fig. 43B); figs 47D, (*nec* fig. 47E); fig. 49A. Holotype ♂ (coll. H. Inoue in BMNH; examined), [Japan]: Tsushima Is(land), Shirikubiyama (synonymised in Mironov 2003).

Eupithecia fujisana Inoue, 1980, *Bull. Fac. domest. Sci. Otsuma Wom. Univ.* **16**: 204, fig. 55W; fig. 57B. Holotype ♀ (coll. H. Inoue in BMNH; examined), [Japan]: Shizuoka Pref(ecture), Mt. Fuji, Shin 5-gomé, 2400 m (synonymised in Mironov 2003).

Eupithecia kunashiriensis Viidalepp & Mironov, 1988, *Eesti NSV Tead. Akad. Toim.* (Biol.) 37 (3): 212, figs 3e-h. Holotype ♂ (IAET; examined), [Russia]: Kurile Islands, Kunashir Island, The Pioneer Camp "Dozorny" (synonymised in Mironov 2003).

Description. Forewing unicolorous smoky grey; transverse lines inconspicuous, except fine, dark grey medial and postmedial; medial sharply angled around discal dot and forming a small dark grey costal spot; postmedial evenly curved, forming a series of narrow, elongate, black dashes on veins and a larger dark grey costal spot; terminal area darker with chequered veins, and a fine whitish wavy line; discal dot relatively large. Hindwing slightly paler, with only faint postmedial; discal spot distinct. Abdomen grey with narrow black lateral stripes.

Male genitalia. Uncus stout, biapical. Valve broad, unmodified. Vinculum broad, semicircular. Papillae elongate, narrow. Aedeagus stout, slightly shorter than length of valve. Vesica armed with three horn-like cornuti,

a flat U-shaped cornutus and a smaller curved cornutus at ductus ejaculatorius base. Sternite A8 very elongate, expanded at apices, with two pincer-like horns.
Female genitalia. Bursa elongate, covered with small spines, with a longitudinal fold along one side and a membranous diverticulum at base. Ductus bursae with longitudinal striae. Ductus seminalis broadened at base, attached to ductus bursae on left side. Colliculum elongate and relatively narrow. Antrum narrow, membranous. Tergite A8 elongate. Apophyses slightly broadened. Papillae anales narrow and elongate.
Range. China (Qinghai); Japan (Hokkaido, Honshu, Tsushima), Russia (Primorje, Kurile Islands, Urals, NW of European part), Norway, Finland.
Phenology. Univoltine: late July to late September. Larva: mid-August to early October. Overwintering as pupa.
Biology. Larva monophagous, found on the flowers and seeds of *Solidago virgaurea* (Asteraceae) (Finland: Peltonen & al. 1970; Seppänen 1970; Mikkola & al. 1989; Russia: Mironov 2001, 2003). In the Far East of Russia, Japan and China possibly on other species of *Solidago*.
Altitude. From sea level to about 2400 m in Honshu (Japan: Inoue 1980, as *E. fujisana*).
Similar species. A rather plain grey species without conspicuous markings, easily confused with *E. virgaureata*, which has more distinct transverse lines and the veins reddish chequered, and *E. selinata,* which has broader wings, a more rounded postmedial, and larger fore wing discal spots. The male genitalia are easily recognised by the three large curved cornuti in the vesica, and the distinctive shape of sternite A8. The female genitalia are also distinct by the long fold on the elongate bursa, and membranous basal diverticulum.
Material. ♂, Suomi [Finland], EH, Pälkane, e. p. 1967, E.O. Peltonen leg. (holotype, ZMUH); ♂, [Japan], Tsushima, Kōnokiyama, 28.ix.1973, T. Watanabe [leg.], 25, Inoue slide no. 6508♂, BM Geom. gen. slide no. 16376♂, Inoue Coll. B.M. 1992–71, Type status verified K. Buckmaster 1992 (holotype of *E. tsushimensis*, coll. H. Inoue in BMNH); ♀, [Japan], Shin 5-gome, 2400 m, Mt. Fuji, 29.vii.1971, H. Inoue leg., 32, Inoue slide no. 3954♀, BM Geom. gen. slide no. 16380♀, Inoue Coll. B.M. 1992–71 (holotype of *E. fujisana*, coll. H. Inoue in BMNH); ♂, [Russia], Kurile Islands, Kunashiri Isl., Pioneer Camp “Dozorny”, 11–15.viii.1976, Ruben leg. (holotype of *E. kunashiriensis*, IAET); 1♀, Qinghai, Huzhu Beishan Linchang, Langshidang, 2600 m, 36°55′ N, 102°23′ E, 7.viii.2005, Han Hongxiang, slide no. ZL691♀ (IZCAS).
Remarks. Not previously recorded from China.

151. *Eupithecia selinata* Herrich-Schäffer, 1861

Eupithecia selinata Herrich-Schäffer, 1861, *Corresp.-Bl. Sammler Ins., insbesondere Schmett.* 17: 131. Syntype(s). [Germany]: Mombach.
Eupithecia divina Vojnits, 1979, *Acta zool. Acad. Sci. hung.* 25 (1–2): 205, fig. 16. Holotype ♂ (ZFMK; examined), [China]: Chekiang [Zhejiang], West Tien-mu-shan, 1600m (synonymised in Mironov 2003).
Eupithecia selinata fusei Inoue, 1980, *Bull. Fac. domest. Sci. Otsuma Wom. Univ.* **16**: 161, fig. 37B; figs 39O-Q; fig. 41A; fig. 42G. Holotype ♂ (BMNH; examined), [Japan]: Tokyo, Takao-san.

Unavailable names (incorrect subsequent spelling): *silenata*: Dietze (1971) *nec* Assmann, 1848 (identity: *Eupithecia silenata*).

Description. Fore wing dark grey or brownish grey, with darker costal margin; transverse lines inconspicuous, medial sinuate, crossing discal dot, postmedial finely waved, evenly curved; pale wavy subterminal line ill-defined; discal dot conspicuous, dark brown, usually touching dark medial costal spot. Hind wing slightly paler than fore wing, all transverse lines rather obsolete; discal dot distinct but small. Abdomen shiny grey, with small dark dorsal tufts and black lateral stripes.
Male genitalia. Uncus elongate, biapical. Valve broad and rather pointed. Vinculum semicircular. Papillae long and narrow. Aedeagus stout, shorter than length of valve. Vesica armed with two short, slightly curved horn-like cornuti, an irregular, folded cornutus, numerous small spines and a folded, V-shaped cornutus at ductus ejaculatorius base; the base of the ductus ejaculatorius (perhaps uniquely) has one to three small, spindle-shaped cornuti. Sternite A8 long and very narrow, the base broadened and apically with two very small lobes.
Female genitalia. Bursa long-oval, strongly sclerotized, with a basal field of small spines edged with larger spines, and a broad longitudinal sclerite bearing large spines reaching from spiniferous area to colliculum on left side; numerous scattered small spines and longitudinal striations in the ductus bursae. Ductus seminalis broad, sclerotized, evenly curved ventrad. Colliculum short and broad. Antrum broad, membranous. Tergite A8 almost quadrate, with sinuate anterior margin and broad, shallow medial hollow in posterior margin. Apophyses short and narrow. Papillae anales relatively small, rounded.
Range. China (Zhejiang, Sichuan); Europe, Caucasus, Transcaucasus (Georgia), Asian Russia (western and southern Siberia, Altai Mts., Sajan Mts., Transbaikalia, Priamurje, Primorje, Sakhalin, Kuril Islands: Kunashiri), Japan (Hokkaido, Honshu).
Phenology. Bivoltine (in Europe): mid-April to mid-June; late June to late August.

Biology. Larva of the first generation on the leaves and the second generation on the flowers and ripening seeds of various species of Apiaceae in Europe.
Altitude. From sea level up to about 1600 m in Tien-mu-shan (China: Vojnits 1979a, as *E. divina*).
Similar species. See under previous species; also similar to *E. addictata*, but distinguished by broader wings, less distinct and more evenly curved postmedial lines, and usually larger discal dots on the fore wings. The two small cornuti in the aedeagus vesica, taken with the narrow eighth sternite, are diagnostic in the male genitalia, as is the long spine-bearing sclerite in the female.
Material. ♂, Pz. Chekiang [Zhejiang], West Tien-Mu-Shan (1600 m), 12. iv.1932, H. Höne, Vojnits slide no. 11485♂ (holotype of *E. divina*, ZFMK); 1♂, same loc. and alt., 28.iv.1932, H. Höne, de Laever slide no. 329♂ (ZFMK); 1♂, same loc., alt. and date, H. Höne, Mironov slide no. 793♂ (SMNK); ♂, [Japan], Tanigawa Spa, Gumma Pref., 15.ix.1974, H. Inoue (holotype of *E. selinata fusei*, BMNH); 1♂, China, Zhejiang, Lin'an, West Tianmushan, 400 m, 26–27. vii.2003, Han Hongxiang, slide no. ZL688♂ (IZCAS); 1♀, China [Sichuan], Kwanhsien [Guanxian], 28.vii.1926, G.M.Franck, slide no. 23876 (BMNH).

152. *Eupithecia tamara* Mironov & Pekarsky, 2011

Eupithecia tamara Mironov & Pekarsky, 2011, *Lep. Science* **62**: 26, figs 13, 30, 32. Holotype ♀ (coll. O. Pekarsky; examined), China: Yunnan.

Description. Fore wing ground colour greyish brown with ochreous tinge; transverse lines distinct, sharply angled onto costa, median forming a distinct arrowhead shape distal of discal spot, and postmedial twice sharply angled onto costa and forming an oblique, black costal spot; terminal area darker, brownish, with indented, pale subterminal line and a series of black, elongate dashes between veins along terminal margin; discal dot large, elongate. Hind wing paler, blackish transverse lines distinct along anal margin only, this area sharply differentiated from paler remainder of wing.
Male genitalia. Uncus stout, biapical. Valve unmodified. Vinculum broad with shallow medial hollow. Papillae elongate and clavate. Aedeagus stout, slightly shorter than length of valve. Vesica armed with one long and one shorter, slightly curved, horn-like cornuti, and an elongate, V-shaped cornutus near ductus ejaculatorius base. Sternite A8 with two narrow, apically pointed rods, heavily sclerotized along their outer margins.
Female genitalia. Bursa small, almost globular, membranous, and almost completely covered with sparse spines, coalescing into a dense elongate

patch of larger spines on the right extending to the base of the colliculum; area near ductus seminalis spineless. Ductus bursae not distinguishable. Ductus seminalis attached to posterior part of corpus bursae near base of colliculum. Colliculum plate-like, relatively broad. Antrum short, membranous. Tergite A8 transverse, with sclerotized anterior margin. Apophyses relatively short and narrow. Papillae anales broad and rounded.
Range. China (Sichuan).
Phenology. All three type specimens were taken in mid July.
Altitude. Found from about 3500 to 4000 m above sea-level in W. Sichuan (China: coll. O. Pekarsky).
Similar species. Externally similar to *E. repentina* and *E. fatigata*. However, in *E. repentina* the transverse lines, especially the ante- and postmedial are less sharply angled onto the costa, the discal dot on the fore wing is usually narrower, and the hind wing has a less rounded terminal margin and rather inconspicuous transverse lines along the anal margin. *E. fatigata* has narrower, more elongate wings, the antemedial less sharply angled onto the costa, the postmedial more oblique, the subterminal line forming a small, whitish tornal spot, and the discal dot longer and narrower on the fore wing; the hind wing is also narrower and more elongate, with a distinct white tornal spot. The male genitalia are primarily distinctive because of the shape and size of the cornuti in the aedeagus vesica. The female genitalia resemble those of *E. nepalata* and *E. eximia,* but *E. nepalata* has a smaller spiniferous area in the corpus bursae with slimmer spines, a different disposition of the ductus seminalis and a shorter colliculum, while *E. eximia* also has a smaller spiniferous area in the corpus bursae with larger spines and a large, membranous spineless area, a different disposition of the dense patch of larger spines and also of the ductus seminalis.
Material. ♀, China, W. Sichuan, Xindugiao evn 3500 m, 30°04′ N / 101°26′ E, 14.vii.2009, I. & A. Floriani, Saldaitis leg., Pekarsky slide no. OP 0514♀ (holotype, coll. O. Pekarsky); 1♂, same data, Pekarsky slide no. OP 0692♂ (paratype, coll. O. Pekarsky); 1♂, China, W. Sichuan, Litang, 4000 m, 29°50′ N / 100°21′ E, 16.vii.2009, I. & A. Floriani, Saldaitis leg., Pekarsky slide no. OP 0735♂ (paratype, coll. O. Pekarsky).

153. *Eupithecia arenosa* Vojnits, 1984

Eupithecia arenosa Vojnits, 1984, *Acta zool. Acad. Sci. hung.* **30** (1–2): 236, fig. 30. Holotype ♂ (TTM; examined), [China]: North Yunnan, A-tun-tse [Dêqên], Talsohle, 3000 m.

Description. Pattern hardly distinguishable on available material. Fore wing narrow, triangular, with pointed apex; ground colour brownish grey;

transverse lines broadened near costa, postmedial angled onto costa; a broad transverse band between medial and terminal areas reddish tinged near costa; terminal area with long dark dashes on veins and a dark tornal spot. Hind wing paler, dark transverse lines marked along anal margin; terminal area darker near tornus; discal dot well marked.

Male genitalia. Uncus stout, biapical. Valve sharply tapered to apex, with convex dorsal margin, and a semicircular process on ventral side. Vinculum trapezoid, with shallow medial hollow. Papillae very broad, clavate, with rounded apices. Aedeagus short and stout, equal to valve length. Vesica armed with two apical and one smaller medial horn-like cornutus, a folded cornutus and a smaller, irregular cornutus near ductus ejaculatorius base. Sternite A8 flask-shaped, with a broad bilobed base and very narrow neck and two very small apical lobes.

Female unknown.

Range. China (Yunnan).

Phenology. late May.

Altitude. Unique holotype from 3000 m above sea-level (China: Vojnits 1984a).

Similar species. The male genitalia are most similar to those of *E. inopinata*, but readily distinguished by the semicircular ventral process of the valve, which is unlike that of any other species.

Material. ♂, Nord-Yünnan, A-tun-tse [Dêqên], Talsohle *ca.* 3000 m, 25.v.1937, H. Höne, Vojnits slide no. 14108♂ (holotype, TTM).

Remarks. The unique holotype of *E. arenosa* is a very worn, and its pattern almost indistinguishable; the genitalia slide is unstained, poorly mounted, and difficult to interpret. The valves appear assymetrical, the left having a deeper hollow behind the ventral process than the right. More material of this species is needed to allow proper assessment.

154. *Eupithecia levata* Mironov & Galsworthy, 2004

Eupithecia ***levata*** Mironov & Galsworthy, 2004, *Trans. lepid. Soc. Japan* 55 (2): 118, figs 2, 14. Holotype ♂ (ZFMK; examined), [China]: North Yunnan, Likiang [Lijiang].

Description. Fore wing ground colour pale brown, with a reddish tinge along costa, and more narrowly, on inner margin; all transverse lines faint except near costa; medial angled and postmedial twice angled near costa, bordered externally by a pale fascia; terminal area darker, with a submarginal line of dark dots on veins. Hind wing ground colour as fore wing; discal dot faint; anal margin shaded with brown scales extending to tornal area.

Male genitalia. Uncus long, biapical. Valve rather pointed. Vinculum short and broad. Papillae large, relatively broad. Aedeagus broad and short. Vesica armed with one long and one shorter, claw-like cornutus, an elongate plate, a large irregular plate near base of ductus ejaculatorius, and an elongate patch of very small spines. Sternite A8 gradually tapered, with two short apical rods, heavily sclerotized at tips.
Female unknown.
Range. China (Yunnan).
Similar species. The single specimen is too worn to allow external comparison with other species. The male genitalia resemble those of the Mediterranean species *E. pauxillaria* Boisduval, 1840 (= *euphrasiata* Herrich-Schäffer, 1861), but can be distinguished by the longer horn-like cornutus on the vesica and the shape of sternite A8.
Material. ♂, China, Provinz Nord-Yuennan, Li-kiang [Lijiang], 30.iii.1935, H. Höne, de Laever slide no. 349♂ (holotype, ZFMK).

155. *Eupithecia molestissima* Vojnits, 1981

Eupithecia molestissima Vojnits, 1981, *Acta zool. Acad. Sci. hung.* 27 (3–4): 425, fig. 46, 48–50. Holotype ♀ (ZFMK; examined), China: North Yunnan, Li-kiang.

Description. Fore wing ground colour dark greyish brown; basal, ante- and postmedial lines straight, right angled onto costa; double medial sinuate, sharply angled onto costa and touching discal dot; terminal area slightly darker, veins covered with black scales and a small blackish blotch near tornus; discal dot large, intensely black with orange mark beneath. Hind wing pale greyish brown; anal margin dark blackish grey; terminal area narrow, slightly darker, with a small whitish tornal spot; discal dot relatively large, distinct.
Male genitalia. Uncus elongate, narrow, biapical. Valve with ventral margin produced at base. Vinculum short and broad. Papillae very broad. Aedeagus stout, shorter than length of valve. Vesica armed with two long horn-like cornuti, a folded, bifurcate cornutus, and a rounded and folded cornutus near ductus ejaculatorius base. Sternite A8 short, almost triangular, with two narrow apical rods like giraffe horns.
Female genitalia. (Described here for the first time) Bursa pear-shaped, more strongly sclerotized in posterior half, basal half covered with spines, with two longitudinal rows of large spines along left and right sides and a patch of small spines near colliculum. Ductus seminalis broad, attached to posterior half of corpus bursae on right. Colliculum short and narrow. Antrum membranous, covered with numerous pores. Tergite A8 with

posterior margin produced centrally, and a small tubercle in centre of process. Apophyses short and narrow. Papillae anales short and angulate.
Range. China (Yunnan, Sichuan).
Phenology. Bivoltine (?): mid-April to late July.
Altitude. Found from about 3000 to 4000 m above sea-level in North Yunnan (China: coll. ZFMK).
Similar species. Rather similar externally to *E. placida* but with less pointed fore wings, larger discal dots, and a paler hind wing. The male genitalia are distinct, particularly in the ventral projection of the valve, the very stout papillae, the two large cornuti and the shape of the eighth sternite. The female genitalia resemble those of the *propagata* group, but the large spines are arranged in rows rather than a round patch.
Material. ♀, China, Provinz Nord-Yuennan, Li-kiang [Lijiang], 23.iv.1935, H. Höne, Vojnits slide no. 13173♀ (holotype, ZFMK); 4♀, same loc., 17, 23, 27.iv., 27.v.1935, de Laever slides nos. 357♀, 267♀, Vojnits slides nos. 10930♀, 14205♀ (ZFMK); 1♀, Nord Yünnan, A-tun-tse [Dêqên], Mittlere Höhe (*ca.* 4000 m), 21.vii.1936, H. Höne, Vojnits slide no. 12483♀ (ZFMK); 2♀, same loc., Talsohle *ca.* 3000 m, 17, 31.v.1937, H. Höne, Vojnits slides nos. 13757♀, 14110♀ (ZFMK); 5♂, 14♀, China, North Yunnan, Diqing Tibetan Aut. Pref., 8 km NNE of Shangri La at Nairi village, 3300 m, 14.vi.2009, leg. B. Benedek, Pekarsky slides nos. OP 0529♀, OP 0822♂, OP 0823♂, OP 0825♂, OP 0826♀, OP 0827♀, OP 0829♀, OP 0830♀, OP 0831♀, OP 0832♀, OP 0833♀, OP 0837♂, OP 0838♂, OP 0839♀, OP 0840♀, OP 0841♀, OP 0842♀, OP 0843♀, OP 0844♀, OP 0845♀ (coll. O. Pekarsky); 1♀, China, North Yunnan, Diqing Tibetan Aut. Pref., 5 km SE of Deqin, 3356 m, 16–17.vi.2009, leg. Balázs Benedek, Pekarsky slide no. OP 0847♀ (coll. O. Pekarsky); 1♂, 2♀, same locality, 19. vi.2009, leg. B. Benedek, Pekarsky slides nos. OP 0542♂, OP 0635♀, OP 0849♀ (coll. O. Pekarsky); 1♀, China, W. Sichuan, road Xinngcheng / Zhongdian, 3300 m, 17.vii.2009, 28°46′ N / 99°51′ E, I. & A. Floriani, Saldaitis leg., Pekarsky slide no. OP 0806♀ (coll. O. Pekarsky).

156. *Eupithecia actaeata* Walderdorff, 1869

Eupithecia actaeata Walderdorff, 1869, *CorrespBl. zool.-min. Ver. Regensburg* **23** (6): 82. Holotype ♂, [Germany].
Eupithecia bergunensis Dietze, 1875, *Stettin. ent. Ztg.* **36** (4–6): 248, pl. 2, fig. 1. Syntypes 1♂, 2♀ (coll. K. Dietze in MNHU; examined), [Switzerland]: Kanton Graubünden, Bergün (treated as var. of *E. actaeata* Walderdorff, 1869 in Bohatsch, 1893).
Eupithecia praenubilata Inoue, 1958, *Tinea* **4** (2): (252), fig. 9; pl. 34, fig. 22. Holotype ♂ (BMNH; examined), [Japan]: Kushiro, Shibecha (treated as subsp. of *E. actaeata* Walderdorff, 1869 in Inoue, 1963).

Unavailable names (incorrect subsequent spelling): *acteata*: Matsumura (1905).

Description. Notably broad-winged for a *Eupithecia*. Fore wing grey; transverse lines usually indistinct; postbasal and antemedial oblique, angled near costa; postmedial slightly sinuate, well defined as dark dashes on the veins; ante- and postmedial lines usually forming two dark costal spots; terminal area with wavy subterminal line and white tornal spot; discal dot large. Hind wing slightly paler, transverse lines usually ill-defined; terminal area darker with small white tornal spot; discal dot distinct. Abdomen grey with black lateral stripes.

Male genitalia. Uncus narrow, biapical. Valve rather narrow. Vinculum unusually long. Papillae elongate, narrow. Aedeagus stout, slightly shorter than valve length. Vesica armed with a broad plate-like apical cornutus, a large, smaller horn-like cornutus and a twisted plate near ductus ejaculatorius base. Sternite A8 very large, rectangular, with two broad basal lobes and two pointed and diverging apical arms.

Female genitalia. Bursa ovate, with an irregular pattern of spines on the right from base to ductus seminalis. Ductus bursae broad, very short. Ductus seminalis narrow, attached to posterior part of corpus bursae from right side. Colliculum short and broad, with irregular longitudinal and transverse wrinkles. Antrum membranous. Tergite A8 transverse. Apophyses short and narrow. Papillae anales short and rounded.

Range. China (Qinghai, Shaanxi); Europe, European Russia, Urals, Asian Russia (Priamurje, Primorje, Magadan region, Kamchatka peninsula, Sakhalin, Kuril Islands: Kunashiri), Korea, Japan (Hokkaido, Honshu).

Phenology. Bivoltine: late April to mid-August.

Biology. Larva found on the leaves of *Actaea* sp. (Ranunculaceae) (Far East Russia: Mironov, original data).

Altitude. Found from 1600 m in Qinling Mts. to about 3000 m in Qinghai (China: coll. ZFMK; coll. O. Pekarsky; coll. IZCAS).

Similar species. Most similar to *E. selinata* and *E. kobayashii*. The former is smaller, usually with smaller, rounded discal dots on the fore wings, and always lacks whitish tornal spots; the latter has smaller, less elongate discal dots on the fore wings and usually on the hind wings. The male genitalia are distinctive, particularly the elongate vinculum and the large sternite A8. The female genitalia are similar to those of *E. kobayashii* but lack the membranous diverticulum.

Material. ♂, [Japan], Shibecha, Kushiro, 3.vi.1957, K. Ijima, No. 911, BM Geom. slide no. 22005♂ (holotype of *E. praenubilata*, coll. H. Inoue in BMNH); 1♂, C. China, prov. Shaanxi, Foping Natural Reserve, Panda area, 1600 m, 20.iv.-11.v.1999, 33°45′ N, 107°48′ E, leg. V. Siniaev & A. Plutenko, Mironov slide no. 732♂ (ZFMK); 1♂, China, Shaanxi prov., Qinling Mts.,

S from Baoji, h – 1600 m, N 34°13.932′ / E 106°58.080′, 31.v.-01.vi.2009, leg. Floriani & Saldaitis, Pekarsky slide no. OP 0679♂ (coll. O. Pekarsky); 1♂ Qinghai, Tongren, Maixiu, 2950 m, 30.7–1.viii.1992, Xue Dayong, slide no. ZL265♂ (IZCAS).
Remarks. Not previously recorded from China. East Asian specimens are usually paler than European specimens with larger discal dots on all wings.

157. *Eupithecia actrix* Mironov & Galsworthy, 2006

Eupithecia actrix Mironov & Galsworthy, 2006, *Trans. lepid. Soc. Japan* 57 (4): 349, figs 25. Holotype ♀ (ZFMK; examined), C. China: prov. Shaanxi, Mt. Tai Bei Shan, 33°35′ N 107°43′ E, 1300–1500 m.

Description. Fore wing elongate, ground colour pale brownish grey; three medial transverse lines inconspicuous, forming oblique dark costal spots; discal dot small, rounded, black. Hind wing rounded triangular, terminal margin slightly sinuous, concolorous with fore wing; transverse lines inconspicuous; discal dot very small, circular.
Female genitalia. Bursa small, heavily sclerotized in large irregular folds, with a long curved chain of different sized spines near base and a short chain of small spines near base of ductus seminalis. Ductus bursae relatively narrow, heavily sclerotized, with some longitudinal striations. Ductus seminalis narrow, attached to membranous swelling in the central part of the bursa on the right side. Colliculum not present. Antrum heavily sclerotized, covered with numerous small pores. Tergite A8 elongate. Apophyses short and narrow. Papillae anales small, slightly tapered.
Male unknown.
Range. China (Shaanxi).
Phenology. The unique specimen was captured between 10 August and 4 September.
Altitude. Found at altitude about 1500 m on Mt. Tai Bei Shan (China: coll. ZFMK).
Similar species. The unique specimen is too worn to allow meaningful comparison with other species, but the small size is helpful. The female genitalia are similar to those of *E. quadripunctata*, but can be distinguished by the heavily sclerotized antrum: the adult of *E. quadripunctata* also has much paler hind wings.
Material. ♀, Central China, prov. Shaanxi, Mt. Tai Bei Shan Mts., 1300–1500 m, 33°35′ N, 107°43′ E, 10.viii.-4.ix.1998, leg. V. Murzin & V. Siniaev (holotype, ZFMK).

158. *Eupithecia uliata* Staudinger, 1897

Eupith[*ecia*] *uliata* Staudinger, 1897, *Dt. ent. Z., Iris* (1896) **9** (2): 281. Lectotype ♀, herewith designated to stabilize nomenclature (MNHU; examined), [Mongolia]: Uliassutai [= Uliastay or Javhlant].

Eup[*ithecia*] *recens* Dietze, 1904, *Dt. ent. Z., Iris* [1903] **16** (2): 349, pl. 3, fig. 26. Syntypes 1♂, 1♀ (coll. K. Dietze and R. Püngeler in MNHU; examined), Central Asia [China]: Kuku-Noor [Qinghai Hu]; [Russia]: Sidimi. **New junior synonym.**

Eup[*ithecia*] *recens* var. *creta* Dietze, 1908, *Dt. ent. Z., Iris* **21** (2): 172, pl. 2, fig. 29. Syntypes 1♂, 4♀ (coll. K. Dietze in MNHU; examined), [Central Asia]: [Kirghizstan]: [Issyk-Kul]; [China]: [Kuldja], [Juldus Ili]. **New junior synonym.**

Eupithecia extrinseca Vojnits & de Laever, 1978, *Acta zool. Acad. Sci. hung.* **24** (1–2): 247, fig. 24. Holotype ♀ (ZFMK; examined), [China]: Sued-Shensi, Tapaishan in Tsinling, 3000 m. **New junior synonym.**

Misidentification: *Eupithecia innotata* sensu Shin, 1996 (*nec* Hufnagel, 1767).
Unavailable names (incorrect subsequent spelling): *Eupithecia ulicada*: Dietze (1910).

Description. Fore wing ground colour grey; transverse lines oblique, basal and antemedial sharply angled onto costa, postmedial oblique, often curved inwards, twice angled onto costa; terminal area with whitish wavy subterminal line forming a small tornal spot; discal dot small, elongate. Hind wing paler than fore wing; transverse lines well marked along anal margin; terminal area slightly darker, with whitish wavy subterminal line; discal dot small and rounded.

Male genitalia. Uncus short, biapical. Valve broad, with a pointed ventral process. Vinculum elongate and broad. Papillae long and stout. Aedeagus stout, longer than valve length. Vesica armed with three broad cornuti, pointed at apices, and a curved cornutus with forked apex connecting with an irregular cornutus near ductus ejaculatorius base. Sternite A8 a convex triangle, with two asymmetrical, pointed horns.

Female genitalia. Bursa large, transverse, with a broad sinuous chain of spines. Ductus bursae heavily sclerotized, with an oblique band of paler sclerotization near posterior end. Ductus seminalis broadened basally and recurved, attached to sclerotized base of ductus bursae from right side. Colliculum not distinguishable. Antrum sclerotized, with numerous minute striations and pores. Tergite A8 transverse. Apophyses short and narrow. Papillae anales relatively small, rounded.

Range. China (Xinjiang, Tibet, Qinghai, Inner Mongolia, Gansu, Shanxi, Shaanxi, Hebei, Hubei); Kazakhstan (south and south eastern), Kirghizstan (Tien-Shan Mts.), Asian Russia (southern Siberia: Altai Mts., Dahuria, Transbaikalia, Priamurje, Primorje), Mongolia, Korea.

Phenology. Univoltine (?): late May to late August.

Biology. Larva found on the flowers of *Seseli libanotis* (Apiaceae) in late August in Zailijsky Alatau Mts. (Kazakhstan, Tien-Shan: Mironov, original data).

Altitude. From about 700 m in Altai Mts. (Russia: coll. M. Čemila and O. Pekarsky) up to about 4000 m in Tibet (China: Vojnits & de Laever 1978; Vojnits 1978; colls. ZFMK, IZCAS).
Similar species. Similar to *E. repentina* Vojnits & de Laever, 1978 and some other allied Asian species of *fletcherata* group. Distinguishable with difficulty from *E. repentina* by the narrower and more pointed fore wings, the very sharply angled postmedial lines and the smaller discal dots on the fore wings. Both the male and female genitalia are very distinctive and easily recognized.
Material. ♀, [Mongolia], Uliassutai, 94. Led., Origin., 52, *Uliata* Stgr., Vojnits slide no. 19026♀ (lectotype, coll. O. Staudinger in MNHU); 1♂, [Russia], *Eup.* sp. ♂. 1889, Sidemi, *recens* n. sp. Original, Sidimi, 1889, Typus, Geom. slide no. 1950–593A (syntype of *E. recens*, coll. K. Dietze in MNHU); 1♀, [Qinghai], Tibet (Kuku-Noor), Rückbeil 98, *recens* Dietze ♀, Original, v. R. Tancré 2.02, *Eupithecia recens* Dietze ♀, Tibet Ku Ku Nor, Cotype, prep. № 1656.1 E. de Laever det. 1957 (syntype of *E. recens*, coll. R. Püngeler in MNHU); 1♂, [Kazakhstan / Kirghizstan], ? *recens* n. sp. aber. Alex.-Gebirge (Bohatsch gesehen), K. Dietze (von Tancré) (coll. K. Dietze in MNHU); 1♀, [Kirghizstan], Jssyk-Kul, *creta* n. sp. Jssyk-Kul. Type ♀, K. Dietze (von Graeser), ? *recens* ? n. spec., Typus, Geom. slide no. 1950–592A (syntype of *E. recens* var. *creta*, coll. K. Dietze in MNHU); 1♂, [Xinjiang], ? *creta* n. sp. ♂, (nahe bei *corroborata*!) Kuldja, K. Dietze, Geom. slide no. 1951–10A (syntype of *E. recens* var. *creta*, coll. K. Dietze in MNHU); 3♀, [Xinjiang], Asia centr. Turkestan, Ili-Gebiet, Coll. Wagner, *creta* n. sp. det. Dietze, Geom. slide no. 1951–11A (syntypes of *E. recens* var. *creta*, coll. K. Dietze in MNHU); 1♀, [Xinjiang], Juldus, Ili (coll. K. Dietze in MNHU); 2♀, [Xinjiang], Asia centr., Ost-Turkestan (Urumtschi) [Urumqi], Rückbeil 1904, v. R. Tancré 05 (coll. R. Püngeler in MNHU); 1♀, [Xinjiang], Juldus Ili (coll. K. Dietze in MNHU); ♀, Sued-Shensi [Shaanxi], Tapaishan [Taibaishan] im Tsinling [Qinling], *ca.* 3000 m, 19.viii.1936, H. Höne, Vojnits slide no. 10932♀ (holotype of *E. extrinseca*, ZFMK); 1♀, same loc. and alt., 12.viii.1936, H. Höne, Vojnits slide no. 10931♀ (paratype of *E. extrinseca*, TTM); 8♂, 9♀, same loc. and alt., 8, 9, 10, 11, 12, 15, 25.viii.1936, H. Höne, Vojnits slides nos. 10908♂, 10968♂, 10894♂, 12938♀, 10906♀, 10965♂, 12354B♂, 12358♀, 10967♂, 10916♀, 10875♂, de Laever slide no. 203♂ (ZFMK); 4♂, 1♀, same loc. and alt., 9, 11, 12.viii.1936, H. Höne, Vojnits slides nos. 10971♂, 12047♀ (TTM); 18♂, 7♀, same loc, *ca.* 1700 m, 14, 25.vii., 6, 7, 8, 9, 10, 12.viii.1936, H. Höne, Vojnits slides nos. 12043♂, 12354A♀, 10893♂, 10895♂, 12937♀, 10877♀, 10910♂, 10915♂, 10964♂, 10969♀, 12318♀ (an incorrectly labelled slide belonging to a female from the *satyrata* group), 10874♂, 10907♂, 10963♂, 12355♂, 12935♂, 10935♂, 10970♂, 12042♂ (ZFMK); 4♂, 5♀,

same loc. and alt., 5, 7, 9, 10.viii.1936, H. Höne, Vojnits slides nos. 10876♂, 10921♂, 11471♀, 11524♀, 12044♀, 12045♀, 12046♀ (TTM); 2♀, same loc. (without alt.), 26.vi., 28.viii.1935, H. Höne, Vojnits slides nos. 10905♀, 10892♀ (ZFMK); 1♂, same loc., 6.viii.1935, H. Höne (abdomen missing, TTM); 1♀, Chasseurs indigènes de Tá-tsien-lou, Récolte de 1910, Vojnits slide no. 13502♀ (TTM); 1♀, Prov. Shansi [Shanxi], Mien-shan, Obere Höhe *ca.* 2000 m, 2.viii.1937, H. Höne, de Laever slide no. 104♀ (paratype of *E. extrinseca*, ZFMK); 3♂, 27♀, same loc. and alt., 14, 27, 29, 30, 31.vii., 1, 2, 3, 5, 6, 7, 9, 10, 14.viii.1937, H. Höne, de Laever slides nos. 294♀, 293♀, 106♀, 428♀, Vojnits slides nos. 10933♀, 10920♀, 12193♀, 10911♂, 10872♀, 10936♀, 11496+, 10914♀, 10871♀, 10938♀, 10873♀, 11520♀, 10957♀, 10913♀, 13118♀ (ZFMK; slides nos. 10914♀, 10957♀, 11496+ absent in ZFMK); 2♂, 6♀, same loc. and alt., 8, 9.vii., 7, 9, 11, 14.viii.1937, H. Höne, Vojnits slides nos. 10870♂, 10912♂, 11521♀, 12050♀, 12051♀, 12052♀, 12053♀ (TTM); 1♂, 1♀, China, Gansu, Xiahe, 2800–3000 m, 27.vii.1992, leg. J. Verhulst, Mironov slides nos. 646♀, 647♂ (ZFMK); 1♀, China, Inner Mongolia, 100 km SW of Hohhot, Mo-Tien-Ling, 12.viii.2007, [leg.] A. Bidzilya (ZISP); 1♀, China, Inner Mongolia, 80 km SW of Hohhot, Nantianmen, 13.viii.2007, [leg.] A. Bidzilya (ZISP); 1♀, Tibet, Jomda, 3400 m, 27.vii.1976; 1♀, Tibet, Zogang, 3800 m, 1.vii.1976, Zhang Xuezhong; 2♀, Tibet, Chajian Nanxue, 3900 m, 28.vi.1973, Huang Fusheng; 5♂, 8♀, Qinghai, Menyuan, Xianmi, 2800 m, 23.vii.1992, slides nos. ZL159♂, 160♀; 1♂, 5♀, Qinghai, Tongren, Maixiu, 2950 m, 30.vii.1992, slide no. ZL298♀; 1♀, Qinghai, Huzhu Beishan, 2300 m, 30.vii.1991, coll Xue Dayong; 1♂, 2♀, Shanxi, Wutaishan, 1992; 2♀, Inner Mongolia, Oragxan, 11.viii.1983, slides nos. ZL121, 191♀; 1♀, Hebei, Xiao Wutaishan, 1200 m, 15.viii.1964, coll Han Yinheng; 1♂, Hubei, Shennongjia, 1640 m, 9.vii.1981, coll Han Yinheng (IZCAS).

159. *Eupithecia detritata* Staudinger, 1897

Eupithecia detritata Staudinger, 1897, *Dt. ent. Z., Iris* **10** (1): 119, pl. 3, fig. 77. Lectotype ♂ (design. by Mironov, coll. O. Staudinger in MNHU; examined), [Russia]: Ask(old).
Eupithecia kamedai Inoue, 1987, *Akitu N.S.* **90**: 1, figs 1a-e, 2a-b. Holotype ♂ (BMNH; examined), [Japan]: Hokkaido, Kasai-gun, Nanasatsunai, Kamisatsunai (synonymised in Mironov, 1990).
Eupithecia amplicornuta Viidalepp & Mironov, 1988, *Eesti NSV Tead. Akad. Toim.* (*Biol.*) 37 (3): 207, figs 2d-g. Holotype ♂ (IAET; examined), [Russia]: Primorsky kraj, "Kedrovaya Pad'" Nature Reserve (synonymised in Mironov, 1990).

Misidentification: *Eupithecia jinboi* Inoue, 1980, fig. 36B (*nec* Inoue, 1976).

Description. Fore wing conspicuously narrow and elongate, ground colour pale ash grey; inconspicuous transverse lines oblique, double medial and postmedial sharply angled onto costa; terminal area with indistinct paler wavy subterminal line; discal dot small, slightly elongate. Hind wing slightly paler, especially basally, with inconspicuous transverse lines and small pale discal dot.

Male genitalia. Uncus narrow, biapical. Valve broad, apex rather narrow, with a pointed ventral process. Vinculum relatively long and broad. Papillae long and stout. Aedeagus very stout, equal to length of valve. Vesica armed with two cornuti, one short, broad, and apically pointed, and the other stout, long and very broad at base, sharply tapered at apex, together with a broad, U-shaped cornutus and a small, folded cornutus near ductus ejaculatorius base. Sternite A8 peg-like, bilobed at base, with two pointed apical horns.

Female genitalia. Bursa very large, membranous with a chain of small spines. Ductus bursae broad and curved, heavily sclerotized, with longitudinal wrinkles and a chain of spines. Ductus seminalis attached to central part of corpus bursae on the dorsal side. Colliculum not distinguishable. Antrum broad, heavily sclerotized, covered with numerous pores. Tergite A8 quadrate. Apophyses short and narrow. Papillae anales short and broad, rounded.

Range. China (Shaanxi); Asian Russia (southern Priamurje, southern Primorje), Korea, Japan (Hokkaido, Honshu).

Phenology. Bivoltine (?): late April to mid-July.

Altitude. From sea level (Far East Russia: coll. ZISP) up to about 1600 m in Qinling Mts. (China: coll. ZFMK).

Similar species. Similar to *E. jinboi*, but usually paler with less distinct transverse lines, especially the postmedial, and with less elongate discal dots on the fore wings. The male valves are similar to those of *E. uliata*, but the cornuti in the vesica and eighth sternite are distinctive. The female genitalia are typical of the group, but lack the straight ductus bursae seen in other species with large, heavily sclerotized bursae.

Material. ♂, [Russia], Askold Bork., 25, *Detritata* Stgr., Origin., 77, bon. spec bei *Altenaria* (Boh.), Geom. slide no. 1950–591A (lectotype, design. by Mironov, coll. O. Staudinger in MNHU); ♂, [Japan], Hokkaido, Kasai-gun, Nanasatsunai, Kamisatsunai, 26.vi.1983, M. Kameda, BM Geom. slide no. 16535 (holotype of *E. kamedai*, coll. H. Inoue in BMNH); ♂, [Russia], Primorsky kraj, "Kedrovaja Pad" Nature Reserve, 26–31.v.1974, leg. Ruben & Viidalepp (holotype of *E. amplicornuta*, IZBE); 2♂, 3♀, C. China, prov. Shaanxi, Foping Natural Reserve, Panda area, 33°45′ N, 107°48′ E, 1600 m, 20.iv.-11.v.1999, leg. V. Siniaev & A. Plutenko, Mironov slide no. 663♂ (ZFMK).

160. *Eupithecia jinboi* Inoue, 1976

Eupithecia jinboi Inoue, 1976, *Tinea* **10** (2): 17, figs 27, 28. Holotype ♀ (coll. Inoue in BMNH; examined), Japan: Nagano Prefecture, Iide Range, Eboshidake, 2018 m

Description. Relatively large with elongate, grey or brownish grey fore wings and, usually, distinct oblique ante- and postmedial lines on the fore wing, sharply angled onto costa behind discal dot and often double; discal dot small, usually narrow and elongate, and with a whitish tornal spot.
Male genitalia. Uncus slim, elongate, biapical. Valve broad, with an obtuse, folded ventral projection. Vinculum broad and elongate, with very shallow medial hollow. Papillae long and very broad. Aedeagus very stout, almost equal to valve length. Vesica armed with three large cornuti, strongly broadened at bases, and an elongate, curved cornutus connected with a smaller, irregular cornutus near ductus ejaculatorius base. Sternite A8 peg-like, with broad bilobed base and two narrow, often asymmetrical, apical horns.
Female genitalia. Bursa very large, ovoid, membranous, with broad curved band of large spines in medial part. Ductus bursae elongate, heavily sclerotized, spineless. Ductus seminalis narrow, attached to posterior part of corpus bursae on dorsal side. Colliculum not present. Antrum short and broad, heavily sclerotized, with numerous pores and minute wrinkles. Tergite A8 almost quadrate, with shallow medial hollow in posterior margin. Apophyses short and narrow. Papillae anales relatively small, short and rounded.
Range. China (Jilin); Japan (Hokkaido, Honshu, Shikoku).
Phenology. Bivoltine (?): mid-June to mid-July; late July to mid-August.
Biology. Larva found on *Tripetaleia bracteata* (Ericaceae), an alpine plant distributed in Hokkaido and the alpine zone of Honshu (Inoue 1980).
Altitude. Found from about 1800 to 2600 m in Japan (Inoue 1980; coll. BMNH).
Similar species. Easily confusable with *E. uliata* and *E. detritata*. *E. uliata* has a narrower and more pointed fore wing with a more strongly angled postmedial; *E. detritata* has paler, light grey fore wings with less distinct transverse lines. The male genitalia of *E. jinboi* are distinguished from both by the shape of the ventral process on the valve and the cornuti in the vesica, and the female by the shape of the sclerotized ductus bursae, which lacks either a slanting pale band or a chain of spines.
Material. ♀, [Japan], Iide Range, Eboshidake, 2018 m, 1.viii.1968, K. Jinbo (holotype, coll. H. Inoue in BMNH; 1♀, Jilin, Changbai mountains, 2.vii.1982, coll Zhang Baolin, slide no. ZL341♀ (IZCAS).

Remarks. Not previously recorded from China. Inoue 1980, page 155, fig. 36B, which is labelled *E. jinboi*, is in fact *E. detritata*. Fig 36C illustrates true *E. jinboi*.

161. *Eupithecia pellicata* Mironov & Galsworthy, 2007

Eupithecia pellicata Mironov & Galsworthy, 2007, *Trans. lepid. Soc. Japan* 58 (3): 354, figs 7, 17. Holotype ♂ (NSMTA; examined), Taiwan, Nantou, Meifeng.

Description. Fore wing elongate, uniform pale brownish grey, with white subtornal dot, and faint signs of crenulate white submarginal; very lightly marked post-medial, darker brown, curved inward towards costa at level of discal dot; discal dot black, elongate. Hind wing short, paler than fore wing, with all but marginal area dingy white; area along anal margin and marginal band suffused with dark scales; discal dot well marked.
Male genitalia. Uncus elongate, biapical. Valve broad, with a spine-like process on ventral margin. Vinculum broad with shallow medial hollow. Papillae elongate, fairly stout. Aedeagus very stout, but shorter than length of valve. Vesica covered with oblique rows of wrinkles, armed with two stout, basally broadened cornuti, an elongate cornutus with longitudinal striations and a smaller, folded cornutus near ductus ejaculatorius base. Sternite A8 peg-like, broadened basally, with two short, pointed apical rods.
Female unknown.
Range. China (Taiwan).
Phenology. Univoltine (?): late February.
Similar species. Very similar to *E. uliata*, and probably a sister species, but somewhat darker and slightly longer-winged. The male valves are virtually identical to those of *E. uliata*, but the cornuti in the aedeagus are substantially different, as is the eighth sternite, which in *E. uliata* is strongly asymmetrical.
Material. ♂, Taiwan, Nantou, Meifeng, 23.ii.1993, leg. C.S. Lin, 1660–75, Galsworthy slide no. ACG TW7 (holotype, NSMTA).

162. *Eupithecia irreperta* Vojnits & de Laever, 1978

Eupithecia irreperta Vojnits & de Laever, 1978, *Acta zool. Acad. Sci. hung.* 24 (1–2): 242, figs 19, 21. Holotype ♂ (ZFMK; examined), [China]: North Yunnan, A-tun-tse [Dêqên], 4000 m.

Description. Fore wing elongate, ground colour brownish grey; transverse lines oblique, curved from anal margin to stalk of Cu and sharply angled from Cu to costa; postmedial twice angled onto costa; double medial touching discal dot; terminal area with inconspicuous whitish subterminal

line; discal dot linear, black. Hind wing paler, transverse lines indistinct, marked along anal margin; terminal area slightly darker with wavy inner border and a series of dark dashes on the veins; discal dot rounded, faint.
Male genitalia. Uncus elongate, biapical. Valve broad, the ventral corner almost square, with 2–3 digitate processes. Vinculum long, trapezoid. Papillae broad and long. Aedeagus very stout, strongly swelling at anterior end, equal to valve length. Vesica armed with three stout, basally broadened horn-like cornuti, an elongate, laterally folded cornutus and a smaller, irregular cornutus near ductus ejaculatorius base. Sternite A8 peg-like, with two short apical horns.
Female genitalia. Bursa membranous, spineless. Ductus bursae heavily sclerotized, long and broad, further broadened at base. Ductus seminalis narrow; attached to corpus bursae near base of ductus bursae from right side. Colliculum not present. Antrum heavily sclerotized, covered by numerous pores. Tergite A8 transverse. Apophyses short and narrow. Papillae anales relatively small, rounded.
Range. China (Tibet, Yunnan).
Phenology. Bivoltine: late April to mid-May; early August to early September.
Altitude. From about 2800 to 4500 m above sea-level in Tibet and Yunnan (China: Vojnits & de Laever 1978).
Similar species. Easily confusable externally with *E. uliata*, *E. vivida*, *E. quadripunctata*, *E. repentina* and *E. fatigata* and best separated by examination of genitalia, which are quite distinct in both male and female.
Material. ♂, Nord Yünnan, A-tun-tse [Dêqên], Mittlere Höhe (*ca.* 4000 m), 11.viii.1936, H. Höne, Vojnits slide no. 10882♂ (holotype, ZFMK); 1♂, 2♀, same loc. and alt., 11, 16.viii., 3.ix.1936, H. Höne, Vojnits slides no. 11000♀, de Laever slides nos. 227♀, 225♂ (paratypes, ZFMK); 4♂, 2♀, same loc. and alt., 8, 11, 12, 15, 16, 30.viii.1936, H. Höne, Vojnits slides nos. 10880♂, 10887♀, 10998♂, 11001♂, 11019♂, 11023♀ (paratypes, TTM); 1♀, same loc., Obere Höhe (*ca.* 4500 m), 9.viii.1936, H. Höne, Vojnits slide no. 10991♀ (paratype, TTM); 2♂, Tibet [Sichuan], Batang, Im Tal des Yangtze (*ca.* 2800 m), 20.iv., 15.v.1936, H. Höne, de Laever slides nos. 99♂, 97♂ (paratypes, ZFMK); 1♂, same loc. and alt., 23.iv.1936, H. Höne, Vojnits slide no. 10879♂ (paratype, TTM); 1♀, same loc. and alt., 13.v.1936, H. Höne, de Laever slide no. 98♀ (ZFMK); 3♀, same loc., Untere Urwaldzone (*ca.* 3800 m), 12.v., 2, 11.viii.1936, H. Höne, de Laever slides nos. 466♀, 105♀, 118♀ (paratypes, ZFMK); 1♂, same loc. and alt., 2.viii.1936, H. Höne (paratype, abdomen missing, TTM); 1♂, same loc. and alt., 10.v.1936, H. Höne, de Laever slide no. 481♂ (ZFMK); 1♀, Tibet Bomi, Yi'ong, 2300 m, 15.viii.1983, coll Han Yinheng, slide no. ZL408♀; 1♀, Tibet, Qamdo, 23.viii.1984, coll Jiang Ba, Galsworthy slide no. ZL417♀ (IZCAS).

163. *Eupithecia vivida* Vojnits & de Laever, 1978

Eupithecia vivida Vojnits & de Laever, 1978, *Acta zool. Acad. Sci. hung.* **24** (1–2): 238, figs 18, 20. Holotype ♀ (ZFMK; examined), China: North Yunnan, Li-kiang.
Eupithecia producta Vojnits, 1981, *Annls hist.-nat. Mus. natn. hung.* **73**: 222, fig. 3. Holotype ♀ (SMNK; examined), India (north): Kumaon Bhimtal (Nainital), 1450 m. [Junior primary homonym of *Eupithecia producta* Bastelberger, 1911; Peru.]
Eupithecia wolfi Vojnits, 1985, *Acta zool. hung.* **31** (1–3): 272. [Replacement name for *Eupithecia producta* Vojnits, 1981] (synonymised in Mironov et al., 2008b).

Description. All wings narrow and elongate. Fore wing ground colour ash grey; dark transverse lines fine and oblique; antemedial curved; double medial crossing discal dot; postmedial sharply angled onto costa behind discal dot; terminal area slightly darker, with trace of wavy, light subterminal line, broadened near tornus; discal dot oblique, linear. Hind wing whitish grey; transverse lines well marked along anal margin; terminal area narrow, darker; discal dot distinct.
Male genitalia. Uncus narrow, biapical. Valve with a blunt medial process ventrally. Vinculum elongate and narrow. Papillae broad. Aedeagus slim, equal to length of valve. Vesica armed with a small apical patch of heavily sclerotized spines, a large broad cornutus with pointed tip, a row of smaller, slim cornuti, and an elongate longitudinally twisted plate-like cornutus, connected with an irregular cornutus near ductus ejaculatorius base. Sternite A8 with a broad bilobed base, sharply tapered to pointed apex.
Female genitalia. Bursa ovate, sclerotized, with left side wrinkled, and almost completely covered with chaotic large spines. Ductus bursae spineless, short, with large chaotic wrinkles. Ductus seminalis narrow, attached to ductus bursae near short and wide colliculum. Antrum heavily sclerotized with numerous minute pores. Tergite A8 transverse, with shallow medial hollow in posterior margin. Apophyses relatively stout. Papillae anales slightly elongate and tapered.
Range. China (Tibet, Yunnan, Sichuan); Pakistan, Nepal.
Phenology. Bivoltine: early March to late April and late August to early October.
Altitude. From about 1750 m up to about 2800 m above sea-level (Nepal: coll. C. Herbulot in ZSM; China: Vojnits & de Laever 1978; Vojnits 1983).
Similar species. Broadly similar to *E. quadripunctata*, but with narrower wings and a more linear discal dot. The male genitalia are superficially similar to those of *E. acutangula* Hampson, 1895, but the ventral process of the valve is rounded rather than pointed. The female genitalia are easily distinguished by the sclerotized transverse antrum.
Material. ♀, China, Provinz Nord-Yuennan, Li-kiang [Lijiang], X, 17.iii.1935, H. Höne, Vojnits slide no. 11057♀ (holotype, ZFMK); 13♂, 12♀, same loc., 26,

27, 30.viii., 24.ix.1934, 9, 11, 13, 14, 17, 18, 19, 21, 22, 23, 24, 25, 28, 31.iii., 2, 3, 5, 20.iv.1935, H. Höne, Vojnits slides nos. 10901♂, 10884♂, 11012♂, 11011♀, 11010♀, 10996♂, 11025♀, 10881♂, 10995♀, 11056♀, 10904♀, 11058♀, 11024♀, 11003♂, 11051♂, 11050♀, de Laever slides nos. 212♀, 371♂, 222♂, 228♀, 344♂, 221♂, 345♀, 210♀, 209♂ (paratypes, ZFMK); 11♂, 12♀, same loc., 8, 9, 11, 12, 13, 14, 19, 20, 21, 23, 24, 25, 28.iii., 3.iv., 6.ix.1935, H. Höne, Vojnits slides nos. 10885♀, 10929♂, 10961♂, 10985♂, 10986♀, 10987♀, 10988♀, 10993♀, 10994♀, 11005♀, 11006♀, 11008♀, 11009♀, 11013♀, 11018♂, 11020♂, 11026♀, 11052♂, 11053♂, 11054♂, 11055♂ (paratypes, two specimens without abdomens, TTM); 3♀, same loc., 21, 27.iii.1935, H. Höne, Vojnits slides nos. 1202♀, 1200♀, 1201♀ (ZFMK); 2♀, Tibet [Sichuan], Batang, Im Tal des Yangtze (*ca.* 2800 m), 5, 6.iv.1936, H. Höne, Vojnits slides nos. 10878♀, 10902♀ (paratypes, TTM); 2♀, same loc. and alt., 10.iv., 4.ix.1936, H. Höne, de Laever slides nos. 100♀, 468♀ (ZFMK); ♀, Nord-Indien/Kumaon Bhimtal (Nainital), 1450 m, 23. ix.1973, Lichtfang, leg. S. Richter, Vojnits slide no. 12124♀ (holotype of *E. producta* Vojnits, 1981, SMNK); 1♀, same loc., 6.ix.1973, Lichtfang, leg. S. Richter, Vojnits slide no. 12127♀ (paratype of *E. producta* Vojnits, 1981, TTM); 2♀, Yunnan, Tsekou, P. Dubernard 1898, Vojnits slides nos. 13453♀, 14005♀ (TTM).

Remarks. The adults of the second generation of this species are considerably smaller than the first generation.

164. *Eupithecia quadripunctata* Warren, 1888

Eupithecia quadripunctata Warren, 1888, *Proc. zool. Soc. Lond.* **1888** (3): 331. Holotype ♀ (BMNH; examined), Western India [Pakistan]: Thundiani.
Eupithecia tricornuta Inoue, 1980, *Bull. Fac. domest. Sci. Otsuma Wom. Univ.* **16**: 153, fig. 35A; fig. 36A; figs 39A-D; fig. 42A. Holotype ♂ (BMNH; examined), [Japan]: Tokyo, Nippara (synonymised in Mironov et al., 2008b).
Eupithecia abiecta Vojnits, 1980, *Acta zool. Acad. Sci. hung.* **26** (4): 439, figs 7, 8, 20. Holotype ♂ (ZFMK; examined), [China]: Sued-Shensi, Tapaishan in Tsinling, 3000 m (synonymised in Mironov et al., 2008b).

Misidentification: *Eupithecia actaeata praenubilata* sensu Inoue, 1988 (*nec* Inoue, 1958).

Description. Fore wing ground colour dark grey or brownish grey with narrow dark, oblique transverse lines, the postmedial more distinct and sharply angled onto the costa; discal dot large, ovoid, conspicuous; hind wings paler, transverse lines clearly marked only along anal margin; discal dot small. Abdomen pale, reddish or ochreous. Moths of the first generation usually considerably larger than those of the second.

Male genitalia. Uncus slim, biapical. Valve broadened medially, with a short, blunt medial process on the ventral margin. Vinculum broad, trapezoid. Papillae relatively broad. Aedeagus broadened at anterior end, almost equal to length of valve. Vesica armed with three sinuate cornuti, an elongate folded cornutus and a small triangular cornutus near ductus ejaculatorius base. Sternite A8 rectangular, bilobed at base, and with two spiniform apical horns.

Female genitalia. Bursa ovate, heavily sclerotized, with a broad membranous diverticulum near base, and armed with spines of varying sizes forming a broad circular area basally and centrally; also with an elongate, transverse patch of small spines at base of the ductus bursae, the latter sclerotized with short longitudinal striations. Ductus seminalis sharply curved around broad colliculum. Antrum lightly sclerotized. Tergite A8 quadrate. Apophyses narrow. Papillae anales broad and rounded.

Range. China (Heilongjiang, Hebei, Beijing, Gansu, Yunnan, Sichuan, Shanxi, Shaanxi, Taiwan); Asian Russia (Priamurje, Primorje), Korea, Japan (Hokkaido, Honshu, Shikoku, Kyushu), Nepal, N Thailand.

Phenology. Bi- or polyvoltine: mid-February to early October.

Altitude. Found from about 1300–1500 m in Tai-Bei Shan (China: coll. ZFMK) up to about 2400 m in Nepal and Taiwan (coll. C. Herbulot in ZSM; TTM).

Similar species. Similar to *E. acutangula* Hampson, 1895 and *E. vivida*, but can usually be distinguished by the ochreous colour of the abdomen. The discal dot is also rounder. In the male genitalia, the valve shape is similar to that of *E. vivida* but the shape of the eighth sternite and the cornuti in the aedeagus vesica are distinctive. The female genitalia are best distinguished by the rather square sclerotized ductus bursae.

Material. ♂, Thundiani 30.iv.[18]87, 87.162, *Eupithecia quadripunctata* n. sp. Type, BM Geom. slide no. 20119 (holotype; BMNH); ♂, [Japan], Tokyo-To, Nippara, 12.v.1962, T. Ebato, BM Geom. slide no. 16373 (holotype of *E. tricornuta*, coll. H. Inoue in BMNH); ♂, Sued-Shensi [Shaanxi], Tapaishan [Taibaishan] im Tsinling [Qinling], *ca.* 3000 m, 11.viii.1936, H. Höne, Vojnits slide no. 12336♂ (holotype of *E. abiecta*, ZFMK); 8♂, 3♀, same loc. and alt., 11, 12, 25.viii.1936, H. Höne, Vojnits slides nos. 12049♂, 12337♂, 12341♂, 12344♂, 12347♀, 12349♂, 12350♂, 12359♂, 12334♂, 12345♂, 12377♀ (paratypes of *E. abiecta*, ZFMK); 1♂, 1♀, same loc. and alt., 11, 25. viii.1936, H. Höne, Vojnits slides nos. 12338♀, 12343♂ (paratypes of *E. abiecta*, TTM); 2♂, 1♀, same loc. and alt., 11, 12.viii.1936, H. Höne, de Laever slides nos. 275♂, 276♀, Vojnits slide no. 12960♂ (ZFMK); 1♂, 1♀, same loc. and alt., 11.viii, 7.ix.1936, H. Höne, Mironov slide no. 830♂, Vojnits slide no. 10972♀ (TTM); 1♂, same loc., *ca.* 1700 m, 17.v.1936, H. Höne, Vojnits slide no. 12351♂ (paratype of *E. abiecta*, ZFMK); 1♂, 1♀, same loc. and alt., 6.v, 9.viii.1936,

H. Höne, Vojnits slides nos. 12219♂, 12333♀ (paratypes of *E. abiecta*, TTM); 1♀, same loc. and alt., 17.v.1936, H. Höne, Vojnits slide no. 12391♀ (ZFMK); 3♀, same loc. and alt., 13, 20.v.1936, H. Höne, Vojnits slides nos. 11480♀, 11481♀, 11490+ (last slide is absent in TTM); 1♂, same loc. (without alt.), L, 20.vi.1935, H. Höne, Vojnits slide no. 12340♂ (paratype of *E. abiecta*, TTM); 3♀, same loc., L, 30.viii.1935, H. Höne, Vojnits slides nos. 12939♀, 12940♀, 12950♀ (ZFMK); 1♀, China, Provinz Nord-Yuennan, Li-kiang [Lijiang], 16.v.1935, H. Höne, de Laever slide no. 213♀ (ZFMK); 1♀, [Taiwan], Mt. Hou Fuan, Nantow, Formosa, 8.viii.[19]74, leg. H. Nakajima, Inoue slide no. 9383♀ (*E. actaeata praenubilata*, det. H. Inoue, BMNH); 1♀, Taiwan, Hualien Hsien, Dayuling, 28–29.viii.[19]83, leg. H. Yoshimoto, Inoue slide no. 13144♀ (*E. actaeata praenubilata*, det. H. Inoue, BMNH); 1♂, 1♀, Taiwan, Prov. Hualien, Taroko N.P., Kuanyuan, 2400 m, 4.iv.1997, leg. Csorba & Ronkay (TTM); 2♀, same data, Galsworthy slides nos. ACG GL5, ACG GL24 (coll. GL); 4♀, Central China, prov. Shaanxi, Mt. Tai Bei Shan, 1300–1500 m, 33°35′ N, 107°43′ E, 10.viii.-4.ix.1998, leg. V. Murzin & V. Siniaev (ZFMK); 1♀, Taiwan, Prov. Nantou, Hohuachi, 1950 m, 24°13′ N, 121°16′ E, 31.iii.2000, leg. A. Kun & L. Peregovits (ZSM); 1♀, Heilongjiang, Dailing, 390 m, 11.viii.1963, coll Bai Jiuwei, slide no. ZL119♀; 1♂, Hebei, Xiao Wutaishan, Nantai, 1700 m, 7.vii.1964, coll Wang Chunguang, slide no. ZL206♂; 1♀, Sichuan, Emeishan, Qingyinge, 800–1000 m, 19.ix.1957, coll Zhu Fuxing, slide no. ZL192♀; 1♂, Gansu, Yongdeng, Tulugou, 1.ix.[19]89, slide no. ZL293♂; 1♂, 2♀, Beijing, Mentougou, Xiaolongmen, 1100 m, 24.ix.1998, Han Hongxiang, slide no. ZL680♀ (IZCAS).

Remarks. The synonymy of this species was dealt with extensively in Mironov et al., 2008b. Inoue (2000) confused it with *E. vivida*.

165. *Eupithecia eximia* Vojnits & de Laever, 1978

Eupithecia eximia Vojnits & de Laever, 1978, *Acta zool. Acad. Sci. hung.* **24** (1–2): 245, figs 22, 23. Holotype ♂ (ZFMK; examined), [China]: North Yunnan, A-tun-tse [Dêqên], 4500 m.

Eupithecia infortunata Vojnits, 1984, *Acta zool. Acad. Sci. hung.* **30** (1–2): 227, fig. 18. Holotype ♂ (TTM; examined), [China]: North Yunnan, A-tun-tse [Dêqên], 4000 m. **New junior synonym.**

Description. A large species with narrow, elongate fore wings and greyish ochreous ground colour; all transverse lines narrow, oblique, sharply angled onto costa, the postmedial twice so; often with a semicircular mark on costa above discal dot; terminal area usually darker; discal dot small, ovoid-oblong, black; hind wing paler, whitish with dark anal margin; third abdominal segment dark brown, forming a band.

Male genitalia. Uncus triangular, biapical. Valve elongate, relatively narrow, broadened medially. Vinculum rounded. Papillae elongate, slightly clavate. Aedeagus stout, shorter than length of valve. Vesica armed with a stout, sinuate cornutus and a very small, folded, irregular cornutus near ductus ejaculatorius base. Sternite A8 elongate, with almost parallel lateral margins, bilobed at base, and with two narrow diverging apical horns.
Female genitalia. Bursa short, spongy, and membranous, about one half of base and whole left side covered with small spines, and with a long patch of spines centrally on ventral side. Ductus bursae not distinguishable. Ductus seminalis slightly broadened at base, attached to central part of corpus bursae on dorsal side. Colliculum short and broad. Antrum membranous, short and broad. Tergite A8 transverse. Apophyses rather narrow. Papillae anales broad and rounded.
Range. China (Tibet, Qinghai, Sichuan, Yunnan).
Phenology. Univoltine: mid-April to mid-August.
Altitude. From about 2800 to 4500 m in SW China (Vojnits & de Laever 1978; coll. Pekarsky; IZCAS).
Similar species. Externally very similar to *E. nepalata*, but distinguished from it by the larger size, more elongate and pointed fore wings, greyish ochreous colour and also the paler, whitish hind wings. The additional apical cornutus on the vesica in the male genitalia distinguishes the male from that species, but the female genitalia are very similar.
Material. ♂, Nord Yünnan, A-tun-tse [Dêqên], Obere Höhe (*ca.* 4500 m), 7.vii.1936, H. Höne, Vojnits slide no. 11015♂ (holotype, ZFMK); 3♂, 2♀, same loc. and alt., 8, 9, 15, 18.vii., 18.viii.1936, H. Höne, de Laever slides nos. 57♂, 218♂, 219♀, Vojnits slides nos. 10886♂, 11022♀ (paratypes, ZFMK); 1♂, 2♀, same loc. and alt., 14, 31.vii., 11.viii.1936, H. Höne, Vojnits slides nos. 10999♀, 11014♀, 11016♀ (paratypes, TTM); ♂, same loc. and alt., 13.vi.1936, H. Höne, Vojnits slide no. 14196♂ (holotype of *E. infortunata*, TTM); 1♂, 2♀, same loc. and alt., 9.vii., 23.viii.1936, 25.vii.1937, H. Höne, Vojnits slides nos. 12466♂ (an incorrectly labelled slide of *E. impavida*), 13125♀, 13132♀ (ZFMK); 1♀, same loc., Mittlere Höhe (*ca.* 4000 m), 13.viii.1936, H. Höne, de Laever slide no. 58♀ (paratype of *E. eximia*, ZFMK); 3♀, same loc. and alt., 12, 13.vi., 17.viii.1936, H. Höne, Vojnits slide no. 10997♀, 11017♀ (paratypes of *E. eximia*, TTM); 1♀, same loc., Talsohle (*ca.* 3500 m), 18.vi.1936, H. Höne, Vojnits slide no. 11021♀ (paratype of *E. eximia*, TTM); 1♂, Szetschwan [Sichuan], Sungpan-ting, Exped. Stötzner, coll. Dr. Wehrli, Basel, Vojnits slide no. 10890♂ (paratype of *E. eximia*, TTM); 1♀, Tibet [Sichuan], Batang, Untere Urwaldzone (*ca.* 3800 m), 11.viii.1936, H. Höne, de Laever slide no. 101♀ (paratype of *E. eximia*, ZFMK); 2♀, same loc. and alt., 4, 5.viii.1936, H. Höne,

de Laever slides nos. 96♀, 108♀ (ZFMK); 1♀, Ta-tsien-Loû, Chasseurs du P. Déjean 1904, Vojnits slide no. 13437♀ (an incorrectly labeled gen. slide of a species from the "*innotata*" group) (TTM); 1♀, China, Sichuan, Daxue Shan Mts., 80 km W Kangding, 30°02′ N, 101°40′ E, ~3800 m, 27–29.vii.1999, leg. V. Siniaev & A. Plutenko (ZFMK); 1♀, China, W. Sichuan, Xindugiao evn 3500 m, 30°04′ N / 101°26′ E, 14.vii.2009, I. & A. Floriani, Saldaitis, Pekarsky slide no. OP 0688♀ (coll. O. Pekarsky); 1♂, China, N. Yunnan, road Xinngcheng / Zhongdian, 3500 m, 28°32′ N / 99°49′ E, 18.vii.2009, I.& A. Floriani, Saldaitis, Pekarsky slide no. OP 0487♂ (coll. O. Pekarsky); 4♀, Tibet, Qamdo, 3900 m, 7–9.viii.1976, coll Han Yinheng, slide no. ZL236♀; 1♀ Qamdo, 3300 m, 13.viii.1976, coll Zhang Xuezhong; 2♂, 1♀, Tibet, Markam, Rongxi, 4000 m, 23–4.vi.1976, slide no. ZL170♀; 10♀, Tibet, Jomda, 3400 m, 22–31.vii.1976, slides nos. ZL55, 221♀; 2♀ same loc, 29.viii.1976, coll Zhang Xuezhong and Han Yinheng; 1♀, Tibet, Zhag'yab Gyitang, 3600 m, 5. vii.1976, coll Han Yinheng; 24♀, Tibet, Riwoqê, 3750 m, 17–29.viii.1976, coll. Zhang Xuezhong, slides nos. ZL340, 460♀; 1♀, Tibet, Nyingchi, Bayi, 3040 m, 2–3.vii.2002, Han Hongxiang, slide no. ZL645♀; 4♂, 10♀, Qinghai, Menyuan Xianmi, 2800 m, 23.vii.1992, slides nos. ZL308, 382♂, ZL241♀; 2♂, 15♀, Qinghai, Tongren, Maixiu, 2950 m, 30.vii.1992, slides nos. ZL245♂, ZL83, 239, 306, 378♀; 1♀, Qinghai, Huzhu Beishan, 2300 m, 30.vii.1991, all coll Xue Dayong; 1♀, Qinghai, Huangzhong, 6.viii.1977, coll Zhang Baolin; 1♀, Sichuan, Batang, 4600 m, 20.viii.1982, coll Wang Shuyong (IZCAS).

166. *Eupithecia nepalata* Schütze, 1961

Eupithecia nepalata Schütze, 1961, *Veröff. zool. StSamml. Münch.* 6: 179, pl. 29, figs 1, 6; pl. 32, figs 1, 1a, 1b, 2. Holotype ♂ (ZSM; examined), [Nepal]: Manangbhot, Kehami, 29°11′ n, Br. 83°58′ ö.L., 3700 m.

Eupithecia abundeli Vojnits, 1988, *Annls hist.-nat. Mus. natn. hung.* 80: 83, pl. 2, figs 13, 14; pl. 6, figs 41–44. Holotype ♀ (ZISP; examined), [Tajikistan]: Pamir, Schach-Darensis Mts., Badjond-Dara River, 3500 m (synonymised in Mironov et al., 2008b).

Eupithecia petrensis Mironov, 1989, *Ent. obozrenie* 68 (2): 371, figs 9–11. Holotype ♂ (ZMMU; examined), Tajikskaya SSR [Tajikistan]: Khorog, Botanical Garden (synonymised in Mironov et al., 2008b).

Description. Fore wing ground colour ash-grey; transverse lines fine, sharply angled onto costa; basal, postbasal and antemedial broken, with three angles; medial very sharply angled onto costa behind discal dot, often touching it; postmedial sinuate, twice angled, forming a sharp angle on the R vein; basal medial and postmedial forming three dark spots on costa; terminal area slightly darker with pale apical streak and weak subterminal line; discal dot ovoid, black. Hind wing whitish grey, paler than fore wing; dark transverse lines well marked along anal margin; discal dot minute.

Male genitalia. Uncus broad, biapical. Valve narrow, unmodified. Vinculum semicircular. Papillae elongate. Aedeagus stout, shorter than valve length. Vesica armed apically with a sinuate cornutus with a hooked tip, centrally with a stout cornutus with a broadened base, and with a small, irregular cornutus near ductus ejaculatorius base. Sternite A8 rectangular, with broadened basal lobes, and with two sharp 'Viking' horns at apex.
Female genitalia. Bursa small, irregularly shaped, right side membranous and spineless, base and left side covered with small spines, and with a distinct elongate patch of small spines on ventral side. Ductus bursae not distinguishable. Ductus seminalis basally broad, attached dorsally to centre of corpus bursae. Colliculum short, very broad. Antrum membranous, broad. Tergite A8 transverse. Apophyses narrow. Papillae anales broad and rounded.
Range. China (Sichuan, Yunnan); Nepal, Tajikistan (Pamir Mts.).
Phenology. Bivoltine (?): mid-April to mid-August.
Altitude. From about 3500 m in Pamir Mts. (Tajikistan: Vojnits 1988) up to about 4500 m in the Himalayas (Nepal: Schütze 1961).
Similar species. Similar externally to *E. relaxata*, which has a narrower fore wing with three fine medial lines and a linear discal dot. Also very similar, including in the genitalia, to the preceding species. In the male, the presence of the apical cornutus in the vesica is decisive. The female genitalia are hard to distinguish, but usually have a smaller spiniferous area and longer spines in the medial patch in the corpus bursae.
Material. ♂, Nepal, Mustangbhot, 29°11′ n. Br., 83°58′ ö. L., Kehami, 3700 m, 12.viii.[19]55. leg. F. Lobbichler, ZSM slide no. G 3892♂ (holotype, ZSM); ♂, Nepal, Mustangbhot, 29°11′ n. Br., 83°58′ ö. L., Penga, 3800 m, 8.viii.[19]55. leg. F. Lobbichler (paratype, ZSM); ♀, [Tajikistan], Pamir, Mts. Schach-Darensis, cliv. sept., fl. Badjond-Dara, 3500 m, 31/VII/1961, Bundel, Vojnits slide no. 17689♀ (holotype of *E. abundeli*, ZISP); ♂, Tajikskaya SSR [Tajikistan], Khorog, Botanical Garden, 24.vii.1958, leg. A. Tzvetajev (holotype of *E. petrensis*, ZMMU); 1♀, China, Sichuan, Daxue Shan Mts., 80 km W Kangding, 30°02′ N, 101°40′ E, 3800 m, 27.-29.vii.1999, leg. V. Siniaev & A. Plutenko, Mironov slide no. 728♀ (ZFMK).
Remarks. The holotype of this species consists only of the genitalia slide and four wings stuck to a piece of card.

167. *Eupithecia fatigata* Vojnits & de Laever, 1978

Eupithecia fatigata Vojnits & de Laever, 1978, *Acta zool. Acad. Sci. hung.* **24** (1–2): 249, figs 25–27. Holotype ♀ (ZFMK; examined), [China]: North Yunnan, A-tun-tse [Dêqên], 4500 m.

Description. Fore wing very elongate, ground colour ash-grey; transverse lines distinct, broadened near costa; basal oblique, sharply angled onto costa; postbasal and antemedial twice obtusely angled; medial touching discal dot and well marked between discal dot and costa only; postmedial double, twice angled near costa; stalk of Cu vein covered with black scales; terminal area with black marginal streaks between veins and wavy, whitish subterminal line broadened near tornus, where bordered by blackish shading; discal dot broadly linear, black. Hind wing paler, darker along anal margin; transverse lines well marked along anal margin; discal dot relatively large.

Female genitalia. Bursa small, ovate-oblong, membranous, covered with small spines in the posterior part on the left side, which form a circular spiniferous area near ductus seminalis base. Ductus bursae not distinguishable. Ductus seminalis narrow, attached to anterior part of corpus bursae. Colliculum short and broad. Antrum membranous. Tergite A8 transverse. Apophyses short and narrow. Papillae anales relatively large and broad.

Male unknown.

Range. China (Tibet, Sichuan, Yunnan).

Phenology. Bivoltine (?): from mid-May to mid-August.

Altitude. Found from 3000 to 4500 m in Sichuan and Yunnan (China: coll. Pekarsky; Vojnits & de Laever 1978).

Similar species. Similar to *E. repentina*, *E. eximia*, and *E. nepalata*, but fore wings narrower and more elongate, and fore wing discal dot larger and longer. The female genitalia are very similar to those of all three species, but are distinguishable by the membranous base of the corpus bursae, and the very narrow ductus seminalis.

Material. ♀, Nord Yünnan, A-tun-tse [Dêqên], Obere Höhe (*ca.* 4500 m), 8.vii.1936, H. Höne, Vojnits slide no. 11004♀ (holotype, ZFMK); 2♀, same loc. and alt., 7, 9.vii.1936, H. Höne, Vojnits slides nos. 10903♀, 11047♀ (paratypes, ZFMK; slide no. 11047♀ not found in ZFMK); 4♀, same loc. and alt., 8.vii., 11, 15.viii.1936, H. Höne, Vojnits slides nos. 10889♀, 10992♀, 11007♀, 11049♀ (paratypes, TTM); 1♀, same loc., Mittlere Höhe (*ca.* 4000 m), 28. vii.1936, H. Höne, Vojnits slide no. 10900♀ (paratype, slide absent in ZFMK); 1♀, Tibet [Sichuan], Batang, Im Tal des Yangtze (*ca.* 2800 m), 17.v.1936, H. Höne, de Laever slide no. 107♀ (ZFMK); 3♀, China, Szetschwan [Sichuan], Ta-tsien-lu, Exped. Stötzner, de Laever slide no. 406♀, Vojnits slide no. 1203♀ (ZFMK); 1♀, same data, Coll. Dr. Wehrli, Basel, Vojnits slide no. 658♀ (TTM); 6♀, China, W. Sichuan, Kangding, evn 3000 m, 29°53′ N / 101°55′ E, 13.vii.2009, leg. I. & A. Floriani, Saldaitis, Pekarsky slides

nos. OP 0488♀, OP 0490♀, OP 0491♀, OP 0492♀, OP 0493♀, OP 0494♀ (coll. O. Pekarsky); 1♀, Tibet, Markam, Rongxi, 4000 m, 26.vi.1976, Zhang Xuezhong, slide no. ZL263♀ (IZCAS).
Remarks. The worn paratype male assigned to this taxon by Vojnits, 1984 is a specimen of *E. egena*.

168. *Eupithecia repentina* Vojnits & de Laever, 1978

Eupithecia repentina Vojnits & de Laever, 1978, *Acta zool. Acad. Sci. hung.* **24** (1–2): 237, figs 15, 17. Holotype ♂ (ZFMK; examined), China: Prov. Hunan, Hoeng-Shan, 900 m.

Misidentification: *Eupithecia recens* Inoue, 1955 (*nec* Dietze, 1904).

Description. Fore wing ground colour rich grey; transverse lines oblique, blackish, broadened near costa, especially ante- and postmedial; basal and antemedial sharply angled onto costa; postmedial right angled, often twice angled, onto costa; terminal area slightly darker, with weak trace of pale wavy subterminal line; discal dot oblique, linear, intensely black. Hind wing paler, light grey, darker along anal margin; transverse lines inconspicuous; terminal area darker near tornus with very weak trace of whitish indented subterminal line; discal dot small.
Male genitalia. Uncus narrow, biapical. Valve with a very narrow spiniform ventral process. Vinculum short and round. Papillae elongate, rather broad. Aedeagus stout, shorter than valve length. Vesica armed with three small stout cornuti, an elongate, broad cornutus and an irregular, curved cornutus at ductus ejaculatorius base. Sternite A8 rectangular, with broadened lobes at base and two small apical horns.
Female genitalia. Bursa small, pear-shaped, densely covered with small spines from base to colliculum on left side. Ductus bursae not distinguishable. Ductus seminalis broadened at base, attached to corpus bursae near its base on right side. Colliculum short and narrow. Antrum membranous. Tergite A8 very transverse. Apophyses relatively long and narrow. Papillae anales large, slightly elongate and rounded at apices.
Range. China (Tibet, Yunnan, Sichuan, Qinghai, Gansu, Hunan, Shaanxi, Beijing); European Russia (Tula oblast'), Asian Russia (southern Primorje), Mongolia, Korea, Japan (Hokkaido, Honshu, Sado, Shikoku, Kyushu, Tsushima).
Phenology. Bivoltine: early April to mid July; late August to mid October.
Altitude. From about 900 m in Huangshan up to about 3000 m in Tapaishan (China: Vojnits & de Laever 1978).
Similar species. Very similar to *E. uliata* Staudinger, 1897, but with broader and less pointed fore wings with postmedial lines less sharply angled onto

costa and larger, linear discal dots. The pattern resembles the three preceding species, but with the fore wing broader, and less pointed. The male genitalia are recognisable by the narrow ventral spine on the valve, and the combination of cornuti in the aedeagus vesica. The female genitalia resemble those of the preceding three species, but the pattern of spining in the corpus bursae is unique, and invariable.

Material. ♂, China, Provinz Hunan, Hoeng-Shan (900 m), 13.iv.1933, H. Höne, Vojnits slide no. 10919♂ (holotype, ZFMK); 1♂, 5♀, same loc. and alt., 13, 15, 16.iv.1933, H. Höne, Vojnits slides nos. 10866♀, 10868♀, 10917♀, 10939♀ (paratypes, ZFMK); 2♂, 2♀, same loc. and alt., 12, 19.iv., 18.x.1933, H. Höne, Vojnits slides nos. 10867♂, 10869♀, 10918♀, 10940♂ (paratypes, TTM); 4♂, 1♀, same loc. and alt., 10, 18, 22, 25.x.1933, H. Höne, Vojnits slides nos. 12930♀, 12934♂, 12931♂, 12932♂, 12933♂ (ZFMK); 1♂, China, Provinz Nord-Yuennan, 28.viii.1934, H. Höne (paratype, ZFMK); 1♂, Sued-Shensi [Shaanxi], Tapaishan [Taibaishan] im Tsinling [Qinling], *ca.* 3000 m, 25. viii.1936, H. Höne, Vojnits slide no. 10922♂ (paratype, ZFMK); 1♂, 2♀, same loc. and alt., 9, 11, 25.viii.1936, H. Höne (ZFMK); 3♂, same loc., *ca.* 1700 m, 25.viii.1936, H. Höne, Vojnits slides nos. 11523♂, 12342♂ (ZFMK); 1♂, same loc. (without alt.), L, 31.viii.1935, H. Höne, Vojnits slide no. 12346♂ (ZFMK); 1♀, Tibet [Sichuan], Batang, Im Tal des Yangtze (*ca.* 2800 m), 22.iv.1936, H. Höne, de Laever slide no. 462♀ (ZFMK); 1♀, China, Shaanxi, Tsinling Mts. [Qinling], Foping Nature Reserve, 33°51′ N, 107°57′ E, 1400 m, 6.-11.iv.1999, leg. V. Siniaev & A. Plutenko, *ex.* coll. Dr. A. Schintlmeister (ZSM); 1♀, C. China, prov. Shaanxi, Mt. Tai bei shan, Hao Djenzy village, 33°48′ N, 107°43′ E, 1400 m, 1.-5.iv.1999, leg. V. Siniaev & A. Plutenko (ZFMK); 1♀, C. China, prov. Shaanxi, Foping Natural Reserve, Panda area, 33°45′ N, 107°48′ E, 1600 m, 15.-20.iv.1999, leg. V. Siniaev & A. Plutenko, Mironov slide no. 662♀ (ZFMK); 1♀, Tibet, Nyingchi, Bayi, 3040 m, 2–3. vii.2002, Han Hongxiang, slide no. ZL647♀; 1♀, Qinghai, Menyuan, Xianmi, 2800 m, 23–24.vii.1992, coll Xue Dayong, slide no. ZL296♀; 1♀, Qinghai, Huzhu, Beishan Linchang, Langshidang, 2600 m, 36°55′ N 102°23′ E, 7. viii.2005, Xue Dayong; 1♀, Sichuan, Batang, Yidun, 3370 m, 16.viii.1982, coll Wang Shuyong, slide no. ZL280♀; 1♀, Gansu, Zhouqu Shatan forestry reserve, 2400 m, 14.vii.1999, coll Yao Jian, slide no. ZL563♀; 1♀, Gansu, Yongdeng, Liancheng Linchang, Tulugou 2600 m 36°41′ N 102°42′ E, 7. viii.2005, Xue Dayong, slide no. ZL705♀;1♀, Beijing, Mentougou, Liyuanling, 1100 m, 20–21.viii.2002, slide no. ZL658♀; 3♀, same loc., 22.viii.2002, slides nos. ZL668, 717♀; 3♀, same loc. and alt., but Xiaolongmen, 24.ix.1998, slide no. ZL669♀, all Han Hongxiang (IZCAS).

169. *Eupithecia virgaureata* Doubleday, 1861

Eupithecia virgaureata Doubleday, 1861, *The Zoologist* **19**: 7566. Syntypes, at least 3 (BMNH), [United Kingdom].
Eupithecia altenaria Staudinger, 1861, *Stettin. ent. Z.* **22** (10–12): 401. Syntypes 5♂, 7♀ (coll. O. Staudinger in MNHU; examined), [Norway]: [near Bossekop] (treated as var. of *E. virgaureata* in Bohatsch, 1893).
Eupithecia offirmata Speyer, 1869, *in* Grentzenberg, *Schr. phys.-ökon. Ges. Königsb.* **10**: 120. Syntypes (?), [Poland]: Johannisberge near Danzig [= Gdansk] (synonymised in Staudinger & Rebel, 1901).
Eupithecia invisa Butler, 1878, *Ann. Mag. nat. Hist.* (5) **1** (6): 444. Holotype ♀ (BMNH), Japan: Hakodaté (treated as subsp. to *E. virgaureata* in Inoue, 1980).
[*Eupithecia*] *albipunctata* f. *privata* Dietze, 1913, *Biologie der Eupithecien* **2**: 87; *ibidem* (1910), **1**, pl. 85, fig. 941. Syntypes 1♂, 3♀ (coll. K. Dietze in MNHU; examined), Italy: Tyrol, Waidbruck. **New junior synonym.**

Misidentification: *Eupithecia pimpinellata* Doubleday, 1856 (*nec* Hübner, 1813).
Unavailable names (infrasubspecific): *nigronotata*: Dietze (1910); *nigrofasciata*: Dietze (1910); *aestiva*: Dietze (1910); *notata*: Dietze (1913); *nigra*: Lempke (1951). – (incorrect subsequent spelling): *virgaurearia*: Morris (1861).

Description. A very variable species in size. Fore wing fuscous grey; transverse lines indistinct, slightly sinuous, often reduced to a series of short black dashes on the veins, especially on the stalk of Cu beyond the discal dot; costa usually with five blackish grey spots; medial usually conspicuous, sharply angled near costa beyond discal dot or touching discal dot; terminal area with pale subterminal line inconspicuous, broken or often obsolete, but with small whitish tornal spot; discal dot distinct, ovate or rounded, black. Hind wing paler, with darker terminal area and minute pale tornal spot; transverse lines usually ill-defined but in some specimens more or less distinct; discal dot small, rounded.
Male genitalia. Uncus narrow, biapical. Valve with centrally prominent ventral margin. Vinculum semicircular. Papillae narrow. Aedeagus narrow, shorter than valve length. Vesica with an apical field of very small spinules, and armed with a narrow apical cornutus and a larger flat and folded medial cornutus. Sternite A8 narrow with broadly bilobed base, terminating in two sharp asymmetric horns curved towards the left side.
Female genitalia. Bursa small, globular, with numerous spinules around perimeter medially. Ductus bursae membranous, narrowed to colliculum, with a small patch of spinules near ductus seminalis base. Ductus seminalis narrow, attached to ductus bursae near colliculum on left side. Colliculum short and narrow, lightly sclerotized. Antrum membranous. Tergite A8 elongate. Apophyses rather narrow. Papillae anales relatively broad and rounded.

Range. China (Zhejiang); Europe, Caucasus, Transcaucasus (Georgia), Asian Russia (Urals, western and southern Siberia, Priamurje, Primorje, Kamtchatka peninsula), Mongolia, Korea, Japan (Hokkaido, Honshu, Sado, Shikoku, Kyushu, Tsushima, Yakushima).
Phenology. Bivoltine: April to June; early (late) July to early September.
Biology. Larva found on the flowers of *Solidago japonica, Kalimeris yomena, Dendranthema japonicum* and *Farfugium japonicum* (Asteraceae) (Japan: Inoue 1980; Murase 1996, 2000), and also on the flowers and ripening seeds of *Heracleum moellendorffi* (Apiaceae) (Far East Russia: Mironov 2003). In Europe the larvae of the first generation are found on the leaves of various Rosaceae; the larvae of the second generation are polyphagous on the flowers and seeds of herbaceous plants, especially on the Asteraceae, and also on some Ericaceae, Primulaceae, Apiaceae and Gentianaceae.
Altitude. Found at about 1600 m above sea-level (Mongolia: Vojnits 1975; China: coll. ZFMK).
Similar species. Variable in the distinctness of the wing pattern; in some specimens the maculation is almost wholly obsolete, except the black discal dot, dark veins along the postmedial and whitish tornal spot on the fore wing. Distinguished with difficulty from *E. subfuscata*, but usually separable by the larger, rounded discal dot and the chequered veins on the fore wing. The male and female genitalia, and the shape of sternite A8 in the male easily distinguish this species from all other Palaearctic *Eupithecia* species.
Material. 1♂, [Latvia], Livon[ia], *undosata* Dietze = *egenaria* H. S ♂ von Frau Pastor Lienig, Livland, Typ., K. Dietze, Typus (syntype of *E. undosata*, coll. K. Dietze in MNHU); 2♀, Pz. Chekiang [Zhejiang], West-Tien-Mu-Shan, (1600 m), 25.iii, 4.iv.1932, H. Höne, Vojnits slide nos. 12577♀, 12531♀ (ZFMK).
Remarks. Not previously recorded from China.

One male syntype of *E. undosata* Dietze, 1875 belongs to this species, but a further female syntype belongs to *E. egenaria* Herrich-Schäffer, 1848 (coll. K. Dietze in MNHU). We have therefore not formally synonymised this species, pending a review of all the syntypes.

170. *Eupithecia taiwana* Wileman & South, 1917

Eupithecia taiwana Wileman & South, 1917, *The Entomologist* **50** (646): 56. Holotype ♂ (BMNH; examined), Formosa [Taiwan]: Arizan, 7300ft.

Description. An easily recognized large, narrow winged species with elongate and pointed fore wings. The combination of clear reddish brown basal and terminal areas, a pale ash grey medial area with distinct, narrow black postmedial, sharply angled onto costa, and an elongate and narrow black discal dot on the fore wing is unique.

Male genitalia. Uncus short, biapical. Valve with a small pointed ventral process. Vinculum short and broad, with shallow medial hollow. Papillae elongate, curved. Aedeagus very stout, almost equal to length of valve. Vesica armed with two very large, heavily sclerotized horn-like cornuti, a bunch of about ten lightly sclerotized apical cornuti, an elongate and twisted medial cornutus, and a folded, irregular cornutus near ductus ejaculatorius base. Sternite A8 peg-like, broadened and bilobed basally, with two very small rounded apical horns.

Female genitalia. Bursa large, ovoid, heavily sclerotized, with a small medial projection on dorsal side, covered with dense spines from base to ductus bursae on ventral side, and with a short curved chain of large spines near base. Ductus bursae with numerous longitudinal striations. Ductus seminalis narrow, arising from a broad transverse tube adhering to corpus bursae on dorsal side. Colliculum broad. Antrum heavily sclerotized, wrinkled, covered with numerous pores. Tergite A8 slightly transverse. Apophyses narrow. Papillae anales broad, angulate at apices.

Range. China (Taiwan).

Phenology. Bivoltine: late February to mid May; mid September to November.

Altitude. From about 2000 to 2800 m above sea-level (Taiwan: coll. TTM, ZSM).

Similar species. Externally easily distinguished. Male and female genitalia also both readily distinguished from related species.

Material. ♂, [Taiwan] Formosa, Arizan, 7,300 ft., 15.ix.1906, A.E. Wileman, *Eupithecia taiwana* sp. n. Type ♂, 830 F, Wileman coll. B.M. 1929–261 (holotype, BMNH); 1♂, Taiwan, Prov. Tai-Tung, 2 km E of Hsiangyang, 2200 m, 11–13.iii.1996, Gy. Fábián & L. Németh (TTM); 1♀, Taiwan, Prov. Nan-Tou, 10 km SE of Shenmu, Yu-Shan Nat. Park, 2200 m, 14.iii.1996, Gy. Fábián & L. Németh (TTM); 2♀, Taiwan, Prov. Chiayi, 3 km S of Alishan, 1950 m, 11.iv.1997, leg. Csorba & Ronkay (TTM); 1♀, Taiwan, Prov. Nantou, 6 km SW Hohuan Pass, Yuanfeng, 2760 m, 10.v.1997, leg. Gy.M. László & G. László (TTM); 1♀, Taiwan, Prov. Nantou, Hohuachi, 1950 m, 24°13′ N, 121°16′ E, 31.iii.2000, leg. A. Kun & L. Peregovits (ZSM); 1♀, Taiwan, county Hualien, Taroko Nat. Park, at sacred Pilu tree, 2000 m, 24°14′ N, 121°24′ E, 10.iv.2002, leg. Gy. Fábián (TTM); 1♀, Taiwan, Kaohsiung county, Shanping Forest Recreation Area near Liukuei, 22°58′16″ N, 120°41′15″ E, 19–21.xi.2002, leg. L. Ronkay & O. Merkl (TTM); 1♂, Central Taiwan, Nantou Co., Meifeng, *ca.* 2100 m NN, 24°05′55″ N / 121°10′48″

E, LF, 29. March 2003, leg. Mei-Yu Chen & U. Buchsbaum (ZSM); 1♂, Central Taiwan, near Puli, Nantou Co., Meifeng, *ca.* 2200 m NN, 24°05′25″ N / 121°10′21″ E, LF, 19. March 2004, leg. Mei-Yu Chen (ZSM).
Remarks: The holotype of this species has a gummed abdomen.

171. *Eupithecia matrona* Mironov & Galsworthy, 2004

Eupithecia matrona Mironov & Galsworthy, 2004, *Trans. lepid. Soc. Japan* 55 (4): 291, figs 6, 19. Holotype ♂ (ZFMK; examined), China: Gansu, Xiahe, 2600–3000 m.

Description. Fore wing narrow and elongate, ground colour light greyish brown; costal margin slightly darker with ochreous tinge; transverse lines oblique, brownish, faintly marked; discal dot narrow and elongate. Hind wing narrow and elongate, paler, light greyish brown; transverse lines inconspicuous, brownish, weakly marked along dorsum; discal dot invisible. Abdomen ochreous brown; first segment dark brown.
Male genitalia. Uncus narrow, biapical. Valve short and narrow, ventral margin unmodified, but sacculus slightly bulging and sclerotized. Vinculum large, rounded. Papillae elongate, broad. Aedeagus narrow, sinuate, equal to length of valve. Vesica armed with one medial and two apical cornuti and an elongate folded V-shaped cornutus near ductus ejaculatorius base. Sternite A8 rectangular, heavily sclerotized apically, with two diverging pointed horns at apex.
Female unknown.
Range. China (Gansu, Qinghai).
Phenology. Univoltine: June.
Altitude. From about 2600 to 3800 m above sea-level (China: coll. ZFMK, IZCAS).
Similar species. Externally, very similar to the Central Asian *E. nobilitata* in shape, colour and pattern, but distinguished by the structure of the male genitalia, in particular the shapes of valve, vinculum, eighth sternite and cornuti on the vesica. The male genitalia are most similar to those of three Himalayan species *E. coccinea* Vojnits, 1981, *E. concinna* Vojnits, 1983, and *E. lata*, especially in the shape of the eighth abdominal sternite.
Material. ♂, China, Gansu, Xiahe, 2600–3000 m, 26–30.vi.1986, Coll. Nr. 246, leg. Görgner (holotype, ZFMK); 2♂, same data, Mironov slides nos. 521♂, 522♂ (paratypes, ZFMK); 2♂, Qinghai, Guinan, 3300–3800 m, 4 and 15. vi.1957, coll. Zhang Yiran, Galsworthy slides nos. ZL73, 500♂ (paratypes, IZCAS).
Remarks. All known specimens are worn: markings are likely to be stronger in fresh specimens.

172. *Eupithecia rhadine* Mironov & Galsworthy, 2007

Eupithecia rhadine Mironov & Galsworthy, 2007, *Trans. lepid. Soc. Japan* 58 (3): 349, figs 5, 19. Holotype ♂ (TTM; examined), Taiwan: Prov. Nantou, Hohuachi, 24°13′ N, 121°16′ E, 1950 m.

Description. Fore wing ground colour unicolourous brownish grey with darker costal margin; transverse lines narrow, indistinct; antemedial slightly curved, postmedial oblique, straight, obtusely angled near costa; discal dot small, dark brown, circular. Hind wing elongate and narrow with more or less narrow apex, termen with shallow medial hollow; ground colour and discal dot as fore wing; transverse lines indistinct.
Male genitalia. Uncus short, biapical. Valve with broad medial hollow on the ventral margin. Vinculum short and broad. Papillae short. Aedeagus slim, almost equal to length of valve. Vesica armed with two stout horn-like cornuti joined basally, and an elongate plate. Sternite A8 peg-like, bilobed at base, and with two asymmetrical pointed apical arms. Tergite A8 broad and elongate, with a narrow sclerotized anterior margin and, most unusually in the genus, a narrow pointed, hook-shaped, posterior process.
Female unknown.
Range. China (Taiwan).
Altitude. Only known specimen taken at 1950 m above sea-level (coll. TTM).
Similar species. Externally similar to *E. addictata*, but smaller, with narrower wings. The male genitalia are unique. The process on the male eighth tergite is known from very few other species of the *Eupithecia* world fauna, and none with this shape.
Material. ♂, Taiwan, Prov. Nantou, Hohuachi, 1950 m, 24°13′ N, 121°16′ E, 31.iii.2000, leg. A. Kun & L. Peregovits (holotype, TTM).
Remarks. The single holotype male is worn, with the fringes almost completely worn off on all wings.

173. *Eupithecia cotidiana* Vojnits, 1979

Eupithecia cotidiana Vojnits, 1979, *Acta zool. Acad. Sci. hung.* 25 (3–4): 437, fig. 15. Holotype ♀ (ZFMK; examined), [China]: Sued-Shensi, Tapaishan [Taibaishan] im Tsinling, 3000 m.

Description. Ground colour unicolourous dark grey. Fore wing with darker costal margin, indistinct transverse lines, two clear white subterminal spots, the tornal larger than the medial, and a conspicuous ovoid black discal dot. Hind wing paler in anterior half, also with two subterminal white spots, inconspicuous transverse lines and a small rounded discal dot.
Male genitalia. (not previously described) Uncus bell-shaped, biapical. Valve with a pointed medial process on the ventral margin. Vinculum slightly elongate, trapezoid. Papillae elongate. Aedeagus stout, almost

equal to valve length. Vesica armed with four large cornuti, a narrow V-shaped cornutus and an irregular cornutus at ductus ejaculatorius base. Sternite A8 broad, triangular, with two short blunt apical horns.
Female genitalia. Bursa very long, heavily sclerotized, with a large area of small spines at base and a patch of long, stout spines medially on left side. Ductus bursae elongate, with a large patch of short spines near base of ductus seminalis, the latter broadened and sclerotized at base, attached to base of ductus bursae on right side. Colliculum short and narrow. Antrum membranous, short and broad. Tergite A8 almost square. Apophyses short and narrow. Papillae anales short and broad, rounded.
Range. China (Gansu, Shaanxi, Sichuan).
Phenology. Univoltine: mid-July to mid-August.
Altitude. From about 1700 to 3200 m in Tapai-shan and Daxue Shan (China: Vojnits 1979b; coll. ZFMK).
Similar species. Very similar to *E. tripunctaria*, but darker, less brightly coloured, with more pointed fore wings. Both male and female genitalia are very distinctive.
Material. ♀, Sued-Shensi [Shaanxi], Tapaishan [Taibaishan] im Tsinling [Qinling], *ca.* 3000 m, 11.viii.1936, H. Höne, Vojnits slide no. 12295♀ (holotype, ZFMK); 1♀, same loc., *ca.* 1700 m, 11.viii.1936, H. Höne, Vojnits slide no. 12287♀ (paratype, TTM); 2♂, China, Sichuan, Daxue Shan Mts., Gongga Shan, NW Moxi, 29°41′ N, 101°58′ E, 2850 m, 14.-19.vii.1999, leg. V. Siniaev & A. Plutenko (ZFMK; 1♂ in ZISP); 2♂, same loc. and collectors, 29°34′ N, 101°59′ E, 3200 m, 20–21.vii.1999 (ZFMK); 4♂, Gansu, Wenxian, Qiujiaba, 2350 m, 22.vii.1999, coll Yao Jian, slides nos. ZL545, 555♂; 1♀, Gansu, Zhouqu, Shatan forestry farm, 2400 m, 14.vii.1999, coll Zhu Chaodong, slide no. ZL544♀; 1♀, same loc. and date, coll Yao Jian.

174. *Eupithecia krampli* Vojnits, 1979

Eupithecia krampli Vojnits, 1979, *Acta zool. Acad. Sci. hung.* 25 (3–4): 437, fig. 16. Holotype ♀ (ZFMK; examined), China: North Yunnan, Li-kiang.

Description. Fore wing dark grey; transverse lines oblique, sharply angled onto costa, indistinct, except for a pale double medial, sharply angled onto costa behind discal dot; terminal area unicolourous, with trace of narrow white subterminal line forming distinct medial and tornal white spots; discal dot small. Hind wing dark grey in tornal half, with more or less distinct transverse lines, with veins chequered black and white, two distinct white medial and tornal subterminal spots and a small blackish discal dot. Fringes distinctly chequered white and dark grey.

Male genitalia. (not previously described) Uncus stout, biapical. Valve with heavily sclerotized sacculus terminating in a hook-shaped process. Vinculum elongate. Papillae long and narrow, broadened at tips. Aedeagus narrow, shorter than valve length. Vesica armed with two slightly curved horn-like cornuti, an elongate, H-shaped plate and a smaller, irregular plate near ductus ejaculatorius base. Sternite A8 large and very broad, lozenge-shaped, with two stout apical horns. Tergite A8 with a sinuate base and almost parallel lateral margins tapering to a spiniform apical projection.

Female genitalia. Bursa ovoid, with sparse spinules in basal third, a large patch of small spines posteriorly and a patch of large, stout spines centrally. Ductus bursae not distinguishable. Ductus seminalis narrow, attached to centre of corpus bursae on dorsal side. Colliculum short and narrow. Antrum short and broad, membranous, covered with numerous pores. Tergite A8 almost quadrate. Apophyses narrow. Papillae anales short and rounded.

Range. China (Sichuan, Yunnan, Tibet); Myanmar.

Phenology. Bivoltine (?): early May to mid-July.

Altitude. From about 2800 to 3600 m (China: coll. ZFMK; coll. O. Pekarsky).

Similar species. A species showing little variation, similar to *E. tenuisquama*, but distinguished by the distinct whitish medial transverse lines. Both the male and female genitalia are highly distinctive.

Material. ♀, China, Provinz Nord-Yuennan, Li-kiang [Lijiang], 22.vii.1935, H. Höne, Vojnits slide no. 12243♀ (holotype, ZFMK); 1♀, Tibet [Sichuan], Batang, Im Tal des Yangtze, *ca.* 2800 m, 10.v.1936, H. Höne, Vojnits slide no. 12418♀ (ZFMK); 1♂, China, Sichuan, Daxue Shan Mts., Gongga Shan, NW Moxi, 29°41′ N, 101°58′ E, 2850 m, 14.-19.vii.1999, leg. V. Siniaev & A. Plutenko (ZISP); 2♂, same loc. and collectors, 29°34′ N, 101°59′ E, 3200 m, 20.-21. vii.1999 (ZFMK); 1♀, China, W. Sichuan, Kangding, 3000 m, 29°53′ N / 101°55′ E, 13.vii.2009, I. & A. Floriani, Saldaitis leg., Pekarsky slide no. OP 0681♀ (coll. O. Pekarsky); 1♀, China, W. Sichuan, road Yajiang / Litang, 3600 m, 15.vii.2009, 29°59′ N / 100°52′ E, I. & A. Floriani, Saldaitis leg., Pekarsky slide no. OP 0656♀ (coll. O. Pekarsky); 1♀, Sichuan, Pingwu, Wanglang, 2830 m, 1.viii.1999, Zhou Xin, slide no. ZL662♀ (IZCAS).

175. *Eupithecia inopinata* Vojnits, 1984

Eupithecia inopinata Vojnits, 1984, *Acta zool. Acad. Sci. hung.* **30** (3–4): 531, figs 12, 20, 28, 29. Holotype ♂ (ZFMK; examined), China: North Yunnan, Li-kiang.

Misidentification: *Eva petulans* Vojnits, 1981 (1♂, 1♀ paratypes in ZFMK).

Description. Fore wing ground colour grey to brownish grey; Cu vein covered with black scales; transverse lines inconspicuous; basal and antemedial straight, oblique, sharply angled onto costa; double medial slightly paler, right angled onto costa beyond discal dot; postmedial obtusely angled near costa, with darker costal shade and black wedge-shaped marks on the veins; terminal area darker, with usually broken, wavy whitish subterminal line and distinct medial and tornal white spots; discal dot rounded. Hind wing ground colour paler; posterior half darker, with distinct transverse lines and a small white tornal spot; discal dot small.
Male genitalia. Uncus elongate, biapical. Valve with a sclerotized sacculus and a large pointed ventral process. Vinculum short with a shallow medial hollow. Papillae rather short. Aedeagus narrow, slightly shorter than length of valve. Vesica armed with two slightly curved cornuti, an elongate, V-shaped cornutus, and a small irregular cornutus near ductus ejaculatorius base. Sternite A8 flask-shaped, with convex lateral margins and two short blunt apical horns.
Female genitalia. Bursa spherical, covered with dense small spines. Ductus bursae elongate, broad, with longitudinal striations and a small patch of minute spines on right side near colliculum. Ductus seminalis broadened near base, attached to ductus bursae from left side. Colliculum short and narrow. Antrum membranous. Tergite A8 trapezoid, with narrowly sclerotized anterior margin and rounded posterior corners. Apophyses short and narrow. Papillae anales narrow and rounded.
Range. China (Yunnan, Sichuan).
Phenology. Univoltine (?): early July to early September.
Altitude. Found from about 4000–4500 m in mountains of Yunnan (China: coll. ZFMK, TTM).
Similar species. Externally similar to *E. tripunctaria* on account of the two white spots on the terminal area of the fore wing and the similar tornal spot on the hind wing, but has narrower, more pointed wings and entirely different male and female genitalia, both of which are very distinctive.
Material. ♂, China, Provinz Nord-Yuennan, Li-kiang [Lijiang], 1.ix.1934, H. Höne, Vojnits slide no. 13069♂ (a wrongly labelled gen. slide of *E. cuneata*, subsequently used as holotype of *E. proba*, ZFMK); 2♂, 1♀, same loc., 19, 28.vii., 1.ix.1934, H. Höne, Vojnits slides nos. 13063♂, 13237♂, 13068♀ (paratypes, ZFMK, 13237♂ earlier cited by Vojnits as paratype of *E. intolerabilis*); 3♂, same loc., 7, 25.viii.1934, 3.ix.1935, H. Höne, Vojnits slides nos. 13040♂, 13042♂, 13260♂ (paratypes, TTM); 1♂, 1♀, same loc., 2.ix.1934, 28.vii.1935, H. Höne, Vojnits slide no. 13163♀ (paratype of *Eva petulans*, with a wrongly labelled gen. slide of *Eva flexa*), de Laever slide no. 361♀ (ZFMK); 1♂, same

loc., 16.viii.1935, H. Höne, Vojnits slide no. 13307♀ (with a wrongly labelled gen. slide of *E. refertissima*, TTM); 1♂, same loc., 23.vii.1935, H. Höne, Vojnits slide no. 1085♂ (with a wrongly labelled gen. slide of *E. coniurata*, TTM); 1♂, same loc., 18.viii.1934, H. Höne, Vojnits slide no. 13178♂ (TTM); 1♂, Nord Yünnan, A-tun-tse [Dêqên], Talsohle (*ca.* 3500 m), 21.vi.1936, H. Höne, Vojnits slide no. 13061♂ (slide not found in ZFMK); 1♀, same loc., Mittlere Höhe (*ca.* 4000 m), 18.viii.1936, H. Höne, Vojnits slide no. 13306♂ (unmarked paratype, TTM); 1♂, same loc., Obere Höhe (*ca.* 4500 m), 8.vii., 13.viii.1936, H. Höne, Vojnits slides nos. 11547♂, 13044♀ (the last a wrongly labelled slide of the male of *E. coniurata*, ZFMK); 1♀, same loc. and alt., 23.vii.1936, H. Höne, Vojnits slide no. 1084♀ (with a wrongly labelled gen. slide of *E. coniurata*, TTM); 1♂, China, Szetschwan [Sichuan], Ta-tsien-lu, Exped. Stötzner, de Laever slide no. 409♂ (ZFMK).

Remarks. A long series of labelling errors on Vojnits' slides has caused great confusion over this species. According to Vojnits (1984), the specimen and genitalia in Vojnits slide no. 13069♂ belong to the holotype of *E. inopinata*. However, neither this specimen nor the gen. slide were marked by him as "holotype", nor do the genitalia on the slide correspond to Vojnits' drawing of the male genitalia of *inopinata*. This specimen (no. 13069♂) was subsequently used by Mironov & Galsworthy, 2004c, unaware at the time that it had been published as the holotype of *E. inopinata*, as the holotype of *E. proba*, which however has turned out to be a synonym of *E. cuneata* Vojnits, 1984, described by him earlier in the same year. We think that the slide number published in the description of *inopinata* is probably a mistake or misprint (there are many such in Vojnits' papers). We have not found any specimen marked by Vojnits as the holotype of *E. inopinata*. In the circumstances, we treat the holotype as having been lost, at least for the time being, and base our concept of the species on those paratypes, listed above, whose genitalia slides match the genitalia drawings published by Vojnits as those of *E. inopinata*.

176. *Eupithecia dealbata* Inoue, 1988

Eupithecia dealbata Inoue, 1988, *Bull. Fac. domest. Sci. Otsuma Wom. Univ.* **24**: 368, figs 28a-b. Holotype ♀ (BMNH; examined), [Taiwan]: Chiai Hsien, Fenchihu, 1600 m.

Description. Fore wings broad, ground colour when fresh a darkish grey-brown; transverse lines indistinct, but pale angled medial and twice angled postmedial usually discernible; a dark streak on medial part of Cu vein, and on all veins close to postmedial; terminal area always with white medial and tornal spots, and usually with similar spots in other spaces between veins; discal dot small. Hind wing concolorous, with white terminal dots and usually a double row of white dots proximal to it; postmedial marked by dark dashes on veins, and often several similar dark marks in basal area; discal dot small.

Male genitalia. Uncus plate-like with short sharply pointed tip, uniapical. Valve with a narrow spinelike ventral process. Vinculum semicircular. Papillae elongate, very narrrow. Aedeagus slim, considerably shorter than valve length. Vesica covered with numerous denticules, and armed with a small, dense patch of short straight cornuti, an elongate, U-shaped cornutus and an irregular cornutus near ductus ejaculatorius base. Sternite A8 flask-shaped, with two short apical horns along lateral margins.

Female genitalia. Bursa small and globular, covered with dense, small spines. Ductus bursae elongate and broad, with lightly sclerotized, curved central section covered with numerous small spines running from corpus bursae to colliculum. Ductus seminalis very narrow, attached to base of ductus bursae from right side. Colliculum short and narrow. Antrum membranous. Tergite A8 elongate, with a shallow medial hollow in the anterior. Apophyses relatively long and narrow, the anterior stouter. Papillae anales broad and slightly elongate, rounded.

Range. China (Shaanxi, Guangxi, Taiwan).

Phenology. Univoltine: March to mid-April.

Altitude. From about 500 to 2200 m (Taiwan: Inoue 1988; China: colls. ZFMK, NSMT).

Similar species. Generally similar to *E. tripunctaria* but with a more chequered appearance and usually more white spots in the terminal area. The male valve resembles that of *E. repentina*, but the uncus and cornuti in the vesica immediately distinguish the male genitalia. The female genitalia are highly distinctive.

Material. ♀, Taiwan, Chiayi County, Fenchihu, 2.iii.1982, S. Hashimoto, BM Geom. slide no. 16555♀ (holotype, coll. H. Inoue in BMNH); 1♀, C. China, prov. Shaanxi, Foping Natural Reserve, Panda area, 33°45′ N, 107°48′ E, 1600 m, 6.-11.iv.1999, leg. V. Siniaev & A. Plutenko, Mironov slide no. 666♀ (ZFMK); 1♀, same loc. and alt., 15.-20.iv.1999, leg. V. Siniaev & A. Plutenko (ZFMK); 2♂, 4♀, China, Guangxi, Guilin, Maoershan, 500–1400 m, 26–29.iii.2005, M. Owada, Galsworthy slides nos. NSMT3♀, NSMT16♂ (NSMTO).

177. *Eupithecia lata* Vojnits, 1981

Eupithecia lata Vojnits, 1981, *Acta zool. Acad. Sci. hung.* 27 (1–2): 228, figs 12, 13. Holotype ♂ (ZSM; examined), Nepal: Prov. Nr. 3, East Junbesi, 2750 m.

Description. Fore wing dark grey, medial area paler; transverse lines inconspicuous; postmedial wavy, twice angled onto costa, with a series of narrow blackish dashes on the veins; terminal area narrow, slightly darker with paler medial blotch and trace of a light subterminal line; discal dot very small, rounded. Hind wing ground colour paler; transverse lines well marked along dorsum; terminal area narrow, darker especially near tornus; discal dot very small.
Male genitalia. Uncus stout, biapical. Valve with a shallow ventral hollow and a short process behind it. Vinculum short and broad. Papillae very long and narrow. Aedeagus stout, shorter than valve length. Vesica armed with two stout cornuti with broadened bases, and a small, flat, lightly sclerotized plate. Sternite A8 broad, with sinuate outer margins and two strongly sclerotized 'viking' horns.
Female genitalia. Bursa small, heavily sclerotized, with a membranous, thick-walled diverticulum at anterior end, a patch of stout spines in anterior half and a smaller patch of dense small spines posteriorly. Ductus bursae not distinguishable. Ductus seminalis attached to posterior half of corpus bursae from left side. Colliculum plate-like, short and broad. Antrum short and broad, membranous but with sclerotized transverse ring covered with numerous pores. Tergite A8 small, almost triangular, with strongly sclerotized posterior margin. Apophyses short and narrow. Papillae anales short and broad.
Range. China (Tibet, Sichuan); Nepal.
Similar species. Superficially similar to *E. krampli*, but discal dots smaller, the pale medial area less prominent, and terminal white spots absent. In the male genitalia the characteristic features are the valve shape, the long narrow papillae, and the eighth sternite. The female genitalia are closest to those of *E. quadripunctata*, which has a smaller diverticulum, lacks the transverse patch of small spines near the colliculum, and has a differently shaped eighth tergite.
Material. ♂, Nepal, Prov. Nr. 3, East Junbesi, 2750 m, 25.–31.vii.1964, leg. W. Dierl, Vojnits slide no. 11721♂ [= ZSM slide no. G 10611♂] (holotype, ZSM); 1♀, same loc., alt. and date, Vojnits slide no. 11717♀ [= ZSM slide no. G 10633♀] (paratype, ZSM); 1♂, Tibet, Bomi Yigong, 2300 m, 22.viii.1983, coll Han Yinheng, slide no. ZL665♂; 1♂, Sichuan, Pingwu, Lao He Gou, 1630 m, 16.viii.[19]99, leg Zhou Xin, slide no. ZL333♂ (IZCAS).

178. *Eupithecia karapinensis* Wileman & South, 1917

Eupithecia karapinensis Wileman & South, 1917, *The Entomologist* **50** (646): 56. Holotype ♀ (BMNH; examined), Formosa [Taiwan]: [Chiai Hsien], Karapin [Chaoliping], 3000ft.

Description. A large broad winged species. Fore wing brownish grey with fine transverse lines; medial sinuate, sharply angled onto costa behind discal dot; postmedial triple, the outer line stronger, very wavy and twice angled near costa; medial area paler; terminal area with relatively broad wavy whitish subterminal line forming a whitish tornal spot near tornus; discal dot elongate. Hind wing paler, with distinct indented transverse lines; discal dot very small or invisible.
Male genitalia. Uncus elongate, biapical. Valve with heavily sclerotized sacculus bearing two broad rounded ventral processes divided by a shallow hollow. Vinculum broad and tapering. Papillae long and narrow. Aedeagus stout, shorter than valve length. Vesica with a small patch of spinules, and armed with four cornuti, one long and sinuate, and a folded, irregular cornutus near base of longest cornutus. Sternite A8 broad, tapered, with two narrow apical horns.
Female genitalia. Bursa pear-shaped, about $^2/_3$ covered with small spines. Ductus bursae broad, with numerous wrinkles. Ductus seminalis broadened basally, attached to ductus bursae on left side. Colliculum short and broad. Antrum membranous, with numerous pores. Tergite A8 almost quadrate. Apophyses narrow. Papillae anales rather narrow and elongate, tapered to apices.
Range. China (Taiwan).
Phenology. Polyvoltine (?): early January to early November.
Altitude. From about 900 to 3100 m above sea-level (Taiwan: Wileman & South 1917; Inoue 1988).
Similar species. Easily distinguished from other Taiwanese species by the broad wings with wavy postmedial lines on the fore wings and distinct wavy transverse lines on the hind wings. The shape of the male valve is distinct from any other Chinese species. The female genitalia have less clear distinguishing features, but do not resemble any other Taiwanese species.
Material. ♀, [Taiwan] Formosa, Karapin, 3,000ft., 1.viii.1908, A.E. Wileman, 4678, *Eupithecia karapinensis* sp. n. Type ♀, BM Geom. slide no. 11854 (holotype, BMNH); 1♂, [Taiwan], Karapin [= Chaoliping, Chiai Hsien], January 1908, BM Geom. slide no. 6162 (? paratype, BMNH); 1♀, [Taiwan], Inoue slide no. 9382♀ (BMNH); 1♂, 1♀, Taiwan, Nantou, Meifeng, *ca.* 2300 m, 29.x-3.xi.1998, leg. Mey & Speidel, Zool. Mus. Berlin, Mironov slide no. 811♂ (MNHU).

179. *Eupithecia kobayashii* Inoue, 1958

Eupithecia kobayashii Inoue, 1958, *Tinea* 4 (2): (250), text-fig. 7. Holotype ♂ (BMNH; examined), [Japan]: Nagano Pref[ecture], Karuizawa.

Eupithecia hundamoi Vojnits & de Laever, 1978, *Acta zool. Acad. Sci. hung.* 24 (1–2): 227, fig. 3. Holotype ♀ (TTM; examined), Korea: Prov[ince] Ryang-gang, Hyesan. **New junior synonym.**
Eupithecia importuna Vojnits, 1981, *Acta zool. Acad. Sci. hung.* 27 (3–4): 426, fig. 47. Holotype ♀ (ZFMK; examined), [China]: Prov. Shansi [Shanxi], Mien-shan, 2000 m. **New junior synonym.**
Eupithecia nonferenda Vojnits, 1981, *Acta zool. Acad. Sci. hung.* 27 (3–4): 427, fig. 51. Holotype ♀ (ZFMK; examined), [China]: Prov. Shansi [Shanxi], Mien-shan, 2000 m. **New junior synonym.**

Description. A relatively large species with broad, almost unicolourous, dark grey or brownish grey wings, the fore wing with inconspicuous, slightly curved transverse lines, and a usually distinct, rounded or ovoid black discal dot and very small whitish tornal spot. Hind wing similarly coloured, with smaller discal dot and without whitish tornal spot.
Male genitalia. Uncus elongate, narrow, biapical. Valve with a short blunt ventral process. Vinculum short and broad. Papillae elongate and broad. Aedeagus very stout, shorter than valve length. Vesica armed with three very large horn-like cornuti and a small, irregular plate near ductus ejaculatorius base. Sternite A8 narrow, peg-like, with a blunt apex and two narrow basal lobes.
Female genitalia. Bursa almost spherical with two membranous diverticula, the smaller near the base and the larger centrally on the right side, the latter with a sclerotized base bearing a patch of stout spines; corpus bursae two thirds densely covered with small spines on left side. Ductus seminalis attached to corpus bursae near colliculum on dorsal side. Colliculum narrow and very short. Antrum membranous. Tergite A8 transverse, with a shallow medial hollow in the posterior margin. Apophyses short and narrow. Papillae anales short and very broad.
Range. China (Shanxi, Shaanxi); Asian Russia (Priamurje, Primorje), Korea, Japan (Honshu, Shikoku).
Phenology. Bivoltine (?): mid-May to early September.
Altitude. Up to about 2000 m in Mien-shan (China: Vojnits 1981; as *E. importuna*).
Similar species. Easily confused with *E. actaeata* and *E. consortaria*. The former is usually paler, with more distinct ante- and postmedial lines and more elongate discal dots on the fore wings; the latter is usually smaller with slightly wavy margins to all wings. The male and female genitalia are similar to those of the following species, but the male has longer cornuti in the vesica and a longer eighth sternite, and the female has a much larger spiniferous area in the corpus bursae.
Material. ♂, [Japan], Nagano, Karuizawa, 5.vii.1957, H. Kobayashi, BM Geom. slide no. 16072♂ (holotype, coll. H. Inoue in BMNH); ♀, Korea, Prov.

Ryang-gang Hyesan, Hotel Hyesan, No. 276, 22.vii.1975, leg. J. Papp et A. Vojnits, Vojnits slide no. 10794♀ (holotype of *E. hundamoi*, TTM); ♀, Prov. Shansi [Shanxi], Mien-shan, Obere Höhe *ca.* 2000 m, 12.viii.1937, H. Höne, *E. hundamoi* det. Dr. A. Vojnits, Vojnits slide no. 13120♀ (holotype of *E. importuna*, ZFMK); ♀, same loc. and alt., 3.viii.1937, H. Höne, *hundamoi* det. Dr. A. Vojnits, Vojnits slide no. 13119♀ (holotype of *E. nonferenda*, ZFMK); 2♂, 1♀, same loc. and alt., 7, 28.vii., 1.viii.1937, H. Höne, Vojnits slide no. 11517♂, de Laever slide no. 305♀, Mironov slide no. 657♂ (ZFMK); 4♀, China, Sued-Shensi [Shaanxi], Tapaishan [Taibaishan] im Tsinling [Qinling], L, 19, 24, 29.vi., 1.vii.1935, H. Höne, Mironov slide no. 656♀, Vojnits slides nos. 11506♀, 11453♀, 11452♀ (*E. hundamoi* det. Dr. A. Vojnits, ZFMK); 2♂, 2♀, same loc., 26, 28.vi., 2.vii.1935, 29.vii.1937, H. Höne, Vojnits slides nos. 12191♂, 12220♀, 12223♀, Mironov slide no. 831♂ (TTM); 1♀, Central China, prov. Shaanxi, Mt. Tai Bei Shan, 33°35′ N, 107°43′ E, 1300–1500 m, 10.viii.-4.ix.1998, leg. V. Murzin & V. Siniaev, Mironov slide no. 648♀ (ZFMK).

180. *Eupithecia incorrupta* Vojnits, 1984

Eupithecia incorrupta Vojnits, 1984, *Acta zool. Acad. Sci. hung.* **30** (1–2): 233, fig. 28. Holotype ♂ (TTM; examined), [China]: North Yunnan, A-tun-tse [Dêqên], 3000 m.

Description. A large broad-winged species. Fore wing grey with yellowish tinge; basal and postbasal transverse lines oblique, sharply angled onto costa; antemedial, medial and postmedial oblique, sinuate, right angled onto costa and forming blackish costal spots; postmedial with a series of dark dashes on the veins; terminal area with dark apical, medial and tornal blotches and a wavy, whitish subterminal line; discal dot small. Hind wing paler, whitish grey; basal, ante- and postmedial transverse lines distinct; terminal area darker, with distinct inner border; discal dot small.
Male genitalia. Uncus long and narrow, biapical. Valve with a very small, pointed medial process on ventral margin. Vinculum semicircular. Papillae elongate, broadened at apices. Aedeagus stout, shorter than length of valve. Vesica armed with three stout horn-like cornuti, a small plate-like cornutus and an irregular cornutus near ductus ejaculatorius base. Sternite A8 peg-like, sharply broadened at base, with an obtuse apex, more strongly sclerotized along its margins.
Female genitalia. Bursa almost globular, with a large membranous diverticulum at base and a second smaller one on the right side, covered with small spines in posterior half, and with a heavily sclerotized area with some stout spines near base of larger diverticulum. Ductus seminalis

broadened at base, attached to central part of corpus bursae on ventral side. Colliculum short and narrow. Antrum very short, broad, covered with pores. Tergite A8 transverse. Apophyses short and narrow. Papillae anales slightly pointed.
Range. China (Sichuan, Yunnan, Shanxi).
Phenology. Univoltine (?): mid- May to July.
Altitude. From about 2000 to 3000 m above sea-level in mountains of Yunnan, Sichuan, and Shanxi (China: Vojnits 1984; coll. TTM).
Similar species. Similar externally to the preceding species (qv) but with more distinct transverse lines and darker terminal areas on all wings, and lighter hind wings. For differences in the genitalia, see under preceding species.
Material. ♂, Nord-Yünnan, A-tun-tse [Dêqên], Talsohle *ca.* 3000 m, 4.vi.1937, H. Höne, Vojnits slide no. 13759♂ (holotype, TTM); 1♂, same loc. and alt., 18.v.1937, H. Höne, Vojnits slide no. 13762♂ (paratype, slide absent in ZFMK); 1♂, same loc. and alt., 30.v.1937, H. Höne, Vojnits slide no. 13294♂ (slide absent in ZFMK); 1♂, 1♀, China, Provinz Nord-Yuennan, Li-kiang [Lijiang], 22, 28.vi.1935, H. Höne, de Laever slides nos. 247♂, 248♀ (ZFMK); 1♂, same loc. and alt., 21.vi.1937, H. Höne, Vojnits slide no. 13140♂ (paratype, TTM); 1♂, same loc. and alt., 31.v.1937, H. Höne, Vojnits slide no. 13761♂ (slide absent in TTM); 1♂, Prov. Shansi [Shanxi], Mien-shan, Obere Höhe *ca.* 2000 m, 1.viii.1937, H. Höne, Vojnits slide no. 11528♂ (incorrectly marked as holotype of *E. incorrupta*, TTM); 1♂, 1♀, China, Szechuen [Sichuan], Sungpan, July [c1930], DC Graham coll., Galsworthy slides nos. USNM59♂, 56♀ (AMNH).
Remarks. According to the original label, the holotype of *E. incorrupta* was collected on 4.vi.1937, not 9.vi.1938 as in Vojnits 1984.

181. *Eupithecia leucenthesis* Prout, 1926

Eupithecia leucenthesis Prout, 1926, *J. Bombay nat. Hist. Soc.* **31** (2): 318. Syntypes 1♂, 2♀ (BMNH; examined), [Burma (upper)]: Hparé.
Eupithecia albicans Vojnits, 1981, *Acta zool. Acad. Sci. hung.* 27 (1–2): 225, figs 10, 11. Holotype ♂ (ZSM; examined), Nepal: Dudh Kosi Tal, 3500 m (synonymised in Mironov & Galsworthy, 2010).

Description. Fore wing white; costal margin rusty brown to centre and with large rusty brown subapical costal blotch; transverse lines inconspicuous; postmedial oblique, composed of small dark dots on veins, angled onto costa; posterior half of fore wing from Cu vein to anal margin pale rusty brown; terminal area rich rusty brown with wavy white subterminal line; apex whitish; discal dot very small, blackish. Hind wing white;

transverse lines brown, broad, well marked along dorsum only; terminal area narrow, rusty brownish with wavy white subterminal line; discal dot small.

Male genitalia. Uncus short, triangular. Valve relatively small, unmodified. Vinculum narrow, rounded. Papillae short and broad. Aedeagus slim, equal to length of valve. Vesica covered with numerous denticules, armed centrally with a patch of large jagged spines, and more basally with an irregular folded plate. Sternite A8 rectangular, slightly tapering, with two diverging apical horns.

Female genitalia. Bursa globular, membranous with two large lateral patches of sparse spines. Ductus bursae sclerotized, elongate, with a characteristic triangular projection at attachment of ductus seminalis on right side. Colliculum narrow, oblique. Antrum very short, membranous. Tergite A8 almost square. Apophyses narrow. Papillae anales short and broad, rounded.

Range. China (Tibet, Sichuan, Yunnan); Nepal, Myanmar.

Phenology. Univoltine (?): mid-July to early September.

Altitude. From about 2600 to 3900 m in the Himalayas (Nepal: Vojnits 1981).

Similar species. Only potentially confusable with some members of the *haworthiata* group which have a similar pattern of extensive white and brown. The presence of a hooked uncus in the male genitalia immediately distinguishes this species from them. The central group of spines in the male vesica is unusual, and shared only with *E. asempiterna*, a very different looking species. The shape of the ductus bursae in the female genitalia is distinctive.

Material. ♂, Hparé, Burma, ix.1923, A.E. Swann, *Eupithecia leucenthesis* Prout type, Rothschild bequest 1939–1, BM Geometrid slide no. 20307 (holotype, BMNH); ♀, data as holotype, *Eupithecia leucenthesis* Prout paratype, BM Geometrid slide no. 20308 (paratype, BMNH); ♂, Nepal, Dudh kosi Tal, 3500 m, 22.-23.vii.1962, leg. G. Ebert u. H. Falkner, Vojnits slide no. 10648♂ [= ZSM slide no. G 10628♂] (holotype of *E. albicans*, ZSM); 1♂, [China, Sichuan], Bet. Pin-E-Pu & Den Shiang vin DC Graham, Galsworthy slide no. USNM20♂ (AMNH); 1♂, Yatung [Yadong], Tibet, A. E. Hobson, 98–201, BM Geom. genit. slide no. 21518♂ (BMNH); 1♂, China, N. Yunnan, road Zhongdian / Lijiang, 3000 m, 27°28′ N / 99°53′ E, 19.vii.2009, I.& A. Floriani, Saldaitis leg., Pekarsky slide no. OP 0815♂ (coll. O. Pekarsky).

Remarks. Prout's holotype and paratype of *E. leucenthesis* are in very worn condition, but, though very smudged, show a distribution of white and brown markings corresponding well with the much fresher holotype of

E. albicans Vojnits, 1981. Both male and female genitalia of *E. leucenthesis* and *E. albicans* are identical in all important respects.

The three known Chinese specimens are smaller than Nepalese specimens of this species.

182. *Eupithecia pimpinellata* Hübner, [1813]

Geometra pimpinellata Hübner, [1813], *Samml. Eur. Schmett.* 5 Geometrae (2), pl. 86: 443, 444. Syntype(s) lost. [Europe].

Eupithecia altaicata Guenée, [1858], *in* Boisduval & Guenée, *Hist. nat. Insectes* (Spec. gén. Lépid.) **10** (2): 330. Syntypes 5♂, 2♀ (BMNH; coll. O. Staudinger and K. Dietze in MNHU; examined), [Russia]: Altai, Ustk[amenogorsk] (treated as var. of *E. pimpinellata* in Staudinger & Rebel, 1901).

Eupithecia lantoscata Millière, 1873, *Cat. Rais. Lépid. dept. Alpes-Maritimes* (**2**): 201. Syntype(s) (MNHN?), [France]: Alpes-Maritimes, vallée de Lantosque (treated as var. of *E. pimpinellata* in Bohatsch, 1893).

Eupithecia pimpinellata var. *cinerascens* Tengström, 1875, *Notis. Sällsk. Faun. Fl. fenn. Förh.* (N.S.) **11**: 32. Holotype ♀ (ZMH), [Russia]: Karelia.

Eup[ithecia] vellicata Dietze, 1908, *Dt. ent. Z., Iris* (1908) **21** (2/3): 173; *ibidem* (1906), **19** (3), pl. 2, fig. 16. Syntypes 2♀ (coll. K. Dietze and R. Püngeler in MNHU; examined), [China]: Urumtschi (treated as subsp. of *E. pimpinellata* Hübner, [1813] in Mironov, 1990).

[*Eupithecia*] *pimpinellata* f. *assimilis* Dietze, 1913, *Biologie der Eupithecien* **2**: 118; *ibidem* (1910), **1**, pl. 76, figs 554–556; pl. 85, fig. 949. Syntypes 7♂, 2♀ (coll. K. Dietze in MNHU; examined), [China]: [Xinjiang], Juldus [Yining] region; [Kazakhstan/China]: Ili region.

[*Eupithecia*] *pimpinellata* f. *elongata* Dietze, 1913, *Biologie der Eupithecien* **2**: 118; *ibidem* (1910), **1**, pl. 76, figs 558, 559. Syntypes 5♂, 1♀ (coll. K. Dietze in MNHU; examined), Italy: Pescocostanzo.

Eupithecia pimpinellata variata Schwingenschuss, 1936, *in* Zerny, *Mém. Soc. Sci. nat. Phys. Maroc* **42**: 85. Syntype(s) (at least 1 syntype in NHRM), [Morocco]: [Great Atlas], Tachdirt; Tizi n'Tachdirt.

Eupithecia pimpinellata retheli Prout, 1938, *in* Seitz, *The Macrolepidoptera of the World* **4** (Suppl.): 200. Syntype(s) (BMNH), Italy.

Misidentification: *Eupithecia denotata* Doubleday, 1856 (*nec* Hübner, 1813).
Unavailable names (infrasubspecific): *grisea*: Lempke (1969). – (incorrect subsequent spelling): *pimpinellaria*: Boisduval (1840).

Description. Fore wing ash-grey; transverse lines (sometimes inconspicuous, represented by a series of small blackish dots on veins) forming from three to five rather weak dark costal spots; basal, postbasal and antemedial indented, sharply angled near costa; triple medial fine, the inner line crossing discal dot and sharply angled towards costa; postmedial sinuate, obtusely angled near costa, usually with a broad pale fascia outside it; terminal area darker, with wavy creamy white subterminal line; discal dot obliquely elongate. Hind wing paler; transverse lines well indicated along dorsum; terminal area darker with faint pale subterminal line; discal dot large, very distinct. Abdomen grey with a soft ochreous tinge, usually without dark lateral stripes but with darker, ochreous brown second segment.

Male genitalia. Uncus stout, biapical. Valve unmodified. Vinculum broad, with shallow medial hollow. Papillae elongate, narrow. Aedeagus shorter than valve length. Vesica armed with two short, hooked cornuti, and a folded plate-like cornutus connected at base with an irregular cornutus at ductus ejaculatorius base. Sternite A8 narrowly rectangular, broadened and bilobed at base with two narrow apical horns.
Female genitalia. Bursa globular, covered with small spines. Ductus bursae elongate, sclerotized, with a small appendix at base on left side. Ductus seminalis broadened at base, attached to ductus bursae from right side. Colliculum narrow, slightly inclined to right. Antrum membranous. Tergite A8 almost square with narrowly sclerotized anterior margin. Apophyses narrow, the posterior with short spurs basally. Papillae anales short and rounded.
Range. China (Xinjiang). Widely distributed in Palaearctic region from Portugal eastwards throughout Europe and Siberia to Kirghizstan, Altai Mts. and Mongolia; an isolated population in North Africa (Morocco: Great Atlas).
Phenology. Bivoltine: late April to late June; mid-July to mid-September. In northern areas univoltine: mid-July to late August. Larva: late July to late October. Overwinters as pupa in insecure cocoon in the ground.
Biology. Larva found on the flowers and ripening seeds of *Pimpinella saxifraga, P. major* (= *magna*)*, Angelica sylvestris, Aegopodium podagraria, Bupleurum falcatum, B. spinos.um, Pastinaca sativa, Peucedanum oreoselinum, P. cervaria, Seseli annuum* (Apiaceae), *Galium verum* (Rubiaceae), *Solidago virgaurea, Achillea ptarmica, Senecio vulgaris, Artemisia campestris* (Asteraceae) in Europe (Mironov 2003). In captivity feeds also on *Senecio jacobaea, S. congestus* (Asteraceae) (United Kingdom: Crewe 1861c). Often heavily parasitized.
Altitude. From sea level up to about 2400 m in Kirghizsky Range (Kirghizstan: coll. ZMKU).
Similar species. Remotely similar to *E. absinthiata* (Clerck, 1759) but paler, with no reddish brown tinge on fore wings. Broader winged than most Chinese species: the pale postmedial fascia is a helpful distinguishing mark. The male genitalia are distinguished by the cornuti on the vesica and the shape of sternite A8. The female genitalia are very distinct.
Material. 1♂, [Kazakhstan], Ustk. 1852, ? *Altaicata* Gn., Zell. Coll. 1884 (syntype of *E. altaicata*, BMNH); 1♂, [Kazakhstan], Ustkam., Altai Kinderm., *altaicata*, Origin., Coll. Led. *Egenaria* HS., Zool. Mus. Berlin, typus, Vojnits slide no. 15434♂, photo 84 C 20 det. A. Vojnits (syntype of *E. altaicata*, coll. O. Staudinger in MNHU); 1♀, [Kazakhstan], Ustkam.,

altaicata, Origin., Coll. Led. *Egenaria* HS. (syntype of *E. altaicata*, coll. O. Staudinger in MNHU); 1♂, [Kazakhstan], Ustkam., Coll. Led. *Egenaria* HS. (syntype of *E. altaicata*, coll. O. Staudinger in MNHU); 1♂, [Kazakhstan], Ustkamenogorsk, Coll. Led. *Egenaria* HS. (syntype of *E. altaicata*, coll. O. Staudinger in MNHU); 1♀, [Kazakhstan], Ustk. 1852, *altaicata* Gn. = *pimpinellata* ♀ Type v. Lederer, K. Dietze, Original. Ustk. 1852 (syntype of *E. altaicata*, coll. K. Dietze in MNHU); 1♂, [Kazakhstan], *altaicata* Gn. = *pimpinellata* ♂. Type v. Lederer, K. Dietze, Original, Ustk. 1852 (syntype of *E. altaicata*, coll. K. Dietze in MNHU); 1♀, [Xinjiang], Asia centr., Ost-Turkestan (Urumtschi) Rückbeil 1904, ♀, *vellicata*. Typ. ? *euphrasiata*-Form ♀, Urumtschi, Ost-Turk. (553), Zool. Mus. Berlin, *vellicata* Dtze, präp. Nr. A 828♀ det. E. Schütze, Kassel (syntype of *E. vellicata*, coll. K. Dietze in MNHU); 1♀, [Xinjiang], Asia centr., Ost-Turkestan (Urumtschi [Urumqi]), Rückbeil 1904, v. R. Tancré 05, 8., *euphrasiata* forma: *vellicata*, K. Dietze (syntype of *E. vellicata*, coll. R. Püngeler in MNHU); 1♀, [Xinjiang], Asia centr., Thian-Schan, Juldus Geb., coll. Wagner, ? *pimpinellata* forma: *affinis* ♀ Typ. Juldus Gebirge (syntype of *E. pimpinellata* f. *affinis*, coll. K. Dietze in MNHU); 1♂, [Xinjiang], Juldus, forma: *affinis* Typ ♂. Ili-Gebiet (syntype of *E. pimpinellata* f. *affinis*, coll. K. Dietze in MNHU); 5♂, [Xinjiang], Juldus, Juldus Ili (coll. K. Dietze in MNHU); 1♀, [Xinjiang], Asia centr., Turkestan, Ili-Gebiet, Coll. Wagner, 109., ♀, bei *pimpinell.* (coll. K. Dietze in MNHU).

Remarks. Usually variable in coloration from light grey to light ochreous brown.

183. *Eupithecia denotata* Hübner, [1813]

Geometra denotata Hübner, [1813], *Samml. Eur. Schmett.* 5 Geometrae (1), pl. 88: 455. Syntype(s) lost. [Europe].

Eupithecia atraria Herrich-Schäffer, 1848, *Syst. Bearb. Schmett. Eur.* 3 (32): 121, 132; *ibidem* (1846), pl. 25, figs 154, 155 (non binominal). Syntype(s), Austria: Schneeberg Mt. (treated as var. of *E. denotata* in Bohatsch, 1893).

Eupithecia campanulata Herrich-Schäffer, 1861, *Corresp.-Bl. Sammler Insect., insbesondere Schmett.* 17: 130. Syntypes, [Germany]: Frankfurt; Braunschweig [= Brunswick] (synonymised as *E. campanularia* [sic] in Heinemann, 1859 before publication of Herrich-Schäffer).

Eupithecia primulata Millière, 1874, *Revue Mag. d'Zool.* (3) 2: 244. Syntype(s) (MNHN ?), Switzerland: Célérina (synonymised in Bohatsch, 1893).

Eupithecia jasioneata Crewe, 1881, *The Entomologist* 14 (220): 198; pl. 1, figs 4, 5. Syntype(s) (BMNH), [United Kingdom]: North Devon (treated as form of *E. denotata* in Dietze, 1910).

Eupithecia denotata var. ? *ferreata* Fuchs, 1901: *in* Pagenstecher, *Jahrb. nassau. Ver. Naturkde* 54: 57. Syntype(s) (NSMW ?), Italy: Trentino, Trafoi.

Eup[ithecia] ? *denotata* f. *livida* Dietze, 1910, *Biologie der Eupithecien* 1, pl. 72, figs 278, 279; *ibidem* (1913), 2: 92. Syntypes 3♂, 4♀ (coll. K. Dietze in MNHU; examined), [China]: (Xinjiang), Juldus (Yining) region; Ili.

Eup[ithecia] pimpinellata f. *limbosignata* Dietze, 1910, *Biologie der Eupithecien* 1, pl. 72, fig. 285; *ibidem* (1913), 2: 119. Holotype ♀ (coll. K. Dietze in MNHU; examined), [China]: Kuldja. **New junior synonym.**

Eupithecia denotata hellenata Schütze, 1959, *Mitt. Münchn. Ent. Ges.* **49**: 37, pl. 1, figs. 3, 4. Holotype ♂ (ZSM), Greece: Olympus Mt., Enipefs-Tal, Kataphygion.

Unavailable names (infrasubspecific): *solidaginis*: Fuchs (1902). – (incorrect subsequent spellings): *denotaria*: Boisduval (1840); *campannlata*: Reisen (1891); *jasionata*: Petersen (1910); *denonata*: Bolshakov et al. (2008).

Description. A broad-winged species. Fore wing ground colour dark ash-grey; dark transverse lines fine, usually wavy; basal, postbasal and antemedial evenly curved, double medial usually touching discal dot and sharply angled onto costa; postmedial right angled onto costa; terminal area with pale subterminal line; discal dot small, intensely black. Hind wing paler; dark transverse lines, especially postmedial, conspicuous; terminal area darker with whitish subterminal line and often a small whitish tornal spot; discal dot small. Abdomen grey, usually with a dorsal row of small black dots and black lateral stripes or with rows of small black lateral spots.

Male genitalia. Uncus short and stout, biapical. Valve with a broad blunt ventral process (when flattened on slide, can appear as an obtuse angle on ventral edge). Vinculum broad, semicircular. Papillae short, slightly curved. Aedeagus stout, shorter than length of valve. Vesica armed with a slightly curved horn-like cornutus, a U-shaped flat cornutus and a smaller, curved cornutus at ductus ejaculatorius base. Sternite A8 narrow and elongate, broadened and bilobed at base, with two small apical horns.

Female genitalia. Bursa elongate pear-shaped, covered with small spines at base and along left side. Ductus bursae tapered, with longitudinal striations and transverse patch of small spines. Ductus seminalis broad at base, attached to centre of corpus bursae. Colliculum narrow, slightly elongate. Antrum membranous. Tergite A8 almost quadrate. Apophyses narrow, the posterior with minute spurs near short and rounded papillae anales.

Range. China (Xinjiang); almost all Europe (from Pyrenees to Urals), Asia Minor, Caucasus, Transcaucasus, Southern Kazakhstan (Tien-Shan), Kirghizstan, Southern Siberia, Mongolia.

Phenology. Univoltine: early April to late August in Europe.

Biology. Larva found in the ripening seed capsules of *Codonopsis clematidea* (Campanulaceae) (Kirghizstan, Tien-Shan Mts.: Mironov 2003); also upon the unripe seeds and seed capsules, and rarely on the flowers of *Campanula trachelium, C. rotundifolia, C. barbata, C, persicifolia, C. medium, C. latifolia* (= *urticaefolia*), *C. collina, C. patula, C. carpatica, C. rapunculus, C. rapunculoides, C. glomerata, Phyteuma campanuloides, Ph. ovatum,*

flower heads of *Jasione montana* (Campanulaceae), and probably occasionally on *Ballota nigra* (Lamiaceae), *Primula integrifolia* (Primulaceae) and *Vaccinium uliginos.um* (Ericaceae) (Europe: Mironov 2003).
Altitude. From sea level in Europe up to about 2300 m in Tien-Shan Mts. (Kirghizstan: coll. ZISP).
Similar species. Distinguished from other broad-winged Asian species by the dark ash-grey ground colour, the distinct wavy transverse lines and relatively small, rounded discal dots. The male genitalia are distinguished by the shape of the ventral process on the valve, by the presence of one slightly curved horn-like cornutus on the vesica, and by the shape of sternite A8, the female by the limited area of spines and the separate patch of spines in the ductus bursae.
Material. 1♀, [Xinjiang], ? *denotata*, form. *livida*, Juldus, Ili (syntype of *E. denotata* f. *livida*, coll. K. Dietze in MNHU); 1♀, [Xinjiang], forma *livida-reducta*, Ili-Gebiet (coll. K. Dietze in MNHU); 1♀, [Xinjiang], form. *reducta*, Juldus, Ili (coll. K. Dietze in MNHU); 1♀, [Xinjiang], ? *pimpinellata*, ? *denotata* form. *limbosignata* Typ ♀, Kuldja, Sicht aus, wie ein Hybride von *pimpinellata* x *denotata* (holotype of *E. denotata* f. *limbosignata*, coll. K. Dietze in MNHU); 1♂, [Xinjiang], 1886, Kuldja, *denotata*!, Kuldja (coll. K. Dietze in MNHU); 1♂, [Xinjiang], ? *denotata*, Juldus, Ili ♂ (coll. K. Dietze in MNHU); 2♂, 2♀, [Xinjiang], Juldus, Ili (coll. K. Dietze in MNHU); 1♀, [Xinjiang], Urumtschi [Urumqi], Tian-shan, 45, Graeser (coll. R. Püngeler in MNHU); 1♀, [Xinjiang], Asia centr., Ost-Turkestan (Urumtschi), Rückbeil 1904, v. R. Tancré 1905 (coll. R. Püngeler in MNHU).
Remarks. All three syntypes of *E. denotata* f. *difficilis* Dietze, 1910 (see *Biologie der Eupithecien* 1, pl. 85, figs 905–907; *ibidem* (1913), **2**: 92) from Tajikistan/Kirghizstan (Alai Mts. and Ispajran [= Isfara]) in the collection of K. Dietze in MNHU belong to *E. nigrilinea* (Warren, 1896), **syn. n.**

184. *Eupithecia chesiata* Dietze, 1904

Eup[ithecia] chesiata Dietze, 1904, *Dt. ent. Z. Iris* [1903] **16** (2): 340, pl. 3, figs 13, 14. Syntypes 5♂, 9♀ (coll. K. Dietze, R. Püngeler and O. Staudinger in MNHU; ZSM; examined), [China]: Korla, Sai-chin, Aksu, Chines. West Turkestan.

Description. A relatively large species with characteristic narrow, elongate and pointed pale brownish ochreous fore wings. Fore wing with narrow, oblique, sinuate brownish transverse lines sharply angled onto costa; postmedial brownish ochreous, evenly curved from A to stalk of Cu and sharply angled onto costa forming a narrow longitudinal black stripe; discal dot invisible. Hind wing paler, with transverse lines visible only at anal margin; discal dot invisible.

Male genitalia. Uncus narrow, biapical. Valve small, unmodified. Vinculum short, rounded. Papillae shortish. Aedeagus narrow, shorter than length of valve, with ductus ejaculatorius attached centrally. Vesica covered with minute blunt spinules, and armed with two broad cornuti and a narrow, weakly sclerotized plate. Sternite A8 broad and bilobed at base, with two narrow apical rods.

Female genitalia. Bursa small, oval, with membranous spineless base and an oblique area of small spines from above base extending up left side; ductus bursae not clearly separated, with longitudinal striations and small oblique patch of spines between colliculum and ductus seminalis base, the latter narrow, attached to central part of corpus bursae on right. Colliculum narrow, slightly elongate. Antrum membranous. Tergite A8 transverse, with a deep central notch on posterior margin. Apophyses short and narrow. Papillae anales rounded.

Range. China (Xinjiang); SE Kazakhstan, Kirghizstan, Tajikistan.

Phenology. Univoltine: early May to mid-July.

Altitude. Found from 2400 m on Tomürt Peak (China: coll. IZCAS) up to about 4000 m above sea-level in Tien-Shan (Kirghizstan: coll. IAET).

Similar species. Similar in shape to *E. vicariata* and *E. nobilitata*, but easily distinguished by the black streak on the fore wing postmedial. The pattern of cornuti and shape of the eighth sternite distinguish the male genitalia. The female genitalia have a characteristically shaped corpus bursae.

Material. 1♀, [Xinjiang], *chesiata* n. sp., Ak-ssu, Maklan-Wüste, Anf. Mai, Original (K. Dietze), Typus, Zool. Mus. Berlin, Gen.-Unters. Nr. [Schmidt slide no.] Eu 117a, Zool. Mus. Berlin (syntype, coll. K. Dietze in MNHU); 1♂, [Xinjiang], *chesiata* Aksu 1903 ♂, Anfang Mai, Typ, K. Dietze (v. Tancré), Zool. Mus. Berlin, Gen.-Unters. Nr. [Schmidt slide no.] Eu 117, Zool. Mus. Berlin (syntype, coll. K. Dietze in MNHU); 1♀, [Xinjiang], *chesiata* n. sp., Original, Aksu, K. Dietze (syntype, coll. K. Dietze in MNHU); 1♀, [Xinjiang], *chesiata* n. sp., Original, Aksu, Type, K. Dietze, 222. (syntype, coll. K. Dietze in MNHU); 1♀, [Xinjiang], *chesiata* n. sp., Type ♀, Aksu, Mai 1900, Tancré, K. Dietze, Typus, Zool. Mus. Berlin, *Eupithecia chesiata* ♀, Ak-su Mai, Type, E. de Laever det. 1957, prep. № 1661.2 (syntype, coll. K. Dietze in MNHU); 1♂, [Xinjiang], *chesiata* Aksu 03, K. Dietze, Type, Ventral Platte freigelegt (syntype, coll. K. Dietze in MNHU); 1♀, [Xinjiang], *chesiata* Type, Aksu, Anfg. Mai, K. Dietze (syntype, coll. K. Dietze in MNHU); 1♀, [Xinjiang], *chesiata* n. sp., Aksu (Maklan-Wüste), Original., K. Dietze, ♀, 220. (syntype, coll. K. Dietze in MNHU); 1♀, [Xinjiang], *chesiata* ♂, Anfg. Mai, Aksu, K. Dietze (syntype, coll. K. Dietze in MNHU); 1♂, [Xinjiang], *chesiata* Typ ♂,

Ak-su, Makan-Wüste, Anfang Mai, Zool. Mus. Berlin, *Eupithecia chesiata* ♂ Ak-su, Type, E. de Laever det. 1957, Prep. № 1661.1 (syntype, coll. K. Dietze in MNHU); 1♂, 2♀, [Xinjiang], Asia centr., Aksu, Rückbeil 1900, *chesiata* Dietze, Type, v. Tancré 3.01 (syntypes, coll. R. Püngeler in MNHU); 1♂, [Xinjiang], Korla 97. Hbhr., 27/6, 541., 537., n. sp. bei *Nobilitata* Stgr. (Bob.), *chesiata* n. sp. original (K. Dietze), Vojnits slide no. 19027♂ (syntype, coll. O. Staudinger in MNHU); 1♀, [Xinjiang], Aksu, Mai 1903, *Eupithecia chesiata* n. sp., Coll. L. Lhomme 1957 (syntype, ZSM); 1♂, [Xinjiang], Ak-su, Makan-Wüste ♂, *chesiata* ♂, K. Dietze (coll. K. Dietze in MNHU); 1♂, [Xinjiang], Asia centr., Aksu, Rückbeil 1900, *chesiata* Dietze ♂, von R. Tancré 11.03 (coll. R. Püngeler in MNHU); 7♀, Tomürt Peak, Xinjiang, 2400 m, 19–20.vi.1977, slides nos. ZL258, 498♀ (IZCAS).

The *rajata* species group

This is a South and East Asian species group of *Eupithecia*, which is here formally introduced for the first time. It is represented in the territory of China by the following fourteen species. The Nepalese species *E. xerophila* Inoue, 2000, and *E. spilocyma* Prout, 1931 from the Philippines also belong to this group. From the morphology, the group may be related to the *undata* and *venosata* groups.

In the male genitalia, the uncus is usually slightly elongate, and the tip biapical, except in *E. mundiscripta*. The valve lacks a ventral process, but in some species the ventral edge protrudes slightly. The papillae are elongate and relatively narrow. In almost all cases, the aedeagus vesica has a pair of apical cornuti of varying sizes and shapes (lacking in *E. raniata*), a U-shaped folded cornutus at centre and a smaller, irregular cornutus near ductus ejaculatorius base. Sternite A8 consists of two almost parallel arms connected with each other by a broad basal band.

The female bursa copulatrix is rounded or ovoid, membranous, and completely or partially covered with small spines. The ductus bursae is usually elongate with a membranous diverticulum of varying sizes and shapes on one side with a patch of spines in its base. The ductus seminalis is narrow, attached to the corpus bursae on the left side. The colliculum is usually short and narrow, but absent in some species.

185. *Eupithecia mundiscripta* Warren, 1907

Tephroclystia mundiscripta Warren, 1907, *Novit. zool.* **14**: 161. Syntypes 2♂ (BMNH, examined), British New Guinea [Papua New Guinea]: Upper Mambaré River, Biagi, 5000ft.

Eupithecia mundiscripta commundata Prout, 1958, *Bull. Br. Mus. nat. Hist.* (Ent.) **6** (12): 393. Holotype ♀ (BMNH; examined), [Borneo]: Mt. Kinabalu.

Eupithecia mundiscripta larutensis Prout, 1958, *Bull. Br. Mus. nat. Hist.* (Ent.) **6** (12): 394. Holotype ♀ (BMNH; examined), Malaya [West Malaysia]: Perak, Larut Hills, 3700ft.
Eupithecia sekkongensis Galsworthy, 1999, *Trans. lepid. Soc. Japan* **50** (3): 232, figs 7, 22. Holotype ♂ (BMNH; examined), Hong Kong: Shek Kong, New Territories (synonymised in Mironov & Galsworthy 2009b).

Description. A small pale species. Fore wing in terminal half pale brown with weak reddish tinge, basal half off-white, costal margin blackish, discal dot small, linear, black. Hind wing rather angulate, white, with narrow, darker terminal area and small, oblique discal dot.
Male genitalia. Uncus elongate, narrow, with tip relatively short, hook-like, uniapical. Valve very broad. Vinculum rather square. Papillae slightly tapered to apices. Aedeagus narrow, shorter than length of valve, with two lateral, apical lobes covered with numerous setae. Vesica armed with two small apical cornuti, a short heavily sclerotized band and an elongate plate-like cornutus. Sternite A8 peg-like, elongate and narrow, with two short pointed apical arms.
Female genitalia. Bursa globular, covered with dense spines. Ductus bursae membranous, with a small diverticulum on right side carrying a small elongate patch of spines. Ductus seminalis narrow, attached to diverticulum. Colliculum not present. Antrum lightly sclerotized. Tergite A8 almost quadrate. Apophyses narrow, slightly expanded and flattened at apices. Papillae anales tapered and pointed.
Range. China (Taiwan, Hong Kong), Malaysia, Sumatra, Borneo, Sundaland, Sulawesi, Seram, New Guinea.
Phenology. Apparently polyvoltine in the tropics. A winter species in Taiwan and Hong Kong.
Altitude. From sea level in Hong Kong up to about 1930 m in Borneo (Holloway 1997).
Similar species. Externally similar to representatives of the *haworthiata* species group, especially *E. jezonica*, but distinguished by the more washed-out coloration and the white hind wing. The male genitalia are distinguished by the uniapical uncus, the apical lobes on the aedeagus, and the shape of sternite A8. The female genitalia are similar to those of several species in the group, but are distinguished by the flattened lobes of the apophyses.
Material. ♂, British New Guinea [Papua New Guinea]: Upper Mambaré R., Biagi, 5000ft., B.N.G., Feb. [19]06 (A.S. Meek.), Rothschild Bequest, B.M. 1931–1, Geom. gen. slide no. 18890♂ (type, BMNH); ♀, Nord Borneo: Mont Kina Balu, 5.8.1903, John Waterstradt, Ex. Oberthur Coll., Brit. Mus. 1927–3 (holotype of *E. mundiscripta commundata*, BMNH); ♀, Malaya

[West Malaysia], Perak F.M.S., Larut Hills, at light, 3700 ft., 13 Feb.: 1932, H.M. Pendlebury, 1892, Brit. Mus. 1939–220 (holotype of *E. mundiscripta larutensis*, BMNH); ♂, Hong Kong, Shek Kong, N: T, Kadoorie Agric. Research Centre, UTM: 50Q KV 020833, 12 Jan. 1997, 125 W MBF, coll. R.C. Kendrick, BMNH(E) 1998–126, Geom. gen. slide no. 19636♂ (holotype of *E. sekkongensis*, BMNH), 3♂, 1♀, Taiwan, Taipei County, Taipei, Nanshih Chiao, Han-Lo-Dae, 450 m, December 2001, leg. A. Kun & L. Ronkay, Mironov slide no. 634♂ (TTM); 1♀, Taiwan, Nantou County, Hui-Sun Exp. Forest Area, 15 km N of Puli, 950 m, 24°04′49″ N, 121°02′08″ E, 3–4.i.2002, leg. A. Kun & L. Ronkay, Mironov slide no. 635♀ (TTM); 1♂, Taiwan, Taoyuan Co., Tungyenshan Recr. Ar., *ca.* 24°50′ N/121°24′ E, 900 m, 21.–24.ix.2002, leg. W. Schacht et al. (ZSM).

Remarks. The inclusion of this species in the *rajata* species group is tentative, but we think justified on the basis of the close similarity of the female genitalia to those of other members of the group.

186. *Eupithecia robiginascens robiginascens* Prout, 1926

Eupithecia robiginascens Prout, 1926, *Novit. zool.* **33** (1): 9. Holotype ♂ (BMNH; examined), Bhutan: Buxa.

Eupithecia captiosa Vojnits, 1984, *Acta zool. Acad. Sci. hung.* **30** (1–2): 225, figs 17. Holotype ♂ (TTM; examined), China: North Yunnan, Li-kiang (synonymised in Mironov et al., 2004a).

Description. Very variable in size and depth of colour. Fore wing ground colour dark greyish brown, often with reddish tinge; basal area usually reddish ochreous or greyish brown with oblique, reddish ochreous transverse band bordering medial area; transverse lines inconspicuous, oblique; medial curved onto costa behind discal dot; postmedial twice angled onto costa; transverse band between medial and terminal areas reddish ochreous; terminal area with whitish wavy subterminal line and medial and tornal white spots; discal dot large, ovoid, intensely black. Hind wing a little paler than fore wing; dark transverse lines marked along anal margin; terminal area darker with whitish subterminal line and small white tornal spot; discal dot small and pale.

Male genitalia. Uncus elongate, biapical. Valve broad, with slightly prominent medial angle on ventral margin. Vinculum narrow and very short, with shallow medial hollow. Papillae long and narrow, slightly curved. Aedeagus stout, slightly shorter than length of valve. Vesica armed with two long sinuate cornuti, a plate-like, apically pointed cornutus at their base and a small curved cornutus near ductus ejaculatorius base. Sternite A8 weakly sclerotized, bilobed at base, with two long narrow arms.

Female genitalia. Bursa small, ovoid, almost completely covered with dense spines. Ductus bursae membranous but thick-walled, with a transverse patch of small spines and a membranous diverticulum near colliculum. Ductus seminalis broadened near base, attached to posterior part of corpus bursae on right side. Colliculum short and broad. Antrum membranous but anterior part sclerotized and with numerous pores. Tergite A8 campaniform, with centrally pointed posterior margin. Apophyses elongate, narrow. Papillae anales apically rounded.
Range. China (Tibet, Sichuan, Yunnan, Gansu); India (Jammu and Kashmir, Sikkim, Khasi Hills), Nepal, Bhutan, Burma.
Phenology. Polyvoltine: almost all the year round from mid-March to late December.
Altitude. From about 1750 to 4100 m in the Himalayas (Nepal: coll. C. Herbulot in ZSM; coll. M. Fibiger).
Similar species. Usually easily distinguished by the two reddish ochreous fore wing bands, and terminal white spots. Both male and female genitalia are distinctive and easily recognized.
Material. ♂, Bhutan, Buxa, *Eupithecia robiginascens* Prout type ♂, BM Geom. slide no. 20149 (holotype, BMNH); ♂, China, Provinz Nord-Yuennan, Li-kiang [Lijiang], 22.vii.1935, H. Höne, Vojnits slide no. 14185♂ (holotype of *E. captiosa*, TTM); 1♀, same loc., 4.viii.1935, Vojnits slide no. 14183♀ (paratype of *E. captiosa*, TTM); 1♂, same loc., 8.ix.1935, H. Höne, Vojnits slide no. 14186♂ (paratype of *E. captiosa*, ZFMK); 1♂, 2♀, same loc., 31.vii, 30.viii, 15.ix.1935, H. Höne, de Laever slides nos. 246♀, 249♂, 250♀ (ZFMK); 2♂, 1♀, same loc., 11.vi., 27.viii.1934, 24.vii.1935, H. Höne, Vojnits slides nos. 12977♀, 13142♂, 13143♀ (last gen. slide wrongly labelled, possibly belonging to *E. praecipitata*, TTM); 1♂, Chasseurs indigènes de Tátsien-lou, Récolte de 1910, Vojnits slide no. 866♂ (TTM); 1♀, China, Tibet [Sichuan], Batang, Untere Urwaldzone (*ca.* 3800 m), 7.viii.1936, H. Höne, Vojnits slide no. 13139♀ (wrongly labelled gen. slide of *E. particeps*, TTM); 1♂, China, Sichuan, Daxue Shan Mts., Gongga Shan, NW Moxi, 29°41′ N, 101°58′ E, 2850 m, 14.-19.vii.1999, leg. V. Siniaev & A. Plutenko (ZFMK); 2♀, same loc., W Moxi, 29°34′ N, 101°59′ E, 3200 m, 20.-21.vii.1999, leg. V. Siniaev & A. Plutenko (ZFMK); 1♀, China, N. Yunnan, road Zhongdian / Lijiang, 3000 m, 27°28′ N / 99°53′ E, 19.vii.2009, I.& A. Floriani, Saldaitis, Pekarsky slide no. OP 0816♀ (coll. O. Pekarsky); 2♂, Yatung [Yadong], Tibet, A.E. Hobson, BM slides nos.21477, 21492♂ (BMNH); 1♀, Tibet, Bomi Yi'ong, 2300 m, 17.vii.1983, coll Han Yinheng, slide no ZL59♀; 2♀, Sichuan, Emeishan, Qingyinge, 800–1000 m, 18–19.ix.1957, coll Lu Youcai & Zhu Fuxing, slides nos. ZL93, 129♀; 1♀, Sichuan, Wenchuan, Wolong, 1920 m,

26.vii.1983, coll Wang Shuyong, slide no. ZL127♀; 2♀, Yunnan, Lijiang, Yulong mountain, 23.vii.1962, coll Song Shimei; 1♀, Gansu, Wenxian, Qiujiaba, 2350 m, 22.vii.1999, coll Yao Jian, slide no. ZL569♀ (IZCAS).

186a *Eupithecia robiginascens flexicornuta* Inoue, 1988, stat. n.

Eupithecia flexicornuta Inoue, 1988, *Bull. Fac. domest. Sci. Otsuma Wom. Univ.* **24**: 345, figs 15a-f. Holotype ♀ (BMNH; examined), [Taiwan]: Hualien Hsien, Tayuling, 2600 m

Differential diagnosis. Specimens from Taiwan described by Inoue as *E. flexicornuta* are slightly paler than specimens of *E. robiginascens*, but we could find no consistent differences in external characteristics between the two. In the male genitalia, *E. flexicornuta* has a narrower apex to the valve, and in the female, a slightly different patch of spining in the ductus bursae. We think it is best regarded as an insular subspecies of *E. robiginascens.*.
Range. China (Taiwan).
Phenology. Bivoltine (?): mid-March – late March; late December.
Altitude. From about 2200 to 2600 m above sea-level (Taiwan: Inoue 1988; coll. TTM).
Material. ♀, Taiwan, Sangshan, 29.xii.1985, H. Nakajima, BM Geom. slide no. 21869♀(holotype, BMNH); 1♂, 2♀, Ta-yu-ling, 2600 m, Hualien Hsien, 28–31.iii.1981 and 2–4.v.1984, H. Yoshimoto leg., BM Geom. slides11192♂, 11193, 13140♀; 1♂, Arizan, Formosa, 7300ft, 21.ix.1906, A.E. Wileman (BMNH); 1♂, Taiwan, Prov. Tai-Tung, 2 km E of Hsiangyang, 2200 m, 11–13.iii.1996, leg. Gy. Fábián & L. Németh (TTM).

187. *Eupithecia luctuosa* Mironov & Galsworthy, 2004

Eupithecia luctuosa Mironov & Galsworthy, 2004, *Trans. lepid. Soc. Japan* 55 (1): 53, figs 10, 28, 29. Holotype ♀ (ZFMK; examined), [China]: [Fujian], Kwangtseh-Fukien.

Description. All body parts unicolorous blackish grey. Ground colour of all wings dark grey-brown; fore wing transverse lines darker brown; antemedial broad, doubled at hind margin; discal spot dark brown, slightly elongate; medial sinuous, passing through discal dot; postmedial broad, bordered distally by a broad pale fascia; terminal area with a pale sinuous submarginal line. Hind wing with transverse lines complete but faint, and a substantial discal dot.
Male genitalia. Uncus elongate, biapical. Valve with a slight ventral protrusion. Vinculum broad and square. Papillae narrow, elongate. Aedeagus stout, longer than length of valve. Vesica armed with a sinuate cornutus almost equal to length of aedeagus, a dense patch of spine-like cornuti

and a small plate-like cornutus near ductus ejaculatorius base. Sternite A8 broadened basally, with two narrow apical rods connected by a wide basal band.

Female genitalia. Bursa membranous, ovate-oblong, covered with small spines in posterior part. Ductus bursae broad, tapered, heavily sclerotized, covered with numerous spinules, and on the dorsal side with a long, sclerotized, digitiform process attached at base of colliculum. Ductus seminalis attached to posterior part of corpus bursae on left side. Colliculum narrow and elongate. Antrum very short, membranous. Tergite A8 almost quadrate. Apophyses short and narrow. Papillae anales relatively short and broadly rounded.

Range. China (Fujian, Taiwan).

Altitude. Found at about 2000–2300 m above sea-level (Taiwan: coll. ZSM, MNHU).

Phenology. Univoltine: early October to early November.

Similar species. Externally unmistakable because of its uniform dark grey-brown coloration with a violet tinge. The male genitalia are distinguished by the very long sinuous cornutus in the vesica. The female genitalia are unlike those of any other species in the genus.

Material. ♀, [Fujian], Kwangtseh-Fukien, 4.x.1937, J. Klapperich, O, de Laever slide no. 167♀ (holotype, ZFMK); 1♂, same loc. and date, de Laever slide no. 35♂ (paratype, ZFMK); 1♀, same data, Vojnits slide no. 12257♀ (TTM); 2♀, Taiwan, Nantou, Meifeng, *ca.* 2300 m, 29.x.-3.xi.1998, leg. Mey & Speidel, Zool. Mus. Berlin, Mironov slides nos. 810♀, 816♀ (MNHU); 1♂, Central Taiwan, near Puli, Nantou Co., Meifeng, *ca.* 2200 m NN, 24°05′25″ N/121°10′21″ E, LF, 29. October 2003, leg. Mei-Yu Chen, Mironov slide no. 560♂ (ZSM).

Remarks. Not previously recorded from Taiwan.

188. *Eupithecia qinlingata* Mironov & Galsworthy, 2011

Eupithecia qinlingata Mironov & Galsworthy, 2011, *Lep. Science,* **62** (1): 29, figs 14, 35. Holotype ♂ (ZFMK; examined), Sued-Shensi [Shaanxi], Tapaishan [Taibaishan] im Tsinling [Qinling], 1700 m.

Misidentification: *Eupithecia praecipitata* Vojnits, 1981 (♂) (*nec* Vojnits, 1979).

Description. Fore wing ground colour pale brownish grey; transverse lines inconspicuous; postmedial obtusely angled onto costa, forming a dark costal spot and with dark wedge-shaped marks on the veins; terminal area slightly darker, especially near apex, with light tornal spot; discal dot distinct, rounded. Hind wing paler, transverse lines well marked along anal

margin only; terminal area with a small, whitish tornal spot; discal dot minute, pale, rounded.
Male genitalia. Uncus elongate, biapical. Valve rather broad, unmodified. Vinculum broad. Papillae narrow. Aedeagus stout, shorter than length of valve. Vesica armed with two large cornuti, one curved and the other hook-shaped, and a small basal sclerotization. Sternite A8 two parallel rods connected by a very narrow basal band.
Female unknown.
Range. China (Shaanxi).
Altitude. The single known specimen was captured at about 1700 m in the Qinling Mts. (China: coll. ZFMK).
Similar species. Externally, on the basis of the shape and pattern of wings, *E. qinlingata* is very similar to the holotype of *E. vojnitsi* Inoue, 2000 (= *tenebricosa* Vojnits, 1983, preocc., *nec* Schaus, 1913) from Nepal (Inoue, 2000, pl. 168: 17♂). The very distinctive cornuti in the aedeagus vesica distinguish the two.
Material. ♂, Sued-Shensi [Shaanxi], Tapaishan [Taibaishan] im Tsinling [Qinling], *ca.* 1700 m, 6.viii.1936, H. Höne, Vojnits slide no. 12951♂ (holotype, ZFMK).
Remarks. The holotype of *E. qinlingata* was incorrectly determined by Vojnits (1981) as a male of *E. praecipitata.*

189. *Eupithecia svetlanae* Mironov & Pekarsky, 2011

Eupithecia svetlanae Mironov & Pekarsky, 2010, *In*: Mironov & al., 2011, *Lep. Science,* **62** (1): 29, figs 15, 36. Holotype ♂ (coll. O. Pekarsky; examined), China, Sichuan, Lao Lin Kou, 28°21′ N, 103°26′ E, 1900 m.

Description. Fore wing ground colour unicolorous ash-grey, basal area blackish brown along costa; antemedial inconspicuous, right angled and broadened onto costa, postmedial formed by a series of dark dashes on veins, evenly curved onto costa and forming a dark costal blotch; discal dot distinct, black; terminal area darker, with whitish tornal spot and wavy subterminal line, both shaded internally, and with a large blackish brown blotch near apex; area between postmedial and latter rusty ochreous. Hind wing unicolorous, darker on anal and terminal margins, veins covered with dark scales; transverse lines indistinct; whitish tornal spot large, bordered internally with blackish; discal dot rounded.
Male genitalia. Uncus elongate, biapical. Valve unmodified. Vinculum narrow with deep medial hollow. Papillae slightly curved. Aedeagus stout, shorter than length of valve. Vesica armed with one straight and one strongly curved apical cornutus, and a small, irregular plate-like cornutus

near ductus ejaculatorius base. Sternite A8 with two almost parallel sinuate arms on a broadly bilobed base.
Female unknown.
Range. China (Sichuan).
Similar species. Like the previous species, externally similar to the Himalayan *E. vojnitsi* Inoue, 2000, but distinguished by the presence of distinct whitish tornal spots on all wings. The male genitalia are well characterised by the medial hollow in the vinculum, the shape of the cornuti in the vesica and the shape of the eighth sternite.
Material. ♂, China, Sichuan, Lao Lin Kou, 1900 m, 28°21′ N, 103°26′ E, 26.vi.-12.vii.2008, leg. Viktor Sinyaev, Pekarsky slide no. OP 0244♂ (holotype, coll. O. Pekarsky).

190. *Eupithecia contraria* Vojnits, 1983

Eupithecia contraria Vojnits, 1983, *Acta zool. Acad. Sci. hung.* **29** (1–3): 263, figs 2, 4, 5. Holotype ♂ (ZSM; examined), Nepal: Tanga, 3800 m.
Eupithecia fuscoferruginea Inoue, 1987, *Bull. Fac. domest. Sci. Otsuma Wom. Univ.* **23**: 239, figs 63f, 65a, b. Holotype ♂ (BMNH; examined), [Nepal]: Kalbani, near Kalingandaki, 2400 m (synonymised in Inoue, 2000).

Description. Fore wing broad, ground colour brownish grey; transverse lines fine, wavy, oblique, angled onto costa; postmedial often just a series of dark dashes on the veins, twice angled onto costa; terminal area with fine wavy white subterminal line, forming a small white tornal dot; discal dot very distinct, round. Hind wing off-white, dark grey brown along anal margin where all transverse lines broadly marked; discal dot small, rounded.
Male genitalia. Uncus narrow, biapical. Valve unmodified. Vinculum short, with shallow medial hollow. Papillae elongate, rather narrow. Aedeagus slim, shorter than length of valve. Vesica armed with a long irregularly curved cornutus, a small apical cornutus, a broad plate-like central cornutus, and a small, irregular cornutus near ductus ejaculatorius base. Sternite A8 short, with two almost parallel, broad, apically pointed apical arms.
Female genitalia. Bursa almost globular, membranous but thick-walled, covered with small spines basally and along left side, and with a longitudinal row of larger spines at base of ductus bursae. Latter elongate, with broad diverticulum at base on left side. Ductus seminalis slightly broadened at base, attached at base of corpus bursae. Colliculum not present. Antrum short and broad. Tergite A8 almost quadrate. Apophyses narrow. Papillae anales large and broadly rounded.

Range. China (Tibet); Nepal, Bhutan.
Phenology. Bivoltine (?): late May to early August.
Altitude. From about 3100 to 3900 m above sea-level in the Himalayas (Nepal: Vojnits 1983; coll. ZSM).
Similar species. Similar to *E. robiginascens* but distinguished by the smaller, rounded discal dot, the absence of reddish ochreous coloration, and the paler hind wings. The cornuti in the male vesica and the shape and spining of the female bursa are quite distinctive.
Material. ♂, Nepal, Tanga, 3800 m, 29.vii.1962, leg. G. Ebert u. H. Falkner, Vojnits slide no. 11654♂ [= ZSM slide no. G 4093♂] (holotype, ZSM); 1♀, Nepal, Prov. Nr. 3, East Khumjung, 3800 m, 11.vii.1964, leg. W. Dierl, Vojnits slide no. 11770♀ [= ZSM slide no. G 4106♀] (paratype of *E. contraria*, ZSM); 1♂, Tibet, Chumbitang, 13000ft, 25.vii.1924, Maj. R.W.G. Hingston, BM Geometrid slide no. 21469♂; 1♂, Tibet, Yatung [Yadong], A.E.Hobson, BM Geometrid slide no. 21477♂; 1♂, Tibet, Gantea, Chumba Valley, 11500 ft, 21.vii.1924, Maj. R.W.G.Hingston, BM Geometrid slide no. 20838♂ (BMNH); 13♂, 1♀ Tibet, Zham, 3300 m, 2–8.vii.1975, slides nos. ZL99,162,186♂; 2♂, 1♀ Tibet, Gyirong, 3300 m, 3–4.viii.1975, coll Huang Fusheng and Wang Ziqing, slides nos. ZL94,189♂ (IZCAS).

191. *Eupithecia rajata* Guenée, [1858]

Eupithecia rajata Guenée, [1858], *in* Boisduval & Guenée, *Hist. nat. Insectes* (Spec. gén. Lépid.) **10**: 328. Holotype ♂ (BMNH; examined), India (central).
Eupithecia apparticeps Inoue, 2000, *Tinea* **16** (Suppl. 1): 29, pl. 165, fig. 6; fig. 1327. Holotype ♀ (NSMTO; examined), [Nepal]: [Kathmandu], Godavari, 1600 m (synonymised in Mironov et al., 2008b).

Misidentifications: *Eupithecia profana* Vojnits, 1984 (1♂ paratype in TTM); *Eupithecia particeps* Vojnits, 1988 (1♀ paratype in NSMTO); *Eupithecia bini* Vojnits, 1981(paratype ♀).

Description. A small species. Fore wing ground colour brownish-grey; transverse lines narrow, broadened near costa to form blackish costal spots; basal and antemedial right angled onto costa; postmedial twice angled; terminal area slightly darker with fine, wavy, whitish subterminal line; discal dot distinct, rounded or ovoid. Hind wing slightly paler; postmedial blackish, well marked near anal margin; terminal area slightly darker with narrow, wavy, whitish subterminal line; discal dot small, pale, and rounded.
Male genitalia. Uncus elongate, biapical. Valve unmodified. Vinculum short, rounded. Papillae small, narrow. Aedeagus slim, shorter than length of valve. Vesica armed with two lightly sclerotized, short and broad apical cornuti, an elongate U-shaped cornutus and a smaller, irregular cornutus

near ductus ejaculatorius base. Sternite A8 consisting of two narrow arms connected by a basal band.
Female genitalia. Bursa globular, covered with dense small spines. Ductus bursae elongate, membranous, broadened centrally with a patch of dense small spines. Ductus seminalis narrow, attached to central part of corpus bursae on ventral side. Colliculum not present. Antrum membranous. Tergite A8 small. Apophyses narrow. Papillae anales short and rounded.
Range. China (Tibet, Yunnan); India, Sri Lanka, Nepal, Vietnam.
Phenology. Bivoltine: early June to November.
Altitude. From about 1600 to 3800 m in the Himalayas (Nepal: Vojnits 1988; China: coll. ZFMK).
Similar species. Reasonably distinctive, with uniform fore wing coloration, and prominent discal and costal spots, but confusable with *E. vasta*, from which the genitalia separate it. See also under *E. ebriosa*.
Material. ♂, Typicum Specimen, Ex Musaeo Ach. Guenée, *Rajata* Gn. Inde., Silma, Ex Oberthür Coll. Brit. Mus. 1927–3, Holotypus *Eupithecia rajata* ♀ abdomen on ♂ body, gen. no. 19349, det. A. Vojnits, BM Geom. slide no. 16414 (holotype, BMNH); ♀, Nepal, Kathmandu, Godavari, 1,600 m, 26.ii.1992, Inoue slide no. 15927♀ (holotype of *E. apparticeps*, coll. H. Inoue in BMNH); 3♂, China, Provinz Nord-Yuennan, Li-kiang [Lijiang], 5.vi., 12.vii.1934, H. Höne, Vojnits slides nos. 13357♂, 13358♂, 13347♂ (ZFMK); 1♂, China, Provinz Nord-Yuennan, Li-kiang [Lijiang], 14.viii.1934, H. Höne, Vojnits slide no. 14201♂ (paratype of *E. profana*, TTM); 1♂, Tibet [Sichuan], Batang, Untere Urwaldzone (*ca.* 3800 m), 5.viii.1936, H. Höne, Vojnits slide no. 13831♂ (ZFMK); 1♀ Yunnan, Pingbian, Dayuanshan, 1500 m, 18.vi.1956, coll Huang Keren, slide no.ZL96♀;1♀ Yunnan, Lijiang, Yulong mountain, 21–22.vii.1962, coll Song Shimei, slide no ZL113♀ (IZCAS).
Remarks. Not previously recorded from China. The identity of this species was discussed in Mironov et al., 2008b.

192. *Eupithecia ebriosa* Vojnits, 1979

Eupithecia ebriosa Vojnits, 1979, *Acta zool. Acad. Sci. hung.* **25** (1–2): 209, fig. 21. Holotype ♂ (ZFMK; examined), China: Kuling.
Eupithecia gummaensis Inoue, 1980, *Bull. Fac. domest. Sci. Otsuma Wom. Univ.* **16**: 201, figs 55V; fig. 57A; fig. 60. Holotype ♀ (coll. H. Inoue in BMNH; examined), [Japan]: Gumma Pref(ecture), Mikaboyama, 750 m (synonymised in Mironov & Galsworthy, 2010).

Description. Fore wing ground colour dark grey; transverse lines inconspicuous, ante- and postmedial obtusely angled onto costa where broadened and forming darker costal spots; postmedial often a series of narrow dark dashes on the veins; terminal area broad, slightly darker, with a wavy, whitish subterminal line and small, white tornal spot; discal dot small, ovoid. Hind wing slightly paler with fairly distinct postmedial; terminal area broad, slightly darker, with whitish subterminal line and small, white tornal spot.

Male genitalia. Uncus elongate, biapical. Valve elongate and narrow, unmodified, but with slightly prominent ventral margin near apex. Vinculum narrow, rounded. Papillae broadened at bases. Aedeagus short, narrowed anteriorly. Vesica armed with two short stout apical cornuti, an elongate, U-shaped cornutus alongside them, and a small, irregular cornutus near ductus ejaculatorius base. Sternite A8 two long and narrow rods connected by a broad band at base.

Female genitalia. Bursa spherical, covered with dense small spines except for a small area around base of ductus seminalis. Ductus bursae elongate, relatively narrow, with numerous wrinkles and a row of dense spines along left side. Ductus seminalis narrow, attached to corpus bursae on right side. Colliculum not present. Antrum lightly sclerotized, covered with numerous minute pores. Tergite A8 small, square, with band-like, sclerotized anterior margin. Apophyses rather short. Papillae anales small, rounded.

Range. China (Tibet, Gansu, Sichuan, Yunnan, Jiangxi); Japan (Honshu), India (Jammu & Kashmir, Uttar Pradesh).

Phenology. Bivoltine: late April to mid-May, mid-July to mid-September.

Altitude. From about 2100 m in West Sichuan up to about 3500 m in North Yunnan (China: coll. O. Pekarsky).

Similar species. Very similar to *E. rajata*, but distinguished usually by larger size, more elongate fore wings, more greyish ground colour, relatively smaller discal dots and less distinct dark costal spots, especially the medial costal spot near the discal dot on the fore wing. The male genitalia are extremely similar, but in *E. ebriosa* the tip of the uncus is broader when viewed laterally, the valve narrower and more elongate, and sternite A8 has narrower apical arms. In the female genitalia, the spining in the ductus bursae is different from that in *E. rajata*, and more closely resembles that of *E. raniata*, which however has a well developed colliculum and longer apophyses.

Material. ♂, China, [Jiangxi], Kuling, 2.v.1934, H. Höne, Vojnits slide no. 11450♂ (holotype, ZFMK); 1♂, same data, Vojnits slide no. 11451♂ (paratype, TTM); 1♂, same data, de Laever slide no. 20♂ (ZFMK); 1♀, China, Provinz Nord-Yuennan, Li-kiang [Lijiang], 16.vi.1934, H. Höne, Mironov slide no. 832♀ (TTM); 1♂, 1♀, China, Sichuan, Daxue Shan, Gongga Shan, NW Moxi, 29°41′ E, 101°58′ N, 2600–3200 m, 23.iv.-15.v.2001, leg. V. Siniaev

& A. Plutenko, *ex* coll. Dr. A. Schintlmeister (ZSM); 1♂, 1♀, China, W. Sichuan, road Ya'an / Kangding, Erlang Shan Mt., 2100 m, 29°51′ N / 102°18′ E, 12.vii.2009, I. & A. Floriani, A. Saldaitis leg., Pekarsky slides nos. OP 0643♂, OP 0644♀ (coll. O. Pekarsky); 3♂, 1♀, China, W. Sichuan, Kangding evn 3000 m, 29°53′ N / 101°55′ E, 13.vii.2009, I. & A. Floriani, Saldaitis leg., Pekarsky slides nos. OP 0592♂, OP 0641♀, OP 0686♂, OP 0857♂ (coll. O. Pekarsky); 1♂, 3♀, China, N. Yunnan, road Xinngcheng / Zhongdian, 3500 m, 18.vii.2009, 28°32′ N / 99°49′ E, I. & A. Floriani, Saldaitis leg., Pekarsky slides nos. OP 0581♂, OP 0629♀, OP 0636♀, OP 0639♀ (coll. O. Pekarsky); 2♂, China, N. Yunnan, road Zhongdian / Lijiang, 3000 m, 19.vii.2009, 27°28′ N / 99°53′ E, I. & A. Floriani, Saldaitis leg., Pekarsky slides nos. OP 0583♂, OP 0652♂ (coll. O. Pekarsky); 1♀, Tibet, Nyingchi, Bayi, 3040 m, 2–3.vii.2002, Han Hongxiang, slide no. ZL644♀; 1♀, Gansu, Diebu, Lazikou, 2000 m, 5.ix.1990, slide no. ZL363♀; 1♀, Gansu, Yongdeng Tulugou, 2280 m, 25–29.vii.1991, slide no. ZL379♀, both Xue Dayong (IZCAS).

Remarks. Indian specimens of this species are considerably smaller than specimens from Chinese populations, but the genitalia are virtually identical.

193. *Eupithecia vasta* Vojnits, 1979

Eupithecia vasta Vojnits, 1979, *Acta zool. Acad. Sci. hung.* 25 (1–2): 197, fig. 6. Holotype ♂ (ZFMK; examined), China: North Yunnan, Li-kiang.

Misidentification: *Eupithecia avara* Vojnits, 1984 (1♀ paratype in TTM).

Description. A small species: fore wing ground colour brownish grey, with five dark spots on costa; all transverse lines indistinct; postmedial formed by an evenly curved row of dots on veins; terminal area slightly darker with a fine whitish subterminal line and a white tornal spot bordered internally with brown; discal dot large, intensely black. Hind wing slightly paler, with darker anal margin, a small whitish tornal spot and a small rounded discal dot.

Male genitalia. Uncus elongate, biapical. Valve unmodified, with rather square apex. Vinculum broad, trapezoid. Papillae elongate, narrow. Aedeagus shorter than valve length. Vesica armed with two almost identical long cornuti curved near bases, and a U-shaped, band-like cornutus near ductus ejaculatorius base. Sternite A8 small, somewhat variable, with two parallel apical rods about one half of sternite length.

Female genitalia. Bursa broad, semi-globular, covered with minute spines. Ductus bursae long and broad, heavily sclerotized in large longitudinal

folds, without spines. Colliculum short and broad. Antrum membranous, short and broad. Tergite A8 small, almost square. Apophyses narrow and short. Papillae anales short and rounded.
Range. China (Gansu, Sichuan, Yunnan).
Phenology. Bivoltine (?): mid-July to late September.
Altitude. Found from about 2000–3000 m (China: coll. O. Pekarsky; IZCAS).
Similar species. Very similar to *E. rajata*, but the genitalia are very different. The male of *E. vasta* has much larger cornuti in the vesica. The female genitalia are unlike those of any other Asian *Eupithecia* species.
Material. ♂, China, Provinz Nord-Yuennan, Li-kiang [Lijiang], 6.ix.1935, H. Höne, Vojnits slide no. 11465♂ (holotype, ZFMK); 1♀, same loc., 2.ix.1935, H. Höne, Vojnits slide no. 14191♀ (paratype of *E. avara*, TTM); 1♀, same loc., 21.ix.1934, H. Höne, de Laever slide no. 359♀ (ZFMK); 1♂, Sichuan, Dukou, 13.vi.1981, coll Zhang Baolin, slide no. ZL111♂; 1♂, Gansu, Diebu, Lazikou, 2000 m, 5.ix.1990, coll Xue Dayong, slide no. ZL371♂ (IZCAS); 1♂, China, N. Yunnan, road Zhongdian / Lijiang, 3000 m, 19.vii.2009, 27°28′ N / 99°53′ E, I. & A. Floriani, Saldaitis leg., Pekarsky slide no. OP 0651♂ (coll. O. Pekarsky).
Remarks. We have concluded that the female specimens listed above, which Vojnits had assigned to a separate species, *E. avara*, in fact belong with the male holotype of the present species, from the same locality. In Vojnits slide no. 14191♀ the ductus seminalis is missing; it probably arises from the ductus bursae near the colliculum.

194. *Eupithecia lobbichlerata* Schütze, 1961

Eupithecia lobbichlerata Schütze, 1961, *Veröff. zool. StSamml. Münch.* **6**: 181, pl. 29, fig. 3; pl. 31, figs 2, 2a. Holotype ♂ (ZSM; examined), [Nepal]: Sabzi-Chu, 3500 m
Eupithecia mortua Vojnits, 1984, *Acta zool. Acad. Sci. hung.* **30** (1–2): 229, fig. 21. Holotype ♂ (TTM; examined), [China]: North Yunnan, A-tun-tse [Dêqên], 4500 m. **New junior synonym.**

Misidentifications: *Eupithecia molestissima* Vojnits, 1981 (1♂ paratype in ZFMK); *Eupithecia refertissima* Vojnits, 1984 (1♀ paratype in TTM).

Description. Fore wing ground colour dark, brownish grey; transverse lines oblique, inconspicuous, usually marked with small, black dots on the veins; basal, antemedial and medial sharply angled onto costa; postmedial right angled near costa; terminal area with a series of small black dashes between veins, a weak pale subterminal line and a small, whitish tornal spot; discal dot black, slightly ovoid. Hind wing off-white, dark along anal margin where transverse lines visible; postmedial sharply angled; discal dot small, rounded.

Male genitalia. Uncus narrow, biapical. Valve unmodified. Vinculum short, semicircular. Papillae long and narrow. Aedeagus stout, shorter than length of valve. Vesica armed with two slightly curved apical horn-like cornuti, and two small irregular cornuti near ductus ejaculatorius base. Sternite A8 weakly sclerotized, consisting of two broad blunt apical rods connected by a basal band.

Female genitalia. Bursa oval, almost completely covered with dense small spines. Ductus bursae elongate, with a sclerotized diverticulum on the left and two longitudinal patches of small spines dorsally. Ductus seminalis arising from membranous bulge on central part of corpus bursae. Colliculum lacking. Antrum membranous. Tergite A8 almost quadrate. Apophyses narrow. Papillae anales narrow and elongate.

Range. China (Sichuan, Yunnan, Tibet, Qinghai); Nepal, North India.

Phenology. Univoltine (?): from mid-July to early October.

Altitude. From about 1800 to 4500 m (China: Vojnits 1984a; coll. ZFMK, IZCAS).

Similar species. Difficult to distinguish from *E. albispumata*, which is usually paler rusty brown on the fore wings. Best separated by its very different genitalia. See also under next species.

Material. ♂, Nepal, Manangbhot, Sabzi-Chu, 28°40′ n. Br., 84°1′ ö. L., 3,500[m], 13.vii.[19]55, leg. F. Lobbichler, Schütze slide no. A 711♂ [= ZSM slide no. G 13369] (holotype, ZSM); ♂, Nord Yünnan, A-tun-tse [Dêqên], Obere Höhe (*ca.* 4500 m), 13.viii.1936, H. Höne, Vojnits slide no. 14120♂ (holotype of *E. mortua*, TTM); ♂, same loc. and alt., 15.viii.1936, H. Höne, Vojnits slide no. 12453♂ (holotype of unpubl. *E. effecta* Vojnits, ZFMK); 5♂, 1♀, same loc. and alt., 7, 26, 27, 30.vii., 11, 12.viii.1936, H. Höne, Vojnits slides nos. 12182♀, 12464♂, 12450♂, 12451♂, 12470♂, 12180♂ (paratypes of unpubl. *E. effecta* Vojnits, ZFMK); 23♂, 17♀, same loc. and alt., 1, 6, 7, 9, 10, 11, 13, 14, 16, 17, 19, 20, 22, 24, 25, 27.vii., 2, 7, 10, 11, 13, 21, 23, 25. viii.1936, H. Höne, Vojnits slides nos. 12458♀ (wrongly labelled gen. slide of *E. supercastigata*), 13914♀, 11550♀ (specimen and gen. slide wrongly labelled as holotype of *E. improba* Vojnits, but gen. slide in fact of *E. lasciva*), 14090♀, 14145♂ (absent in ZFMK), 13923♂, 13927♂, 13930♀, 13938♂, 13952♀, 14148♂, 14188♂, 13915♀, 13921♀, 13925♀, 13295♀ (wrongly labelled gen. slide of *E. egena*), 13942♂, 11545♀ (wrongly labelled gen. slide of *E. lasciva*), 14113♂, 13924♂, 13926♂, 13932♂, 13936♂, 13941♂, 13746♂, 14151♂, 14119♀, 13917♂, 14114♂, 13916♂, 13780♀, 14107♂, 14150♀, 13779♀, 14061♂, 13919♂, 14094♀, 14117♂, 14152♂, de Laever slide no. 80♀ (ZFMK); 1♂, 1♀, same loc., Mittlere Höhe (*ca.* 4000 m), 5.vii., 3.viii.1936, H. Höne, Vojnits slides nos. 12167♂, 12457♀ (paratypes of unpubl. *E. effecta* Vojnits, ZFMK); 3♂, 3♀, same loc.

and alt., 12, 13, 25.vi., 4.vii., 2, 18.viii.1936, H. Höne, Vojnits slides nos. 13751♂, 13750♀, 13754♂, 13928♂, 14187♀, 13211♀ (ZFMK); 1♀, same loc., Talsohle (*ca.* 3500 m), 26.iv.1936, H. Höne, Vojnits slide no. 14181♀ (ZFMK); 2♂, 2♀, same loc., Talsohle *ca.* 3000 m, 22.v., 3, 10, 11.vii.1937, H. Höne, Vojnits slides nos. 14104♂, 13920♀, 14068♂, 13321♀ (ZFMK); 1♀, Tibet [Sichuan], Batang, Untere Urwaldzone (*ca.* 3800 m), 2.viii.1936, H. Höne, Vojnits slide no. 13803♀ (paratype of *E. refertissima*, TTM); 16♂, 13♀, same loc. and alt., 10, 12, 13, 16, 18, 20, 22, 23, 26.vi., 1, 2, 6, 12.viii.1936, H. Höne, Vojnits slides nos. 12151♀, 12150♂, 12146♀, 12398♀, 12399♀, 12409♀, 12144♀, 12423♀, 12149♀, 12176♂, 12400♀, 12397♀, 12405♂, 12394♂, 12142♂, 12147♂, 12141♂, 12204♂ (all paratypes of unpubl. *E. effecta* Vojnits, ZFMK); 8♂, 3♀, same loc. and alt., 13, 17, 19.vi., 1, 2, 5, 6, 7.viii.1936, H. Höne, de Laever slides nos. 447♂, 448♀, 451♂, 459♂, 477♂, 440♀, Vojnits slides nos. 13896♂, 13820♀, 13841♂, 13827♂, 13837♂ (ZFMK); 1♂, 2♀, same loc. and alt., 1, 5.viii.1936, H. Höne, Mironov slides nos. 633♂, 823♀ (TTM); 1♂, China, Provinz Nord-Yuennan, Li-kiang [Lijiang], 24.vii.1934, H. Höne, Vojnits slide no. 13187♂ (paratype of *E. molestissima*, ZFMK); 1♂, same loc., 29.vii.1935, H. Höne, Vojnits slide no. 11560♂ (paratype of unpubl. *E. effecta* Vojnits, ZFMK); 21♂, 17♀, same loc., 2, 9, 10, 13, 15, 23, 29.vii., 4, 5, 7, 8, 11, 13, 14.viii.1934, 20, 24, 25, 27.vi., 3, 5, 21, 23, 24, 26.vii., 4, 12.viii.1935, H. Höne, de Laever slides nos. 350♀, 296♂, 364♂, 268♂, 128♀, 356♂, 266♂, 355♀, 353♂, 366♂, Vojnits slides nos. 13355♂, 13773♂, 13166♀, 13171♀, 13360♀, 14192♀, 14200♂, 13356♀, 13060♂, 13039♀, 13064♂, 13770♂, 13080♂, 13037♂, 13774♀, 13035♀, 13362♀, 13257♀, 14182♂, 14184♂, 13361♂, 13348♂, 13364♂, 13067♀, 13359♀, 13343♀, 13772♀, Mironov slide no. 717♂ (ZFMK); 2♂, same loc., 5, 23.vii.1934, H. Höne, Vojnits slides nos. 13079♂, 13213♂ (TTM); 1♂, 1♀, China, Sichuan, Daxue Shan Mts., Gongga Shan, NW Moxi, 29°41′ N, 101°58′ E, 2850 m, 14.-19. vii.1999, leg. V. Siniaev & A. Plutenko, Mironov slides nos. 735♀, 738♂ (ZFMK); 3♂, 5♀, China, Sichuan, Daxue Shan Mts., 80 km W Kangding, 30°02′ N, 101°40′ E, 3800 m, 27–29.vii.1999, leg. V. Siniaev & A. Plutenko, Mironov slide no. 644♀ (ZFMK); 1♂, China, North Yunnan, Diqing Tibetan Aut. Pref., 8 km NNE of Shangri La at Nairi village, 3300 m, 14.vi.2009, leg. B. Benedek, Pekarsky slide no. OP 0828♂ (coll. O. Pekarsky); 1♂, 3♀, China, W. Sichuan, Kangding, 3000 m, 29°53′ N / 101°55′ E, 13.vii.2009, I. & A. Floriani, Saldaitis leg., Pekarsky slides nos. OP 0628♀, OP 0640♀, OP 0654♀, OP 0685♂ (coll. O. Pekarsky); 3♂, 4♀, China, W. Sichuan, road Yajiang / Litang, 3600 m, 29°59′ N / 100°52′ E, 15.vii.2009, I. & A. Floriani, Saldaitis leg., Pekarsky slides nos. OP 0575♂, OP 0578♂, OP 0627♀, OP 0657♂, OP 0693♀, OP 0694♀, OP 0697♀ (coll. O. Pekarsky); 17♂, 13♀, China, W. Sichuan, Litang, 4000 m, 29°50′ N / 100°21′ E, 16.vii.2009, I. & A. Floriani,

Saldaitis leg., Pekarsky slides nos. OP 0504♀, OP 0505♀, OP 0506♂, OP 0540♂, OP 0648♀, OP 0650♂, OP 0705♀, OP 0706♀, OP 0707♀, OP 0708♀, OP 0710♂, OP 0711♂, OP 0713♂, OP 0714♂, OP 0715♂, OP 0716♀, OP 0721♂, OP 0722♂, OP 0728♀, OP 0730♀, OP 0731♀, OP 0732♂, OP 0733♂, OP 0739♀, OP 0741♀, OP 0744♂, OP 0749♂, OP 0750♂, OP 0751♂, OP 0752♂ (coll. O. Pekarsky); 2♀, China, W. Sichuan, road Xinngcheng / Zhongdian, 3300 m, 28°46′ N / 99°51′ E, 17.vii.2009, I. & A. Floriani, Saldaitis leg., Pekarsky slides nos. OP 0804♀, OP 0805♀ (coll. O. Pekarsky); 5♂, 4♀, China, N. Yunnan, road Xinngcheng / Zhongdian, 3500 m, 28°32′ N / 99°49′ E, 18.vii.2009, I. & A. Floriani, Saldaitis leg., Pekarsky slides nos. OP 0579♂, OP 0580♂, OP 0582♂, OP 0630♂, OP 0637♀, OP 0638♀, OP 0811♀, OP 0813♀, OP 0814♂ (coll. O. Pekarsky); 1♀, Tibet, Jomda, 3400 m, 25.vii.1976, coll Zhang Xuezhong, slide no. ZL202♀; 1♀, Sichuan, Yajiang, 3300 m, 24. viii.1982, slide no. ZL342♀; 3♂, Yunnan, Lijiang, Yulongshan, 21.vii and 30.viii.1962, coll Song Shimei, slides nos. ZL134, 223♂; 1♂, same loc, 2800 m, 15.vii.1984, coll Liu Dajun, slide no. ZL542♂; 1♂, Yunnan, Zhongdian, Geza, 3150 m, 6.viii.1981, coll Liao Suobai, slide no. ZL322♂; 1♂, 1♀, Sichuan, Batang, Yidun, 3500 m, 16.viii.1982, coll Chai Huaicheng, slides nos. ZL440♂, 327♀; 1♀ Sichuan, Kangding, 10.vii.1959, slide no. ZL153♀; 1♂, Sichuan, Yajiang, 3300 m, 24.vii.1982, coll Zhang Xuezhong, slide no. ZL 132♂; 1♂, Sichuan, Wenchuan, 1920 m, 27.vii.1983, coll Chai Huaicheng, slide no. ZL426♂; 1♀, Sichuan, Luding Xinxing, 1800 m, 13.vi.1983, coll Chai Huaicheng; 1♀, Qinghai, Menyuan, Xianmi, 2800 m, 37°16′ N, 102°00′ E, 6. viii.2005, Xue Dayong, slide no. ZL716♀ (IZCAS).

Remarks. Mironov et al. (2008b) dealt with the original confusion of this species with *E. albispumata*.

195. *Eupithecia coniurata* Vojnits, 1979

Eupithecia coniurata Vojnits, 1979, *Acta zool. Acad. Sci. hung.* 25 (1–2): 198, fig. 7. Holotype ♂ (ZFMK; examined), China: North Yunnan, Li-kiang.

Description. Fore wing ground colour pale grey; transverse lines inconspicuous; postmedial usually a series of dark dots on veins; terminal area with fine, wavy, whitish subterminal line and small tornal spot; discal dot small, comma-like. Hind wing whitish grey, darker along anal and terminal margins, with a small white tornal dot; discal dot small.

Male genitalia. Uncus slim, biapical. Valve unmodified. Vinculum short and broad. Papillae elongate, slightly clavate. Aedeagus narrow, equal to length of valve. Vesica armed with two narrow, straight, apical cornuti, and an elongate folded cornutus connected with a smaller, irregular cornutus near ductus ejaculatorius base. Sternite A8 relatively short, with two broad parallel arms connected by a broad basal band.

Female genitalia. Bursa elongate, covered with dense small spines. Ductus bursae relatively short, mostly covered with small spines, and with a curved patch of sclerotization. Ductus seminalis narrow, attached to central part of corpus bursae. Colliculum short and narrow. Antrum membranous, with numerous pores. Tergite A8 almost quadrate. Apophyses narrow, the posterior slightly flattened at apices. Papillae anales small and rounded.

Range. China (Tibet,Yunnan, Sichuan, Qinghai, Shaanxi).

Phenology. Univoltine: mid-July to late August.

Altitude. From 3000 to about 4500 m in mountains of Shaanxi and Yunnan (coll. TTM and O. Pekarsky).

Similar species. Similar to *E. lobbichlerata*, but has a smaller discal dot on the fore wing. In the male genitalia the much narrower aedeagus with straight apical cornuti is diagnostic. The female genitalia are very similar to those of *E. lobbichlerata*, but the absence of the small sclerotized diverticulum on one side of the ductus bursae gives them a more elongate appearance.

Material. ♂, China, Provinz Nord-Yuennan, Li-kiang [Lijiang], 21.viii.1935, H. Höne, Vojnits slide no. 11443♂ (holotype, ZFMK); 1♂, same loc., 27. viii.1934, H. Höne, Vojnits slide no. 12063♂ (paratype, TTM); 1♂, 1♀, same loc., 23.vii., 9.ix.1935, H. Höne, de Laever slide no. 352♀, Vojnits slide no. 11500♂ (ZFMK); 1♂, same loc., 27.viii.1934, H. Höne, Vojnits slide no. 11502♂ (wrongly labelled gen. slide of *E. lobbichlerata*, TTM); 2♂, same loc., 5, 14.ix.1935, H. Höne, Vojnits slide no. 12062♂ (TTM); 2♀, same loc., 30.viii., 26.ix.1934, H. Höne, Vojnits slides nos. 1052♀ (marked by Vojnits as unpubl. *E. nucleatae*), 1055♀ (TTM); 1♂, Sued-Shensi [Shaanxi], Tapaishan [Taibaishan] im Tsinling [Qinling], *ca.* 1700 m, 10.viii.1936, H. Höne, Vojnits slide no. 12331♂ (as unpubl. *E. effecta* Vojnits, ZFMK); 1♀, same loc., *ca.* 3000 m, 7.ix.1936, H. Höne, Vojnits slide no. 11433♀ (paratype of *E. vana*, TTM); 1♀, same loc. and alt., 9.ix.1936, H. Höne, Vojnits slide no. 12373♀ (ZFMK); 1♂, Nord Yünnan, A-tun-tse [Dêqên], Talsohle (*ca.* 3500 m), 25. vi.1936, H. Höne, Vojnits slide no. 14056♂ (ZFMK); 1♂, same loc., Obere Höhe (*ca.* 4500 m), 9.viii.1936, H. Höne, Vojnits slide no. 13239♂ (TTM); 1♂, same loc. and alt., 23.viii.1936, H. Höne, Vojnits slide no. 14062♂ (ZFMK); 1♂, China, W. Sichuan, Kangding evn 3000 m, 29°53′ N / 101°55′ E, 13. vii.2009, I. & A. Floriani, Saldaitis leg., Pekarsky slide no. OP 0683♂ (coll. O. Pekarsky); 1♂, China, W. Sichuan, road Yajiang / Litang, 3600 m, 29°59′ N / 100°52′ E, 15.vii.2009, I. & A. Floriani, Saldaitis leg., Pekarsky slide no. OP 0695♂ (coll. O. Pekarsky); 5♂, 5♀, China, W. Sichuan, Litang, 4000 m, 29°50′ N / 100°21′ E, 16.vii.2009, I. & A. Floriani, Saldaitis leg., Pekarsky slides nos. OP 0550♂, OP 0649♀, OP 0719♀, OP 0720♀, OP 0727♂, OP 0734♂, OP 0743♀, OP 0745♂, OP 0746♂, OP 0753♀ (coll. O. Pekarsky); 3♂, 1♀, China, W. Sichuan, Xinduqiao, 3500 m, 30°04′ N / 101°26′ E, 14.vii.2009, I.& A. Floriani, Saldaitis, Pekarsky slides nos. OP 0687♀, OP 0689♂, OP 0690♂, OP 0691♂ (coll. O. Pekarsky); 4♂, 6♀, Tibet, Jomda, 3400 m, 24–27.vii.1976, coll Zhang Xuezhong and Han Yinheng, slides nos. ZL187, 328♂, 184, 231, 620, 625♀; 1♀, Tibet, Zhag'yab, Gyitang, 3600 m, 16.viii.1976, coll Han Yinheng, slide no.ZL332♀; 1♀, Tibet, Zham, 330 m, 1.vii.1975, coll Wang Ziqing; 2♂,2♀, Qinghai, Menyuan, Xianmi, 2800 m, 23.vii.1992, coll Xue Dayong, slides nos. ZL519♂, 284, 301♀ (IZCAS).

196. *Eupithecia cuneata* Vojnits, 1984

Eupithecia cuneata Vojnits, 1984, *Acta zool. Acad. Sci. hung.* **30** (1–2): 214, fig. 3. Holotype ♂ (TTM; examined), China: North Yunnan, Li-kiang.

Eupithecia proba Mironov & Galsworthy, 2004, *Trans. lepid. Soc. Japan* **55** (3): 229, figs 4, 16. Holotype ♂ (ZFMK; examined), North Yunnan, Likiang [Lijiang], 2000 (synonymised in Mironov & Galsworthy, 2009b).

Description. Fore wing elongate, ground colour pale reddish brown; discal dot raised; except postmedial, transverse lines obsolete, marked only by dark spots on costa; postmedial faint, darker brown, twice angled before costa, and bordered externally by a pale red brown fascia; terminal area darker, crossed by a wavy pale line with a white tornal dot. Hind wing off-white, discal dot faint; anal margin with strongly marked transverse lines, medial and postmedial almost complete; terminal area darker, with transverse wavy pale line.

Male genitalia. Uncus slim, biapical. Valve unmodified. Vinculum short with shallow medial hollow. Papillae elongate and narrow. Aedeagus stout, shorter than valve length. Vesica armed with two slim apical cornuti, a broad H-shaped cornutus and an irregular plate near ductus ejaculatorius base. Sternite A8 elongate, with two recurved apical rods.

Female genitalia. Bursa globular, covered with spines, except small area in posterior part. Ductus bursae long, sinuate, membranous, but with striate sclerotized base on one side, and covered with spinules between base of ductus seminalis and colliculum. Ductus seminalis narrow, attached to basal part of ductus bursae on left side. Colliculum short and narrow. Antrum membranous. Tergite A8 elongate, broadened posteriorly. Apophyses broad, the posterior expanded and flattened at apices. Papillae anales broad, but sharply tapered and pointed.

Range. China (Yunnan); N Thailand, N Vietnam.
Phenology. Univoltine: early September to mid-October.
Altitude. From about 1600–1800 m (Vietnam: coll. ZFMK), and up to about 2000 m in North Yunnan (coll. ZFMK).
Similar species. Similar externally to *E. coniurata*, but more reddish, and with stronger hind wing markings. In the male genitalia the two slim apical cornuti are distinctive. The female genitalia are similar in shape to several other species in the group, but lack any larger spines in the ductus bursae. See also next species.
Material. ♂, China, Provinz Nord-Yuennan, Li-kiang [Lijiang], 12.ix.1934, H. Höne, Vojnits slide no. 13771♂ (holotype, TTM); 9♂, same loc., 13, 18, 21 and 25.ix.(2), 17.x.1934, 3, 14 and 24.ix.1935, H. Höne, respectively Vojnits slide no. 13185♂, de Laever slides nos. 358♂, 381♂, 127♂, Vojnits slides nos. 12509♂, 13167♂, 13172♂, 13180♂, 13181♂ (paratypes of *E. proba*, ZFMK); 1♂, 1♀, same loc., 25.vii.1934, 11.viii.1935, H. Höne, Vojnits slides nos. 13038♂, 12527♀ (wrongly labelled gen. slides of ♂ and ♀ of *E. granata*, ZFMK); 3♂, same loc., 17.vi., 25.ix., 3.x.1935, H. Höne, Vojnits slides nos. 11554♂, 11555♀ (wrongly labelled gen. slide of *E. granata*), 13116♂ (all in TTM); 2♂, same loc., 6.ix.1934, H. Höne, Vojnits slides nos. 11553♂ (wrongly labelled gen. slide of *E. egena*), 11562♀ (wrongly labelled gen. slide of *E. lobbichlerata*, all in TTM); 1♀, same loc., 31.viii.1934, H. Höne, Mironov slide no. 827♀ (TTM); ♂, same loc., *ca.* 2000 m, 17.ix.1934, H. Höne, Vojnits slide no. 13069♂ (holotype of *E. proba*, ZFMK).

197. *Eupithecia particeps* Vojnits, 1988

Eupithecia particeps Vojnits, 1988, *Acta. zool. hung.* **34** (1): 41, figs 7–9; pl. 3, fig. G. Holotype ♂ (NSMTO; examined), Nepal (east): Solukhumbu Sagarmatha, Ringmo, 2,780 m.
Eupithecia orba Mironov & Galsworthy, 2004, *Trans. lepid. Soc. Japan* **55** (4): 287, figs 3, 15. Holotype ♀ (ZFMK; examined), China: Tibet [Sichuan], Batang, Untere Urwaldzone [Lower virgin forest zone], 3800 m. **New junior synonym.**

Misidentification: *Eupithecia lobbichlerata* Inoue, 2000 (male, *nec* Schütze, 1961).

Description. A small species; fore wing brownish grey with a reddish tinge; transverse lines, especially ante- and postmedial, right angled onto costa, postmedial bordered externally by a pale fascia; medial area slightly darker, covered with blackish scales; terminal area reddish with faint pale subterminal line; discal dot prominent, black. Hind wing off-white, transverse lines well marked along anal margin; terminal area with reddish tinge; discal dot small.
Male genitalia. Uncus thickened at base, biapical. Valve unmodified. Vinculum short and rounded. Papillae long and narrow. Aedeagus slim,

shorter than length of valve. Vesica armed with two slim, parallel cornuti, a small plate-like cornutus and a small irregular cornutus near ductus ejaculatorius base. Sternite A8 elongate and narrow, with two parallel apical arms.
Female genitalia. Bursa spherical, covered with small spines, a tongue extending into the ductus bursae. Latter broad, with a membranous diverticulum on one side. Ductus seminalis narrow, attached centrally to corpus bursae. Colliculum short and broad. Antrum weakly sclerotized, with numerous pores. Tergite A8 almost square. Apophyses very short and narrow, the anterior with short basal spurs. Papillae anales short and narrow.
Range. China (Tibet, Sichuan); Pakistan (Himalayas), Nepal.
Phenology. Bivoltine (?): late July and early October.
Altitude. Found from about 1750 m to 3800 m in the Himalayas (Pakistan and Nepal: coll. ZFMK, coll. C. Herbulot in ZSM, and coll. U. Ratzel).
Similar species. Similar to preceding species, but with larger discal dots. The male genitalia are similar to those of *E. coniurata*, but the uncus is stouter, and both the cornuti in the vesica and sternite A8 much longer. The female genitalia are similar to those of *E. ebriosa*, but can be distinguished by the shorter apophyses and the large membranous diverticulum.
Material. ♂, E Nepal, Solukhumbu Sagarmatha, Ringmo, 2780 m, 9.x.1979, M. Owada, Vojnits slide no. 14689♂ (holotype, NSMTO); 1♀, China, Tibet [Sichuan], Batang, Untere Urwaldzone [Lower virgin forest zone], *ca.* 3800 m, 17.vi.1937, H. Höne, Vojnits slide no. 13263♀ (holotype of *E. orba*, ZFMK).

198. *Eupithecia sempiterna* Vojnits, 1984

Eupithecia sempiterna Vojnits, 1984, *Acta zool. Acad. Sci. hung.* **30** (1–2): 230, fig. 22. Holotype ♂ (TTM; examined), China: North Yunnan, Li-kiang.

Description. Fore wing ground colour greyish brown; transverse lines inconspicuous, oblique, right angled onto costa; postmedial crenellate, forming a dark shade near costa; terminal area darker near apex with wavy white subterminal line, broadened at middle and near tornus; discal dot small, ovoid-acuminate. Hind wing paler, brownish grey; transverse lines well marked along anal margin; terminal area distinctly bordered, slightly darker, with fine dentate, whitish subterminal line and small white tornal spot; discal dot small, circular.
Male genitalia. Uncus elongate, biapical. Valve unmodified. Vinculum broad with shallow medial hollow. Papillae long and narrow. Aedeagus stout, slightly longer than length of valve. Vesica armed with two large

parallel cornuti, the shorter with some heavily sclerotized, transverse wrinkles near base, and a smaller irregular cornutus near ductus ejaculatorius base. Sternite A8 broadened basally, with two parallel apical arms.

Female unknown.

Range. China (Yunnan).

Phenology. Univoltine (?): late August to late September.

Similar species. A rather plain facies, a little similar to *E. adoranda*, but with more elongate fore wings and more angulate hind wings. Best separated on genitalia. The cornuti in the male vesica and the long papillae are distinctive.

Material. ♂, China, Provinz Nord-Yunnan, Li-kiang [Lijiang], 27.viii.1934, H. Höne, Vojnits slide no. 13338♂ (holotype, TTM); 1♂, same loc., 21.ix.1935, H. Höne, Vojnits slide no. 13252♂ (paratype, gen. slide not found in ZFMK); 2♂, same loc., 28.viii.1934, 21.ix.1935, H. Höne, Vojnits slides nos. 13036♂, 13066♂ (ZFMK); 3♂, same loc., 27.ix.1934, 12, 28.viii.1935, H. Höne, Vojnits slides nos. 11473♂, 13175♂, 13212♂ (TTM).

Remarks. This species should be distinguished from Vojnits' later use (1987) of the same name for a species from Nepal, now *E. asempiterna* Inoue.

199. *Eupithecia serpentigena* Mironov & Galsworthy, 2006

Eupithecia serpentigena Mironov & Galsworthy, 2006, *Trans. lepid. Soc. Japan* 57 (4): 342, figs 5, 24. Holotype ♀ (IZCAS; examined), China: Sichuan, Pinglu, Wanglang, 1100 m.

Description. Fore wing ground colour pale grey, darker at apex and marginal area; two conspicuous dark patches on costa; inner and medial transverse lines absent, postmedial present as a narrow black crenulate line, sharply angled at level of discal dot and running inwards to costa; a broad grey postmedial fascia outside postmedial; discal dot black, large and square. Hind wing clear white, with grey-brown shading along anal margin; discal dot small, round.

Female genitalia. Bursa globular, covered with spines except near ductus seminalis base. Ductus bursae broad, elongate, slightly curved, with sclerotized band along middle, a prominent diverticulum on right and two small patches of small spines. Ductus seminalis long and narrow, sharply broadened at base, attached to central part of corpus bursae from left side. Colliculum not present. Antrum broad, membranous. Tergite A8 narrow and elongate. Apophyses short and narrow. Papillae anales short and rounded.

Male unknown.

Range. China (Sichuan).

Phenology. Known only from the holotype specimen, which was collected in early August.
Altitude. Found at 1100 m in the Wanglang Nature Reserve north of Chengdu (China, Sichuan: coll. IZCAS).
Similar species. Externally, the combination of a strong discal dot, a strongly crenulate postmedial line, a broad grey postmedial fascia, and a generally grey cast to the fore wing are distinctive. The coloration and pattern is similar to *E. amicula*, but the postmedial is a different shape, and the discal dot stronger. The female genitalia are very distinctive.
Material. ♀, China, Sichuan, Pingwu, Wanglang, 1100 m, 2.viii.1999, leg. Zhou Kan, slide no. ZL655♀ (holotype, IZCAS).

200. *Eupithecia raniata* Prout, 1958

Eupithecia raniata Prout, 1958, *Bull. Br. Mus. nat. Hist.* (Ent.) **6** (12): 391. Holotype ♂ (BMNH; examined), India: Darjeeling.
Eupithecia bini Vojnits, 1981, *Acta zool. Acad. Sci. hung.* **27** (1–2): 231, fig. 15. Holotype ♂ (ZSM; examined), Nepal: Prov. Nr. 3, East Junbesi, 2750 m. (synonymised in Mironov & Galsworthy, 2010a).
Eupithecia darjeelica Inoue, 2000, *Tinea* **16** (Suppl. 1): 32, pl. 165, fig. 16; fig. 1293. Holotype ♂ (BMNH; examined), [NE India]: Darjeeling, 2000 m. (synonymised in Mironov & Galsworthy, 2009b).

Description. A small species with almost unicolorous rusty brown fore wings and lighter rusty brown hind wings. Fore wing transverse lines indistinct, except postmedial, the latter with pale inner border, and right angled onto costa; terminal area with broken whitish subterminal line forming a small, white tornal spot; discal dot relatively very large, intensely black. Hind wing paler, but dark brown along anal margin; transverse lines indistinct; terminal area rusty brown with faint whitish subterminal line and small white tornal spot; discal dot small but clear.
Male genitalia. Uncus subulate, biapical. Valve broad, unmodified. Vinculum very short, with shallow medial hollow. Papillae narrow. Aedeagus slim, about equal to length of valve. Vesica with a single short striate cornutus and a small irregular cornutus near ductus ejaculatorius base. Sternite A8 broadly bilobed at base with two parallel apical rods.
Female genitalia. Bursa globular, densely covered with small spines except small area in posterior part. Ductus bursae elongate and sinuate, broadened centrally into a membranous diverticulum, with an elongate patch of small spines. Ductus seminalis broadened at base, attached to posterior part of corpus bursae on left. Colliculum short and narrow. Antrum elongate and narrow, with numerous pores. Tergite A8 and apophyses elongate and narrow. Papillae anales narrow, rounded at apices.

Range. China (Yunnan, Sichuan); Nepal, Myanmar.
Phenology. Univoltine (?): mid-July to early October.
Altitude. From 1750 m in Nepal to 2850 m in Daxue Shan Mts. (China: coll. C. Herbulot in ZSM; coll. ZFMK).
Similar species. Similar to *E. sacrimontis* and *E. vasta*, but distinguished from both by the larger, ovoid discal dots on the fore wings. The hollow in the vinculum and the reduced armature of the vesica are the best distinguishing marks in the male genitalia. The female are very similar to those of *E. ebriosa*, but the longer apophyses and elongate eighth tergite are helpful.
Material. ♂, Type, Darjeeling, Aug. 1904, *Eupithecia raniata* Prout, ♂ type, Rothschild Bequest B.M. 1939–1, Vojnits slide no. 19357♂ [= BM Geom slide no. 16421♂] (holotype, BMNH); 1♀, [China, Sichuan] Bet. Pin-E-Pu & Den shiang vin, DC Graham, Galsworthy slide no. USNM 50 (AMNH); 1♂, China, Sichuan, Daxue Shan Mts., Gongga Shan, NW Moxi, 29°41′ N, 101°58′ E, 2850 m, 14–19.vii.1999, leg. V. Siniaev & A. Plutenko, Mironov slide no. 636♂ (ZFMK); 1♂, Yunnan, Lijiang, Gan He Ba, 3200 m, 23.vi.2009, A.C. Galsworthy (BMNH).

The *atrisignis* species group

This small group was introduced by Galsworthy & Mironov (2005) with seven South Asian species. Species of the group are characterized by an enlarged discal dot and, generally, a contrasting purplish brown ground colour with grey transverse lines and sometimes larger areas of grey. The male genitalia have a distally narrowed valve, no distinct cornuti on the vesica, but instead a number of ill-defined plates and an amorphous mass of small spines. The eighth sternite is bifid. In the female genitalia the antrum is long and membranous, leading to a stout colliculum and a large oval bursa with various patterns of spining. The colliculum is offset to one side of the apex of the bursa, with the ductus seminalis leaving the bursa close to the colliculum.

With the exception of *E. nuceistrigata* and *E. yazakii* in Taiwan, the group as so far known is confined to the slopes of the Himalayas and adjacent mountain systems on the southern and eastern sides. Six species occur in China.

201. *Eupithecia atrisignis* Butler, 1889

Eupithecia atrisignis Butler, 1889, *Illust. typical Specimens Lepid. Heterocera Colln Br. Mus.* 7: 23: 114, pl. 137, fig. 10. Holotype ♂ (BMNH; examined), [India]: [Kangra district], Dharmsala.

Eupithecia circumacta Prout, 1958, *Bull. Br. Mus. nat. Hist. (Ent.)* **6** (12): 391. Holotype ♀ (BMNH; examined), India: Darjeeling (synonymised in Galsworthy & Mironov, 2005).

Eupithecia forsteri Vojnits, 1983, *Acta zool. Acad. Sci. hung.* **29** (1–3): 266, figs 7, 9. Holotype ♂ (ZSM; examined), Nepal: Dudh Kosi Valley, 3000 m (synonymised in Galsworthy & Mironov, 2005).

Eupithecia profana Vojnits, 1984, *Acta zool. Acad. Sci. hung.* **30** (1–2): 220, figs 10, 13. Holotype ♀ (TTM; examined), China: North Yunnan, Li-kiang [Lijiang] (synonymised in Galsworthy & Mironov, 2005).

Misidentifications: *Eupithecia silenata* Butler, 1889 (*nec* Assmann, 1848); *Eupithecia violacea* Inoue, 2000 (*nec* Vojnits, 1981) (identity: *Eupithecia szelenyica* Vojnits & de Laever, 1974).

Description. Fore wing ground colour purplish brown; discal dot large, black, rounded; transverse lines grey, antemedial curved onto costa, medial curved around outside discal dot, and postmedial twice angled; terminal area broad with white wavy line. Hind wing paler, with indistinct transverse lines and small, rounded discal dot.

Male genitalia. Uncus relatively broad, biapical. Valve narrowed in outer half to narrowly rounded apex. Vinculum rounded. Papillae long and narrow. Aedeagus narrow, considerably shorter than valve length. Vesica with numerous longitudinal striations and minute teeth, and armed with two lightly sclerotized amorphous cornuti near base. Sternite A8 two slightly diverging narrow pointed rods, joined at base by a spreading bilobed laterally pointed band.

Female genitalia. Bursa ovate-oblong, all but left side covered with small spines. Ductus bursae not distinguishable. Ductus seminalis attached near colliculum on left side. Colliculum short and narrow. Antrum elongate and narrow. Tergite A8 trapezoid. Apophyses narrow Papillae anales small, rounded apically.

Range. China (Tibet, Yunnan); N.W. and N.E. India, Nepal, Sikkim, Bhutan, N.E. Myanmar.

Phenology. Bi- or polyvoltine: early June to late September.

Altitude. From about 1800 to 3000 m in the Himalayas (Sikkim and Nepal: coll. ZFMK; ZSM).

Similar species. Darker than other members of the group, with less contrast with paler areas of wing which hardly extend beyond transverse lines. Shares a doubly angled postmedial with *E. tricrossa*, which is absent in other species of the group. Distinguished from most other *Eupithecia* in the region by a combination of the large dark discal dot, and the generally dark coloration. Closest to *E. robiginascens*, from which distinguished by the presence of clear grey transverse lines (buff and faint in *E. robiginascens*). The male genitalia have a broader apex to the valve and base to the eighth sternite than other species of the group, and the female genitalia a longer and more heavily spined bursa.

Material. ♂, [India], Dharmsala 87–59, *Eupithecia atrisignis* Butler type, BM Geom. slide no. 20314 (holotype, BMNH); ♀, [India], Darjieling Aug 1904, *Eupithecia circumacta* Prout ♀ type, BM Geom. slide no. 20150 (holotype of *E. circumacta*, BMNH); 2♀, [India], Darjeeling, August 1904, BM Geom. slide no. 20144 (paratypes, BMNH); 1♀, [India], Sikkim, H. J. Elwes (paratype, BMNH); 1♀, [India], Sikkim, Tendong, 8000 ft., 1.8.[18]86, H. J. Elwes (paratype, BMNH); 1♂, [India], Sikkim, 22.ix.[18]89, J. G. Pilcher, BM Geom. slide no. 21766 (paratype, BMNH); 1♂, Bhutan, J. G. Pilcher (paratype, BMNH); ♂, Nepal, Dudh Kosi Tal, 3000 m, Vojnits slide no. 11796♂ (holotype of *E. forsteri*, ZSM); ♀, China, Provinz Nord-Yuennan, Li-kiang [Lijiang], 11.vi.1934, H. Höne, Vojnits slide no. 13769♀ (holotype of *E. profana*, TTM); 1♀, same loc., 11.vii.1934, H. Höne, Vojnits slide no. 13767♀ (ZFMK); 6♂, 8♀, Yunnan, Tengchong, Danzha forestry farm, 2500 m, 2–4.vi.1992, coll Xue Dayong, slides nos. ZL79, 501♂, ZL80, 502♀; 2♀, Yulong mountain, Lijiang, 27–28.vi.1962, coll. Song Shimei, slide no. ZL95♀; 1♂, Tibet, Mêdog, Hanmi, 2095 m, 29°21′ N, 95°07′ E, 11.viii.2006, Lang Songyun, slide no. ZL695♂ (IZCAS).

202. *Eupithecia pyricoetes* Prout, 1958

Eupithecia pyricoetes Prout, 1958, *Bull. Br. Mus. nat. Hist.* (Ent.) **6** (12): 392. Holotype ♂ (BMNH; examined), [India]: Sikkim.

Description. Fore wing ground colour purplish brown; antemedial and postmedial broad, gently curving grey bands, medial a narrow wavy grey line passing through discal spot; latter very large and oval; median area strongly marked with red along veins; terminal area with apical, medial and tornal dark blotches. Hind wing ash-grey, without distinct transverse lines, but darkened on anal and terminal margins and with small, conspicuous blackish discal dot.

Male genitalia. Uncus biapical. Valve with very pointed apex. Vinculum short. Papillae narrow, slightly curved. Aedeagus stout, slightly shorter than valve length. Vesica with two longitudinal patches of spinules, and two indistinct cornuti, one longer and curved. Sternite A8 peg-like, broadened and basally bilobed with two short apical arms.

Female genitalia. Bursa ovate, with spines concentrated in two lateral longitudinal patches, and a narrow transverse band of single spines running from posterior part of bursa round circumference to meet longitudinal patches. Ductus seminalis shortly broadened at base, attached to

corpus bursae near colliculum on right. Colliculum short and narrow, slightly inclined. Antrum elongate, broadening. Tergite A8 small, almost quadrate with membranous medial hollow in posterior margin. Apophyses short and narrow. Papillae anales broad and rounded.
Range. China (Shanxi); Nepal, India (Sikkim, West Bengal).
Phenology. Univoltine (?): late June to late August.
Altitude. Found from about 1700 to 2650 m (China; coll. TTM; Sikkim: Galsworthy & Mironov 2005).
Similar species. *E. pyricoetes*, *E. nuceistrigata*, and *E.suspiciosa* have elongate discal dots on the fore wings, not present in other species of this group. This species is readily distinguished from other members of the group, except for *E. suspiciosa* (qv.), and from all other *Eupithecia* by the presence of wide unmarked grey areas on the fore wing, consisting of a broad postmedial band, extended in two broad areas to the outer margin, and a broad slanting, slightly zigzag antemedial band. *E. suspiciosa* lacks the antemedial band. The male genitalia are best distinguished by the shape of the eighth sternite, and the female by the shape of the spining in the bursa.
Material. ♂, [India], Sikkim, H. J. Elwes, Collectio H. J. Elwes, *Eupithecia pyricoetes* Prout, ♂ type, Rothschild Bequest B.M. 1939–1, Vojnits slide no. 19331♂ [= BM Geom. genit. slide no. 20145] (holotype, BMNH); ♀, [India], Sikkim, O. Möller, 7000ft., [18]89, Collectio H. J. Elwes, BM Geom. genit. slide no. 20146 (paratype, BMNH); 1 ex., Sued-Shensi [Shanxi], Tapaishan [Taibaishan] im Tsinling [Qinling], *ca.* 1700 m, 9.viii.1936, H. Höne, Vojnits slide no. 11486 (gen. slide absent in TTM).
Remarks. The genitalia preparation of the Shanxi specimen is missing. Externally it appears to belong to this species, but, given the distance from its location to the normal range of the species, its identification as such should be seen as tentative until the genitalia are found.

203. *Eupithecia suspiciosa* Vojnits, 1983

Eupithecia suspiciosa Vojnits, 1983, *Acta zool. hung.* **29** (1–3): 267, fig. 10l Holotype ♀ (ZSM; examined), Nepal: Bhandar, Thodung, 2200 m.

Description. Externally hard to distinguish from the preceding species, but seems to lack a grey antemedial band.
Female genitalia. Bursa copulatrix oval, tapered to colliculum, the basal two thirds covered with large spines, and with a separate patch of larger spines in posterior part of corpus and two patches of small spines near base of colliculum. Ductus seminalis not present on Vojnits' slide.

Colliculum short and narrow. Antrum elongate, membranous. Tergite A8 relatively small, almost quadrate. Apophyses rather short, narrow. Papillae anales small, rounded.

Male unknown.

Range. China (Yunnan), Nepal.

Phenology. Both known specimens collected in August.

Altitude. Both known specimens from 2000–2200 m above sea-level.

Similar species. Can only be reliably distinguished from *E. pyricoetes* by dissection. In the present species, the spining of the bursa is quite different, covering only the basal half of the bursa.

Material. ♀, Nepal, Prov. Nr. 2, East Bhandar unter Thodung, 2200 m, 2. viii.1964, leg. W. Dierl, Holotypus *Eupithecia suspiciosa* Vojnits, det. A. Vojnits, gen. prep. No. 11062♀, ZSM Genitalprp. No. 10616 (holotype, ZSM); 1♀, China, Yunnan, Tengchong Dahaoping, 2020 m, 98,43N 24,58E, 5–7.viii.2007, Xue Dayong, slide no. ZL687♀ (IZCAS).

Remarks. The two known specimens are externally so similar to the distinctive *E. pyricoetes* that we have considered whether this might represent the true female of the species. The only female specimen in the BMNH series of *E. pyricoetes* was dissected in the 1980s by Dr. Vojnits, and we have found many of his genitalia slides to be wrongly labelled. However, the female genitalia on the slide appear to belong to a female in the present species group, and are quite different to those of the present species, so that we would scarcely be justified in concluding that they are the wrong genitalia. The genitalia in slide no. ZL687 are damaged, but enough remains to make it clear that they belong to the present rather than the preceding species. Further material will be needed to clarify the situation.

204. *Eupithecia nuceistrigata* Bastelberger, 1911

Eupithecia nuceistrigata Bastelberger, 1911, *Int. ent. Z.* (Guben) 4 (45): 241. Syntype(s) (DEI), Formosa [Taiwan]: [Arizan].

Description. Fore wing grey, blackish along costa; transverse lines dark brown, double medial sinuate and curved onto costa behind discal spot, postmedial evenly curved; reddish on veins around elongate oval discal dot; a broad ash-grey band outside postmedial, extended subapically and medially to termen; terminal area with apical, medial and tornal dark blotches, and an ash-grey subterminal line. Hind wing paler; basal, medial and postmedial distinct on anal margin; discal dot distinct but small.

Male genitalia. Uncus very elongate, biapical. Valve narrow, with almost pointed apex. Vinculum rounded. Papillae elongate and narrow. Aedeagus stout, shorter than valve length. Vesica with numerous longitudinal striations and minute teeth, forming a dense patch at anterior end of aedeagus, and armed with a small U-shaped cornutus, connected at its base with an irregular cornutus near ductus ejaculatorius base. Sternite A8 bilobed at base with lateral extremities pointed, and with two long narrow rods.

Female genitalia. Bursa oval, membranous, covered with large spines on dorsal, right and ventral sides; posterior end with a short transverse chain of spines dorsally and some small spines near base of colliculum. Ductus seminalis sharply broadened at base, attached to posterior part of bursa close to colliculum. Colliculum short and narrow. Antrum narrow and elongate, with numerous pores. Tergite A8 almost quadrate. Apophyses short and narrow. Papillae anales short and broad, rounded.

Range. China (Taiwan); N Vietnam.

Phenology. Polyvoltine (?): from February to mid-November.

Altitude. From about 1200 to 2400 m above sea-level (Taiwan: Inoue 1988; coll. ZSM).

Similar species. Very similar externally to *E. pyricoetes*, but with smaller discal dot. The male genitalia are similar to those of *E. pyricoetes*, but the shape of the eighth sternite is different. The pattern of spining in the female genitalia is diagnostic. Distinct from any other species in Taiwan.

Material. 11♂♀ exx., Arizan, Taiwan, BM slide no. 6329, which carries a note by D. M. Fletcher to indicate that it had been compared with the type genitalia, and BM slides nos. 21712♀, 21713♂ (BMNH); 1♀, Arizan, Formosa, 25 September 06 (A. E. Wileman), Rothschild Bequest B.M. 1939–1, Vojnits slide no. 19406♀ (BMNH, now in TTM); 1♂, [Taiwan], Arizan, Formosa, 7300ft., 14.ix.1906, A. E. Wileman, Wileman Coll. B.M. 1929–261, 828F, *Eupithecia nuceistrigata* Bstlbgr., Vojnits slide no. 19404♂ (BMNH, now in TTM); 1♀, Central Formosa, Arrizan, vis-a-vis Mt. Morrisson, Kagé district, 8000ft, Juin Juillet 1908, ex. Moltrecht (ZFMK); 1♀, Taiwan, Prov. Tai-Tung, Hsiangyang, Police station, 2320 m, 19–20.x.1996, leg. Gy. Fábián & F. Nemes (TTM); 2♂, Taiwan, Prov. Hualien, Taroko N.P., on the road 8, 1400 m, 2.iv.1997, leg. Csorba & Ronkay (TTM); 1♀, Taiwan, Prov. Taitung, Hsiangyang, 2200 m, 10.iv.1997, leg. Csorba & Ronkay (TTM); 1♂, Taiwan, Prov. Chiayi, 3 km S of Alishan, 1950 m, 11.iv.1997, leg. Csorba & Ronkay (TTM); 1♀, Taiwan, Prov. Hualien, Taroko N.P., Kuanyuan, 2380 m, 23.v.1997, leg. Gy. M. László & G. László (ZSM); 1♂, 9♀, Taiwan, Nantou, Meifeng, *ca.* 2300 m, 29.x.-3.xi.1998, leg. Mey & Speidel (MNHU); 3♂, 6♀, Taiwan, Prov. Nantou, Hohuachi, 1950 m, 24°13′ N, 121°16′ E, 31.iii.2000, leg. A. Kun & L. Peregovits (ZSM); 1♂, Taiwan, Nantou county, Kao-Leng Dyi,

18 km W Wushe, 24°4.605′ N, 121°7.583′ E, 2100 m, 18–19.iv.2002, leg. D. Anstine, Gy. Fábián & O. Merkl (TTM); 2♀, Central Taiwan, Nantou Co., Meifeng, *ca.* 2100 m NN, 24°05′55″ N / 121°10′48″ E, LF, 04, 08. September 2002, leg. M.-Y. Chen & U. Buchsbaum (ZSM, ZISP); 2♀, Central Taiwan, near Puli, Meifeng, *ca.* 2200 m NN, 24°05′25″ N / 121°10′21″ E, LF, 05, 10. Nov. 2003, leg. M.-Y. Chen (ZSM); 1♂, same loc. and alt., 07. Mai 2004, leg. U. Buchsbaum (ZISP); 4♂, 2♀, same loc. and alt., 14, 20. Mai, 13, 15, 20. October 2006, leg. Mei-Yu Chen & U. Buchsbaum (ZSM).

205. *Eupithecia tricrossa* Prout, 1926

Eupithecia tricrossa Prout, 1926, *J. Bombay nat. Hist. Soc.* **31** (2): 316. Holotype ♂ (BMNH; examined), Burma (upper): Hpimaw Fort.
Eupithecia multa Vojnits, 1981, *Acta zool. Acad. Sci. hung.* 27 (1–2): 234, figs 19, 20. Holotype ♂ (ZSM; examined), Nepal: Prov. Nr. 2, East Bhandar under Thodung, 2200 m (synonymised in Galsworthy & Mironov, 2005).

Misidentifications: *Eupithecia albibaltea* Inoue, 2000 (*nec* Prout, 1958); *Eupithecia circumacta* Inoue, 2000 (*nec* Prout, 1958) (identity: *Eupithecia atrisignis* Butler, 1889).

Description. A small species with pale grey medial areas and large black discal dots on fore wings. Fore wing with brownish grey basal area to discal dot; medial pale grey, curved outside discal dot; postmedial sinuate, angled onto costa, shaded internally with grey; terminal area dark brown with pale medial blotch and narrow apical streak; pale subterminal line fine. Hind wing ground colour brownish white; basal, medial and postmedial distinct; terminal area narrowly darker, with narrow indented subterminal line; discal dot small.
Male genitalia. Uncus elongate, biapical. Valve rather narrow, especially at apex. Vinculum short, rounded. Papillae short and narrow. Aedeagus shorter than valve, very narrow, with two apical lobes covered with dense spicules. Vesica granulate, armed with one small cornutus. Sternite A8 small, with laterally pointed basal lobes, and two narrow apical rods.
Female genitalia. Bursa oval, covered with small spines along one side. Ductus seminalis narrow, attached to bursa near short narrow colliculum. Antrum elongate, slightly sclerotized and covered with numerous pores. Tergite A8 almost square. Apophyses narrow. Papillae anales short, relatively broad and rounded.
Range. China (Tibet, Yunnan); Nepal, Sikkim, Bhutan, N.E. India, Northern Myanmar.
Phenology. Bivoltine (?): mid-June to late August.
Altitude. From about 1800 to 3800 m in the Himalayas (Sikkim and Nepal: Galsworthy & Mironov 2005).

Similar species. Smaller than *E. atrisignis*, and distinguishable from that species and from other *Eupithecia* except the Himalayan *E. albibaltea* Prout, 1958 by the broader and more prominent pale transverse lines in the medial and postmedial areas of the fore wing, contrasting strongly with the dark ground colour. *E. albibaltea* has a pale subapical streak running from the outer margin to the discal spot. The male genitalia are very similar to those of *E. albibaltea*, but distinguished from it by the much narrower aedeagus. The female genitalia are rather similar to those of *E. atrisignis*, but bursa copulatrix and colliculum usually smaller, and all apophyses shorter.

Material. ♂, Upper Burma, Hpimaw Fort, Nr. Myitkyina, 8000ft., 9–13. viii.1923, Capt. A.E. Swann, *Eupithecia tricrossa* Prout ♂ type, Brit. Mus. 1923–488, BM Geom. slide no. 20147 (holotype, BMNH); 1♀, same loc., BM Geom. slide no. 20748 (paratype, BMNH); ♂, Nepal, East Bhandar below Thodung, 2200 m, 2.viii.1964, coll. W. Dierl, ZSM slide no. G 10622, together with paratype series as listed by Vojnits (holotype of *E. multa*, ZSM); 1♀, China, Provinz Nord-Yuennan, Li-kiang [Lijiang], X, 10.viii.1935, H. Höne, Mironov slide no. 821♀ (TTM); 1♂, China, Tibet, Gyirong, Xiao Gyirong, 2030 m, 23.vii.1971, Huang Fusheng, slide no. ZL406♂ (IZCAS).

206. *Eupithecia yazakii* Inoue, 1988

Eupithecia yazakii Inoue, 1988, *Bull. Fac. domest. Sci. Otsuma Wom. Univ.* **24**: 343, figs 13a-e. Holotype ♂ (BMNH; examined), [Taiwan]: Chiai Hsien, Alishan, 2200 m.

Description. Fore wing ground colour reddish brown; medial and postmedial faint and very fine, angled onto costa behind discal dot, and enclosing a blackish brown area, very dark at costa; terminal area blackish brown, with pale wavy subterminal line and medial and tornal white spots; discal dot large, rounded, intensely black. Hind wing paler than fore wing, greyish; transverse lines inconspicuous; anal and terminal areas usually darker, latter with small white tornal spot; discal dot relatively large but pale.

Male genitalia. Uncus stout, biapical. Valve unmodified. Vinculum short and narrow, rounded. Papillae long and narrow. Aedeagus slim, shorter than valve length. Vesica armed with a medial elongate cornutus and a smaller, irregular one at ductus ejaculatorius base, both lightly sclerotized, and with a patch of short obliquely longitudinal wrinkles nearby. Sternite A8 roundly bilobed basally, with two parallel apical arms.

Female unknown.

Range. China (Taiwan).

Phenology. Bivoltine (?): mid-May to late September.
Altitude. Found at about 2100–2300 m above sea-level (Taiwan: Inoue 1988; coll. ZSM).
Similar species. Externally similar to *E. conterminata* because of its small size, and large fore wing discal dots, but with rounder wings, and white tornal spots on all wings. The rounder valve and different armature of the vesica distinguish this species from any other with similarly large discal dots.
Material. ♂, Taiwan, Mt. Alishan, 2200 m, Chiayi Hsien, 2–5.viii.1984, Coll. K. Yazaki, BM Geom. slide no. 16545 (holotype, coll. H. Inoue in BMNH); 1♂, Taiwan, Nantou, Meifeng, *ca.* 2300 m, 29.x.-3.xi.1998, leg. Mey & Speidel (MNHU); 1♂, Taiwan, Prov. Nantou, 2250 m, Meifong, 24°05′ N, 121°10′ E, 25.ix.1999, leg. G. Csorba & B. Herczig (ZSM); 2♂, Central Taiwan, near Puli, Tsuifeng, *ca.* 2150 m NN, 24°04′17″ / 121°12′12″ E, LF, 11. May 2001, leg. U. Buchsbaum (ZSM; ZISP); 5♂, Central Taiwan, Nantou Co., Meifeng, *ca.* 2100 m NN, 24°05′55″ N / 121°10′48″ E, LF, 04, 10. September 2002, leg. M.-Y. Chen & U. Buchsbaum (ZSM, 1♂ in ZISP); 5♂, same loc. and alt., 24°05′00″ N / 121°10′45″ E, LF, 15, 20. Sept. 2002, leg. M.-Y. Chen & U. Buchsbaum (ZSM); 2♂, Taiwan, N-Nantou Co., Road No. 14, NE Puli, Reyen Shi-Reg., Meifeng, *ca.* 24°06′ N / 121°10′ E, 2200 m, 9.-11.ix.2002, leg. W. Schacht et al. (ZISP).
Remarks. We have seen 17 specimens of this species in different collections, but all are males. This species is placed very tentatively in this species group. The female genitalia, when known, may help to determine its affinities.

The *suboxydata* species group

This group was described as a separate genus *Catarina* by Vojnits & de Laever, 1973, but Viidalepp & Mironov (1988b) synonymised this name with *Eupithecia*, showing how the features regarded as distinct by those authors were probably derived from the normal *Eupithecia* pattern.

The group is best defined by the eighth sternite in the male genitalia, which takes the form of two separate latero-caudal sclerites joined by membranous bands to the central part. The sacculus of the male valve is heavily sclerotized. The female genitalia are also distinctive, having a spined globular bursa with a heavily wrinkled ductus, and strong sclerotization to the antrum and the eighth sternite. The following three species belong to the group, as does a further Himalayan species *E. incurvaria* Hampson, 1903 (= *propoxydata* Schütze, 1961) from Kashmir and Nepal.

207. *Eupithecia suboxydata* Staudinger, 1897

Eup[ithecia] suboxydata Staudinger, 1897, *Dt. ent. Z., Iris* **10** (1): 112, pl. 3, fig. 74. Syntypes, at least 3♂, 6♀ (coll. O. Staudinger in MNHU; examined), [Russia]: Vlad(ivostok), Ask(old), Suif(un), Sutschan.

Misidentifications: *Larentia minutata* Eversmann, 1844 (*nec* Denis & Schiffermüller, 1775) (identity: *Eupithecia absinthiata*); *Eupithecia hannemanni* Vojnits & de Laever, 1973 (1♀ paratype in ZFMK).
Unavailable names (incorrect subsequent spelling): ***suboxydota***: Djakonov (1927); (infrasubspecific): *minorata*: Dietze (1913).

Description. Fore wing ground colour fuscous greyish brown; antemedial triple, obtusely angled near costa; medial double, passing through discal dot; postmedial almost straight, obtusely angled near costa; medial and terminal areas separated by a wide pale double stripe; terminal area broad, whitish wavy line forming an ovate tornal spot, edged inwardly by darker scales; discal spot fairly large, round. Hind wing hardly paler than fore wing; traces of transverse lines along anal margin; terminal area usually slightly darker without wavy line; discal spot small.
Male genitalia. Uncus elongate, biapical. Valve with sacculus expanded, abruptly narrowing at apex. Vinculum broad, with large medial hollow. Papillae short. Aedeagus large, almost equal to valve length. Vesica armed with a pointed dentate apical cornutus with transverse wrinkles at its base, and a large, elongate cornutus centrally. Central part of sternite A8 like the figure of a man, with two very large, ovate-oblong latero-caudal sclerites.
Female genitalia. Bursa globular, covered with small spines. Ductus bursae short, with longitudinal striations and spineless. Ductus seminalis stout, with a short row of spinules on inner side near base, attached to ductus bursae near short and narrow colliculum. Antrum membranous with sclerotized poriferous swelling. Tergite A8 square with convex anterior margin. Sternite A8 very large, heavily sclerotized and poriferous. Apophyses narrow and unusually short, the anterior especially so, and with narrow spurs. Papillae anales short and broad, rounded.
Range. China (Xinjiang, Shaanxi, Shanxi, Gansu, Qinghai, Sichuan, Beijing); Russia (south-east of European part, S Urals, W and S Siberia: Tjumen'oblast', Altai Mts., Dahuria, Transbaikalia, Priamurje, Primorje), Kazakhstan, Kirghizstan (Tien-Shan Mts.), Mongolia, Korea, Japan (Hokkaido).
Phenology. Probably bivoltine: late March to early August. Larvae in August and September. Overwintering as pupa.
Biology. Larva found on flowers of *Artemisia santolinifolia* (Asteraceae) (Kirghizstan: Mironov, original data).

Altitude. From sea level to about 2200 m in Tien-Shan Mts. (Kirghizstan: Mironov, original data).

Similar species. Similar in pattern to *E. subbrunneata* but lacks reddish tinge. Also to *E. formosa* (qv.), and to *E. hannemanni*, which has very different genitalia. In the male genitalia, valve shape is the best distinction from other members of the group. The female genitalia can be distinguished from *E. subbrunneata* by the absence of spines in the ductus bursae, and from both this and *E. formosa* by the short row of spines in the ductus seminalis.

Material. 1♀, [Russia], Wladiwstk Chr., 1/7, *Eupith. Vulgata* Hw., *Suboxydata* Stgr., Origin. (syntype, coll. O. Staudinger in MNHU); 1♂, [Russia], *Suboxydata* ♂ Askold, Origin., Gen. vérifiés 1451[1] *Eupithecia suboxydata* ♂ Askold, dét. E. de Laever 1954 (syntype, coll. O. Staudinger in MNHU); 1♀, [Russia], Suifun 89 D., 158., *suboxydata* ♀, Origin., Gen. vérifiés 1451[2] *Eupithecia suboxydata* ♀, dét. E. de Laever 1954 (syntype, coll. O. Staudinger in MNHU); 1♂, [Russia ?], Origin., *Eupithecia subfulvata* ab. *oxydata* Tr. (syntype, coll. O. Staudinger in MNHU); 1♀, [Russia], Sutschan 90 Dörr., Origin., 189., nicht Aberrat. v. *Mandschurica* aber. *Suboxyd.* ab. *Bruneata* (Boh.) (syntype, coll. O. Staudinger in MNHU); 1♀, [Russia ?], *Eupithecia pimpinellata* Hb., 20/6, Origin. (syntype, coll. O. Staudinger in MNHU); 1♀, [Russia], Askold Dörr. 83, *Tephr. suboxydata* Stgr., Origin., *Catarina suboxydata* Stgr. det. Vojnits A., Vojnits slide no. 10045♀ (syntype, coll. O. Staudinger in MNHU); 1♂, 1♀, [Russia ?], Origin. (syntypes, coll. O. Staudinger in MNHU); 1♂, [Xinjiang], Asia centr., Turkestan, Ili-Gebiet, Coll. Wagner, ? *suboxydata* form. *minorata* Typ., Juldus, Ili, *Catarina suboxydata* Stgr., det. Vojnits A., Vojnits slide no. 10015♂ (syntype of *E. suboxydata minorata*, coll. K. Dietze in MNHU); 1♀, same loc., *minorata* Form Typ., Juldus, Ili, *Catarina suboxydata* Stgr., det. Vojnits A., Vojnits slide no. 10028♀ (syntype of *E. suboxydata minorata*, coll. K. Dietze in MNHU); 2 exx., Peking [Beijing], Westberge, Exped. Stötzner (TTM); 2♂, West-China, Szetschwan [Sichuan], Exped. Stötzner (ZFMK); 2♀, same loc., Vojnits slide no. 1151♀ (TTM); 2♂, Szetschwan [Sichuan], Kwansien [Guanxian], Exped. Stötzner, Vojnits slide no. 1158♂ (TTM); 3♂, 1♀, China, Sued-Shensi [Shaanxi], Tapaishan [Taibaishan] im Tsinling [Qinling], 20, 21, 22, 26. vi.1935, H. Höne, de Laever slides nos. 39♂, 40♀, Vojnits slide no. 10891♂ (ZFMK); 2♂, same loc., 20.vi.1935, H. Höne, Vojnits slides nos. 10001♂, 10023♂ (TTM); 1♀, same loc., *ca.* 1700 m, 14.viii.1936, H. Höne (paratype of *E. hannemanni*, ZFMK); 3♀, same loc. and alt., 13, 20, 22.v.1936, H. Höne, de Laever slide no. 285♀, Vojnits slide no. 12936♀ (ZFMK); 1♂, 1♀, same loc. and alt., 12, 15.vi.1936, H. Höne, Vojnits slide no. 1154♀ (TTM); 2♂, same loc., *ca.* 3000 m, 16.vi.1936, H. Höne, de Laever slide no. 184♂ (ZFMK); 1♂, same loc. and alt., 15.vi.1936, H. Höne, Mironov slide no. 824♂ (TTM); 2♂, 5♀, Prov. Shansi [Shanxi], Mien-shan, Mittlere Höhe *ca.* 1500 m, 3, 7, 10, 12. vi.1937, H. Höne, Vojnits slides nos. 1149♂, 1157♀, 1168♀, 10008♀, 10009♀, 10010♂ (TTM); 2♂, 2♀, same loc., Obere Höhe *ca.* 2000 m, 1, 29.vii. 1viii.1937, H. Höne, de Laever slides nos. 19♀, 44♀, Vojnits slide no. 11513♂ (ZFMK); 1♂, 1♀, same loc. and alt., 1, 5.vii.1937, H. Höne, Vojnits slide no. 10011♂ (TTM); 1♀ Qinghai, Tongren, Maixiu, 2950 m, 30.vii.1992, coll Xue Dayong; 1♂, 3♀, Gansu, Dangchang, 1800 m, 7.vii.1998, coll. Liu Dajun and Yuan Decheng; 1♀, Gansu, Wenxian, Qiujiaba, 2350 m, 21.vii.1999, coll Zhu Chaodong, slide no. ZL564♀; 1♀, Wenxian, Bikou, Zhongmiao, 24.vi.1998, coll Chen Jun; 1♀, Wenxian, Liujiaping, 1300 m, 27.vi.1998, coll Yao Jian; 1♀, Gansu, Kangxian, Baiyunshan, 1250–1750 m, 12.vii.1998, coll Yao Jian; 1♀ Gansu,Yongdeng Tulugou, 2280 m, 25.vii.1991, coll Xue Dayong; 3♂, 3♀, Shanxi, Wutaishan, 1992, slides no. ZL398♀, 1♀, Shanxi, Pangquangou, 26. vi.1988, slide no. ZL289♀; 1♀ Beijing, Badaling, 700 m, 29.vi.1962, coll Wang Chunguang, slide no. ZL118♀; 1♀, Beijing, Baihuashan, 7.vii.1973, slide no. 282♀; 1♂, same loc, 29.v.1973, coll Han Yinheng, slide no. ZL299♂; 2♀, Beijing, Mentougou, Xiaolongmen, 1100 m, 23.vi.2001, Xue Dayong, slide no ZL706♀ (IZCAS).

Remarks. Not previously recorded from China.

208. *Eupithecia formosa* Vojnits & de Laever, 1973

Catarina formosa Vojnits & de Laever, 1973, *Acta zool. Acad. Sci. hung.* **19** (3–4): 430, figs 1B, 5B, 9C. Holotype ♂ (ZFMK; examined), [China]: North Yunnan, A-tun-tse [Dêqên], 3000 m.

Catarina formosa nova Vojnits & de Laever, 1973, *Acta zool. Acad. Sci. hung.* **19** (3–4): 432. Holotype ♂ (ZFMK; examined), [China]: Tibet [Sichuan], Batang, Yangtse Valley, 2800 m.

Description. Fore wing ground colour greyish brown; transverse lines brown, basal and postbasal right angled, and antemedial very sharply angled onto costa; postmedial double, pale, slightly sinuate, but almost straight near costa; medial area slightly darker, with a long transverse, blackish brown line between CuA_2 and A; terminal area with two medial and tornal blackish blotches, and two blackish dashes between M_1 and M_3; whitish subterminal line inconspicuous; discal dot relatively large, ovoid or circular. Hind wing much paler, transverse lines well marked along anal margin; terminal area darker near tornus, with indistinct whitish subterminal line; discal dot conspicuous.

Male genitalia. Uncus short, biapical. Valve with sacculus only slightly broadened. Vinculum broad, with shallow medial hollow. Papillae short and broad. Aedeagus stout, almost equal to length of valve. Vesica with numerous oblique wrinkles, armed with two cornuti, one slim and horn-like, the other large and plate-like. Sternite A8 a cup-shaped central part with two diverging basal lobes and two basally concave latero-caudal sclerites.

Female genitalia. Bursa globular, covered with small spines. Ductus bursae short, with numerous longitudinal striations and without spines. Ductus seminalis broadened at base, attached to ductus bursae near relatively short colliculum. Antrum membranous with sclerotized poriferous swelling near colliculum. Tergite A8 broad, with convex and sclerotized anterior margin. Sternite A8 transverse, heavily sclerotized and poriferous, with broad and deep medial hollow in posterior margin. Apophyses short and narrow, the anterior with narrow spurs. Papillae anales short and rounded.

Range. China (Tibet, Sichuan, Yunnan).

Phenology. Univoltine: mid-May to mid-June.

Altitude. From about 2800 to 3800 m above sea-level in Tibet (China: Vojnits & de Laever 1973).

Similar species. Externally similar to *E. suboxydata* and *E. hannemanni*, but with more sharply angled antemedial transverse lines and larger discal dots on the fore wings. *E. hannemanni* has a more sharply angled postmedial. The male genitalia resemble those of *E. subbrunneata*, but the apical cornutus on the vesica is smaller, and the eighth sternite has a slightly different shape. The female genitalia are similar to those of other members of the group, but lack spines in the ductus bursae or ductus seminalis.

Material. ♂, Nord-Yünnan, A-tun-tse [Dêqên], Talsohle *ca.* 3000 m, 19.v.1937, H. Höne, Vojnits slide no. 10019♂ [Genital präp. Nr. ZFMK 331♂] (holotype, ZFMK); 4♂, 3♀, same loc. and alt., 17, 24, 26, 31.v., 1, 3, 6.vi.1937, H. Höne, de Laever slides nos. 45♂, 46♀, 113♂ (paratypes, ZFMK); 1♂, 6♀, same loc. and alt., 24, 25, 30.v., 3.vi.1937, H. Höne, Vojnits slides nos. 1152♀, 1153♀, 1164♀, 1165♀, 1169♀, 10018♂, 10020♀ (paratypes, TTM); 1♂, 1♀, same loc. and alt., 18, 30.v.1937, H. Höne, Vojnits slide no. 14109♂, Mironov slide no. 754♀ (ZFMK); ♂, Tibet [Sichuan], Batang, Im Tal des Yangtze (*ca.* 2800 m), 12.v.1936, H. Höne, de Laever slide no. 298♂ (holotype of *C. formosa nova*, ZFMK); 1♀, same loc., Untere Urwaldzone (*ca.* 3800 m), 11.vi.1936, H. Höne, de Laever slide no. 89♀ (paratype of *E. formosa nova*, TTM); 1♂, Thibet [Sichuan], Tà-tsien-loû, Mai, Juin 1896, Chasseurs Thibétains, Vojnits slide no. 868♂ (TTM); 1♂, 1♀, China, North Yunnan, Diqing Tibetan Aut. Pref., 5 km SE of Deqin, 3356 m, 16–17.vi.2009, leg. Balázs Benedek (coll. O. Pekarsky); 1♂, 2♀, same locality and altitude, 19.vi.2009, leg. B. Benedek, Pekarsky slide no. OP 0484♂ (coll. O. Pekarsky); 1♂, 1♀, Tibet, Nyingchi, Bayi, 3040 m, 2–3.vii.2002, Han Hongxiang, slide no. ZL648♀ (IZCAS).

209. *Eupithecia subbrunneata* Dietze, 1906

Eup[ithecia] ab. *subbrunneata* Dietze, 1904, *Dt. ent. Z., Iris* [1903] **16** (2): 371. Syntypes 15, at least 2♂, 6♀ (coll. O. Staudinger in MNHU; examined), [China]: Korla; Aksu; Nord Tibet.

[Eupithecia] (ab.) *subbrunneata* ab. *amita* Dietze, 1904, *Dt. ent. Z., Iris* [1903] **16** (2): 372, pl. 3, fig. 24 (non binominal). Syntypes 10, at least 4♂, 2♀ (coll. K. Dietze and R. Püngeler in MNHU; examined), [China]: North Tibet.

[Eupithecia] subbrunneata f. *patruelis* Dietze, 1906, *Dt. ent. Z., Iris* **19** (2): pl. 2, fig. 22. Syntypes 2♂ (coll. K. Dietze in MNHU; examined), [China]: Korla.

Eup[ithecia] suboxydata f. *amita pallida* Dietze, 1910, *Biologie der Eupithecien* **1**: pl. 73, fig. 391; pl. 85, fig. 979; *ibidem* (1913), **2**: 109. Holotype (coll. K. Dietze in MNHU, not traced), As(ien) [China]: Aksu.

[Eupithecia] subbrunneata f. *cinnamomeata* Dietze, 1910, *Biologie der Eupithecien* **1**: pl. 73, figs 392, 393 (as *Eup. suboxydata* f. *subbrunneata*), pl. 85, figs 980, 981; *ibidem* (1913), **2**: 109. Syntypes 12♂, 1♀ (coll. K. Dietze in MNHU; examined), [China]: Juldus, Ili-Gebiet, Urumtschi (Ost-Turkestan).

Catarina carissima Vojnits & de Laever, 1973, *Acta zool. Acad. Sci. hung.* **19** (3–4): 434, figs 2A-B, 3A, 6A-B, 7A, 10A. Holotype ♀ (coll. O. Staudinger in MNHU; examined), [Russia]: Suifun (synonymised in Inoue, 1998).

Catarina dolosa Vojnits, 1977, *Acta zool. Acad. Sci. hung.* **23** (3–4): 467, fig. 4. Holotype ♂ (TTM; examined), Mongolia: Central aimak, SE of Somon Bajanzogt, 1600 m (synonymised in Mironov, 2003).

Catarina necessaria Vojnits, 1977, *Acta zool. Acad. Sci. hung.* **23** (3–4): 469, fig. 6. Holotype ♀ (TTM; examined), Mongolia: Uburchangaj aimak, Arc Bogdo ul., 20 km S of Somon Chovd, 1760 m (synonymised in Mironov, 2003).

Misidentification: *Eupithecia hannemanni* Vojnits & de Laever, 1973 (1♀ paratype in ZFMK).

Description. Fore wing ground colour strongly rusty brown; transverse lines doubled, oblique, curving or obscurely angled onto costa; postmedial in part shaded paler; terminal area darker brown at centre and near tornus with whitish subterminal line and a small white tornal spot; discal dot linear, black. Hind wing darker brownish grey; transverse lines obscurely marked along anal margin; terminal area without light subterminal line; discal dot small, inconspicuous.

Male genitalia. Uncus stout, biapical. Valve with sacculus unmodified, but rather squarely angled at base. Vinculum broad with deep medial hollow. Papillae short and narrow. Aedeagus stout, shorter than valve length. Vesica with a stout horn-like cornutus with wrinkled base, and two twisted plates, one central and one basal. Sternite A8 a lozenge-shaped central part with broadly diverging basal lobes and two small, heavily sclerotized latero-caudal sclerites.

Female genitalia. Bursa globular, completely covered with dense small spines. Ductus bursae short, with longitudinal striations and a patch of sparse spines near base of colliculum. Ductus seminalis broad, attached to ductus bursae near colliculum on right. Colliculum short and wide. Antrum lightly sclerotized. Tergite A8 almost square with a small membranous medial hollow on posterior margin. Sternite A8 broad, heavily sclerotized and poriferous. Apophyses narrow and short, especially the broadly spurred anterior ones; papillae anales broad and rounded.
Range. China (Xinjiang, Tibet, Qinghai, Sichuan, Shanxi, Shaanxi); Asian Russia (Altai Mts., Tuva, Dahuria, Transbaikalia, Priamurje, Primorje), Kazakhstan, Mongolia, Korea.
Phenology. Univoltine (?): mid-May to early August.
Altitude. From sea level up to about 1700 m in Tapaishan (China: Vojnits & de Laever 1973) and Zailijsky Alatau mts. (coll. ZMH) and 3300 m in Tibet (coll. IZCAS).
Similar species. Usually instantly recognisable by the rusty brown coloration, but also with longer and more pointed fore wings than other species in the group. In the male genitalia the straight ventral margin of the valve is distinctive. The female can be distinguished by the patch of spines in the ductus bursae and the lack of sclerotization of the antrum.
Material. 1♂, [Xinjiang], Korla, Sai-chin, ? *subbrunneata* ab. *patruelis* Type, Korla, K. Dietze, Vojnits slide no. 10025♂ (syntype of *E. subbrunneata patruelis*; paratype of *Catarina carissima*, coll. K. Dietze in MNHU); 1♂, [Xinjiang], Korla, Sai-chin, *subbrunneata* ab. *patruelis*, Original, Korla, K. Dietze, Palpen Rürger als bei *vulgata* aus Korla, Fransen gescheckt (syntype of *E. subbrunneata patruelis*, coll. K. Dietze in MNHU); 1♀, [Russia], Wladiwstk Chr., 9/6, ab. *Subbrunneata* Stgr., Origin. (syntype of *E. suboxydata* ab. *subbrunneata*, coll. O. Staudinger in MNHU); 1♀, [Russia], 9/6, *Eupithecia subfulvata* Hw., Origin. (syntype of *E. suboxydata* ab. *subbrunneata*, coll. O. Staudinger in MNHU); 1♂, [Russia], Sutschan 90 Dörr. (syntype of *E. suboxydata* ab. *subbrunneata*, coll. O. Staudinger in MNHU); 1♂, [Russia], Askold Dörr., 74., Origin. (syntype of *E. suboxydata* ab. *subbrunneata*, coll. O. Staudinger in MNHU); 4♀, [Russia ?], small violet circular label, Origin. (syntypes of *E. suboxydata* ab. *subbrunneata*, coll. O. Staudinger in MNHU); 1♂, [Xinjiang], Korla 97. Hbhr., 18/6, 538., n. sp. zehr nahe *Suboxydata* Stgr. (Boh.) (coll. O. Staudinger in MNHU); 1♀, [Xinjiang], Korla 97. Hbhr., var. bei *suboxydata*, K. Dietze (coll. O. Staudinger in MNHU); 3♂, 2♀, [Qinghai], Tibet (Kuku-Noor), Rückbeil 98, *suboxydata* var. *amita* Dietze, Original, v. R. Tancré 2.00 (syntypes of *E. suboxydata* f. *amita*, coll. R. Püngeler in MNHU); 1♀, [Qinghai], Tibet

(Kuku-Noor), Rückbeil 98, *suboxydata* ♀, v. R. Tancré 2.00 (coll. R. Püngeler in MNHU); 1♂, [Qinghai], Tibet (Kuku-Noor) 1898, v. Tancré 2/1900, *subbrunneata* ab. *amita* Typ., Nord-Tibet., Vojnits slide no. 10014♂ (syntype of *E. subbrunneata amita*; paratype of *Catarina carissima*, coll. K. Dietze in MNHU); 2♂, [Qinghai], Tibet (Kuku-Noor) 1898, R. Püngeler, ab. *amita* Dietze (coll. K. Dietze in MNHU); 1♀, *subbrunneata* ab. *amita*, Nord-Tibet, K. Dietze (coll. K. Dietze in MNHU); 1♂, 4♀, ab. *amita*, Nord-Tibet (coll. K. Dietze in MNHU); 1♂, [Xinjiang], ? *subbrunneata* Stgr. forma: *amita* ab. *minor* Typ. ♂, Aksu (coll. K. Dietze in MNHU); 1♂, [Xinjiang], forma: *amita* ab. *minor*, Aksu (coll. K. Dietze in MNHU); 1♂, forma: *amita* ab. *minor*, Nord-Tibet (coll. K. Dietze in MNHU); 1♂, [Xinjiang], *cinnamomeata* (bei *subbruneata*) Typ., Juldus, Ili, Vojnits slide no. 10013♂ (syntype of *E. subbrunneata cinnamomeata*; paratype of *Catarina carissima*, coll. K. Dietze in MNHU); 1♂, [Xinjiang], Juldus, *cinnamomeata* Typ., Juldus, Ili (syntype of *E. subbrunneata cinnamomeata*, coll. K. Dietze in MNHU); 2♂, [Xinjiang], Asia centr., Ost-Turkestan, Urumtschi, Rückbeil 1904, *subbrunneata* ♂, R. Püngeler, *amita* form: *cinnamomeata* (syntypes of *E. subbrunneata cinnamomeata*, coll. K. Dietze in MNHU); 7♂, [Xinjiang], Juldus, Ili (syntypes of *E. subbrunneata cinnamomeata*, coll. K. Dietze in MNHU); 1♂, [Xinjiang], *cinnamomeata*-Form, Juldus, Ili (coll. R. Püngeler in MNHU); 1♂, [Xinjiang], Asia centr., Juldus-Gebiet 1911, *cinnamomeata* Dietze ♂, 5/1912 v. K. Rolla (coll. R. Püngeler in MNHU); 2♂, [Xinjiang], Asia centr., Juldus, ab. *cinnamomeata* Dietze 4/1912, v. B.-Haas Püng. (coll. R. Püngeler in MNHU); 4♂, [Xinjiang], Asia centr., Ost-Turkestan (Urumtschi) Rückbeil 1904, v. R. Tancré 1905, bei *amita* und *cinnamomeata* Dietze, i. l. 2.1913 (coll. R. Püngeler in MNHU); ♀, [Russia], Suifun 85 D., *Tephr. subbrunneata* Stgr., Origin., Vojnits slide no. 10044♀ (syntype; holotype of *Catarina carissima*, coll. O. Staudinger in MNHU); 1♂, West China, Szetschwan [Sichuan], Exped. Stötzner, Vojnits slide no. 1150♂ (paratype of *Catarina carissima*, ZFMK); 1♂, same data (paratype of *Catarina carissima*, TTM); 1♂, same data, de Laever slide no. 407♂ (ZFMK); 1♀, same data, Coll. Dr. Wehrli, Basel, Vojnits slide no. 10021♀ (TTM); ♂, Mongolia, Central aimak, SO von Somon Bajanzogt, 1600 m, Exp. Dr. Z. Kaszab, 1966, Nr. 521, 11.vi.1966, Vojnits slide no. 10539♂ (holotype of *Catarina dolosa*, TTM); ♀, Mongolia, Uburchangaj aimak, Arc Bogdo ul., *cca* 20 km S von Somon Chovd, 1760 m, Exp. Dr. Z. Kaszab, 1964, No. 168, 21.vi.1964, Vojnits slide no. 10518♀ (holotype of *Catarina necessaria*, TTM); 1♀, Sued-Shensi [Shaanxi], Tapaishan [Taibaishan] im Tsinling [Qinling], 3.vi.1935, H. Höne (paratype of *E. hannemanni*, ZFMK); 1♂, same loc., 30.v.1935, H. Höne, Vojnits slide no. 10511♂ (ZFMK); 1♀, same loc., *ca.* 1700 m, 18.v.1936, H. Höne, Vojnits slide no. 1163♀ (paratype of *Catarina carissima*, ZFMK); 2♂, 1♀, same loc. and alt., 19, 20, 22.v.1936, H. Höne, Vojnits slides nos. 10510♀ [slide with this number belongs to the paratype specimen of *E. breviculata georgi*(*ca*) Vojnits, 1977 in ZFMK (!)], 11509♂, de Laever slide no. 310♂ (ZFMK); 1 ex., same loc. and alt., 17.v.1936, H. Höne (TTM); 4♂, 2♀, Prov. Shansi [Shanxi], Mien-shan, Mittlere Höhe *ca.* 1500 m, 4, 5, 10.vi.1937, H. Höne, Vojnits slides nos. 13112♂, 11526♂ (ZFMK); 1♀, same loc. and alt., 10.vi.1937, H. Höne, Vojnits slide no. 10012♀ (TTM); 1♂, 1♀, Ta-tsien-Loû, 1899, Chasseurs indigènes, Vojnits slides nos. 13438♂, 13439♀ (TTM); 1♀, [Xinjiang], Kuldja, Mai, ♀, Vojnits slide no. 13520♀ (TTM); 1♂, [Xinjiang], Juldus, Schmidt slide no. Eu 197 (TTM); 1♂, China, Lin si hien, Chizgan mont., Schmidt slide no. Eu 198a (TTM); 1♂, Tibet, Qamdo, 3300 m, 25.v.1976, coll Han Yinheng, slide no. ZL128♂; 1♂, Qinghai, Tongren, Maixiu, 2950 m, 30.vii.1992, coll Xue Dayong, slide no. ZL372♂; 1♂ 5♀, Shanxi Wutaishan, 1992, slides nos. ZL399♂, 243, 249, 291, 397, 531♀ (IZCAS).

The *innotata* species group

This species group was introduced by McDunnough (1949) with three North American species.

The male valve is unmodified, often with pointed apex, and an angle on the ventral edge. The vesica usually has two characterisitic small apical cornuti. Sternite A8 is usually a narrow peg-like shape. The females have a large pyriform bursa, heavily spined at base and striate at the posterior end; the antrum is broadened and sclerotized. The genitalia are generally very uniform in this group, and determination by external pattern is normally most effective.

The group is very rich in Central Asia, including the two western Chinese provinces of Xinjiang and Tibet, where at least the nine following species occur. Three other species, *E. tshimganica* Viidalepp, 1988 from the Tien-Shan and Pamir ranges (Uzbekistan, Kirghizstan and Tajikistan), *E. myoma* Vojnits, 1988 from Kirghizstan, and *E. efferata* Mironov & Ratzel, 2008 from the Western Himalayas (Pakistan and Jammu & Kashmir) might also occur in China.

210. *Eupithecia innotata* Hufnagel, 1767

Phalaena innotata Hufnagel, 1767, *Berlin. Mag.* 4 (6): 616. Syntype(s) lost, [Germany]: Berlin region.

Phalaena innotata Thunberg, 1788, *D. D. Mus nat. Acad. Upsal.* 6: 74. Lectotype ♀ (DEUU), [Europe]. [Junior primary homonym and junior synonym of *Phalaena innotata* Hufnagel, 1767].

Larentia tamarisciata Freyer, 1836, *Neuere Beitr. Schmettkde* **2** (32): 153, pl. 192, figs 1a-e. Syntype(s), Germany: Wolfratshausen (synonymised in Herrich-Schäffer, 1848).
Eupithecia fraxinata Crewe, 1863, *The Entomologist's Annual* **1863**: 116. Syntypes (BMNH ?), United Kingdom: England (treated as var. of *E. innotata* in Staudinger & Rebel, 1901).
Eupithecia innotata f. *meridionalis* Mabille, 1868, *Annls Soc. ent. France* (4) **9**: 73. Syntype(s), [France]: Corsica.
Eupithecia suspectata Dietze, 1871, *Stett. Ent. Z.* **32** (4–6): 210. Syntypes, at least 2♀ (coll. K. Dietze in MNHU; examined), France: Paris; Germany: Warthe, Wiesbaden, Frankfurt a.M. (synonymised in Staudinger & Rebel, 1901).
[*Eupithecia*] *innotata* f. *perturbatrix* Dietze, 1906, *Dt. ent. Z., Iris* **19** (2): pl. 2, fig. 20 (non binominal); *ibidem* (1908) **21** (2): 168 (*Eup*[*ithecia*]). Lectotype ♂ (design. by Mironov, coll. K. Dietze in MNHU; examined), [China]: Korla.
E[*upithecia*] *innotata* var. *grisescens* W. Petersen, 1910, *Dt. ent. Z., Iris* [1909] **22** (4): 271. Syntype(s) (IAET; examined), [Estland].
Eupithecia petersenaria Wnukowsky, 1929, *Zool. Anz.* **83** (9/10): 223. [Replacement name for *Eupithecia innotata grisescens* W. Petersen, 1910] (synonymised in Prout, 1939).
E[*upithecia*] *nanata kozhantschikovi* Wehrli, 1929, *Ezheg. gosud. Muz. N.M. Mart'yanova* **6** (1): 19 (non binominal). Lectotype ♂, published here to stabilize nomenclature (ZFMK; examined), [Russia]: environs of Minussinsk. **New junior synonym.**
Eupithecia alexandriana Wardikjan, 1972, *Dokl. Acad. Nauk Arm. SSR* **55** (4): 253, figs 1, 2. Holotype ♂ (IZAR; examined; the slide of holotype included genitalia of 3 females, 5 armature of male genitalia, 3 aedeaguses and 1 sternum A8), Armenia: Razdan region, Tzakhkadzor (synonymised in Mironov, 1991).
Eupithecia sergiana Wardikian, 1972, *Dokl. Acad. Nauk Arm. SSR* **55** (4): 255, figs 3, 4. Holotype ♀ (IZAR; examined), Armenia: Erivan (synonymised in Mironov, 1991).
Eupithecia inexplicabilis Vojnits, 1982, *Annls hist.-nat. Mus. natn. hung.* **74**: 224, fig. 7. Holotype ♀ (coll. K. Dietze in MNHU; examined), [Kirghizstan]: Karagai-tau, Issyk-Kul. **New junior synonym.**
Eupithecia mystica Vojnits, 1988, *Annls hist.-nat. Mus. natn. hung.* **80**: 81, pl. 1, figs 6, 7; pl. 4, figs 29–32. Holotype ♂ (ZISP; examined), Tajikistan: Badakhshan, Horog, 2300 m. [Junior primary homonym of *Eupithecia gemellata* f. *mystica* Dietze, 1910]. **New junior synonym.**

Misidentification: *Eupithecia omnigera* Vojnits, 1988 (*nec* Vojnits, 1982) (at least 8♂ and 1♀ in ZISP).
Unavailable names (infrasubspecific): *rotundata*: Bastelberger (1908); *melanotica*: Warnecke (1962). – (incorrect subsequent spellings): ***innotaria***: Boisduval (1840); *tamarisciaria*: Boisduval (1840); *innonata*: Kostrowicki (1953).

Description. Fore wing narrow, elongate, ground colour grey to brownish in medial area, veins darker; transverse lines inconspicuous, except postmedial, which is pale, double, parallel to termen, and twice sharply angled near costa; whitish faintly wavy subterminal line fine, sometimes indistinct; discal dot narrow, obliquely elongate. Hind wing paler, flecked with brown scales along anal margin; discal dot minute.
Male genitalia. Uncus narrow, biapical. Valve with a small bulge, or blunt process in the middle of ventral margin. Vinculum short and broad. Papillae large, slightly curved. Aedeagus stout, slightly shorter than valve length. Vesica armed with two slim apical cornuti, one broad, folded and one smaller, curved plate at ductus ejaculatorius base and an elongate

patch of dense spinules, like a fox-brush. Sternite A8 peg-like with broad bilobed base.

Female genitalia. Bursa almost 2/3 covered with small spines in anterior part except short ductus bursae and near base of ductus seminalis. Ductus bursae tapered, striate. Ductus seminalis broadly attached to centre of bursa on dorsal side. Colliculum short and narrow. Antrum transverse, sclerotized. Tergite A8 trapezoid, broadened to posterior margin, with medially pointed anterior margin. Apophyses short and narrow, not expanded at apices. Papillae anales rounded, short and broad.

Range. China (Beijing, Hebei, Gansu, Xinjiang, Qinghai); Europe (except north), North Africa, Asia Minor, Georgia, Armenia, Azerbajan, Kazakhstan, Uzbekistan, Kirghizstan, Tajikistan, Pakistan, Russia (Caucasus, Urals, western and southern Siberia).

Phenology. Bivoltine: early April to early June; mid-July to late October.

Altitude. From sea level up to about 3400 m in Pamirs Mts. (Tajikistan: coll. ZISP).

Similar species. A widespread and common species, very variable in size, depth of coloration and distinctness of markings, especially in the various biotopes of Central Asia. Specimens of the second generation are usually considerably smaller than those of the first. Often very difficult to separate from other Central Asian species of this group. Points of difference from them are given under each species.

Material. ♂, [Xinjiang], Korla, Sai-chin, *innotata*-Gruppe, Forma: *perturbatrix*, Type, Korla, K. Dietze (von Tancré), *innota perturbatrix* Dtze, präp. Nr. A 824♂, det. E. Schütze, Kassel, Zool. Mus. Berlin, Vojnits slide no. 13556♀ (lectotype of *E. innotata* f. *perturbatrix* Dietze, 1910, pl. 75, fig. 489, design. by Mironov, coll. K. Dietze in MNHU); 1♀, [Kirghizstan], *innotata* form. *concoloris*, Karagai-tau, Issyk-Kul, 3, Zool. Mus. Berlin, Vojnits slide no. 13554♀ (holotype of *E. inexplicabilis*, coll. K. Dietze in MNHU); 1♀, [Xinjiang], Mus-tag-ata Yarkend, "*omniparens*", Zool. Mus. Berlin, Vojnits slide no. 13553♀ (coll. K. Dietze in MNHU); 1♂, [Qinghai], Tibet (Kuku-Noor) Rückbeil 1898, *innotata* ab. ♂ nach Bohatsch 3.01, 4/00 v. Tancré, Zwischen *innot.* var. u. *uliata* (Dietze 2.06), Zool. Mus. Berlin, *uliata* Stgr., *Eupithecia ochridata* Pk., det. A. Vojnits, Vojnits slide no. 13589♂ (coll. R. Püngeler in MNHU); ♂, [Russia], environs of Minusinsk, Tagarsk. Isl., 17.v.1924, Kozhanchikov (leg.), 3097♂ (lectotype of *E. nanata kozhantschikovi*, design. V. Mironov; ZFMK); ♂, Tadžikistan, Badakhshan, Horog, 2300 m, 30.v-4.vi.1978 L, Metsaviir, Ratviir & Viidalepp leg., Vojnits slide no. 14952♂ (holotype of *E. mystica* Vojnits, 1988, ZISP); 7♂, 5♀, Gansu, Wudu, 1000 m, 14.ix.1990, Xue Dayong, slide no. ZL124♂; 1♂, Gansu,

Yongdeng, Liancheng Linchang, Tulugou, 2600 m 36° 41′ N 102° 42E, 7–8.viii.2005, Han Hongxiang, slide no. ZL708♂; 2♂, 3♀, Shanxi, Wutaishan, 1992, slide no. ZL352♂; 1♀, Beijing, Yanqing, 29.viii.1964, ZL123♀; 1♂, 1♀, Beijing, Changping, 31.viii.1964; 1♂, Beijing, Mentougou, Xiaolongmen, 1100 m, 24.ix.1998, Han Hongxiang, slide no. ZL679♂; 3♂, 2♀, Hebei, Xiaowutaishan, 10.ix.1980; 1♀, Xinjiang, Baicheng, Talieke Kuotan, 2300 m, 8.vi.1978; 1♂, 1♀, Xinjiang, Yanerwo, 2.v.1985, slides nos.ZL453♂, 437♀; 1♂, Xinjiang, Akto Wuyitake, 2650 m, 6.viii.1989, coll Zhang Xuezhong, slide no.ZL470♂ (IZCAS).

Remarks. The taxonomy of this species, especially in its Asian range, remains very uncertain.

211. *Eupithecia parallelaria* Bohatsch, 1893

Eupithecia unedonata var. *parallelaria* Bohatsch, 1893, *Dt. ent. Z., Iris* **6** (1): 22 (non binominal). Lectotype ♀ (coll. R. Püngeler in MNHU; examined), no locality designated.

Eupithecia innotata contracta Vojnits, 1982, *Annls hist.-nat. Mus. natn. hung.* **74**: 228, figs 11–13. Lectotype ♀ (coll. K. Dietze in MNHU; examined), [Uzbekistan/Tajikistan]: Syr-Darja. **New junior synonym.**

Eupithecia emittens Inoue, 1996, *Trans. lepid. Soc. Japan* **47** (4): 241, figs 5, 6, 10, 13. Holotype ♂ (BMNH; examined), [Pakistan]: Baltistan, Route de Shigar, 17 km Nord de Skardu. **New junior synonym.**

Misidentification: *Eupithecia unedonata* f. *famelica* Dietze, 1910 (1♀ paralectotype in coll. K. Dietze in MNHU); *Eupithecia mystica* Vojnits, 1988 (nec *Eupithecia gemellata* f. *mystica* Dietze, 1910) (at least 3♀ paratypes from Turkmenistan in ZISP).

Description. Fore wing narrow, elongate, ground colour grey-brown, usually with ochreous tinge; transverse lines very distinct, oblique, dark grey-brown; basal sharply angled near costa; postbasal and antemedial triply angled; double medial sharply angled behind discal dot; postmedial doubly angled; terminal area with ochreous apical streak and whitish subterminal line; discal dot small, oblique; terminal line black, interrupted by veins. Hind wing off-white with slightly darker terminal area; anal margin with brown blotch basally and transverse lines marked to Cu vein; discal dot very small. Adults of second generation much smaller.

Male genitalia. Uncus narrow, biapical. Valve with an angle or small medial process on ventral margin. Vinculum narrow. Papillae broad, clavate. Aedeagus short and stout, slightly shorter than valve length. Vesica armed with two slim apical cornuti, a broad, plate-like cornutus and a dense patch of small spines like a fox-brush. Sternite A8 peg-like.

Female genitalia. Bursa 2/3 covered with small spines. Ductus bursae sclerotized and striate. Ductus seminalis evenly curved across bursa, broadly attached dorsally to its centre. Colliculum short and narrow. Antrum broad, sclerotized and bilobed. Tergite A8 broadened posteriorly, with medially convex anterior margin, and small point in centre of concave posterior margin. Apophyses short and narrow. Papillae anales short, rounded, and broad.

Range. China (Xinjiang); Iran, Turkmenistan, Uzbekistan, Kirghizstan, Tajikistan, Pakistan, Kazakhstan, Mongolia.

Phenology. Bivoltine: late March to mid-May; mid-September to late October.

Biology. Found on seeds of *Artemisia turanica* (Asteraceae) in Uzbekistan (original data of M. I. Falkovitch; ZISP).

Altitude. From sea level up to about 3300 m in the Himalayas (Pakistan: Mironov et al., 2008c).

Similar species. Similar in shape to *E. innotata*, but fore wings longer and narrower with more oblique terminal margins, more richly coloured, and all transverse lines more oblique and more sharply angled onto costa. The male genitalia have a slightly narrower valve and vinculum. The female genitalia are very similar to those of *E. innotata*, but with slight differences in the shape of the eighth tergite.

Material. ♀, Type ♀: Iris 1893, v. *Parallelaria* Bohatsch, Origin., Zool. Mus. Berlin, Vojnits slide no. 13574♀ (lectotype, design. by Vojnits, coll. R. Püngeler in MNHU); ♀, [Uzbekistan/Tajikistan], *innotata*-Form. *contracta*. Type Syr-Darja, K. Dietze, A-Ata, Zool. Mus. Berlin, *E. innotata* Hufn. det. Vojnits A., Vojnits slide no. 13533♀ (lectotype of *E. innotata contracta* Vojnits, coll. K. Dietze in MNHU); 1♀, [Xinjiang], "*famelica*", Aksu 1903 (paralectotype of *E. unedonata* f. *famelica* Dietze, 1910, pl. 75, fig. 544, design. by Mironov, coll. K. Dietze in MNHU); 1♀, [Xinjiang], "*famelica*", Aksu 1900 (coll. K. Dietze in MNHU); ♂, Pakistan, Baltistan, 17 kms Nord de Skardu, Route de Shigar, 18–10.1989, Alt, 2300 m, F. Aulombard et J. Plante, Type status verified K. Buckmaster 1997, Inoue Coll. B.B. 1992–71, Inoue slide no. 15822♂ [= BM Geom. genit. slide no. 19216] (holotype of *E. emittens*, BMNH); 1♂, [Xinjiang], Culdscha, Alph[éraky], Coll. Vel. Knjazja Nikolaja Mikhajlovicha (ZISP); 1♀, same data, 12.iv.[18]79 (ZISP); 2♀, [Xinjiang], Aksu 1900 (coll. K. Dietze in MNHU); 1♀, [Xinjiang], var. *parallelaria*, Mitte April, Ak-ssu (coll. K. Dietze in MNHU).

212. *Eupithecia corroborata* Dietze, 1908

[*Eupithecia*] *innotata* f. *corroborata* Dietze, 1908, *Dt. ent. Z., Iris* 21 (2): 170, pl. 3, figs 5, 6 (non binominal). Lectotype ♀ (design. by Vojnits, coll. K. Dietze in MNHU; examined), [China]: Kuldja [Yining (Gulja)].

Eupithecia fulgurata Vojnits, 1982, *Annls hist.-nat. Mus. natn. hung.* 74: 223, fig. 6. Holotype ♀ (coll. K. Dietze in MNHU; examined), [China]: [Xinjiang], Juldus [Yining] region, Ili. **New junior synonym.**

Eupithecia hysterica Vojnits, 1988, *Annls hist.-nat. Mus. natn. hung.* 80: 82, pl. 1, fig. 8; pl. 2, fig. 9; pl. 5, figs 33–36. Holotype ♂ (ZISP; examined), Tajikistan: Vantsh, Gutshevast, 2200 m. **New junior synonym.**

Misidentification: *Eupithecia recentissima* Vojnits, 1982 (2♀ paratypes in coll. K. Dietze in MNHU); *Eupithecia omnigera* Vojnits, 1988 (*nec* Vojnits, 1982) (at least 1♂ and 11♀ in ZISP).

Description. Fore wing narrow and pointed, ground colour typically dark brown, sometimes greyer; transverse lines oblique; basal and antemedial sinuous, sharply angled; medial less sharply angled beyond discal dot and touching it; postmedial twice angled onto costa; medial and terminal areas divided by a narrow, double pale band; terminal area with conspicuous pale apical streak and a wavy whitish subterminal line; discal dot narrow and slightly elongate. Hind wing pale grey with darker terminal area, transverse lines well marked along anal margin; discal dot small.

Male genitalia. Uncus elongate, biapical. Valve narrow, with almost pointed apex and angulate ventral margin. Vinculum short, semicircular. Papillae broad, slightly clavate. Aedeagus shorter than valve length. Vesica armed with two small sinuate apical cornuti, an elongate patch of spinules, and one large and one smaller plate near ductus ejaculatorius base. Sternite A8 peg-like.

Female genitalia. Bursa not clearly differentiated from ductus, almost completely covered with small spines except striate area near ductus seminalis base. Latter broadened at base, attached to centre of bursa. Colliculum short and narrow. Antrum broadening, lightly sclerotized. Tergite A8 identical to previous species. Apophyses narrow and short. Papillae anales rounded, short and broad.

Range. China (Xinjiang); Kirghizstan, Tajikistan.

Phenology. Bivoltine (?): late April to early August.

Altitude. From about 2200 to 3250 m above sea level in the Pamirs (Tajikistan: coll. ZISP).

Similar species. Darker than other members of the group, and quite well differentiated by the pale double fascia between medial and terminal areas. The male genitalia are not well differentiated from other species, but the angled ventral margin of the valve is helpful. In the female genitalia the spining comes closer to the colliculum, and the antrum is smaller than in the two preceding species.

Material. ♀, [Xinjiang], *corroborata* ♀, Type, *innotata* Formen Gruppe, Kuldja, 15.6.03, K. Dietze, Original (Djarkent), Zool. Mus. Berlin, Vojnits slide no. 13539♀ (lectotype of *E. innotata* f. *corroborata*, design. by Vojnits, coll. K. Dietze in MNHU); 1♀, [Kirghizstan], anf. Juli, 14., *corroborata* Cotype, *innotata* Formen-Gruppe, Alex.-Gebirge, K. Dietze, Original, Alexg., Juli, Zool. Mus. Berlin, Vojnits slide no. 13540♀ (paralectotype of *E. innotata* f. *corroborata*, design. by Vojnits, coll. K. Dietze in MNHU); 1♂, [Xinjiang], "*corroborata*", *Innotata*-Formen Gruppe, K. Dietze, cotype, Kuldja, *innotata* var., Kuldja, Zool. Mus. Berlin, Vojnits slide no. 13543♂ (paralectotype of *E. innotata* f. *corroborata*, design. by Vojnits, coll. K. Dietze in MNHU); 1♀, [Xinjiang], forma *corroborata*, Kuldja, Mitte Juni 1903 (coll. K. Dietze in MNHU); 1♀, [Xinjiang], Korla, v. *parallelaria* forma. *rotundata* ♀ Typ., Korla (see *Dietze, 1910, pl. 75, fig. 539*; coll. K. Dietze in MNHU); ♀, [Xinjiang], Forma *fulgurata*, *corroborata* – Form, Juldus, Ili, Zool. Mus. Berlin, Vojnits slide no. 13544♀ (holotype of *E. fulgurata*, coll. K. Dietze in MNHU); 1♀, [Xinjiang], Thian. or. 96, Hbhr., Zool. Mus. Berlin, Vojnits slide no. 13572♀ (paratype of *E. recentissima*, coll. K. Dietze in MNHU); 1♀, [Xinjiang], Ak-ssu, *perturbatra*, Zool. Mus. Berlin, *perturbata*, Vojnits slide no. 13538♀ (paratype of *E. recentissima*, coll. K. Dietze in MNHU); 1♀, [Xinjiang], v. *parallelaria*, (Aksu), Bohatsch gesehen, 79. (coll. K. Dietze in MNHU).

213. *Eupithecia praesignata* Bohatsch, 1893

[***Eupithecia***] ***praesignata*** Bohatsch, 1893, *Dt. ent. Z., Iris* **6** (1): 21 (non binominal). Lectotype ♀ (design. by Vojnits, coll. O. Staudinger in MNHU; examined), [Central Asia]: type locality not designated [Replacement name for *Eupithecia insignata* Bohatsch, 1893].

[*Eupithecia*] *insignata* Bohatsch, 1893, *Dt. ent. Z., Iris* **6** (1): 21 (non binominal). [Junior secondary homonym of *Phalaena Geometra insignata* Hübner, 1789].

Eupithecia innotata f. *omniparens* Dietze, 1908, *Dt. ent. Z., Iris* **21** (2): 171, pl. 3, figs 9, 10 (non binominal). Lectotype ♀ (coll. K. Dietze in MNHU; examined), Central Asia [Tajikistan]: Sarawschan [= Zerafshan], Sary-ob. **New junior synonym.**

[*Eupithecia*] ? *unedonata* f. *famelica* Dietze, 1908, *Dt. ent. Z., Iris* **21** (2): 171, pl. 3, figs 21, 22 (non binominal). Syntypes 2♀ (coll. K. Dietze in MNHU; examined), [China]: Aksu (synonymised to *E. costisignata* Dietze, 1904 in Prout, 1915). **New junior synonym.**

Eupithecia omnigera Vojnits, 1982, *Annls hist.-nat. Mus. natn. hung.* 74: 218, fig. 1. Holotype ♂ (coll. K. Dietze in MNHU; examined), [China]: [Sinkiang], Korla. **New junior synonym.**

Eupithecia recentissima Vojnits, 1982, *Annls hist.-nat. Mus. natn. hung.* 74: 219, fig. 2. Holotype ♀ (coll. K. Dietze in MNHU; examined), [China]: [Xinjiang], Juldus [Yining] region, Ili (synonymised to *E. omnigera* Vojnits, 1982 in Vojnits, 1988). **New junior synonym.**

Misidentification: *Eupithecia parallelaria* Bohatsch, 1893 (2♂ paralectotypes in coll. O. Staudinger in MNHU).

Description. Fore wing elongate, ground colour yellowish grey; transverse lines conspicuous, narrow, oblique, sharply angled onto costa, postmedial twice so; a pale transverse band between medial and terminal areas, and usually a substantial unmarked pale area across centre of wing below discal dot; terminal area with pale wavy subterminal line, very dentate near apex; discal dot small, slightly elongate. Hind wing pale grey; transverse lines visible along anal margin; terminal area slightly darker; discal dot small and pale.

Male genitalia. Uncus stout, biapical. Valve with a very small, blunt medial projection on ventral margin. Vinculum short, trapezoid. Papillae very broad, slightly curved. Aedeagus stout, shorter than length of valve. Vesica with a brush of spines in centre, and armed with two apical cornuti, one larger with a broadened base, a large, broad, folded, H-shaped plate-like cornutus, and a small, irregular cornutus near ductus ejaculatorius base. Sternite A8 peg-like.

Female genitalia. Bursa with basal half covered with small spines, with a protruding apex in centre; posterior half with longitudinal striations and strongly wrinkled area near colliculum. Ductus bursae not distinguishable. Ductus seminalis broadened at base, sharply curved backwards, attached to central part of bursa on right. Colliculum short and narrow. Antrum short and very broad, sclerotized. Apophyses relatively short, narrow. Tergite A8 quadrate. Papillae anales short, rather narrow and rounded.

Range. North western China (Xinjiang); Kirghizstan, Tajikistan, Jammu & Kashmir, India (Ladakh Mts.).

Phenology. Bivoltine (?): mid-April to early August.

Altitude. From about 2000 to 3000 m above sea level in Alai Mts. (Kirghizstan: coll. ZISP).

Similar species. Mainly distinguished from other species of this group by the paler pattern and paler areas below discal dot and between antemedial and medial transverse lines. The male genitalia are similar to those of *E. innotata*, but the papillae are broader, the two apical cornuti slightly larger, and the eighth sternite broader at apex. The female genitalia are very similar to other species in the group, but have a slightly smaller spined area in the bursa.

Material. ♀, Type ♀, Iris 93 *praesignata* Btoch, 55., Origin., Zool. Mus. Berlin, Vojnits slide no. 13571♀ (lectotype, design. by Vojnits, coll. O. Staudinger in MNHU); 1♂, [Kirghizstan], Jssyk-Kul, *innotata* form: *praesignata* Boh., Jssyk-Kul (In Coll. Staudgr. Typen unter sich verschieben), Zool. Mus. Berlin, Vojnits slide no. 13545♂ (paralectotype, coll. K. Dietze in MNHU); 1♂, Type ♂: Iris 1893, v. *Parallelaria* Bohatsch, Origin., Zool. Mus. Berlin (paralectotype of *E. parallelaria*, coll. O. Staudinger in MNHU); 1♂, Asia, Turan Stgr. 1884, Type ♂: Iris 1893, v. *Parallelaria* Bohatsch

(paralectotype of *E. parallelaria*, coll. K. Dietze in MNHU); ♀, [Tajikistan], Saraw., Sary-ob, 18/7.1900, Funke, *Eupithecia* ♀, *innotata*-Form *omniparens*, Type, Sary-ob, Sarawschan, Original, K. Dietze, Zool. Mus. Berlin, Vojnits slide no. 13550♀ (lectotype of *E. omniparens* Dietze/Vojnits, coll. K. Dietze in MNHU); 1♂, [Tajikistan], Saraw., Sary-ob, 18/7.1900, Funke, *Eupithecia* ♂, *innotata*-Form *omniparens*, Sarawschan, Original, K. Dietze (paralectotype of *E. innotata forma omniparens* Dietze, design. by Mironov, coll. K. Dietze in MNHU); ♂, [Xinjiang], *innotata* form. *omnigera*; übergang zu *parallelaria*, Korla, Zool. Mus. Berlin, Vojnits slide no. 13549♂ (holotype of *E. omnigera* Vojnits, 1982, coll. K. Dietze in MNHU); ♀, [Xinjiang], ? *costisignata* form. *famelica* Ak-su (lectotype of *E. unedonata* f. *famelica* Dietze, 1910, pl. 75, fig. 543, design. by Mironov, coll. K. Dietze in MNHU); 1♂, [Xinjiang], Aksu 1900 (coll. K. Dietze in MNHU); ♀, [Xinjiang], No. 486, Übergang, Juldus, Ili, *perturbatra*, Zool. Mus. Berlin, *perturbata*, Vojnits slide no. 13537♀ (holotype of *E. recentissima*, coll. K. Dietze in MNHU); 1♂, 1♀, [Xinjiang], v. *parallelaria*, Ak-ssu, Mitte April, 15. (coll. K. Dietze in MNHU); 1♂, [Xinjiang], Korla, Sai-chin, № 516 Korla (coll. K. Dietze in MNHU); 1♂, [Xinjiang], Korla 97. Hbh., *Unedonata* v. *Parallelaria* (coll. O. Staudinger in MNHU); 1♂, [Xinjiang], Korla 97. Hbhr., 16/6, *Unedonata* v. *Parallelaria* (Boh.) (coll. O. Staudinger in MNHU); 1♀, [Xinjiang], Korla 97. Hbhr., 20/6 (coll. O. Staudinger in MNHU); 1♂, 1♀, [Xinjiang], Mus-tag-ata Yarkend, "*unedonata* var. *parallelaria*" (coll. K. Dietze in MNHU); 1♀, Xinjiang, Zhaosu, Alasan, 2430 m, 29.vii.1978, coll. Zhang Xuezhong, slide no. ZL286♀; 1♀, Xinjiang, Tomürt Peak, 2400 m, 20.vi.1977, slide no. ZL496♀; 3♀ same loc, 2600 m, 21.vii.1977, slides nos. ZL321, 512♀; 1♀, Xinjiang, Akto, 2650 m, 6.viii.1989, coll Zhang Xuezhong; 1♀, Xinjiang, Zhaosu, Xiata, 2400 m, 28.vii.1978, slide no. ZL320♀ (IZCAS).
Remarks. *E. unedonata* f. *famelica* Dietze, 1908 was synonymised with *E. costisignata* by Prout (1915). However, both syntype females of this taxon belong to *E. praesignata*.

214. *Eupithecia paupera* Dietze, 1910, stat. n.

[*Eupithecia*] *innotata* f. *paupera* Dietze, 1910, *Biologie der Eupithecien* 1: pl. 85, fig. 955; *ibidem* (1913), 2: 127. Lectotype ♀ (here design., coll. K. Dietze in MNHU; examined), [China]: Juldus (Ili-Gebirge).
Eupithecia paupera Vojnits, 1982, *Annls hist.-nat. Mus. natn. hung.* 74: 221, fig. 3. Holotype ♂ (coll. K. Dietze in MNHU; examined), [Kirghizstan]: Gouldscha Ferghana [Junior objective homonym of *E. innotata* f. *paupera* Dietze, 1910]. **New junior synonym.**

Description. Fore wing elongate, and narrow, ground colour rather unicolorous blackish grey; transverse lines usually invisible, sometimes with weak pale transverse band between medial and terminal areas; discal dot small; terminal line narrow, black; fringe almost unicolorous dark grey. Hind wing slightly paler grey, transverse lines indistinct, but anal margin shaded dark brown; discal dot very small.

Male genitalia. Uncus slim, biapical. Valve relatively short, with pointed apex and angulate ventral edge. Vinculum tapered. Papillae long and broad. Aedeagus stout, shorter than length of valve. Vesica armed with two narrow, curved apical cornuti, a large plate-like cornutus centrally and a small, irregular cornutus near ductus ejaculatorius base. Sternite A8 peg-like.

Female genitalia. Bursa about 2/3 covered with small spines basally. Ductus bursae not distinguishable. Ductus seminalis broadened at base, curved along corpus bursae, attached to posterior part dorsally. Colliculum short and narrow. Antrum very broad, sclerotized. Apophyses short. Tergite A8 quadrate, with a membranous posterior margin with a small central tooth. Papillae anales short, broad, and rounded.

Range. China (Xinjiang, Qinghai); Kirghizstan, southern Kazakhstan.

Phenology. Univoltine (?): late May to late July.

Altitude. From about 1500 m (Kazakhstan, Ketmen Mts.: coll. ZISP) up to about 2700 m (Kirghizstan, Sary-Djaz: coll. ZISP).

Similar species. Externally similar to *E. mitigata*, but larger, darker, more uniform, without light costal spot between fore wing medial and terminal areas. The male genitalia resemble those of *E. corroborata*, but the valve has a less sharply angled ventral margin, and the two apical cornuti are more curved. The female genitalia cannot easily be distinguished from those of other species in the group which lack a bilobed antrum, but have the attachment of the ductus seminalis slightly more posterior than other species.

Material. ♀, [Xinjiang], *paupera* ♀ Juldus Ili, Zool. Mus. Berlin, Vojnits slide no. 13555♀ (lectotype, here design.; paratype of *E. paupera* Vojnits, 1982, coll. K. Dietze in MNHU); 1♂, [Kirghizstan], Gouldscha, Ferghana 1905, Korb, 8.vi.05, *innotata* form: *praesignata* ab. *paupera*, Gouldja, Ferghana, 8.6.05, Fast genau so, etwas langfransiger ist verglichene Type der *praesignata*!, Zool. Mus. Berlin, Vojnits slide no. 13548♂ (paralectotype; holotype of *E. paupera* Vojnits, 1982, coll. K. Dietze in MNHU); 1♂, [Xinjiang], *innotata* form *praesignata* ab. *paupera* Typ, Juldus Ili, Zool. Mus. Berlin, Vojnits slide no. 13546♂ (paralectotype; paratype of *E. paupera* Vojnits, 1982, coll. K. Dietze in MNHU); 1♂, [Xinjiang], Juldus, forma *paupera* ♂, Zool. Mus. Berlin, Vojnits slide no. 13547♂ (paralectotype; paratype of *E. paupera* Vojnits, 1982, coll. K. Dietze in MNHU); 1♂, [Xinjiang], *paupera* [hand-writing of K. Dietze], Asia centr., Thian-Schan, Juldus Geb., coll. Wagner (coll. K. Dietze in MNHU); 2♀, Qinghai, Tongren, Maixiu, 2950 m, 30.vii-1.viii.1992, Galsworthy slides nos. ZL374♀, ZL581♀ (IZCAS).

Remarks. Contra Vojnits (1982), the name *Eupithecia innotata* f. *paupera* Dietze, 1910 is available, having subspecific (not infrasubspecific) rank according to the International Code of Zoological Nomenclature (article 45g). We designated as lectotype a specimen of *E. paupera* which was first illustrated by K. Dietze under this name (1910, pl. 85, fig. 955). The male type-specimen illustrated by Dietze (1910: pl. 75, fig. 488) was treated as *E. innotata* f. *praesignata*. Dietze included this specimen in the type series of *E. innotata* f./ab. *paupera* in the second volume of the folio in 1913. However, he had earlier used the name *paupera* for a different female specimen (Dietze, 1910; pl. 85, fig. 955).

215. *Eupithecia mitigata* Dietze, 1906

[*Eupithecia*] *mitigata* Dietze, 1906, *Dt. ent. Z., Iris* 19 (2): pl. 2, figs 18, 19 (non binominal); *ibidem* (1908), 21 (2): 168 (*Eup*[*ithecia*]). Lectotype ♂ (design. by Vojnits, coll. R. Püngeler in MNHU; examined), [China]: Lob-Noor.

Eupithecia sclerata Vojnits, 1982, *Annls hist.-nat. Mus. natn. hung.* 74: 220, fig. 4. Holotype ♂ (coll. O. Staudinger in MNHU; examined), [China]: Lob-Noor. **New junior synonym.**

Misidentification: *Eupithecia omnigera* Vojnits, 1988 (*nec* Vojnits, 1982) (at least 1♂ in ZISP).

Description. Fore wings narrow and elongate, almost unicolorous dark to ash grey; all transverse lines inconspicuous, oblique, running parallel with wider space between ante- and postmedial, the latter forming a pale costal spot; terminal area with faint pale subterminal line; discal dot small. Hind wing paler than fore wing, anal margin scattered with darker scales; discal dot minute.

Male genitalia. Uncus elongate, biapical. Valve with a strong angle on ventral margin. Vinculum short, rounded. Papillae very broad. Aedeagus stout, almost equal to length of valve. Vesica armed with one substantial and one tiny apical cornutus, a large plate-like cornutus and a smaller curved, irregular cornutus near ductus ejaculatorius base. Sternite A8 peg-like.

Female genitalia. Bursa covered with small spines in anterior half, and with a patch of larger spines in the centre. Ductus bursae striate. Ductus seminalis broadened at base, sharply recurved, attached to ductus bursae on right. Colliculum short. Antrum very wide, with a deep indentation on anterior edge. Tergite A8 transverse, with pointed medial processes on anterior and posterior margins. Apophyses short and narrow. Papillae anales rounded.

Range. China (Xinjiang); S. Kazakhstan, Kirghizstan, Tajikistan.
Phenology. Bivoltine: late May to early August.
Altitude. From about 1200 m in Zailijsky Alatau Mts. (Kazakhstan) up to about 3250 m in the Pamirs (Tajikistan: coll. ZISP).
Similar species. Most similar externally to *E. innotata*, but distinguished by faint pattern of parallel transverse lines and lack of a pale apical streak on the fore wing. The male genitalia are best distinguished by the tiny size of the second apical cornutus, and the female by the characteristic shape of the antrum.
Material. ♂, [Xinjiang], Asia centr., Lob-Noor, E. Mai 1904 Rückbeil sen. 4/05 v. Tancré, *mitigata* cotype, K. Dietze, Zool. Mus. Berlin, Vojnits slide no. 13587♂ (lectotype, design. by Vojnits, coll. R. Püngeler in MNHU); 2♂, [Xinjiang], same data, *mitigata* cotype, K. Dietze (coll. R. Püngeler in MNHU); 1♂, [Xinjiang], *mitigata* n. sp. ♂, Type, Lob-noor, Asia centr. (Ende Mai 1904. Rückbeil), K. Dietze, *mitigata* Dtze, präp. Nr. A 827♂, det. E. Schütze, Kassel (paralectotype, coll. K. Dietze in MNHU); 1♂ (*nec* ♀), *mitigata* ♀, Type, Lob-noor, Asia centr., Rückbeil, Ende Mai 1901t., Original., K. Dietze (paralectotype, coll. K. Dietze in MNHU); ♂, Asia centr., Lob-Noor, t. Mai 1904 Rückbeil sen., Zool. Mus. Berlin, Vojnits slide no. 13588♂ (holotype of *E. sclerata*, coll. O. Staudinger in MNHU); 1♂, [Xinjiang], *unedon.* v. *relaxata* Dietze, Lob-Noor, Nob-Noor [on the underside of label] (coll. R. Püngeler in MNHU); 2♂, [Xinjiang], Asia centr., Lob-Noor, E. Mai 1904, Rückbeil sen, *E. relaxata* Dietze ♂, v. R. Tancré 5.05 (coll. R. Püngeler in MNHU); 2♂, 1♀, [Xinjiang], Lob-Noor, E. Mai 1904, Rückbeil sen., *mitigata* Dietze, 10/1910 v. Tancré (coll. R. Püngeler in MNHU); 1♂ Xinjiang, Baicheng Taliekekuotan, 2300 m, 8.vii.1978, slide no. ZL534♂; 3♂, Xinjiang Tomürt peak, 2400 m, 4, 6.vii.1977, and 2600 m, 21.vii.1977; 2♂ Xinjiang, Zhaosu, Alasan, 2400 m, 24, 28.vii.1978, slide no. ZL536♂; 1♀ Xinjiang, Tomürt Peak, 2400 m, 21.vi.1977, slide no. ZL497♀ (IZCAS).

216. *Eupithecia vicariata* Dietze, 1904

Eup[ithecia] vicariata Dietze, 1904, *Dt. ent. Z., Iris* [1903] **16** (2): 343, pl. 3, fig. 17. Syntypes 3, at least 1♂, 1♀ (coll. K. Dietze in MNHU; examined), [Tajikistan]: Sarafscham [= Zerafshan], 2500 m.

Eup[ithecia] vicariata var. *adjunctata* Dietze, 1904, *Dt. ent. Z., Iris* [1903] **16** (2): 344, pl. 3, fig. 18. Syntypes 1♂, 2♀ (coll. R. Püngeler and K. Dietze in MNHU; examined), [China]: Korla, Sai-Chin.

Eup[ithecia] ?sp./f. *privata* Dietze, 1910, *Biologie der Eupithecien* **1**: pl. 75, fig. 545; *ibidem* (1913), **2**: 137. Holotype ♀ (coll. R. Püngeler in MNHU; examined), Cent(ral) As(ia): Tura. **New junior synonym.**

Eupithecia costisignata f. *lutulenta* Dietze, 1913, *Biologie der Eupithecien* **2**: 137. Syntypes 1♂, 1♀ (coll. K. Dietze in MNHU; examined), [Kazakhstan]: Yarkend, Mus-tag-ata. **New junior synonym.**

Eupithecia proprivata Vojnits, 1982, *Annls hist.-nat. Mus. natn. hung.* 74: 224, fig. 8. Type ♀ (coll. R. Püngeler in MNHU; examined), Centr(al) Asia [Turkmenistan]: Tura. [Objective synonym of *Eupithecia privata* Dietze, 1910]. **New junior synonym.**

Misidentification: *Eupithecia omnigera* Vojnits, 1988 (*nec* Vojnits, 1982) (at least 1♂ and 2♀ in ZISP).

Description. Fore wings elongate and pointed, pale ash-grey, often with a faint or even quite strong yellowish tinge; transverse lines varying from strong to sometimes almost obscured, strongly oblique; medial usually more distinct, broadened at anal margin, sharply angled onto costa behind discal dot; postmedial twice angled near costa; terminal area slightly darker; discal dot small. Hind wing paler, usually with well marked dark transverse lines along anal margin; discal dot invisible or very small.

Male genitalia. Uncus elongate, biapical. Valve with a strong angle on ventral margin, and sometimes a tiny tooth. Vinculum narrow, rounded. Papillae very large and stout, slightly curved. Aedeagus stout, just shorter than length of valve. Vesica armed with two curved apical cornuti, a large elongate patch of spinules, a broad, folded and a smaller, curved, hook-like cornutus at ductus ejaculatorius base. Sternite A8 peg-like with narrow apex.

Female genitalia. Bursa densely covered with small spines in basal half. Ductus bursae broad and striate, with a small longitudinal patch of small spines. Ductus seminalis sharply recurved, broadly attached to base of ductus bursae from right side. Colliculum slightly elongate, narrow. Antrum broad, covered with numerous pores. Tergite A8 transverse. Apophyses short and narrow. Papillae anales broad and rounded.

Range. China (Xinjiang, Qinghai); SE Kazakhstan, Kirghizstan, Tajikistan.

Phenology. Univoltine: early June to late July.

Altitude. From about 2000 to 2800 m in Hissar Mts. and Serafshan Mts. (Tajikistan: coll. A. Tzvetaev in ZMMU).

Similar species. Specimens with stronger transverse lines easily distinguished from related species (except *E. magica*, which has different genitalia) by the strongly marked medial line on the fore wing. More faintly marked specimens can be very similar to *E. mitigata*, but the genitalia of both sexes lack the features highlighted under that species.

Material. 1♂, [Tajikistan], Asia centr., Sarafschan, 2500 m, 30.vii.1900, v. E. Funke 10/00, *vicariata* n. sp., Typ. ♂, Sarafschan, K. Dietze, *vicariata*, *vicariata* Dtze., Präp. Nr. A 819♂ det. E. Schütze, Kassel (syntype, coll. K. Dietze in MNHU); 1♀, [Tajikistan], Saraw. 10, Sary-ob 7, 2500 m, Funke 1900, *Eupithecia* ♀, 33, *vicariata* n. sp. Original (K. Dietze), *vicariata* Dtze., Präp. Nr. A 818♀ det. E. Schütze, Kassel (syntype, coll. K. Dietze in MNHU);

1♀, [Xinjiang], Korla, Sai-chin, *adjunctata*, Korla, Sai-chin, *adjunctata* Dtze., Präp. Nr. A 822♀ det. E. Schütze, Kassel (syntype of *E. vicariata* var. *adjunctata*, coll. K. Dietze in MNHU); 1♀, [Xinjiang], Korla, Sai-chin, *adjunctata* Original ♀, Korla, Sai-chin, *adjunctata* Dtze., Präp. Nr. A 823♀ det. E. Schütze, Kassel (syntype of *E. vicariata* var. *adjunctata*, coll. K. Dietze in MNHU); 1♂, [Xinjiang], Asia centr., Saichin-Kette, nördl. von Korla: Rückbeil 1902, *vicariata* Dietze ♂, Type, v. R. Tancré 11.02, *vicariata* Original (Sattel characteristisch für ditte Art.), Cotype *adjunctata* Dietze ♂ (syntype of *E. vicariata* var. *adjunctata*, coll. R. Püngeler in MNHU); 1♂, [Xinjiang], ? *vicariata* form. *lutulenta* ♂, Mus-tag-ata Yarkend, *E. vicariata* Dtze., Präp. Nr. A 820♂ det. E. Schütze, Kassel (syntype of *E. vicariata* f. *lutulenta*, coll. K. Dietze in MNHU); 1♀, [Xinjiang], ? *vicariata* forma: *lutulenta*, Mus-tag-ata Yarkend, ♀, *E. vicariata* Dtze., Präp. Nr. A 821♀ det. E. Schütze, Kassel (syntype of *E. vicariata* f. *lutulenta*, MNHU); 1♀, Asia centr. (Tura), *Eup. praesignata* Bohatsch ♀ (als *insignata* Stgr. von Stgr. 3.93), Type *privata* Dietze ♀, Püng., Zool. Mus. Berlin, Vojnits slide no. 13586♀ (holotype of *E.* ?sp./f. *privata* Dietze, 1910; type of *E. proprivata* Vojnits, 1982, coll. R. Püngeler in MNHU); 1♀, [Xinjiang], Korla 97. Hbhr., n. sp. bei *Biornata* or *Aequata*; vielleicht nur ab. des letztem (Boh.), möchle ich als Verwandte des *unedonata* ansehen, Dietze (coll. O. Staudinger in MNHU); 1♂, [Xinjiang], Asia centr., Ost-Turkestan (Urumtschi), Rückbeil 1904, v. R. Tancré 1905 (coll. R. Püngeler in MNHU); 2♀ Xinjiang, Akto, Wuyitake, 2650 m, 6.viii.1989, coll Zhang Xuezhong, slides nos. ZL474, 582♀; 1♀, Xinjiang, Baicheng, Talieke Kuotan,1.vi.1978, coll Han Yinheng, slide no. ZL535♀; 1♂, Qinghai, Tongren, Maixiu, 2950 m, 30.vii–1.viii.1992, coll Xue Dayong, slide no. ZL383♂ (IZCAS).

217. *Eupithecia magica* Mironov & Galsworthy, 2006

Eupithecia magica Mironov & Galsworthy, 2006, *Trans. lepid. Soc. Japan* 57 (4): 351, figs 20, 31. Holotype ♂ (ZFMK; examined), Tibet [Sichuan], Batang, Valley of Yangtze River, 2800 m.

Description. Fore wing ground colour pale brownish grey; transverse lines dark brown, oblique, sharply angled near costa; antemedial doubled below angle; postmedial strongest near anal margin, bordered outside by a broad pale fascia; terminal area darker with an inconspicuous, pale wavy subterminal line; discal dot small, slightly elongate. Hind wing paler; antemedial and medial strongly marked along anal margin only; discal dot indistinct.

Male genitalia. Uncus broad, biapical. Valve short, ventral margin only slightly angled. Vinculum short. Papillae broad, slightly clavate. Aedeagus

shorter than valve length. Vesica armed with a small anterior plate and a smaller, irregular plate near ductus ejaculatorius base. Sternite A8 strongly broadened basally, with two narrow apical rods.
Female genitalia. Bursa completely covered with small spines. Ductus bursae relatively short, with longitudinal striations and two small patches of larger spines. Ductus seminalis attached to ductus bursae near colliculum on right. Colliculum short and narrow. Antrum broad, sclerotized. Tergite A8 almost square with heavily sclerotized anterior margin, and deep, narrow medial hollow on posterior margin. Apophyses short and moderately stout. Papillae anales broad and rounded.
Range. China (Sichuan, Yunnan).
Phenology. Univoltine (?): late August to early September.
Altitude. Found at about 2800 m above sea-level in southern Tibet (China: coll. ZFMK).
Similar species. Very similar to more heavily marked examples of *E. vicariata*. Distinction has to be by genitalia. This and the next species have similar genitalia, which differ from those of the rest of this species group. The best distinction in the male genitalia is the much longer papillae in the present species. The female genitalia are best distinguished by the presence of two patches of larger spines in the ductus bursae.
Material. ♂, Tibet [Sichuan], Batang, Im Tal des Yangtze, *ca.* 2800 m, 28. viii.1936, H. Höne; de Laever slide no. 437♂ (holotype, ZFMK); 1♀, same loc. and alt., 4.ix.1936, H. Höne; de Laever slide no. 438♀ (paratype, ZFMK); 2♂, Nord Yünnan, A-tun-tse [Dêqên], Mittlere Höhe (*ca.* 4000 m), 19.vii., 6.viii.1936, H. Höne, Vojnits slides nos. 1204♂, 1205♂; 1♀, same loc., Obere Höhe (*ca.* 4500 m), 15.viii.1936, H. Höne, Vojnits slide no. 10990♀ (ZFMK).

218. *Eupithecia relaxata* Dietze, 1904

[*Eupithecia*] *unedonata* var. *relaxata* Dietze, 1904, *Dt. ent. Z., Iris* [1903] **16** (2): 367, pl. 5, fig. 17 (non binominal). Holotype ♀ (coll. K. Dietze in MNHU; examined), [Iran]: Schakuh [= Shahkuh].
[*Eupithecia*] *costisignata* Dietze, 1904, *Dt. ent. Z., Iris* [1903] **16** (2): 367, pl. 5, figs 18, 19 (non binominal). Syntypes 11, at least 4♂, 3♀ (coll. K. Dietze and R. Püngeler in MNHU; examined), Central Asia [Iran]: Schakuh [= Shahkuh]; [China]: Korla. [Originally as ab. of (*Eupithecia*) *unedonata* var. *relaxata* Dietze, 1904.] (synonymised in Dietze, 1913).
Eupithecia pamiri Vojnits, 1988, *Annls hist.-nat. Mus. natn. hung.* **80**: 81, pl. 1, figs 3–5; pl. 4, figs 25–28. Holotype ♂ (ZISP; examined), [Tajikistan]: Pamir merid. occid., Schachdarensis Mts., Seiojdara River, 3250 m (synonymised in Mironov et al., 2008c).

Misidentification: *Eupithecia emittens* Inoue, 1996 (at least 1♀ paratype in ZISP).

Description. Fore wing ground colour ash-grey; transverse lines fine, dark brown; medial sharply angled onto costa behind discal dot; postmedial curved, then sharply angled back onto costa; antemedial, medial and postmedial forming dark patches at costa, the latter two forming a characteristic semicircle; terminal area darker; discal dot oblique, linear. Hind wing slightly paler, transverse lines well marked on anal margin; terminal area darker; discal dot small, rounded.

Male genitalia. Uncus short, uniapical. Valve unmodified. Vinculum relatively narrow. Juxta with two forked, uniquely broadened posterior lobes with rounded apices (the anellus lobes of Pierce, 1914). Papillae elongate and broad. Aedeagus slim and elongate, about equal to valve length. Vesica armed with an elongate Y-shaped plate, and a large, irregular plate near ductus ejaculatorius base. Sternite A8 short, broadened basally, with two narrow apical rods.

Female genitalia. Bursa completely covered with small spines. Ductus bursae short, narrow. Ductus seminalis evenly curved around corpus bursae, attached to ductus bursae near short and narrow colliculum on right. Antrum broad, triangular, slightly sclerotized. Tergite A8 with a deep medial hollow on posterior margin. Apophyses short and narrow. Papillae anales broad and rounded.

Range. China (Xinjiang); Turkmenistan (Kopet-Dagh Mts.), Iran (Golestan), Afghanistan (north-east), Kazakhstan and Kirghizstan (Tien-Shan Mts.), Tajikistan (Pamirs Mts.), Pakistan (Baltistan), India (Ladakh Range, Jammu and Kashmir), Mongolia (Mongol Altai Mts.).

Phenology. Bivoltine: early April to late October.

Biology. Larva found on *Prunus* (*Amygdalus*) *spartioides* (= *Prunus arabica*) (Rosaceae) (Iran: Wiltshire 1952).

Altitude. From about 1100 m in Tien-Shan Mts. (Kirghizstan: coll. ZISP) up to about 4500 m in Jammu and Kashmir (Pakistan/India: coll. ZFMK).

Similar species. Externally similar to *E. nepalata* but with finer, narrower and more oblique transverse lines and narrower discal spots on fore wings. The genitalia resemble those of the previous species (qv.).

Material. ♀, [Iran], Schakuh, 31/7.98, Funke, *costisignata* form. *relaxata*, Typ, Schahkuh ♀, 31.vii.98, Zool. Mus. Berlin, Typus, Gen.-Unters. Nr. [Schmidt slide no.] Eu 104, Zool. Mus. Berlin (syntype of *E. relaxata*, coll. K. Dietze in MNHU); 1♂, [Xinjiang], *costisignata*, *costisignata* Korla, Type ♂, Zool. Mus. Berlin, Typus, *Eupithecia costisignata* ♂, Korea, Type, Prep. № 1659.1 E. de Laever det. 1957 (syntype of *E. costisignata*, coll. K. Dietze in MNHU); 1♀, [Xinjiang], *costisignata* Korla, *costisignata* ♀ Typ., Korla, *Eupithecia costisignata* ♀, Kor(e)a, Type, Prep. № 1659.2 E. de Laever det. 1957 (syntype of *E. costisignata*, coll. K. Dietze in MNHU); 1♀, [Xinjiang],

Korla, Sai-chin, 372, *costisignata* [hand writing by Dietze] (coll. K. Dietze in MNHU); 1♀, [Xinjiang], Ak-ssu, *famelica* (coll. K. Dietze in MNHU); 3♂, [Xinjiang], Asia centr., Saichin-Kette nördl. v. Korla, Rückbeil 1902, *relaxata* var. *costisignata* Dietze, Type (v. R. Tancré 11.02), subspec: *relaxata* var. *costisignata* Original (Dietze), Zool. Mus. Berlin, Vojnits slide no. 13584♂ (syntypes of *E. relaxata* var. *costisignata*, coll. R. Püngeler in MNHU); 2♂, 5♀, [Xinjiang], Mus-tag-ata Yarkend (coll. K. Dietze in MNHU); 1♀, [Xinjiang], *costisignata*, Mus-tag-ata Yarkend (coll. K. Dietze in MNHU); ♂, [Tajikistan], Pamir merid. occid., Mts. Schach-Darensis, Cliv septr., fl. Seidj-Dara, 3250 m, 27.vii.[1]957, A. Bundel leg., Vojnits slide no. 17686♂ (holotype of *E. pamiri*, ZISP).

Remarks. The identity of this taxon was dealt with extensively in Mironov et al. (2008c). Mr. de Laever (1960) referred to type specimens of this species from Korea, having apparently misread 'Korla' (NW China).

The *centaureata* species group

This species group was introduced by Vojnits (1977), and elaborated by Mironov (2003). The male genitalia are characterised by heavy modifications to the ventral margin of the valves and an H-shaped eighth sternite. The female genitalia are fairly uniform, with a globular spined bursa, and attachment of the ductus seminalis on the ductus bursae, but also share a sclerotized funnel-shaped antrum with two distinctive lobes.

The following three Chinese species are included in the group.

219. *Eupithecia centaureata* [Denis & Schiffermüller], 1775

Geometra centaureata [Denis & Schiffermüller], 1775, *Syst. Verz. Schmett. Wien*: 114. Syntype(s) lost, [Austria]: Vienna district.

? *Phalaena signata* Scopoli, 1763, *Ent. Carniol.,Ordo* 3: 231. Syntype(s) lost, [Slovenia and NE. Italy]: Carniolia. Senior synonym ? (synonymised in Treitschke, 1828).

Geometra oblongata Thunberg, 1784, *Diss. ent. sistens Insecta Suecica* 1: 14, fig. 1. Lectotype ♂ (DEUU), Sweden (synonymised in Stephens, 1829a).

Phalaena boloniensis Fourcroy, 1785, *Ent. Parisiensis* 2: 282. Syntype(s) in Mus. Autun (?), France: Paris (synonymised in Prout, 1915).

Eupithecia oblongata var. *centralisata* Staudinger, 1892, *Dt. ent. Z., Iris* 5 (1): 250. Syntypes 13, at least 5♂, 6♀ (coll. O. Staudinger in MNHU; examined), Central Asia: [Uzbekistan]: Margelan; Namangan; Samarkand; [Kirghizstan]: Alexander Mts.; [Kazakhstan]: Saisan.

Eup[ithecia] oblongata f. *obscura* Dietze, 1910, *Biologie der Eupithecien* 1: pl. 70: 132; *ibidem* (1913), 2: 63. Syntypes 6, at least 4♀ (coll. K. Dietze in MNHU; examined), [Kirghizstan]: Alai Mts.; [Kazakhstan]: Uralsk; [Italy]: southern Tyrol.

Eupithecia centaureata dagestani Vojnits, 1977, *Acta zool. Acad. Sci. hung.* 23 (1–2): 229, fig. 2. Holotype ♂ (ZFMK; examined), [Russia]: Dag(h)estan, Petrovsk-port [= Makhachkala].

Eupithecia centaureata dsharkendi Vojnits, 1977, *Acta zool. Acad. Sci. hung.* 23 (1–2): 230, fig. 3. Holotype ♀ (ZFMK; examined), [Kazakhstan]: Dsharkend, Semiretschensk.

Eupithecia chinae Vojnits, 1977, *Acta zool. Acad. Sci. hung.* 23 (1–2): 231, figs 4, 5b. Holotype ♂ (ZFMK; examined), China: Shantung Province, Tai-shan, 1550 m (synonymised in Mironov & Galsworthy, 2007).

Unavailable name (incorrect subsequent spelling): *centaurearia*: Boisduval (1840).

Description. Fore wing white with a dark costal blotch between crescent-shaped discal dot and crenellate postmedial line; terminal area darker ochreous with wavy white subterminal line. Hind wing white with transverse lines and termen marked in darker ochreous.

Male genitalia. Uncus elongate, narrow, biapical. Valve with heavily modified and sclerotized ventral margin. Vinculum elongate and broad, trapezoid. Papillae stout and short, clavate. Aedeagus slim, shorter than valve length. Vesica minutely granulate, armed with a broad, H-shaped cornutus with two anterior arms. Sternite A8 broad, with diverging basal and apical lobes.

Female genitalia. Bursa ovate, completely covered with small spines. Ductus bursae elongate and broad, sclerotized. Ductus seminalis strongly broadened basally. Colliculum short and narrow. Antrum broad, funnel-shaped, strongly sclerotized, with two broad ventral lobes. Tergite A8 almost rectangular with a small, pointed medial tip on posterior margin. Apophyses narrow. Papillae anales slim and rounded.

Range. China (Xinjiang, Inner Mongolia, Beijing, Heilongjiang, Shanxi, Shandong, Guangdong, Taiwan); Europe, North Africa (Morocco, Algeria, Tunisia), Cyprus, Asian Turkey, Lebanon, Israel, Jordan, Syria, Georgia, Armenia, Azerbaijan, Iran, Kazakhstan, Turkmenistan, Uzbekistan, Tajikistan, Afghanistan, Kirghizstan, Pakistan/India (Jammu and Kashmir), Asian Russia (Urals, western and southern Siberia: Altai Mts., Sajan Mts., Transbaikalia, Dahuria, Priamurje), Mongolia.

Phenology. Polyvoltine: early March to mid-October.

Altitude. From sea level to about 3300 m in the Himalayas(Kashmir: coll. ZFMK).

Similar species. Distinctive, and not likely to be confused with any other species, externally or in genitalia.

Material. 1♂, [Xinjiang], *oblongata* Aksu 03 (coll. K. Dietze in MNHU); 2♂, [Xinjiang], Aksu 1900 (coll. K. Dietze in MNHU); 2♂, 1♀, [Xinjiang], Ak-ssu (coll. K. Dietze in MNHU); 1♂, [Xinjiang], Ili, Tancré 1902 (coll. K. Dietze in MNHU); 2♂, [Xinjiang], Juldus Ili (coll. K. Dietze in MNHU); 1♂, 1♀, [Xinjiang], Korla (coll. K. Dietze in MNHU); 1♀, [Xinjiang], Mus-tag-ata Yarkend (coll. K. Dietze in MNHU); ♂, [Russia], Dagestan, Petrovsk-port [now Makhachkala], 29.v.[19]26, M. Rjabov, Vojnits slide no. 743♂

(holotype of *E. centaureata dagestani*, ZFMK); ♀, [Kazakhstan], Dsharkend [= Djarkent], Semiretschensk, Vojnits slide no. 1110♀ (holotype of *E. centaureata dsharkendi*, ZFMK); ♂, China, Prov. Shantung, Tai-shan, 1550 m, 16.viii.1934, H. Höne, Vojnits slide no. 1103♂ (holotype of *E. chinae*, ZFMK); 1♀, same loc. and alt., 11.viii.1934, H. Höne, de Laever slide no. 424♀ (paratype of *E. chinae*, TTM); 1♀, Manchuria [Heilongjiang], Harbin, Juni 1923, V. Tolmachov, Vojnits slide no. 1101♀ (paratype of *E. chinae*, TTM); 1♂, China, [Guangdong], Canton, Mironov slide no. 545♂ (ZFMK); 1♀, Taiwan, Prov. Ping-Tung, 10 km SE of Mntan, 470 m, 7–10.iii.1996, leg. Gy. Fábián & L. Németh (TTM); 1♀, China, Inner Mongolia, 80 km SW of Hohhot, Nantianmen, 13.viii.2007, A. Bidzilya (ZISP); 1♂, Xinjiang, Tomürt Peak, 3200 m, 11.vii.1977; 1♀, same loc, 2400 m, 4.vii.1977, both coll Li Changjiang; 1♀, Inner Mongolia, Shihezi, 3.v.1981, slide no. ZL543♀; 2♀, Inner Mongolia, Yakeshi, Wuerquhan, 6, 12.viii.1983; 1♂, Inner Mongolia, Yakeshi, Oragxan, Miaoguo, 31.vii.1987, slide no. ZL541♂ (IZCAS).

Remarks. The taxonomy of this species was dealt with in detail in Mironov and Galsworthy (2007). We regard the various forms which have been described across Asia as belonging to the same widespread species.

220. *Eupithecia exquisita* Vojnits, 1979

Eupithecia exquisita Vojnits, 1979, *Acta zool. Acad. Sci. hung.* 25 (1–2): 210, fig. 22. Holotype ♂ (ZFMK; examined), [China]: Sued-Shensi, Tapaishan in Tsinling, 1700 m.

Description. Very small. Fore wing ground colour pale grey with a soft ochreous tinge; transverse lines distinct, blackish, broadened near costa; basal, postbasal, antemedial, double medial and postmedial gently curved onto costa; terminal area slightly darker with weak trace of pale wavy subterminal line; discal dot distinct, linear. Hind wing paler, dark transverse lines well marked only on anal margin; terminal area slightly darker, with weak trace of dentate light subterminal line; discal dot distinct, linear.

Male genitalia. Uncus stout, biapical. Valve rather square, sacculus heavily sclerotized, with a blunt process on ventral corner. Vinculum small, circular. Papillae elongate, tapered. Aedeagus narrow, shorter than length of valve. Vesica armed with two blunt plate-like cornuti. Sternite A8 with broad apical and basal lobes.

Female genitalia. Bursa oval, completely covered with small spines. Ductus bursae spineless, elongate, more strongly sclerotized and tapered to base of colliculum. Ductus seminalis broad at base, attached at base of ductus bursae on right. Colliculum elongate and narrow. Antrum funnel-shaped,

heavily sclerotized, with numerous minute pores. Apophyses short and narrow. Papillae anales elongate and narrow, tapered.
Range. China (Hebei, Shanxi, Shaanxi).
Phenology. Univoltine: early July to early August.
Altitude. Found at about 1700–2000 m above sea-level (China: Vojnits, 1979a; coll. ZFMK).
Similar species. Very similar to European and Central Asian *E. extraversaria* Herrich-Schäffer, 1852, but smaller, with more rounded transverse lines. The male and female genitalia are also close to those of *E. extraversaria*, but differ in the pattern of cornuti in the male vesica and the much longer ductus bursae in the female.
Material. ♂, Sued-Shensi [Shaanxi], Tapaishan [Taibaishan] im Tsinling [Qinling], *ca.* 1700 m, 6.viii.1936, H. Höne, Vojnits slide no. 12065♂ (holotype of *E. equisita* [sic], ZFMK); 1♂, same loc. and alt., 6.viii.1935, H. Höne, Vojnits slide no. 12064♂ (paratype, TTM); 1♂, same loc. and alt., 9.viii.1936, H. Höne, de Laever slide no. 277♂ (ZFMK); 1♀, same loc., 6.viii.1935, H. Höne, Vojnits slide no. 12271♀ (ZFMK); 1♀, Prov. Shansi [Shanxi], Mienshan, Obere Höhe *ca.* 2000 m, 5.viii.1937, H. Höne, Vojnits slide no. 13737♀ (ZFMK); 1♀, China, [Hebei], 100 km W of Beijing, 6–8.vii.2006, [leg.] V. Krivokhatsky, Mironov slide no. 806♀ (ZISP).

221. *Eupithecia brunneilutea* Mironov & Galsworthy, 2004

Eupithecia brunneilutea Mironov & Galsworthy, 2004, *Trans. lepid. Soc. Japan* 55 (1): 53, figs 9, 27. Holotype ♂ (ZFMK; examined), [China]: North Yunnan, Li-kiang [Lijiang].

Description. Fore wing ground colour creamy white, transverse markings dark brown; postbasal and double antemedial very finely zigzag; discal dot small, round; medial curving from costa through discal dot, then angled, doubled below cell; double postmedial curved round cell; a strongly zigzag submarginal line; costa, inner margin and terminal area suffused yellow, with some dark brown scales; terminal line lacking; fringes pale yellow. Hind wing ground colour, subterminal line and terminal area as fore wing, with a small discal dot; transverse lines marked on anal margin.
Male genitalia. Uncus long, narrow, biapical. Valve rather square, with a blunt ventral process. Vinculum broad, with shallow medial hollow. Papillae short and stout. Aedeagus elongate and narrow. Vesica armed with one large U-shaped plate. Sternite A8 with spreading apical and broader basal lobes.
Female unknown.
Range. China (Yunnan); Nepal.

Phenology. Bivoltine (?): early June to mid-August.
Altitude. From about 2100 to 3000 m above sea-level in the Himalayas and Tibet (China: coll. ZFMK).
Similar species. Uniquely coloured and patterned, and unmistakeable.
Material. ♂, China, Provinz Nord-Yuennan, Li-kiang [Lijiang], 8.viii.1934, H. Höne, Mironov slide no. 508♂ (holotype, ZFMK); 1♂, same data (abdomen missing; paratype, ZFMK); 1♂, same loc., *ca.* 3000 m, 14.viii.1934, H. Höne, Mironov slide no. 509♂ (paratype, ZFMK); 1♂, Nepal, Annapurna Himal, valley of Kali Gandaki, near Ghasa, 2080 m, 83°39.5′ E, 28°36′ N, 4.vi.1996, László and Ronkay (paratype; coll. Gy.M. László, Budapest).

The *despectaria* species group

This small Palaearctic group of *Eupithecia* was introduced and described by Mironov (2003), and is probably close to the preceding group, with which it shares an H-shaped eighth sternite. The male valve however is unmodified. The female ductus bursae is usually sclerotized, but lacks the antrum lobes of the *centaureata* group. It is represented in China by one species only.

222. *Eupithecia despectaria* Lederer, 1853

Eupithecia despectaria Lederer, 1853, *Verh. zool.-bot. Ver. Wien* 3 (Abh.): 382, pl. 6, fig. 7. Holotype ♂ (coll. O. Staudinger in MNHU; examined), [Russia]: [Siberia or.], [Altai, Ustkamenogorsk].

Description. Fore wing ground colour whitish yellow to ochreous; transverse lines narrow, black, wavy, forming blackish brown costal spots; a paler medial crossing discal dot and sharply angled towards costa; postmedial broadly curved beyond discal dot; terminal area darker, with costal and medial brownish spots, and pale wavy subterminal line; discal dot elongate and narrow. Hind wing paler, suffused with blackish brown in rear half up to pale wavy subterminal line, beyond that ochreous; discal dot elongate and narrow.
Male genitalia. Uncus elongate, biapical. Valve short, broadened at middle, but unmodified. Vinculum short and narrow, rounded. Papillae rather stout. Aedeagus slim, almost equal to length of valve. Vesica armed with a small, irregular, plate at ductus ejaculatorius base and a band-like, V-shaped cornutus at middle. Sternite A8 with narrow rounded apical and broader basal lobes.
Female genitalia. Bursa globular, covered with small spines. Ductus bursae elongate, sclerotized. Ductus seminalis slightly broadened at base,

attached to base of ductus bursae on right. Colliculum narrow and short. Antrum membranous, elongate. Tergite A8 transverse. Apophyses short and narrow. Papillae anales short, broad and rounded.
Range. China (Xinjiang); Uzbekistan, Kirghizstan, Tajikistan, NW. Pakistan, S. Kazakhstan, Russia (Altai Mts., Tuva).
Phenology. Univoltine: mid-April to late June.
Altitude. From about 1100 m in Hissar Mts. (Tajikistan: Viidalepp 1988) up to about 2600 m in the Northern Himalayas (Pakistan, Chitral: coll. TTM).
Similar species. Small size is a helpful guide. Similar to the Central Asian *E. graciliata* Dietze, 1908, but with darker ground colour, less distinct transverse lines, and longer discal dots. The shape of the eighth sternite in the male genitalia, and longer ductus bursae in the female distinguish the present species from *E. graciliata.*
Material. ♂, [Kazakhstan], Ustkam[enogorsk], Altai Kinderm[ann], *Despectaria* Led. Sib. or., Origin., Vojnits slide no. 19032♂ (holotype, coll. O. Staudinger in MNHU); 1♂, [Xinjiang], Kuldja Rükb., 10., *Despectaria* ♂ (Boh.) (coll. O. Staudinger in MNHU).

The *gueneata* species group

The *gueneata* species group was introduced by Schütze (1961) as subgenus *Delaeveria*, and developed in Vojnits (1978) and Mironov (1989, 2003). Many members of the group have a strikingly coloured facies. The male genitalia have an unmodified valve, but usually with a strong angle on the ventral margin, armature of the vesica consisting of a number of strongly toothed cornuti, and a broad-based tapering eighth sternite, usually with a bifid apex. The female genitalia are fairly uniform, with a large partially spined bursa, normally with a number of isolated very large spines in the posterior part.

This group is primarily Mediterranean and Central Asian, and is represented in China by the following three species.

223. *Eupithecia molybdaena* Mironov & Galsworthy, 2004

Eupithecia ***molybdaena*** Mironov & Galsworthy, 2004, *Trans. lepid. Soc. Japan* **55** (2): 127, figs 9, 27. Holotype ♀ (ZFMK; examined), [China]: Sued-Shensi [South Shaanxi], Tapaishan [Taibaishan] im Tsinling [Qinling Mts.], 1700 m.
Misidentification: *Eupithecia finitima* Vojnits, 1979 (1♀ paratype in TTM).

Description. Fore wing ground colour pale brown, rather uniform, transverse lines reduced, visible only at costa as short brown streaks, one antemedial, two medial, and one postmedial; postmedial just visible as a curved

line parallel to termen, without angle; discal dot small, dark brown. Hind wing ground colour as fore wing, discal dot faintly visible.
Female genitalia. Bursa elongate, sclerotized, with prominent broad diverticulum on right, covered with large spines in anterior half. Ductus bursae with a discrete patch of spines and longitudinal striations. Ductus seminalis narrow, attached to corpus bursae on right. Colliculum short but relatively broad. Antrum membranous. Tergite A8 slightly elongate. Apophyses narrow. Papillae anales short, rounded.
Male unknown.
Range. China (Shaanxi).
Phenology. First half of August.
Altitude. Found between about 1700 and 3000 m above sea-level in Tapaishan (China: Vojnits 1979a; Mironov & Galsworthy 2004b; coll. TTM).
Similar species. Similar to the Euro-Siberian *E. trisignaria* Herrich-Schäffer, 1848. The female genitalia are very similar to those of *E. trisignaria* but the spines, especially the marginal spines in the corpus bursae, are larger than in the compared species, and the patch of spines in the ductus bursae is separated from those in the bursa.
Material. ♀, Sued-Shensi [Shaanxi], Tapaishan [Taibaishan] im Tsinling [Qinling], *ca.* 1700 m, 9.viii.1936, H. Höne, Vojnits slide no. 12382♀ (holotype, ZFMK); 1♀, same loc., *ca.* 3000 m, 19.viii.1936, H. Höne, Vojnits slide no. 11437♀ (paratype of *E. finitima*, TTM).

224. *Eupithecia gratiosata* Herrich-Schäffer, 1861

Eupithecia gratiosata Herrich-Schäffer, 1861, *Neue Schmett. Eur. angr. Ländern* 3: 27, pl. 135. Syntypes ♂ ♀ lost, [Europe?].
Eupithecia gueneata var. *subseparata* Christoph, 1885, *in* Romanoff, *Mém. Lépid.* 2: 134, pl. 7, fig. 3. Holotype ♀ (ZISP, not traced). [Turkmenistan]: Kisil-Arwat (synonymised in Prout, 1915).
Eupithecia gratiosata var. *lavata* Fuchs, 1901, *Stett. Ent. Z.* **62** (1–2): 124. Holotype ♀ (coll. K. Dietze in MNHU; examined), [Sicily].
Eup[ithecia] gratiosata f. *pallida* Dietze, 1910, *Biologie der Eupithecien* 1, pl. 70, fig. 127; *ibidem* (1913), 2: 67. Holotype ♀ (coll. K. Dietze in MNHU, not traced), Russia: Orenburg.

Description. Fore wing ground colour white; transverse lines present, but obscured by prominently coloured fasciae; costal part of basal area and much angled medial and postmedial fasciae ash-grey; a fainter ochreous antemedial fascia; postmedial marked by black dots on veins; terminal area shaded internally with a grey to ochreous fascia with two subapical crenellations, and often further greyish shading on terminal line; discal dot inconspicuous. Hind wing white with ash-grey transverse lines near dorsum, and blackish dots on veins marking postmedial; discal dot greyish.

Male genitalia. Uncus biapical. Valve broad, with angulate ventral margin. Vinculum short and broad, with shallow medial hollow. Papillae elongate and narrow. Aedeagus very stout. Vesica armed with two large dentate cornuti with a patch of dense, heavily sclerotized wrinkles at their base, and an elongate, U-shaped cornutus at ductus ejaculatorius base. Sternite A8 almost triangular, with two short diverging apical arms.
Female genitalia. Bursa spherical, sclerotized, basal half covered with spines, and with numerous large spines in the heavily sclerotized central part on the ventral side, and a broad, transverse patch of small spines near colliculum. Ductus seminalis narrow, attached to posterior part of corpus bursae near short and very broad colliculum on right. Antrum membranous. Tergite A8 transverse. Apophyses robust. Papillae anales broad and rounded.
Range. China (Xinjiang); Southern Europe (northward to southern Poland), North Africa, Asia Minor, Caucasus, Transcaucasus, Lebanon, Syria, Iraq, Iran, Turkmenistan (Kopet-Dagh Mts.), southern Urals, Kazakhstan, Uzbekistan.
Phenology. Univoltine (?): mid-April to mid-July. Larva: mid-June to mid-July (mid-August). Overwintering as pupa.
Altitude. From sea level up to about 2300 m in Asia Minor (coll. ZSM).
Similar species. An unmistakable species.
Material. 1♀, [Xinjiang], m. Juni, var. *lavata*!, *gratiosata*, Mitte Juni, Kuldja, v. Tancré 03, K. Dietze (coll. K. Dietze in MNHU); 1♀, [Xinjiang], Ili, ♀ *gratiosata*, Vojnits slide no. 757♀ (ZFMK).

225. *Eupithecia subpulchrata* Alphéraky, 1883

Eup[*ithecia*] *subpulchrata* Alphéraky, 1883, *Horae Soc. ent. Ross.* 17 (3/4): 221, pl. 8, fig. 75. Holotype ♂ (ZISP, not traced), [China]: Ili (region), Kouldjà [= Kuldja] [Yining (Gulja)].

Description. Fore wing basal and medial areas blackish brown, area between basal and terminal area rusty; transverse lines pale, basal and antemedial indented; medial sharply angled onto costa beyond discal dot; postmedial rounded; terminal area with two darker areas divided by a paler medial blotch and a pale subterminal line; discal dot inconspicuous. Hind wing paler, brownish along anal margin and with reddish terminal area crossed by a pale crenellate submarginal; pale transverse lines almost complete; discal dot minute. Fringes of both wings strongly chequered.
Male genitalia. Uncus stout, biapical. Valve broadened medially, with obtusely angled ventral margin. Vinculum short and broad. Papillae shortish. Aedeagus stout, shorter than valve length. Vesica armed with

two dentate cornuti, an elongate patch of dense small spines and heavily sclerotized wrinkles, and a U-shaped plate near ductus ejaculatorius base. Sternite A8 almost triangular, with two narrow apical arms.

Female genitalia. Bursa globular, with two spiniferous areas at anterior end and close to colliculum, and a heavily sclerotized medial area with some separate stout spines. Ductus seminalis broadened at base, attached to posterior part of bursa near short and broad colliculum. Antrum short. Tergite A8 almost quadrate. Apophyses very narrow. Papillae anales rather short and narrow.

Range. China (Xinjiang); Turkmenistan, S. Kazakhstan, Uzbekistan and Afghanistan.

Phenology. Univoltine: late April to early June.

Altitude. From about 1700 to 2100 m above sea-level (Afghanistan: coll. SMNK; TTM).

Similar species. Externally distinctive, but very similar to *E. albertiata* Schütze, 1961, than which it is slightly larger and paler, with broader and less pointed wings. The male has a broader valve and broader apical arms to the eighth sternite, but there are no significant structural differences in the female genitalia.

Material. 2♀, [Xinjiang], *subpulchrata*, Ili-Gebiet (coll. K. Dietze in MNHU); 1♀, [Xinjiang], anf. Juni, *subpulchrata* Alph., Kuldja, Anf. Juni, typische Form, Original verglichen, Der Type von ganz ähnlich, Original verglichen, Karl Dietze, *subpulchrata* Dtze., Präp. Nr. A 850♀, det. E. Schütze, Kassel (coll. K. Dietze in MNHU).

Remarks. We have been unable to find type material of this species.

The *graphata* species group

The *graphata* species group was described by Schütze (1958) as subgenus *Petersenia*. Mironov (2003) developed it further. We here include in it the *simpliciata* species group, which was introduced by Mironov (1991), since differences between the genitalia of the two groups are minimal. The male genitalia are characterised by an unmodified valve, a vesica armed with a large field of spines and an elongate U-shaped cornutus, sometimes with one or more dentate cornuti, and a peg-like eighth sternite. In the female the bursa is oval, with a large basal spine field, and nearly always a separate small field near the colliculum; the attachment of the ductus seminalis is high on the right of the bursa.

The group is rich in the Mediterranean and Central Asia, but represented in China only by the four species below.

226. *Eupithecia perendina* Vojnits, 1980

Eupithecia perendina Vojnits, 1980, *Acta zool. Acad. Sci. hung.* **26** (4): 441, figs 9, 10, 21. Holotype ♂ (ZFMK; examined), [China]: North Yunnan, A-tun-tse [Dêqên], Talsohle, 3000 m.

Eupithecia garrula Vojnits, 1984, *Acta zool. Acad. Sci. hung.* **30** (1–2): 222, figs 12, 15. Holotype ♂ (ZFMK; examined), [China]: [North Yunnan], A-tun-tse [Dêqên], 4000 m. **New junior synonym.**

Description. Fore wing grey brown, area from base to medial line darker; latter running through discal dot and sharply angled onto costa; postmedial a series of short dark dashes on the veins obtusely curved onto costa; terminal area usually darker with light wavy subterminal line; discal dot small, black. Hind wing paler, with darker terminal area, transverse lines marked along anal margin; discal dot small and rounded.
Male genitalia. Uncus short, biapical. Valve with a very slight broadening at tip of sacculus. Vinculum short, rounded. Papillae long and narrow, slightly curved. Aedeagus stout, almost equal to length of valve. Vesica with numerous tiny teeth in a longitudinal band, a small dentate cornutus at apex, and one long and two small plate-like cornuti. Sternite A8 peg-like, with two very short, pointed apical horns.
Female genitalia. Bursa oval, about two thirds covered with spines at base and up left side, and with a transverse row of larger spines near colliculum. Ductus bursae short, with longitudinal striations. Ductus seminalis broadened at base, attached to bursa high on right side. Colliculum short and narrow. Antrum membranous; its posterior part slightly sclerotized, with numerous pores. Tergite A8 trapezoid, with shallow medial hollows in both margins. Apophyses short and narrow, the anterior stouter. Papillae anales short and relatively broad.
Range. China (Sichuan, Yunnan).
Phenology. Univoltine: early July to mid-September.
Altitude. From about 2000 to 4000 m above sea level in Yunnan (China: Vojnits 1984).
Similar species. Usually easily distinguished externally by the dark basal half of the fore wing. The genitalia of both sexes are unlike any other Chinese species.
Material. ♂, Nord-Yünnan, A-tun-tse [Dêqên], Talsohle *ca.* 3000 m, 29.v.1937, H. Höne, Vojnits slide no. 12500♂ (holotype, ZFMK); 1♂, 4♀, same loc. and alt., 14.vi.1936, 25, 29.v., 1.vi.1937, H. Höne, Vojnits slides nos. 12541♀, 12523♀, 12501♂, 12503♀, 12521♀ (paratypes, ZFMK); 1♂, 2♀, same loc. and alt., 25, 26.v., 3.vi.1937, H. Höne, Vojnits slides nos. 12496♀, 12504♀, 12524♂ (paratypes, TTM); 1♂, same loc. and alt., 28.vii.1936, H. Höne,

Vojnits slide no. 13286♂ (paratype of *E. subplacida*, wrongly linked to gen. slide of *E. refertissima*, TTM); 1♂, same loc. and alt., 18.viii.[19]36, H. Höne, Vojnits slide no. 13285♂ (paratype of *E. garrula*, TTM); 3♀, same loc. and alt., 17, 19.vii.1936, H. Höne, Vojnits slides nos. 13023♀, 13025♀, 13085♀ (TTM); 8♂, 1♀, same loc. and alt., 19, 24, 25, 26, 31.v., 2, 4, 5.vi.1937, H. Höne, Vojnits slides nos. 12994♂, 12996♂, de Laever slides nos. 84♂, 283♂, Mironov slide no. 643♂ (ZFMK); 3♂, same loc. and alt., 18.v., 3.vi.1937, H. Höne, Vojnits slides nos. 13083♂, 13086♂, 13088♂ (TTM); 5♂, 2♀, same loc., Talsohle (*ca.* 3500 m), 18, 21, 23.vi., 13.vii.1936, H. Höne, Vojnits slides nos. 12510♂, 12511♂, 12512♂, 12494♀, 12513♂, 12498♀, 12525♂ (paratypes, ZFMK); 1♀, same loc. and alt., 27.vi.1936, H. Höne, Vojnits slide no. 12499♀ (paratype, TTM); 3♂, 4♀, same loc. and alt., 19.vi., 9, 13, 15.vii.1936, H. Höne, Vojnits slides nos. 13087♂, 13024♂, Mironov slide no. 642♀ (ZFMK); 5♂, 1♀, same loc., Mittlere Höhe (*ca.* 4000 m), 16, 17, 26.vii., 2.viii.1936, H. Höne, Vojnits slides nos. 12526♀, 11531♂, 11532♂, 12529♂, 12516♂ (labelled as paratype of unpubl. *E. suavissima* Vojnits), 12522♂ (all paratypes, ZFMK); 1♂, same loc. and alt., 27.vii.1936, H. Höne, Vojnits slide no. 12518♂ (paratype, TTM); 1♂, 7♀, same loc. and alt., 19.vi., 16, 17, 19, 23.vii.1936, H. Höne, Vojnits slides nos. 13958♀, 12997♀, 13082♀, 13089♀, 13084♀, 13956♀ (ZFMK); 1♂, same loc., Obere Höhe (*ca.* 4500 m), 10.vii.1936, H. Höne, Vojnits slide no. 12507♂ (paratype, ZFMK); 2♂, same loc. and alt., 11.viii.1936, H. Höne, de Laever slide no. 114♂, Vojnits slide no. 12505♂ (specimen labelled as paratype ♂ of *E. depressa* Vojnits, 1984 but gen. slide labelled as paratype ♂ of *E. perendina*, ZFMK); 1♂, Tibet [Sichuan], Batang, Im Tal des Yangtze (*ca.* 2800 m), 11.vii.1936, H. Höne, Vojnits slide no. 11474♂ (paratype, ZFMK); 1♂, same loc. and alt., 3.vi.1936, H. Höne, Vojnits slide no. 11463♂ (paratype, TTM); 3♂, 1♀, same loc. and alt., 6.iv., 15.vi., 5, 11.vii.1936, H. Höne, Vojnits slide no. 11462♀, de Leaver slide no. 449♂ (ZFMK); 3♂, 2♀, Tibet [Sichuan], Batang, Untere Urwaldzone (*ca.* 3800 m), 10, 12, 19, 24.vi., 6.viii.1936, H. Höne, Vojnits slides nos. 12424♂, 12444♀, 12443♀, 12442♂, 12440♂ (paratypes, ZFMK); 1♂, same loc. and alt., 1.viii.1936, H. Höne, Vojnits slide no. 13852♂ (paratype of *E. garrula*, TTM); 1♂, same loc. and alt., 7.viii.1936, H. Höne, Vojnits slide no. 13822♂ (paratype of *E. garrula*, ZFMK); 1♂, 1♀, same loc. and alt., 21.vi., 3.viii.1936, H. Höne, Vojnits slide no. 13022♀ (ZFMK); 1♀, China, Provinz Nord-Yuennan, Li-kiang [Lijiang], 29.vi.1934, H. Höne, Vojnits slide no. 12540♀ (paratype, ZFMK); 1♀, same loc., 3.vi.1934, H. Höne, de Laever slide no. 365♀ (ZFMK); ♂, same loc., *ca.* 2000 m, 14.vii.1934, H. Höne, Vojnits slide no. 13271♂ (holotype of *E. garrula*, ZFMK); 2♀, China, Sichuan Prov., Batang, 2500 m, 10.vii.2001, Li Houhun leg. (coll. P. Skou); 1♂, China, North Yunnan, Diqing Tibetan Aut. Pref.,

5 km SE of Deqin, 3356 m, 19.vi.2009, leg. B. Benedek, Pekarsky slide no. OP 0854♂ (coll. O. Pekarsky); 2♀, China, W. Sichuan, road Xinngcheng / Zhongdian, 3300 m, 17.vii.2009, 28°46′ N / 99°51′ E, I. & A. Floriani, Saldaitis leg., Pekarsky slides nos. OP 0653♀, OP 0807♀ (coll. O. Pekarsky).

227. *Eupithecia lamata* Mironov & Galsworthy, 2004

Eupithecia ***lamata*** Mironov & Galsworthy, 2004, *Trans. lepid. Soc. Japan* 55 (2): 130, figs 10, 28. Holotype ♀ (IZCAS; examined), China: Qinghai, Menyuanxianmi, 2800 m.

Description. Fore wing ground colour pale buff, transverse lines dark brown; antemedial, medial and triple postmedial curved onto costa, medial just outside dark brown, elongate discal dot; terminal area darker. Hind wing ground colour as fore wing; discal dot large but faint; antemedial and postmedial lines complete, mostly parallel to termen but angled close to anal margin.
Female genitalia. Bursa oval, covered with numerous small spines except immediately adjacent to colliculum. Ductus bursae very short, with short longitudinal striations. Ductus seminalis broadened at base, attached to corpus bursae near short and narrow colliculum. Antrum funnel-shaped, slightly sclerotized. Tergite A8 almost square. Apophyses narrow, rather short. Papillae anales rounded.
Male unknown.
Range. China (Qinghai).
Phenology. late July.
Altitude. Found at about 2800 m above sea level in Qinghai (China: coll. IZCAS).
Similar species. Known material is worn, but the rather heavy transverse lines are distinctive, and should enable recognition. The female genitalia, particularly the antrum, are reminiscent of the West Mediterranean *E. cooptata* Dietze, 1904, from which it differs in the lack spines at the entrance of the ductus seminalis.
Material. ♀, China, Qinghai, Menyuan, Xianmi, 2800 m, 23–24.vii.1992, coll. Xue Dayong, slide no. ZL513♀ (holotype, IZCAS); 1♀, same data, Galsworthy slide no. ZL273♀ (paratype, IZCAS).
Remarks. Attribution to this species group is tentative in the absence of the male.

228. *Eupithecia ancillata* Mironov & Galsworthy, 2004

Eupithecia ancillata Mironov & Galsworthy, 2004, *Trans. lepid. Soc. Japan* 55 (3): 233, figs 7, 21, 24. Holotype ♂ (ZFMK; examined), C[entral] China, prov. Shaanxi, Foping Natural Reserve, Panda area, 33°45′ N, 107°48′ E, 1600 m.

Description. Fore wing ground colour pale brownish grey, all transverse lines relatively broad, dark brown, broadened near costa; basal and antemedial evenly curved; narrow medial triply curved; postmedial oblique, straight until sharply curved near tornus; medial area relatively broad with veins Cu, CuA_2 and A covered with black scales; terminal area darker brown with pale wavy line; discal spot large, elongate. Hind wing paler, greyish; transverse lines visible along dorsum; discal spot small, indistinct. Abdomen light brownish grey, with a row of small dark brown dorsal dots; third segment brown; the last segment pale grey with a brown tip.
Male genitalia. Uncus elongate, biapical. Valve unmodified. Vinculum short and narrow. Papillae relatively broad. Aedeagus narrow, shorter than length of valve. Vesica granulate, armed with an elongate V-shaped plate, and a narrow string of tooth-like cornuti. Sternite A8 very small, peg-like, broadened basally, with two short pointed apical lobes.
Female genitalia. Bursa oval, about ¾ covered with spines on base and left side. Ductus bursae short with small longitudinal striations. Ductus seminalis broadened at base, attached to bursa near colliculum on right. Colliculum short and narrow. Antrum membranous. Tergite A8 relatively large, square. Apophyses short and narrow. Papillae anales broad and rounded.
Range. China (Shaanxi, Shanghai); Korea (Chuncheon); Japan (Tsushima).
Phenology. Univoltine: early March to early May.
Altitude. Found at about 1600 m above sea level in Shaanxi (China: coll. ZFMK).
Similar species. Readily recognized and easily distinguished from all other Asian *Eupithecia* by the distinctive fore wing maculation. The long chain of small cornuti on the vesica in the male genitalia, and the disposition of the spiniferous area in the bursa in the female genitalia are distinctive.
Material. ♂, C. China, Shaanxi prov., Foping Natural Reserve, Panda area, 33°45′ N, 107°48′ E, 1600 m, 6.-11.iv.1999, leg. V. Siniaev & A. Plutenko (holotype, ZFMK); 2♂, same data (paratypes, ZFMK); 1♀, same data, Mironov slide no. 512♀ (paratype, ZISP); 1♂, same loc. and alt., but date 15.–20. iv.1999, Mironov slide no. 511♂ (paratype, ZISP); 1♀, [Korea], Chuncheon-dam, Chuncheon GW, 2.v.1996, leg. Y.M. Park (paratype, coll. Oh); 1♂, China, Provinz Kiangsu [Jiangsu], Schanghai, 6.iii.1946, H. Höne (TTM).

229. *Eupithecia simpliciata* Haworth, 1809

Phalaena simpliciata Haworth, 1809, *Lepid. Br.* (2): 359. Syntype(s) (BMNH), [United Kingdom].

Geometra subnotata Hübner, 1813, *Samml. Eur. Schmett.* 5 Geometrae (2), pl. 89, fig. 458. Syntype(s) lost. [Europe] (synonymised as *E. subnotataria* [sic] in Herrich-Schäffer, [1848]).
Eupithecia subnotata f. ***collustrata*** Dietze, 1910, *Biologie der Eupithecien* 1, pl. 85, fig. 947; *ibidem* (1913), **2**: 107. Syntypes 2♂, 3♀ (coll. K. Dietze in MNHU; examined), [Kirghizstan]: Karagai-tau; Naryn; [China]: Altyn-dagh.

Misidentification: *Eupithecia scabiosata*: Duponchel (1830) *nec* Borkhausen, 1794 (identity: *Eupithecia subumbrata*).
Unavailable names (infrasubspecific): *brunnea*: Lempke (1951); *impuncta*: Lempke (1951). Incorrect subsequent spellings: *subnotaria*: Boisduval (1840); *subnotataria*: Herrich-Schäffer (1848).

Description. Fore wing ochreous; transverse lines fine; postmedial brownish, finely waved, slightly curved but not angled near costa, bordered externally by a ribbon-like whitish line; terminal area with wavy whitish subterminal line; discal dot small, brown. Hind wing light greyish with darker brownish anal margin and fine, indistinct transverse lines; terminal area ochreous, wavy whitish line distinct; discal dot minute, sometimes absent.
Male genitalia. Uncus slim, biapical. Valve unmodified. Vinculum broad, sometimes with shallow medial hollow. Papillae short and broad. Aedeagus stout, slightly shorter than length of valve. Vesica covered with tiny teeth and with an elongate patch of dense spinules, and armed with a broad, V-shaped cornutus near ductus ejaculatorius base. Sternite A8 peg-like, elongate and narrow, with two blunt, heavily sclerotized apical horns.
Female genitalia. Bursa pear-shaped, with numerous small spines at base. Ductus bursae broad, spineless, with longitudinal striations. Ductus seminalis broadened at base, attached to bursa near middle on right. Colliculum short and wide. Antrum membranous, short and broad. Apophyses short. Tergite A8 almost square. Papillae anales short and rounded.
Range. China (Xinjiang); Europe, Asia Minor, Caucasus, Transcaucasus, western Siberia, Kazakhstan, Turkmenistan, Uzbekistan, Tajikistan, Kirghizstan.
Phenology. Univoltine: early July to late August. Larvae from late July to early September. Overwintering as pupa.
Biology. Larvae on the flowers and ripening seedheads of *Chenopodium* sp. (Kirghizstan: original data of Mironov), also *Chenopodium album*, *C. vulvaria*, *C. hybridum*, *C. bonus-henricus*, *C. glaucum*, *C. rubrum*, *Atriplex patula*, *A. prostrata* (= *hastata*), *A. littoralis*, *A. rosea*, *A. oblongifolia* (Chenopodiaceae), and occasionally (?) on *Mercurialis annua* (Euphorbiaceae) and *Artemisia maritima* (Asteraceae) (Europe: Mironov 2003).
Altitude. From sea level to about 3400 m in Pamirs (Tajikistan: coll. ZISP; Viidalepp 1988, as *E. inculta*).

Similar species. Best distinguished by smooth pale ochreous coloration, and fine and slightly curved postmedial bordered externally by whitish ribbon-like line. The cornuti on the vesica, and the shape of the eighth sternite in the male, and the spiniferous area in the bursa in the female genitalia are distinctive.
Material. 1♀, [Kirghizstan], Karagai tau, *subnotata* v. *collustrata* v. *issyka* Bang-Hs., Karagai tau (syntype, coll. K. Dietze in MNHU); 1♀, [Xinjiang], Asia centr., Kuldja 1886, *subnotata* ♀ (Stgr. det.), R. Tancré 5.05 (coll. R. Püngeler in MNHU).

The *sinuosaria* species group

This species group was introduced by Vojnits (1976), and elaborated by Mironov (2003). Moths of the group are characterised externally by a strongly marked median band. In the male genitalia the valve is unmodified, and the eighth sternite broad, but deeply bifid. In the female genitalia the bursa is large and oval, strongly spined, and with a striate ductus bursae and broad colliculum. The following two species occur in China.

230. *Eupithecia sinuosaria* Eversmann, 1848

Larentia sinuosaria Eversmann, 1848, *Bull. Soc. imp. Nat. Moscou* **21** (3): 230. Lectotype ♂ (ZISP; examined), [Russia]: Irkuzk [= Irkutsk]
Eupithecia obliquaria Leech, 1897, *Ann. Mag. nat. Hist.* (6) **20**: 70. Lectotype ♂ (BMNH; examined), China (western): Tachien-lu [Kangding] (treated as form of *E. sinuosaria* in Prout, 1915).
Eupithecia sinuosaria tenella Vojnits, 1976, *Acta zool. Acad. Sci. hung.* **22** (3–4): 417, fig. 2b. Holotype ♀ (ZFMK; examined), [China]: Prov. Shansi [Shanxi], Mien-shan, 2000 m.

Unavailable names (incorrect subsequent spellings): ***sinnos.aria***: Dietze (1913); *sinuoaria*: Valle (1946); *sinnuosaria*: Bielewicz (1966); *sinousaria*: Lastukhin (1998).

Description. Fore wing yellowish brown, costal margin ash-grey; basal area with dark basal line and thrice sharply angled postbasal line; antemedial slightly sinuate, right angled, black and widened near costa; medial curved and obtusely angled near discal dot, straight, black and widened near costa; basal part of medial area between antemedial and medial narrow, constricted in centre, broadened towards costa, dark reddish brown; postmedial inconspicuous, sinuate and evenly curved; terminal area with large blackish brown apical blotch, sometimes with small similar tornal blotch and always with slightly sinuate pale wavy subterminal line; discal dot linear, oblique. Hind wing paler, dark transverse lines marked only on anal margin; discal dot small, rounded.
Male genitalia. Uncus narrow, biapical. Valve unmodified. Vinculum broad, semicircular. Papillae short, quite broad. Aedeagus shorter than

length of valve, vesica armed with a curved cornutus at apex and with a flat V-shaped cornutus at ductus ejaculatorius base. Sternite A8 elongate, bilobed basally, narrowed medially, with two diverging blunt rods.

Female genitalia. Bursa oval, basal half covered with small spines. Ductus bursae broad, with numerous longitudinal striations and a large ovoid patch of spines on ventral side. Ductus seminalis narrow, attached to central part of ductus bursae. Colliculum elongate and broad. Antrum membranous. Tergite A8 almost quadrate with rounded posterior corners. Apophyses short and narrow. Papillae anales short and rounded.

Range. China (Xinjiang, Gansu, Inner Mongolia, Shanxi, Hebei, Sichuan, Tibet, Qinghai, Yunnan); Central, northern and eastern Europe, Urals, northern Kazakhstan, Asian Russia (western and southern Siberia, Altai Mts., Burjatia, Transbaikalia, Dahuria; Priamurje, Sakhalin), Mongolia and Korea.

Phenology. Univoltine (?): late May to late August.

Biology. Larvae found on the leaves, flowers and seeds of *Chenopodium album, C. pratericola, C. hybridum, C. glaucum, C. rubrum, C. bonushenricus, Atriplex patula, A. littoralis, A. laciniata, A. oblongifolia* (Chenopodiaceae) and *Polygonum aviculare* (Polygonaceae) in northern and central Europe.

Altitude. From sea level up to about 3950 m in Sichuan (China: coll. Pekarsky).

Similar species. Only likely to be confused with the following species; see there for distinctions.

Material. ♂, [Russia], Irkuzk [= Irkutsk], 273e, *Larentia sinuosata* Evm., coll. Eversmann (lectotype, ZISP); 3♂, 1♀, same loc. (paralectotypes, ZISP); ♂, [Sichuan], Ta-Chien-Lu, 8300ft., Pratt. Coll., May & June 1890, Leech Coll. 1900–64, *Eupithecia obliquaria* sp. n. Type ♂ (lectotype of *E. obliquaria*, BMNH); 1♀, [Sichuan], Tà-tsien-Loû, Chasseurs indigènes 1893, Vojnits slide no. 967♀ (TTM); 2♂, [Sichuan], Tà-tsien-Lou, 1898, Chasseurs indigènes, Vojnits slides nos. 10468♂, 10469♂ (ZFMK); 1♀, [Sichuan], Siao-Lou, 1900, Chasseurs indigènes, Vojnits slide no. 976♀ (ZFMK); 1♂, [Sichuan],Yarégong, P. Soulié 1902 (ZFMK); 1♀, [Sichuan], Ta-tsien-Lou, Chasseurs du P. Déjean 1904 (abdomen lost; ZFMK); 1♂, [Sichuan], Chasseurs indigènes de Tà-tsien-lou, Récolte de 1910, Vojnits slide no. 10470♂ (ZFMK); 1♀, Thibet, Tàtsiénloû, Mgr. F. Biet, Vojnits slide no. 966♀ (ZFMK); 1♀, [Yunnan], Atuntse [Dêqên], 9/57, ex coll. R. Oberthür (ZFMK); 1♂, Szetschwan [Sichuan], Kwanhsien [Guanxian], Exp. Stötzner (ZFMK); 1♂, 2♀, Szetschwan [Sichuan], Sunpanting, Exp. Stötzner (both females without abdomens, ZFMK); 1♂, 1♀, Szetschwan [Sichuan],

Tatsienlu, Exp. Stötzner (female without abdomen, ZFMK); 1♂, Szetschwan [Sichuan], Tschengtufu, Exp. Stötzner (ZFMK); 2♂, China, Szetschwan [Sichuan], Ta-tsien-lu, Exped. Stötzner, Vojnits slides nos. 10464♂, 10467♂ (TTM); 1♀, Szetschwan [Sichuan], Ta-tsien-lu, Exped. Stötzner, Coll. Dr. Wehrli, Basel, Vojnits slide no. 10471♀ (ZFMK); 1♀, China, Szetschwan [Sichuan], Ta-tsien-lu, Exped. Stötzner (ZFMK); 1♂, 1♀, Szetschwan [Sichuan], Tatsienlu, Exp. Stötzner, Vojnits slide no. 10465♀ (ZFMK); 1♀, Szetschwan [Sichuan], Omisien, Exped. Stötzner, Vojnits slide no. 975♀ (ZFMK); 1♀, [Sichuan], Sung-pan-ting, Szetschwan-China, Exped. Stötzner, Vojnits slide no. 10454♀ (ZFMK); 1♀, [Sichuan],Wassekou, Szetschwan-China, Exped. Stötzner, Vojnits slide no. 10455♀ (ZFMK); 1♀, Tibet [Sichuan], Batang, Im Tal des Yangtze (*ca.* 2800 m), 22.v.1936, H. Höne, Vojnits slide no. 10436♀ (ZFMK); 1♂, 2♀, same loc., Untere Urwaldzone (*ca.* 3800 m), 12, 13.vi., 30.vii.1936, H. Höne, Vojnits slide no. 10437♂ (as *E. sinuosaria obliquaria*, ZFMK); 1♀, same loc. and alt., 4.viii.1936, H. Höne, Vojnits slide no. 10456♀ (TTM); 1, 2♀, Nord Yünnan, A-tun-tse [Dêqên], Obere Höhe (*ca.* 4500 m), 7, 9, 23.vii.1936, H. Höne, Vojnits slide no. 10432♀ (as *E. sinuosaria obliquaria*, ZFMK); 2♂, 8♀, same loc., Mittlere Höhe (*ca.* 4000 m), 5, 21, 23, 28.vii., 2, 12, 14, 16, 17.viii.1936, H. Höne, de Laever slide no. 50♀, Vojnits slide no. 10433♂ (one male without abdomen, ZFMK); ♀, Prov. Shansi [Shanxi], Mien-shan, Obere Höhe *ca.* 2000 m, 8. viii.1937, H. Höne, Vojnits slide no. 10458♀ (holotype of *E. sinuosaria tenella*, ZFMK); 1♂, 6♀, same loc. and alt., 2, 3, 5, 7.viii.1937, H. Höne (paratypes of *E. sinuosaria tenella*; one specimen without abdomen; ZFMK); 1♂, 12♀, same loc. and alt., 28, 29, 30.vii., 1, 2, 3, 7, 10.viii.1937, H. Höne, de Laever slides nos. 47♂, 48♀, Vojnits slides nos. 10438♀, 10439♀, 10440♀, 10459♀, 10476♀, 10477♀, 19340♀, 19341♀ (paratypes of *E. sinuosaria tenella*, TTM); 1♀, same loc. and alt., 6.viii.1937, H. Höne, de Laever slide no. 90♀ (ZFMK); 2♂, China, Gansu, Xiahe, 2600–3000 m, 26.-30.vi.1986, Coll. Nr. 246, leg. Görgner, Mironov slide no. 725♂ (ZFMK); 1♂, 1♀, China, Sichuan, Daxue Shan Mts., 80 km W Kangding, 30°01′ N, 101°40′ E, 3800 m, 26.-27.vii.1999, lg. Siniaev & A. Plutenko, *ex* coll. Viktor Siniaev, Mironov slide no. 724♀ (ZFMK); 1♂, China, Yunnan Prov., Zhongdian, 3100 m, 14.vii.2001, Li Houhun leg. (coll. P. Skou); 1♀, China, North Yunnan, Diqing Tibetan Aut. Pref., 5 km SE of Deqin, 3356 m, 19.vi.2009, leg. B. Benedek, Pekarsky slide no. OP 0513♀ (coll. O. Pekarsky); 1♂, China, Sichuan, 3950 m, Garzé Tibetan Aut. Pref., Litang, 23.vi.2009, leg. Balázs Benedek, Pekarsky slide no. OP 0512♂ (coll. O. Pekarsky); 10♂, 8♀, Qinghai, Menyuan, Xianmi, 2800 m, 23.vii.1992; 4♀, Qinghai, Tongren, Maixiu, 2950 m, 30.vii.1992; 1♀, Qinghai, Huzhu, Beishan, 2300 m, 30.vii.1991, coll Xue Dayong; 2♀, Qinghai, Qilian,

28.vii.1977, coll Zhang Baolin; 8♀, Yunnan, Deqin, 3250 m, 15.vii.1982, coll Wang Shuyong, slide no. ZL89♀; 1♂, 17♀, Tibet, Qamdo, 3900 m, 7–9. viii.1976, 1♀, same loc., 23.v.1976; 6♀, Tibet, Zhag'yab Gyitang, 3600 m, 5–23. vii.1976, slide no. ZL169♀; 2♀, Tibet, Zogang, 3800 m, 29.vi.and 1.viii. 1976; 1♂, 10♀, Tibet, Riwoqê, 3750 m, 17–25.viii.1976; 1♂, 15♀, Tibet, Jomda, 3400 m, 22–31.vii.1976, slide no. ZL344♀; 1♂, Tibet, Markam, Rongxi, 4000 m, 26.vi.1976; 1♀, Tibet, Bomi, 2700 m, 19.viii.1983, all coll Zhang Xuezhong and Han Yinheng; 1♂, Tibet, Shigatse, 5.viii.1983, coll Yan Zhaoxing; 2♂, 5♀, Tibet, Mainling, 2950 m, 17.viii.1974, coll Huang Fusheng; 3♂, 7♀, Tibet, Renbu, 1.viii.1983, slide no. ZL90♂, coll Basang; 1♂, ♀, Tibet, Gyirong, 2–5. vii.1984, coll Yan Yaoxing etc; 1♂, Tibet, Chaqiansangjiu, 3400 m, 25.vi.1973; 1♂, Tibet, Namling, 1982, slide no. ZL177♂; 1♀, Tibet, Lhasa, 3600 m, 4. vii.1974, coll Huang Fusheng; 1♀, Inner Mongolia, Jiwen, 7.viii.1982; 1♀, Inner Mongolia, Oragxan; 2♀, Inner Mongolia, Wumeng, Xinghe banner, 15.vii.1987, coll Bao Yongqing; 1♀, Inner Mongolia, Humeng, Eyou banner, 10.vii.1986, 1♀, same loc 5.vii.1986, slide no. ZL634♀; 1♀, Inner Mongolia, Tumuertai, 1400 m, 29.vii.1971; 3♀, Xinjiang, Zhaosu, Alasan, 28.vii.1978, coll Han Yinheng; 2♀, Xinjiang, Tomürt Peak, 2600 m, 21.vii.1977, 2♀, same loc, 2400 m, 23.vi and 4.vii.1977, coll Li Changjiang; 1♂, Gansu, Zhuoni, 2500 m, 4.ix.1990, coll Xue Dayong; 1♂, Shanxi, Huoguo, 1550 m, 28.vii.1972, coll Wang Shuyong; 1♂, Hebei, Xiao Wutaishan, 12.vii.1980 (IZCAS).

231. *Eupithecia pallescens* Dietze, 1910, stat. n.

Eup[ithecia] sinuosaria f. *pallescens* Dietze, 1910, *Biol. Eupithecien* 1, pl. 73, fig. 385; *ibidem* (1913), 2: 110. Holotype ♀ (coll. K. Dietze in MNHU; examined), ([China]: Aksu, Takla-Makan.

[*Eupithecia*] *sinuosaria* f. *modesta* Dietze, 1913, *Biol. Eupithecien* 2: 110 (non binominal). Holotype ♂ (coll. K. Dietze in MNHU; examined), ([Kirghizstan]: Alai Mts., Ferghana. **New junior synonym.**

Eupithecia morosa Vojnits, 1976, *Acta zool. Acad. Sci. hung.* 22 (3–4): 420, figs 4a-c, 5b. Holotype ♂ (ZFMK; examined), Oriental Turkestan [Kirghizstan]: Semirechgensee Province, Fort Narine [= Naryn]. **New junior synonym.**

Misidentifications: *Eup[ithecia] sinuosaria*: Staudinger, 1900 (*nec* Eversmann, 1848); *Eupithecia hannemanni*: Vojnits, 1977 (*nec* Vojnits & de Laever, 1973).

Description. Fore wing basal and medial areas a soft reddish brown; area between basal and medial areas blackish grey; transverse lines broadened, black near costa; basal evenly curved; postbasal and antemedial twice angled; postmedial curved inwards, touching discal dot, where obtuse angled onto costa; terminal area darker, blackish brown or rusty brown near apex with two black marginal dashes; pale subterminal line very distinct; discal dot oblique, linear, intensely black. Hind wing pale greyish brown, dark transverse lines well marked near anal margin only; terminal area broad, slightly darker with weak trace of indented pale subterminal line; discal dot small, rounded or comma-like.

Male genitalia. Uncus narrow, biapical. Valve unmodified. Vinculum broad, semicircular. Papillae long and broad. Aedeagus stout, shorter than valve length. Vesica armed with two dentate cornuti at apex, and a longitudinally folded cornutus with a connected irregular cornutus at base. Sternite A8 elongate, bilobed at base, with parallel lateral margins and two long blunt apical horns.

Female genitalia. Bursa large, covered with spines except for large area on right side. Ductus bursae broad, with longitudinal striations and on right a continuation of the spine-field on the ductus bursae. Ductus seminalis narrow, attached to posterior part of bursa. Colliculum elongate, only moderately broad. Antrum membranous. Tergite A8 transverse. Apophyses short and narrow. Papillae anales short and rounded.

Range. China (Xinjiang, Gansu); Uzbekistan, Kirghizstan, Kazakhstan (SE), Mongolia.

Phenology. Univoltine: late June to mid-August. Larva: late July to early September. Overwintering as pupa in flimsy cocoon in the ground.

Biology. Larva found on the flowers and unripened seeds of *Atriplex tatarica* (Chenopodiaceae) in Tien Shan Mts. (Kirghizstan: original data of Mironov).

Altitude. From about 1900 m (Mongolia: Vojnits 1977, as *E. hannemanni*) up to about 3500 m above sea-level in Tien Shan mountains (Kirghizstan: Viidalepp 1988; coll. ZISP; China, Xinjiang: IZCAS).

Similar species. Very similar to the preceding species, from which no significant differences in size and wing maculation could be found. *E. pallescens* appears only slightly paler than *E. sinuosaria*. In the male genitalia the cornuti in the aedeagus vesica and shape of sternite A8 are characteristic. The female can be distinguished by the spining of the bursa running up as far as the narrower colliculum.

Material. ♀, [Xinjiang], ♀, *sinuosaria* Ev. forma: *pallescens* Typ. (385), Aksu, Takla-Makan, Vojnits slide no. 10241♀, Holotypus *Eupithecia sinuosaria* Ev. *pallescens* Dietze [Vojnits design.] (holotype of *E. sinuosaria pallescens*; paratype of *E. morosa* Vojnits, 1975/76, coll. K. Dietze in MNHU); ♂, [Kirghizstan], Alai mont. Fergana Juli, *sinuosaria* forma: *modesta* Bang-Haas, Ferghana, Alai mont., Vojnits slide no. 10243♂, Holotypus *Eupithecia sinuosaria* Ev. *modesta* Dietze [Vojnits design.] (holotype of *E. sinuosaria modesta*, coll. K. Dietze in MNHU); 1♀, [Xinjiang], Asia centr., Saichin-Kette nördl. von Korla, Rückbeil 1902, *sinuosaria* ♀, v. R. Tancré 12/02 (coll. R. Püngeler in MNHU); 1♀, [Xinjiang], Korla 97 Hbh., 2/7 (coll.

O. Staudinger in MNHU); 1♂, 2♀, [Xinjiang], Mus-tag-ata Yarkend (coll. K. Dietze in MNHU); ♂, [Kirghizstan], Turkestan Oriental, Province Semirechgensee, Fort Narine, G. S. Akulin, 1907, Vojnits slide no. 972♂ (holotype of *E. morosa*, ZFMK; slide absent in ZFMK); 1♂, 2♀, same data, but 1♀ was captured in 1906, Vojnits slides nos. 971♀, 10449♀ (paratypes of *E. morosa*, ZFMK; slide no. 10449♀ absent in ZFMK); 1♀, [Kirghizstan], Issyk Kul, *Sinuosaria* Ev. ♀, 1886, Vojnits slide no. 10472♀ (paratype of *E. morosa*, ZFMK); 1♂, 1♀, [Kirghizstan], Kysyl Jart, *Sinuosaria* Ev., ♂, End Juni (other specimen ♀ marked 5.vii.[18]89), Vojnits slides nos. 10473♂, 10474♀ (paratypes of *E. morosa*, ZFMK; slide no. 10473♂); 1♀, [Mongolia], Bajan-Ölgij aimak, im Tal des Flusses Charcalyn gol, 24 km O von Somon Cagannuur, 1890 m, 29.vi.1968, No. 1042, Vojnits slide no. 10517♀ (as *E. hannemanni*, TTM); 1♂, China, Gansu, Xiahe, 2600–3000 m, 26.-30. vi.1986, Coll. Nr. 246, leg. Görgner (ZFMK); 3♂, 9♀, Xinjiang, Akto, 2650 m, 16.viii.1989, slides nos. ZL87♂, 88♀; 2♀, Xinjiang, Taxkorgan, 4500 m, 13. viii.1989; 1♀ same loc., 3550 m, 12.viii.1989, coll Zhang Xuezhong (IZCAS).

The *irriguata* species group

This species group was introduced and described by Mironov (2001, 2003). It is probably closely related to the *satyrata* group, sharing with it the small dentate cornuti in the vesica and general shape of the eighth sternite in the male, but having in addition a large field of small spines in the vesica, and generally much smaller male and female genitalia. It is represented in China, if at all, only by the following species.

232. *Eupithecia conterminata* Lienig & Zeller, 1846

Larentia conterminata Lienig & Zeller, 1846, *Isis* 3: 197. Syntypes 1♂, 1♀ (BMNH; examined), [Latvia]: ?Koknese [= Kokenhusen].
Eupithecia manniaria Herrich-Schäffer, 1848, *Syst. Bearb. Schmett. Eur.* 3 (32): 122, 136; ibidem (1846), pl. 25, fig. 156 (non binominal). Syntype(s), [Czech Republic], Böhmen (synonymised in Guenée, 1857 [1858]).
Eupithecia idiopusillata Inoue, 1979, *Bull. Fac. domest. Sci. Otsuma Wom. Univ.* 15: 213, figs 22M, N; fig. 23K; fig. 31B; fig. 33A. Holotype ♂ (BMNH; examined), [Japan]: Toyama Prefecture, Kanetsuri Spa, Kurobe (treated as subsp. of *E. conterminata* Lienig & Zeller, 1846 in Viidalepp & Mironov, 1988b).

Unavailable names (incorrect subsequent spellings): *maniaria*: Sievers (1852); *conterminanta*: Garthe (1976).

Description. Very small, with triangular pointed fore wings. Fore wing pale smoky-grey, darker along costa and in terminal area; transverse lines distinct, medial and postmedial right angled onto costa; terminal area with trace of pale subterminal line; discal dot very large, ovoid, intensely

black. Hind wing paler, with curved postmedial and paler and smaller discal dot. Abdomen pale grey, with a series of black dorsal dots.
Male genitalia. Uncus narrow, biapical. Valve very small, unmodified. Vinculum short, semicircular. Papillae relatively broad, elongate. Aedeagus narrow and short. Vesica armed with an elongate patch of spinules, three slim dentate cornuti, and a twisted, U-shaped plate near base of ductus ejaculatorius. Sternite A8 small, peg-like, tapered to apex, with two short and blunt apical horns, varying from short to longer and diverging.
Female genitalia. Bursa pear-shaped, with a large squarish patch of rather sparse spines in basal area, the most posterior considerably larger. Ductus bursae short, with an elongate transverse patch of small spines. Ductus seminalis broadened at base, attached to ductus bursae on right. Colliculum short and narrow. Antrum membranous. Tergite A8 very transverse, with a shallow medial hollow on posterior margin. Apophyses very short and narrow. Papillae anales short and broad, rounded.
Range. China (?); Central and Eastern Europe (incl. Scandinavia), Great Caucasus Mts., Urals, Asian Russia (Transbaikalia, Priamurje, Primorje), Mongolia, Japan (Honshu).
Phenology. Univoltine: late May to mid-July. Larva: mid-June to early August in Europe. Overwintering as pupa.
Biology. Larva found on the young needles of *Picea abies* (= *excelsa*) (Pinaceae) in Europe.
Altitude. From sea level up to about 1500–1600 m (Japan: Inoue 1979, as *E. idiopusillata*).
Similar species. A very distinctive species, due to small size and large discal dots. Asian specimens usually have darker fore wings with a more distinct pattern, but paler hind wings than European specimens.
Material. ♂, [Japan], Toyama Pref., Kanetsuri, 2.vi.1973, H. Yamanaka (holotype of *E. idiopusillata*, coll. H. Inoue in BMNH).
Remarks. There are two Vojnits male genitalia slides both numbered 12139, the first belonging to a male of *E. fenita* in TTM, and the other in ZFMK belonging to a male of *E. conterminata*, and marked as a holotype of the unpublished *E. sacrilega* Vojnits: no corresponding specimen was found in ZFMK, but in the context the origin is likely to have been in China. Hence we include this species, particularly since its distribution in Japan makes its occurrence in northern China likely.

The *nevadata* species group

This species group was introduced by McDunnough (1949) to contain five North American species of *Eupithecia*, and elaborated by Mironov (2003).

Moths in the group have distinctive purplish brown maculation on the wings, and usually a prominent ventral process on the male valve and distinctive shape to the male eighth sternite. We place the following two species in this group.

233. *Eupithecia insignioides* Wehrli, 1923

Eupithecia insignioides Wehrli, 1923, *Dt. ent. Z., Iris* 37 (3/4): 65, pl. 1, figs 11, 22. Holotype ♀ (ZFMK; examined), [China]: Schanghai.

Misidentifications: *Eupithecia venosata*: Inoue, 1955 (*nec* Fabricius, 1787); *Eupithecia insigniata*: Inoue, 1958 (*nec* Hübner, 1790).

Description. Fore wing pale ash-grey, with three long dark brown costal spots; transverse lines fine, blackish grey; basal and double antemedial strongly angled, double medial sinuate, touching discal dot; postmedial double, sinuate with a series of black dashes on veins; terminal area narrow, broadened near apex, darker, with trace of a pale subterminal line; discal dot narrow and elongate, black. Hind wing concolorous with brownish grey basal and terminal areas, usually with black dashes on veins; discal dots clear but small.
Male genitalia. Uncus biapical. Valve with a short, pointed ventral process. Vinculum narrow, tapered. Papillae straight, membranous. Aedeagus stout, shorter than valve length. Vesica armed with two small dentate cornuti, one elongate, minutely spined area and a V-shaped cornutus with apically broadened arms. Sternite A8 weakly sclerotized, consisting of two narrow rods, connected by a broad basal band.
Female genitalia. Bursa large, oval, completely covered with spines except small rounded area at left 'shoulder'. Ductus bursae elongate, sclerotized, with longitudinal striations. Ductus seminalis broad, sharply recurved, attached to base of ductus bursae on right. Colliculum short, but broad. Antrum membranous. Tergite A8 quadrate. Apophyses short and narrow. Papillae anales relatively small, rounded.
Range. China (Gansu, Shaanxi, Shanghai); Asian Russia (Dahuria, Transbaikalia, Priamurje, southern Primorje, Sakhalin), Korea, Japan (Hokkaido, Honshu, Shikoku, Kyushu).
Phenology. Bivoltine: mid-March to early July; September.
Altitude. From sea level (Far East Russia: coll. ZISP) up to about 1700 m in Qinling Mts. (China: coll. ZFMK, ZISP and TTM).
Similar species. Could only be confused with the closely related European *E. insigniata* Hübner, 1790. The male genitalia are distinguished by the much shorter, and symmetrical ventral processes on the valves, and the

shape of the eighth sternite. In the female, the ductus bursae is much longer, and the eighth tergite quadrate rather than transverse.
Material. ♀, Shanghai China, *insignioides* Wehrli, *Eupithecia*, abgebildet Seitz IV, Suppl. fig., Type, Vojnits slide no. 670♀ (holotype, ZFMK); 2♂, China, Sued-Shensi [Shaanxi], Tapaishan [Taibaishan] im Tsinling [Qinling], *ca.* 1700 m, 18, 20.v.1936, H. Höne (TTM); 2♂, C. China, prov. Shaanxi, Mt. Tai bei shan, Hao Djenzy village, 33°48′ N, 107°43′ E, 1400 m, ix.1998, leg. loc. coll. (ZFMK); 2♀, same loc. and alt., iv.1999, leg. loc. coll. (ZFMK); 6♂, same loc. and alt., 1.-5.iv.1999, leg. V. Siniaev & A. Plutenko (ZFMK, ZISP); 1♂, 3♀, C. China, prov. Shaanxi, Foping Natural Reserve, Panda area, 33°45′ N, 107°48′ E, 1600 m, 6.-11.iv.1999, leg. V. Siniaev & A. Plutenko (ZFMK); 1♂, 6♀, same loc. and alt., 15.-20.iv.1999, leg. V. Siniaev & A. Plutenko (ZFMK); 1♂, Gansu, Yongdeng, Tulugou, 2280 m, 20–26. vii.1991, coll Meng Feng, slide no. ZL274♂ (IZCAS).

234. *Eupithecia costimacularia* Leech, 1897

Eupithecia costimacularia Leech, 1897, *Ann. Mag. nat. Hist.* (6) **20**: 69. Holotype ♀ (BMNH; examined), Japan: ?Yokohama.

Description. Fore wing ground colour white to ochreous, with four brown costal blotches, a brownish anal margin to the medial area, and a darker terminal area with a distinct white subterminal line broadened near tornus; discal dot narrow and elongate. Hind wing concolorous with brown basal area along anal margin and brown to ochreous tornal blotch. Specimens from Taiwan are darker.
Male genitalia. Uncus short and narrow, uniapical. Valve small, apically rather pointed, with a sclerotized and sinuate sacculus. Vinculum broad and elongate, with shallow medial hollow. Papillae short and relatively broad. Aedeagus stout, equal to length of valve. Vesica with several longitudinal wrinkles and an elongate patch of small teeth, armed with a U-shaped plate-like cornutus and an irregular cornutus near ductus ejaculatorius base. Sternite A8 with two broad basal lobes, apical half very narrow with two apically diverging rods.
Female genitalia. Bursa ovate, covered with dense spines. Ductus bursae sclerotized, transverse, with short longitudinal striations. Ductus seminalis slightly broadened basally, sharply recurved, attached to central part of bursa on right. Colliculum long and tapering, broadest at base. Antrum membranous. Tergite A8 almost quadrate. Apophyses short and narrow. Papillae anales broad and rounded.
Range. China (Guangxi, Guangdong, Hong Kong, Taiwan); Japan (Honshu, Shikoku, Kyushu, Tsushima).

Phenology. Univoltine: late February to late March.
Altitude. Found from about 500 to 2600 m above sea-level (NSMT, Taiwan: Inoue 1988).
Similar species. Not readily confusable with any other Chinese *Eupithecia* species. The eighth sternite in the male is very distinctive. Although externally superficially resembling other members of this species group, the male genitalia differ substantially, so that attribution to this species group is tentative.
Material. ♀, H. Preyer Coll. Japan, Leech Coll. 1909–64, *Eupithecia costimacularia* sp. n. Type ♀ (holotype, BMNH); 1♀, Taiwan, Prov. Nan-Tou, 15 km N of Puli, 500 m, 4.iii.1996, leg. Gy. Fábián & L. Németh (TTM); 2♀, Taiwan, Prov. Nantou, Hohuachi, 1950 m, 24°13′ N, 121°16′ E, 31.iii.2000, leg. A. Kun & L. Peregovits (ZSM); 1♂, 1♀, China, Guangxi, Guilin, Maoershan, 500–1400 m, 26–29.iii.2005, M. Owada; 1♀, Guangdong, Shaoguan, Nanling, 600–1400 m, 11–14.iii.2004, M. Owada (NSMTO); 1 specimen, Kadoorie Farm Botanic Garden, Hong Kong, 570 m, 22.iii.2009, R. Kendrick (coll. KFBG).
Remarks. Not previously recorded from mainland of China.

The *russeliata* species group

Bolte (1990) introduced this group as a segregate from the *satyrata* species group (see McDunnough 1949). Mironov (1991) introduced the similar Palaearctic *vulgata* species group, and (2003) combined it with the *russeliata* group. In the male genitalia the valve is usually unmodified, the aedeagus with several dentate cornuti, and the eighth sternite peg-like, sometimes weakly bifid at the apex. The female genitalia have a large rounded bursa, nearly always covered in spines, the ductus seminalis attached close to the colliculum, and often a bifid eighth tergite. The group is represented in China by the following thirteen species.

235. *Eupithecia assimilata* Doubleday, 1856

Eupithecia assimilata Doubleday, 1856, *The Zoologist* 14: 5140, 5142. Syntype(s) (BMNH), [United Kingdom].
Tephroclystis fumosa Hulst, 1896, *Trans. Am. ent. Soc.* 23 (3): 269. Lectotype ♂ (AMNH), [U.S.A.]: Massachusetts, South Abington (synonymised in Bolte, 1990).

Misidentifications: *Geometra minutata* Hübner, 1813 (*nec* Denis & Schiffermüller, 1775) (identity: *Eupithecia absinthiata*); *Eupithecia absynthiata* Packard, 1876 (*nec* Fischer von Röslerstamm, 1829) (identity: *Eupithecia absinthiata*); *Eupithecia coagulata* McDunnough, 1949 (*nec* Guenée, [1858]).
Unavailable names (incorrect subsequent spelling): *assimilaria*: Morris (1861).

Description. Fore wing grey-brown, often with reddish tinge; costal margin mottled and irrorate with white scales, with two to five darkish costal spots in place of transverse lines; latter inconspicuous, except postmedial, which is right-angled beyond discal dot and well marked on veins with a series of dark dashes; medial fine, angled around discal dot; wavy subterminal line broken, forming a series of small triangular white dots and a larger, often double, white tornal spot, shaded on inner margin; discal dot narrow, elongate. Hind wing paler, transverse lines indicated along anal margin; terminal area with ill-defined subterminal line, and a small white tornal spot; discal dot small.
Male genitalia. Uncus narrow, biapical. Valve unmodified. Vinculum short, semicircular. Papillae relatively short. Aedeagus stout, narrowing anteriorly, slightly shorter than length of valve. Vesica armed with three dentate cornuti and an elongate folded cornutus. Sternite A8 almost triangular, bilobed at base.
Female genitalia. Bursa oval, almost completely covered with small spines. Ductus bursae with dense patch of longitudinal striations centrally and a large patch of small spines near base of colliculum. Ductus seminalis attached to ductus bursae on right. Colliculum short and relatively broad. Antrum membranous. Tergite A8 transverse, with shallow medial hollow in posterior margin. Apophyses short and narrow. Papillae anales relatively small and rounded.
Range. China (Xinjiang); almost all Europe, Caucasus, Russia to southern Siberia (Altai Mts., Sajan Mts. and Transbaikalia), and to Primorje and Sakhalin Island, Kirghizstan (Tien-Shan Mts.); North America (Bolte 1990).
Phenology. Bivoltine: from mid-April to early October.
Biology. Larva disjunct oligophagous. Found on the leaves of *Ribes nigrum* (Grossulariaceae) and also *Humulus lupulus* (Cannabaceae), occasionally on *Ribes rubrum* and *Urtica dioica* (Urticaceae) in Europe (Mironov 2003). In North America has been found on cultivated *Ribes rubrum* (Grossulariaceae) (Canada: Bolte 1990).
Altitude. From sea level to about 1850 m in Caucasus (coll. ZISP).
Similar species. Very similar to *E. absinthiata* but with distinctly rounded costa and more prominent white subterminal dots and tornal spot on fore wing. The genitalia in both sexes are quite distinct from those of *E. absinthiata*.
Material. 1♂, [Xinjiang], *assimilata*, Juldus, Prov. Ili, K. Dietze (coll. K. Dietze in MNHU).
Remarks. The only known Chinese specimen of this species was illustrated in Dietze 1910, pl. 72, fig. 295.

236. *Eupithecia vulgata* Haworth, 1809

Phalaena vulgata Haworth, 1809, *Lepid. Br.* (2): 359. Syntype(s) (BMNH), [United Kingdom].
Geometra clusterata Hübner, [1813], *Samml. Eur. Schmett.* 5 Geometrae (2), pl. 89, fig. 457. Syntype(s) lost. [Europe] (synonymised in Prout, 1939).
Tarachia austerata Hübner, [1825], *Verz. bek. Schmett.* (21): 324. Syntype(s) lost. [Europe] (synonymised as *E. austeraria* [sic] in Herrich-Schäffer, 1848).
Eupithecia lepsaria Staudinger, 1882, *Ent. Z., Stettin* **43** (1–3): 74. Syntypes 2 ♂ (coll. O. Staudinger in MNHU; examined), [Kazakhstan]: Lepsa (treated as subsp. of *E. vulgata* in Mironov, 1989a).
Eup[ithecia] vulgata f. *montium* Dietze, 1910, *Biologie der Eupithecien* **1**, pl. 72, figs. 299, 300; *ibidem* (1913), **2**: 91. Syntypes 2♂ (coll. K. Dietze in MNHU; examined), Switzerland: Zermatt.
Eup[ithecia] vulgata f. *atropicta* Dietze, 1910, *Biologie der Eupithecien* **1**, pl. 72, fig. 301; *ibidem* (1913), **2**: 91. Syntype(s) (coll. K. Dietze in MNHU; examined), Austria: Tirol.
Eupithecia vulgata var. *cyrneata* Schawerda, 1933, *Z. Öst. Ent. Ver.* **18** (6): 48, pl. 8, figs. 1, 4, 7, 8. Syntypes (NHMW), Corsica: Vizzavona, Ponte de Fulminato.
Eupithecia vulgata scotica Cockayne, 1951, *The Entomologist* **84** (1058): 154. Holotype ♂ (BMNH), [United Kingdom]: [Scotland], Aviemore.
Eupithecia vulgata clarensis Huggins, 1962, *The Entomologist* **95** (1185): 45. Holotype ♀ (BMNH), [Ireland]: Co(untry) Clare, the Burren, Ballyvalackan.

Unavailable names (infrasubspecific): *nigrofasciata*: Dietze (1913); *unicolour*: Lempke (1951); *impuncta*: Lempke (1951). – (incorrect subsequent spellings): *austeraria*: Boisduval (1840); *vulgaria*: Morris (1861).

Description. Fore wing ground colour pale greyish brown, veins often chequered dark and light; transverse lines slightly sinuate, sharply angled, broadened and slightly darker near costa; basal, postbasal and antemedial rather inconspicuous; postmedial usually straight, twice angled onto costa; pale wavy subterminal line fine, forming a whitish tornal spot; discal dot minute, circular to ovate. Hind wing slightly paler, greyish brown, darker towards termen; transverse lines weakly marked; pale subterminal line fine, indented, but sometimes ill-defined; discal dot minute.
Male genitalia. Uncus short and stout, biapical. Valve unmodified. Vinculum broad, semicircular. Papillae short and stout. Aedeagus tapered to anterior end, shorter than length of valve. Vesica armed with three small dentate cornuti and a small basal folded cornutus. Sternite A8 triangular with bilobed base.
Female genitalia. Bursa globular, almost completely covered with small spines and with a separate small patch of spinules near colliculum. Ductus seminalis attached to posterior part of bursa near colliculum on right. Colliculum short and narrow. Antrum membranous. Tergite A8 transverse, with a deep medial hollow in posterior margin. Apophyses short and narrow. Papillae anales, short, broad and rounded.
Range. China (Xinjiang); Europe, Northern Africa (Morocco), Asia Minor, Lebanon, Caucasus, Transcaucasia, northern Iran, Kazakhstan, Uzbekistan,

Kirghizstan, Tajikistan, Afghanistan, Russia (European part, Urals, western and southern Siberia [Altai Mts., Priamurje]), Mongolia, Korea.
Phenology. Uni- or bivoltine (?): early June to early August.
Biology. Larva found on the leaves of a numerous mainly herbaceous plants in Europe.
Altitude. From sea level to about 3900 m in Pamirs. (Tajikistan: Viidalepp 1988).
Similar species. Very variable in intensity of ground colour, but usually distinguishable by reduced discal spot, clear, doubly angled postmedial, and white tornal spot. Asian specimens differ little from those from Europe. In the male genitalia, the short and broad uncus, coupled with the three dentate cornuti in the vesica and the shape of the eighth sternite are diagnostic. The pattern of spining in the female bursa and the deeply cleft eighth tergite are helpful.
Material. 1♂, [Kazakhstan], Lepsa Hbhr., *lepsaria* ♂, Origin., Gen. vérifiés 1454[1] *Eupithecia lepsaria* Stgr. ♂, Lepsa, dét. E. de Laever 1954 (syntype, coll. O. Staudinger in MNHU); 1♂, *lepsaria* ♂ Lepsa, Origin., bon. sp. bei *Vulgata* nicht *gueneata* (Boh.), Vojnits slide no. 19034♂ (syntype, coll. O. Staudinger in MNHU); 1♂, [Xinjiang], Korla, Sai-Chin, ? *vulgata*, Korla, Sai-chin, Palpen länger als bei *patruelis* aus Korla. Fransen nicht gescheckt (coll. K. Dietze in MNHU); 1♂, [Xinjiang], Kuldja, v. *Oxydata* Tr. ♂, 1886, Mironov slide no. 679♂ (ZFMK).

237. *Eupithecia andrasi* Mironov & Galsworthy, 2004

Eupithecia ***andrasi*** Mironov & Galsworthy, 2004, *Trans. lepid. Soc. Japan* 55 (2): 117, figs 1, 13. Holotype ♂ (ZFMK; examined), [China]: Sued-Shensi [South Shaanxi], Tapaishan [Taibaishan] im Tsinling [Qinling Mts.], 3000 m.

Description. Fore wing ground colour pale buff; discal dot small, round; antemedial almost straight; postmedial doubly angled; area between speckled with darker brown scales, perhaps forming darker medial fascia in fresh specimens; terminal area slightly darker, with wavy, pale submarginal line. Hind wing concolourous, with scattered darker scales between anal margin and discal area.
Male genitalia. Tip of uncus short and narrow, main part broadened and bulbous. Valve unmodified. Vinculum short, tapered. Papillae small and narrow. Aedeagus short and stout. Vesica armed with two large dentate cornuti, one elongate and one short irregular plate at ductus ejaculatorius base, and a patch of minute spines. Sternite A8 peg-like, with a pointed, sclerotized and curved apical rod.
Female unknown.

Range. Central China (Shaanxi).
Phenology. Univoltine (?): mid-June.
Altitude. Found at about 3000 m above sea-level in Taibaishan (China: coll. ZFMK).
Similar species. The only known specimen is too worn to permit comparison with other species. The male genitalia are very similar to those of *E. vulgata*, but the aedeagus is stouter, and the eighth sternite ends in a strong asymmetric curved process.
Material. ♂, South Shensi [Shaanxi], Tapaishan [Taibaishan] im Tsinling [Qinling], *ca.* 3000 m, 16.vi.1936, H. Höne, Vojnits slide no. 12375♂ (holotype, ZFMK).

238. *Eupithecia nigristriata* Mironov & Galsworthy, 2004

Eupithecia nigristriata Mironov & Galsworthy, 2004, *Trans. lepid. Soc. Japan* 55 (4): 295, figs 9, 17, 22. Holotype ♂ (IZCAS; examined), China: Tibet, Jomda military post, 3400 m.

Description. Fore wing ground colour uniform pale brown, transverse lines entirely absent; discal dot jet black, lengthened into a broad streak, with two costal spots on either side; terminal line brown. Hind wing ground colour as fore wing; anal margin with a band of brighter colouring, slightly yellowish, with brown scales scattered near base, forming beginnings of double medial and single postmedial lines.
Male genitalia. Uncus narrowly bell-shaped, biapical. Valve with angulate ventral margin. Vinculum broad and short. Papillae small, clavate. Aedeagus narrow, equal to valve length. Vesica armed with three small dentate cornuti, a V-shaped cornutus and an irregular cornutus near ductus ejaculatorius base. Sternite A8 peg-like with bilobed base.
Female genitalia. Bursa globular, covered with small spines. Ductus bursae short, with longitudinal striations and an elongate transverse patch of small spines. Ductus seminalis broadened at base, attached to upper part of ductus bursae. Colliculum short and narrow. Antrum membranous. Tergite A8 small, almost square. Apophyses short and narrow. Papillae anales short and rounded.
Range. China (Tibet, Qinghai).
Phenology. Univoltine: late July and early August.
Altitude. From about 2800 to 3400 m above sea-level (China: coll. IZCAS).
Similar species. Very distinct because of heavy discal streak. Male genitalia similar to those of *E. assimilata*, but with angled ventral margin to valve. Female genitalia resemble those of *E. vulgata*, but have a differently shaped patch of spines near the colliculum, and no membranous medial hollow in the posterior margin of the eighth tergite.

Material. ♂, China, Tibet, Jomda military post, 3400 m, 25.vii.1976, coll. Zhang Xuezhong (holotype, IZCAS); 1♂, China, Tibet, Jomda, 3400 m, 27. vii.1976, coll. Zhang Xuezhong; 4♂, 3♀, Qinghai, Tongren, Maixiu, 2950 m, 30.vii–1.viii.1992, slides nos. ZL70, 366, 370, 525♂, 71, 358, 525♀; 2♂, Qinghai, Menyuan, Xianmi, 2800 m, 31.vii–1.viii.1991, slides nos. 375, 381♂ all coll. Xue Dayong (IZCAS).

239. *Eupithecia sacrimontis* Prout, 1938

Eupithecia sacrimontis Prout, 1938, *in* Seitz, *The Macrolepidoptera of the World* 4 (Suppl.): 199, pl. 18, fig. c. Holotype ♂ (BMNH; examined), [China]: Mt. Omei, 4000ft.
Eupithecia finitima Vojnits, 1979, *Acta zool. Acad. Sci. hung.* 25 (1–2): 201, figs 10, 11. Holotype ♂ (ZFMK; examined), [China]: Sued-Shensi, Tapaishan in Tsinling, 1700 m. **New junior synonym.**
Eupithecia deformis Vojnits, 1979, *Acta zool. Acad. Sci. hung.* 25 (3–4): 426, fig. 1. Holotype ♂ (ZFMK; examined), [China]: Sued-Shensi, Tapaishan in Tsinling, 1700 m. **New junior synonym.**

Description. Fore wing dark brownish grey, with inconspicuous transverse lines, slightly curved near costa and forming three small blackish costal spots; narrow double medial running through discal dot; terminal area with indistinct white wavy subterminal line and small white tornal spot; discal dot relatively large, ovoid, intensely black. Hind wing very slightly paler, darker along anal margin, with small white tornal dot and small dark discal dot.
Male genitalia. Uncus longish, biapical. Valve with a short, angulate medial process on ventral margin. Vinculum rather short, semicircular. Papillae elongate. Aedeagus stout, shorter than valve length. Vesica granulate, armed with a long and narrow cornutus, slightly broadened and twisted at anterior end. Sternite A8 peg-like, broadly bilobed at base, narrow at middle and slightly broadened near apex, with two short apical lobes.
Female genitalia. Bursa large, globular, almost completely covered with small spines. Ductus bursae not clearly distinguished, with transverse patch of spines near base of colliculum. Ductus seminalis broadened at base, attached to ductus bursae near colliculum on right. Colliculum short, moderately broad, and strongly oblique. Antrum membranous. Tergite small, almost quadrate. Apophyses short and narrow. Papillae anales short and rounded.
Range. China (Shaanxi, Sichuan, Hubei, Gansu).
Phenology. Univoltine: early July to mid-August.
Altitude. From about 1220 m in Mt. Omei up to about 1700 m in Tapaishan (China: Vojnits 1979a).
Similar species. Externally similar to *E. raniata* and *E. vasta*, but distinguished from both by the smaller discal dots and less elongate fore wings

with more obtuse apices. In the male genitalia the valve shape is distinctive, as are the ornamentation of the vesica and the eighth sternite. The female genitalia are very similar to those of *E. vulgata*, but can be distinguished by the entire eighth tergite.

Material. ♂, [Sichuan], Mt. Omei, 4000ft., 4.7.[19]31, G.M. Franck, *Eupithecia sacrimontis* Prout type ♂, L.B. Prout coll., B.M. 1939–643, BM Geom. slide no. 20205♂ (holotype, BMNH); ♂, Sued-Shensi [Shaanxi], Tapaishan [Taibaishan] im Tsinling [Qinling], *ca.* 1700 m, 8.viii.1936, H. Höne, Vojnits slide no. 11435♂ (holotype of *E. finitima*, ZFMK); ♂, same loc. and alt., 6.viii.1936, H. Höne, Vojnits slide no. 12299♂ (holotype of *E. deformis*, ZFMK); 1♀, same loc. and alt., 9.viii.1936, H. Höne, Vojnits slide no. 12211♀ (paratype of *E. deformis*, ZFMK); 1♂, 2♀, same loc. and alt., 5, 10, 11.viii.1936, H. Höne, Vojnits slides nos. 12385♀, 12361♀, 12945♂ (ZFMK); 1♂, same loc. and alt., 7.viii.1936, [H. Höne], Mironov slide no. 829♂ (TTM); 1♂, same loc., *ca.* 3000 m, 11.viii.1936, H. Höne, Vojnits slide no. 12298♂ (paratype of *E. deformis* TTM); 1♂, 1♀, same loc. and alt., 11, 13.viii.1936, H. Höne, *deformis* det. Vojnits A., Vojnits slides nos. 12378♂, 12363♀ (ZFMK); 2♀, Sichuan, Emeishan, 1800–1900 m, 28.vii.1957,coll Huang Keren, slide no. ZL92♀; 1♀ same loc and alt, 31.vii.1957,coll Lu Youcai; 1♀ same loc, 800–1000 m 22.ix.1957, coll Zhu Fuxing; 1♂, same loc, 1800–1900 m, 12.vii.1957, coll Zhu Fuxing; 1♀ Hubei, Shennongjia, Hongping forestry reserve, 1660 m, 16.vii.1981, coll Han Yinheng, slide no. ZL195♀; 2♀ Gansu, Wenxian, Qiujiaba, 2350 m, 21.vii.1999, coll Yao Jian, slides nos. ZL548, 553♀ (IZCAS).

240. *Eupithecia citraria* Mironov & Galsworthy, 2004

Eupithecia ***citraria*** Mironov & Galsworthy, 2004, *Trans. lepid. Soc. Japan* 55 (1): 49, figs 7, 25 (*nec* 24), 26. Holotype ♂ (ZFMK; examined), [China]: North Yunnan, Likiang [Lijiang].

Description. Fore wing ground colour reddish brown, transverse lines purplish brown; antemedial double, curved; postmedial almost straight, slightly angled close to costa, bordered externally by a pale fascia; discal spot black, elongate, touching postmedial; terminal area with a comma-shaped dark subtornal spot, and a dark subapical blotch. Hind wing brownish white, with a small discal dot; transverse lines marked in brown along anal margin only; terminal area slightly darker; discal dot very small.

Male genitalia. Uncus short and stout, biapical. Valve unmodified, but centrally broadened. Vinculum trapezoid, with shallow medial hollow. Papillae elongate and narrow. Aedeagus short and stout, tapered anteriorly. Vesica armed with two dentate cornuti and a V-shaped plate. Sternite A8 peg-like with broadly bilobed base, and two narrow apical rods.

Female genitalia. Bursa oval, densely covered with small spines and with a separate patch of small spines in the short ductus bursae near the colliculum. Ductus seminalis attached to bursa near colliculum on right. Colliculum moderately broad, short. Antrum slightly sclerotized. Tergite A8 quadrate. Apophyses narrow, the posterior broadened near elongate and narrow papillae anales, with very small spurs.
Range. China (Yunnan).
Phenology. Univoltine (?): late May to late June.
Similar species. Similar externally to *E. subbrunneata*, but ground colour of fore and hind wings paler, and with less dark costal spots on fore wing. In spite of this external similarity, *E. citraria* belongs to the *russeliata* species group on the base of the genitalia structure. In this group it is closest externally to *E. rubeni*, but the wing pattern is different. The genitalia in both sexes are distinct from those of *E. rubeni* (see diagrams), though typical generally of this species group.
Material. ♂, China, Provinz Nord-Yuennan, Li-kiang [Lijiang], 26.v.1935, H. Höne, de Laever slide no. 348♂ (holotype, ZFMK); 1♀, Yunnan, Yulongshan, Lijiang, 28.vi.1962, coll. Song Shimei, slide no. ZL529♀ (paratype, IZCAS).

241. *Eupithecia nobilitata* Staudinger, 1882

Eupithecia nobilitata Staudinger, 1882, -*Ent. Z. Stettin* **43** (1–3): 76. Holotype ♂ (coll. O. Staudinger in MNHU; examined), [Kazakhstan]: Lepsa.
Tephroclystia medionotata Warren, 1899, *Novit. zool.* **6** (1): (40). Syntypes 2♂ (BMNH; examined), [China]: Thibet, Kuku-nor (synonymised in Prout, 1915).

Description. Fore wing narrow, elongate; ground colour grey to ochreous grey, often with brown shading around disc; transverse lines straight, sharply angled onto costa, darkest near dorsal margin; postmedial broadened, dark grey or brown between M_1 and M_3 veins, right or sharply angled from M_1 to costa; apex with an oblique paler streak; terminal area narrow, with weakly defined whitish subterminal line; discal dot small, elongate, usually touching medial. Hind wing paler, unicolourous; transverse lines weakly marked along anal margin only; discal dot minute, rounded or absent.
Male genitalia. Uncus narrow, biapical. Valve small, unmodified. Vinculum short but broad. Papillae shortish. Aedeagus narrowed anteriorly, vesica armed with three small dentate cornuti and two small flat cornuti near ductus ejaculatorius base. Sternite A8 peg-like, basally bilobed with two short and blunt apical rods.
Female genitalia. Bursa spherical, covered with short spines except longitudinal membranous gap under ductus seminalis base. Ductus bursae

short with some longitudinal striations and a small patch of spines. Ductus seminalis attached to ductus bursae on right. Colliculum very short. Antrum membranous. Tergite A8 square with very deep medial cleft. Apophyses stout, the anterior slightly expanded at apices. Papillae anales broad and tapered.

Range. China (Tibet, Qinghai, Gansu); Kazakhstan, Russia (Altai & Sajan Mts., Transbaikalia, Dahuria), Mongolia.

Phenology. Univoltine: early June to July.

Altitude. From about 700 m in Altai Mts. (Russia: coll. O. Pekarsky) to about 2200 m (Mongolia: Vojnits 1975) and 4000 m in Tibet (China: IZCAS).

Similar species. Generally recognizable from the distinctive shape and coloration, but somewhat variable in size, shape of fore wings and depth of colour. Similar to *E. chesiata* but lacks the strong dark fore wing streak. The male and female genitalia will immediately distinguish the two species.

Material. ♂, [Kazakhstan], Lepsa Hbhr., *nobilitata* ♂, Origin., bon. spec. bei *Lanceata* von Christ. & Tancré aus Mongolei, die abordnen (Boh.), Gen. vérifiés 1445[2] *Eupithecia nobilitata* Stgr., dét. E. de Laever 1954 (holotype, coll. O. Staudinger in MNHU); 1♂, [Qinghai], *nobilitata* Stdgr. Tibet ♂, Kuku-Noor, Tibet (Kuku-Noor) 98., *nobilitata* ♂, Püngeler, Zool. Mus. Berlin, Gen.-Unters. Nr. [Schmidt slide no.] Eu 103b, Zool. Mus. Berlin (coll. K. Dietze in MNHU); 1♀, [Qinghai], *nobilitata* Stgr. ♀, Kuku-Noor, Mitte Juni, von Tancré, Rückbeil (K. Dietze), Zool. Mus. Berlin, Gen.-Unters. Nr. [Schmidt slide no.] Eu 103a, Zool. Mus. Berlin (coll. K. Dietze in MNHU); 1♀, [Qinghai], *Nobilitata* Stg. ♂ Rükb. Kuku N., Zool. Mus. Berlin, Gen.-Unters. Nr. [Schmidt slide no.] Eu 103, Zool. Mus. Berlin (coll. K. Dietze in MNHU); 1♂, [Qinghai], *nobilitata* Stgr., *Nobilitata* Stg. ♂, R. Kuku Noor (coll. K. Dietze in MNHU); 1♂, [Qinghai], Tibet 1894, *Eupithecia nobilitata* Stgr. ♂, Tancré 2.97, *Nobilitata* Stgr. (coll. K. Dietze in MNHU); 1♂, [Qinghai], Tibet (Kuku-Noor), Rückbeil 94, *chesiata* ♂, v. R. Tancré 4.95 (coll. R. Püngeler in MNHU); 1♂, [Qinghai], same data, *nobilitata* ♂, v. R. Tancré 4.95 (coll. R. Püngeler in MNHU); 1♀, [Qinghai], Tibet (Kuku-Noor), Rückbeil 1894, *nobiliaria* Stgr. ♀ (Stgr. det.), 10/1910 v. Tancré (coll. R. Püngeler in MNHU); ♂, [Qinghai], Type, 165, Kuku-nor, Thibet, *Tephroclystia medionotata* Warr., Type ♂, Rothschild Bequest B.M. 1939–1 (holotype of *E. medionotata*, BMNH); 1♂, China, Gansu, Xiahe, 2600–3000 m, 26.-30.vi.1986, Coll. Nr. 246, leg. Görgner, Mironov slide no. 730♂ (ZFMK); 15♂, 5♀, Qinghai, Menyuan, Xianmi, 2800 m, 23–24.vii.1992, slides nos. ZL86♂, 463, 499♀; 1♂, same loc., 31.vii–1.viii.1991; 1♂, Qinghai, Tongren, Maixiu, 2950 m, 30.vii–1.viii.1992, slide no. ZL168♂, all coll Xue Dayong; 1♀,

Tibet, Markam, Rongxi, 4000 m, 26.vi.1976, slide no. ZL461♀, coll Zhang Xuezhong (IZCAS).

242. *Eupithecia sutiliata* Christoph, 1876

Eupithecia sutiliata Christoph, 1876, *Horae Soc. Ent. Ross.* **12**: 262, pl. 7, fig. 34. Lectotype ♂ (ZISP; examined), [Iran]: Schahrud.
Eupithecia scopariata var.? *orientata* Staudinger, 1901: in Staudinger & Rebel, *Cat. Lepid. Palaearct. Faunengeb.* (Edn 3) **1**: 314. Lectotype ♂ (designated in Mironov 2003; ZISP; examined), Arm(enia); Tura (synonymised in Mironov, 1991).
Eupithecia rjabovi Vojnits, 1975, *Acta zool. Acad. Sci. Hung.* **21** (1–2): 237, fig. 3. Holotype ♂ (TTM; examined), Russia: Daghestan, district Petrovsk [= Makhachkala] (synonymised in Mironov, 1991).
Eupithecia kasahstanica Viidalepp, 1988, *Fauna pjadenitz gor Srednej Azii*: 120, figs 26 (2–4), pl. 2, fig. 17. Holotype ♂ (ZISP; examined), Kazakhstan: Kzyl-Orda (synonymised in Mironov, 1991).

Misidentification: *Eupithecia satyrata* Lvovsky, 1971 (*nec* Hübner, 1813).

Description. Fore wing ground colour pale greyish ochreous, with numerous closely placed, fine and oblique transverse lines, sharply angled near costa; postmedial often more distinct, right angled onto costa; indistinct pale wavy subterminal line; discal dot very small, rounded. Hind wing paler than fore wing, with terminal shade; numerous transverse lines well marked along anal margin; discal dot inconspicuous or absent.
Male genitalia. Uncus elongate, biapical. Valve unmodified. Vinculum narrow and tapered. Papillae long, broadened at apices. Aedeagus long and narrow, but shorter than length of valve. Vesica armed with an elongate patch of granulation and a narrow V-shaped cornutus at ductus ejaculatorius base. Sternite A8 triangular, bilobed at base.
Female genitalia. Bursa oval, almost completely covered with dense spines. Ductus bursae short, membranous, with two small separate patches of spines. Ductus seminalis broadened basally, attached to ductus bursae on right. Colliculum short and narrow. Antrum membranous. Tergite A8 almost quadrate, with a deep central hollow. Apophyses narrow. Papillae anales broad and rounded.
Range. China (Xinjiang); Europe (Greece, S. Russia [Lower Volga, Kalmykia and Daghestan]), Armenia, Azerbaijan, northern Iran, Turkmenistan, Kazakhstan, Uzbekistan, Kirghizstan, Tajikistan, Mongolia, Korea ? (Shin 1996).
Phenology. Probably bi- or polyvoltine: early May to late September. Overwintering as pupa.
Biology. Larva found on *Suaeda* (Chenopodiaceae) (Russia (Daghestan): Mironov 2003), *Kalidium foliatum* and *Climacoptera* (Chenopodiaceae) (SE Kazakhstan: Mironov 2003).

Altitude. From sea level up to about 2000 m (Mongolia: coll. ZISP).
Similar species. Generally distinct on the basis of pale coloration, but darker specimens may be confused with *E. vulgata*: it is distinguished by the narrow and elongate fore wing with less arched costa, the numerous fine oblique transverse lines, and the absence of a pale tornal spot on the fore wing. In the male genitalia it is separated by the absence of dentate cornuti in the vesica, and in the female by the absence of larger marginal spines in the bursa.
Material. ♂, [Iran], Schahrud, *Eupithecia sutiliata*, 2/7, *ochrovittata* Chr. aberr. det. M. Rjabov, coll. Erschov (lectotype, ZISP); ♂, [Armenia], Erivan, 12. 7.[18]81, Chr., 13, *Scopariata* ♂♀ *Orientata* Stgr. Origin., Coll. Vel. Kn. Nikolaja Mikhajlovicha (lectotype of *E. scopariata orientata*, ZISP); 1♂, ? spec. ? *aequata* form. *leuca* Typ, [Xinjiang], Aksu. 03, Gen.-Unters. Nr. [Schmidt slide no.] Eu 71b, Zool. Mus. Berlin (coll. K. Dietze in MNHU); 1♀, ? spec. ? *aequata/diffidata*, Ili (coll. K. Dietze in MNHU); 1♀, *aequata* Stgr., [Xinjiang], Juldus Ili-Gebiet, Gen.-Unters. Nr. [Schmidt slide no.] Eu 71a, Zool. Mus. Berlin (coll. K. Dietze in MNHU); 1♀, ? spec. ? *aequata* form. *leuca* Typ, [Xinjiang], Aksu. 03 (coll. K. Dietze in MNHU); ♂, Dagestan, distr. Petrovsk, vi.[1]926, M. Rjabov, *T. ochrovittata* Chr. det. M. Rjabov, Vojnits slide no. 831♂ (holotype of *E. rjabovi*, TTM); ♂, [Kazakhstan], st. Kzyl-Orda, 18.vii.[19]50, [A. Bundel leg.], Viidalepp slide no. 2317♂, ZISP slide no. 17086 (holotype of *E. kasahstanica*, ZISP).

243. *Eupithecia biornata* Christoph, 1867

Eupithecia biornata Christoph, 1867, *Ent. Z., Stettin* **28** (4–6): 238. Syntypes 2♀ (MNHU, BMNH; examined), [Russia]: Sarepta.

Description. Fore wing elongate with a series of narrow parallel transverse lines on a pale, ash grey background, and a broad longitudinal pale yellow stripe from base to apex; discal dot minute; hind wing white with darker terminal area and narrow parallel transverse lines at anal margin only.
Male genitalia. Uncus long and narrow. biapical. Valve unmodified, very broad. Vinculum short and narrow. Papillae expanded at apices. Aedeagus narrowed anteriorly, vesica with three dentate cornuti and a narrow, folded V-shaped cornutus. Sternite A8 elongate with two long inwardly curved apical rods.
Female genitalia. Bursa large, globular, completely covered with small spines. Ductus bursae very short, with longitudinal striations. Ductus seminalis broadened at base, attached to ductus bursae near colliculum on right. Colliculum short and narrow. Tergite A8 square, with deep medial

cleft in posterior margin. Apophyses narrow. Papillae anales relatively small, rounded.

Range. China (Xinjiang, Heilongjiang); South eastern Europe (north-eastern Greece, Bulgaria, Romania, southern Ukraine, southern European Russia, S Urals, Daghestan), Turkey (north-east), Kazakhstan and Asian Russia (south-western and southern Siberia, Dahuria and Transbaikalia), Mongolia.

Phenology. Univoltine: mid-June to mid-August (mid-September).

Biology. Larva found on *Artemisia* sp. (Asteraceae) (Russia: Christoph 1867).

Altitude. Up to about 900–1000 m above sea-level in eastern Turkey (Mironov 2003).

Similar species. Shares yellow fore wing streak only with the next two species. See under those for differences. The male genitalia are similar to those of *E. rubeni*, but the dentate cornuti are smaller and the eighth sternite longer. The female bursa is best distinguished by the almost complete spining, with no separate patch of spines by the colliculum.

Material. 1♀, [Russia], Sarepta Chr., ♀ *Biornata*, Origin., (syntype, coll. O. Staudinger in MNHU); 1♀, [Russia], Sarepta, 22.7.60, *ex.* coll. Christoph, Sarepta 22.vii.[18]60, Ex. Coll. H. J. Elwes, 1920 (syntype, BMNH); 1♂, [Xinjiang], Asia centr., Turkestan, Ili-Gebiet, Coll. Wagner (coll. K. Dietze in MNHU); 1♂, [Xinjiang], Asia centr., Aksu, Chin. Turk., 320, *biornata* Chr., K. Rosen det., Vojnits slide no. 15411♂ [= ZSM slide no. G 13382] (ZSM); 1♂, Mandschuria [Heilongjiang], Maoreschan (= Mauschan), Juli (ZFMK); 1♂, 2♀, [Heilongjiang], Harbin, Manchuria, 9.viii.1935 (1), 15.vii.1937 (2), M.I. Nikitin, Inoue slides nos. 4028♀, 4029♀, BM Geom. slide no. 20947♂ (BMNH).

Remarks. Not previously recorded from China.

244. *Eupithecia kuldschaensis* Staudinger, 1892

Eup[*ithecia*] *kuldschaensis* Staudinger, 1892, *Dt. ent. Z., Iris* 5 (1): 255. Syntypes 3♀ (coll. O. Staudinger in MNHU, ZISP; examined), [China]: Kuldja [= Yining (Gulja)].

Eupithecia aporia Vojnits, 1975, *Acta zool. Acad. Sci. hung.* 21 (3–4): 452, fig. 4. Holotype ♂ (TTM; examined), Mongolia: Chovd aimak, 10 km SSW of Somon Bulgan, 1200 m. **New junior synonym.**

Unavailable name (incorrect subsequent spelling): *kuldshaensis* Viidalepp, 1988.

Description. Fore wing pale ochreous with weak grey tinge, costal margin pale grey; transverse lines marked as a series of small dark dots on veins; terminal area darker, greyish with weakly wavy white subterminal line; discal dot very small. Hind wing slightly paler than fore wing, transverse lines inconspicuous, weakly marked along anal margin only; terminal area darker, pale grey with weak wavy whitish subterminal line; discal dot lacking.

Male genitalia. Uncus narrow, biapical. Valve unmodified. Vinculum short and narrow. Papillae narrow, slightly clavate. Aedeagus narrow, shorter than valve length. Vesica armed with a V-shaped cornutus and a smaller elongate cornutus, both near ductus ejaculatorius base. Sternite A8 entire, triangular.

Female genitalia. Bursa globular, completely covered with small spines. Ductus bursae very short, membranous, with a few separate spines near colliculum on left. Ductus seminalis broad, attached to ductus bursae near colliculum on right. Colliculum short and narrow, inclined. Tergite A8 square with deep medial cleft. Apophyses narrow and rather short. Papillae anales broad, rounded.

Range. China (Xinjiang); SE Kazakhstan, Mongolia.

Phenology. Univoltine: mid-June to mid-July.

Altitude. Found from about 1200 to 2000 m above sea-level (Mongolia: Vojnits 1975; coll. ZISP).

Similar species. Very similar to preceding species, but fore wing streak paler, much broader on rear margin, and crossed by transverse lines formed by minute dots on the veins. The male genitalia lack dentate cornuti on the vesica and have sternite A8 entire. The female bursa has the spiniferous area not reaching the base of the smaller colliculum.

Material. 1♀, [Xinjiang], Kuldja 89 Rckbl., *kuldschaensis* ♀, Origin., Gen. vérifiés 1455[2] *Eupithecia kuld(gah)ensis* Stgr. ♀, [Xinjiang], Kuldja, dét. E. de Laever 1954 (syntype, coll. O. Staudinger in MNHU); 2♀, [Xinjiang], Culdscha, v. *Kuldschaensis* Stgr., Origin., Coll. Vel. Knjazja Nikolaja Mikhajlovicha (syntypes, ZISP); ♂, Mongolia, Chovd aimak, 10 km SSW of Somon Bulgan, 1200 m, Exp. Dr. Z. Kaszab, 1966, Nr. 632, 4–6.vii.1966, Vojnits slide no. 10291♂ (holotype of *E. aporia*, TTM).

245. *Eupithecia rubeni* Viidalepp, 1976

Eupithecia rubeni Viidalepp, 1976, *Nasekomye Mongolii* 4: 390, figs 16, 30–33. Holotype ♂ (IAET; examined), USSR [Russia]: Tuvinskaya ASSR [Tuva], near Kyzyl town.

Eupithecia propria Vojnits, 1977, *Acta zool. Acad. Sci. hung.* 23 (3–4): 465, figs 3, 5. Holotype ♂ (TTM; examined), Mongolia: Uvs aimak, Chöndlön gol river, 32 km NW of Ulaangom, 1200 m (synonymised in Viidalepp & Mironov, 1990).

Description. Wings narrow and elongate. Fore wing ground colour rusty ochreous, blackish brown along costa, with a broad yellowish longitudinal medial stripe; transverse lines inconspicuous, curved onto costa and forming blackish brown costal spots; terminal area brownish black near tornus, with small pale tornal spot; discal dot minute. Hind wing paler, darker along anal and terminal margins with indistinct transverse lines; postmedial crenulate; discal dot minute.

Male genitalia. Uncus stout, biapical. Valve broad, unmodified. Vinculum short and curved. Papillae narrow. Aedeagus broad, shorter than length of valve, vesica armed with three large dentate cornuti connected at base and a twisted V-shaped cornutus at base; surface of vesica granulate. Sternite A8 peg-like, strongly broadened basally.

Female genitalia. Bursa very large, globular, sclerotized close to colliculum only, covered with an abruptly delimited field of spines on one side of posterior part, and with short longitudinal striations and small separate patch of spinules near base of colliculum. Ductus seminalis broadened basally, attached to corpus bursae near colliculum on right. Colliculum short and very wide. Antrum short and broad. Tergite A8 elongate, with deep membranous medial cleft in posterior margin. Apophyses narrow. Papillae anales slightly elongate, rounded apically.

Range. China (Inner Mongolia, Heilongjiang); Asian Russia (southern Siberia: Tuva, Burjatia, Dahuria, southern Primorje), Mongolia.

Phenology. Univoltine: late July to mid-August.

Altitude. Found at about 1200 m above sea-level (Mongolia: Vojnits 1977).

Similar species. Only likely to be confused with the preceding two species. Distinguished from both by the rich rusty ochreous ground colour and darker costal margin on the fore wing. The male genitalia can be distinguished by the much larger dentate cornuti on the vesica and the shorter and broader eighth sternite. The female bursa has a smaller and abruptly delimited spine field and a very broad colliculum.

Material. ♂, U.S.S.R. [Russia], Tuvinskaya ASSR [Tuva], environs of Kyzyl, at light, 23.vii.1972, leg. T. Ruben & J. Viidalepp (holotype, IAET); 1♀, [Heilongjiang], Harbin, Manchuria, Aug. 1923, [leg.] V. Tolmachov, Vojnits slide no. 12258♀ (TTM); 1♀, same loc., 4.viii.1937, [leg.] M. I. Nikitin, Inoue slide no. 4027♀ (coll. H. Inoue in BMNH); 2♀, China, Inner Mongolia, 80 km N of Hohhot, Zhao-he, 10.viii.2007, (leg.) A. Bidzilya (ZISP); 1♀, Inner Mongolia, Bazhong, 9.viii.1985, slide no. ZL259♀ (IZCAS).

Remarks. Not previously recorded from China.

246. *Eupithecia rebeli* Bohatsch, 1893

[*Eupithecia*] *rebeli* Bohatsch, 1893, *Dt. ent. Z., Iris* **6** (1): 23 (non binominal). Syntypes 2♂, 4♀ (coll. O. Staudinger in MNHU; examined), [Uzbekistan]: Samarkand; Namangan; [Kirghizstan]: Transalai (Mts.); Alai (Mts.); [China]: Kuldja [Yining (Gulja)].

Description. Fore wing elongate, with a characteristic darker, blackish grey costal margin, numerous fine, oblique transverse lines, and a pale, usually yellowish, triangular blotch stretching from the base of CuA_2 to the apex; terminal area greyish ochreous with distinct whitish subterminal line; discal dot very small often invisible. Hind wing greyish white with multiple wavy transverse lines usually marked with small dark dashes on the veins; terminal area darker, narrow, with crenellate inner border and an indistinct whitish wavy subterminal line; discal dot small.
Male genitalia. Uncus biapical. Valve narrow, with a pointed ventral process. Vinculum short and narrow. Papillae rather small. Aedeagus slim, shorter than valve length. Vesica armed with two small dentate cornuti and two small striate plates near ductus ejaculatorius base. Sternite A8 triangular, with two short pointed apical rods.
Female genitalia. Bursa oval, almost completely covered with very small spines. Ductus seminalis narrow, attached near base of colliculum on right. Colliculum elongate and narrow. Antrum elongate. Tergite A8 transverse. Apophyses short and narrow. Papillae anales relatively large, rounded.
Range. China (Xinjiang); Uzbekistan, Kirghizstan, Tajikistan.
Phenology. Univoltine: late May to early July.
Altitude. From about 1800 m in the Ghissar Mts. to about 3600 m in the Pamirs (Tajikistan: Viidalepp 1988).
Similar species. Although slightly reminiscent of the three previous species, easily distinguished by the dark leading half of the forewing and the regular transverse lines. The presence of a ventral process on the valve immediately distinguishes the male genitalia. The female are less distinctive, but can be distinguished by the longer colliculum and the complete coverage of tiny spines on the bursa.
Material. 1♂, [Kirghizstan], Transalai 86 Maur., *Rebeli* Bohatsch Type. Iris 1893♂, Origin., Zool. Mus. Berlin, Vojnits slide no. 13576♂ (syntype, coll. K. Dietze in MNHU); 1♀, *Rebeli* Bohatsch Type ♀ Iris 1893, Origin., Zool. Mus. Berlin, Vojnits slide no. 13577♀ (syntype, coll. K. Dietze in MNHU); 1♂, [Uzbekistan], Prov. Samarkand Hbhr. 81, Origin., Zool. Mus. Berlin (syntype, coll. K. Dietze in MNHU); 2♀, [Kirghizstan], Alai 85 Hbh., Origin., Zool. Mus. Berlin (syntypes, coll. K. Dietze in MNHU); 1♀, [Xinjiang], Kuldja Rükb., Origin., Zool. Mus. Berlin (syntype, coll. K. Dietze in MNHU).

The *addictata* species group

This species group was introduced by Krampl (1994). In the male genitalia the valve is broad, and thickened, sometimes protruding on the ventral edge. The vesica carries, usually, one large and one or more smaller, dentate cornuti. The female bursa is pear-shaped, spined at the base, with a sclerotized ductus bursae, often with a row of separate spines. The morphology appears close to the preceding group. It is represented in China by the following three species.

247. *Eupithecia thalictrata* Püngeler, 1902

Tephroclystis thalictrata Püngeler, 1902, *Dt. ent. Z., Iris* 15 (1): 137, pl. 6, figs 21, 22. Syntypes, at least 6♂, 5♀ (coll. R. Püngeler in MNHU; examined), [Switzerland]: Wallis.
Eupithecia ijimai Inoue, 1963, *Tinea* 6 (1/2): (33), text-fig. 5; pl. 7, figs 11, 12. Holotype ♂ (BMNH; examined), Japan: Hokkaido, Kushiro, Shibecha (treated as subsp. of *E. thalictrata* in Viidalepp & Mironov, 1988b).
Eupithecia trita Vojnits, 1977, *Acta zool. Acad. Sci. hung.* 23 (3–4): 476, fig. 12. Holotype ♂ (TTM; examined), Mongolia: Central aimak, SE of Somon Bajanzogt, 1600 m. [Junior secondary homonym of *Tephroclystia trita* Turati, 1926.].
Eupithecia inveterata Vojnits, 1987, *Articulata* 3 (1): 37. [Replacement name for *Eupithecia trita* Vojnits, 1977]. **New junior synonym.**

Misidentification: *Eupithecia indecora* Vojnits, 1981 (1♀ paratype in ZFMK).

Description. Fore wing unicolourous brownish grey, usually with paler band between medial and terminal areas, well indicated only near costa; transverse lines indistinct, weakly sinuate, slightly angled and darker near costa; postmedial evenly curved near costa; discal dot brownish, weakly discernible or absent. Hind wing paler, transverse lines indistinct, faintly marked only at posterior margin; discal dot absent. Head, notum and abdomen grey.
Male genitalia. Uncus narrow, biapical. Valve unmodified, sacculus slightly thickened. Vinculum short and broad. Papillae fairly stout. Aedeagus slim, shorter than length of valve. Vesica granulate, armed with one large and two small dentate cornuti, and a flat cornutus at ductus ejaculatorius base. Sternite A8 bilobed at base with two broad apical rods.
Female genitalia. Bursa subglobular, about $^2/_3$ densely covered with small spines at base and with a separate transverse row of sparse spines between spiniferous area and colliculum. Ductus bursae short, sclerotized. Ductus seminalis broad, attached near relatively short colliculum on right. Antrum membranous. Tergite A8 slightly transverse. Apophyses narrow. Papillae anales short and rounded.
Range. China (Qinghai, Heilongjiang, Shanxi); Europe to Urals, Asian Russia (southern Siberia: Transbaikalia, Dahuria, Priamurje, Primorje, Sakhalin, Kuril Islands: Kunashiri), Mongolia, Japan (Hokkaido).
Phenology. Univoltine: late May to late July.

Biology. Larva found on the leaves, flowers and young seeds of *Thalictrum foetidum* (Switzerland: Püngeler 1902; Favre 1905; Dietze 1913) and also *T. minus* (Ranunculaceae) (Italy: Weigt 1990).
Altitude. From sea level (Russia, Primorje: coll. ZISP) up to about 2000 m in the Mien-Shan Mts. (China: coll. ZFMK).
Similar species. Very similar to the two following species, and also to *E. severa*. See under those species, but best distinguished by examination of genitalia.
Material. 1♂, [Switzerland], Valesia, Saasthal, *Thalictrum foet.*, 7.01. e. l. 1.6.02, *thalictrata* Püng. ♂, Original, abgebildet Iris 1902, R. Püngeler (type, coll. R. Püngeler in MNHU); ♂, [Japan], Shibecha, Kushiro, 30.vi.1951, K. Ijima (holotype of *E. ijimai*, coll. H. Inoue in BMNH); ♂, Mongolia, Central aimak, SO von Somon Bajanzogt, 1600 m, Exp. Dr. Z. Kaszab, 1966, Nr. 521, 11.vi.1966, Vojnits slide no. 10535♂ (holotype of *E. trita* Vojnits, 1977, nec *Tephroclystia trita* Turati, 1926, TTM); 1♀, Prov. Shansi [Shanxi], Mien-shan, Obere Höhe *ca.* 2000 m, 13.vii.1937, H. Höne, Vojnits slide no. 13113♀ (paratype of *E. indecora*, ZFMK); 1♀, same loc. and alt., 6.vii.1937, H. Höne, Vojnits slide no. 13739♀ (ZFMK); 1♀, same loc. and alt., 9.vii.1937, H. Höne, Vojnits slide no. 11518♀ (TTM); 1♀, same loc. and alt., 10.vii.1937, H. Höne, Vojnits slide no. 13165♂ (incorrectly associated gen. slide of *E. robiginascens*, TTM); 1♀, same loc., Mittlere Höhe *ca.* 1500 m, 10.vi.1937, H. Höne, Vojnits slide no. 13164♀ (incorrectly associated gen. slide of *E. robiginascens*, TTM); 1♂, Qinghai, Menyuan, Xianmi, 2800 m, slide no. ZL315♂; 1♀, Heilongjiang Dailing, 390 m, 6.7.1962, coll Bai Jiuwei, slide no. ZL222♀; 1♀, Shanxi, Wutaishan, 1992, slide no. ZL385♀ (IZCAS).
Remarks. The male holotype of *E. indecora* belongs to *E. impolita*.

248. *Eupithecia dura* Vojnits, 1981

Eupithecia dura Vojnits, 1981, *Acta zool. Acad. Sci. hung.* 27 (3–4): 421, figs 32, 36–38. Holotype ♂ (ZFMK; examined), China: North Yunnan, Li-kiang.

Misidentification: *Eupithecia pseudoplumbeolata* Vojnits, 1973 (1♀ paratype in ZFMK).

Description. Fore wings dark brownish grey. Fore wing with inconspicuous pale transverse lines, postmedial slightly more distinct, straight, twice sharply angled onto costa; terminal area darker with weak trace of pale wavy subterminal line; discal dot rounded, blackish. Hind wing only slightly paler, broad transverse lines well marked along anal margin only; discal dot small but distinct.
Male genitalia. Uncus elongate, biapical. Valve broad, with a large obtuse ventral process. Vinculum rounded. Papillae elongate and narrow.

Aedeagus slim, shorter than length of valve. Vesica granulate, armed with a small, dentate apical cornutus and an elongate, plate-like cornutus near ductus ejaculatorius base. Sternite A8 broadened basally, with parallel lateral margins, and with two slightly expanded, more sclerotized, blunt apices.

Female genitalia. Bursa almost globular, completely covered with uniform spines, except for a narrow area along right side. Ductus bursae more strongly sclerotized, short and broad, tapered to colliculum, with short longitudinal striations, a small patch of larger spines on right side, and a transverse chain of small spines dorsally. Ductus seminalis broadened basally; attached near colliculum on left. Colliculum short and narrow. Antrum membranous, with numerous pores. Tergite A8 almost quadrate. Apophyses narrow, the anterior slightly stouter. Papillae anales broad and rounded.

Range. China (Yunnan).

Phenology. Univoltine: late May to mid July.

Altitude. From about 1700 to 3000 m (China: Vojnits 1981b; coll. ZFMK).

Similar species. Very similar externally to *E. addictata*, from which we have found no reliable and constant external differences. The male genitalia are easily distinguished by the prominent medial process on the ventral margin of the valve and the cornuti on the vesica, and the female by the presence of a transverse chain of small spines near the base of the ductus seminalis and by the narrower colliculum.

Material. ♂, China, Provinz Nord-Yuennan, Li-kiang [Lijiang], 20.v.1935, H. Höne, Vojnits slide no. 13214♂ (holotype, ZFMK); 1♂, same loc., 2.vi.1934, H. Höne, Vojnits slide no. 12969♂ (paratype, TTM); 1♀, same loc., 8.vi.1934, H. Höne, Vojnits slide no. 13209♀ (paratype, ZFMK); 1♀, same loc., 2. vii.1934, H. Höne, Vojnits slide no. 1069♀ (paratype of *E. pseudoplumbeolata*, ZFMK); 2♂, 1♀, same loc., 12, 17.vi.1934, H. Höne, Vojnits slides nos. 11472♂, 18762♂ (TTM); 1♂, same loc., 21.vi.1934, H. Höne, Vojnits slide no. 13186♂ (ZFMK); 1♀, same loc., 9.vii.1934, H. Höne, Mironov slide no. 637♀ (ZFMK); 1♂, Nord-Yünnan, A-tun-tse [Dêqên], Talsohle *ca.* 3000 m, 25.v.1937, H. Höne, Vojnits slide no. 12970♂ (paratype, ZFMK).

249. *Eupithecia addictata* Dietze, 1908

Eup[ithecia] ? *selinata* f. *addictata* Dietze, 1908, *Dt. ent. Z., Iris* **21** (3): 194, pl. 2, figs 5, 6. Lectotype ♂ (design. by Krampl, 1994, coll. K. Dietze in MNHU; examined), [Russia]: Ussuri, Kasakewitsch. Raised to species rank by Prout (1930).

Eup[ithecia] ? *selinata* f. *tenebricosa* Dietze, 1910, *Biologie der Eupithecien* 1: pl. 74, fig. 405. Lectotype ♂ (coll. K. Dietze in MNHU; examined), [Russia]: Amur, Radde. **New junior synonym.**

Eupithecia danielata Schütze, 1959, *Mitt. Münch. ent. Ges.* **49**: 35, pl. 1, figs 1, 2, 5, 6. Syntypes 4♀ (ZSM; examined), Greece (northern): Olympos, Enipefs Valley, Ag(ios) Dionyssios, 800–1200 m (synonymised in Krampl, 1994).

Eupithecia rudniki Vojnits, 1973, *Annls hist.-nat. Mus. natn. hung.* **65**: 253, fig. 3b. Holotype ♀ (ZFMK; examined), [Russia]: Ussuri, Sutschanski Rudnik (synonymised in Mironov, 1991).

Eupithecia pseudoplumbeolata Vojnits, 1973, *Annls hist.-nat. Mus. natn. hung.* **65**: 254, figs 5a, b. Holotype ♂ (ZFMK; gen. slide examined), [Russia]: near Minusinsk (synonymised in Mironov, 1991).

Eupithecia falkovitshi Viidalepp, 1976, *Nasekomye Mongolii* **4**: 386, figs 14, 19–23. Holotype ♂ (IAET; examined), USSR [Russia]: Sajan (Mts.), Khakasskaya AO [Khakassia], Sartaban Valley (synonymised in Viidalepp & Mironov, 1988b).

Eupithecia koreaica Vojnits & de Laever, 1978, *Acta zool. Acad. Sci. hung.* **24** (1–2): 225, figs 1, 2. Holotype ♂ (TTM; examined), Korea: Ryang-gang Prov(ince), Hyesan (synonymised in Mironov, 1991).

Misidentifications: *Eupithecia selinata* Inoue, 1953 (*nec* Herrich-Schäffer, 1861); *Eupithecia submelanochroa* Vojnits, 1973 (1♀ paratype in TTM); *Eupithecia Peking [Beijing]iana* Vojnits, 1973 (1♂ paratype in TTM).

Unavailable names (incorrect subsequent spelling): *addictaria*: Prout (1914).

Description. Fore wing brownish grey, transverse lines mostly indistinct, obtuse angled and slightly broadened near costa; medial with two pronounced sinuate curves, crossing discal dot; postmedial wavy, curved near costa; usually a prominent pale double stripe beyond postmedial; wavy subterminal line inconspicuous; discal dot variable from large to very small or obsolete. Hind wing faintly paler, with inconspicuous transverse lines, except a paler, evenly curved stripe between medial and terminal areas; discal dot minute or absent.

Male genitalia. Uncus elongate, biapical. Valve broadened medially with obtuse angled ventral margin. Vinculum short, sometimes with shallow medial hollow. Papillae elongate, sightly clavate. Aedeagus stout, shorter than valve length. Vesica granulate, armed with a dentate cornutus at apex and a U-shaped cornutus at ductus ejaculatorius base. Sternite A8 broadened basally, with apically sclerotized lateral margins and two very small apical lobes.

Female genitalia. Bursa ovate, covered with small spines, with a longitudinal spineless gap on the dorsal side to the right. Ductus bursae truncate-conical, sclerotized, with one or two large spines on the oblique wrinkle on the right. Ductus seminalis broad, attached to ductus bursae near colliculum on left. Colliculum simple, circular. Antrum membranous. Tergite A8 almost quadrate. Apophyses short and narrow. Papillae anales rounded.

Range. China (Heilongjiang, Beijing, Shandong, Shaanxi, Gansu, Sichuan); Europe, Russia (Udmurtia, North Caucasus, W Siberia, Altai Mts., Priamurje, Primorje, Kuril Islands: Kunashiri), Korea, Japan.

Phenology. Univoltine: late June to early September.
Biology. Larvae found on the flowers and ripening seeds of *Thalictrum thunbergii* (Japan: Inoue 1979) and on the flowers of *Thalictrum minus* and *T. amurense* (Ranunculaceae) (Russia, southern Primorje: Mironov 2003).
Altitude. From sea level to about 3000 m in the Qinling range(China, Shaanxi: coll. TTM).
Similar species. Difficult to separate from *E. thalictrata*, but darker, with a more distinct postmedial line, a larger discal dot on the fore wing, and less spotted fringes. *E. selinata* has a paler, shiny, silky grey ground colour and usually larger discal dots, especially on the hind wings. The genitalia are easily distinguished by the combination of the ventrally obtusely angled valve, the presence of only one elongate dentate cornutus on the vesica, the characteristic shape of sternite A8 in the male, and by the sclerotized, truncate-conical ductus bursae without a separate patch of spines in the female.
Material. ♂, [Russia], Ussuri, Kasakewitsch 1907 Korb, *addictata* Typ. ♂, Zool. Mus. Berlin, Typus, Lectotypus ♂ *Eupithecia addictata* Dietze, 1908, F. Krampl det. 1993, Figured: *Eur. J. Entomol.*, 91 (1994): 335–348 (lectotype, design. by Krampl, coll. K. Dietze in MNHU); 1♀, same loc., *addictata* Typ. ♀, Zool. Mus. Berlin, Typus, Paralectotypus ♀ *Eupithecia addictata* Dietze, 1908, F. Krampl det. 1993, Figured: *Eur. J. Entomol.*, 91 (1994): 335–348 (paralectotype, coll. K. Dietze in MNHU); 1♀, same loc., *addictata* (syntype, paralectotype ?, coll. K. Dietze in MNHU); 1♂, [Russia], Radde, m. Amur, 25 Juli 03 Korb, *addictata*; an *zibellinata* und *selinata* erinnernd (syntype, paralectotype ?, coll. K. Dietze in MNHU); 1♀, [Russia], Ussuri, Kasakewitsch 1907 Korb, *addictata* Typ. ♀, Typus, Geom. slide no. 1950–595A (syntype, paralectotype ?, coll. K. Dietze in MNHU); ♂, [Russia], Radde, m. Amur, 6. Juli 03 Korb, *selinata*, form: *tenebricosata* Amur, ähnliches Tier hat Bartels dort ist Querzeichnung noch schärfer betont, hier ist mittelfeld Begrenzung näher am Zellfleck, als bei *selinata*, Typus. (lectotype of *E. selinata tenebricosa,* design. by Mironov, 2009, coll. K. Dietze in MNHU); ♀, [Greece], Nordgriechenland, Olympos, Enipefs-Tal, Ag. Dionyssios, 800 bis 1200 m, Waldgebiet, 24.vi.-12.vii.1957, F. Daniel leg., Schütze slide no. A 815♀ (holotype of *E. danielata*, ZSM); ♀, [Russia], Sutschanski-Rudnik, Ussuri, 461, Vojnits slide no. 523♀ (holotype of *E. rudniki*, TTM); 1♂, [Russia], okr. Minusinska, 27.vi.1928, [leg.] Kozhanchikov, Vojnits slide no. 551♂ (holotype of *E. pseudoplumbeolata*, ZFMK); 1♂, 2♀, China, Sued-Shensi [Shaanxi], Tapaishan [Taibaishan] im Tsinling [Qinling], 20.vi., 23.vii.1935, H. Höne, de Laever slides nos. 161♀, 162♂, Vojnits slide no. 1126♀ (paratypes of *E. pseudoplumbeolata*, TTM); 3♂, same loc., *ca.* 1700 m, 8,

24.vii., 10.viii.1936, H. Höne, Schmidt slide no. Eu 125♂, de Laever slides nos. 181♂, 196♂ (paratypes of *E. pseudoplumbeolata*, ZFMK); 2♀, same loc. and alt., 7, 12.vii.1936, H. Höne, Vojnits slides nos. 1057♀, 1125♀ (paratypes of *E. pseudoplumbeolata*, TTM); 2♂, 1♀, same loc. and alt., 11, 12.vii.1936, H. Höne, Vojnits slides nos. 12374♀, 12379♂ (ZFMK); 2♂, same loc. and alt., 12.vii.1936, H. Höne, Vojnits slides nos. 12225♂, 12974♂ (TTM); 1♀, same loc., *ca.* 3000 m, 13.vii.1936, H. Höne, Vojnits slide no. 1075♀ (paratype of *E. pseudoplumbeolata*, TTM); 1♀, same loc. and alt., 13.vii.1936, H. Höne (ZFMK); 3♂, West-China, Szetschwan [Sichuan], Exped. Stötzner, Coll. Dr. Wehrli, Basel, Schmidt slide no. Eu 125a♂, Vojnits slides nos. 1061♂, 1097♂ (paratypes of *E. pseudoplumbeolata*, ZFMK); 1♂, same data, Vojnits slide no. 1046♂ (paratype of *E. pseudoplumbeolata*, TTM); 1♂, West-China, Szetschwan [Sichuan], Exped. Stötzner, Vojnits slide no. 1060♂ (paratype of *E. pseudoplumbeolata*, TTM); 1♀, [Russia], Ussuri, Sutschanski-Rudnik, 463., Vojnits slide no. 1100♀ (paratype of *E. submelanochroa*, TTM); ♂, U.S.S.R. [Russia], Khakasskaya AO [Khakassia], Sajan Mts., valley of Sartaban river, 19.vii.1972, (leg.) Yu. P. Korshunov, Viidalepp slide no. K-294 (holotype of *E. falkovitshi*, IAET); ♂, Korea, Prov. Ryang-gang, Hyesan, room of Hotel Hyesan, No. 294, 26.vii.1975, leg. J. Papp et A. Vojnits, Vojnits slide no. 10823♂ (holotype of *E. koreaica*, TTM); 1♀, Korea, Prov. Ryang-gang, Hyesan, room of Hotel Hyesan, No. 294, 26.vii.1975, leg. J. Papp et A. Vojnits, Vojnits slide no. 10824♀ (paratype of *E. koreaica*, TTM); 1♂, Szetschwan [Sichuan], Yahothal, Exp. Stötzner (paratype of *E. Peking [Beijing]iana*, TTM); 1♂, 7♀, Heilongjiang, Dailing, 390 m, 6–16.vii.1962, coll Jiu Shixiu and Bai Jiuwei, slides nos. ZL211, 215♀; 1♂, Beijing, Sanpu, 23.vii.1972, coll Zhang Baolin, slide no. ZL217♂; 1♀, Beijing, Badachu, 650 m, 2.viii.1961, coll Zhang Xuezhong, slide no. ZL207♀; 1♂, Shandong, Taishan, 18.vii.1981, slide no. ZL349♂; 3♀, Shaanxi, Foping, 950 m, 23–24.vii.1998, coll Yao Jian, slides nos. ZL558, 579, 612♀; 1♀, Gansu, Zhouqu Shatan forestry reserve, 2350 m, 4.vii.1998, coll Zhang Xuezhong, slide no. ZL560♀; 1♀, Gansu, Wenxian, 700 m, 24.vi.1998, coll Chen Jun, slide no. ZL585♀ (IZCAS).

Remarks. The holotype of *E. pseudoplumbeolata* Vojnits, 1973 was not found either in ZFMK (Bonn) or TTM (Budapest), but Vojnits slide no. 551♂, belonging to this holotype was examined by Mironov in ZFMK.

The *druentiata* species group

This small species group was introduced by Mironov (2003). In the male genitalia the valve is usually angled ventrally; the aedeagus is armed with one to four dentate cornuti and an elongate V-shaped cornutus; sternite

A8 is peg-like, varying in shape. The female bursa is heavily sclerotized posteriorly, and spined basally. The group is represented in China by the following four species.

250. *Eupithecia exiguata* Hübner, [1813]

Geometra exiguata Hübner, [1813], *Samml. Eur. Schmett.* 5 Geometrae (2), pl. 73, fig. 379. Syntype(s) lost. [Europe].

Phalaena trimaculata Haworth, 1809, *Lepid. Br.* (2): 362. Syntype(s) (BMNH, not traced), [United Kingdom]. Junior primary homonym of *Phalaena Geometra trimaculata* De Villers, 1789 (identity: *Stegania trimaculata,* Ennominae) (synonymised in Curtis, 1829 and Stephens, 1829b).

Eupithecia ochreata Stephens, 1831, *Illustr. Br. Ent.* (Haustellata) **3**: 283. Syntype(s) (BMNH, not traced), [United Kingdom]: London (synonymised in Curtis, 1829).

Eupithecia lanceolaria Wood, 1854, *Index Ent.* (Edn **2**): 261. Syntypes (BMNH, not traced), [United Kingdom]: [England], Sudbury, Suffolk; Isle of Wight. Junior primary homonym of [*Eupithecia*] *lanceolaria* Boisduval, 1840 (treated as var. of *E. exiguata* in Doubleday, 1856).

Eupithecia exiguata muricolor Prout, 1938, *in* Seitz, *The Macrolepidoptera of the World,* **4** (Suppl.): 187. Syntype(s) (BMNH; examined), [Scotland]: E Aberdeenshire.

Eupithecia subexiguata Vojnits, 1974, *Acta zool. Acad. Sci. hung.* **20** (1–2): 224, figs 4a-e. Holotype ♂ (TTM; examined), China: Sued-Shensi, Tapaishan [Taibaishan] im Tsinling. **New junior synonym.**

Unavailable names (incorrect subsequent spellings): *exiguaria*: Boisduval (1840); *exiquata*: Ivinskis (1993).

Description. Fore wing ground colour pale brownish grey; transverse lines light brown, obscure, almost straight or slightly sinuate, widened and right-angled near costa; postmedial angled near costa, with brownish shade and small black wedge-shaped marks on veins; terminal area brownish with slightly sinuate pale wavy subterminal line, a quadrate dark blotch between M_1 and $M_{3,}$ and weaker blotches on costa and at tornus; discal dot conspicuous, black, elongate. Hind wing pale brownish grey, dark transverse lines marked only along posterior margin; discal dot brownish, rounded. Abdomen brownish grey usually with darker, brown first and second segments.

Male genitalia. Uncus elongate, narrow, uniapical. Valve strongly broadened medially, with a broad rounded right angle on ventral margin. Vinculum broad, with shallow medial hollow. Papillae long, clavate. Aedeagus stout, shorter than valve length. Vesica armed with two large, dentate cornuti, and a large, basal V-shaped cornutus. Sternite A8 peg-like, broadened at base, with narrow apex, breadth variable.

Female genitalia. Bursa pear-shaped, sclerotized, covered with small spines basally, except for a longitudinal area along left side. Ductus bursae tapered, spineless, with large longitudinal striations. Ductus seminalis recurved, attached to ductus bursae on right. Colliculum short and broad. Antrum slightly sclerotized, covered with numerous pores. Tergite A8 very broad, with medially convex anterior margin, and slightly concave posterior margin. Apophyses short and narrow. Papillae anales broad and rounded.

Range. China (Shaanxi, Gansu, Yunnan); Europe, Urals, Asian Russia (western and southern Siberia, Altai Mts., Sajan Mts., Transbaikalia, Priamurje, Primorje), Mongolia.

Phenology. Univoltine: late May to late July.

Biology. Larva polyphagous, found on leaves of a number of deciduous trees and shrubs.

Altitude. From sea level to about 1700 m in Kuraisky Range (Russia, Altai Mts.: coll. ZISP).

Similar species. Generally separable by the combination of grey ground colour, medial and postmedial strongly angled near costa, large discal dots, wedge-shaped marks in postmedial area and brownish grey abdomen, lacking dark lateral stripes. In the male genitalia, the valve shape is highly distinctive. The female is recognizable by the disposition of spines in the bursa and the striate ductus bursae.

Material. ♂, China, Sued-Shensi [Shaanxi], Tapaishan [Taibaishan] im Tsinling [Qinling], F, 19.vi.1935, H. Höne, Vojnits slide no. 1093♂ (holotype of *E. subexiguata*, TTM); 3♂, 3♀, same loc., 20, 24, 26.vi.1935, H. Höne, Vojnits slide no. 1095♂, Schmidt slide no. Eu 147 (paratypes of *E. subexiguata*, ZFMK); 2♂, 5♀, same loc., 19, 20, 22, 26.vi.1935, H. Höne, Vojnits slides nos. 1092♀, 1094♀, 10030♀, 10031♂ (paratypes of *E. subexiguata*, TTM); 1♂, same loc., 19.vi.1935, H. Höne, Vojnits slide no. 12273♂ (TTM); 1♂, China, Provinz Nord-Yuennan, Li-kiang [Lijiang], 20.vi.1934, H. Höne (paratype of *E. subexiguata*, ZFMK); 6♀, Gansu, Yongdeng, Tulugou, 1.vi.1988, slide no. ZL109♀, 6.vi.1988,, slide no. ZL343♀, 19.vi.1988, 1.viii.1989, slide no. ZL618♀, 25.iv.1992 (IZCAS).

Remarks. We have not found the external differences between Chinese *E. subexiguata* and European *E. exiguata* as listed by Vojnits (1974). Moths of the western Palaearctic population tend to have a yellowish tinge with brownish pattern elements; the Asian population tends to have a greyer ground colour and narrower and more pointed fore wings with more oblique terminal margins in proportion. This seems to be normal geographical variability for a widespread Palaearctic species. There are no clear differences in the male and female genitalia across the range. The shape of the eighth sternite in the males is very variable everywhere, including among European and Asian populations.

251. *Eupithecia hannemanni* Vojnits & de Laever, 1973

Eupithecia hannemanni Vojnits & de Laever, 1973, *Acta zool. Acad. Sci. hung.* **19** (3–4): 442, figs 4B, 8B, 10C. Holotype ♂ (ZFMK; examined), China: Sued-Shensi [Shaanxi], Tapaishan [Taibaishan] im Tsinling [Qinling].

Eupithecia manca Vojnits, 1979, *Acta zool. Acad. Sci. hung.* **25** (3–4): 433, fig. 10. Holotype ♀ (ZFMK; examined), China: Sued-Shensi [Shaanxi], Tapaishan [Taibaishan] im Tsinling [Qinling]. **New junior synonym.**

Description. Fore wing broad, ground colour pale greyish brown, basal and especially medial and terminal areas darker; transverse lines pale, all slightly sinuate, antemedial oblique, sharply angled onto costa; medial passing through discal dot; postmedial weakly angled onto costa; terminal area relatively broad, pale wavy subterminal line broadened near tornus into a tornal spot; discal dot large, ovoid. Hind wing pale greyish brown, transverse lines well marked on anal margin only, except postmedial, which faint but complete; terminal area darker; discal dot relatively large.

Male genitalia. Uncus stout, biapical. Valve broad medially with blunt ventral process; dorsal margin slightly prominent medially. Vinculum short, broad. Papillae short and rather broad. Aedeagus long and stout, broadened anteriorly, longer than valve length. Vesica armed with an elongate dentate cornutus, and a flat and twisted plate connected with a smaller irregular plate near ductus ejaculatorius base. Sternite A8 peg-like with broadened and rounded apex and two tapered basal lobes.

Female genitalia. Bursa globular, sclerotized with a large medial, heavily sclerotized blotch on ventral side, covered with large spines at base. Ductus bursae sclerotized, short and broad, with a short row of minute spines near base of ductus seminalis. Latter broad at base, curved forward, attached to posterior part of corpus bursae on right. Colliculum short and narrow. Antrum membranous, covered by numerous pores. Tergite A8 very transverse, broadest at posterior margin. Apophyses short and narrow. Papillae anales short and broad.

Range. China (Shaanxi, Gansu, Qinghai, Sichuan, Yunnan); Mongolia.

Phenology. Univoltine: mid-May to mid-July.

Altitude. From about 2350 to 3600 m above sea-level in Gansu and Yunnan (China: coll. O. Pekarsky).

Similar species. Externally similar to *E. suboxydata* but distinguishable by paler ground colour, the more oblique and sharply angled antemedial, and the broader postmedial. The male genitalia are unmistakable. The female are also immediately recognisable by the characteristic shape of the ductus bursae.

Material. ♂, China, Sued-Shensi [Shaanxi], Tapaishan [Taibaishan] im Tsinling [Qinling], 25.vi.1935, H. Höne, Vojnits slide no. 1162♂ [Genital präp. Nr. ZFMK 338♂] (holotype, ZFMK); 16♂, 5♀, same loc., 20, 21, 22, 24, 25, 26, 27, 29.vi., 2.vii.1935, H. Höne, de Laever slides nos. 55♂, 56♀, Vojnits slides nos. 1146♂ [Nr. ZFMK 336], 1147♂ [Nr. ZFMK 337], 1166♂ [Nr. ZFMK 339] (paratypes, 3♂ without abdomens, ZFMK); 14♂, 3♀, same loc., 20, 21, 22, 24, 25, 26, 27, 28, 29, 30.vi., 2.vii.1935, H. Höne, Vojnits slides nos. 1155♂, 1156♀, 1161♀, 1167♂, 10002♂, 10003♂, 10004♂, 10005♂, 10006♂, 10007♂, 10022♂, 10024♂ (paratypes, TTM); 1♂, 1♀, same loc., 24, 30.vi.1935, H. Höne (coll. R. Püngeler [!] in MNHU); ♀, same loc., 19.vi.1935, H. Höne, Vojnits slide no. 12208♀ (holotype of *E. manca*, ZFMK); 1♂, [Sichuan], Chasseurs Indigènes des Missionnaires de Ta-tsien-Loû 1906, Vojnits slide no. 869♂ (TTM); 1♂, Westchina, Stötzner, Coll. Dr. Wehrli (ZFMK); 1♂, Szetschwan [Sichuan], Yahothal, Exp. Stötzner (ZFMK); 1♂, China, Gansu Prov., Wenxian, 2350 m, 3.–4.vii.2001, Li Houhun leg. (coll. P. Skou); 1♀, China, Yunnan Prov., Zhongdian, 3100 m, 14.vii.2001, Li Houhun leg. (coll. P. Skou); 1♀, China, W. Sichuan, road Yajiang/Litang, 3600 m, 15.vii.2009, 29°59′ N / 100°52′ E, I. & A. Floriani, Saldaitis leg., Pekarsky slide no. OP 0503♀ (coll. O. Pekarsky); 3♂, 1♀, Gansu, Tulugou, 2280 m, 25–29.vii.1991, coll Xue Dayong, slides nos. ZL268, 377, 455♂, 456♀; 1♀, Qinghai, Huzhu, Beishan, 2300 m, 30.vii.1991, coll Xue Dayong, slide no. ZL367♀ (IZCAS).

252. *Eupithecia bella* Staudinger, 1897

Eup[*ithecia*] *bella* Staudinger, 1897, *Dt. ent. Z., Iris* **10** (1): 115, pl. 3, fig. 75. Syntypes 4♀ (coll. O. Staudinger in MNHU; examined), [Russia]: Sutschan.

Description. Fore wing rich rusty ochreous; transverse lines inconspicuous, but forming three narrow brownish costal spots; postmedial more distinct, straight, obtusely angled onto costa; a dark brown streak between M_1 and M_3 with medial and postmedial white blotches crossing it; terminal area with narrow wavy white subterminal line; a white tornal spot with brown inner shade; discal dot small, blackish. Hind wing paler, whitish grey with ochreous tinge; transverse lines well marked along anal margin only; terminal area darker, rusty ochreous with fine, dentate, white subterminal line, broadened near tornus; tornus with brownish blotch; discal dot distinct, rounded.
Male genitalia. Uncus elongate, narrow, biapical. Valve long and narrow, with slightly bulging sacculus. Vinculum tapered, with shallow medial hollow. Papillae small. Aedeagus stout, shorter than length of valve. Vesica armed with four dentate cornuti and a V-shaped cornutus near ductus

ejaculatorius base. Sternite A8 peg-like, broadened basally, with digitiform apical lobe.
Female genitalia. Bursa pear-shaped, densely covered with small spines anteriorly. Ductus bursae sclerotized with a large patch of spines. Ductus seminalis long, broadened basally, attached to corpus bursae on right. Colliculum short and broad. Antrum heavily sclerotized, short and broad. Sternite A8 very large, heavily sclerotized and poriferous. Tergite A8 large, almost square; its anterior margin convex medially. Apophyses relatively stout, slightly expanded at apices. Papillae anales broad and rounded.
Range. China (Hebei); Far East Russia (Priamurje, southern Primorje).
Phenology. Univoltine: late June to late July.
Similar species. Easily recognized by its rich coloration. The male and female genitalia are close to those of the following species: in the male the uncus is broader at base, the valve longer and without ventral angle, and the cornuti in the vesica much broader. In the female there is a separate patch of spines in the ductus bursae.
Material. 1♀, *bella* ♂ ♀ Sutschan, Origin., Gen. vérifiés 1438[2] *Eupithecia bella* Stgr. ♀, dét. E. de Laever 1954 (syntype, coll. O. Staudinger in MNHU); 1♀, [Russia], Sutschan 90 Dörr., *bella*, Origin. (syntype, coll. O. Staudinger in MNHU); 1♀, 75., Origin. (syntype, coll. O. Staudinger in MNHU); 1♀, bon. sp. bei *Subnotata* (Boh.), Origin. (syntype, coll. O. Staudinger in MNHU); 1♀, Lidahaeuzy [in Russian], 10.vii.1940, 8/1957, *bella* Stgr. (ZFMK).
Remarks. All four syntypes of this species in coll. O. Staudinger (MNHU) are females (see Staudinger 1897; Scoble 1999). We believe the collection locality of the single known Chinese specimen of this species to have been Liudaohezi in Hebei province, about 100 km NE of Beijing.

253. *Eupithecia insolita* Vojnits & de Laever, 1973

Eupithecia insolita Vojnits & de Laever, 1973, *Acta zool. Acad. Sci. hung.* **19** (3–4): 440, figs 4A, 8A. Holotype ♂ (ZFMK; examined), China (west): [Szechwan (Sichuan)], Sung-pan-ting.

Description. Fore wing ground colour ochreous brown; transverse lines pale; antemedial sharply angled onto costa; postmedial broad, double, very sharply angled twice before costa, and shaded inside darker brown; terminal area with three blackish brown blotches; discal dot large, ovoid. Hind wing white, brown along anal margin, with inconspicuous postmedial line and darker terminal area near tornus; discal dot small, rounded.
Male genitalia. Uncus elongate and narrow, biapical. Valve strongly broadened medially, with angulate ventral margin. Vinculum short and broad,

with shallow medial hollow. Papillae short and broad, clavate. Aedeagus long and stout, equal to length of valve. Vesica armed with four narrow dentate cornuti, an elongate, Y-shaped cornutus, and a smaller, pointed cornutus near ductus ejaculatorius base. Sternite A8 peg-shaped, bilobed at base, with a digitate process at apex.

Female genitalia. Bursa pear-shaped, the basal half covered with spines. Ductus bursae spineless, more strongly sclerotized on left. Ductus seminalis broad, curved. Colliculum short and broad. Antrum lightly sclerotized and covered with numerous pores. Tergite A8 transverse, broadest posteriorly, and with medially pointed anterior margin. Apophyses stout and very short. Papillae anales short and tapered.

Range. China (Tibet, Sichuan).

Phenology. Univoltine: late June to mid-July.

Altitude. Found from about 3250 to 3400 m above sea-level in Tibet (China: coll. IZCAS).

Similar species. The broad pale postmedian fascia recalls *E. hannemanni* or *E. suboxydata*, but the very white hind wing distinguishes it from both. The ornamentation of the aedeagus vesica and the shape of the eighth sternite are distinctive in the male genitalia. The partial sclerotization of the ductus bursae is helpful in the female genitalia.

Material. ♂, China, Szetschwan [Sichuan], Sung-pan-ting, Exped. Stötzner, Vojnits slide no. 1159♂ (holotype, ZFMK); 2♀, same loc., Vojnits slides nos. 13130♀, 13131♀ (paratypes, but not marked, TTM); 1♂, 3♀, China, Tibet, Jomda, 3400 m, 31.vii.1976, Galsworthy slides nos. ZL339♂, ZL633♀ (IZCAS).

The *cretaceata* species group

This species group was introduced by Bolte (1990), and developed by Mironov (2003). It appears to be closely related to the *satyrata* group, but is distinguished by the characteristic basally bilobed bifid eighth sternite in the male. The female genitalia closely resemble those of the *satyrata* group, but are more elongate, with striate wrinkles in the membranous part of the bursa. The group is represented in China by the following three species. It is likely that *Eupithecia veratraria* Herrich-Schäffer, 1848, will also be found to occur in NE China.

254. *Eupithecia duplex* Sterneck, 1931

Eupithecia duplex Sterneck, 1931, *Dt. ent. Z., Iris* 45 (2): 86. Holotype ♀ (ZFMK; examined), China.

Eupithecia pollens Vojnits, 1981, *Acta zool. Acad. Sci. hung.* 27 (3–4): 418, fig. 30. Holotype ♀ (ZFMK; examined), China: North Yunnan, Li-kiang. **New junior synonym.**

Description. Fore wing very broad, ground colour ash-grey; costal margin darker, blackish grey, especially at base; transverse lines marked as a series of black dashes on veins, sometimes connected by faint wavy brown lines; terminal area darker, with wavy pale subterminal line and sometimes whitish tornal spot bordered by blackish grey on inner side; discal dot large, smudgy. Hind wing slightly paler, transverse lines inconspicuous except faint postmedial; terminal area slightly darker.
Male genitalia. Uncus sharply tapered, biapical. Valve unmodified. Vinculum broad, with very shallow medial hollow. Papillae elongate, broadened basally. Aedeagus stout, slightly shorter than length of valve. Vesica with numerous small teeth, three large dentate cornuti, a complex elongate patch of linked teeth, a broad U-shaped cornutus and a small irregular cornutus near ductus ejaculatorius base. Sternite A8 basally bilobed, with two apically rounded arms.
Female genitalia. Bursa pear-shaped, more strongly sclerotized near colliculum, covered with small spines at base and with a separate transverse patch of spines near colliculum; spiniferous area on ventral side with a large spineless gap along left wall. Ductus seminalis broadened near base, sharply recurved, attached to bursa near colliculum on right. Colliculum short and narrow. Antrum short, membranous. Tergite A8 transverse. Apophyses short and narrow, broadened at apices. Papillae anales broad and rounded.
Range. China (Sichuan, Yunnan).
Phenology. Univoltine: late June to mid-July.
Altitude. Only altitude record is of 2850 m in Daxue Shan (China: coll. ZFMK).
Similar species. Similar in shape to the following species, but distinguished by more uniform darker colour and blackish grey costal margin and basal area of the fore wing. The ornamentation of the male vesica and the disposition of spines in the female bursa readily distinguish the genitalia of the two species.
Material. ♀, Typus *Eupithecia duplex* Sterneck, *Tephroclyst(h)ia duplex* Ster., 5/1930, 94, Staatl. Museum für Tierkunde Dresden (holotype, ZFMK); ♀, China, Provinz Nord-Yuennan, Li-kiang [Lijiang], 24.vi.1935, H. Höne, de Laever slide no. 133♀ (holotype of *E. pollens*, ZFMK); 1♀, same loc., 25.vi.1935, H. Höne, Vojnits slide no. 13123♀ (paratype of *E. pollens*, TTM); 1♀, same loc., 24.vi.1935, H. Höne, Vojnits slide no. 12123♀ (paratype of *E. pollens*, TTM); 1♀, same loc., 27.vi.1935, H. Höne (paratype of *E. pollens*, abdomen missing, ZFMK); 1♂, 4♀, China, Sichuan, Daxue Shan Mts., Gongga Shan, NW Moxi, 29°41′ N, 101°58′ E, 2850 m, 14.–19.vii.1999, leg. V. Siniaev & A. Plutenko (ZFMK).

Remarks. The male genitalia of this species are described here for the first time.

255. *Eupithecia tectaria* Mironov & Galsworthy, 2011

Eupithecia tectaria Mironov & Galsworthy, 2011, *Lep. Science* **62** (1): 30, figs 16, 37, 39. Holotype ♀ (ZFMK; examined), [China]: [Fujian], Kuatun [Guadun], 2300 m.

Description. A large species. Fore wing ground colour pale brown, slightly darker in terminal area; transverse lines inconspicuous; postmedial obtusely angled onto costa; discal dot small, rounded, black. Hind wing also pale brown with inconspicuous transverse lines; discal dot small, rounded, dark brown.
Male genitalia. Uncus elongate, biapical. Valve unmodified, costal margin sinuate. Vinculum broad, semicircular. Papillae relatively short, with parallel sides. Aedeagus stout, with three stout toothed cornuti at centre, a small irregular plate closer to apex, and a twisted knot near base. Sternite A8 broadly bilobed at base, with two sinuous apical arms.
Female genitalia. Bursa almost globular, sclerotized in posterior half, about 2/3 covered with large spines; a transverse row of small spines near base of ductus seminalis. Ductus bursae broad, sclerotized, with numerous striations, and a short transverse row of spines near base of colliculum. Colliculum short and broad. Antrum broad, membranous. Tergite A8 transverse, with medially concave posterior margin. Anterior apophyses short, broader then posterior, the latter slightly expanded at apices. Papillae anales broad and rounded.
Range. China (Fujian).
Altitude. Found at about 2300 m above sea-level in Fujian (China: coll. ZFMK).
Similar species. Very similar externally to *E. veratraria* Herrich-Schäffer, 1848, but can be distinguished by the darker ground colour, and the narrower fore wing with oblique termen and rather pointed apex. The male genitalia are closer to those of *E. duplex*, but both the aedeagus and eighth sternite are broader. The female bursa has a larger spine field than in *E. veratraria*, and is considerably larger than in *E. duplex*.
Material. ♀, [Fujian], Kuatun [Guadun] (2300 m), 27, 40 n. Br. 117, 40 ö. L., J. Klapperich, 6.v.1938 (Fukien), Mironov slide no. 826♀ (holotype, ZFMK); 1♂, [Fujian], Kuatun [Guadun] (2300 m), 27, 40 n. Br. 117, 40 ö. L., J. Klapperich, 21.iv.1938 (Fukien), Vojnits slide no. 12253♂ (paratype, ZFMK).

256. *Eupithecia mandschurica* Staudinger, 1897

Eup[ithecia] absinthiata var. ? *mandschurica* Staudinger, 1897, *Dt. ent. Z., Iris* **10** (1): 118. Syntypes, at least 1♂, 4♀ (coll. O. Staudinger in MNHU; examined), [Russia]: Sutschan, Ussuri, Askold, Suifun.

Eup[ithecia] helveticaria f./ab. *robusta* Dietze, 1910, *Biologie der Eupithecien* 1: pl. 74, fig. 418; *ibidem* (1913), 2: 76. Holotype ♀ (coll. K. Dietze in MNHU; examined), [Russia]: Amur, Radde. **New junior synonym.**

Eupithecia mandschurica japonica Inoue, 1979, *Bull. Fac. domest. Sci. Otsuma Wom. Univ.* **15**: 201, fig. 22C; fig. 23C; fig. 24B; fig. 25B; fig. 26B. Holotype ♀ (*nec* ♂) (BMNH; examined), [Japan]: Niigata Pref[ecture], Mt. Yahiko.

Eupithecia depressa Vojnits, 1979, *Acta zool. Acad. Sci. hung.* **25** (3–4): 430, fig. 6. Holotype ♀ (ZFMK; examined), [China]: Prov. Shansi [Shanxi], Mien-shan, 2000 m. **New junior synonym.**

Misidentification: *Eupithecia certa* Vojnits, 1981 (1♂ paratype in ZFMK).
Unavailable names (incorrect subsequent spelling): *mandschuria*: de Laever (1956).

Description. Fore wing ground colour dark reddish brown; transverse lines usually indistinct, well marked along costa only; postmedial more distinct, obtusely angled onto costa; terminal area slightly darker with inconspicuous pale wavy subterminal line and tornal spot; discal dot elongate, black. Hind wing concolorous, transverse lines faintly marked; terminal area darker with weak trace of pale subterminal line and tornal spot; discal dot small, sometimes invisible.

Male genitalia. Uncus elongate, biapical. Valve unmodified. Vinculum tapered. Papillae of average size. Aedeagus stout, shorter than length of valve. Vesica armed with three dentate cornuti, a long narrow patch of dense small teeth, a folded, U-shaped cornutus and a smaller irregular cornutus near ductus ejaculatorius base. Sternite A8 peg-like, broadly bilobed at base, with two narrow, pincer-like apical rods.

Female genitalia. Bursa pear-shaped, covered with large spines except for a membranous gap on left. Ductus bursae sclerotized, with a transverse patch of spines opposite ductus seminalis and longitudinal striations. Ductus seminalis broadened at base, recurved, attached to ductus bursae near colliculum on right. Colliculum short and narrow. Antrum membranous. Tergite A8 quadrate or elongate. Apophyses short and narrow. Papillae anales broad and rounded.

Range. China (Heilongjiang, Shandong, Shanxi, Shaanxi, Gansu, Henan, Hubei, Hunan, Shanghai, Jiangsu); Asian Russia (southern Priamurje, Primorje, Sakhalin, Kuril Islands: Kunashiri), Korea, Japan (Hokkaido, Honshu, Sado).

Phenology. Univoltine: late July to early September.

Altitude. From sea level to about 3000 m in Tapai-shan (China: Vojnits 1979b).

Similar species. Rather uniform dark colour separates from *E. veratraria*. Also similar to *E. absinthiata*, which has narrower wings and clearer cut markings. Genitalia in both sexes are similar to those of *E. veratraria*, but the dentate cornuti in the male vesica are stouter, and the female bursa is narrower and more elongate.

Material. 1♂, [Russia], Sutschan 90 Dörr., *mandschuri(c)a* ♂, Origin., 188., 18., Zool. Mus., Berlin, Gen. vérifiés 1453[3] *Eupithecia mandshuri(c)a* Stgr. ♂, Sutschan, *Eupithecia* photo 84 C 21 det. A. Vojnits (syntype of *E. absinthiata* var. *mandschurica*, coll. O. Staudinger in MNHU); 1♀, [Russia], Ussuri, Dörr. 82 (syntype of *E. absinthiata* var. *mandschurica*, coll. O. Staudinger in MNHU); 2♀, [Russia], Askold, Dörr. 83 (syntypes of *E. absinthiata* var. *mandschurica*, coll. O. Staudinger in MNHU); 1♀, [Russia], Suifun, 85 D. (syntype of *E. absinthiata* var. *mandschurica*, coll. O. Staudinger in MNHU); ♀, [Russia], Radde, m. Amur, 6.Juli 63, Korb, forma: *robusta* ? *mandschurica*, Radde, Amur, 6.7.03 (holotype of *E. helveticaria* f. *robusta*, design. by Mironov, 2009 (coll. K. Dietze in MNHU); ♀ [*nec* ♂], [Japan], Mt. Yahiko, Niigata Pref., Aug. 18, 1961, Coll. S. Sakurai, No. 1006 (holotype of *E. mandschurica japonica*, coll. H. Inoue in BMNH); 1♀, [Heilongjiang], Mandschurei, Tschang-kwan-zailing [Zhang Guang Cai Ling], 500–700 m, Juni-Juli (ZFMK); ♀, Prov. Shansi [Shanxi], Mien-shan, Obere Höhe *ca.* 2000 m, 29.vii.1937, H. Höne, Vojnits slide no. 12190♀ (holotype of *E. depressa* Vojnits, 1979, ZFMK); 1♀, same loc. and alt., 2.viii.1937, H. Höne, Vojnits slide no. 12189♀ (paratype of *E. depressa* Vojnits, 1979, ZFMK); 1♂, same loc. and alt., 1.viii.1937, H. Höne, Vojnits, slide no. 13122♂ (paratype of *E. certa*, ZFMK); 1♂, 1♀, same loc. and alt., 31.vii., 3.viii.1937, H. Höne, de Laever slide no. 154♀, Mironov slide no. 711♂ (ZFMK); 1♀, Sued-Shensi [Shaanxi], Tapaishan [Taibaishan] im Tsinling [Qinling], *ca.* 3000 m, 11.viii.1936, H. Höne, Vojnits slide no. 12207♀ (paratype of *E. depressa* Vojnits, 1979, TTM); 2♀, China, Kiangsu [Jiangsu], Vojnits slides nos. 819♀, 821♀ (TTM); 1♀, China, Shanghai, Vojnits slide no. 820♀ (TTM); 1♂, Henan, Funiushan, 7–9. viii.1993, Wang Zhiguo, slide no. ZL667♂; 1♀, Hubei, Shennongjia, Dajiuhu, 1800 m, 4.viii.1981, coll Han Yinheng, slide no. ZL200♀; 1♀, Shandong, Laoshan, c800 m, slide no. ZL566♀; 1♀ Hunan, Zhangjiajie, 8.x.1988, coll Zhao Zhongling, slide no. ZL234♀; 2♀, Gansu, Yongdeng, Tulugou, 2280 m, 25–29.vii.1991,slides nos. ZL149, 267♀; 2♀, Gansu,Wenxian, Qiujiaba, 2350 m, 21.vii.1999, coll Zhu Chaodong, slides nos. ZL547, 565♀ (IZCAS).

The *satyrata* species group

Introduced by McDunnough (1949), and one of the largest Holarctic groups in the genus. The genitalia in both sexes are unusually uniform, making identification of species often difficult, especially in the females. The male genitalia are characterized by an unmodified valve, an aedeagus vesica with a number of small dentate cornuti, and a peg-shaped eighth sternite. The female bursa is usually large and robust-walled, well spined at the base and normally with a separate group of spines close to the colliculum, with the intervening membranous area often heavily sclerotized and striate. The group is represented in China by the following thirteen species.

257. *Eupithecia intricata* Zetterstedt, 1839

Larentia intricata Zetterstedt, 1839, *Insecta Lapponica* 4 (6): 962. Syntype(s) (MZLU): [Sweden]: Lapponia.

Eupithecia helveticaria Boisduval, 1840, *Gen. Index meth. Eur. Lepid.*: 208. Syntype(s), [Switzerland]: [Wallis], Valesia (synonymised in Staudinger, 1871).

Larentia arceuthata Freyer, 1842, *Neuere Beitr. Schmett.* 4 (62): 145, pl. 372. Syntype(s) (1♂ syntype in coll. K. Dietze, MNHU; examined), [Germany]: Augsburg (synonymised to *Eupithecia helveticaria* Boisduval, 1840 in Staudinger, 1871).

Eupithecia helveticaria var. *anglicata* Millière, 1870, *Iconogr. Descr. Chenilles Lépid. inédits* 3 (25): 102, pl. 110, figs 20, 21. Syntypes 8♂♀ (MNHN?), United Kingdom. Junior primary homonym of *Eupithecia anglicata* Herrich-Schäffer, 1863.

Eupithecia taylorata Swett, 1907, *The Can. Ent.* 39 (11): 377. Lectotype ♂ (MCZ), U.S.A.: Massachusetts, Sherborn; Maine, Monmouth (synonymised to *Eupithecia arceuthata* Freyer, 1842 in McDunnough, 1949).

Eupithecia gibsonata Taylor, 1910, *The Can. Ent.* 42 (3): 78. Holotype ♀ (USNM), Canada: Ontario, Ottawa (synonymised to *Eupithecia helveticaria* Boisduval, 1840 in McDunnough, 1929).

Eup[ithecia] helveticaria f. *septemtrionalis* Dietze, 1910, *Biologie der Eupithecien* 1, pl. 74, figs 423, 424, 434, 435; *ibidem* (1913), 2: 75. Syntypes 13, at least 10♂ (coll. K. Dietze, MNHU; examined), Finland: Kuusamo.

Eupithecia chagnoni Swett, 1911, *The Can. Ent.* 43 (7): 255. Holotype ♂ (MCZ), Canada: Quebec, Montreal (synonymised to *Eupithecia helveticaria* Boisduval, 1840 in McDunnough, 1929).

Eup[ithecia] absinthiata sp.?/f. *lunata* Dietze, 1910, *Biologie der Eupithecien* 1: pl. 71, fig. 190; *ibidem* (1913), 2: 88. Holotype ♂ (coll. K. Dietze in MNHU; examined), [China]: [Xinjiang], Ili, Juldus [Yining region]. **New junior synonym.**

[*Eupithecia*] *helveticaria* f. *suffusa* Dietze, 1913, *Biologie der Eupithecien* 2: 76. Syntypes 5♂♀, at least 1♂ (coll. K. Dietze in MNHU; examined), Germany: Mittelrhein; Austria: Stainz.

[*Eupithecia*] *helveticaria* f. *mediofasciata* Dietze, 1913, *Biologie der Eupithecien* 2: 76. Syntypes 3♂ (coll. K. Dietze in MNHU, not traced), Finland; Germany: Erfurt.

Eupithecia helveticaria millieraria Wnukowsky, 1929, *Zool. Anz.* 83 (9/10): 223 [Replacement name for *Eupithecia helveticaria* var. *anglicata* Millière, 1870].

Eurithecia [sic!] *intricata hibernica* Mere, 1964, *Entomologist's Gaz.* 15: 73, pl. 3, fig. 2. Holotype ♂ (BMNH; examined), Ireland: Co(unty) Clare, The Burren, Doughbranneen.

Unavailable names (incorrect subsequent spellings): *helveticata*: Duponchel (1842); *arceutharia*: Herrich-Schäffer (1848).

Description. Fore and hind wings unicolorous brownish grey. Fore wing transverse lines indistinct; basal, ante- and postmedial weakly visible, evenly curved onto costa; terminal area unicolorous, without pale subterminal line; discal dot relatively small, black. Hind wing with a paler transverse band between medial and terminal areas; dark transverse lines inconspicuous; terminal area without pale subterminal line; discal dot small, comma-like.

Male genitalia. Uncus narrow, uniapical. Valve with rather pointed apex. Vinculum short, rounded. Papillae rather broad. Aedeagus stout, shorter than length of valve. Vesica armed with four dentate cornuti, an elongate U-shaped cornutus and a heavily sclerotized, curved irregular cornutus near ductus ejaculatorius base. Sternite A8 triangular, with bilobed base and blunt, rounded apex.

Female genitalia. Bursa pear-shaped, with spines separated by a wide, membranous gap on left side and base, and with broader marginal spines. Ductus bursae tapered, with two small patches of spines. Ductus seminalis curved, broadened near base, attached to posterior half of corpus bursae. Colliculum short and broad. Antrum membranous with elongate shagreened dorsal patch. Tergite A8 transverse. Anterior apophyses short and stout, posterior narrower. Papillae anales broad and rounded.

Range. North western China (Xinjiang); almost all Europe from Spain to Urals, western Caucasus, north-eastern Turkey, SE Kazakhstan, all southern Siberia to Amur region and Mongolia, Korea (Shin 1996, as *lunata*); North America (subsp. *taylorata* Swett, 1907).

Phenology. Univoltine: late May to mid-July.

Biology. Larva feeding on species of *Juniperus.*

Altitude. From about 1700 to 2300–2800 m in Zailijsky Alatau Mts. and Dzhungarsky Alatau Mts. (SE Kazakhstan).

Similar species. Closely related to *E. satyrata*. Central Asian specimens are distinguishable from it by the brownish colour, darker ante- and postmedial lines on the fore wings, faintly larger oblique discal spots on all wings, more chequered fringes and darker second abdominal segment. The male is the only species in the group with a uniapical uncus. The female is well characterized by the very broad marginal spines in the bursa.

Material. ♂, [Xinjiang], ? n. sp. *lunata* Type, Juldus, Ili, K. Dietze, Zool. Mus. Berlin, Vojnits slide no. 1183♂ (holotype of *E. lunata* wrongly associated with gen. slide of *E. leptogrammata*, coll. K. Dietze in MNHU); 1♂, USSR, Kazakhstan, Zailiskiy Alatau, Alma-Atinskij Nat. P., 43°5′ N, 77°15′ E, 1700 m, Picea/meadow, 22.vi.1990, ad luc., leg. K. Mikkola & L. Kaila (ZISP); 1♂, [Kazakhstan], Dzhungarsky Alatau, 40–42 km ENE of Tekeli, near Semjenov-Tianshansky Pass, 2300–2800 m, 21.vi.1993, V. Dubatolov & O. Kosterin (ZISP); 1♂, 1♀, Kazakhstan, Prov. Almaty,

Uzunbulak, Mt. Kuluktau, 1800 m, 79°02′ E, 43°08′ N, 29.v.1994, leg. Gy. Fábián & I. Retezár (coll. M. Sommerer, Munich).

Remarks. This species is known from China only from Dietze's holotype of *E. lunata*. We have compared this with four very similar specimens from the Dzhungarsky Alatau Mts. and Zailijsky Alatau Mts. in south-eastern Kazakhstan. They belong to *E. intricata* and are more similar to its Scottish form *millieraria* Wnukowsky, 1929 (= *anglicata* Millière, 1870) than to the nominate subspecies, but distinguished by larger size and larger discal spots on all wings. We are satisfied that the holotype specimen of *E. lunata* belongs to this species. However, the male genitalia slide associated with this specimen (Vojnits slide no.1183), although very badly prepared and damaged, appears to belong to a species of very different appearance, *E. leptogrammata* Staudinger, 1882 (= *melanochroa* Wehrli, 1927). We assume that this is yet another instance of a mistake in the slide preparation process.

258. *Eupithecia lupa* Mironov & Galsworthy, 2007

Eupithecia lupa Mironov & Galsworthy, 2007, *Trans. lepid. Soc. Japan* 58 (3): 353, figs 6, 16, 22. Holotype ♂ (TTM; examined), Taiwan: Prov. Nantou, Hohuachi, 24°13′ N, 121°16′ E, 1950 m.

Description. Fore wing ground colour brownish grey mottled with whitish grey scales; basal area and a broad median fascia darker brown, the latter broadest at the costa; basal line sharply angled near costa; postmedial weakly angled towards costa; terminal area mottled pale and dark brown, with paler apical and subtornal blotches; discal dot elongate, black. Hind wing ground colour whitish grey with brownish tinge; basal and postmedial lines marked near anal margin; postmedial well marked, evenly curved; terminal area darker with wavy inner border; discal dot small, pale.

Male genitalia. Uncus longish, biapical. Valve unmodified. Vinculum short and broad. Papillae very broad, rather clavate. Aedeagus stout, shorter than length of valve. Vesica armed with an elongate patch of small spines, three large dentate cornuti, a folded, U-shaped cornutus near ductus ejaculatorius base and a small curved cornutus near anterior end of aedeagus. Sternite A8 rectangular, with saw-toothed lateral margins near apex, and two short pincer-like apical horns.

Female genitalia. Bursa pear-shaped, with two patches of large spines near base. Ductus bursae with longitudinal striations and an obliquely transverse patch of small spines on ventral side. Ductus seminalis attached to ductus bursae from right. Colliculum relatively broad. Antrum

membranous. Tergite A8 square with deep membranous medial hollow at posterior margin. Apophyses narrow and pointed. Papillae elongate, rounded apically.
Range. China (Taiwan).
Phenology. Univoltine (?): mid-February to late March.
Altitude. Found at about 1950 m above sea-level in Taiwan (coll. TTM).
Similar species. Easily distinguished from other Taiwanese *Eupithecia* by the pointed triangular fore wings and distinctive maculation. The shape of the eighth sternite in the male genitalia is unique. The female bursa is broadly similar to that of other members of the group, but unlike any other species in Taiwan.
Material. ♂, Taiwan, Prov. Nantou, Hohuachi, 1950 m, 24°13′ N, 121°16′ E, 15–19.ii.1999, leg. L. Peregovits & G. Ronkay (holotype, TTM); 3♀, same loc., 31.iii.2000, leg. A. Kun & L. Peregovits (paratypes, TTM).

259. *Eupithecia satyrata* Hübner, [1813]

Geometra satyrata Hübner, [1813], *Samml. Eur. Schmett.* 5 Geometrae (1), pl. 85, fig. 439. Syntype(s) lost. [Europe].
Eupithecia grammaria Boisduval, 1840, *Gen. Index meth.Eur. Lepid.*: 208. Syntype(s), France: Alp. Gall (treated as var. of *Eupithecia satyrata* in Guenée, [1858]).
Eupithecia callunaria Doubleday, 1850, *The Zoologist* 8 (App., Art. 8): 55. Syntypes (BMNH), [United Kingdom]: Scotland; northern England (synonymised in Doubleday, 1856).
Eupithecia italicata Guenée, [1858]: in Boisduval & Guenée, *Hist. nat. Insectes* (Spec. gén. Lépid.) **10** (2): 316. Holotype ♂ (BMNH ?), Italy: Lombardie, Domo-Dossola (synonymised in Staudinger & Rebel, 1901).
Eupithecia satyrata ab./var. *subatrata* Staudinger, 1871: *in* Staudinger & Wocke, *Cat. Lepid. Eur. Faunengeb.* **1**: 197. Syntypes, at least 1♂, 1♀ (coll. O. Staudinger in MNHU; examined), Central Asia: Ferghana, Alai Mts., Ili region.
Eupithecia rivosulata Dietze, 1875, *Ent. Z., Stettin* **36** (4–6): 254, pl. 2, fig. 6. Syntypes 1♂, 1♀ (MNHU, not traced), [Russia]: Altai Mts. (treated as var. of *Eupithecia satyrata* in Staudinger & Rebel, 1901).
Eupithecia curzoni Gregson, 1884, *The Entomologist* 17 (257): 230. Syntype(s) (BMNH ?), [United Kingdom]: Shetland Islands, Balta Sound (treated as var. of *Eupithecia satyrata* in Staudinger & Rebel, 1901).
Eupithecia satyrata var. *fagicolouria* Robson & Gardner, 1886, *Young Naturalist* 7 (Suppl.): (part 84) 7, 41. Syntype(s) (BMNH). Type locality not stated.
Eupithecia satyrata var. *serenata* Staudinger, 1896, *Dt. ent. Z., Iris* 8 (2): 340. Syntypes 4♀ (coll. O. Staudinger, K. Dietze and R. Püngeler in MNHU; examined), [China]: Tibet; [Russia]: Altai.
Eupithecia dodata Taylor, 1906, *The Can. Ent.* **38** (3): 103. Lectotype ♀ (USNM), Canada: Calgary, Head of Pine Creek (synonymised in McDunnough, 1929).
Eupithecia slocanata Taylor, 1908, *The Can. Ent.* **40** (2): 58. Lectotype ♂ (USNM), Canada: British Columbia, Kaslo (synonymised to *Eupithecia terminata* Taylor, 1908 in McDunnough, 1929).
Eupithecia terminata Taylor, 1908, *The Can. Ent.* **40** (2): 58. Lectotype ♂ (USNM), Canada: British Columbia, Kaslo (synonymised in Bolte, 1990).
Eupithecia satyrata intimata Pearsall, 1908, *Jl N. Y. ent. Soc.* **16** (2): 100. Lectotype ♀ (AMNH), U.S.A.: North Carolina, Black Mts.

Eupithecia fumata Taylor, 1910, *The Can. Ent.* **42** (3): 82. Holotype ♀ (USNM), Canada: Ontario, Ottawa (synonymised in McDunnough, 1929).

Eup[ithecia] satyrata f. *juldusi* Dietze, 1910, *Biologie der Eupithecien* **1**, pl. 71, figs 196, 197. Syntypes 6♂, 3♀ (coll. K. Dietze in MNHU; examined), [China]: Ili region, [Xinjiang], Juldus [Yining] Mountains.

Eup[ithecia] satyrata f. *medionotata* Dietze, 1910, *Biologie der Eupithecien* **1**, pl. 71, figs 206, 207; *ibidem* (1913), **2**: 85. Syntypes 6♂, 8♀ (coll. K. Dietze in MNHU; examined), [Russia]: Altai Mts., Sajan Mts.

[Eupithecia] satyrata f. *concolour* Dietze, 1913, *Biologie der Eupithecien* **2**: 85. Unnecessary replacement name for *juldusi* Dietze.

Eupithecia divinula Cassino & Swett, 1924, *Lepidopterist* **4** (4): 26. Holotype ♂ (MCZ), Canada: Alberta, Nordegg (synonymised in McDunnough, 1932).

Eupithecia mackieata Cassino & Swett, 1925, *Lepidopterist* **4** (6–7): 45. Holotype ♂ (MCZ), Canada: Alberta, Edmonton (synonymised in McDunnough, 1929).

Eupithecia satyrata f. *zermattensis* Wehrli, 1928: *in* Vorbrodt, *Dt. ent. Z., Iris* **42** (2): 85. Syntype(s), [Switzerland]: Zermatt.

Eupithecia submelanochroa Vojnits, 1973, *Folia ent. Hung.* **26** (1): 211, figs 1c, d; 3c, d. Holotype ♂ (ZFMK, now in TTM; examined), [Kirghizstan]: Turkestan Oriental, Province Semirechgensee, Fort Narine [= Naryn] (synonymised in Mironov, 1991).

Eupithecia inculta Vojnits, 1975, *Acta zool. Acad. Sci. Hung.* **21** (3–4): 450, fig. 2. Holotype ♂ (TTM; examined), Mongolia: Central aimak, SE of Somon Bajanzogt, 1600 m. **New junior synonym.**

Misidentifications: *Eupithecia aggregata* Vojnits, 1975 (*nec* Guenée, [1858]); *Eupithecia fuscicostata* Vojnits, 1975 (*nec* Christoph, 1887); *Eupithecia innotata* Stanescu, 1997 (*nec* Hufnagel, 1767).

Unavailable names (infrasubspecific): *strandi*: Fuchs (1901); *limbopunctata*: Dietze (1910); *caeca*: Dietze (1913); *nigrofasciata*: Dietze (1913); *transversa*: Dietze (1913); *bistrigata*: Dietze (1913); *marginata*: Lempke (1969). – (incorrect subsequent spellings): *satirata*: Alexinschi (1942); *satyraria*: Boisduval (1840).

Description. A broad winged species. Fore wing almost unicolorous, usually light brown or light greyish brown; transverse lines pale, indistinct; terminal area with weak trace of pale subterminal line; discal dot small, inconspicuous, rounded. Hind wing unicolorous pale brown, slightly paler than fore wing; transverse lines and discal dot usually invisible.

Male genitalia. Uncus narrow, biapical. Valve unmodified. Vinculum short and rounded. Papillae parallel-sided. Aedeagus rather stout, shorter than valve length. Vesica armed with four dentate cornuti, the apical one smaller and narrower than the others, and a flat, U-shaped cornutus, connected with a smaller irregular, curved cornutus at ductus ejaculatorius base. Sternite A8 peg-like with slightly bifid apex.

Female genitalia. Bursa sclerotized, pear-shaped, mostly covered with uniform spines. Ductus bursae sclerotized, with a longitudinal patch of spines near base of ductus seminalis and a transverse patch near colliculum, and with longitudinal striations. Ductus seminalis broadened at base; attached to central part of ductus bursae on right. Colliculum short and broad. Antrum broad, membranous. Tergite A8 transverse. Apophyses short and narrow, the posterior with small spurs near broad and rounded papillae anales.

Range. China (Xinjiang, Qinghai); Europe, Asian Turkey, SE Kazakhstan, Kirghizstan, Asian Russia (western and southern Siberia to Transbaikalia, Dahuria, Magadan region and Priamurje), Mongolia, North America (subsp. *dodata* Taylor, 1906).

Phenology. Univoltine: late May to early August. Overwintering as pupa in flimsy cocoon in the ground.

Biology. Larva polyphagous on the flowers and seeds of various, mainly herbaceous plants in Europe and North America.

Altitude. From sea level to about 2300–2800 m in Dzhungarsky Alatau Mts. (Kazakhstan: coll. ISEN).

Similar species. Central Asian specimens are distinguished from nominate European by the almost unicolourous brown ground colour on all wings, and resemble the Asian species *E. leptogrammata* Staudinger, 1882. They can be distinguished by the paler ground colour and weakly visible pale transverse lines, which are usually distinct and wavy in *E. leptogrammata*. The genitalia are quite distinct from those of *E. leptogrammata*, but not easily distinguished from other members of this species group.

Material. 1♀, [Qinghai], Kuku-Noor 94, Rückb., *Satyrata* var. *Serenata* Stgr., Form welche stark zu *aggregata* Gn. hinüberneigt (Dietze gescheckt) (syntype of *E. satyrata* var. *serenata*, coll. O. Staudinger in MNHU); 1♀, [Qinghai], *Satyr.* v. *Serenata* Stgr., ♀, Kuku-Noor, *serenata* Stgr. (? *satyrata*) Kuku-Noor (syntype of *E. satyrata* var. *serenata*, coll. K. Dietze in MNHU); 1♀, [Qinghai], Tibet (Kuku-Noor), 1892–93, *satyrata* var. *serenata* Stgr. ♀, Original, v. R. Tancré 5.05, *Satyrata* var. *Serenata* Stgr. (syntype of *E. satyrata* var. *serenata*, coll. R. Püngeler in MNHU); 1♀, [Russia], Altai, Altai Kinderm., *Satyrata* var. *Serenata* Stgr., Origin., Coll. Led., Zool. Mus. Berlin, Gen. vérifiés 1455[3] *Eupithecia* ..., dét. E. de Laever 1954 (syntype of *E. satyrata* var. *serenata*, coll. O. Staudinger in MNHU); 1♂, [Xinjiang], Juldus, *satyrata*, form: *juldusi* Bang-Hs, Typ., Juldus Ili (syntype of *E. satyrata juldusi*, coll. K. Dietze in MNHU); 1♂, 2♀, [Xinjiang], Juldus, ab. *juldusi* Bang-Haas (syntypes of *E. satyrata juldusi*, coll. K. Dietze in MNHU); 4♂, 1♀, [Xinjiang], Juldus Ili (syntypes of *E. satyrata juldusi*, coll. K. Dietze in MNHU); 1♂, 1♀, [Xinjiang], Asia centr., Juldus, als *satyr. juldusi*, 2/1911 v. B.-Haas, Püng. (coll. R. Püngeler in MNHU); ♂, [Kirghizstan], Turkestan Oriental, Province Semirechgensee, Fort Narine (= Naryn), G. S. Akulin 1906, Vojnits slide no. 716♂ (holotype of *E. submelanochroa*, TTM); ♂, Mongolia, Central aimak SO von Somon Bajanzogt 1600 m,

Exp. Dr. Z. Kasab, 1966, Nr. 521, 11.vi.1966, *E. collustrata* det. Vojnits A., Vojnits slide no. 10192♂, HNHM Lepidoptera No. 00106515 (holotype of *E. inculta*, TTM).

260. *Eupithecia pernotata* Guenée, [1858]

Eupithecia pernotata Guenée, [1858], *in* Boisduval & Guenée, *Hist. nat. Insectes* (Spec. gén. Lépid.) **10** (2): 316. Syntypes 2♂ 1♀ (coll. M. Feisthamel, not traced), [Italy]: Mont-Rose; Mont-Ossolano.
Eupithecia aggregata Guenée, [1858], *in* Boisduval & Guenée, *Hist. nat. Insectes* (Spec. gén. Lépid.) **10**: 315. Holotype ♀ (coll. O. Staudinger in MNHU; examined), [Russia]: Altai [Ustkamenogorsk] (synonymised in Mironov & Galsworthy, 2010).
Eupithecia antaggregata Inoue, 1977, *Bull. Fac. domest. Sci. Otsuma Wom. Univ.* **13**: 321, fig. 39. Holotype ♂ (coll. H. Inoue in BMNH; examined), [Japan]: Hokkaido, Tokachi, Nukabira (synonymised to *Eupithecia aggregata* Guenée, [1858] in Mironov, 2003).
Eupithecia aggregata enictata Pellmyr & Mikkola, 1984, *Notulae Ent.* **64** (2): 54, fig. 2. Holotype ♂ (ZMUH), Finland: E-K, Hamina.

Misidentification: *Eupithecia fuscicostata* Vojnits, 1975 (*nec* Christoph, 1887).
Unavailable names (incorrect subsequent spelling): *pernotaria*: Morris (1861).

Description. Fore wing ground colour whitish grey with delicate light ochreous suffusion; transverse lines numerous, fine, alternately dark and pale, slightly sinuate, rather oblique, right-angled, slightly broader and darker near costa; whitish wavy line sometimes broken; small whitish tornal spot present; discal spot minute, frequently ill-defined. Hind wing paler with inconspicuous transverse lines except double pale band between medial and terminal areas; discal dot minute or obsolete.
Male genitalia. Uncus long and narrow, biapical. Valve unmodified, apex rather narrow. Vinculum broad and rounded. Papillae parallel-sided. Aedeagus shorter than length of valve. Vesica armed with one larger and two smaller dentate cornuti, a short broad apical horn-like cornutus, an elongate U-shaped cornutus and a small, irregular cornutus. Sternite A8 elongate, peg-like, broadly bilobed at base, with almost parallel margins in apical half and a broad indented apex.
Female genitalia. Bursa sclerotized, pear-shaped, mostly covered with large uniform spines. Ductus bursae sclerotized, with prominent bulge on left side, two small patches of spines near base of ductus seminalis and longitudinal striations. Ductus seminalis narrow, slightly broadened at base, attached to central part of ductus bursae on right. Colliculum relatively short and broad. Antrum membranous. Tergite A8 transverse, with shallow medial hollow in posterior margin. Apophyses short and narrow. Papillae anales relatively small, rounded.

Range. China (Tibet, Gansu, Shaanxi, Shanxi, Heilongjiang); Europe, Urals, northern Kazakhstan, southern Siberia (Altai Mts., Minusinsk region, Transbaikalia, Dahuria, Priamurje, Primorje, Kuril Islands), Mongolia, Korea, Japan.
Phenology. Univoltine: late May to early August.
Biology. Larvae on the leaves of *Scabiosa japonica* (Dipsacaceae) and *Serratula wolffii* (= *coronata*) (Asteraceae) (Japan: Inoue 1979, as *E. antaggregata*). In Europe larvae found on the leaves of *Tanacetum vulgare, Solidago virgaurea, Artemisia absinthium, A. campestris, A. vulgaris* var. *coarctata* (Asteraceae) and *Pimpinella saxifraga* (Apiaceae) (Dietze 1910, 1913; Kontuniemi 1941, Valle 1946; Seppänen 1970; Pellmyr & Mikkola 1984).
Altitude. Found from sea level up to about 3000 m in Qinling Mts. (China: coll. ZFMK).
Similar species. Asian specimens have a clear yellowish ochreous ground colour, which makes them more distinct than European ones. The male genitalia have a longer uncus and papillae, larger, longer and broader valve, two larger and broader dentate cornuti on the vesica, and more parallel margins to sternite A8 than European specimens. The apical cornutus distinguishes the male genitalia from those of other members of the group. The female have a characteristic sclerotized rounded swelling on the left side of the corpus bursae beyond the spiniferous area.
Material. ♀, [Kazakhstan], Ustkam[enogorsk], *aggregata*, Origin., 41, Coll. Led. (holotype, design. by Mironov, 2009, coll. O. Staudinger in MNHU); ♂, [Japan], Hokkaido, Nukabira, 27.vi.1962, H. Ono, No. 1005 (holotype of *E. antaggregata*, coll. H. Inoue in BMNH); 1♀, Westchina, Stötzner, Coll. Dr. Wehrli (ZFMK); 1♂, 4♀, Sued-Shensi [Shaanxi], Tapaishan [Taibaishan] im Tsinling [Qinling], 19, 20, 21.vi.1935, H. Höne, Vojnits slides nos. 12226♀, 12227♀, 12228♀ (ZFMK); 1♀, same loc., *ca.* 1700 m, 16.vi.1936, H. Höne (ZFMK); 1♂, same loc., *ca.* 3000 m, 16.vi.1936, H. Höne (ZFMK); 1♂, 1♀, China, Shaanxi prov., Qinling Mts., S. from Baoji, h-1600 m, N 34°13.932′, E 106°58.080′, 31.v.-01.vi.2009, leg. Floriani & Saldaitis, Pekarsky slides nos. OP 0516♂, OP 0521♀ (coll. O. Pekarsky); 2♀, Heilongjiang, Dailing, 390 m, 4–6.vii.1962, coll Bai Jiuwei, slide no. ZL226♀; 1♀, Shanxi, Wutaishan, 1992, slide no. ZL396♀; 1♂, Tibet, Jomda military station, 3400 m, 26.vii.1976, coll Zhang Xuezhong, slide no. ZL331♂;1♂ Gansu, Wenxian, Qiujiaba, 2350 m, 22.vii.1999, coll Yao Jian, slide no. ZL546♂; 1♀, Gansu, Yongdeng, Tulugou, 2280 m, slide no. ZL389♀; 1♀, Shaanxi, Ningshan crater, 1580 m, 2.vii.1999, coll Yuan Decheng. slide no. ZL554♀ (IZCAS).
Remarks. Not previously recorded from China.

261. *Eupithecia amplexata* Christoph, 1880

Eupithecia amplexata Christoph, 1880, *Bull. Soc. imp. Nat. Moscou* 55 (3): 119. Syntypes, at least 1♂, 3♀ (BMNH; coll. O. Staudinger in MNHU; examined), [Russia]: Wladiwostok.
Eupithecia pryeriaria Leech, 1897, *Ann. Mag. nat. Hist.* (6) **20**: 67. Syntypes 3 ♂, 3 ♀ (BMNH; examined), Japan: Yokohama; Yesso (synonymised in Prout, 1939).

Description. Ground colour white, usually with delicate ochreous tinge. Both wings with brownish ochreous outer margin. Fore wing transverse lines faint, but usually marked with a series of dark spots on veins, and strongly marked on costa; a distinct zigzag white subterminal line, ending in a white tornal spot; discal dot conspicuous but small, black. Hind wing identically coloured with distinct postmedial transverse line; discal dot small and pale. Body white.
Male genitalia. Uncus shortish, biapical. Valve unmodified, with relatively narrow apex. Vinculum broad, rounded. Papillae narrow. Aedeagus almost equal to length of valve. Vesica armed with two prominent dentate cornuti, an elongate U-shaped cornutus and a smaller irregular, curved cornutus at ductus ejaculatorius base. Sternite A8 bilobed at base, peg-like, with slightly concave sides.
Female genitalia. Bursa pear-shaped, mostly covered with large uniform spines. Ductus bursae sclerotized, with longitudinal striations and a small patch of spines near base of colliculum. Ductus seminalis slightly broadened at base, attached to central part of ductus bursae on right. Colliculum short and fairly broad. Antrum membranous. Tergite A8 transverse. Apophyses short and narrow. Papillae anales broad and rounded.
Range. China (Gansu, Shanxi, Qinghai, Shaanxi, Inner Mongolia, Heilongjiang); Asian Russia (Transbaikalia, Priamurje, Primorje, Sakhalin, Kuril Islands, Magadan region), Mongolia, Korea, Japan (Hokkaido, Honshu).
Phenology. Univoltine: late May to early August. Larva: July and August. Overwintering as pupa in flimsy cocoon in the ground.
Biology. Larva polyphagous on the flowers of *Hypericum gebleri* (Clusiaceae), *Valeriana koreana* (Valerianaceae), *Achillea* (= *Ptarmica*) *acuminata* (Asteraceae), unripe seeds of *Heracleum moellendorffii* (Apiaceae) (Far East Russia: Mironov, original data) and on the *Dianthus superbus* var. *speciosus* (Caryophyllaceae) (Japan: Inoue 1979).
Altitude. Up to about 2600–3200 m in Gansu and Shanxi (China: coll. ZFMK, TTM).
Similar species. Variable in the depth of coloration and distinctness of markings, but generally easily recognized by the white ground colour of wings and body and the distinctly darker, ochreous terminal areas on all

wings. The relatively small valve, and the presence of only two dentate cornuti on the vesica are helpful distinctions in the male. The female genitalia are very similar to those of the majority of other species in this group.

Material. ♂, [Russia], ♂ Wladiwost., ex coll. Christoph, Vladivostock 26.6.[18]77, Ex. Coll. H. J. Elwes, 1920, Joicey Bequest. Brit. Mus. 1934–120 (syntype, BMNH); 1♀, [Russia], Wladiwstk Chr., Origin. (syntype, coll. O. Staudinger in MNHU); 1♀, [Russia ?], same loc. ?, 24/6, Origin. (syntype, coll. O. Staudinger in MNHU); 1♀, [Russia], Raddeffskaja Chr., *Eupithecia amplexata* Chr., 166, 122[a], Origin. (syntype, coll. O. Staudinger in MNHU); 1♀, Prov. Shansi [Shanxi], Wu-Tai-Shan, 3200 m, 15.vii.1936, H. Höne, Vojnits slide no. 12255♀ (TTM); 1♂, China, Gansu, Xiahe, 2600–3000 m, 26.-30.vi.1986, Coll. Nr. 246, leg. Görgner, Mironov slide no. 733♂ (ZFMK); 1♀, Qinghai, Menyuan, Xianmi, 2800 m, 23.vii.1992, coll Xue Dayong, slide no. ZL288;1♀, Shaanxi, Liuba, 1020 m, 18.vii.1998, coll Yao Jian, slide no. ZL549♀; 2♀, Shanxi, Wutaishan 1992, slides nos. ZL264, 266♀; 2♂,1♀, Inner Mongolia, Oragxan, slides nos. ZL175, 479♂, 176♀; 1♀, Inner Mongolia, Jiwen, 500 m, 4.vii.1985; 1♀, Heilongjiang, Dailing, 390 m, 12.vii.1962, coll Bai Jiuwei, slide no. ZL174♀ (IZCAS).

Remarks. Not previously recorded from China.

262. *Eupithecia extensaria* Freyer, 1845

Acidalia extensaria Freyer, 1845, *Neuere Beitr. Schmettkde* 5 (73): 96, pl. 438, fig. 3. Lectotype ♂ (design. by Mironov, coll. K. Dietze in MNHU; examined), [Russia]: Sarepta.

Larentia prolongata Lienig & Zeller, 1846, *Isis* 3: 198. Holotype ♂ (BMNH; examined), [Latvia]: ? Koknese [= Kokenhusen] (synonymised in Herrich-Schäffer, 1848).

Eupithecia sydyi Staudinger, 1885, *Ent. Z., Stettin* **46** (7–9): 349. Syntypes 9, at least 4♂, 4♀ (coll. O. Staudinger in MNHU; examined), [China]: Kuldja [= Yining (Gulja)] (treated as var. of *Eupithecia extensaria* in Bohatsch, 1893).

Eup[ithecia] extensaria f. *leuca* Dietze, 1910, *Biologie der Eupithecien* 1, pl. 74, fig. 467; *ibidem* (1913), 2: 121. Holotype ♀ (coll. K. Dietze in MNHU; examined), [Russia]: Ussuri, Kasakewitsch.

Eup[ithecia] extensaria f. *prolongata* Dietze, 1910, *Biologie der Eupithecien* 1, pl. 74, figs 468, 469; *ibidem* (1913), 2: 121. [Junior secondary homonym of *Larentia prolongata* Lienig & Zeller, 1846].

Eupithecia extensaria occidua Prout, 1915, *in* Seitz, *The Macrolepidoptera of the World* 4: 293, pl. 13, fig. O. Syntypes, at least 1♂, 1♀ (BMNH). [Replacement name for *Eupithecia extensaria* f. *prolongata* Dietze, 1910 (*nec* Lienig & Zeller, 1846)].

Description. Fore wing narrow, elongate and pointed, ground colour ash-grey; transverse lines dark grey-brown, oblique, almost straight, sharply angled near costa; three conspicuous broad oblique, chalk-white transverse lines; whitish wavy subterminal line almost straight, curved only near apex; discal dot very small or absent. Hind wing pale grey with slightly darker terminal area and weak indications of transverse lines, and without discal dot.

Male genitalia. Uncus elongate, biapical. Valve broad, unmodified. Vinculum broad, rounded. Papillae relatively large. Aedeagus rather stout, much shorter than valve length. Vesica armed with two horn-like cornuti apically, two dentate cornuti centrally, an elongate U-shaped cornutus and a small, irregular cornutus at ductus ejaculatorius base. Sternite A8 almost triangular, broad at base with pointed apex.

Female genitalia. Bursa oblong, its base entirely spined and with longer marginal spines. Ductus bursae with longitudinal striations and two patches of large spines. Ductus seminalis sharply recurved, much broadened at base, attached to central part of corpus bursae. Colliculum broad and fairly long. Tergite A8 with centrally toothed anterior margin and emarginate posterior margin. Apophyses short and narrow, the anterior spurred near apices. Papillae anales broadly rounded.

Range. China (Xinjiang, Inner Mongolia, Shaanxi); Eastern Europe, Asian Turkey, Georgia, Armenia, Azerbaijan, Kazakhstan, Kirghizstan, Asian Russia (western and southern Siberia, Altai Mts., Sajan Mts., Transbaikalia, Dahuria, Primorje), Mongolia, Korea, Japan (Honshu). In Western Europe (UK, Spain) subsp. *occidua* Prout, 1914.

Phenology. Univoltine: mid-May to late July.

Biology. Larvae found on *Artemisia santalinifolia* (Asteraceae) in Tien-Shan Mts. (Kirghizstan: Mironov 2003); in Europe on flowers and leaves of *Artemisia campestris* and *A. maritima* (Asteraceae).

Altitude. From sea level up to about 2500 m in the Tien-Shan Mts. (Kirghizstan: coll. ZISP).

Similar species. One of the most easily recognized species among Chinese representatives of the genus. In the genitalia, the triangular sternite A8, the cornuti on the vesica, and the shape of the bursa should distinguish this species.

Material. ♂, [Russia], *G. Extensaria, extensaria* Fr. Type. Sarepta. collect. Freyer, Type aus Freyers Samlung im Lepid. verein Frankfurt gewesen, Karl Dietze, Gen.-Unters. Nr. [Schmidt slide no.] Eu 62a, Zool. Mus. Berlin (syntype, coll. K. Dietze in MNHU); 3♂, [Xinjiang], Juldus, Ili, Gen.-Unters. Nr. [Schmidt slide no.] Eu 62b, Zool. Mus. Berlin (coll. K. Dietze in MNHU); 1♂, [Xinjiang], Dscharkent (Kuldja) (coll. K. Dietze in MNHU); 1♂, [Xinjiang], Kuldja Rkbl. 84, *Sydyi* Stgr., Origin., Zool. Mus. Berlin, Vojnits slide no. 13579♂ (syntype of *E. sydyi*, coll. O. Staudinger in MNHU); 2♂, 4♀, [China ?], Origin., Zool. Mus. Berlin, Vojnits slide no. 13580♀ (syntypes of *E. sydyi*, coll. O. Staudinger in MNHU); 1♂, [Xinjiang], Juldus, Kuldscha,

669., *Sydyi* Stdgr. (syntype of *E. sydyi*, ZFMK); 1♂, [Xinjiang], Kuldja 86 Rckbl. (coll. O. Staudinger in MNHU); 1♂, [Xinjiang], anf. Mai, v. *Sydyi* Anf. Mai ♀, Kuldja (coll. K. Dietze in MNHU); 1♂, [Xinjiang], *extensaria* Fr. v. ! *Sydyi* Stgr.! Kuldscha, K. Dietze (coll. K. Dietze in MNHU); 1♂, [Xinjiang], ubergang von *extensaria* zu *sydyi*-Form, Juldus. Ili (coll. K. Dietze in MNHU); ♀, [Russia], Ussuri, Kasakewitsch 1907 Korb, *extensaria* Fr. forma: *leuca* Typ. Ussuri (467) (holotype of *E. extensaria* f. *leuca*, coll. K. Dietze in MNHU); 1♀, [Xinjiang], Juldus, forma. *leuca*, Juldus, Ili (coll. K. Dietze in MNHU); 1♂, [Xinjiang], Juldus Ili (coll. K. Dietze in MNHU); 1♂, [Xinjiang], anf. Mai, *extensaria*, Anfang Mai, Kuldja (coll. K. Dietze in MNHU); 1♂, 1♀, [Xinjiang], Dscharkent (Kuldja) (coll. K. Dietze in MNHU); 1♂, [Xinjiang], Asia centr., Ost-Turkestan (Urumtschi), Ruckbeil 1904, *extensaria*, v. R. Tancré 1905, Praep. Gen. Nr. 2247♂, *Eupithecia extensaria* Frr., det. Błeszyński 1955 (coll. R. Püngeler in MNHU); 2♀, [Xinjiang], Asia centr., Ost-Turkestan (Urumtschi), Rückbeil 1904, *extensaria* ♀, v. R. Tancré 1905 (coll. R. Püngeler in MNHU); 1♂, [Xinjiang], Aksu (ZFMK; specimen without abdomen); 1♂, [Xinjiang], Juldus, Kuldscha (TTM); 1♂, [Xinjiang], Kuldsha (ZISP); 1♀, Sued-Shensi [Shaanxi], Tapaishan [Taibaishan] im Tsinling [Qinling], *ca.* 1700 m, 7.vii.1936, H. Höne (ZFMK); 1♀, Inner Mongolia,Oragxan (IZCAS).

Remarks. The pale whitish form *leuca* seems to predominate in Far East Asia. The Central Asian and South Siberian form *sydyi* is similar externally to European specimens, and may not be distinguishable from the nominate form.

263. *Eupithecia absinthiata* Clerck, 1759

[*Phalaena*] *absinthiata* Clerck, 1759, *Icon. Insect. rariorum* **1**, pl. 8, fig. 9. Lectotype ♂ (design. by Mikkola, 1985, NHRM), [Sweden].

Geometra minutata Denis & Schiffermüller, 1775, *Syst. Verz.Schmett. Wien.* (**2**): 110. Syntype(s) lost. [Austria]: Vienna district (synonymised to *Eupithecia absinthiaria* [sic] in Herrich-Schäffer, 1848).

Phalaena elongata Haworth, 1809, *Lepid. Br.* (**2**): 358. Syntype(s) (BMNH), [United Kingdom] (synonymised to *Eupithecia absinthiaria* [sic] in Herrich-Schäffer, 1848).

Eupithecia notata Stephens, 1831, *Illustr. Br. Ent.* (Haustellata) **3**: 286. Syntype(s) (BMNH), [United Kingdom]: [England], near London (synonymised to *Eupithecia absinthiaria* [sic] in Herrich-Schäffer, 1848).

Eupithecia coagulata Guenée, [1858], in Boisduval & Guenée, *Hist. nat. Insectes* (*Spec. gén. Lépid.*) **10** (2): 339. Holotype ♂ (MNHN), [U.S.A.]: Pennsylvania (synonymised in Bolte, 1990).

Eupithecia absinthiata var. *callunae* Speyer, 1867, *Verh. naturh. Ver. preuss. Rheinl.* **24**: 249. Syntype(s), Germany: Westfallen, Waldeck.

Eupithecia goossensiata Mabille, 1869, *Annls Soc. ent. France* (4) **9** (1): 78. Lectotype ♀ (coll. C. Herbulot in ZSM; examined), [France] (treated as form of *Eupithecia absinthiata* in Petersen, 1910; synonymysed to *Eupithecia absinthiata* in Riley, 1986).

Eupithecia geminata Packard, 1873, 5th *Ann. Rep. Trustees Peab. Acad. Sci.*: 58. Lectotype ♀ (MCZ), [U.S.A.]: Massachusetts (synonymised in Packard, 1876 or 1874?).
Eupithecia knautiata Gregson, 1874, *The Entomologist* 7 (135): 255. Syntype(s) (BMNH), [United Kingdom]: Lancashire, Bolton, Ball Hill Lodges (synonymised to *Eupithecia goossensiata* Mabille, 1869 in Staudinger & Rebel, 1901).
Eup[ithecia] packardata Taylor, 1907, *Can. Ent.* **39** (8): 277 [Unnecessary replacement name for *Eupithecia geminata* Packard, 1873].
Eupithecia meritata Pearsall, 1908, *Ent. News* **19** (5): 195. Lectotype ♂ (AMNH), [U.S.A.]: [New York], Catskill Mts., Big Indian Valley (synonymised to *Eupithecia geminata* Packard, 1873 in Barnes & McDunnough, 1916, Contributions to the natural history of the Lepidoptera of North America 3 (3): 178 or in McDunnough, 1949).
Eup[ithecia] absinthiata f. *obscura* Dietze, 1910, *Biologie der Eupithecien* **1**: pl. 71, fig. 182; *ibidem* (1913), **2**: 88. Holotype ♀ (coll. K. Dietze in MNHU; examined), Ross(ia) centr.: Kassimow.
Eupithecia catharinae Vojnits, 1969, *Acta zool. Acad. Sci. hung.* **15** (1–2): 232, figs 1a, 2a. Holotype ♂ (TTM; examined), [Hungary]: Kunfehértó (synonymised in Mironov, 2003).

Misidentifications: *Eupithecia innotata* Wood, 1835 (*nec* Hufnagel, 1767); *Eupithecia innotata* Matsumura, 1925 (*nec* Hufnagel, 1767); *Eupithecia mandschurica* Staudinger, 1897 (1♀ syntype in coll. O. Staudinger in MNHU).
Unavailable names (infrasubspecific): *nigrescens*: Lempke (1969); *obscura*: Lempke (1969). – (incorrect subsequent spellings): *absynthiata*: Fischer von Röslerstamm (1829); *minutaria*: Boisduval (1840); *absinthiaria*: Herrich-Schäffer (1848); *elongaria*: Doubleday (1849); *absynthiaria*: De La Harpe (1855); *goosensiata*: Sterneck (1929); *goossentiata*: Schmidlin (1964); *gooseniata*: Bielewicz (1966); *goosentiata*: Garthe (1976).

Description. Fore wing reddish grey brown; costal margin mottled and irrorate with white scales; transverse lines inconspicuous, usually well marked on veins as blackish dots and always as larger black costal spots, especially ante- and postmedial; two weak medials close to postmedial, curved around discal dot; a series of dark dots in place of postmedial usually margined outside by white scales; whitish subterminal line usually broken, forming a series of minute white dots and always with larger white tornal spot, margined by dark scales on inner side; discal dot more or less elongate, intensely black. Hind wing brownish grey without reddish tinge; transverse lines inconspicuous, except on anal margin; a small whitish tornal spot; discal dot minute.
Male genitalia. Uncus elongate, narrow, biapical. Valve unmodified. Vinculum broad, rounded. Papillae elongate and narrow. Aedeagus relatively slim, slightly shorter than valve. Vesica armed with four prominent dentate cornuti, an elongate, U-shaped cornutus and a smaller curved cornutus at ductus ejaculatorius base. Sternite A8 peg-like, bilobed at base, sides strongly concave, very weakly bilobed at apex.
Female genitalia. Bursa strongly sclerotized, pear-shaped, covered with large uniform spines, the marginal ones longest. Ductus bursae usually with prominent bulge on left, and with two small patches of spines and longitudinal striations. Ductus seminalis rather narrow, attached to

central part of ductus bursae on right. Colliculum short and fairly narrow. Antrum membranous. Tergite A8 slightly transverse Apophyses short and narrow. Papillae anales small and rounded.
Range. China (Heilongjiang, Beijing, Shanxi); Europe, Caucasus, Transcaucasia (Georgia, Armenia), Kazakhstan, Kirghizstan, Tajikistan, Russia (Urals, western and southern Siberia, Priamurje, Primorje, Sakhalin), Korea, Japan (Hokkaido, Honshu, Tsushima).
Phenology. Univoltine: late June to late September.
Biology. Larva polyphagous on the flowers and seeds of various herbaceous plants, especially on the Asteraceae, and also on the Campanulaceae, Lamiaceae, Dipsacaceae, Apiaceae, Myricaceae and Ericaceaein Europe (Mironov 2003).
Altitude. From sea level to about 2700–3000 m in Anzob Pass (Tajikistan: coll. ZFMK).
Similar species. Variable in depth of coloration and distinctness of markings, with transverse lines variably expressed. East Asian specimens are less rich reddish brown on the fore wings than European and Central Asian specimens. See under *E. interpunctaria* and *E. korbi* for distinction from these very similar species.
Material. 1♀, [Russia], *mandschuria* ♀ Ussuri, Origin., Zool. Mus. Berlin, 19., Gen. vérifiés 1453[4] *Eupithecia mandshuria* ♀ ussuri, dét. E. de Laever 1954, *Eupithecia* photo 84 C17 det. A. Vojnits, "This ♀ was placed by Staudinger only doubtfully with *mandschurica*. It is probably a ♀ *absinthiata* Cl. det. D. S. Fletcher 1965" (coll. O. Staudinger in MNHU); 1♂, [Heilongjiang], Mandshuria or., Maoershan, Vojnits slide no. 543♂ (ZFMK); 1♂, Shanxi, [loc. illegible], 1550 m, 29.vii.1972, coll Wang Shuyong, slide no. ZL434♂; 1♂, Heilongjiang, Dailing, 390 m, 25.vii.1958, coll. He Zhong, slide no. ZL205♂; 1♀, Beijing, Baihuashan, 1.viii.1972, coll Zhang Baolin, slide no. L-0348♀ (IZCAS).
Remarks. Not previously recorded from China.

264. *Eupithecia interpunctaria* Inoue, 1979

Eupithecia interpunctaria Inoue, 1979, *Bull. Fac. domest. Sci. Otsuma Wom. Univ.* 15: 202, figs 22D, E; fig. 23D; fig. 24C; fig. 27A. Holotype ♂ (BMNH; examined), [Japan]: Shizuoka Pref[ecture], Sakumacho, Nada.

Misidentifications: *Eupithecia lunata* Vojnits & de Laever, 1978 (*nec* Dietze, 1910); *Eupithecia mandschurica* Vojnits & de Laever, 1978 (*nec* Staudinger, 1897).

Description. Fore wing with ground colour grey, almost unicolorous; transverse lines inconspicuous, well marked near costa only as blackish basal, ante- and postmedial costal dashes; terminal area broad, often slightly darker with narrow, indented whitish subterminal line, broadened near tornus; discal dot large, broadly ovoid, intensely black. Hind wing unicolorous grey, slightly paler than fore wing; transverse lines indistinct, discal dot small, rounded.

Male genitalia. Uncus narrow, biapical. Valve unmodified. Vinculum short, rounded. Papillae elongate. Aedeagus stout, slightly shorter than length of valve. Vesica armed with four dentate cornuti, a flat, U-shaped cornutus and a smaller curved cornutus at ductus ejaculatorius base. Sternite A8 peg-like, bilobed at base, slightly concave at sides and weakly bifid at apex.

Female genitalia. Bursa sclerotized, pear-shaped, mostly covered with large uniform spines. Ductus bursae sclerotized, with two patches of spines and longitudinal striations. Ductus seminalis broadened at base, attached to central part of ductus bursae on right. Colliculum relatively elongate and narrow. Antrum membranous. Tergite A8 almost quadrate. Apophyses short and narrow. Papillae anales rather small, rounded.

Range. China (Beijing, Jiangsu, Taiwan); Asian Russia (southern Priamurje, Primorje, Sakhalin), Korea, Japan (Hokkaido, Honshu, Sado, Shikoku, Kyushu, Tsushima).

Phenology. Univoltine: late July to October.

Altitude. From about 250 m (Korea: coll. TTM) up to about 2600 m (Taiwan: Vojnits 1978; Inoue 1988).

Similar species. Very similar to *E. absinthiata* and *E. korbi* Dietze, 1908. Distinguished from both by the larger, broadly ovoid fore wing discal dots and from *E. absinthiata* by the shorter, more obtuse, and greyer fore wings. The shorter uncus, and the straighter sides to the eighth sternite are helpful in distinguishing the male genitalia from both. No significant structural differences from *E. absinthiata* and *E. korbi* can be found in the female genitalia.

Material. ♂, [Japan], Shizuoka Pref., Sakuma-cho, Nada, 28.viii.1957, Coll. Y. Saitoh (holotype, coll. H. Inoue in BMNH); 3♂, 3♀, China, Prov. Kiangsu [Jiangsu], Lungtan bei Nanking [Nanjing], 29.viii., 6, 10, 28.ix., 7.x.1933, H. Höne, respectively Vojnits slides nos. 12368♂ (marked as holotype of unpubl. *E. qualtuorversiba* Vojnits, ZFMK), 12366♀, 13269♂, 12364♀, 12365♂ (marked as paratype of unpubl. *E. qualtuorversiba* Vojnits), 12367♀ (ZFMK); ♀, [Korea], Prov. Gan-von, district On-dzong, Kum-gang san, near Hotel Go-sang, 250 m, 6.viii.1975, No. 322, leg. J. Papp et A. Vojnits, Vojnits slide no. 10796♀ (*nec* 10976♀) (as *E. lunata*, TTM); 2♀, same loc., 5.viii.1975, No. 319, Vojnits slides nos. 10797♀, 10798♀ (as *E. mandschurica*,

TTM); 2♂, Beijing, Mentougou, Xiaolongmen, 1100 m, 24.ix.1998, slide no. ZL682♂; 4♂, 3♀ same data, 22.viii.2002, slides nos. ZL660, 663, 682♂, 659, 683♀.

Remarks. Not previously recorded from the mainland of China.

265. *Eupithecia korbi* Dietze, 1908

Eupithecia korbi Dietze, 1908, *Dt. ent. Z., Iris* **21** (3): 193; *Biologie der Eupithecien* 1, pl. 71, figs 191, 192. Syntypes 8♀ (coll. K. Dietze in MNHU; examined), [Russia]: Ussuri.

*Eupithecia **infensa*** Vojnits, 1979, *Acta zool. Acad. Sci. hung.* **25** (1–2): 202, fig. 13, (*nec* 14). Holotype ♂ (ZFMK; examined), [China]: Sued-Shensi, Tapaishan [Taibaishan] im Tsinling, 3000 m (synonymised in Mironov & Galsworthy, 2010).

Eupithecia memorata Mironov, 1988, *in* Viidalepp & Mironov, *Eesti NSV Tead. Akad. Toim.* (Biol.) 37 (4): 282, figs 2a-c. Holotype ♂ (ZISP; examined), [Russia]: Ussuri, Sutchansky rudnik (synonymised in Mironov & Galsworthy, 2010).

Misidentification: *Eupithecia absinthiata* Inoue, 1979 (part., *nec* Clerck, 1759); *Eupithecia ultrix* Mironov & Galsworthy, 2004 (1♂ paratype in IZCAS).

Description. Fore wing ground colour grey with a weak brownish tinge; transverse lines rather weak, but ante- and postmedial well marked on costa; antemedial perpendicular to anal margin, slightly curved onto costa; crenellate double medial between discal dot and postmedial; terminal area with a distinct, rather straight border, and darker with a slight rusty brownish tinge; pale subterminal line slightly wavy, broadened near tornus into a small, whitish spot; discal dot circular to ovoid, and prominent. Hind wing only slightly paler than fore wing; transverse lines inconspicuous, sometimes stronger close to anal margin; tornal area often darker; discal dot invisible or very small.

Male genitalia. Uncus elongate, biapical. Valve unmodified. Vinculum short, rounded. Papillae elongate. Aedeagus stout, slightly shorter than length of valve. Vesica armed with three dentate cornuti, a flat, U-shaped cornutus and a smaller irregular cornutus near base of ductus ejaculatorius. Sternite A8 peg-like, elongate, evenly narrowed to a slightly asymmetric bifid apex.

Female genitalia. Bursa pear-shaped, mostly covered with large spines. Ductus bursae sclerotized, with one longitudinal and one transverse patch of spines and longitudinal striations. Ductus seminalis curved, slightly broadened at base, attached to central part of bursa on right. Colliculum short and broad. Antrum membranous. Tergite A8 almost square. Apophyses markedly short, especially anterior. Papillae anales relatively small, rounded.

Range. China (Gansu, Shaanxi, Sichuan, Hubei, Zhejiang); Far East Russia (S Primorje), Korea, Japan.

Phenology. Univoltine: late June to early September.
Altitude. Found from about 2000–3000 m above sea-level (China: Vojnits 1979a; IZCAS).
Similar species. Very similar to *E. absinthiata* and *E. interpunctaria*; ground colour is greyer than in *E. absinthiata* and the fore wing discal dots are smaller than in *E. interpunctaria*. The rather regular inner edge to the darker border is often a good pointer. The hind wing ground colour is also deeper than in the other two species, but not as uniform as in *E. mandschurica*, another potential confusion species. The male genitalia are the only reliable distinction of this species, having only three rather small dentate cornuti on the vesica and an eighth sternite without concave sides and with a very shallow and asymmetrical medial hollow at the apex. The female genitalia cannot be reliably distinguished.
Material. ♀, [Russia], Ussuri, Kasakewitsch 1907 Korb, *korbi* n. sp. Typ. Ussuri ♀, K. Dietze, Type, *Eupithecia korbi* Dietze, Holotype ♀, Zool. Mus. Berlin, Gen. slide Geom. 1951–8A (holotype, design. by Fletcher [possibly], coll. K. Dietze in MNHU); 2♀, same loc., *korbi* Typ., Ussuri, Zool. Mus. Berlin, Vojnits slides nos. 14967♀, 14972♀ (syntypes [paratypes ?], coll. K. Dietze in MNHU); 1♀, [Russia], *korbi* n. sp., Ussuri Type, K. Dietze, Zool. Mus. Berlin, Typus, Geom. slide 1950–588A (syntype [paratype ?], coll. K. Dietze in MNHU); 4♀, [Russia], Ussuri, Kasakewitsch 1907 Korb, *korbi* Uss., Zool. Mus. Berlin, Vojnits slides nos. 14968♀, 14969♀, 14970♀, 14971♀ (syntypes [paratypes ?], coll. K. Dietze in MNHU); ♂, Sued-Shensi [Shaanxi], Tapaishan [Taibaishan] im Tsinling [Qinling], *ca.* 3000 m, 12. viii.1936, H. Höne, Vojnits slide no. 11522♂ (holotype of *E. infensa*, ZFMK); 1♂, same loc. and alt., 10.viii.1936, H. Höne, Vojnits slide no. 11482♂ (paratype of *E. infensa*, ZFMK); 2♂, 1♀, same loc. and alt., 8, 10, 12.viii.1936, H. Höne, Vojnits slides nos. 12216♂, 12307♂, 12321♀ (ZFMK); 4♂, 2♀, same loc. and alt., 10, 11, 12, 13.viii.1936, H. Höne, Vojnits slides nos. 11475♂, 12206♀, 12222♂, 12941♂, 12955♂, 12965♀ (TTM); 1♂, same loc., *ca.* 1700 m, 9.viii.1936, H. Höne, Vojnits slide no. 11479♂ (paratype of *E. infensa*, TTM); 3♂, 3♀, same loc. and alt., 5, 6, 7.viii.1936, H. Höne, Vojnits slides nos. 11370♀, 12319♀, 12316♂, 12317♂, 12313♀, 12339♂ (ZFMK; slides nos. 11370♀, 12319♀ absent in ZFMK); 1♂, 3♀, same loc. and alt., 5, 8, 9, 10.viii.1936, H. Höne, Vojnits slides nos. 10896♀, 12303♀, 12304♀, 12964♂ (TTM); 1♂, 2♀, same loc. but without alt., 6, 26.viii., 1.ix.1935, H. Höne, Vojnits slides nos. 12305♀, 12954♀, 12956♂ (TTM); 1♀, Pz. Chekiang [Zhejiang], West-Tien-Mu-Shan, 1600 m, 1.x.1932, H. Höne, Vojnits slide no. 12957♀ (TTM); ♂, [Russia], Ussur. kr[aj], Sutchansky rudnik, 28.vi.[1]935, leg. Palshkov (holotype of *E. memorata*, ZISP); 1♀, Hubei, Shennongjia, 920 m, 28.

viii.1981, coll Han Yinheng, slide no. ZL97♀; 2♀, Gansu, Yongdeng, Tulugou, 2280 m, 25–29.vii.1991,slides nos. ZL117, 362♀; 1♂, Gansu,Wenxian, Qiujiaba, 2350 m, 22.vii.1999, coll Yao Jian, slide no. ZL571♂; 1♂, Sichuan, Yajiang, 3300 m, 24.viii.1982, coll Zhang Xuezhong, slide no. ZL562♂ (IZCAS).
Remarks. Prout (1915) synonymised this species with *Eupithecia emanata* Dietze, 1908.

266. *Eupithecia kurilensis* Bryk, 1942

Eupithecia kurilensis Bryk1942, *Dt. ent. Z., Iris* **56** (1/2): 77, pl. 2, fig. 19. Holotype ♂ (NHRS), [Russia]: Kurile Islands, Urup (Isl.), Tokotanbai.
Eupithecia angustipunctaria Inoue, 1979, *Bull. Fac. domest. Sci. Otsuma Wom. Univ.* **15**: 204, figs 22F, G; fig. 23E; fig. 24D; fig. 27B. Holotype ♂ (BMNH; examined), [Japan]: Hokkaido, Kamikawa, Fukiage Spa, 1100 m (synonymised in Viidalepp & Mironov, 1988b).
Eupithecia kurilensis mironovi Beljaev, 2002, *Entomol. Fennica* **13** (4): 228, figs 5 H, 8 A–C. Holotype ♂ (IBSS), Russia: Kamchatka Peninsula, 17 km NE of Malki (Poperechnaya River).

Misidentifications: *Eupithecia absinthiata* Djakonov, 1929, 1931 (*nec* Clerck, 1759); *Eupithecia castigata* Konovalova, 1970 (*nec* Hübner, 1813); *Eupithecia absinthiata* Sedykh, 1979 (*nec* Clerck, 1759); *Eupithecia castigata* Sedykh, 1979 (*nec* Hübner, 1813).

Description. Fore wing fairly broad, ground colour uniform grey; antemedial, double medial and postmedial weakly visible, slightly darker grey, medial passing outside discal dot, and all angled to meet costa, medial most sharply so; postmedial forming a costal spot. Terminal area slightly darker, without subterminal line. Discal dot small, narrow, slightly elongate. Hind wing slightly paler, transverse lines weakly marked, but usually complete. Discal dot small.
Male genitalia. Uncus narrow, elongate, bifid. Valve unmodified. Vinculum broad and rounded. Aedeagus stout, approximately equal to valve length. Vesica armed with four dentate plates, three long and narrow, and the fourth much broader, and a flat U-shaped cornutus near base. Sternite A8 peg-like, broadly bilobed at base, with apical part parallel-sided, and slightly bifid at apex.
Female genitalia. Bursa sclerotized, pear-shaped, all but a small posterior area covered with large uniform spines. Ductus bursae short, robust, with longitudinal striations. Ductus seminalis broadened at base, attached dorsally to posterior part of bursa. Colliculum broad and short. Antrum membranous, triangular. Tergite A8 quadrate. Apophyses posteriores narrow and rather short, anteriores shorter and stouter. Papillae anales broad and rounded.
Range. China (Jilin); Russia (Kamchatka Peninsula, Sakhalin, Kuril Islands); Japan (Hokkaido).

Phenology. Univoltine. Taken in mid July in Jilin. In Russia from late June to mid-August (Viidalepp & Mironov 1988b). In Hokkaido from late June to early September (Inoue 1979).
Altitude. Found to about 1100 m above sea-level in Hokkaido (Japan: Inoue 1979, as *E. angustipunctaria*).
Similar species. Broadly similar to *E. absinthiata, E. interpunctaria* and *E. korbi*, but with a more uniform and greyer fore wing and much smaller discal dots. The male genitalia are very similar to those of *E. absinthiata*, but the three smaller cornuti in the vesica are smaller and narrower in contrast to the larger cornutus. The female genitalia are indistinguishable from other species in this group.
Material. ♂, [Japan], Fukiage Spa, 1100 m, Hokkaido, 25.vii.1958, K. Ishiangka, Inoue slide no. 2000=BM Geometrid slide no. 16361♂ (holotype of *E. angustipunctaria*); 5♀, China, Jilin, Changbaishan, 42°01′ N, 128°04′ E, 13.vii.2001, A. C. Galsworthy (BMNH).

267. *Eupithecia ensifera* Mironov & Galsworthy, 2004

Eupithecia ***ensifera*** Mironov & Galsworthy, 2004, *Trans. lepid. Soc. Japan* 55 (4): 298, figs 11, 24, 27. Holotype ♀ (IZCAS; examined), China: Qinghai, Tongren, Maixiu, 2950 m.

Description. Fore wing ground colour pale brown; antemedial double, angled very close to costa, medial running through discal dot, angled at dot, postmedial double, angled outside discal dot, all three marked with darker spots along costa; discal dot brown, faint, elongate; a broader pale fascia outside postmedial, terminal area darker brown, crossed by wavy pale subterminal line. Hind wing ground colour much paler than fore wing, discal dot prominent, transverse lines visible only at anal margin, dark brown; terminal area darker.
Male genitalia. Uncus relatively elongate, slim, biapical. Valve short and broad, unmodified. Vinculum short, slightly tapered. Papillae rather short and stout. Aedeagus slim, shorter than valve length. Vesica armed with three elongate dentate cornuti, a small elongate patch of small spines, an elongate U-shaped cornutus and a shorter irregular cornutus near ductus ejaculatorius base. Sternite A8 peg-like, broadly bilobed at base, parallel-sided and blunt at apex.
Female genitalia. Bursa pear-shaped, about $^2/_3$ covered with large spines. Ductus bursae with two small separate patches of spines, one near base of ductus seminalis. Ductus seminalis long, curved, attached to central part of corpus bursae on left. Colliculum short, rather broad. Antrum triangular, covered by numerous pores. Tergite A8 transverse, with small shallow

medial hollow on the posterior margin. Apophyses narrow. Papillae broad and rounded.
Range. China (Qinghai).
Similar species. More variegated in tone than other species in this group, and with contrastingly pale hind wing. In the male genitalia the short rounded valve, and the presence of only three dentate cornuti in the vesica are distinctive, as well as the shape of the eighth sternite. The female genitalia are very similar to those of other species in the group.
Material. ♀, China, Qinghai, Tongren, Maixiu, 2950 m, 30.vii–1.viii.1992, coll. Xue Dayong, slide no. ZL247♀ (holotype, IZCAS); 2♂, 1♀, same loc., slides nos. ZL539, 386♂, ZL373♀ (paratypes, IZCAS).

268. *Eupithecia ultrix* Mironov & Galsworthy, 2004

*Eupithecia **ultrix*** Mironov & Galsworthy, 2004, *Trans. lepid. Soc. Japan* 55 (4): 299, figs 12, 25, 28. Holotype ♂ (IZCAS; examined), China: Yunnan, Lijiang, Yulongshan.

Description. Fore wing ground colour light brown, very uniform over entire wing surface, except for slight darkening in terminal area; except for postmedial, transverse lines absent, basal and antemedial marked only by small brown spots on costa; postmedial marked on costa by an elongate brown spot, with a pale fascia and a dark subapical spot visible outside it, then faint and scalloped, curved around discal dot; discal spot large, dark brown, ovate. Hind wing concolourous with fore wing; discal dot prominent; postmedial and darkened subterminal zone visible at anal margin.
Male genitalia. Uncus short and narrow, biapical. Valve rather narrow, unmodified. Vinculum short, rounded. Papillae elongate and narrow. Aedeagus stout, slightly shorter than length of valve. Vesica armed with four prominent dentate cornuti, one larger than the rest, a flat cornutus and a smaller irregular cornutus at ductus ejaculatorius base. Sternite A8 peg-like, broadly bilobed at base and with two short, blunt apical horns.
Female genitalia. Bursa pear-shaped, basal half covered with large spines, interrupted by a longitudinal membranous gap. Ductus bursae with longitudinal striations, a small patch of spines and a row of spines near base of ductus seminalis. Ductus seminalis broadened at base, attached to middle of corpus bursae on right. Colliculum short and quite broad. Antrum membranous, short and broad. Tergite A8 relatively small, quadrate. Apophyses short and narrow. Papillae anales small, rounded.
Range. China (Sichuan, Yunnan).
Phenology. Univoltine (?): late July to late August.

Altitude. The only altitude data is of 3000 m in Sichuan (China: coll. IZCAS).
Similar species. Very similar externally to *E. absinthiata*, particularly in the uniform fore wing colour. The male eighth sternite has two small apical rods like the horns of a giraffe, which do not occur in *E. absinthiata*. The female genitalia are probably not distinguishable. However, *E. absinthiata* is not known from SW China.
Material. ♂, China, Yunnan, Lijiang, Yulongshan, 23.vii.1962, coll. Song Shimei, slide no. ZL561♂ (holotype, IZCAS); 1♂, China, Sichuan, Yajiang, 3300 m, 24.viii.1982, coll. Zhang Xuezhong, slide no. ZL91♂ (paratype, IZCAS); 1♀, China, Sichuan, Wushuiyuan, 20.vii.1961, coll. Li Shengfu, slide no. ZL416♀ (paratype, IZCAS).
Remarks. The third male paratype originally published as *E. ultrix* is in fact probably a specimen of *E. korbi*.

269. *Eupithecia cervina* Mironov & Galsworthy, 2004

*Eupithecia **cervina*** Mironov & Galsworthy, 2004, *Trans. lepid. Soc. Japan* 55 (4): 296, figs 10, 23, 26. Holotype ♂ (IZCAS; examined), China: Tibet, Zogang military outpost, 3800 m.

Description. Fore wing narrow and very elongate, ground colour uniform mid brown, almost without transverse or other markings; discal dot small, faint, only slightly darker than ground colour; one short oblique streak on costa at position of antemedial line, and three short faint lines on costa between discal dot and apex, the inner one forming the beginning of a postmedial, and sharply angled around discal dot. Hind wing elongate, well rounded; concolorous with fore wing, and slightly darker along anal margin.
Male genitalia. Uncus short, biapical. Valve unmodified, sharply narrowed to apex. Vinculum short, rounded. Papillae relatively long and broad. Aedeagus slim, approximately valve length. Vesica armed with four small dentate cornuti, a small elongate U-shaped cornutus and a smaller V-shaped cornutus. Sternite A8 peg-like, bilobed at base, with irregular lateral margins and blunt apex.
Female genitalia. Bursa ovate, about $^2/_3$ covered with sparse spines, the marginal spines longer than the others. Ductus bursae with a patch of sparse spines near colliculum. Ductus seminalis long and recurved, attached to ductus bursae near colliculum. Colliculum short and rather broad. Antrum membranous. Tergite A8 quadrate. Apophyses narrow. Papillae anales short and rounded.
Range. China (Tibet).

Phenology. July.
Altitude. From about 3400 to 4000 m in Tibet (coll. IZCAS).
Similar species. Externally, the adults are rather similar to species of the *innotata* group. However, the male and female genitalia are typical for representatives of the *satyrata* group. In the male genitalia, the apically very sharply narrowed valve, and the shape of the eighth sternite are distinctive. In the female, the sparsely spined and rather thin-walled bursa is a helpful distinction.
Material. ♂, China, Tibet, Zogang military outpost, 3800 m, 1.vii.1976, coll. Zhang Xuezhong, slide no. ZL262♂ (holotype, IZCAS); 5♀, Tibet, Markam, Rongxi, 4000 m, 24 and 26.vii.1976, coll. Zhang Xuezhong and Han Yinheng, slides nos. ZL334, 335, 411, 469♀ (paratypes, IZCAS); 1♀, Tibet, Jomda, 3,400 m, 27.vii.1976, coll. Zhang Xuezhong, slide no. ZL431♀ (paratype, IZCAS).

270. *Eupithecia impolita* Vojnits, 1980

Eupithecia impolita Vojnits, 1980, *Acta zool. Acad. Sci. hung.* **26** (4): 433, figs 1, 2, 17. Holotype ♂ (ZFMK; examined), [China]: Prov. Shansi [Shanxi], Mien-shan, 1500 m.
Eupithecia indecora Vojnits, 1981, *Acta zool. Acad. Sci. hung.* **27** (3–4): 429, figs 53–55. Holotype ♂ (ZFMK; examined), [China]: Prov. Shansi [Shanxi], Mien-shan, 1500 m. **New junior synonym.**

Description. Fore wings rather elongate. All wings unicolourous, rusty brown, without any visible transverse lines or discal dots; in fresher specimens, some very weak mottling with paler parallel lines is visible.
Male genitalia. Uncus narrow, biapical. Valve unmodified, rather short. Vinculum short and rounded. Papillae relatively stout and long, slightly clavate. Aedeagus narrow, shorter than length of valve. Vesica armed with one large and two smaller dentate cornuti, an elongate, U-shaped cornutus and a smaller irregular, curved cornutus at base. Sternite A8 peg-like, broadly bilobed at base, otherwise very narrow with a rounded apex.
Female genitalia. Bursa ovoid, basal half covered with uniform spines. Ductus bursae short, with longitudinal striations and a patch of large spines near base of colliculum. Ductus seminalis attached to ductus bursae near colliculum on right. Colliculum short and narrow. Antrum membranous. Tergite A8 transverse, with medially concave posterior margin. Apophyses short and narrow. Papillae anales short, broad, and rounded.
Range. China (Shanxi, Beijing).
Phenology. Univoltine: early June to mid-July.
Altitude. Found at about 1500 m above sea-level in Mien-shan (China: Vojnits 1981b, 1984a).

Similar species. Externally distinct because of lack of pattern, and closest to *E. thalictrata*, which has a broad and less pointed fore wing, a paler, ash grey colour and more distinct pale transverse lines. The genitalia differ strongly from those of *E. thalictrata*. The pattern of cornuti in the vesica and the shape of the eighth sternite distinguish the male genitalia fairly well from other species of the present group. The female bursa is more rounded and thinner walled than in most other species of the group.
Material. ♂, Prov. Shansi [Shanxi], Mien-shan, Mittlere Höhe *ca.* 1500 m, 3.vi.1937, H. Höne, Vojnits slide no. 11514♂ (holotype, ZFMK); 2♀, same loc. and alt., 5.vi.1937, H. Höne, Vojnits slides nos. 11499♀, 11516♀ (paratypes, ZFMK); 1♂, 1♀, same loc. and alt., 4, 10.vi.1937, H. Höne, Vojnits slides nos. 11512♀, 11515♀ (paratypes, TTM); ♂, same loc. and alt., 5.vi.1937, H. Höne, Vojnits slide no. 13110♂ (holotype of *E. indecora*, ZFMK); 1♂, same loc. and alt., 7.vi.1937, H. Höne, Vojnits slide no. 13111♂ (paratype of *E. indecora*, TTM); 1♂, 2♀, same loc. and alt., 3, 10.vi.1937, H. Höne, Vojnits slides nos. 13733♀, 13734♀, 13735♂ (TTM); 2♂, 1♀, same loc., Mittlere Höhe *ca.* 1500 m, 5, 10.vi.1937, H. Höne, Vojnits slides nos. 11493, 11494, 11497 (slides not found in TTM); 3♀, Shanxi, Wutaishan, 1992, slides nos. ZL388, 390, 403♀; 1♂, Beijing, Baihuashan, 31.v.1973, coll Han Yinheng, slide no. ZL120♂ (IZCAS).
Remarks. The shapes of the male eighth sternites of *E. impolita* and *E. indecora* were incorrectly illustrated by Vojnits (1980, 1981). The sternite lacks apical horns. The sternite on Vojnits' slides is so transparent as to be almost invisible.

271. *Eupithecia neosatyrata* Inoue, 1979

Eupithecia neosatyrata Inoue, 1979, *Bull. Fac. domest. Sci. Otsuma Wom. Univ.* **15**: 214, figs 22O-P; fig. 23L; fig. 31C; fig. 33D. Holotype ♂ (BMNH; examined), [Japan]: Kagawa Pref[ecture], Mt Shiramine, Sakaide City.
Eupithecia semicalva Vojnits, 1979, *Acta zool. Acad. Sci. hung.* **25** (1–2): 204, fig. 15. Holotype ♂ (ZFMK; examined), China (west): Szetschwan [Sichuan]. **New junior synonym.**

Description. Fore wings narrow, elongate and pointed; ground colour greyish brown with very distinct blackish brown transverse lines; basal and antemedial evenly curved; double medial sharply curved onto costa behind discal dot; postmedial sinuously curved onto costa; terminal area almost unicolourous with trace of light subterminal line and with small light tornal spot; discal dot elongate and narrow, intensely black. Hind wing slightly paler, transverse lines more or less marked along anal margin; discal dot very small and pale.
Male genitalia. Uncus elongate, biapical. Valve unmodified, rather long and narrow. Vinculum short, tapering. Papillae relatively short. Aedeagus

narrowed to anterior end. Vesica armed with four small dentate plates in apical and medial pairs and an elongate plate connecting with a small plate near ductus ejaculatorius base. Sternite A8 peg-like, with very broadly spreading basal lobes, and a very narrow, tapered apex.

Female genitalia. Bursa rounded, almost completely covered with dense, long spines. Ductus bursae short and broad, tapered, completely covered with dense, long spines. Ductus seminalis broad, attached ventrally to central part of corpus bursae. Colliculum broadish. Antrum membranous. Tergite A8 transverse, with matching concave and convex profiles on anterior and posterior margins. Apophyses narrow. Papillae anales short and broad, rounded.

Range. China (Sichuan, Shaanxi); Asian Russia (Transbaikalia, Dahuria), Korea, Japan (Shikoku).

Phenology. Univoltine (?): late April to mid-June.

Similar species. Similar externally to *E. carpophilata*, but distinguished from it by narrower, more pointed fore wings, sharply angled medial lines on the fore wing and almost unicolourous hind wings without conspicuous transverse lines. In the male genitalia the shape of the eighth sternite is characteristic. In the female, the bursa has far more spines at the posterior end than other species in this group.

Material. ♂, [Japan], Mt. Iwaseo, Takamatsu, 29.iv.1972, T. Masui, BM Geom. slide no. 16362♂ (holotype, coll. H. Inoue in BMNH); ♂, West-China, Szetschwan [Sichuan], Exped. Stötzner, Coll. Dr. Wehrli, Basel, Vojnits slide no. 11484♂ (holotype of *E. semicalva*, ZFMK); 3♀, Sued-Shensi [Shaanxi], Tapaishan [Taibaishan] im Tsinling [Qinling], *ca.* 1700 m, 16, 17, 22.v.1936, H. Höne (TTM).

272. *Eupithecia indissolubilis* Vojnits, 1979

Eupithecia indissolubilis Vojnits, 1979, *Acta zool. Acad. Sci. hung.* 25 (1–2): 202, fig. 12. Holotype ♂ (ZFMK; examined), [China]: Sued-Shensi, Tapaishan [Taibaishan] im Tsinling.

Eupithecia subita Vojnits, 1979, *Acta zool. Acad. Sci. hung.* 25 (3–4): 431, figs 8, 9. Holotype ♀ (ZFMK; examined), [China]: Sued-Shensi, Tapaishan [Taibaishan] im Tsinling, 1700 m. **New junior synonym.**

Eupithecia assa Mironov, 1989, *Annales Ent. Fennici* 55 (4): 149, figs 1–5. Holotype ♀ (ZISP; examined), USSR [Russia]: SW Altai (Mts.), 10 km W of Katanda, 1200 m. **New junior synonym.**

Description. Fore wing ground colour dark brownish grey, unmarked except for three blackish costal spots; transverse lines slightly darker, only just visible; discal dot large, ovoid or rounded, black. Hind wing ground colour similar, fuscous grey; darker medial transverse line evenly curved; discal dot large, but very pale.

Male genitalia. Uncus shortish, biapical. Valve unmodified, with narrow apex. Vinculum short, rounded. Papillae short with rather pointed apices. Aedeagus short and stout, slightly curved. Vesica armed with four small dentate plates, one larger and broader than the rest, a twisted U-shaped plate and a larger, irregular plate near ductus ejaculatorius base. Sternite A8 peg-like, with broad basal lobes and narrow blunt apex.
Female genitalia. Bursa ovate, covered with small spines over half of base, leaving a membranous gap on left. Ductus bursae very short, with a small transverse patch of small spines. Ductus seminalis broadened to base, attached to ductus bursae near base of colliculum on right. Colliculum short and narrow, strongly inclined. Antrum membranous. Tergite A8 square with shallow medial hollow on anterior margin. Apophyses short and narrow, the anterior stouter. Papillae anales broad and rounded.
Range. China (Shaanxi); Russia (Southern Siberia: Altai Mts., Tuva, Dahuria, Transbaikalia).
Phenology. Univoltine: early June to early July.
Altitude. From about 1100 m in Altai Mts. (Russia: coll. ISEN) up to about 1700 m in Tapai-shan (China: Vojnits 1979a).
Similar species. Externally similar to *E. interpunctaria* and *E. korbi*, but with larger, distinct, but rather fuzzy discal dots and larger, more conspicuous, dark costal spots marking ante- and postmedial transverse lines on the fore wings. The curved aedeagus, and the shape of the eighth sternite are helpful in the male. In the female the rather small bursa, with only partial spining at base, and the inclined colliculum are diagnostic.
Material. ♂, Sued-Shensi [Shaanxi], Tapaishan [Taibaishan] im Tsinling [Qinling], F, 24.vi.1935, H. Höne, Vojnits slide no. 11436♂ (holotype, ZFMK); 1♂, same loc., *ca.* 1700 m, 8.vi.1936, H. Höne, Vojnits slide no. 11438♂ (paratype, ZFMK); 1♂, same loc. and alt., 22.vi.1935, H. Höne, Vojnits slide no. 12061♂ (paratype, TTM); 3♀, same loc. and alt., 7, 11.vi.1936, H. Höne, Vojnits slides nos. 12360♀, 12308♀, 12311♀ (ZFMK); ♀, same loc. and alt., 7.vi.1936, H. Höne, Vojnits slide no. 12297♀ (holotype of *E. subita*, ZFMK); 3♀, same loc. and alt., 3, 11, 12.vi.1936, H. Höne, Vojnits slides nos. 12212♀, 12217♀, 12210♀ (paratypes of *E. subita*, ZFMK); 2♀, same loc., 19.vi.1935, L, H. Höne, Vojnits slides nos. 12221♀, 12300♀ (paratypes of *E. subita*, TTM); 1♂, 1♀, same loc. and date, H. Höne, Vojnits slides nos. 12301♂, 12302♀ (paratypes of *E. subita*, ZFMK); 3♂, 2♀, same loc., 19, 22, 25, 26, 30.vi.1935, H. Höne, de Laever slide no. 425♀, Vojnits slide no. 12320♀ (ZFMK); 2♂, same loc., 21.vi.1935, H. Höne, Vojnits slide no. 12958♂ (TTM); 1♀, same loc., *ca.* 3000 m, 15.vi.1936, H. Höne, Vojnits slide no. 12348♀ (ZFMK); ♀, USSR [Russia], SW Altai, 10 km W of Katanda, 1200 m, 22.vi.-5.vii.1983, Exp.

Mikkola, Hippa & Jalava leg. (holotype of *E. assa*, ZISP); 5♂, 4♀, same loc., alt. and date (paratypes of *E. assa*, ZISP, ZMH); 1♂, 2♀, same loc., alt. and date, Zolotarenko & Dubatolov leg. (paratypes of *E. assa*, ISEN); 3♀, China, Shaanxi prov., Qinling Mts., Hauzherze vill., N 33°52.423′, E 107°48.779′, 2–3.vi.2009, h-1600 m, leg. Floriani & Saldaitis, Pekarsky slide no. OP 0515♀ (coll. O. Pekarsky).

The *millefoliata* species group

This small Palaearctic species group was introduced and described by Mironov (2003). The most obvious common feature in the male genitalia is the peg-like eighth sternite with two small sclerotized apical horns. The aedeagus vesica always has one or more longish cornuti, and often several dentate plates in addition. The female bursa is large and oval, always strongly spined, and with a characteristic anteriorly orientated ductus seminalis. The group is represented in China by the following three species.

273. *Eupithecia nephelata* Staudinger, 1897

Eupith[*ecia*] *nephelata* Staudinger, 1897, *Dt. ent. Z., Iris* [1896] 9 (2): 282. Syntypes 1♂, 1♀ (coll. O. Staudinger in MNHU; examined), [Mongolia]: Uliassutai, Tumartin-Gol.

Unavailable names (incorrect subsequent spelling): *nephalata*: de Laever (1956).

Description. Fore wing broad, ground colour brownish yellow; transverse lines pale brownish ochreous; basal and postbasal evenly curved; antemedial sharply angled onto costa; medial passing through discal dot; postmedial usually evenly curved onto costa; area beteen ante- and postmedial often darker, especially near costa; terminal area also darker near apex; subterminal line yellowish white, dentate; discal dot oblique, narrow and linear, brown. Hind wing concolorous, transverse lines marked in posterior half; terminal area broad, slightly darker; discal dot small, brownish.

Male genitalia. Uncus narrow, biapical. Valve broadened basally, narrowing to rather pointed apex. Vinculum narrow, rounded. Papillae elongate and narrow. Aedeagus stout, shorter than valve length, ductus ejaculatorius attached near anterior end. Vesica armed with two large (and sometimes two very small apical) dentate cornuti, and a long and narrow plate connecting with an irregular, curved cornutus near ductus ejaculatorius base. Sternite A8 peg-like, moderately broadened and bilobed at base, elongate and tapering to two short sclerotized apical horns.

Female genitalia. Bursa large, ovate, basal half covered with large, elongate spines, which continue in a narrow field to a transverse patch in the short and striate ductus bursae. A smaller separate patch of spines near ductus seminalis base. Long, anteriorly curving ductus seminalis attached to ductus bursae on right. Colliculum narrow and tapered. Antrum membranous. Tergite A8 slightly transverse. Apophyses narrow, slightly expanded at apices. Papillae anales elongate and narrow, tapered.
Range. China (Xinjiang, Tibet); Kirghizstan, Tajikistan, Afghanistan (NE), Pakistan/India (Jammu and Kashmir), Mongolia.
Phenology. Univoltine: early June to mid-August.
Altitude. From about 2200 m in Tien-Shan Mts. (Kirghizstan: coll. ZISP) up to about 3700 m in Jammu and Kashmir (Pakistan/India: coll. ZFMK) and 4500 m in Xinjiang (coll. IZCAS).
Similar species. Similar to the following species, but larger and paler with broader fore wings and a less distinct pattern on all wings. The broader papillae, presence of only two large dentate cornuti on the vesica, and the longer and narrower eighth sternite distinguish the male genitalia; the female genitalia are easily distinguished from those of *E. exactata* by the much narrower apophyses.
Material. 1♂, [Mongolia], Uliassutai, 94, Led., *Nephelata* Stgr., Origin., Gen. vérifiés 1441[4] *Eupithecia nephalata* Stgr. ♂, dét. E. de Laever 1954 (syntype, coll. O. Staudinger in MNHU); 1♀, [Mongolia], Uliassutai, 94, Led., *nephelata* ♀, Origin., Gen. vérifiés 1441[5] *Eup. nephalata* Stgr. ♀, dét. E. de Laever 1954 (syntype, coll. O. Staudinger in MNHU); 1♂, [Xinjiang], Korla, 97. Hbh., 27/6, *Nephelata* Stgr., ? *Lithographata* Chr. mer. Or. ttt 9 J. I, 5, *Poecilata* Püngl. Stett. e. Z. 1888 p. 349 (Boh.), Vojnits slide no. 19043♂ (coll. O. Staudinger in MNHU); 2♂, 1♀, [Xinjiang], Mus-tag-ata Yarkend, Schmidt slide no. Eu 15c (coll. K. Dietze in MNHU); 1♀, [Xinjiang], Altyntagh, Tancré 906, 58., *Nephelata* Stgr., Gen.-Unters. Nr. [Schmidt slide no.] Eu 15 (coll. K. Dietze in MNHU); 1♀, [Xinjiang], *nephelata* Stgr. Korla, Type verglichen K. Dietze, Rückbeil 1902 durch R. Püngeler, Zool. Mus. Berlin, Gen.-Unters. Nr. [Schmidt slide no.] Eu 15b (coll. K. Dietze in MNHU); 2♀, [Xinjiang], Korla, Sai-chin, 377 and 379, *nephelata* Stgr., K. Dietze (coll. K. Dietze in MNHU); 1♂, [Xinjiang], *nephelata* Stgr., Aksu ♂, K. Dietze (Tancré 03), Zool. Mus. Berlin, Gen.-Unters. Nr. [Schmidt slide no.] Eu 15a (coll. K. Dietze in MNHU); 1♀, [Xinjiang], *nephelata* Stgr., Aksu., Turkest., K. Dietze (coll. K. Dietze in MNHU); 1♀, [Xinjiang], *nephelata* Aksu 1903, K. Dietze (v. Tancré) (coll. K. Dietze in MNHU); 1♀, [Xinjiang], Asia centr., Saichin-Kette, nördl. v. Korla, Rückbeil 1902, *nephelata* Stgr. ♀ (v. R. Tancré 11.02), *nephelata* Stgr. mit Typen verglichen (K. Dietze) (coll. R. Püngeler in

MNHU); 2♂, 2♀, [Xinjiang], Asia centr., Altyn Tagh, Rückbeil 1903, *nephelata* Stgr., v. R. Tancré 5.05 and /04 (coll. R. Püngeler in MNHU); 1♀, [Xinjiang], Altyn Tag [= Altun Shan], Vojnits slide no. 858♀ (TTM); 4♂, 14♀ Xinjiang, Akto, Wuyitake, 2650 m, 16.viii.1989, slide no. ZL630♂; 6♂,3♀ Xinjiang, Yecheng Qipanxiang, 3000 m, 16.vii.1989, slides no. ZL297♂, 464♀, L-2170; 1♀, 1♂, same loc, 2600 m, 18.vii.1989; 2♂,5♀ Xinjiang, Tashkhurgan Hongqilapu, 4500 m, 13.viii.1989, slide no. ZL465♂; 1♀ same loc, 3550 m, 14.viii.1989; 1♂, same loc, 2650 m, 17.viii.1989; 1♂,1♀ Xinjiang, Akto, Wuyitake, 2650 m, 8.viii.1989; all coll. Zhang Xuezhong; 1♀, Tibet, Ngari, Shiquanhe, 4280 m, 4.vi.1976, coll Huang Fusheng, slide no. ZL325♀(IZCAS).

Remarks. Not previously recorded from China.

274. *Eupithecia exactata* Staudinger, 1882

Eupithecia exactata Staudinger, 1882, *Ent. Z., Stettin* **43** (1–3): 78. Syntypes 1♂, 2♀ (coll. O. Staudinger in MNHU; examined), [Kazakhstan]: Saisan.

Eupithecia exactata f. *modesta* Dietze, 1910, *Biologie der Eupithecien* **1**, pl. 73, fig. 380. Holotype ♂ (coll. K. Dietze in MNHU; examined), [China]: Aksu, Makan-Wüste [Korla, Sai-chin on the label]. **New junior synonym.**

Eupithecia opulenta Vojnits, 1982, *Acta zool. Acad. Sci. hung.* **28** (3–4): 420, fig. 21. Holotype ♀ (TTM; examined), East Turkestan [Kirghizstan]: Prov. Semirechgensee, Fort-Naryne (synonymised in Mironov, 1991).

Eupithecia subolivacea Inoue, 1996, *Trans. lepid. Soc. Japan* **47** (4): 241, figs 7, 14. Holotype ♀ (BMNH; examined), Pakistan: Gilgit (synonymised in Mironov et al., 2008c).

Description. Fore wing ground colour brownish ochreous; transverse lines usually fine and distinct; antemedial curved or angled; postmedial oblique, evenly curved onto costa; often double medial line broadly curved outside discal dot; terminal area usually with darker apical and tornal blotches and a fine pale dentate subterminal line; medial area sometimes slightly darker; discal dot linear, relatively small, black. Hind wing paler, transverse lines usually distinct, evenly curved; terminal area darker, with fine pale dentate subterminal line; discal dot very small.

Male genitalia. Uncus narrow, biapical. Valve unmodified, with rather pointed apex. Vinculum short and narrow, rounded. Papillae elongate and stout. Aedeagus stout, equal to valve length; ductus ejaculatorius attached near anterior end. Vesica armed with three heavily sclerotized, dentate cornuti, and a long U-shaped plate connecting with a smaller, irregular, curved plate near ductus ejaculatorius base. Sternite A8 peg-like, broadly bilobed basally, apical part rather short with two small apical horns.

Female genitalia. Bursa large, pear-shaped, basal half covered with large spines which connect on one side with a transverse patch of spines in

short, tapered, striate ductus bursae. Ductus seminalis curved, broadened basally, attached to ductus bursae on right. Colliculum small and narrow, tapered. Antrum membranous. Tergite A8 square with medially concave anterior and posterior margins. Apophyses long and broad, the posterior expanded and flattened at apices. Papillae anales elongate and narrow, tapered.
Range. China (Xinjiang); Kirghizstan, south-eastern Kazakhstan, Tajikistan, Afghanistan (NE), Pakistan/India (Jammu and Kashmir), Mongolia.
Phenology. Uni- or bivoltine: mid-June to late August.
Biology. Larva found on *Eurotia ceratoides* (= *Ceratoides papposa*) (Chenopodiaceae) (SE Kazakhstan, Ili River: original data).
Altitude. From about 1500 (Mongolia: coll. ISEN) to 4500 m above sea-level (Kashmir: coll. ZFMK).
Similar species. A little varying species, which is similar only to the preceding species. For distinctions see thereunder.
Material. 1♂, [Kazakhstan], *exactata* ♂ Saisan, Origin., Gen. vérifiés 1456[1] *Eupithecia exactata* Stgr. ♂, Saisan, dét. E. de Laever 1954 (syntype, coll. O. Staudinger in MNHU); 1♀, [Kazakhstan], Saisan, Origin., gehórt ui den Náhe der *Gratiosata* HS. Boh., *exactata* Stgr. det. E. Schütze, Kassel (syntype, coll. O. Staudinger in MNHU); 1♀, [Kazakhstan], Saisan Hbhr., Origin., *Exactata* Stgr ♀, Boh. Schakuh, *exactata* Stgr. det. E. Schütze, Kassel (syntype, coll. O. Staudinger in MNHU); ♂, [Xinjiang], Korla, Saichin, ? *exactata* forma: *modesta* Typ., Korla. (380), Zool. Mus. Berlin, Typus *exactata* Stgr. det. E. Schütze, Kassel, Gen.-Unters. Nr. [Schmidt slide no.] Eu 113a, Zool. Mus. Berlin (holotype of *E. exactata* f. *modesta*, coll. K. Dietze in MNHU); ♀, [Kirghizstan], Turkestan Oriental, Prov. Semirechgensee, Fort Naryne, 1910–1911, S. Akulin, Vojnits slide no. 891♀ (holotype of *E. opulenta*, TTM); ♀, Pakistan, Gilgit, Alt. 1600 m, 17.vii.1990, F. Aulombard et J. Plante leg., BM Geom. slide no. 19217 (holotype of *E. subolivacea*, coll. H. Inoue in BMNH); 1♀ (*nec* ♂ as on the label), Korla (undissected, TTM); 1♀, Xinjiang, Qiemo Amei, 3200 m, 25.vii.1988, slide no. L-2171♀; 1♂, Xinjiang, Akto, Mei, 3200 m, 25.vii. 1988, both coll Zhang Xuezhong, slide no. L-2164♂ (IZCAS).

275. *Eupithecia omissa* Mironov & Galsworthy, 2004

Eupithecia ***omissa*** Mironov & Galsworthy, 2004, *Trans. lepid. Soc. Japan* 55 (2): 119, figs 3, 15, 19. Holotype ♀ (IZCAS; examined), [China]: Shaanxi, Ningshan Huoditan, 1580 m.

Description. Fore wing ground colour mid brown, transverse lines pale brown; basal line curved, most strongly marked at costa; antemedial double, sharply angled in cell; discal dot dark brown, slightly elongate; two weakly marked postmedial lines, sharply angled close to the costa, with a strongly marked broad parallel pale line outside them. Hind wing ground colour white, dotted throughout with brown scales, most strongly towards the outer margin and on the anal margin, where brown scales are concentrated, marking the beginnings of transverse lines; discal dot small.

Male genitalia. Valve long and rather pointed with a broad sclerotized saccular projection near base. Vinculum elongate and narrow, with shallow medial hollow. Papillae broad, quite long. Aedeagus long and stout. Vesica armed with two apical and one medial stout horn-like cornuti, a long tubulate cornutus, a large irregular plate near ductus ejaculatorius base and an elongate patch of microspines. Sternite A8 almost triangular, bilobed at base, with two small asymmetrical apical rods.

Female genitalia. Bursa elongate oval, covered with large uniform spines on the right side and base. Ductus bursae broad and partly corrugated, with a separate patch of smaller spines. Ductus seminalis broadly attached to corpus bursae near base. Colliculum relatively elongate and broad. Antrum membranous. Tergite A8 elongate. Apophyses fairly narrow. Papillae anales elongate and narrow.

Range. China (Shaanxi, Gansu).

Phenology. Bivoltine: mid-April to late May; late June to early July.

Altitude. From about 1000 to 2350 m above sea-level (China: coll. O. Pekarsky; IZCAS).

Similar species. Externally, the broad pale strongly angled fascia outside the postmedial is a useful distinguishing feature. The male genitalia are similar to those of the Mediterranean *E. weissi* Prout, 1938 and *E. pauxillaria* Boisduval, 1840, but are readily distinguished by the shape of the valves, the cornuti on the vesica and the eighth sternite. The female genitalia are rather similar to those of the West Mediterranean *E. santolinata* Mabille, 1871, but the separate patch of small spines in the ductus bursae and the elongate eighth tergite are distinctive.

Material. ♀, Shaanxi, Ningshan Huoditan, 1580 m, 2.vii.1999, coll. Yuan Decheng, slide no. ZL587♀ (holotype, IZCAS); 1♂, Gansu, Wenxian, Qiujiaba, 2350 m, 28.vi.1988, coll. Yao Jian, slide no. ZL588♂ (paratype, IZCAS); 1♀, Shaanxi, Foping, 1,750 m, 28.vi.1999, coll. Yao Jian, slide no. ZL586♀ (paratype, IZCAS); 1♂, 1♀, C. China, prov. Shaanxi, Foping Natural Reserve, Panda area, 33°45′ N, 107°48′ E, 1600 m, 20.iv.–11.v.1999, leg. V. Siniaev & A. Plutenko, Mironov slides nos. 660♀, 661♂ (ZFMK); 3♀, China, Shaanxi, South Taibai Shan, Tsinling Mts. [Qinling], Houzhensi, 33°53′ N, 107°49′ E, 1500 m, 16.-21.v.2000, leg. Siniaev & Plutenko (ZSM); 1♀,

China, Shaanxi prov., Qinling Mts., N from Taibai, h – 1000 m, N 34°14.124′ / E 107°18.482′, 30.v.2009, leg. Floriani & Saldaitis, Pekarsky slide no. OP 0675♀ (coll. O. Pekarsky).

The *semigraphata* species group

This group was first introduced by Schütze (1956) as subgenus *Dietzea*, and developed in Mironov (2003). The group as developed in the latter publication included species allied to *E. subfuscata*. We now believe that the latter group, although probably closely related to *E. semigraphata* and its allies, should be separated, since the males have a rather differently shaped eighth sternite, and lack dentate cornuti in the vesica, and the females have genitalia of more diverse patterns. The eighth sternite of males in this group is generally broad, with heavily sclerotized portions in the apical part. The aedeagus vesica is armed with several, usually dentate, cornuti. The female bursa is elongate oval, with spines in the basal part and a separate patch close to the attachment of the broad ductus seminalis, which loops around the base of the bursa. We place the following four Chinese species in this group.

276. *Eupithecia succenturiata* Linnaeus, 1758

Phalaena (*Geometra*) *succenturiata* Linnaeus, 1758, *Syst. Nat.* (Ed. 10) 1: 528. Syntype(s) (LSL), [Sweden].
Geometra disparata Hübner, [1799], *Samml. Eur. Schmett.* 5 Geometrae (2), pl. 47: 246 (*nec* 247). Syntypes lost, [Europe] (synonymised in Treitschke, 1828).
Tephroclystia succenturiata var./ab. *exalbidata* Staudinger, 1901, in Staudinger & Rebel, *Cat. Lepid. palaearct. Faunengeb.* (Edn 3) 1: 315. Syntypes, at least 2♂, 4♀ (coll. O. Staudinger in MNHU; examined), [Russia]: Altai; [Kazakhstan]: Ala Tau Mts.; [China]: Ili region; [Kirghizstan]: Issyk-Kul; Germany.
[*Eupithecia*] *succenturiata* f. *extrema* Dietze, 1913, *Biologie der Eupithecien* 2: 98; *ibidem* (1910), 1, pl. 73, fig. 337. Holotype ♀ (coll. K. Dietze in MNHU; examined), [China]: Kuldja.

Unavailable names (infrasubspecific): *obscurata*: Lempke (1951); *bistrigata*: Lempke (1951). – (incorrect subsequent spellings): *succenturiaria*: Boisduval (1840); *subccenturiata*: Vallhonrat (1989).

Description. Fore wing chalky white, costa brown usually with reddish tinge; transverse lines indistinct, brownish, well marked on costa with large dark brown spots; medial area more or less brownish with reddish tinge along posterior margin; terminal area dark brown on inner side, and reddish brown with short dark brown dashes between veins towards margin, usually with whitish apical streak; subterminal line white, distinct, ending in white tornal spot; discal dot large, intensely black. Hind wing chalky white with darker shade and weak indications of transverse lines on posterior margin; terminal area darker with dentate white subterminal line; discal dot minute.

Male genitalia. Uncus fairly broad, biapical. Valve long and broad, unmodified. Vinculum broad, rounded. Papillae elongate and broad. Aedeagus stout, slightly shorter than length of valve. Vesica armed with three large, dentate cornuti, a narrow, U-shaped cornutus and an irregular curved cornutus at ductus ejaculatorius base. Sternite A8 very large, broadened near apex, with two heavily sclerotized digitiform apical lobes.

Female genitalia. Bursa elongate oval, covered with small spines except on right side, with a small, separate patch of small spines near ductus seminalis base. Ductus bursae short, spineless. Ductus seminalis long and broad, attached to bursa near colliculum on right. Colliculum short. Antrum membranous. Tergite A8 transverse, with shallow medial hollow in posterior margin. Apophyses rather short, quite broad. Papillae anales broad, with rounded apices.

Range. China (Xinjiang); Europe, Asia Minor, Caucasus, Transcaucasus, Urals, Uzbekistan (Ferghana), Kirghizstan (Tien-Shan Mts.), SE Kazakhstan, Asian Russia (Siberia: Altai Mts., near Baikal Lake, Kamchatka peninsula).

Phenology. Univoltine: late May to early September. Larva: mid-July to early October. Overwintering as pupa.

Biology. Larva found on flowers, fruits and leaflets of *Artemisia vulgaris, A. absinthium, A. campestris, A. maritima, Achillea millefolium, Tanacetum vulgare* (Asteraceae), *Angelica sylvestris* (Apiaceae), *Vaccinium uliginos. um, Calluna vulgaris* (Ericaceae), *Verbascum* (Scrophulariaceae), *Rubus* (Rosaceae) (Europe: Mironov 2003).

Altitude. From sea level up to about 2400 m in Kyrghyzsky Mts. (Kirghizstan: coll. ZMKU).

Similar species. Easily distinguishable externally from other *Eupithecia* species.

Material. 1♂, 1♀, [Xinjiang], m. Juni, v. *exalbidata*, Kuldja, Mitte Juni 03 (coll. K. Dietze in MNHU); ♀, [Xinjiang], forma: *exalbidata* ab. *extrema* Typ., Kuldja, Anfgs. Juli 03 ♀ (holotype of *E. succenturiata* f. *extrema*, coll. K. Dietze in MNHU); 1♂, 1♀, [Xinjiang], m. Juni, v. *exalbidata*, Kuldja, Mitte Juni 03 (coll. K. Dietze in MNHU); 2♂, 3♀, [Xinjiang], Juldus, Prov. Ili (or Ili or Ili-Gebiet) (coll. K. Dietze in MNHU).

277. *Eupithecia marginata* Staudinger, 1892

Eup[*ithecia*] *marginata* Staudinger, 1892, *Dt. ent. Z.*„ *Iris* 5 (1): 257. Syntypes 2♂, 4♀ (coll. O. Staudinger and K. Dietze in MNHU; examined), [China]: Kuldja [= Yining (Gulja)].

Unavailable names (incorrect subsequent spelling): *marginita*: de Laever (1956).

Description. Rather small, variable in size and depth of colour. Fore wing ground colour rusty ochreous to blackish grey with rusty tinge; costal margin darker, blackish grey; transverse lines inconspicuous, except brownish postmedial, bluntly angled onto costa; terminal area darker, blackish grey with indistinct wavy whitish subterminal line; discal dot small, sometimes invisible. Hind wing usually less rusty, blackish grey; terminal area darker; transverse lines indistinct except evenly curved, double pale postmedial band; discal dot small or invisible.
Male genitalia. Uncus short and narrow, biapical. Valve unmodified. Vinculum short, with shallow medial hollow. Papillae rather narrow. Aedeagus slim, about equal to length of valve. Vesica armed with a fusiform, smooth cornutus, an elongate, longitudinally bifurcate cornutus, a ribbed cornutus and an elongate, irregular cornutus near ductus ejaculatorius base. Sternite A8 rectangular, bilobed at base, sides slightly bulging in apical half, and with two small, rounded lobes at apex.
Female genitalia. Bursa small, pear-shaped, over $^2/_3$ densely covered with a field of small spines with a diagonal margin. Ductus bursae short and striate, with a separate patch of small spines near ductus seminalis base. Ductus seminalis evenly curved around base of bursa, attached to ductus bursae on right. Colliculum short and narrow, sharply tapered. Antrum membranous. Tergite A8 elongate. Apophyses narrow, slightly expanded at apices. Papillae anales small, relatively narrow and slightly elongate with rounded apices.
Range. China (Xinjiang); Russia (Caucasus), Armenia, Iran, Uzbekistan, Kirghizstan, Tajikistan, Afghanistan, Kazakhstan (Tien-Shan Mts.).
Phenology. late June to early September.
Altitude. From about 1500 m in Zailijsky Range (Kazakhstan, Tien-Shan: coll. ZISP) up to about 3000 m above sea-level in Pamirs Mts. (Tajikistan: coll. ZMMU).
Similar species. Unlikely to be confused with any other eastern Palaearctic species. In the male the shape of sternite 8 is distinctive.
Material. 1♂, [Xinjiang], Kuldja Rkbl. 84, *marginata* Stgr., Origin., Gen. vérifiés 1451[3] *Eupithecia marginata* Stgr. ♂, Kuldja 84, dét. E. de Laever 1954 (syntype, coll. O. Staudinger in MNHU); 1♀, *marginata* ♀ [Xinjiang] Kuldja, Origin., Gen. vérifiés 1451[4] *Eupithecia marginata* ♀, Kuldja, dét. E. de Laever 1954 (syntype, coll. O. Staudinger in MNHU); 3♀, small brownish circular label, Origin., (syntypes, coll. O. Staudinger in MNHU); 1♂, [Xinjiang] Kuldja, Stgr. 98, Origin. (syntype, coll. K. Dietze in MNHU); 1♀, [Xinjiang] Asia centr., Turkestan, Ili-Gebiet, Coll. Wagner (coll. K. Dietze in MNHU).

Remarks. The cornuti in the aedeagus of this species are usually well visible only on the everted vesica.

278. *Eupithecia masculina* Inoue, 1988

Eupithecia masculina Inoue, 1988, *Bull. Fac. domest. Sci. Otsuma Wom. Univ.* **24**: 364, figs 25a-e. Holotype ♂ (BMNH; examined), [Taiwan]: Hualien Hsien, Tayuling, 2600 m.

Description. A small species with pale ash-grey ground colour to fore and hind wings. Fore wing transverse lines very narrow, sharply angled onto costa; antemedial, medial and postmedial forming dark costal spots; postmedial otherwise formed by a series of short dark dashes on veins; terminal area with trace of pale subterminal line; discal dot narrow and elongate, black. Hind wing with relatively conspicuous transverse lines, postmedial formed by a series of dark dashes on veins; discal dot very small. Abdomen whitish grey with black lateral spots on first two segments.
Male genitalia. Uncus short and narrow, biapical. Valve unmodified, apically rather narrow. Vinculum short and narrow. Papillae elongate and narrow. Aedeagus shorter than length of valve. Vesica armed with a medial slightly dentate cornutus, an H-shaped plate, and a small irregular cornutus near ductus ejaculatorius base. Sternite A8 peg-like, broadly bilobed at base, with two short apical horns.
Female genitalia. Bursa ovoid, pinched medially on left, with some spiniferous areas on the dorsal side, a large spiniferous area on the ventral side and a row of larger spines between base of ductus seminalis and colliculum. Ductus bursae not distinguishable. Ductus seminalis joining corpus bursae at centre, strongly broadened basally and sharply recurved, attached to process of bursa on right. Colliculum short and narrow, inclined. Antrum lightly sclerotized. Tergite A8 square. Apophyses short and narrow. Papillae anales small, rounded.
Range. China (Taiwan).
Phenology. Univoltine: early April to early May.
Altitude. From about 1400 to 2600 m above sea-level (Taiwan: Inoue 1988; coll. TTM).
Similar species. Externally similar to the Taiwanese *E. assulata*, especially to narrow winged males of the latter. However, *E. masculina* has a smaller discal dot and inconspicuous white subterminal line and tornal spot on the fore wing, and darker transverse lines and discal dot on the hind wing. The male and female genitalia are easily distinguished from other Asian species.
Material. ♂, Taiwan, Tayuling, Hualien Hsien, 2600 m, 2–4.v.1984, leg. H. Yoshimoto, BM Geom. slide no. 16549 (holotype, coll. H. Inoue in BMNH);

1♀, [Taiwan], Shyrsen, Chiai Hsien, 2600 m, 20.iv.1991, leg. B.S. Chang, 1282–8584, Galsworthy slide no. ACG TW14♀ (NSMTA); 1♂, 1♀, Taiwan, Prov. Hualien, Taroko N.P., on the road 8, 1400 m, 2.iv.1997, leg. Csorba & Ronkay (TTM).

279. *Eupithecia subumbrata* [Denis & Schiffermüller], 1775

Geometra subumbrata [Denis & Schiffermüller], 1775, *Syst. Verz. Schmett. Wien.*: 110. Syntype(s) lost. [Austria]: Vienna district.

Phalaena (*Geometra*) *scabiosata* Borkhausen, 1794, *Natur. Eur. Schmett.* 5: 336. Syntype(s). [Europe] (synonymised in Staudinger & Rebel, 1901).

Eupithecia piperata Stephens, 1831, *Illustr. Br. Ent.* (Haustellata) 3: 288. Syntypes, at least 1♂ (coll. O. Staudinger in MNHU; examined). [United Kingdom]: [England], [Surrey], Riddlesdown, near Croydon (synonymised to *Eupithecia obrutaria* in Herrich-Schäffer, 1848).

Eupithecia obrutaria Herrich-Schäffer, 1848, *Syst. Bearb. Schmett. Eur.* 3 (31): 119, 126, pl. 24, figs 145, 146. Syntypes. [Germany]: Regensburg (synonymised in Doubleday, 1856).

Eupithecia majoraria De La Harpe, 1855, *Fauna Suisse. Lépidoptères.* 4: 149. Syntypes 2♀, [Switzerland] (synonymised in Staudinger & Rebel, 1901).

[*Eupithecia*] *D*[*ietzea*] *subumbrata iliata* Schütze, 1956, *Z. Wien. Ent. Ges.* 41 (11): 318, pl. 34 (D), figs 36–40, text-fig. 6: C1. Holotype ♂ (coll. K. Dietze in MNHU; examined), [Kirghizstan]: Ferghana, Gouldscha.

Eup[*ithecia*] *scabiosata* f. *limbofasciata* Dietze, 1910, *Biologie der Eupithecien* 1: pl. 76, figs 570, 571; *ibidem* (1913), 2: 103. Lectotype ♂ (coll. K. Dietze in MNHU; examined), [Kirghizstan]: North Tibet, Alexander Mts. (elevated to species rank in Schütze, 1956; synonymised in Prout, 1939 and independently in Mironov, 1991).

[*Eupithecia*] *scabiosata* f. *bistrigata* Dietze, 1913, *Biologie der Eupithecien* 2: 103; *ibidem* (1910), 1, pl. 76, fig. 578; pl. 85, fig. 993. Syntype(s) (coll. K. Dietze in MNHU; examined), [Austria]: Styria, Stainz, ex. l.

Misidentification: *Eupithecia residuaria* Herrich-Schäffer, 1856 (*nec* Boisduval, 1840) (identity: *Eupithecia lariciata*).

Unavailable names (infrasubspecific): *aequistrigata*: Staudinger (1871); *obscurata*: Lempke (1951); *impuncta*: Lempke (1951). – (incorrect subsequent spellings): *piperitata*: Stephens (1831); *subumbraria*: Boisduval (1840); *piperaria*: Doubleday (1849); *pipeperata*: Millière (1869); *subumrata*: Bielewicz (1973); *subbumbrata*: Vives Moreno (1994).

Description. Fore wing ground colour greyish white, darker grey along costa and termen; stalks of R and Cu often ochreous; transverse lines numerous, fine, oblique, wavy, and broadened, darker, sharply angled onto costa; first medial line crossing discal dot; postmedial shaded on inner side and usually well marked by a series of dark grey dashes on veins, twice angled near costa; a whitish subterminal line; discal dot minute, sometimes obsolete. Hind wing paler, smoky white; transverse lines inconspicuous or well marked along dorsum; terminal area distinctly darker; a whitish dentate subterminal line; discal dot small.

Male genitalia. Uncus narrow, biapical. Valve unmodified, rather long. Vinculum broad and rounded. Papillae elongate, strongly narrowed near base. Aedeagus shorter than length of valve. Vesica armed with two dentate cornuti medially and an irregular, bowed cornutus at ductus ejaculatorius base. Sternite A8 elongate, narrowing slightly before apex, and with heavily sclerotized lateral margins and two short apical rods.

Female genitalia. Bursa pear-shaped, spined in basal half. Ductus bursae broad, tapered to colliculum, with a large triangular patch of small spines on ventral side. Ductus seminalis long and curved round base of bursa, attached to ductus bursae from right side. Colliculum elongate and broad. Antrum membranous. Tergite A8 elongate with shallow medial hollow on posterior margin. Apophyses short and narrow. Papillae anales broad and rounded.

Range. China (Xinjiang); Europe, Urals, Uzbekistan (Ferghana), Tajikistan (west Pamirs), Kirghizstan (Tien-Shan Mts.), S and SE Kazakhstan, Asian Russia (southern Siberia), Mongolia.

Phenology. Uni- and probably bivoltine in southern areas: early May to early August.

Biology. Larva found on the flowers of *Scabiosa songarica* (Dipsacaceae), *Gypsophila altissima* (Caryophyllaceae), *Gentiana turkestanorum* (Gentianaceae), *Codonopsis clematidea* (Campanulaceae), *Ziziphora clinopodioides* (Lamiaceae), *Echinops tianschanicus* and *Artemisia* (Asteraceae) (Kirghizstan, Tien-Shan Mts.: Mironov 2003); also found on the flowers, rarely on the seeds and leaves of numerous, mainly herbaceous plants in Europe.

Altitude. From sea level up to about 3000 m (Kirghizstan: coll. ISEN).

Similar species. Varies from pale greyish white with distinct transverse lines to darker grey with inconspicuous markings on fore wings. In spite of its variability, the pale greyish white colouration with characteristic distinct dark costal margin and termen on the fore wing and along termen on the hind wing, should usually distinguish this species from any other Chinese congeners. Both male and female genitalia are distinct.

Material. 1♂, [United Kingdom], Anglia Dbld., 1866., *Su(m)umbrata* Gn., *Piperata* Stph., Origin. (syntype of *E. piperata*, coll. O. Staudinger in MNHU); 1♀, small white circular label, Origin. (syntype of *E. piperata*, coll. O. Staudinger in MNHU); ♂, [Kirghizstan], Gouldscha, Ferghana 1905 Korb, *scabiosata*, Holotypus *Eupithecia subumbrata iliata* Schü., E. Schütze, Kassel (holotype of *E. subumbrata iliata*, coll. K. Dietze in MNHU); 1♂, [Xinjiang], *scabiosata*, Dscharkend (Kuldja), *scabiosata*, *subumbrata iliata* Schü., Präp. Nr. 523♂ det. E. Schütze, Kassel (paratype of *E. subumbrata iliata*, coll. K. Dietze in MNHU); 3♀, [Xinjiang], Juldus Ili, *scabiosata* (paratypes of *E. subumbrata iliata*, coll. K. Dietze in MNHU); 1♀, [Xinjiang], Juldus, v. *juldussi* Bang-Haas, *scabiosata* (paratype of *E. subumbrata iliata*,

coll. K. Dietze in MNHU); 1♀, [Xinjiang], Juldus, Ili, 1116, Paratypus *Eupithecia subumbrata iliata* Schü., E. Schütze, Kassel, präp. Nr. 523b♀, det. E. Schütze, ZSM slide no. G 13398♀ (paratype of *E. subumbrata iliata*, ZSM); ♂, [Kirghizstan], Alex. Geb. *scabiosata, scabiosata limbofasciata* Typ. Alex.-Gebirge, Nord-Tibet, *E. limbofasciata* Dietze, Präp. Nr. 525b♂, det. E. Schütze, Kassel (holotype of *E. limbofasciata*, coll. K. Dietze in MNHU); 1♂, 1♀, [Xinjiang], Asia centr., Ost-Turkestan (Urumtschi), Rückbeil 1904, v. R. Tancré 1905 (coll. R. Püngeler in MNHU); 1♂, 1♀, [Xinjiang], Asia centr., Saichin-Kette, nördl. von Korla, Rückbeil 1902, *succentur.* v. *exalbidata*, v. R. Tancré 12/02 and 1/2.03 (coll. R. Püngeler in MNHU).
Remarks. This species is represented by subsp. *iliata* Schütze, 1956 in the mountains of Central Asia (Kirghizstan, south-eastern Kazakhstan and north-western China).

The *subfuscata* species group

Our reasons for placing *E. subfuscata* and its allies in a separate species group outside the *semigraphata* group are explained in the introduction to that group. This species group is therefore introduced here for the first time, and genitalia characteristics are set out below. It is represented in China by the following fourteen species.
Male genitalia. Uncus usually short and stout, with pointed tip, uni- or biapical. Valve unmodified, except in *E. tibetana*, where there is a small, blunt ventral projection. The aedeagus is usually stout, and shorter than the length of the valve. The vesica is usually covered with numerous minute teeth, and armed with a U-shaped medial cornutus and an irregular cornutus near ductus ejaculatorius base. Some species have a horn-like cornutus in addition. Sternite A8 normally consists of two parallel arms, connected with each other by basal band, but in a few cases the space between the arms is also sclerotized, giving a peg-like appearance.
Female genitalia. Bursa globular to oval, with a large spine field of variable size. Ductus bursae usually well developed, tapered, often striate, and in some species with a separate patch of spines. Ductus seminalis usually long, curved around bursa, and attached to ductus bursae near to colliculum, but short and attached to corpus bursae in *E. praecipitata*. Colliculum present, usually narrow. Antrum membranous, covered with numerous pores in some species. Tergite A8 very variable in size and shape. Its posterior margin is covered with short setae in *E. egena, E. lunatica, E. acutipapillata* and *E. supercastigata*. Apophyses varying from short and narrow to extremely long and stout with expanded and flattened apices. Papillae

anales also variable, from broad and rounded to narrow, and very elongate. Two Chinese species, *E. lasciva* and *E. egena*, have heavily sclerotized, pointed papillae.

280. *Eupithecia subfuscata* Haworth, 1809

Phalaena subfuscata Haworth, 1809, *Lepid. Br.* (2): 360. Syntype(s) (BMNH), [United Kingdom].

Phalaena singulariata Haworth, 1809, *Lepid. Br.* (2): 360. Syntype(s) (BMNH), [United Kingdom] (synonymised to *Eupithecia castigata* Hübner, 1813 in Wood, 1854).

Geometra castigata Hübner, [1813], *Samml. Eur. Schmett.* 5 Geometrae (2), pl. 89, fig. 456. Syntype(s) lost. [Europe] (synonymised in Prout, 1915).

Eupithecia compressata Guenée, 1858, in Boisduval & Guenée, *Hist. nat. Insectes* (*Spec. gén. Lépid.*) **10** (2): 324. Holotype ♀. (Type locality not stated, [Europe ?]) (synonymised in Staudinger & Rebel, 1901).

Eupithecia implicata Walker, 1862, *List Specimens lepid. Insects. Colln. Br. Mus.* **24**: 1241. Syntypes 8♀ (BMNH), [Canada]: Hudson's Bay, Albany River, St Martins Falls; Nova Scotia (synonymised to *Eupithecia castigata* Hübner, 1813 in Ringe, 1952).

Eupithecia blancheata Cooke, 1881, *The Entomologist* **14** (213): 43. Holotype ♀ (BMNH?), [United Kingdom]: Isle of Man (synonymised in Staudinger & Rebel, 1901).

Tephroclystis latipennis Hulst, 1898, *The Can. Ent.* **30** (5): 114. Lectotype ♂ (AMNH), [Canada]: Quebec, Quebec City (synonymised to *Eupithecia castigata* Hübner, 1813 in McDunnough, 1929).

Eup[*ithecia*] *castigata* sp./f. *ussuriensis* Dietze, 1910, *Biologie der Eupithecien* **1**, pl. 72, figs 314–316; *ibidem* (1913), **2**: 94. Syntypes 21, including ♂♀ (coll. K. Dietze in MNHU; examined), [Russia]: Ussuri, Kasakewitsch.

Eupithecia latimarginata Matsumura, 1925, *J. Coll. Agric. Hokkaido Imp. Univ.* **15** (3): 171, pl. 10, fig. 17. Holotype ♀ (EIHU), [Japan/ Russia]: South-Saghalien, (Ichinos.awa) (synonymised in Inoue, 1977).

Eupithecia ichinosawana Matsumura, 1925, *J. Coll. Agric. Hokkaido Imp. Univ.* **15** (3): 172, pl. 11, fig. 23. Holotype ♀ (EIHU), [Japan/Russia]: South-Saghalien, (Ichinos.awa) (synonymised in Inoue, 1980).

Eupithecia verecunda Vojnits, 1980, *Acta zool. Acad. Sci. hung.* **26** (4): 437, figs 5, 6, 19. Holotype ♂ (ZFMK; examined), China: Sued-Shensi, Tapaishan in Tsinling. **New junior synonym.**

Eupithecia perpetua Vojnits, 1984, *Acta zool. Acad. Sci. hung.* **30** (1–2): 214, fig. 2. Holotype ♂ (TTM; examined), [China]: North Yunnan, A-tun-tse [Dêqên], Talsohle, 3000 m. **New junior synonym.**

Misidentifications: *Eupithecia residuata* Eversmann, 1844 (*nec* Hübner, 1817) (identity: *Eupithecia lariciata*); *Eupithecia virgaureata* Petersen, 1910: 244–245, pl. 9: 42A, (*nec* Doubleday, 1861).

Unavailable names (infrasubspecific): *obscura*: Dietze (1910); *obscurissima*: Prout (1915). – (incorrect subsequent spellings): *castigaria*: Boisduval (1840); *eastigata*: Fassnidge (1926).

Description. Fore wing broad, ground colour grey or pale grey; transverse lines variable, usually indistinct, slightly sinuate, broadened near costa, dark grey; medial often more conspicuous, sharply angled near costa beyond discal dot or just touching discal dot; postmedial usually also well defined, slightly sinuate and evenly curved near costa, sometimes with dark dashes on veins; terminal area with ill-defined, unbroken, pale subterminal line and with a weak pale tornal spot; discal dot small but distinct. Hind wing slightly paler, with more or less conspicuous dark and light transverse lines, especially dark postmedial and paler double stripe beyond it; terminal area always darker with more or less distinct, pale subterminal line and minute tornal spot; discal dot minute.

Male genitalia. Uncus elongate, narrow, biapical. Valve large, with narrow, rather pointed apex. Vinculum broad and rounded. Papillae fairly long, slightly clavate. Aedeagus elongate and narrow, shorter than length of valve. Vesica armed with a small U-shaped cornutus and an irregular, curved cornutus at ductus ejaculatorius base. Sternite A8 broadly bilobed at base, with two narrow, parallel rods with apices blunt or often slightly expanded.

Female genitalia. Bursa spherical, covered with spines apart from rounded area on right. Ductus bursae membranous, narrow, with a patch of small spines near ductus seminalis base. Ductus seminalis long and broad, evenly curved around bursa, attached to ductus bursae near colliculum on right. Colliculum short and narrow, tapered. Antrum membranous. Tergite A8 rectangular, short and broad, with convex anterior, and concave posterior margins. Apophyses short and narrow. Papillae anales broad and rounded.

Range. China (Xinjang, Tibet, Gansu, Inner Mongolia, Heilongjiang, Beijing, Sichuan, Shanxi, Shaanxi, Yunnan); Europe, Turkey, Caucasus, Transcaucasus (Georgia, Armenia), Russia (Urals, western and southern Siberia, Altai Mts., Transbaikalia, Dahuria, Priamurje, Primorje, Sakhalin, Kuril Islands), Kazakhstan, Kirghizstan (Tien-Shan Mts.), Mongolia, Korea, Japan (Hokkaido, Honshu), North America.

Phenology. Univoltine: mid-May to late July.

Biology. Larva polyphagous on the flowers, seeds and leaves of various mainly herbaceous plants in Europe and North America.

Altitude. From sea level to about 3400 m in Yunnan and Tibet (China: coll. O. Pekarsky; IZCAS).

Similar species. A variable species which could be confused with *E. lariciata* or *E. subumbrata*. However, *E. lariciata* has a darker area between antemedial and medial lines, a more sharply angled postmedial line, a larger fore wing discal dot, and usually a distinct white spot on the end of the notum; *E. subumbrata* has more elongate, narrowed and pointed fore wings, usually with whitish coloration and with the stalks of R and Cu covered with reddish scales. Both male and female genitalia will easily distinguish the present species from either.

Material. ♂, [Russia], Ussuri, Kasakewitsch 1907 Korb, ♂, Typ. *ussuriensis*, Typus, Holotypus *Eupithecia ussuriensis* Dietze [label of Vojnits], Vojnits slide no. 10242♂ (lectotype of *E. castigata* f. *ussuriensis*, coll. K. Dietze in MNHU); 1♂, [Xinjang], Juldus, Ili, ? *castigata* var., Juldus Ili (coll. K. Dietze in MNHU); 2♂, 2♀, [Xinjang], Juldus, Ili (coll. K. Dietze in MNHU); 1♀, [Xinjang], Urumtschi, Tian-schan, 44, Graeser (coll. R. Püngeler in MNHU); 1♂, 1♀, Yarégong, P. Soulié 1902, Vojnits slides nos. 867♂, 876♀ (TTM); 1♀, Szetschwan-China [Sichuan], Wassekou, Exped. Stötzner, de Laever slide no. 400♀ (ZFMK); 1♂, Szetschwan [Sichuan], Ta-tsien-lu, Exped. Stötzner, Coll. Dr. Wehrli, Basel, Vojnits slide no. 12145♂ (paratype of *E. improba*, ZFMK); 1♀, [Sichuan], Ta-tsien-Loû, Chasseurs du P. Déjean 1904, Vojnits slide no. 875♀ (TTM); 1♂, China, Provinz Nord-Yuennan, Li-kiang [Lijiang], 29.vi.1934, H. Höne (ZFMK); ♂, Sued-Shensi [Shaanxi], Tapaishan [Taibaishan] im Tsinling [Qinling], F, 19.vi.1935, H. Höne, Vojnits slide no. 12381♂ (holotype of *E. verecunda*, ZFMK); 6♀, same loc., 20, 21, 26.vi., 2.vii.1935, H. Höne, Vojnits slides nos. 11458♀, 12224♀, 12265♀, 12264♀, 11507♀, 11456♀ (paratypes of *E. verecunda*, ZFMK); 1♂, 2♀, same loc., 20, 21, 24.vi.1935, H. Höne, Vojnits slides nos. 12266♀, 12267♀, 12380♂ (paratypes of *E. verecunda*, TTM); 2♂, 3♀, same loc., 20, 21, 24, 27, 28.vi.1935, H. Höne, Vojnits slide no. 12387♀ (incorrectly associated gen. slide of *E. stomachosa*), de Laever slides nos. 188♀, 403♀ (ZFMK); 2♀, same loc., 22, 24.vi.1935, H. Höne, Vojnits slides nos. 12943♀, 12952♀ (TTM); 1♀, same loc., L, 22.vi.1935, H. Höne, Mironov slide no. 797♀ (SMNK); ♂, Nord-Yünnan, A-tun-tse [Dêqên], Talsohle *ca.* 3000 m, 26.v.1937, H. Höne, Vojnits slide no. 14102♂ (holotype of *E. perpetua*, TTM); 1♂, same loc. and alt., 28.v.1937, H. Höne, Vojnits slide no. 14103♂ (paratype of *E. perpetua*, ZFMK); 1♀, same loc. and alt., 14.vi.1936, H. Höne, Vojnits slide no. 12263♀ (paratype of *E. verecunda*, ZFMK); 3♂, 14♀, same loc. and alt., 14, 21, 22, 28, 31.v., 1, 4, 6, 12, 16, 17, 19, 25.vi.1937, H. Höne, Vojnits slides nos. 14105♀, 13763♀, 13293♂, 13323♂, 13765♀, 13760♀, 14101♀, 14106♀, 13758♀, 14190♀, 12539♀, 13292♀, 13931♀, 13168♂, 13764♀, 12538♀, 13766♀ (ZFMK); 1♀, same loc. and alt., 31.v.1937, H. Höne, Mironov slide no. 819♀ (TTM); 1♀, Prov. Shansi [Shanxi], Mien-shan, Obere Höhe *ca.* 2000 m, 7.vii.1937, H. Höne, Vojnits slide no. 11519♀ (paratype of *E. verecunda*, ZFMK); 1♀, same loc. and alt., 13.vii.1937, H. Höne, Vojnits slide no. 13743♀ (ZFMK); 1♀, China, Yunnan Prov., Zhongdian, 3100 m, 14.vii.2001, Li Houhun leg. (coll. P. Skou); 1♂, 1♀, China, W. Sichuan, Kangding, evn 3000 m, 29°53′ N / 101°55′ E, 13.vii.2009, I. & A. Floriani, Saldaitis leg., Pekarsky slides nos. OP 0631♂, OP 0682♀ (coll. O. Pekarsky); 1♂, China, North Yunnan, Diqing Tibetan Aut. Pref., 8 km NNE of Shangri La at Nairi village, 3300 m, 14.vi.2009, leg. B. Benedek, Pekarsky slide no. OP 0824♂ (coll. O. Pekarsky); 2♀, China, North Yunnan, Diqing Tibetan Aut. Pref., 5 km SE of Deqin, 3356 m, 19.vi.2009, leg.

B. Benedek, Pekarsky slides nos. OP 0852♀, OP 0856♀ (coll. O. Pekarsky); IZCAS collection: Heilongjiang: 2♀, Dailing, 390 m, 3 and 9.vii.1962, coll Bai Jiuwei, slides nos. ZL199, 212♀; Inner Mongolia: 2♂, 3♀, Oragxan, 11.viii.1983, slides nos. ZL173♂, 139♀, and 28.vi.1985, slide no. ZL429♂; 1♀, Giwen, 500 m, 4.vii.1985, coll Xue Dayong; Beijing: 1♀, Baihuashan, 8.vii.1973, coll Han Yinheng; Shanxi: 1♂, 5♀, Wutaishan, 1992, slide no. ZL353♂, 252♀; 1♀, Pangquangou, 7.vii.1998, slide no. ZL285♀; Gansu: 3♀, Wenxian, Qiujiaba, 2350 m, 28.vi.1998 and 21–22.7.1999, coll Yao Jian and Zhu Chaodong; 1♀, Wenxian, Bikou, Zhongmiao, 700 m, 24.vi.1998, coll Yang Xingke; 1♂, 4♀, Zhouqu, Shatan forestry reserve, 2400 m, 14.vii.1999, coll Zhu Chaodong, slide no. ZL572♂; Yunnan: 1♀, Lijiang, Yulongshan, 28.vi.1962, coll Song Shimei, slide no. ZL143♀; Tibet, 1♀, Jomda, 3400 m, 26.vii.1976, coll Zhang Xuezhong, slide no. ZL347♀; 3♂, 1♀, Nyingchi, Bayi, 3040 m, 2–3.vii.2002, Han Hongxiang, slides nos. ZL 641, 642, 651♂, 649♀.

281. *Eupithecia costipicta* Warren, 1893

Eupithecia costipicta Warren, 1893, *Proc. zool. Soc. Lond.* **1893** (2): 383, pl. 30, fig. 21. Holotype ♀ (BMNH; examined), [India]: Sikkim, 8000ft.
Eupithecia gibbosa Mironov & Galsworthy, 2004, *Trans. lepid. Soc. Japan* 55 (4): 285, figs 1, 13. Holotype ♀ (IZCAS; examined), China: Hubei, Shennongjia, Dajiuhu, 1800 m. **New junior synonym.**

Description. Fore wing elongate, ground colour pale buff; transverse lines indistinct, marked by six to seven dark brown spots along costa; double postmedial faintly visible as undulating lines and dark dashes on veins; discal dot dark brown, elongate; terminal area shaded brown. Hind wing ground colour off-white; discal dot fairly large, but faint; only postmedial present, double or triple, pale brown, crenellate.
Male genitalia. Uncus bell-shaped, biapical. Valve unmodified. Vinculum broad, rounded. Papillae relatively short and narrow. Aedeagus broad, about valve length. Vesica covered with numerous minute spicules, and armed with a broad plate-like cornutus, and a small irregular cornutus at ductus ejaculatorius base. Sternite A8 bilobed at base, with two narrow parallel arms.
Female genitalia. Bursa globular to oval, with two large rounded lateral patches of spines. Ductus bursae not distinguishable. Ductus seminalis attached near colliculum on left side. Colliculum very elongate, tapered towards posterior end. Antrum membranous, poriferous. Tergite A8 elongate. Apophyses rather stout. Papillae anales elongate.
Range. China (Hubei, Hunan); Pakistan/India (Jammu & Kashmir), Nepal.
Phenology. Univoltine: early August to early September.

Altitude. From about 1800 to 2300–2700 m (China: coll. IZCAS; Jammu & Kashmir: coll. U. Ratzel).
Similar species. Externally similar to *E. ruficorpus* in the *haworthiata* group, but larger, with narrower elongate fore wings, darker transverse lines and more distinct discal dots, the ground colour lacks a reddish tinge and the abdomen is not covered by pink or reddish scales as in *E. ruficorpus*. The genitalia in both sexes easily distinguish the two species. Externally not confusable with other species of the present group.
Material. ♀, [India], Sikkim, O. Möller, 8000, 89, *Eupithecia costipicta* Warr. Type ♀, Collectio H.J. Elwes, BM Geom. slide no. 20130♀ (holotype, BMNH); 1♀, China, Hubei, Shennongjia, Dajiuhu, 1800 m, 1.viii.1981, coll. Han Yinheng, slide no. ZL201♀ (holotype of *E. gibbosa*, IZCAS); 2♀, China, Hunan, Sangzhi, Tianping shan, 7.ix.1981, coll. Tong Xinwang, slide no. ZL337♀ (paratypes, IZCAS).
Remarks. There is a large geographical gap between the Chinese and Himalayan records. There are slight differences between the female genitalia from the two areas, principally in the length of the colliculum and shape of the antrum, but given the strong external similarity we do not think that there is sufficient justification for treating the Chinese specimens as a separate species, and therefore synonymise them here. Discovery of the male from central China might resolve the issue.

282. *Eupithecia mediocincta* Mironov & Galsworthy, 2004

*Eupithecia **mediocincta*** Mironov & Galsworthy, 2004, *Trans. lepid. Soc. Japan* 55 (2): 130, figs 11, 24, 29. Holotype ♂ (IZCAS; examined), [China]: Gansu, Yongdeng, Tulugou, 2280 m.

Description. Fore wing ground colour reddish brown; basal line curved, close to body, basal area darker brown; antemedial oblique to discal dot, then acutely angled onto costa; medial wavy, bluntly angled at discal dot onto costa; postmedial sinuate, hardly angled; area between antemedial and postmedial darker brown, forming a strong transverse band; discal dot large, dark brown; terminal area with a pale wavy submarginal line. Hind wing concolorous, discal dot small and faint; postmedial well marked, parallel to termen, whole area basad darker brown.
Male genitalia. Uncus long and stout, biapical. Valve unmodified. Vinculum short and narrow, rounded. Papillae elongate, narrow. Aedeagus narrow, shorter than valve length. Vesica armed with a U-shaped plate and a small irregular plate near ductus ejaculatorius base. Sternite A8 broadly bilobed at base and with two narrow apical rods.

Female genitalia. Bursa pear-shaped, about half covered with small spines. Ductus bursae membranous. Ductus seminalis broad, curved around bursa, attached to ductus bursae on right. Colliculum short and narrow, tapered. Antrum membranous. Tergite A8 square. Apophyses short and narrow. Papillae anales large and broad.
Range. China (Gansu, Sichuan, Shaanxi).
Phenology. Bivoltine: late April; late July to early September.
Altitude. From about 1700 to 2400 m in Sichuan and Shaanxi (China: coll. ZFMK, IZCAS).
Similar species. The dark central fascia separates this species from all other Asian *Eupithecia* except *E. latimedia* Hampson, 1895. The latter can be distinguished from the present species by the different shape of the dark medial fascia, and the antemedial being well separated from the discal dot. The genitalia are entirely different.
Material. ♂, Gansu, Yongdeng, Tulugou, 2280 m, 25–29.vii.1991, coll. Xue Dayong, slide no. 29.iv.1992, slide no. ZL494♀, coll. Meng Feng (paratypes, IZCAS); 1♂, same loc., 2270 m, 25–29.vii.1991, coll. Xue Dayong, slide no. ZL632♂ (paratype, IZCAS); 1♀, Sichuan, Nanping, Jiuzhaigou, 2400 m, 7.ix.1983, coll. Chai Huaicheng, slide no. ZL409♀ (paratype, IZCAS); 1♀, Sued-Shensi [Shaanxi], Tapaishan [Taibaishan] im Tsinling [Qinling], *ca.* 1700 m, 7.viii.1936, H. Höne, Vojnits slide no. 12270♀ (paratype, ZL62♂ (holotype, IZCAS); 3♀, same loc., 20–26.vii.1991, slide no. ZL257♀, 11–16. viii.1991, and 24-ZFMK); 1♀, same loc. and alt., 6.viii.1936, H. Höne, Vojnits slide no. 11477♀ (TTM); 1♂, 2♀, Gansu, Yongdeng Liancheng Linchang, Tulugou, 2600 m, 36°41′ N 102°42E, 8.viii.2005, slide no. ZL693♂, Li Jing, Han Hongxiang (IZCAS).

283. *Eupithecia chui* Inoue, 1970

Eupithecia chui Inoue, 1970, *Bull. Fac. domest. Sci. Otsuma Wom. Univ.* **6**: 3, pl. 1, figs 3, 4; pl. 2, figs 14, 15; pl. 3, figs 19–22. Holotype ♂ (BMNH; examined), Taiwan: Chiayi, Alishan, 2200 m.

Description. Fore wing ground colour pale grey-brown, transverse lines slightly darker, with dark dots on veins; antemedial double, sharply angled onto costa; medial and postmedial indistinguishable in at least four wavy lines curved onto costa outside discal dot; terminal area slightly darker with faint pale wavy line; discal dot black, elongate. Hind wing white with well marked brownish transverse lines along anal margin only; discal dot small.
Male genitalia. Uncus elongate, fairly broad, biapical. Valve unmodified. Vinculum short, rounded. Papillae elongate, very narrow. Aedeagus slim, shorter than length of valve. Vesica granulate at apex, armed with a

V-shaped cornutus and a smaller irregular cornutus near ductus ejaculatorius base. Sternite A8 very narrow, with two short pointed apical horns, and broadly bilobed at base.

Female genitalia. Bursa small, pear-shaped with two patches of dense small spines. Ductus bursae narrow, tapered. Ductus seminalis broad, curved around corpus bursae, attached to bursa centrally on left. Colliculum narrow and slightly elongate. Antrum narrow with numerous pores. Tergite A8 almost quadrate. Apophyses narrow. Papillae anales narrow, apically rounded.

Range. China (Taiwan).

Phenology. Univoltine: July to September.

Altitude. Found at about 2200 m above sea-level (Taiwan: Inoue 1970, 1988).

Similar species. Externally closest among Taiwan species to *E. masculina*, but easily separated by the white hind wings. The genitalia of both sexes are similar to those of *E. nishizawai*, but the male is distinguished by the differing shape of the eighth sternite, and the female by the presence of two patches of spines and the narrower colliculum.

Material. ♂, Formosa [Taiwan], Alishan, 2200 m, Chiayi, ix.1964, leg. Y.I. Chu, BM Geom. slide no. 16312♂ (holotype, coll. H. Inoue in BMNH); 1♂, 2♀, Taiwan, Alishan, Chiayi, 2200 m, ix.1967, Y.I. Chu, Inoue slides nos. 2035♂, 2036♀, 11205♀ (BMNH); 1♀, [Taiwan], Alishan, Chiayi, 7.viii.1974, Y. Shibata (BMNH); 1♂, Taiwan, Chiayi, 2–5.viii.1984, K. Yazaki, Inoue slide no. 13133♂ (BMNH); 2♀, Taiwan, Alishan, Chiayi, 11.viii.1963, R. Sato, BM Geom. slide no. 21824♀ (BMNH).

284. *Eupithecia tibetana* Mironov & Galsworthy, 2004

Eupithecia tibetana Mironov & Galsworthy, 2004, *Trans. lepid. Soc. Japan* 55 (1): 47, figs 6, 21, 23. Holotype ♂ (IZCAS; examined), [China]: Tibet, Chaqiansangjiu, 3000 m.

Description. Fore wing ground colour pale brown; basal area as far as medial line darker brown, basal and antemedial represented by short oblique lines on costa; medial passing through discal dot, sharply angled onto costa; discal dot elongate; postmedial double, the inner line consisting of dots on veins, hardly angled near costa; terminal area darker with traces of pale submarginal line. Hind wing off-white; discal dot small but distinct; transverse lines visible only on anal margin.

Male genitalia. Uncus rather short, biapical. Valve rather square, with a blunt asymmetrical ventral process, more pronounced on right valve. Vinculum short, rounded. Papillae elongate and narrow. Aedeagus slim. Vesica armed with a U-shaped cornutus and a small irregular plate near

ductus ejaculatorius base. Sternite A8 peg-like, bilobed at base, and with a small membranous apical hollow.
Female genitalia. Bursa elongate, basal half covered with small spines except for a small membranous area at left of base. Ductus seminalis attached centrally to bursa on right. Colliculum short and narrow. Antrum membranous. Apophyses narrow. Tergite A8 nearly quadrate. Papillae anales relatively large and broad.
Range. China (Tibet).
Phenology. Bivoltine (?): late June to late September.
Altitude. From about 2750 to 3100 m above sea-level in Tibet (China: coll. IZCAS).
Similar species. Slightly similar in general pattern to E. pusillata, but lacking strong transverse lines. The combination of valve shape and ornamentation of the aedeagus vesica will distinguish the male. The female genitalia are similar to those of the preceding species, but distinct from any other species in the Tibetan region.
Material. ♂, Tibet, Chaqiansangjiu, 3000 m, 25.vi.1973, coll. Huang Fusheng, slide no. ZL629♂ (holotype, IZCAS); 1♀, same loc., 3100 m, 1.viii.1973, slide no. ZL219♀; 1♀, Tibet, Bomi, Jiamu, 2750 m, 24.ix.1973, slide no. ZL418♀, both coll. Huang Fusheng (paratypes, IZCAS).

285. *Eupithecia tepida* Mironov & Galsworthy, 2004

Eupithecia ***tepida*** Mironov & Galsworthy, 2004, *Trans. lepid. Soc. Japan* 55 (2): 131, figs 12, 25. Holotype ♂ (ZFMK; examined), China: Sued-Shensi [South Shaanxi], Tapaishan [Taibaishan] im Tsinling [Qinling Mts.], 1700 m.

Description. Fore wing ground colour pale fawn, transverse lines very faintly marked; antemedial regularly curved from costa to inner margin; postmedial double, angled near costa; discal dot small and faint; terminal area darker. Hind wing ground colour as fore wing; discal dot faint; postmedial running from anal margin to edge of cell; a faint submarginal line running parallel to termen.
Male genitalia. Uncus broad, biapical. Valve unmodified. Vinculum narrow and rounded. Papillae rather short. Aedeagus slim, rather short. Vesica granulate, and armed with one small U-shaped cornutus. Sternite A8 peg-shaped, very narrow, broadly bilobed at base.
Female unknown.
Range. China (Shaanxi).
Altitude. The unique specimen is from 1700 m above sea-level in South Shaanxi (China: coll. ZFMK).

Similar species. The unique holotype is too worn to permit comparison with other species. The genitalia are very close to those of *E. chui* from Taiwan, except for the lack of a bifid apex to the eighth sternite.
Material. ♂, Sued-Shensi [Shaanxi], Tapaishan [Taibaishan] im Tsinling [Qinling], *ca.* 1700 m, 19.viii.1936, H. Höne, Vojnits slide no. 12388♂ (holotype, ZFMK).
Remarks. The unique holotype male is in very worn condition. It is possible that this species might prove to be conspecific with *E. chui*, but there is a large geographical separation, and further material would be needed to resolve this issue.

286. *Eupithecia nishizawai* Inoue, 1988

Eupithecia nishizawai Inoue, 1988, *Bull. Fac. domest. Sci. Otsuma Wom. Univ.* **24**: 364, figs 24a-e. Holotype ♂ (BMNH; examined), [Taiwan]: Kaohsiung Hsien, Paiyunshanchuang, Yushan, 3528 m.

Description. Fore wing broad, ground colour greyish brown, sometimes with weak reddish tinge; transverse lines black; basal and antemedial oblique, sharply angled onto costa; double medial sinuate, curved onto costa behind discal dot; postmedial more distinct, right angled on to costa, forming a series of black dashes on veins; terminal area with blackish shades near apex, at middle and near tornus, and a narrow, pale wavy subterminal line forming a small whitish tornal spot, discal dot relatively large, black. Hind wing paler; transverse lines marked along anal margin only; terminal area darker with weak trace of dentate subterminal line; discal dot small, sometimes invisible.
Male genitalia. Uncus broad, biapical. Valve unmodified. Vinculum broad, rounded. Papillae short. Aedeagus slim, shorter than valve length. Vesica armed with a U-shaped cornutus, and a smaller irregular cornutus near ductus ejaculatorius base. Sternite A8 peg-like, broadly bilobed at base, with two pointed apical horns.
Female genitalia. Bursa pear-shaped, covered basally with spines, which extend posteriorly on left. Ductus bursae short. Ductus seminalis broad, attached to ductus bursae near colliculum on right. Colliculum short and narrow. Antrum membranous. Tergite A8 transverse, with widened anterior margin. Apophyses narrow. Papillae anales rounded.
Range. China (Taiwan).
Phenology. Univoltine: early August to late September.
Altitude. From about 3000 to 3500 m above sea-level (Taiwan: Inoue 1988; coll. ZSM).

Similar species. Similar externally to *E. acutipapillata*, which has a more evenly curved fore wing postmedial and narrower terminal area, and a broader hind wing. The male and female genitalia are closer to those of *E. chui*. In the male the shape of the eighth sternite is the best distinction; in the female, the presence of a continuous spine field instead of two separate patches.

Material. ♂, [Taiwan], Mt. Yu Shan (3528 m), Paiyunshanchuang, Kaohsiung Hsien, Aug. 2–3, 1985, leg. M. Nishizawa, Inoue slide no. 12283♂ [= BM Geom. slide no. 16548♂] (holotype, coll. H. Inoue in BMNH); 2♂, 3♀, Taiwan, prov. Nantou, Hohuanshan Exp. Station, 3100 m, 24°09′ N, 121°17′ E, 15–16 and 26–28.ix.1999, leg. G. Csorba and B. Herczig (ZSM); 2♀, Central Taiwan, Taroko Nat. Park, Hohuanshan, 3000 m NN, 24°09′27″ N / 121°17′34″ E, LF, 12. Sept. 2002, leg. U. Buchsbaum (ZSM).

287. *Eupithecia praecipitata* Vojnits, 1979

Eupithecia praecipitata Vojnits, 1979, *Acta zool. Acad. Sci. hung.* **25** (3–4): 428, fig. 3. Holotype ♀ (ZFMK; examined), [China]: Sued-Shensi, Tapaishan in Tsinling, 1700 m.

Eupithecia iterata Vojnits, 1980, *Acta zool. Acad. Sci. hung.* **26** (4): 444, figs 13, 14, 23. Holotype ♂ (ZFMK; examined), [China]: Tibet [Sichuan], Batang, 3800 m. **New junior synonym.**

Eupithecia tempestuosa Vojnits, 1984, *Acta zool. Acad. Sci. hung.* **30** (1–2): 237, figs 31, 33. Holotype ♂ (TTM; examined), [China]: Tibet [Sichuan], Batang, 3800 m. **New junior synonym.**

Eupithecia laudenda Vojnits, 1984, *Acta zool. Acad. Sci. hung.* **30** (3–4): 538, figs 34, 45, 53 (*nec* 42, 50). Holotype ♂ (TTM; examined), [China]: Tibet [Sichuan], Batang, 3800 m. **New junior synonym.**

Misidentification: *Eupithecia ficta* Vojnits, 1984 (2♂ paratypes in TTM).

Description. Fore wing ground colour pale ash-grey; basal, ante- and postmedial very fine, sometimes absent altogether, forming small blackish costal spots, and a larger one on postmedial; latter usually more conspicuous in costal half, right angled onto costa; terminal area darker with relatively broad pale wavy subterminal line; discal dot small, rounded, black. Hind wing concolorous; dark transverse lines lightly marked along anal margin; terminal area slightly darker with weak trace of pale subterminal line; discal dot small, rounded.

Male genitalia. Uncus broadly triangular, biapical. Valve broad with pointed apex. Vinculum short, with medial hollow. Papillae short and narrow. Aedeagus slim, slightly shorter than length of valve. Vesica armed with a U-shaped cornutus and a small, irregular cornutus near ductus ejaculatorius base. Sternite A8 long and narrow, broadly bilobed basally, with thickened and sclerotized lateral margins near apex, and two short curling apical rods.

Female genitalia. Bursa globular, with ovoid patches of small spines medially on dorsal and ventral sides. Ductus bursae long and narrow, with many longitudinal striations. Ductus seminalis narrow, attached to posterior part of corpus bursae. Colliculum elongate and narrow. Antrum membranous. Tergite A8 quadrate, with sculptured anterior margin. Apophyses short and narrow. Papillae anales short and rounded.
Range. China (Tibet, Shaanxi, Sichuan, Yunnan).
Phenology. Univoltine: late July to late August.
Altitude. From about 1700 m in Taibaishan to about 3800 m in Tibet (Vojnits 1979b, 1984a, 1984b).
Similar species. In good condition, easily recognizable by the very fine transverse lines terminating in three blackish costal spots and the small, neat discal dot; worn specimens appear pale all over with a slightly darker terminal margin and the usual spots. In the male genitalia the unusual uncus shape and the eighth sternite are diagnostic. The female genitalia are unmistakeable.
Material. ♀, Sued-Shensi [Shaanxi], Tapaishan [Taibaishan] im Tsinling [Qinling], *ca.* 1700 m, 22.viii.1936, H. Höne, Vojnits slide no. 12306♀ (holotype, ZFMK); 1♀, same loc. and alt., 19.viii.1936, H. Höne, Vojnits slide no. 12383♀ (ZFMK); 1♂, same loc. (without alt.), F, 22.vi.1935, H. Höne, Vojnits slide no. 12315♂ (incorrectly associated with gen. slide of ♂ *E. robiginascens*, ZFMK); ♂, Tibet [Sichuan], Batang, Untere Urwaldzone (*ca.* 3800 m), 4.viii.1936, H. Höne, Vojnits slide no. 12396♂ (holotype of *E. iterata*, ZFMK); 2♂, same loc. and alt., 30.vii., 6.viii.1936, H. Höne, Vojnits slides nos. 12403♂, 12404♂ (paratypes of *E. iterata*, TTM); 2♂, 1♀, same loc. and alt., 2, 3, 6.viii.1936, H. Höne, Vojnits slides nos. 12436♂, 12415♂, 12413♀ (paratypes of *E. iterata*, ZFMK); 1♂, same loc. and alt., 1.viii.1936, H. Höne, Vojnits slide no. 13844♂ (holotype of *E. tempestuosa*, TTM); 1♀, same loc. and alt., 3.viii.1936, H. Höne, Vojnits slide no. 13835♀ (paratype of *E. tempestuosa*, TTM); 1♂, same loc. and alt., 4.viii.1936, H. Höne, Vojnits slide no. 13830♂ (holotype of *E. laudenda*, TTM); 5♂, 1♀, same loc. and alt., 4, 5, 6.viii., 17.ix.1936, H. Höne, Vojnits slides nos. 13217♂, 13219♀ (slide not found), 13225♂, 13821♂, 13829♂, 13836♂ (paratypes of *E. laudenda*, TTM); 2♂, same loc. and alt., 1, 2.viii.1936, H. Höne, Vojnits slides nos. 12414♂, 13222♂ (paratypes of *E. ficta*, TTM); 4♂, same loc. and alt., 5, 7.viii.1936, H. Höne, Vojnits slides nos. 13229♂, 13230♂, 13828♂ (incorrectly associated with gen. slide of *E. coniurata*), 13825♂ (ZFMK); 3♂, same loc. and alt., 2, 3, 7.viii.1936, H. Höne, Vojnits slides nos. 12420♂ (incorrectly associated with gen. slide of *E. coniurata*, TTM), 13834♂ (gen. slide absent in TTM and ZFMK also), 13840♂ (TTM); 1♀, Thibet, Ta-Ho, Chasseurs Indigènes,

Printemps 1895, ex-Oberthür coll., Brit. Mus. 1927–3, BM Geometrid slide no. 21971♀ (BMNH); 1♀, Sichuan, Batang Yidun, 3370 m, 17.viii.1982, slide no. ZL235♀, coll Wang Shuyong,; 1♂, Yunnan, Lijiang, Yulongshan, 21.7.1962, slide no. ZL224♂, coll Song Shimei (IZCAS).
Remarks. The holotypes of *E. praecipitata*, *E. iterata*, *E. tempestuosa* and *E. laudenda* are all very worn. According to the original label, the holotype of *E. tempestuosa* was collected on 1.viii.1936 (*nec* 1.viii.1931; Vojnits 1984). The shape of the eighth sternite of the male of *E. tempestuosa* published by Vojnits (1984a) is not visible on the slide.

288. *Eupithecia lasciva* Vojnits, 1980

Eupithecia lasciva Vojnits, 1980, *Acta zool. Acad. Sci. hung.* **26** (4): 446, figs 15, 16, 24. Holotype ♂ (ZFMK; examined), China: Sued-Shensi, Tapaishan in Tsinling.
Eupithecia fragmentaria Vojnits, 1984, *Acta zool. hung.* **30** (1–2): 227, fig. 19. Holotype ♂ (TTM; examined), [China]: North Yunnan, A-tun-tse [Dêqên], 4000 m. **New junior synonym.**
Eupithecia depressa Vojnits, 1984, *Acta zool. Acad. Sci. hung.* **30** (3–4): 536, figs 37, 43, 51 (*nec* 45, 53). Holotype ♂ (TTM; examined), [China]: North Yunnan, A-tun-tse [Dêqên], 4500 m. [Junior primary homonym of *Eupithecia depressa* Vojnits, 1979]. **New junior synonym.**
Eupithecia depressa disiuncta Vojnits, 1984, *Acta zool. Acad. Sci. hung.* **30** (3–4): 537, figs 36, 44, 52, 57. Holotype ♂ (TTM; examined), China: North Yunnan, Li-kiang. **New junior synonym.**
Eupithecia benigna Vojnits, 1984, *Acta zool. Acad. Sci. hung.* **30** (3–4): 541, figs 40, 48, 56, 61. Holotype ♂ (TTM; examined), [China]: Prov. Shansi [Shanxi], Mien-shan, 2000 m. **New junior synonym.**

Misidentifications: *Eupithecia seminuda* Vojnits, 1984 (5♀ paratypes in ZFMK; 4♀ paratypes in TTM); *Eupithecia subplacida* Vojnits, 1984 (1♂ paratype in TTM); *Eupithecia sublasciva* Vojnits, 1984 (1♂ paratype in ZFMK); *Eupithecia ficta* Vojnits, 1984 (5♂ paratypes in ZFMK; 7♂ paratypes in TTM).

Description. Fore wing relatively narrow, ground colour blackish grey, medial area under discal dot ochreous or reddish ochreous; transverse lines narrow, oblique, right angled onto costa, very indistinct; terminal area with inconspicuous pale subterminal line and small white tornal spot; discal dot small, brown, rounded. Hind wing off-white, darker along anal margin and near tornus, with indistinct transverse lines and a very small discal dot.
Male genitalia. Uncus bell-shaped, biapical. Valve unmodified. Vinculum rounded. Papillae short and narrow, tapered. Aedeagus slim, about valve length. Vesica armed with a V-shaped cornutus, and a compact, rounded cornutus at base. Sternite A8 broadly bilobed at base, with two broad, blunt lyre-shaped apical arms.
Female genitalia. Bursa pear-shaped, covered with small spines except small area on one or both sides. Ductus bursae elongate, tapered, with

numerous longitudinal striations and a large spine field. Ductus seminalis broad, curved around bursa, attached to ductus bursae on right. Colliculum elongate and narrow. Antrum membranous, covered by numerous pores. Tergite A8 transverse. Apophyses very stout, both pairs apically expanded and flattened. Papillae anales elongate and narrow, pointed apically, and heavily sclerotized.

Range. China (Tibet, Qinghai, Gansu, Shanxi, Shaanxi, Sichuan, Yunnan).

Phenology. Two generations: April and mid-June to mid-September.

Altitude. From about 1700 m in Qinling to about 4500 m in mountains of Yunnan (China: Vojnits 1984; coll. ZFMK, TTM).

Similar species. Similar externally to *E. batangi* in the reddish blotch under the discal dot on the fore wing, but the present species has ovoid or circular rather than linear discal dots on the fore wings and paler hind wings. The male genitalia are readily distinguished by the shape of the eighth sternite, combined with the uncus and ornamentation of the vesica. The female genitalia are best distinguished by the pointed papillae anales, which it shares with the next species (qv).

Material. ♂, China, Sued-Shensi [Shaanxi], Tapaishan [Taibaishan] im Tsinling [Qinling], L, 6.viii.1935, H. Höne, Vojnits slide no. 11423♂ (holotype, ZFMK); 1♂, same loc., 6.viii.1935, H. Höne, Vojnits slide no. 12372♂ (paratype, TTM); ♀ [*nec* ♂], same loc., 28.viii.1935, H. Höne, Vojnits slide no. 11424♂ (holotype of *E. sacrosancta*, wrongly associated with gen. slide of *E. sacrosancta*, ZFMK: Mironov et al. 2011); 2♀, same loc., 6.viii.1935, H. Höne, Vojnits slides nos. 12953♀, 12963♀ (TTM); 9♂, 1♀, same loc., *ca.* 1700 m, 5, 6, 7, 8, 9, 10.viii.1936, H. Höne, Vojnits slides nos. 11504♂, 11505♂, 12353♂, 11460♂, 11420♂, 11503♂, 12357♂, 11461♂, 12328♂, 11421♀ (paratypes, ZFMK); 1♂, 1♀, same loc. and alt., 8, 10.viii.1936, H. Höne, Vojnits slides nos. 11459♂, 11508♀ (paratypes, TTM); 1♂, 1♀, same loc. and alt., 9, 10.viii.1936, H. Höne, Vojnits slides nos. 13156♀, 13160♀ (paratypes of *Eva petulans*, both gen. slides belonging to females of *Eva flexa*, ZFMK); 7♂, 2♀, same loc. and alt., 8, 9, 10, 11, 12, 19.viii.1936, H. Höne, Vojnits slides nos. 12946♂, 12972♂, 12973♀, 13081♂, 13106♂, 13157♂ (wrongly associated gen. slide of *E. refertissima*), 13732♂ (TTM); 1♂, same loc. and alt., 9.viii.1936, H. Höne, Vojnits slide no. 13158♀ (paratype of *Eva flexa* with wrongly associated gen. slide, TTM); 1♂, same loc. and alt., 9.viii.1936, H. Höne, Vojnits slide no. 13155♂ (paratype of *Eva petulans* with wrongly associated gen. slide, TTM); 2♂, same loc. and alt., 8.viii.1936, H. Höne, Vojnits slides nos. 11487, 11488 (neither slide found in TTM); ♂ [*nec* ♀], same loc., *ca.* 3000 m, 10.viii.1936, H. Höne, Vojnits slide no. 13154♀ (holotype of *Eva petulans* with wrongly associate gen. slide of *Eva flexa*, ZFMK); 9♂, 1♀, same loc. and alt., 9, 11, 12,

25.viii.1936, H. Höne, Vojnits slides nos. 11439♂, 12352♂, 11432♂, 12323♂, 12324♂, 12325♂, 12329♂, 12332♂, 11422♀ (absent in ZFMK), 12376♂ (paratypes, ZFMK); 2♂, same loc. and alt., 11.viii.1936, H. Höne, Vojnits slides nos. 12326♂, 12327♂ (paratypes, TTM); 1♂, same loc. and alt., 11.viii.1936, H. Höne, Vojnits slide no. 12322♂, *liquida* Vojnits, det. Vojnits A. (with wrongly associated gen. slide of *E. albispumata*, ZFMK); 1♂, same loc. and alt., 19.viii.1936, H. Höne, Vojnits slide no. 12389♂ (with wrongly associated gen. slide of *E. amicula*, ZFMK); 3♂, 3♀, same loc. and alt., 10, 11, 12, 19. viii.1936, H. Höne, Vojnits slides nos. 10966♀, 12942♀, 12947♂, 13107♂, 13108♂, 13109♀ (TTM); ♂, Nord Yünnan, A-tun-tse [Dêqên], Mittlere Höhe (*ca.* 4000 m), 17.viii.1936, H. Höne, Vojnits slide no. 13308♂ (holotype of *E. fragmentaria*, with wrongly associated gen. slide of *E. sinicaria*, TTM); 1♂, same loc. and alt., 16.viii.1936, H. Höne, Vojnits slide no. 13302♂ (paratype of *E. subplacida*, TTM); 1♂, same loc. and alt., 2.ix.1936, H. Höne, Vojnits slide no. 12168♂ (paratype of *E. depressa* Vojnits, 1984, ZFMK); 2♂, same loc. and alt., 13.vi., 16.viii.1936, H. Höne, Vojnits slides nos. 11533♂, 12463♂ (paratypes of *E. depressa* Vojnits, 1984, TTM); ♂, same loc., Obere Höhe (*ca.* 4500 m), 14.vii.1936, H. Höne, Vojnits slide no. 12153♂ (holotype of *E. depressa* Vojnits, 1984, TTM); 6♂, same loc. and alt., 18, 26, 28.vii., 10, 13.viii.1936, H. Höne, Vojnits slides nos. 12157♂, 12134♂, 14096♂, 13950♂, 12154♂, 12184♂ (paratypes of *E. depressa* Vojnits, 1984, ZFMK); 6♂, 1♀, same loc. and alt., 7.vii., 2, 6, 11, 13.viii.1936, H. Höne, Vojnits slides nos. 12133♂, 12155♂, 12166♂, 12177♂, 12506♂, 13027♀, 13937♂ (paratypes of *E. depressa* Vojnits, 1984, TTM); 2♀, same loc. and alt., 9.viii.1936, 31.vii.1937, H. Höne, Vojnits slides nos. 13319♀, 13745♀ (paratypes of *E. seminuda*, ZFMK); 6♂, 2♀, same loc. and alt., 7.vii., 6, 7, 13, 14.viii.1936, 17.vii.1937, H. Höne, Vojnits slides nos. 12492♂(wrongly associated gen. slide of *E. refertissima*), 13243♂, 12173♀, 13246♂, 12488♂ (wrongly associated gen. slide of *E. refertissima*); 13245♂, 12519♀, de Laever slide no. 300♂ (ZFMK); 1♀, same loc. and alt., 9.viii.1936, H. Höne, Vojnits slide no. 11544♀ (TTM); 1♂, same loc., Talsohle (*ca.* 3500 m), 10.vii.1936, H. Höne, Vojnits slide no. 13943♂ (paratype of *E. E. depressa* Vojnits, 1984, ZFMK); 1♂, same data, Vojnits slide no. 12152♂ (paratype of *E. depressa* Vojnits, 1984, TTM); 1♂, same loc. and alt., 27.vi.1936, H. Höne, de Laever slide no. 85♂ (ZFMK); 1♂, [Sichuan], Sungpan-ting, Szetschwan-China, Exped. Stötzner, Vojnits slide no. 12530♂ (paratype of *E. sublasciva*, ZFMK); 1♂, same data, Coll. Dr. Wehrli, Basel, Vojnits slide no. 13207♂ (ZFMK); 5♂, Tibet [Sichuan], Batang, Untere Urwaldzone (*ca.* 3800 m), 5, 6, 7, 9.viii., 5.x.1936, H. Höne, Vojnits slides nos. 13905♂, 13885♂, 13903♂, 13904♂, 13845♂ (paratypes of *E. ficta*, ZFMK); 6♂, same loc. and alt., 10, 12, 13.vi., 4, 6.viii.1936, H. Höne, Vojnits slides nos. 12199♂, 12201♂, 12406♂, 12408♂, 12434♂, 13797♂ (paratypes of *E. ficta*, TTM); 3♀, same loc. and alt., 2, 10, 12.viii.1936, H. Höne, Vojnits slides nos. 13264♀, 13851♀, 13863♀ (paratypes of *E. seminuda*, TTM); 2♂, 3♀, same loc. and alt., 19.vi., 1, 4, 14.viii., 1.x.1936, H. Höne, Vojnits slides nos. 12428♀, 13824♀, de Laever slides nos. 129♂, 443♂, 119♀ (ZFMK); 1♂, same loc., Im Tal des Yangtze (*ca.* 2800 m), 30.viii.1936, H. Höne, Vojnits slide no. 13220♂ (paratype of *E. ficta*, TTM); 1♀, same loc. and alt., 25.ix.1936, H. Höne, Vojnits slide no. 12392♀ (ZFMK); 1♂, Prov. Shansi [Shanxi], Mien-shan, Obere Höhe *ca.* 2000 m, 29.vii.1937, H. Höne, Vojnits slide no. 12192♂ (paratype, TTM); ♂, same loc. and alt., 3.viii.1937, H. Höne, Vojnits slide no. 13194♂ (holotype of *E. benigna*; gen. slide designated as holotype by Mironov, TTM); 4♂, 3♀, same loc. and alt., 29.vii., 7, 10, 14.viii.1937, H. Höne, Vojnits slides nos. 13203♂, 13198♂, 13202♀, 13740♂, 13200♂, 13199♀, 13201♀ (paratypes of *E. benigna*, ZFMK); 4♂, 2♀, same loc. and alt., 31.vii., 7, 8, 10, 14.viii.1937, H. Höne, Vojnits slides nos. 11529♀, 13191♀, 13192♂, 13193♂, 13195♂, 13196♂ (paratypes of *E. benigna*; gen. slides nos. 13191♀, 13192♂ designated as paratypes by Mironov, TTM); 3♀, same loc. and alt., 7, 8.viii.1937, H. Höne, Vojnits slides nos. 13738♀, 13742♀, 13267♀ (paratypes of *E. seminuda*, ZFMK); 1♀, same loc. and alt., 14.viii.1937, H. Höne, Vojnits slide no. 13736♀ (paratype of *E. seminuda*, TTM); 1♀, same loc. and alt., 7.viii.1937, H. Höne, Vojnits slide no. 12194♀ (ZFMK); 1♂, 1♀, same loc. and alt., 31.vii., 14.viii.1937, H. Höne, Vojnits slides nos. 10934♀, 13197♂ (TTM); ♂, China, Provinz Nord-Yuennan, Li-kiang [Lijiang], 1.ix.1934, H. Höne, Vojnits slide no. 13787♂ (holotype of *E. depressa disiuncta*, TTM); 4♂, same loc., 28.viii., 13.ix.1934, 25.viii., 5.ix.1935, H. Höne, Vojnits slides nos. 12161♂, 13169♂, 13183♂, 13072♂ (paratypes of *E. depressa disiuncta*, ZFMK); 4♂, same loc., 28, 29.viii., 3.ix.1934, 15.viii.1935, H. Höne, Vojnits slides nos. 12976♂, 12978♂, 13028♂, 13071♂ (paratypes of *E. depressa disiuncta*, TTM); 2♂, 2♀, same loc., 29.viii., 16, 18.ix.1934, H. Höne, de Laever slides nos. 229♂, 134♂, 262♀, 132♀ (ZFMK); 1♀, same loc., 31.viii.1934, H. Höne, Mironov slide no. 828♀ (TTM); 1♂, same loc., *ca.* 3000 m, 5.ix.1934, H. Höne, Vojnits slide no. 12984♂ (paratype of *E. depressa disiuncta*, TTM); 1♀, China, Sichuan, Daxue Shan Mts., 80 km W. Kangding, 30°01′ N, 101°40′ E, H, 3800 m, 26.–27.vii.1999, lg. V. Siniaev & A. Plutenko, ex coll. Viktor Siniaev, Mironov slide no. 727♀ (ZFMK); 1♂, Szetschwan [Sichuan], Kunkalaschan, *Eupithecia endolampera* Wrli. typ, Type [on the separate red label], *Eupithecia endolampera* Wrli., Type, Prep. № 1494.2 [unpublished name], de Laever det. 1954 (specimen in TTM; its genitalia preparation in ZFMK); 1♂, China, W. Sichuan, road Yajiang / Litang, 3600 m, 15.vii.2009, 29°59′ N / 100°52′ E, I. & A. Floriani, Saldaitis leg., Pekarsky slide no. OP 0530♂ (coll. O. Pekarsky);

IZCAS material from China: Tibet: 3♂, Jomda, 3400 m, 26–27.vii.1976, coll Zhang Xuezhong and Han Yinheng, slide no. ZL628♂; 1♂, Bomi, 2300 m, 22.viii.1983, coll Han Yinheng, slide no. ZL225♂; Qinghai: 6♂, 5♀, Tongren, Maixiu, 2950 m, 30.vii.1992, slides nos. ZL304, 365, 537♂; 7♂, 2♀, Huzhu, Beishan, 2300 m, 30.vii.1991, coll Xue Dayong, slides nos. ZL68♂, 69, 481♀; 3♂, 1♀, Huzhu, Beishan Linchang, Langshidang, 2600 m, 36°55′ N, 102°23′ E, 7.viii.2005, Xue Dayong and Han Hongxiang, slides no. ZL699, 714♂, 704♀; Gansu: 13♂, 2♀, Yongdeng, Tulugou, 2280 m, 25–29.vii.1991, coll Xue Dayong, slides nos. ZL394, 480, 517♂; 2♂, 2♀, same loc 24–29.iv.1992, slide no. ZL312♂, 516♀; 1♂, 2♀, same loc, 20–26.vii.1991, slide no. ZL616♂; 1♂, same loc, 11–16.viii.1991, all coll Meng Feng; 1♂, Wenxian, Qiujiaba, 2350 m, 21.vii.1999, coll Yao Jian; 1♂, Tianshui, Dongfen Linchang, Taohuaping, 990 m, 15–16.iv.2005, Wang Xuejian; 2♂, 20♀, Yongdeng, Liancheng Linchang, Tulugou, 2600 m, 36°41′ N 102°42′ E, 7–8.viii.2005, Xue Dayong, slide no.709♂; Shaanxi: 1♀, Ningshan Huoditang, 1580 m, 30.vi.1999, Yuan Decheng; Sichuan: 1♂, Pingwu, Wanglang, 2560 m, 9–11.viii.1999, Zhou Xin, slide no. ZL681♂.

Remarks. The holotype of *E. sacrosancta* published by Vojnits in 1979 consists of a female specimen of the present species linked to a male genitalia slide. Mironov et al. (2011) designated the male genitalia slide as the holotype of *E. sacrosancta*, thus removing the need to synonymize *E. lasciva* with it. The ductus bursae shown in Vojnits' drawing of the female genitalia is in the wrong place, and entirely imaginary.

There is great confusion in the material of this species worked on by Dr. Vojnits; for the convenience of future researchers, we list here the problems we have found:

Vojnits slides nos. 12201♂, 12406♂, 12434♂, which belong to paratypes of *E. ficta*, were marked by Dr. Vojnits as paratypes of *E. sublasciva*.

Vojnits slides nos. 12463♂, 12506♂, which belong to paratypes of *E. depressa* Vojnits, 1984, were marked by him as paratypes of *E. sublasciva*. Slide no. 11533♂ was not published as as a paratype of *E. sublasciva* but was marked by him as such.

Vojnits slide no. 12161♂, which belongs to a paratype of *E. depressa disiuncta* Vojnits, 1984, was marked by him as a paratype of *E. sublasciva*.

Vojnits slides nos. 13191♀, 13192♂, 13193♂, 13195♂, 13196♂ (TTM), which were published as paratypes of *E. benigna*, were marked by Dr. Vojnits as paratypes of *E. lasciva*. He had earlier (1980b) published determinations of all but 13193 as *E. lasciva*, but not as paratypes. Slide no. 11529♀ was not published as a paratype of *E. lasciva* but was so marked, presumably in error for 12529♀, which was.

Vojnits slides nos. 13198♂, 13199♀, 13200♂ were published as paratypes of *E. benigna*, but not marked by Vojnits; all were marked by him as *E. lasciva*, and had been published by him as such.

Vojnits slides nos. 11487, 11488, 12408♂, 13191♀, 13192♂, 13197♂ were not found in TTM (Budapest).

289. *Eupithecia egena* Vojnits, 1984

Eupithecia egena Vojnits, 1984, *Acta zool. Acad. Sci. hung.* **30** (1–2): 213, fig. 1. Holotype ♂ (TTM; examined), [China]: North Yunnan, A-tun-tse [Dêqên], 4000 m.

Eupithecia seminuda Vojnits, 1984, *Acta zool. Acad. Sci. hung.* **30** (1–2): 217, figs 5, 8. Holotype ♂ (TTM; examined), [China]: Tibet [Sichuan], Batang, 3800 m. **New junior synonym.**

Eupithecia sublasciva Vojnits, 1984, *Acta zool. Acad. Sci. hung.* **30** (3–4): 533, figs 30, 33, 41, 49. Holotype ♂ (TTM; examined), [China]: North Yunnan, A-tun-tse [Dêqên], 4000 m. **New junior synonym.**

Eupithecia ficta Vojnits, 1984, *Acta zool. Acad. Sci. hung.* **30** (3–4): 535, figs 31, 35, 43, 51. Holotype ♂ (TTM; examined), [China]: Tibet [Sichuan], Batang, Tal des Yangtze, 2800 m. **New junior synonym.**

Eupithecia anteacta Vojnits, 1984, *Acta zool. Acad. Sci. hung.* **30** (3–4): 543, figs 63–65. Holotype ♂ (TTM; examined), [China]: North Yunnan, A-tun-tse [Dêqên], 4000 m. **New junior synonym.**

Misidentifications: *Eupithecia fatigata* Vojnits & de Laever, 1978 (1♂ paratype in ZFMK); *Eupithecia inexhausta* Vojnits, 1984 (1♀ paratype in ZFMK); *Eupithecia profana* Vojnits, 1984 (1♂ paratype in TTM); *Eupithecia depressa disiuncta* Vojnits, 1984 (1♂ paratype in ZFMK).

Description. Fore wing ground colour grey-brown; transverse lines distinct; antemedial double, right angled onto costa; medial double, sinuate, sharply curved onto costa behind discal dot; postmedial separated by a darker fascia, double and twice angled near costa; terminal area with vague pale wavy subterminal line; discal dot small, slightly elongate, sometimes with a slightly reddish patch below it. Hind wing off-white; transverse lines distinct on anal margin; discal dot very small.

Male genitalia. Uncus bell-shaped, biapical. Valve unmodified. Vinculum rounded. Papillae rather short, tapered. Aedeagus narrow, shorter than length of valve. Vesica armed with a weak apical cornutus, a U-shaped cornutus and an irregular cornutus near ductus ejaculatorius base. Sternite A8 broadly bilobed at base, with two broad apical arms, each terminating in a sclerotized cap.

Female genitalia. Bursa globular, covered with small spines except left side and posterior part. Ductus bursae membranous, with a short diverticulum on left. Ductus seminalis broadened at base, attached to ductus bursae on right. Colliculum short and narrow. Antrum membranous. Tergite A8 trapezoid, covered with short setae at posterior margin. Apophyses

long and very stout, broadened and flattened at apices. Papillae anales long and pointed at apices.
Range. China (Tibet, Gansu, Sichuan, Yunnan); India (Himachal Pradesh), Nepal.
Phenology. Univoltine (?): mid-July to mid-September.
Altitude. From about 1800 to 4000 m (Nepal: coll. W. Schäfer, Stuttgart; China: Vojnits 1984).
Similar species. Very similar to the preceding species, but distinguished by larger size, much more distinct transverse lines, and either a smaller or no reddish patch on the fore wing. The presence of a weakly sclerotized apical cornutus, sometimes hidden behind the sclerotized apical section of the vesica, and the different shape of the eighth sternite distinguish the male genitalia. The female genitalia have a much smaller bursa, and a shorter, narrower ductus seminalis.
Material. ♂, Nord Yünnan, A-tun-tse [Dêqên], Mittlere Höhe (*ca.* 4000 m), 6.ix.1936, H. Höne, Vojnits slide no. 13755♂ (holotype, TTM); ♂, same loc. and alt., 11.viii.1936, H. Höne, Vojnits slide no. 14197♂ (holotype of *E. sublasciva*, TTM); ♂, same loc. and alt., 15.viii.1936, H. Höne, Vojnits slide no. 14099♂ (holotype of *E. anteacta*, TTM); 1♂, same loc. and alt., 6.ix.1936, H. Höne, Vojnits slide no. 10962♂ (paratype of *E. fatigata*, ZFMK); 1♀, same loc. and alt., 30.vii.1936, H. Höne, Vojnits slide no. 14095♀ (paratype of *E. inexhausta*, ZFMK); 7♂, same loc. and alt., 3, 17, 18, 30.viii., 1, 3, 4.ix.1936, H. Höne, Vojnits slides nos. 13333♂, 13331♂, 13329♂, 13752♂, 13749♂, 13781♂, 13778♂ (paratypes of *E. sublasciva*, ZFMK); 9♂, same loc. and alt., 11, 12, 13, 14, 16, 17, 30.viii.1936, H. Höne, Vojnits slides nos. 13280♂, 13301♂, 13304♂, 13317♂, 13320♂, 13322♂, 13918♂, 14054♂, 14156♂ (paratypes of *E. sublasciva*, TTM); 3♂, 9♀, same loc. and alt., 31.vii., 5, 11, 15, 16, 17, 18.viii., 6.ix.1936, H. Höne, Vojnits slides nos. 13334♀, 13287♀, 13955♀, 14179♀, 13300♀, 11537♂, 11566♂, 14178♀, 11002♀, 12455♂, 13784♀, 14198♀ (ZFMK); 1♂, same loc. and alt., 15.viii.1936, H. Höne, Mironov slide no. 822♂ (TTM); 1♂, same loc., Talsohle (*ca.* 3500 m), 27.vi.1936, H. Höne, Vojnits slide no. 13756♂ (paratype of *E. sublasciva*, ZFMK); 1♂, same loc. and alt., 12. vii.1936, H. Höne, Vojnits slide no. 12449♂ (ZFMK); 1♂, same loc., Obere Höhe (*ca.* 4500 m), 9.viii.1936, H. Höne, Vojnits slide no. 13782♂ (paratype of *E. sublasciva*, ZFMK); 1♂, same loc. and alt., 6.viii.1936, H. Höne, Vojnits slide no. 13327♂ (paratype of *E. sublasciva*, TTM); 1♀, same loc. and alt., 29.vii.1937, H. Höne, Vojnits slide no. 13124♀ (ZFMK); ♂, Tibet [Sichuan], Batang, Untere Urwaldzone (*ca.* 3800 m), 2.ix.1936, H. Höne, Vojnits slide no. 13858♂ (holotype of *E. seminuda*, TTM); 1♂, same loc. and alt., 3. viii.1936, H. Höne, Vojnits slide no. 13899♂ (paratype of *E. ficta*, ZFMK);

4♂, 9♀, same loc. and alt., 1, 14.viii., 2, 10, 12, 15, 17, 22, 28, 29.ix., 3, 5.x.1936, H. Höne, Vojnits slides nos. 12427♂, 13889♀, 13843♀, 13799♀, 14050♀, 12431♂, 13860♀, 13866♀, 13891♀, 12426♀, de Laever slides nos. 441♂, 444♀, 453♂ (ZFMK); ♂, same loc., Im Tal des Yangtze (*ca.* 2800 m), 27.viii.1936, H. Höne, Vojnits slide no. 13906♂ (holotype of *E. ficta*, TTM); 10♂, same loc. and alt., 25, 26, 29, 30.viii., 1, 3, 18, 22.ix.1936, H. Höne, Vojnits slides nos. 13880♂, 13801♂, 13862♂, 13898♂, 13913♂, 13826♂, 13881♂, 13884♂, 13886♂, 13876♂ (paratypes of *E. ficta*, ZFMK); 5♂, same loc. and alt., 24, 28.viii., 2, 3.ix.1936, H. Höne, Vojnits slides nos. 12412♂, 12198♂, 13798♂, 13804♂ 13792♂ (paratypes of *E. ficta*, TTM); 4♂, 32♀, same loc. and alt., 25, 26, 27, 28, 29, 30, 31.viii., 1, 2, 3, 4, 12, 13, 16, 18, 21, 22, 23, 29.ix.1936, H. Höne, Vojnits slides nos. 13909♀, 13871♀, 13794♀, 13853♀, 13901♀, 13809♀, 13807♀, 14044♀, 13795♀, 13868♀, 13910♀, 12435♀, 13791♀, 13808♀, 14045♀, 13776♀, 13813♀, 13911♀, 13865♀, 13867♀, 13793♀, 13870♀, 14041♀, 13800♀, 13859♀, 13895♀, 12433♀, 12429♂, 13833♀, de Laever slides nos. 435♂, 436♀, 450♀, 458♀, 434♂, 439♂, 197♀ (ZFMK); 1♂, China, Provinz Nord-Yuennan, Li-kiang [Lijiang], 14.vi.1934, H. Höne, Vojnits slide no. 14201♂ (paratype of *E. profana*, TTM); 1♂, same loc., 11.viii.1935, H. Höne, Vojnits slide no. 13256♂ (paratype of *E. depressa disiuncta*, ZFMK); 1♀, same loc., 3.ix.1935, H. Höne, Vojnits slide no. 12475♀ (ZFMK); 1♂, same loc. and date, H. Höne, Vojnits slide no. 13258♂ (TTM); 1♂, same loc., 21.viii.1935, H. Höne, Mironov slide no. 791♂ (SMNK); 1♀, Tibet, 72) Auf Wiesen im Buschland bei Jekundo, 97 öL, 33 nB, 4300 m, 7.viii.1935, H. Höne, Vojnits slide no. 13127♀ (ZFMK); 4♂, 3♀, Tibet, Jomda, 3400 m, 22–31.vii.1976, coll Zhang Xuezhong and Han Yinheng, slides nos. ZL115♂, 102, 336♀; 2♂, 3♀, Tibet, Zhag'yabGyitang, 3600 m, 20 and 27–28.viii.1976, coll Han Yinheng, slides nos. ZL326, 330♂, 181,188, 621♀; 1♂, Gansu, Yongdeng Tulugou, 25.vii.1991, coll Xue Dayong; 2♀, Gansu, Diebu, Lazikou, 2000 m, 5.ix.1990, coll Xue Dayong, slide no. ZL384♀ (IZCAS).

Remarks. The apical cornutus in the vesica is important in the identification of this species, but may be no more than a sclerotized edge to a patch of light sclerotization near the apex. In one specimen (Vojnits slide no. 12429♂) there appear to be two such cornuti.

290. *Eupithecia lunatica* Vojnits, 1984

Eupithecia lunatica Vojnits, 1984, *Acta zool. Acad. Sci. hung.* **30** (3–4): 539, figs 39, 46, 54 (*nec* 45, 55), 59. Holotype ♂ (TTM; examined), [China]: North Yunnan, A-tun-tse [Dêqên], 4000 m.

Misidentification: *Eupithecia epileptica* Vojnits, 1984 (1♂ paratype in TTM).

Description. Fore wing ground colour reddish brown; dark brown transverse lines oblique, sharply angled onto costa where broadened, forming blackish costal spots; postmedial wavy, with dark dashes on veins; terminal area with veins covered with black scales, and a distinct white wavy subterminal line broadening to a large white tornal spot; discal dot large, rounded, black. Hind wing slightly paler, ante- and postmedial distinct, strongest along anal margin; terminal area slightly darker, with white subterminal line and tornal spot; discal dot distinct.
Male genitalia. Uncus bell-shaped, with probably uniapical tip. Valve long and rather rounded. Vinculum narrow, rounded. Papillae broad, tapered, quite long. Aedeagus stout, shorter than length of valve. Vesica granulate near apex, armed with a V-shaped cornutus and an irregular cornutus near ductus ejaculatorius base. Sternite A8 very difficult to see on Vojnits' poorly prepared slides, but apparently consisting of two long arms narrowly joined at base.
Female genitalia. Bursa small, globular, covered with small spines in basal half, except small area on right side. Ductus bursae elongate, tapered, with numerous longitudinal striations. Ductus seminalis broadened at base, attached to basal part of ductus bursae on right. Colliculum long and narrow. Antrum membranous. Tergite A8 very elongate. Apophyses expanded and flattened at apices, the posterior very long. Papillae anales long and very narrow.
Range. China (Sichuan, Yunnan).
Phenology. Univoltine (?): early July to early September.
Altitude. From about 3000 to 4500 m above sea-level (China: coll. O. Pekarsky; Vojnits 1984b).
Similar species. Externally similar both to preceding and following species, but with much stronger pale tornal spots. The male genitalia are very close to those of the following species, but have a much shorter plate on the aedeagus vesica. Like the next three species, the female genitalia have the apophyses and papillae enormously elongate, but in the present species the bursa is much smaller, and there are no spines in the ductus bursae.
Material. ♂, Nord Yünnan, A-tun-tse [Dêqên], Mittlere Höhe (*ca.* 4000 m), 5.vii.1936, H. Höne, Vojnits slide no. 14154♂ (holotype, TTM); 1♂, same data, H. Höne, Vojnits slide no. 14100♂ (paratype, ZFMK); 1♀, same loc. and alt., 22.vii.1936, H. Höne, Vojnits slide no. 13946♀ (a paratype of *E. placida*, wrongly associated with a gen. slide of *E. fenita*, TTM); 2♂, same loc., Obere Höhe (*ca.* 4500 m), 9, 26.vii.1936, H. Höne, Vojnits slides nos. 14153♂, 14116♂

(paratypes, ZFMK); 1♂, 1♀, same loc. and alt., 19.vii.1936, 17.vii.1937, H. Höne, Vojnits slides nos. 13744♂, 13929♀ (paratypes, TTM); 1♀, same loc. and alt., 23.vii.1936, H. Höne, Vojnits slide no. 11549♂ (wrongly associated with gen. slide of *E. caduca*, TTM); 1♂, China, Provinz Nord-Yuennan, Li-kiang [Lijiang], 4.ix.1935, H. Höne, Vojnits slide no. 13783♂ (paratype of *E. epileptica*, TTM); 1♀, China, W. Sichuan, Kangding, 3000 m, 29°53′ N / 101°55′ E, 13.vii.2009, I. & A. Floriani, Saldaitis leg., Pekarsky slide no. OP 0680♀ (coll. O. Pekarsky); 1♀, China, N. Yunnan, road Xinngcheng / Zhongdian, 3500 m, 28°32′ N / 99°49′ E, 18.vii.2009, I.& A. Floriani, Saldaitis, Pekarsky slide no. OP 0812♀ (coll. O. Pekarsky).

291. *Eupithecia otiosa* Vojnits, 1981

Eupithecia otiosa Vojnits, 1981, *Acta zool. Acad. Sci. hung.* **27** (3–4): 414, figs 24–26. Holotype ♂ (ZFMK; examined), [China]: Tibet [Sichuan], Batang, 3800 m.
Eupithecia insana Vojnits, 1984, *Acta zool. Acad. Sci. hung.* **30** (1–2): 232, figs 24, 27. Holotype ♂ (TTM; examined), [China]: Tibet [Sichuan], Batang, Tal des Yangtze, 2800 m. **New junior synonym.**
Eupithecia nodosa Vojnits, 1984, *Acta zool. Acad. Sci. hung.* **30** (3–4): 541, figs 38, 47, 55 (*nec* 46, 53 and 54), 60. Holotype ♂ (TTM; examined), [China]: North Yunnan, A-tun-tse [Dêqên], 4500 m. **New junior synonym.**

Description. Fore wing ground colour fuscous greyish brown. Transverse lines distinct, antemedial sharply angled onto costa; postmedial right or obtuse angled onto costa; medial fascia with discal dot slightly darker; double medial and stronger postmedial evenly curved onto costa outside discal dot; terminal area darker with inconspicuous pale subterminal line; discal dot large, ovoid. Hind wing slightly paler than fore wing, with distinct transverse lines, strongest along anal margin and complete only outside small rounded discal dot.
Male genitalia. Uncus bell-shaped, biapical. Valve unmodified, with rather narrow apex. Vinculum rounded. Papillae unusually long and narrow. Aedeagus very stout, shorter than length of valve. Vesica armed with a long V-shaped cornutus with pointed apex, often with an additional small cornutus in apical half, and an irregular cornutus near ductus ejaculatorius base. Sternite A8 two long parallel rods, with expanded and rounded tips, connected by a narrow band at base.
Female genitalia. Bursa large, ovoid, anterior part and left side covered with spines. Ductus bursae elongate, tapered, longitudinally striate, and with a transverse band of spines at junction with corpus bursae, and a row of spines extending down into ductus seminalis. Latter very broad, curved around corpus bursae, attached to upper part of corpus bursae on right. Colliculum narrow and elongate, tapered. Antrum slightly

sclerotized. Tergite A8 elongate. Apophyses very broad, expanded and flattened at apices, the posterior extremely long. Papillae anales narrow and elongate.

Range. China (Tibet, Qinghai, Gansu, Sichuan, Yunnan).

Phenology. Bivoltine (?): mid-June to early October.

Altitude. From about 2400 to 4500 m above sea-level in Tibet (China: Vojnits 1981b, 1984a, IZCAS).

Similar species. Externally very similar to specimens of *E. egena* which lack the reddish fore wing patch, but with less sharply angled postmedial. The male genitalia are very similar to those of *E. lunatica* and *E. supercastigata*, but have longer papillae, and a much longer plate in the aedeagus vesica than either. The female genitalia are easily recognised by the row of spines extending into the ductus seminalis.

Material. ♂, Tibet [Sichuan], Batang, Untere Urwaldzone (*ca.* 3800 m), 15.vi.1936, H. Höne, Vojnits slide no. 12926♂ (holotype, ZFMK); 3♂, same loc. and alt., 2.viii.1936, H. Höne, de Laever slide no. 457♂, Vojnits slide no. 12927♂ (paratypes, TTM); 1♂, 2♀, same loc. and alt., 2, 7.viii., 6.x.1936, H. Höne, Vojnits slides nos. 13811♂, 13812♀, 13847♀ (paratypes of *E. insana*, ZFMK); 2♀, same loc. and alt., 12.viii., 2.ix.1936, H. Höne, Vojnits slides nos. 13854♀, 14049♀ (paratypes of *E. insana*, TTM); ♂, same loc. and alt., 5.x.1936, H. Höne, Vojnits slide no. 12422♂ (holotype of unpubl. *E. nonsuspecta* Vojnits, ZFMK); 1♂, 2♀, same loc. and alt., 7, 16.viii, 12.ix.1936, H. Höne, de Laever slides nos. 75♀, 25♂, 442♀ (ZFMK; slide no. 25♂ belongs to another species, and is wrongly associated); ♂, same loc., Im Tal des Yangtze (*ca.* 2800 m), 28.viii.1936, H. Höne, Vojnits slide no. 13802♂ (holotype of *E insana*, TTM); 1♀, same loc. and alt., 16.ix.1936, H. Höne, Vojnits slide no. 13861♀ (paratype of *E. insana*, ZFMK); 1♀, same loc. and alt., 26.ix.1936, H. Höne, Vojnits slide no. 12928♀ (TTM); ♂, Nord Yünnan, A-tun-tse [Dêqên], Obere Höhe (*ca.* 4500 m), 25.viii.1936, H. Höne, Vojnits slide no. 14180♂ (holotype of *E. nodosa*, TTM); 1♀, same loc., Mittlere Höhe (*ca.* 4000 m), 31.viii.1936, H. Höne, Vojnits slide no. 14089♀ (paratype of *E. nodosa*, TTM); 1♂, same loc. and alt., 18.viii.1936, H. Höne, Vojnits slide no. 12929♂ (paratype of *E. otiosa*, ZFMK); 1♂, [Yunnan], Tsekou, [Dêqên], P. Dubernard 1898, Vojnits slide no. 874♂ (TTM); 9♂, 1♀, China, W. Sichuan, road Xinngcheng / Zhongdian, 3300 m, 17.vii.2009, 28°46′ N / 99°51′ E, I. & A. Floriani, Saldaitis leg. (coll. O. Pekarsky); 1♀, China, W. Sichuan, road Yajiang / Litang, 3600 m, 15.vii.2009, 29°59′ N / 100°52′ E, I. & A. Floriani, Saldaitis leg., Pekarsky slide no. OP 0528♀ (coll. O. Pekarsky); 2♂ 1♀, Tibet, Jomda, 3400 m, 25–26.vii.1976, coll Zhang Xuezhong, slides no. ZL101♂, 133♀; 1♀, Tibet, Markam, Haitong, 3250 m, 8.viii.1982, coll Wang Shuyong, slide no. ZL116♀; 23♂, 10♀, Qinghai, Tongren, Maixiu, 2950 m, 30.vii.1992, coll Xue Dayong, slides nos. ZL157, 178, 300♂, 240♀; 1♀, Gansu,Yongdeng Tulugou, 2280 m, 20.vii.1991, coll Xue Dayong; 1♀, Gansu, Dangchang, 1800 m, 7.vii.1998, coll Liu Dajun; 1♂, Gansu, Zhouqu Shatan forestry reserve, 2400 m, 14.vii.1999, coll Yao Jian, slide no. ZL578♂; 1♀, Sichuan, Batang, 2600 m, 14.viii.1982, slide no. ZL167♀; 1♀, Sichuan,Yajiang, 3300 m, 29. viii.1982, coll Wang Shuyong, slide no. ZL232♀ (IZCAS).

292. *Eupithecia acutipapillata* Inoue, 1988

Eupithecia acutipapillata Inoue, 1988, *Bull. Fac. domest. Sci. Otsuma Wom. Univ.* **24**: 340, figs 11a–c. Holotype ♀ (BMNH; examined), [Taiwan]: Hualien Hsien, Tayuling, 2600 m.

Description. Fore wing broad, ground colour dark brown; transverse lines inconspicuous; medial double, tinged ochreous, slightly sinuate, angled onto costa outside discal dot; postmedial faint, doubly angled onto costa and with a series of dark dashes on veins; terminal area with narrow pale subterminal line and small dot near tornus; discal dot fairly large, blotchy. Hind wing paler, progressively more brownish towards anal margin; transverse lines marked on veins as dark dashes; terminal area with weakly visible pale subterminal line; discal dot small.
Male genitalia. Uncus bell-shaped, uniapical. Valve unmodified. Vinculum short, evenly rounded. Papillae short and broad. Aedeagus stout, shorter than length of valve. Vesica armed with a broad, lozenge-shaped apical cornutus, a long and narrow, basally curved cornutus and a broad irregular cornutus near ductus ejaculatorius base. Sternite A8 two broad parallel arms, rounded at apices, connected by a broad basal band with two lateral lobes.
Female genitalia. Bursa very long, covered with small spines along left side from base to ductus bursae, and with a belt of smaller spines close to base. Ductus bursae broad and tapered with longitudinal striations and a transverse row of larger spines. Ductus seminalis broad, attached on right side. Colliculum relatively broad. Antrum membranous. Tergite A8 elongate. Apophyses very long and stout, the membrane around anterior part of posterior apophyses sclerotized. Papillae anales elongate and narrow.
Range. China (Taiwan).
Phenology. Univoltine: late August to late November.
Altitude. From about 1000 to 2600 m above sea-level (Taiwan: coll. NSMTA, ZFMK).
Similar species. Similar to the following species (which is not known from Taiwan), but distinguished from it by darker, blackish brown ground colour, especially on the fore wing between base and medial line, the less

distinct transverse lines, the clearer white tornal spot and the darker hind wing with less distinct transverse lines. The genitalia of both sexes are also quite similar to those of the next species, but the male can be distinguished by the apical cornutus on the vesica, and the female by the different pattern of spining in the bursa.

Material. ♀, Taiwan, Tayuling, 2600 m, Hualien Hsien, 28–29.viii.1983, leg. H. Yoshimoto, BM Geom. slide no. 16543 (holotype, coll. H. Inoue in BMNH); 1♂, Taiwan, Nantou, Wantashe, 17–18.xi.1987, leg. C.S. Lin, NMNS ENT 1379–105 (NSMTA); 1♂, Taiwan, Nantou, Wushe, 1000 m, 7–8.xi.1989, leg. C.S. Lin, NMNS ENT 492–213, Galsworthy slide no. ACG TW22 (NSMTA); 2♂, 1♀, Taiwan, Nantou, Meifeng, 15.x.1990, leg. C.S. Lin, NMNS ENT 670–299, 670–317, and 658–184, Galsworthy slides nos. ACG TW-19, ACG TW-23 (NSMTA); 2♀, Taiwan, Nantou, Meifeng, *ca.* 2300 m, 29.x.-3.xi.1998, leg. Mey & Speidel (MNHU); 1♀, Taiwan, Kaohsiung, Yaoyuan, Tengchih, 20–22.xi.2002, leg. W.T. Yang, NMNS ENT 39731029 (NSMTA); 2♂, 1♀, Taiwan, Taichung Co., Wuling Farm, 35 km NE of Lishan, alt. 1750 m, 26.ix.1992, [leg.] F. Aulombard et J. Plante, Mironov slides nos. 747♀, 748♂, 749♂ (ZFMK); 1♂, Taiwan, Prov. Nantou, Meifong, 2250 m, 24°05′ N, 121°10′ E, 25.ix.1999, leg. G. Csorba and B. Herczig (ZSM).

293. *Eupithecia supercastigata* Inoue, 1958

Eupithecia supercastigata Inoue, 1958, *Tinea* 4 (2): (248), text-fig. 2. Holotype ♂ (BMNH; examined), [Japan]: Yokosuka, Funakoshi.

Description. Fore wing ground colour greyish brown; dark transverse lines distinct, slightly sinuate, angled onto costa, forming dark costal blotches; medial touching discal dot and sharply angled onto costa; terminal area slightly darker, with pale wavy subterminal line, and a small white tornal spot; discal dot relatively large, ovoid. Hind wing slightly paler, with oblique antemedial and evenly curved postmedial; terminal area slightly darker, with distinct inner border and fine, dentate, whitish subterminal line, forming a small white tornal spot; discal dot small and pale.

Male genitalia. Uncus bell-shaped, uniapical. Valve unmodified. Vinculum broad and rounded. Papillae short and broad. Aedeagus very stout, rather short. Vesica armed with a broad V-shaped cornutus, and a compact, folded cornutus near ductus ejaculatorius base. Sternite A8 lyre-shaped, bilobed at base, with two broad curved arms, broadened and rounded at apices, connected by a broad basal band.

Female genitalia. Bursa oval, densely covered with spines, except right side under, and a small area opposite, base of ductus seminalis. Ductus bursae tapered, with longitudinal striations and a transverse row of sparse spines. Ductus seminalis very long and broad, attached to ductus bursae on right. Colliculum relatively elongate and broad. Antrum slightly sclerotized. Tergite A8 elongate. Apophyses very long and stout, strongly expanded and flattened at apices, the membrane around the posterior sclerotized. Papillae anales narrow, elongate and tapered, almost without setae.

Range. China (Zhejiang, Shaanxi, Hunan, Yunnan); Korea, Japan (Honshu, Shikoku, Kyushu, Tsushima).

Phenology. Bivoltine (?): June to July; September to mid-November.

Altitude. Found from about 1900 m in Qinling Mts to 3200 m in Yunnan. (China: coll. ZFMK, IZCAS).

Similar species. Externally similar to *E. subfuscata*, but distinguished from it by the more obtuse fore wings, with more distinct transverse lines and larger, ovoid discal dots, and the darker hind wings. The male genitalia resemble those of *E. acutipapillata* and *E. otiosa*, but the aedeagus vesica differs from the former in the lack of an apical cornutus, and from the latter in the much shorter central cornutus; the shape of sternite A8 is different from either. The female genitalia can be distinguished from those of *E. acutipapillata* by the different disposition of spines in the bursa, and from *E. otiosa* by the lack of spines in the entrance of the ductus seminalis.

Material. ♂, [Japan], Yokosuka, Funakoshi, 8.x.1953, H. Inoue (holotype, coll. H. Inoue in BMNH); 1♂, Prov. Chekiang [Zhejiang], West Tien-mu-shan, 18.x.1932, H. Höne, de Laever slide no. 327♂ (ZFMK); 1♂, China, Shaanxi, South Taibashan, Tsinling Mts. [Qinling], near Houzhenzi, 33°53′ N, 107°49′ E, 1900 m, 1.-12.viii.1999, leg. Siniaev & Plutenko, Mironov slide no. 655♂ (ZFMK); 5♂, Yunnan, Dêqên, 3180 m, 15–22.vii.1982, coll Wang Shuyong, slides nos. ZL107, 166, 204, 238, 540♂; 1♂, Hunan, Zhangjiajie, 8.x.1988, coll Fang Chenglai, slide no ZL422♂ (IZCAS).

Not placed in species groups

294. *Eupithecia concremata* Dietze, 1904

Eup[ithecia] concremata Dietze, 1904, *Dt. ent. Z., Iris* [1903] **16** (2): 360, pl. 4, fig. 29. Holotype ♂ (coll. K. Dietze in MNHU; examined), [China]: Sinin [Xining].

Description. Fore wing quite broad, unicolourous brownish grey; transverse lines very indistinct, scalloped and right angled onto costa; discal dot elongate, weakly visible; most veins picked out with alternating brown and white dashes; hind wing pale, off white with transverse lines marked only along anal margin, and traces of a darker postmedial; discal dot invisible; fringes short, white, distinctly spotted with brownish grey at vein ends.

Male genitalia. Uncus slightly elongate and narrow, biapical. Valve with sharply tapered outer half and narrow apex. Vinculum short and narrow, tapered. Papillae broad and rather short. Aedeagus very small and narrow. Vesica armed with three small dentate cornuti, an elongate, folded plate and a small, irregular plate near ductus ejaculatorius base. Sternite A8 square, bilobed at base, with two inwardly curving pincer-like apical arms. Female unknown.

Range. China (Qinghai, Sichuan).

Phenology. Only dated specimen from June.

Altitude. Only altitudinal information is 4600–4800 m in Sichuan (China: coll. BMNH).

Similar species. Both specimens of this rare mountain species are very similar to the unicolourous form of *E. fennoscandica* Knaben, 1949 from Northern Europe, Altai, Sajan Mts. and northern Mongolia, and are difficult to separate on external characters. However, the male genitalia are entirely different, and the shape of the eighth sternite immediately distinguishes the two.

Material. ♂, [Qinghai], *concremata* n. sp., Original, Sinin, Tibet, K. Dietze (Bohatsch und D: Staudinger gesehen.), Vojnits slide no. 1176♂ (holotype, coll. K. Dietze in MNHU); 1♂, China, Sichuan province, Zhedo mountain, 4600–4800 m, 11–20.vi.1989, Native coll., Inoue slide no. 14170♂ (coll. H. Inoue in BMNH).

Remarks. *E. concremata* might perhaps be related to species in the *russeliata* group, members of which have a similar ornamentation in the vesica. However, the shape of the eighth sternite is not typical for this group and is closer to those of the *venosata* group (e.g. *E. alliaria* Staudinger, 1870 and *E. dissertata*) or the *centaureata* group.

295. *Eupithecia coribalteata* Mironov & Galsworthy, 2004

Eupithecia coribalteata Mironov & Galsworthy, 2004, *Trans. lepid. Soc. Japan* 55 (4): 288, figs 4, 16. Holotype ♀ (IZCAS; examined), [China]: Qinghai, Menyuan, Xianmi, 2800 m.

Description. Fore wing ground colour pale grey, with darker dashes on veins; transverse lines faint, forming darker blotches on costa; postbasal and antemedial angled close to costa; medial pale, twice angled beyond discal dot; postmedial brown, sharply angled close to costa, crenellate and

with short dark dashes on veins; discal dot dark brown, elongate; terminal area darker with wavy white subterminal line ending in tornal dot. Hind wing ground colour white, with scattered brownish scales, and transverse lines marked only along anal margin; terminal area shaded pale brown.
Female genitalia. Bursa ovate-oblong, covered with tiny, poorly sclerotized spines apart from a small posterior area; central part of bursa with a long curving sclerotized band, armed with spines of varying sizes. Ductus bursae short with a wrinkled membranous swelling on right. Ductus seminalis broad, broadly attached to corpus bursae below centre on right. Colliculum plate-like, very broad. Antrum short and broad. Tergite A8 trapezoid with a deep medial hollow in posterior margin. Anterior apophyses stout and very short, with two heavily sclerotized areas at their base; posterior apophyses very narrow. Papillae anales short and broad, rounded. Male unknown.
Range. China (Qinghai, Gansu).
Phenology. Possibly univoltine: late June to early August.
Altitude. From about 2600 to 3000 m in mountains of Qinghai and Gansu (China: coll. IZCAS; ZFMK).
Similar species. Externally rather similar to *E. uliata* and other related species. The female genitalia are unique in the heavily sclerotized transverse band around the bursa.
Material. ♀, Qinghai, Menyuan, Xianmi, 2800 m, 23–24.vii.1992, coll. Xue Dayong, slide no. ZL482♀ (holotype, IZCAS); 10♀, same data, slides nos. ZL147, 164♀ (paratypes, IZCAS); 3♀, Qinghai, Tongren, Maixiu, 2950 m, 30.vii–1.viii.1992, coll. Xue Dayong, slide no. ZL277♀ (paratypes, IZCAS); 2♀, Gansu, Xiahe, 2600–3000 m, 26–30.vi.1986, Coll. Görgner, Mironov slide no. 688♀ (paratypes, ZFMK).
Remarks. It is surprising that despite a long series of females, no males have been found which appear to match this species.

296. *Eupithecia lilliputata* Mironov & Galsworthy, 2011

Eupithecia lilliputata Mironov & Galsworthy, 2010, *Lep. Science* **62** (1): 12, figs 1, 17, 20. Holotype ♀ (ZFMK; examined), China, Yunnan, Li-kiang [Lijiang].

Description. Fore wing ground colour rusty ochreous; basal and medial areas along costa brown; transverse lines very narrow, rather inconspicuous; terminal area darker, rusty between postmedial and subterminal transverse lines; subterminal line wavy, whitish; discal dot small, black. Hind wing ground colour off-white, with weak ochreous tinge; transverse lines weakly marked, brownish; discal dot very small and pale.

Male genitalia. Uncus broad, but finely tipped. Valve unmodified. Vinculum rounded. Papillae parallel-sided. Aedeagus shorter than valve, fairly stout, with a single plate-like cornutus, tapering at apex and bifid at base. Sternite A8 bilobed at base, with two narrow arms connected by a broad band.

Female genitalia. Bursa ovoid, covered with small spines except diagonal medial area and areas near ductus seminalis and colliculum. Ductus bursae not distinguishable. Ductus seminalis narrow, attached to central part of bursa on right. Colliculum elongate and narrow. Antrum membranous, narrow and elongate. Tergite A8 and apophyses narrow and elongate. Papillae anales short and tapered.

Range. China (Yunnan).

Similar species. Externally similar to *E. fletcheri* Prout, 1926, but much smaller, the dark medial area of the fore wing well marked between costa and discal dot only, the hind wing whitish, without a dark basal area along anal margin and without a rusty ochreous terminal area as in *E. fletcheri*. The male genitalia are instantly separable from those of *E. fletcheri* by the presence of a hooked uncus. The long colliculum is distinctive in the female.

Material. 1♀, China, Yunnan, Li-kiang [Lijiang], 11.ix.1934, H. Höne, Vojnits slide no. 12249♀ (holotype, ZFMK); 1♂, same loc., date and collector, Vojnits slide no. 12247♂ (paratype, ZFMK).

297. *Eupithecia salubris* Mironov & Galsworthy, 2004

Eupithecia salubris Mironov & Galsworthy, 2004, *Trans. lepid. Soc. Japan* 55 (1): 46, figs 4, 19, 22. Holotype ♂ (ZFMK; examined), [China]: Sued-Shensi [South Shaanxi], Tapaishan [Taibaishan] im Tsinling [Qinling Mts.], 1700 m.

Misidentification: *Eupithecia infecta* Vojnits, 1981 (1♀ paratype in ZFMK).

Description. Ground colour uniform pale brown, dotted in basal, costal, postmedial and terminal areas with darker scales; all transverse lines darker brown, but very faint; a narrow, curved basal line; one postbasal and two antemedials, angled on Cu+M; a single medial passing through discal dot, strongly angled on Cu+M; a faint oblique postmedial, angled below costa. Hind wing uniform pale brown, dotted with darker scales along anal margin.

Male genitalia. Uncus short, stout, probably biapical. Valve unmodified, but with the long basal setae on the ventral margin placed on short processes. Vinculum short and rather broad. Papillae elongate and narrow. Aedeagus very short and stout. Vesica armed with two horn-like cornuti,

one hook-shaped, and a small irregular plate near ductus ejaculatorius base. Sternite A8 small, with two long, narrow rods.
Female genitalia. Bursa oval, with two separate patches of spines, one central and one near the colliculum. Ductus seminalis missing in slide, but possibly attached to corpus bursae near colliculum. Colliculum rather short and narrow. Antrum membranous. Tergite A8 quadrate. Apophyses short and narrow. Papillae anales short and broad.
Range. China (Shanxi, Shaanxi).
Phenology. Univoltine: early August to early September.
Altitude. From about 1700 to 3000 m in mountains of Shanxi and Shaanxi (China: Vojnits, 1981b, as *E. infecta*; Mironov & Galsworthy 2004).
Similar species. Similar to *E. coniurata*, but the distance between the antemedial and medial lines on the fore wing is shorter in the present species. In the male genitalia, distinguished from *E. coniurata* by the small processes carrying long setae on the ventral margin of the valve, and the stouter hook-like cornuti on the vesica; the female genitalia are entirely distinct.
Material. ♂, Sued-Shensi [Shaanxi], Tapaishan [Taibaishan] im Tsinling [Qinling], *ca.* 1700 m, 9.viii.1936, H. Höne, de Laever slide no. 287♂ (holotype, ZFMK); 1♀, same loc. and alt., 9.ix.1936, H. Höne, Vojnits slide no. 12944♀ (paratype, labelled as a paratype of *E. infecta*, ZFMK); 1♂, Prov. Shansi [Shanxi], Mien-Shan, Obere Höhe *ca.* 2000 m, 8.viii.1937, H. Höne, Vojnits slide no. 13268♂ (paratype, ZFMK).
Remarks. All known specimens are very worn.

298. *Eupithecia villica* Mironov & Galsworthy, 2006

Eupithecia villica Mironov & Galsworthy, 2006, *Trans. lepid. Soc. Japan* 57 (4): 352, figs 12, 26, 29. Holotype ♀ (IZCAS; examined), [China]: Tibet, Jomda military post, 3400 m.

Description. Fore wing elongate, ground colour pale yellowish buff. Transverse lines absent, except for a faint and narrow brown postmedial, very sharply angled at the level of the minute discal dot. A series of brown subterminal dots on the veins. Terminal line dark brown; fringes buff. Hind wing white with a minute discal dot; transverse lines present only as obscure buff markings close to anal margin. Abdomen light yellowish ochreous.
Male genitalia. Uncus elongate, biapical, with narrow base. Valve unmodified. Vinculum broad, with shallow medial hollow. Papillae narrow and elongate. Aedeagus slim, shorter than valve length. Vesica armed with one blunt and one sharp elongate plate, and a smaller, irregular plate at base

of ductus ejaculatorius. Sternite A8 peg-like, sharply narrowed to apex, with two short, very close, apical rods.
Female genitalia. Bursa elongate pear-shaped, almost completely covered with small spines except posterior half on dorsal side and along left side. Ductus seminalis broadened at base, attached to centre of bursa on left. Colliculum short and rather broad. Antrum membranous. Tergite A8 almost quadrate. Apophyses moderately stout. Papillae anales short and narrow, tapered.
Range. China (Tibet, Yunnan, Sichuan).
Phenology. Bivoltine (?): mid-July to late August; early October.
Altitude. From about 3400 to 4500 m in the Eastern Himalayas (China: IZCAS, coll. TTM).
Similar species. Very distinctive because of its unusual sandy coloration. In the male, the shape of the eighth sternite is diagnostic. The female genitalia are similar to those of *E. tibetana*, but have a larger spiniferous area from the base of the bursa to the colliculum.
Material. ♀, Tibet, Jomda military post, 3400 m, 24.vii.1976, leg. Zhang Xuezhong, slide no. ZL472♀ (holotype, IZCAS); 1♀, Tibet, Riwoqê, 3750 m, 29.viii.1976, leg. Zhang Xuezhong, slide no. ZL432♀ (paratype, IZCAS); 1♂, China, Provinz Nord-Yuennan, Li-kiang [Lijiang], 1.x.1934, H. Höne, gen. slide no. ACG Bonn 10♂ (paratype, ZFMK); 1♂, Tibet [Sichuan], Batang, Untere Urwaldzone, (*ca.* 3800 m), H. Höne, Vojnits slide no. 13266♂ (paratype, ZFMK); 1♂, Nord Yünnan, A-tun-tse [Dêqên], Obere Höhe (*ca.* 4500 m), 13.viii.1936, H. Höne, Vojnits slide no. 12179♂ (TTM); 1♂, China, W. Sichuan, Litang, 4000 m, 29°50′ N / 100°21′ E, 16.vii.2009, I. & A. Floriani, Saldaitis leg., Pekarsky slide no. OP 0748♂ (coll. O. Pekarsky).

Pareupithecia Mironov & Galsworthy, 2012

Type species: *Eupithecia spadix* Inoue, 1955, *Tinea* 2 (1/2): (81), pl. 6: 11; pl. 7: 15. Holotype ♂ (BMNH), [Japan]: Tokyo, Takao-san.

A monotypic genus introduced by Mironov and Galsworthy (2012) to contain the single species included below. The main differences from *Eupithecia* are as follows: externally the broad wings and very distinct black ante- and postmedial transverse lines bordering the darker medial area on the fore wing resemble species in the genus *Pasiphila*; in the male genitalia the juxta has an unusual lip at its apex supporting the aedeagus, and the papillae on the ends of the anterior arms of the labides are untypical, being situated on the outer rather than the upper surface of the arms, and orientated away from the anellus, and bearing much longer setae than those found normally in *Eupithecia*. The transtillae are unusually sclerotized, and meet rigidly in the centre. Most striking, however, are the processes of the eighth sternite, which are massive, rise clear of the sternite well before its posterior margin before bending inwards, and carry large hook-like processes on their ends. In *Eupithecia*, the sclerotization of the sternite is normally contained within the sternite, or at most slightly protruding at the posterior edge. In the female genitalia, the large membranous bursa copulatrix has several chains of very narrow spines which are not found in any species of *Eupithecia*. The relatively small eighth tergite and papillae anales with very short anterior and posterior apophyses are reminiscent of species in *Micrulia* Warren and *Mariaba* Walker. But the present species has an unusually short antrum and a large, platelike, sclerotized colliculum, also unlike any *Eupithecia* species.

299. *Pareupithecia spadix* Inoue, 1955

Eupithecia spadix Inoue, 1955, *Tinea* **2** (1/2): (81), pl. 6: 11; pl. 7: 15. Holotype ♂ (BMNH; examined), [Japan]: Tokyo, Takao-san.

Misidentification: *Eupithecia abietaria* Shin et Kim, 1993 (*nec* Goeze, 1781).

Description. Fore wing ground colour of basal and terminal areas from pale brownish grey to dark reddish brown; medial area from pale greyish brown through dark reddish brown to blackish brown, usually darker than basal and terminal areas; basal, antemedial and postmedial transverse lines very distinct, fine, black; basal and antemedial oblique, evenly curved; postmedial sinuate, oblique in posterior half and characteristically twice obtusely angled in anterior half; terminal area with apical, medial and tornal brown blotches, but always with a light spot in middle; discal dot small, blackish, usually bordered with paler scales. Hind wing paler than fore wing, transverse lines fine, distinct, evenly curved; terminal area darker with weak trace of pale subterminal line; discal dot small.

Male genitalia. Uncus narrow, uniapical with seta near tip on ventral side. Valve long and narrow. Transtilla connected with dorso-proximal tips of valves and with labides at its posterior ends. Juxta with broad posterior arms, a narrow, semilunar, heavily sclerotized band in the middle, and a large lip at apex. Vinculum broad, semicircular. Papillae elongate, orientated outwards. Aedeagus slim, slightly curved, shorter than length of valve. Vesica granulate, armed with a dense patch of spinules and two small cornuti near ductus ejaculatorius base. Sternite A8 two biflex, broadly diverging rods, heavily sclerotized and connected by a concave band at base.

Female genitalia. Bursa copulatrix very large, ovate, with numerous curved wrinkled striations, three chains of slim spines and a small sclerotized prominence at the centre of corpus bursae; inner surface of corpus bursae almost completely covered with minute spines. Ductus bursae not distinguishable. Ductus seminalis very narrow, attached to corpus bursae at the middle of dorsal side. Colliculum elongate and broad, with two spiniform horns and a convex posterior margin. Antrum membranous, short and broad. Tergite A8 rectangular, small, with narrowly sclerotized anterior margin. Apophyses very short, relatively narrow, the bases of the anterior broadened. Papillae anales small and rounded.

Range. China (Shanxi); Asian Russia (southern Primorje), Korea, Japan (Honshu).

Phenology. Bi- or polyvoltine: late March (late April) to late August.

Altitude. From about 700 m on Mt. Sudosan (Korea: Inoue 1979) to about 2000 m in Mien-shan (China: coll. ZFMK).

Similar species. Very variable in depth of colour, but not likely to be confused with any Asian *Eupithecia* species. Externally resembles species of *Chloroclystis*, *Pasiphila* or *Celaenaclystis* in shape of wings, coloration and maculation, especially in the postmedial line, but the pattern is fairly distinctive and the species should usually be easily recognizable. The unique male and female genitalia and the shape of the eighth sternite in the male immediately distinguish the species.

Material. ♂, [Japan], Tokyo, Takao-san, 28.iv.[19]40, H. Inoue, BM Geom. slide no. 16042 (holotype, coll. H. Inoue in BMNH); 4♂, 1♀, [China], Prov. Shansi [Shanxi], Mien-Shan, Obere Höhe *ca.* 2000 m, 28.vi, 9, 30.vii, 9.viii.1937, H. Höne, de Laever slide no. 28♂ (ZFMK).

Remarks. Not previously recorded from China.

Eva Vojnits, 1981

Eva Vojnits, 1981, *Acta zool. Acad. Sci. hung*, 27 (3–4): 407.

Type species: *Eva flexa* Vojnits, 1981, *Acta zool. Acad. Sci. hung.* 27 (3–4): 408, figs 1–7. Holotype ♂ (ZFMK), China: North Yunnan, Li-kiang [Lijiang].

This genus was erected by Vojnits to include two species. Mironov & Galsworthy (2012) showed that one of these species was based on a chimaera, and that in fact the genus is monotypic. The single species is externally very similar to *Eupithecia*. The genitalia of both sexes however differ significantly as follows: the setation of the valves is quite different from *Eupithecia*: in the latter the valves always have a smooth and regular setation on the ventral surface originating close to the anterior edge, covering most of the surface of the valve, and virtually impossible to remove without tearing the membrane. In *Eva* the setation is confined to the central part of the valve, and is wispy and not strongly attached. The position of the ductus ejaculatorius at the anterior end of the aedeagus is also not found in any *Eupithecia* species we have seen. The female genitalia are also highly atypical, having a pyriform membranous bursa with two bands of linked spines somewhat similar to those seen in *Hybridoneura* Warren; two other features which are not found in *Eupithecia* are the medial longitudinal depression in tergite A8 and the branches of the sinus arising from the apices of the anterior apophyses.

300. *Eva flexa* Vojnits, 1981

Eva flexa Vojnits, 1981, *Acta zool. Acad. Sci. hung.* **27** (3–4): 408, figs 1–7. Holotype ♂ (ZFMK; examined), China: North Yunnan, Li-kiang [Lijiang].

Description. A small species with a distinct emargination in the outer margin of the hind wing. Fore wing dark greyish brown; transverse lines oblique, sharply angled onto costa; medial area with reddish blotch under discal dot between M_3 and CuA_2; terminal area with palesubterminal line and a small pale tornal spot; discal dot large, rounded, intensely black. Hind wing paler with fine transverse lines well marked along anal margin; posterior half much darker than anterior; postmedial double, evenly curved; terminal area narrow with an indistinct, fine, subterminal line; discal dot small. End of abdomen usually broad, obtuse.

Male genitalia. Uncus stout, with broad basal lobes, uniapical. Valve short and broad with large medial transverse wrinkle on ventral margin; inner surface of valve covered with sparse short setae along its central part only; sacculus lightly sclerotized. Transtilla thick. Juxta almost ovoid with two narrow, broadly diverging posterior arms. Vinculum short, with shallow medial hollow. Papillae very short and stout. Aedeagus long and stout, longer than length of valve, the ductus ejaculatorius attached to anterior end of aedeagus. Vesica armed with a plate-like W-shaped cornutus, and a smaller, irregular cornutus near anterior end of aedeagus. Sternite A8 very broad, with two blunt, sclerotized, broadly diverging apical horns. Tergite A8 triangular, tapered to a curved, hook-like apex.

Female genitalia. Bursa membranous, with two long curved chains of sparse small spines. Ductus bursae not distinguishable. Ductus seminalis narrow, attached to ductus bursae near colliculum. Colliculum elongate

and very narrow. Antrum membranous. Tergite A8 almost square, with a medial longitudinal thickening. Apophyses short and narrow, especially the posterior; spurs branching off anterior apophyses close to apices. Papillae anales short and broad, rounded.
Range. China (SE Tibet, Sichuan, Yunnan).
Phenology. Bivoltine (?): early June to mid-August.
Altitude. From about 400 to 4000 m in Yunnan and Tibet (China: coll. ZFMK, IZCAS).
Similar species. Superficially similar to Asian representatives of the genus *Eupithecia* with a reddish subdiscal spot, but generally smaller. Useful distinctions are the emarginate termen of the hind wing, the large, round fore wing discal spot, and the broad apex to the abdomen. The genitalia are very distinctive in both sexes.
Material. ♂, China, Provinz Nord-Yuennan, Li-kiang [Lijiang], 8.vii.1934, H. Höne, Vojnits slide no. 13097♂ (holotype, ZFMK; gen. slide absent in ZFMK); 10♂, 15♀, same loc., 21.vi., 2, 4, 5, 8, 9, 11, 15, 23, 31.vii., 18.viii.1934, 7.vi., 11, 13.viii.1935, H. Höne, Vojnits slides respectively nos 13012♀, 13016♀, 12998♀, 13011♀, 12980♀, 13015♀, 13096♀, 13077♂, 12986♂, 13010♂, 13013♀, 13014♀, 13076♂, 13103♀, 13105♂, 13236♂, 13100♀, 13008♀, 13104♂, 13004♂, 13031♂, 13098♀, 12979♂, 13101♀, 12982♀ (paratypes, ZFMK); 3♂, 3♀, same loc., 5, 10, 11, 26.vi., 4, 8.vii.1934, H. Höne, Vojnits slides nos 12983♀, 12988♀, 12999♂, 13003♂, 13099♀, 13102♂ (paratypes, TTM); 6♂, 5♀, same loc., 2, 20, 21.vi, 1, 8, 9, 10, 12, 23.vii.1934, H. Höne, Vojnits slides nos 13144♂, 13145♂, 13148♀, 13149♂, 13151♂, 13152♀, 13153♀, 13350♂ (TTM); 4♂, 18♀, same loc., 5, 18, 23, 30.vi., 2, 4, 5, 7, 9, 11, 13, 15, 31.vii., 7.viii.1934, 14.viii.1935, H. Höne, Vojnits slides respectively nos 12160♂, 11398♂, 13349♀, de Laever slides nos 123♀, 230♀ (ZFMK); 1♂, 1♀, same loc., 12, 23.vii.1934, H. Höne, Mironov slide no. 792♂ (SMNK); 1♂, 3♀, same loc., *ca.* 2000 m, 9.vii., 8.viii.1934, H. Höne, Vojnits slides respectively nos 13017♀, 13029♀, 13093♂, 13033♀ (paratypes, ZFMK); 2♀, same loc. and alt., 9, 14.vii.1934, H. Höne (ZFMK); 1♂, same loc., *ca.* 3000 m, 2.vi.1934, H. Höne (ZFMK); 1♂, 4♀, China, Tibet, Batang, Untere Urwaldzone (*ca.* 3800 m), 3, 11, 21.vi., 30.vii.1936, H. Höne, Vojnits slides respectively nos 13074♀, 13075♀, 12993♀, 12989♂, 12990♀ (paratypes, ZFMK); 1♂, 2♀, same loc. and alt., 14.vi., 4.viii., 12.ix.1936, H. Höne, de Laever slides respectively nos 120♂, 126♀, 116♀ (ZFMK); 2♀, same loc. and alt., 12, 15.vi.1936, H. Höne, Vojnits slides nos 12401♀, 12407♀ (TTM); 7♂, 7♀, [China], Nord Yünnan, A-tun-tse [Deqing], Mittlere Höhe (*ca.* 4000 m), 16, 18, 19, 21, 24, 26, 27, 28, 31.vii., 8, 17, 18.viii.1936, H. Höne, Vojnits slides respectively nos 13090♀, 13002♀, 13095♀, 13001♀, 13007♂, 13092♀, 13032♂, 12987♂, 13091♀, 13094♂, 13006♀, 13005♂, 13026♂, 13030♂

(paratypes, ZFMK); 6♂, same loc. and alt., 21, 26, 29, 31.vii., 12.viii., 3.ix.1936, H. Höne, Vojnits slides respectively nos 12515♂, 12472♂, 12471♂, 12517♂, 11535♂, 11534♂ (holotype and paratypes of unpubl. *Eupithecia suavissima* Vojnits, ZFMK); 3♂, 7♀, same loc. and alt., 29.vi., 14, 17, 23, 24, 26.vii., 2, 4.viii.1936, H. Höne, de Laever slide no. 124♂, Vojnits slide no. 12508♀ (ZFMK); 1♂, 2♀, same loc. and alt., 23.vii., 1.viii.1936, H. Höne, Vojnits slides nos 12476♀, 12477♀, 12489♂ (TTM); 2♀, same loc., Talsohle *ca.* 3000 m, 17.vi., 3.vii.1937, H. Höne, Vojnits slides nos 13019♀, 13000♀ (paratypes, ZFMK); 5♂, 3♀, same loc. and alt., 4, 23, 27.vi., 2, 3, 4, 11.vii.1937, H. Höne (ZFMK); 2♂, 1♀, same loc. and alt., 1.vi., 4, 11.vii.1937, H. Höne, Vojnits slides nos 12487♀, 13146♂ (TTM); 1♀, same loc., Talsohle *ca.* 3500 m, 15.vii.1936, H. Höne, Vojnits slide no. 13147♀ (TTM); 1♀, China, Yunnan Prov., Lijiang, 2,650 m, 16.vii.2001, Li Houhun leg. (coll. P. Skou); 1♂, 3♀, China, North Yunnan, Diqing Tibetan Aut. Pref., 5 km SE of Deqin, 3356 m, 19.vi.2009, leg. B. Benedek (coll. O. Pekarsky); 11♂, 16♀, China, W. Sichuan, road Xinngcheng / Zhongdian, 3300 m, 17.vii.2009, 28°46′ N / 99°51′ E, I. & A. Floriani, Saldaitis leg. (coll. O. Pekarsky); 6♂, 7♀, KIB/RBGE research station, Lijiang, Yunnan, 15–20 June 2009, A C Galsworthy (BMNH); 1♂ Tibet, Bomi, Yi'ong, 2700 m, 18.viii.1983, coll Han Yinheng; 1♂, Tibet, Zhag'yab, Gyitang, 12.vi.1984, coll Jiang Ba; 2♀, Tibet, same loc, west side, 3600 m, 3 and 5.vii.1976; 1♀, same loc, 27.viii.1976; 1♂,1♀ Tibet, same loc, 3600 m, 7.vii.1976,all coll Han Yinheng; 1♂, Lijiang Yulongshan, 21.vii.1962, 2♀, same loc, 4.vii.1962, coll Song Shimei (IZCAS).

Girida Mironov & Galsworthy, 2012

Type species: *Eupithecia rigida* Swinhoe, 1892, *Trans. ent. Soc. Lond.* **1892** (1): 2, pl. I, fig. 6. Syntypes 2 (BMNH), [India]: [Khasia Hills].

This genus was erected by Mironov & Galsworthy (2012) to contain two aberrant species which had hitherto been included in *Eupithecia*, but in fact lack all the normal defining features of that genus. The two species are extremely similar externally, having a simple very dark pattern, and a wavy terminal margin in the hind wing, both abnormal features for *Eupithecia*. In the male genitalia, the valve has a broadly swollen sacculus, and lacks the universal 'hour-glass' juxta of *Eupithecia*, and the normal papillae on the anterior arms of the labides. The eighth sternite is not specialised. The genus ranges from East and South Asia across the Indian Ocean to much of sub-Saharan Africa.

301. *Girida rigida* Swinhoe, 1892

Eupithecia rigida Swinhoe, 1892, *Trans. ent. Soc. Lond.* **1892** (1): 2, pl. I, fig. 6. Lectotype ♂ (design. by Inoue, 1979, BMNH; examined), [India]: [Khasia Hills].

Description. Fore wing unicolourous dark reddish or chocolate brown with a fine dentate white postmedial line, curved strongly inward near costa to form a white costal spot; discal dot dark brown, poorly marked; terminal area darker, with a distinct white tornal spot and a series of white dots on veins, sometimes linked by a fine zigzag line. Hind wing unicolourous pale brown, darkened near tornus; postmedial white, strongest near anal margin; discal dot lacking.
Male genitalia. Uncus broad with sharply pointed central uniapical tip. Valve short with prominent semicircular sacculus, its apex narrowly rounded. Transtilla absent. Vinculum short and narrow. Labides short and stout, apices of their anterior arms connected with juxta and covered by sparse long setae. Fultura inferior broad, jug-shaped: juxta ovoid; caulis heavily sclerotized, trapezoid with longer posterior margin. Papillae not developed. Aedeagus narrow, longer than length of valve; ductus ejaculatorius attached to central part of aedeagus. Vesica armed with one narrow elongate cornutus. Sternite A8 trapezoid with two rounded and slightly prominent posterior corners covered by sparse short setae.
Female genitalia. Bursa globular, completely covered with small spines. Ductus bursae weakly sclerotized, short, tapered, spineless. Ductus seminalis attached to ductus bursae near base of colliculum on right side. Colliculum narrow. Antrum membranous. Tergite A8 short and broad, with shallow medial hollows in anterior and posterior margins. Apophyses short and stout. Papillae anales rather small, slightly pointed.
Range. China (Guangxi, Hong Kong,Taiwan); Japan (Ishigakijima, Iriomotejima), India (Nilgiri Hills, Khasia Hills), Thailand, Philippines (Luzon), Borneo, New Caledonia.
Phenology. Polyvoltine throughout the year in the tropics.
Altitude. From sea level to about 1900 m in Luzon (Philippines: coll. ZFMK).
Similar species. Not likely to be confused with any other eupitheciine species in the area.
Material. ♂, [India], Khasi Hills 94–66, *Eupithecia rigida* Swinhoe ♂ type (lectotype, BMNH); 1♀, Taiwan, Pingtung County, Kenting, 16.iii.1982, S. Hashimoto, Inoue slide no. 12286♀ (coll. H. Inoue in BMNH); 1♂, Taiwan, County Taichung, Chi Chi, TESRI Botanical Garden, 250 m, 30.ix.2000, leg. L. Papp, L. Peregovits & L. Ronkay (TTM); 1♂, Taiwan, County Pingtung,

Kenting NP, Kenting FRA, 300 m, 21°57′62″ N, 120°48′89″ E, 4.–6.ix.2000, leg. L. Ronkay (TTM); 1♂, 1♀, same loc. and alt., 4.–5.x.2000, leg. L. Papp & L. Peregovits (TTM); 1♀, Taiwan, same loc. and alt., 31.xii.2001–1.i.2002, leg. A. Kun & L. Ronkay (TTM); 1♀, Guangxi, 230 m, 11.vi.1976, Zhang Baolin (IZCAS).

Mesoptila Meyrick, 1891

Mesoptila Meyrick, 1891, *Proc. Linn. Soc. N.S.W.* (2) 5: 793 (key), 794. Type species: *Mesoptila compsodes* Meyrick, 1891.

Mironov & Galsworthy (2012) reviewed the genus *Mesoptila* Meyrick, and transferred the following species to it from *Eupithecia*. The genus differs externally from *Eupithecia* by the central tuft of erect scales in the longitudinally elongate fore wing disc. The male abdomen has long and narrow apodemes on the second tergite, while those in *Eupithecia* are short and broad. The male genitalia are distinguished from *Eupithecia* by the unusually narrow, usually curved, valve with a broadened basal lobe to the sacculus bearing a tuft of long setae, by the replacement of the normal papillae on the anterior arms of the labides by a narrow membranous cushion on the ventral surface bearing a row of setae, and by the more elongate shape of the juxta. The female genitalia have no significant differences from those of representatives of the genus *Eupithecia*. The external pattern in most known species is distinctive, and resembles that of the following species, but in at least one other species it is very similar to *Eupithecia*, and diagnosis can only be by examination of the genitalia. The genus ranges from South and South East Asia, Sulawesi, New Guinea and Australia across the Indian Ocean to East and South Africa.

302. *Mesoptila melanolopha* Swinhoe, 1895

Eupithecia melanolopha Swinhoe, 1895, *Ann. Mag. nat. Hist.* (6) **16**: 296. Holotype ♂ (BMNH; examined), [India]: Cherrapunji.

Eupithecia owadai Vojnits, 1988, *Acta zool. hung.* **34** (1): 47, fig. 24; pl. 3, fig. F. Holotype ♀ (NSMTO), Nepal (east): Solukhumbu Sagarmatha, Junbesi, 2670 m (synonymised in Inoue, 2000).

Unavailable name (incorrect subsequent spelling): *melarolopha*: Murase (1997).

Description. Fore wing brownish ochreous; basal and medial areas darker, blackish brown along costa, often appearing as a dark triangle on costa; antemedial oblique, sharply angled onto costa; postmedial narrow, white, sinuate near costa; terminal area darker, brownish in medial and tornal parts, with distinct, white subterminal line broadened near tornus; discal dot large, elongate, blackish brown. Hind wing pale brown, darker along terminal margin with white double postmedial and single subterminal lines, enclosing between them a characteristic darker brown spot near tornus. Discal dot absent.

Male genitalia. Uncus elongate, bulbous centrally, uniapical. Valve narrow and strap-like. Vinculum short. Papillae reduced to a narrow setose patch on underside of anterior arms of labides. Aedeagus narrow, shorter than length of valve. Vesica with numerous longitudinal striations and closely dentate, armed with a long and narrow cornutus. Sternite A8 very lightly sclerotized, with two narrow and apically blunt arms, connected by a short band at base.

Female genitalia. Bursa globular, covered with small spines apart from a narrow area from colliculum to base on left and from base to about $^2/_3$ on right. Ductus bursae not distinguishable. Ductus seminalis slightly broadened at base, attached near short and narrow colliculum on left. Antrum membranous, long and narrow. Tergite A8 rectangular, elongate. Apophyses thin very narrow. Papillae anales small and rounded.

Range. China (Shaanxi, Shanghai, Hunan, Yunnan, Hong Kong,Taiwan); Japan, Korea, India, Sri Lanka, Nepal, Thailand, Peninsular Malaysia, Philippines (Mindanao), Java, Bali, Borneo.

Phenology. Polyvoltine (possibly all year round in tropics): early February to early December.

Biology. Larva found on the flowers and leaves of *Stephania cephalantha* and *S. japonica* (Menispermaceae) (Japan: Inoue 1963, 1979, 1988; Murase 1997).

Altitude. From sea level to about 2700 m in the Himalayas (Nepal: Vojnits 1988, as *E. owadai*).

Similar species. Readily recognized and easily distinguished from any other Chinese eupitheciine species by the distinctive maculation.

Material. ♂, [India], Cherra Punji, 1901–178, *Eupithecia melanolopha* Swinhoe ♂ type (holotype, BMNH); 1♀, China, Provinz Nord-Yuennan, Li-kiang [Lijiang], 7.viii.1934, H. Höne (TTM); 1♀, [Taiwan], Ile Formosa, Moltrecht 1908 (ZFMK); 1♂, [China], Prov. Hunan, Hoengshan, 16.vii.1933, Höne, de Laever slide no. 324♂ (ZFMK); 1♀, China, Provinz Kiangsu [Jiangsu], Schanghai [Shanghai], 22.vii.1942, H. Höne (ZFMK); 1♂, 1♀, Taiwan, Taoyuan, Ming Chyr Forest Recreation Area, 1160 m, 29.ii.1996, Gy. Fábián & L. Németh (ZSM); 1♂, 1♀, Taiwan, Prov. Ilan, Ming Chyr Forest Recreation Area, 1200 m, 30–31.iii.1997, leg. Csorba & Ronkay (TTM); 1♀, C. China, prov. Shaanxi, Mt. Tai bei Shan, 1400 m, Hao Djenzy village,

33°48′ N, 107°43′ E, xi.1998, leg. loc. coll. (ZFMK); 1♀, Taiwan, Nantou, Meifeng, *ca.* 2300 m, 29.x.-3.xi.1998, leg. Mey & Speidel (MNHU); 1♂, Taiwan, Prov. Nantou, Meifong, 2250 m, 24°05′ N, 121°10′ E, 17–18.ix.1999, leg. G. Csorba and B. Herczig (ZSM); 1♂, Taiwan, Prov. Nantou, Yuanfeng, SW to Hohuan Pass, 2,760 m, 24°07′ N, 121°15′ E, 1.iv.2000, leg. A. Kun & L. Peregovits (ZSM); 1♂, Central Taiwan, Road Nr. 14, Reyen Shi-Region, 2200 m *ca.*, 24°08′ N, 121°10′ E, NE Puli, 28.vi-2.vii.2000, leg. W. Schacht (ZSM); 1♂, Central Taiwan, near Puli, Tsuifeng, *ca.* 2150 m NN, 24°04′17″ N / 121°12′12″ E, LF, 11. May 2001, leg. U. Buchsbaum (ZSM); 3♂, Central Taiwan, Nantou Co., 1300 m NN, Chun-Yang, near Lushan Hot Spring, 24°01′25″ N / 121°09′54″ E, LF, 26, 27. March 2003, leg. M.-Y. Chen & U. Buchsbaum (ZSM; ZISP); 1♂, Central Taiwan, Nantou Co., Meifeng, *ca.* 2100 m NN, 24°05′55″ N / 121°10′48″ E, LF, 20. May 2003, leg. Mei-Yu Chen (ZSM); 1♂, same loc., *ca.* 2200 m NN, 24°05′25″ N / 121°10′21″ E, LF, 01. August 2003, leg. Mei-Yu Chen (ZSM); 1♂, same loc. and alt., 24. March 2004, leg. Mei-Yu Chen (ZISP); 6 specimens, Yunnan, Tengchong, Heinitang, 1930 m, 28–30.v.1992; 1 specimen, Tengchong, Dalaoping, 2000 m, 24–26.v.1992, coll Xue Dayong (IZCAS).

APPENDIX

CHECKLIST OF THE *EUPITHECIA* (AND SOME ALLIED GENERA) OF CHINA AND ADJACENT AREAS

The checklist below includes both species from China and species from areas adjacent to China, many of which may well eventually be found in China. They are assigned where possible to species groups, as defined in the introduction. Species which have not yet been recorded from China are marked with an asterisk, and known ranges are given for them.

Eupithecia Curtis, 1825

The *haworthiata* species group

E. subtacincta Hampson, 1895
E. cichisa Prout, 1939
E. aenigma Mironov & Galsworthy, 2004
E. zibellinata Christoph, 1880
**E. anasticta* Prout, 1926 [NE India (Sikkim), N Myanmar, Laos]
E. olgae Mironov, 1986
E. illaborata Dietze, 1904
**E. bastelbergeri* Dietze, 1910 [Turkey to Kirghizstan]
E. rubellata Dietze, 1904
**E. fletcheri* Prout, 1926 [Southern Himalayas]
E. bohatschi Staudinger, 1897
E. boneta Mironov & Galsworthy, 2011
E. jezonica Matsumura, 1927
E. sophia Butler, 1878
E. likiangi Vojnits, 1976
E. ustata Moore, 1888
E. delaeveri Vojnits, 1976
**E. lactibasis* Inoue, 2000 [Nepal, NE India (Sikkim)]
E. sinicaria Leech, 1897
E. albigutta Prout, 1958
E. rubridorsata Hampson, 1895

*E. *discolor** Vojnits, 1983 [Nepal]
E. ruficorpus (Warren, 1897)
**E. lusoria* Mironov & Galsworthy, 2009 [Nepal]
E. szelenyica Vojnits & De Laever, 1974
E. lineisdistincta Vojnits, 1981
**E. matura* Vojnits, 1981 [Nepal]
**E. idaeoides* Mironov, 2010 [NE India (Sikkim)]

The *kamburonga* species group

E. liberata Inoue, 2000
E. yoshimotoi Inoue, 1988
E. buysseata Mironov & Galsworthy, 2011
**E. schnitzleri* Mironov & Galsworthy, 2009 [N Thailand]
E. kuroshio Inoue, 1980

The *tenuiata* species group

**E. phaea* Mironov & Galsworthy, 2008 [Pakistan]
**E. niveivena* Prout, 1928 [N Thailand, Sri Lanka]
**E. praepupillata* Wehrli, 1927 [Far East Russia, Korea]

The *brevifasciaria* species group

Eupithecia chingana (Wehrli, 1926)
Eupithecia brevifasciaria Leech, 1897
Eupithecia eurytera (Prout, 1938)
**Eupithecia subrubescens* (Warren, 1888) [N Pakistan and India, Nepal]

The *abietaria* species group

E. costalis (Walker, 1863)
E. rufescens Butler, 1878
E. abietaria (Goeze, 1781)
E. gigantea Staudinger, 1897
E. interrubrescens (Hampson, 1902)
**E. dalhousiensis* Mironov & Galsworthy, 2008 [Jammu & Kashmir]

The *linariata* species group

**E. mirificata* Brandt, 1938 [Iran, S Kazakhstan, Uzbekistan, Kyrgyzstan, Tajikistan, Afghanistan]

The *proterva* species group

E. lucigera Butler, 1889
E. jermyi Vojnits, 1976
E. lini Mironov & Galsworthy, 2007
**E. peregovitsi* Mironov & Galsworthy, 2009 [N Vietnam]
**E. laszloi* Mironov & Galsworthy, 2009 [Nepal]
**E. hreblayi* Mironov & Galsworthy, 2009 [N Thailand]
E. subbreviata Staudinger, 1897
E. proterva Butler, 1878
E. megaproterva Inoue, 1988
E. verprota Mironov & Galsworthy, 2006
**E. niphonaria* Leech, 1897 [Japan]
E. wangi Mironov & Galsworthy, 2011
E. signigera Butler, 1879
**E. thaica* Mironov & Galsworthy, 2009 [N Thailand, N Vietnam]
E. flavoapicaria Inoue, 1979
**E. subfumosa* Inoue, 1965 [Japan]
**E. nagaii* Inoue, 1963 [Japan]
E. opicata Mironov & Galsworthy, 2004
E. blenna Mironov & Galsworthy, 2006
E. convexa Inoue, 1988
E. concava Mironov & Galsworthy, 2007
E. russula Mironov & Galsworthy, 2004
**E. antivulgaria* Inoue, 1965 [Japan]
E. clavifera Inoue, 1955
E. costiconvexa Inoue, 1979
**E. superata* Mironov & Galsworthy, 2009 [NE India (Meghalaya)]
**E. skoui* Mironov & Galsworthy, 2009 [N Thailand]
E. missionerata Mironov & Galsworthy, 2011
E. aritai Inoue, 1977
E. admiranda Mironov & Galsworthy, 2011
**E. macrodisca* Mironov & Galsworthy, 2009 [N Thailand]
**E. tropicata* Mironov & Galsworthy, 2009 [N Thailand]
E. hashimotoi Inoue, 1988
E. flavimacula Mironov & Galsworthy, 2007
**E. szaboi* Mironov & Galsworthy, 2009 [N Thailand]

The *irambata* species group

E. amplificata Inoue, 2000
E. wardi Prout, 1958

E. irambata (Warren, 1893) [NE India (Sikkim)]
E. asempiterna Inoue, 2000

The *ultimaria* species group

E. minusculata Alphéraky, 1883
E. mima Mironov, 1989 [Uzbekistan, Kazakhstan, Pakistan, Mongolia]
E. opistographata Dietze, 1906
E. djakonovi Shchetkin, 1956 [S Kazakhstan, Tajikistan]
E. dominaria Shchetkin, 1956 [Turkmenistan, Uzbekistan, Tajikistan]
E. usbeca Viidalepp, 1992 [Russia, Uzbekistan, Turkmenistan]
E. vicina Mironov, 1989 [Kazakhstan, Mongolia]
E. viata Mironov & Galsworthy, 2004

The *leucostaxis* species group

E. mustangata Schütze, 1961
E. solianikovi Viidalepp, 1988 [Kyrgyzstan, Tajikistan]
E. leucostaxis Prout, 1926
E. thomasi Mironov & Galsworthy, 2008 [N India, Nepal]

The *undata* species group

E. ochracea (Warren, 1888)
E. yakushimensis Inoue, 1980 [Japan]
E. plumbeolata (Haworth, 1809)
E. homogrammata Dietze, 1906
E. severa Mironov & Galsworthy, 2006
E. nigrinotata Swinhoe, 1895 [Nepal, E India (Assam), Thailand]
E. spissata Mironov & Galsworthy, 2006
E. pygmaeata (Hübner, [1799]) [Far East Russia, Japan]
E. vinibua Mironov & Galsworthy, 2008 [Jammu & Kashmir]
E. apparatissima Vojnits, 1988 [Nepal]
E. lilliptana Inoue, 2000 [Nepal]
E. cordata Mironov & Galsworthy, 2004
E. holti Viidalepp, 1973 [Mongolia]
E. kostjuki Mironov, 1989 [Uzbekistan, S Kazakhstan]
E. fennoscandica Knaben, 1949 [N Europe, Russia (Siberia), Mongolia]
E. leptogrammata Staudinger, 1882 [Russia (S Siberia)]
E. hemileuca Hampson, 1895 [N India (Punjab), N Thailand]
E. apta Mironov & Galsworthy, 2004

E. antiqua Mironov & Galsworthy, 2004
E. impavida Vojnits, 1979
E. fervida Vojnits, 1984
E. inexhausta Vojnits, 1984
**E. fusca* Vojnits, 1981 [Nepal]
E. hongxiangae Mironov & Galsworthy, 2011
E. hoenehermanni Mironov & Galsworthy, 2011
E. adoranda Vojnits, 1984
E. granata Vojnits, 1979
E. tamara Mironov & Pekarsky, 2011
E. alishana Inoue, 1970
**E. burmata* Mironov & Galsworthy, 2009 [N Myanmar]
E. assulata Bastelberger, 1911
E. amicula Mironov & Galsworthy, 2004
E. extinctata Dietze, 1904
E. consortaria Leech, 1897
E. burselongata Mironov & Galsworthy, 2004
E. minutula Mironov & Galsworthy, 2004
E. vana Vojnits, 1979
E. nirvana Mironov & Galsworthy, 2006
**E. shachdarensis* Viidalepp, 1988 [Tajikistan]
**E. noncoacta* Vojnits, 1988 [Nepal]
E. peguensis Prout, 1958
E. turpis Vojnits, 1981

The *venosata* species group

**E. silenicolata* Mabille, 1867 [S Europe, Asia Minor, Iran, Jammu & Kashmir]
**E. karli* Ratzel & Mironov, 2008 [Jammu & Kashmir]
E. hesperina Mironov & Galsworthy, 2004
E. hilariata Dietze, 1908
**E. kondarana* Viidalepp, 1988 [Tajikistan]
E. dissertata (Püngeler, 1905)
**E. vacuata* Dietze, 1904 [Uzbekistan, Kazakhstan, Kyrgyzstan]
E. carpophilata Staudinger, 1897
**E. bestia* Mironov & Galsworthy, 2008 [Pakistan]
E. albispumata Warren, 1893
**E. claudei* Mironov & Galsworthy, 2009 [Nepal]
E. astricta Inoue, 1988

The *interruptofasciata* species group

E. abbreviata Stephens, 1831
E. pusillata ([Denis & Schiffermüller], 1775)
E. kozlovi Viidalepp, 1973
E. exrubicunda Inoue, 1988

The *tripunctaria* species group

E. tripunctaria Herrich-Schäffer, 1852
E. barteli Dietze, 1908
**E. invicta* Vojnits, 1981 [N India]
**E. stueningi* Mironov & Galsworthy, 2009 [N Thailand]
E. tenuisquama (Warren, 1896)
E. amandae Galsworthy & Mironov, 2011
E. blandula Mironov & Galsworthy, 2007
E. puella Mironov & Galsworthy, 2004
E. alexiae Galsworthy & Mironov, 2008 [Kashmir]

The *propagata* species group

E. werneri Vojnits, 1985
E. placida Vojnits, 1984
E. batangi Vojnits, 1984
E. refertissima Vojnits, 1984
E. laudabilis Vojnits, 1984
**E. coccinea* Vojnits, 1981 [Nepal]
E. intolerabilis Vojnits, 1984
E. mentita Vojnits, 1984
E. sacrosancta Vojnits, 1979
E. fenita Mironov & Galsworthy, 2004
E. avara Vojnits, 1984
**E. persuastrix* Mironov, 1990 [Far East Russia]
E. caduca Vojnits, 1984
E. manifesta Mironov & Galsworthy, 2011
**E. inoueata* Galsworthy & Mironov, 2005 [N India, Nepal, Bhutan]
E. propagata Prout, 1926
E. balintzsolti Vojnits, 1988
**E. karnaliensis* Inoue, 2000 [Afghanistan, Nepal, India]
**E. russeola* Prout, 1926 [India (Sikkim), N Myanmar, N Thailand]
**E. spinibarbata* Mironov & Galsworthy, 2009 [NE India (West Bengal), N Vietnam]

The *lariciata* species group

E. lariciata (Freyer, 1841)
**E. mesodeicta* Prout, 1938 [NE Pakistan, Jammu & Kashmir, India]
**E. pannosa* Mironov & Galsworthy, 2008 [NE Pakistan, N India, Nepal, N Thailand, N Vietnam]
**E. vetula* Mironov & Ratzel, 2008 [NE Pakistan]
**E. firmata* Mironov & Ratzel, 2008 [NE Pakistan]
E. sacrivicae Mironov & Galsworthy, 2004
E. emanata Dietze, 1906
E. honesta Mironov & Galsworthy, 2004
**E. damnosa* Vojnits, 1983 [Nepal]
**E. maenamiella* Inoue, 1980 [Japan]
E. marmorea Vojnits, 1983
E. sola Vojnits, 1983
E. conjunctiva Hampson, 1895
E. pediba Vojnits, 1981
E. argentea Mironov & Galsworthy, 2004
E. horrida Vojnits, 1980
E. stomachosa Vojnits, 1980
E. albimedia Mironov & Galsworthy, 2004
E. hebes Mironov & Galsworthy, 2004
E. primitiva Mironov & Galsworthy, 2004
E. stataria Inoue, 1988

The *lanceata* species group

E. lanceata (Hübner, [1825])
E. kudoi Inoue, 1983
E. daemionata Dietze, 1904
**E. takao* Inoue, 1955 [Far East Russia, Japan]
E. habermani Viidalepp & Mironov, 1988
E. emendata Vojnits, 1983
**E. shikokuensis* Inoue, 1980 [Japan]
**E. filia* Mironov & Galsworthy, 2009 [Nepal]
E. dolichoptera Galsworthy & Mironov, nom. nov.
E. phantastica Mironov & Galsworthy, 2006
E. atuni Mironov & Galsworthy, 2006
E. funerea Inoue, 1988
**E. tantilloides* Inoue, 1958 [Far East Russia, Japan]
E. fulvidorsata Mironov & Galsworthy, 2006

E. okadai Inoue, 1958
E. delicata Mironov & Galsworthy, 2004
E. camilla Mironov & Galsworthy, 2004
E. julia Mironov & Galsworthy, 2004
E. fortis Mironov & Galsworthy, 2004

The *fletcherata* species group

E. hainanensis Mironov & Galsworthy, 2004
E. groenblomi Urbahn, 1969
E. selinata Herrich-Schäffer, 1861
E. arenosa Vojnits, 1984
E. levata Mironov & Galsworthy, 2004
E. molestissima Vojnits, 1981
E. actaeata Walderdorff, 1869
E. actrix Mironov & Galsworthy, 2006
E. uliata Staudinger, 1897
E. detritata Staudinger, 1897
E. jinboi Inoue, 1976
E. pellicata Mironov & Galsworthy, 2007
E. irreperta Vojnits & De Laever, 1978
E. vivida Vojnits & De Laever, 1978
**E. acutangula* Hampson, 1895 [N India]
E. quadripunctata Warren, 1888
E. fatigata Vojnits & De Laever, 1978
E. repentina Vojnits & De Laever, 1978
E. virgaureata Doubleday, 1861
**E. latimedia* Hampson, 1895 [N India]
E. nepalata Schütze, 1961
E. eximia Vojnits & De Laever, 1978
**E. karakasykensis* Viidalepp, 1988 [Kyrgyzstan]
E. taiwana Wileman & South, 1917
E. lata Vojnits, 1981
E. cotidiana Vojnits, 1979
E. krampli Vojnits, 1979
**E. concinna* Vojnits, 1983 [Nepal]
E. dealbata Inoue, 1988
**E. ryukyuensis* Inoue, 1971 [Japan]
**E. mira* Vojnits, 1988 [Nepal, N Thailand]
**E. acyrtoterma* Prout, 1926 [Myanmar]
E. karapinensis Wileman & South, 1917

E. kobayashii Inoue, 1958
E. incorrupta Vojnits, 1984
E. matrona Mironov & Galsworthy, 2004
E. rhadine Mironov & Galsworthy, 2007
E. leucenthesis Prout, 1926
**E. thermosaria* Hampson, 1903 [Kyrgyzstan, Tajikistan, Jammu & Kashmir]
**E. scortillata* Dietze, 1904 [Kyrgyzstan, Tajikistan]
E. pimpinellata (Hübner, [1813])
E. denotata (Hübner, [1813])
**E. assectata* Dietze, 1904 [Uzbekistan, Kyrgyzstan, Tajikistan, Pakistan]
E. chesiata Dietze, 1904
E. inopinata Vojnits, 1984
**E. improba* Vojnits, 1983 [Nepal]
**E. dzhirgatalensis* Viidalepp, 1988 [Kyrgyzstan, Tajikistan, Pakistan, Jammu & Kashmir]
**E. transalaiensis* Viidalepp, 1988 [Kyrgyzstan, Tajikistan]
**E. pamirica* Viidalepp, 1988 [Kyrgyzstan, Tajikistan, Pakistan]
**E. anemica* Viidalepp, 1988 [SE Kazakhstan, Uzbekistan, Tajikistan]

The *rajata* species group

E. mundiscripta (Warren, 1907)
E. robiginascens Prout, 1926
**E. spilocyma* Prout, 1931 [Philippines]
E. luctuosa Mironov & Galsworthy, 2004
E. qinlingata Mironov & Galsworthy, 2011
**E. swanni* Mironov & Galsworthy, 2009 [N Myanmar]
**E. vojnitsi* Inoue, 2000 [Nepal]
E. svetlanae Mironov & Pekarsky, 2011
**E. fessa* Mironov & Galsworthy, 2009 [Nepal, NE India]
**E. fulvipennis* Butler, 1889 [N India, Nepal]
E. contraria Vojnits, 1983
E. lobbichlerata Schütze, 1961
E. rajata Guenée, [1858]
**E. nervosa* Mironov & Galsworthy, 2010
**E. himalayata* Mironov & Galsworthy, 2010 [Nepal]
E. vasta Vojnits, 1979
E. coniurata Vojnits, 1979
E. cuneata Vojnits, 1984
**E. xerophila* Inoue, 2000 [Nepal, NE India]
E. particeps Vojnits, 1988

E. sempiterna Vojnits, 1984
E. serpentigena Mironov & Galsworthy, 2006
E. ebriosa Vojnits, 1979
**E. laoica* Mironov & Galsworthy, 2009 [Laos]
**E. leucospila* (Swinhoe, 1906) [India (Assam), N Thailand]
E. raniata Prout, 1958
**E. danialena* Mironov & Galsworthy, nom. nov. [N Thailand, W Myanmar] (Replacement name for *E. ronkayi* Mironov & Galsworthy, 2009 praeocc.)

The *atrisignis* species group

E. atrisignis Butler, 1889
E. pyricoetes Prout, 1958
**E. garuda* Galsworthy & Mironov, 2005 [Nepal, NE India, N Thailand, Laos]
E. nuceistrigata Bastelberger, 1911
E. tricrossa Prout, 1926
**E. albibaltea* Prout, 1958 [NE India]
E. yazakii Inoue, 1988

The *suboxydata* species group

E. suboxydata Staudinger, 1897
E. formosa Vojnits & De Laever, 1973
**E. incurvaria* Hampson, 1903 [Jammu & Kashmir, Nepal, India (Sikkim)]
E. subbrunneata Dietze, 1906

The *sinuosaria* species group

E. sinuosaria (Eversmann, 1848)
E. pallescens Dietze, 1910
**E. nigrilinea* (Warren, 1896) [Uzbekistan, Kyrgyzstan, Tajikistan, Afghanistan, Pakistan, Jammu & Kashmir, India, Nepal, N Thailand]
**E. acolpodes* Prout, 1938 [Jammu & Kashmir]

The *innotata* species group

E. innotata (Hufnagel, 1767)
E. parallelaria Bohatsch, 1893
E. corroborata Dietze, 1908
E. praesignata Bohatsch, 1893
**E. myoma* Vojnits, 1988 [Kyrgyzstan]

*E. *efferata* Mironov & Ratzel, 2008 [Pakistan, Jammu & Kashmir]
E. paupera Dietze, 1913
E. mitigata Dietze, 1906
*E. *tshimganica* Viidalepp, 1988 [Uzbekistan, Tajikistan]
E. vicariata Dietze, 1904
E. magica Mironov & Galsworthy, 2006
E. relaxata Dietze, 1904

The *graphata* species group

E. perendina Vojnits, 1980
E. ancillata Mironov & Galsworthy, 2004
E. simpliciata (Haworth, 1809)
*E. *tshimkentensis* Viidalepp, 1988 [S Kazakhstan, Tajikistan]
*E. *relictata* Dietze, 1904 [Kyrgyzstan]
*E. *saisanaria* Staudinger, 1882 [SE Kazakhstan (Saisan)]

The *irriguata* species group

*E. *indigata* (Hübner, [1813]) [W Europe to Far East of Russia]
E. conterminata (Lienig & Zeller, 1846)

The *centaureata* species group

*E. *egregiata* Mironov & Ratzel, 2008 [Afghanistan, Pakistan]
*E. *brunneomarginata* Mironov & Galsworthy, 2008 [Pakistan]
E. centaureata ([Denis & Schiffermüller], 1775)
*E. *repetita* Vojnits, 1981 [Pakistan]
*E. *subtilis* Dietze, 1910 [Iran, Uzbekistan, Kyrgyzstan, Tajikistan, Afghanistan]
*E. *decipiens* Petersen, 1910 [Iran, Afghanistan, Kyrgyzstan]
E. exquisita Vojnits, 1979
E. brunneilutea Mironov & Galsworthy, 2004
E. lamata Mironov & Galsworthy, 2004

The *despectaria* species group

E. despectaria Lederer, 1853
*E. *peridonea* Mironov, 1989 [SE Kazakhstan, Kyrgyzstan]

The *variostrigata* species group

*E. *variostrigata* Alphéraky, 1876 [S Europe, Near East to Tajikistan (Pamirs)]

The *nevadata* species group

E. insignioides Wehrli, 1923
E. costimacularia Leech, 1897

The *terrenata* species group

**E. marnoti* Viidalepp, 1988 [Uzbekistan, S Kazakhstan]

The *gueneata* species group

E. molybdaena Mironov & Galsworthy, 2004
E. gratiosata Herrich-Schäffer, 1861
**E. gluptata* Dietze, 1904 [Turkmenistan, Kyrgyzstan]
E. subpulchrata Alphéraky, 1883
**E. albertiata* Schütze, 1961 [Iran, Uzbekistan, Kyrgyzstan, Tajikistan]
**E. dechkanata* Mironov, 1989 [Turkmenistan, Tajikistan]
**E. remmi* Viidalepp, 1988 [Uzbekistan, S Kazakhstan, Kyrgyzstan, Tajikistan]
**E. accurata* Staudinger, 1892 [Iran, Turkmenistan, Uzbekistan, Kyrgyzstan]
**E. kruusi* Viidalepp, 1988 [Tajikistan]
**E. cingulata* Christoph, 1885 [Turkmenistan, Tajikistan]

The *cretaceata* species group

E. duplex Sterneck, 1931
E. veratraria Herrich-Schäffer, 1848
E. tectaria Mironov & Galsworthy, 2011
E. mandschurica Staudinger, 1897

The *satyrata* species group

E. intricata (Zetterstedt, 1839)
E. lupa Mironov & Galsworthy, 2007
E. satyrata (Hübner, [1813])
E. amplexata Christoph, 1881
**E. scribai* Prout, 1938 [Far East Russia, Korea, Japan]
E. extensaria (Freyer, 1845)
E. absinthiata (Clerck, 1759)
E. interpunctaria Inoue, 1979
E. korbi Dietze, 1908
E. cervina Mironov & Galsworthy, 2004
E. ensifera Mironov & Galsworthy, 2004

E. ultrix Mironov & Galsworthy, 2004
**E. lineosa* Moore, 1888 [Jammu & Kashmir, India (West Bengal)]
E. impolita Vojnits, 1980
E. neosatyrata Inoue, 1979
E. indissolubilis Vojnits, 1979

The *russeliata* species group

E. assimilata Doubleday, 1856
E. andrasi Mironov & Galsworthy, 2004
**E. pseudassimilata* Viidalepp & Mironov, 1988 [Far East Russia, Japan]
E. nigristriata Mironov & Galsworthy, 2004
**E. acosmos* Mironov, 1989 [Turkmenistan, Kazakhstan, Kyrgyzstan, Tajikistan]
E. sacrimontis Prout, 1938
E. vulgata (Haworth, 1809)
E. citraria Mironov & Galsworthy, 2004
E. nobilitata Staudinger, 1882
E. rebeli Bohatsch, 1893
E. rubeni Viidalepp, 1976
E. sutiliata Christoph, 1876 [SE Europe, Near East to Tajikistan]
E. biornata Christoph, 1867
E. kuldschaensis Staudinger, 1892
**E. cohorticula* Dietze, 1910 [Kyrgyzstan]

The *addictata* species group

E. thalictrata (Püngeler, 1902)
E. dura Vojnits, 1981
E. addictata Dietze, 1908
**E. caliginea* Butler, 1878 [Japan]

The *druentiata* species group

E. exiguata (Hübner, [1813])
E. hannemanni Vojnits & De Laever, 1973
E. bella Staudinger, 1897
E. insolita Vojnits & De Laever, 1973

The *millefoliata* species group

E. omissa Mironov & Galsworthy, 2004
**E. millefoliata* Rössler, 1866 [W Europe to Kazakhstan, Uzbekistan, Kyrgyzstan, Tajikistan]

E. nephelata Staudinger, 1897
E. exactata Staudinger, 1882

The *semigraphata* species group

E. succenturiata (Linnaeus, 1758)
**E. icterata* (De Villers, 1789) [Europe, Caucasus, SE Kazakhstan]
**E. exicterata* Mironov & Ratzel, 2008 [Jammu & Kashmir]
E. marginata Staudinger, 1892
**E. herczigi* Mironov & Galsworthy, 2008 [N Thailand]
E. subumbrata ([Denis & Schiffermüller], 1775)
**E. orphnata* Petersen, 1909 [Europe, S Kazakhstan, Kyrgyzstan, Tajikistan]
E. masculina Inoue, 1988

The *subfuscata* species group

E. subfuscata (Haworth, 1809)
E. mediocincta Mironov & Galsworthy, 2004
E. chui Inoue, 1970
E. tibetana Mironov & Galsworthy, 2004
E. tepida Mironov & Galsworthy, 2004
E. nishizawai Inoue, 1988
E. praecipitata Vojnits, 1979
E. lasciva Vojnits, 1980
E. egena Vojnits, 1984
**E. phulchokiana* Herbulot, 1984 [Nepal, India (Uttar Pradesh, Assam)]
E. lunatica Vojnits, 1984
E. otiosa Vojnits, 1981
**E. fibigeri* Mironov & Galsworthy, 2009 [Nepal]
**E. maculosa* Vojnits, 1981 [Nepal, NE India]
E. acutipapillata Inoue, 1988
E. supercastigata Inoue, 1958
E. costipicta Warren, 1893

Not assigned to species group

E. concremata Dietze, 1904
E. coribalteata Mironov & Galsworthy, 2004
**E. costinotata* Inoue, 2000 [Nepal]
**E. kuni* Mironov & Galsworthy, 2009 [N Vietnam]
E. lilliputata Mironov & Galsworthy, 2011

*E. *lindti* Viidalepp, 1988 [Uzbekistan, S. Kazakhstan, Tajikistan, Afghanistan, Jammu & Kashmir]
E. salubris Mironov & Galsworthy, 2004
*E. *seditiosa* Vojnits, 1981 [Nepal]
E. villica Mironov & Galsworthy, 2006

Pareupithecia Mironov & Galsworthy, 2012

P. spadix Inoue, 1955

Eva Vojnits, 1981

E. flexa Vojnits, 1981

Girida Mironov & Galsworthy, 2012

G. rigida Swinhoe, 1892

Mesoptila Meyrick, 1891

M. melanolopha Swinhoe, 1895

BIBLIOGRAPHY

Agassiz, D.J.L. & other members of the BENHS, 1981. An identification guide to the British pugs (Lepidoptera: Geometridae). BENHS, London, 56 pp.

Alphéraky, S., 1883. Lépidoptères du district de Kouldjà et des montagnes environnantes, III. Geometrae. *Horae Soc. ent. Ross.* [1882] 17 (3/4): 156–227, pls VIII, IX.

——, 1892. Lépidoptères rapportés de la Chine et de la Mongolie par G.N. Potanine. II. Heterocera. *Mém. Lépid. Romanoff* 6: 1–81, pls. I-III.

——, 1897. Lépidoptères rappotés par M-r Gr. Groum-Grshimaïlo de l'Asie Centrale en 1889–1890. *Mém. Lépid. Romanoff* 9: 1–81.

Bastelberger, S.R., 1911. Neubeschreibungen von Geometriden aus den Hochgebirge von Formosa. *Int. ent. Z., Guben* 4 (45): 241–242.

Bidzilya, A.V. & Z. Ph. Kljuchko, 1994. The Lepidoptera of the Lazo Reserve (An annotated Checklist). Kiev, Zool. Mus. Shevchenko's Univ., 44 pp., 43 photos (in Russian).

Bohatsch, O., 1882. Revision der Eupithecien Treitschke's. *Wien. Ent. Zeitung* 1: 279–282, 308–312.

——, 1893. Mittheilungen über Eupithecien. *Dt. ent. Z., Iris* 6 (1): 1–35.

Bolte, K.B., 1990. Guide to the Geometridae of Canada (Lepidoptera). VI. Subfamily *Larentiinae*. 1. Revision of the genus *Eupithecia*. *Mem. ent. Soc. Can.* **151**: 1–253.

Borkhausen, M.B., 1794. Naturgeschichte der Europäischen Schmetterlinge nach systematischer Ordnung. 5. Theil: der Phalaenen dritte Horde: Spanner. Frankfurt, S. 1–572.

Bryk, F., 1948. Zur Kenntnis der Grossschmetterlinge von Korea. *Arkiv för Zoologi* **41A** (1): 1–225, pls. 1–7.

Buckler, W., 1899. The Larvae of the British Butterflies and Moths, **8**: i-xii, 1–120, pls. 128–147. London.

Butler, A.G., 1878. Descriptions of new Species of Heterocera from Japan, Part III. Geometrites. *Ann. Mag. nat. Hist.* (5) **1** (no. 6): 440–452.

——, 1879. Descriptions of new Species of Lepidoptera from Japan. Geometrites. *Ann. Mag. nat. Hist.* (5) **4** (no. 24): 437–457.

——, 1880. On a second Collection of Lepidoptera made in Formosa by H.E. Hobson, Esq. *Proc. zool. Soc. Lond.* **1880**: 666–691.

——, 1889. Illustrations of typical Specimens of Lepidoptera Heterocera in the Collection of the British Museum 7. London, 124 pp., pls. cxxi–cxxxviii.

Chang, B.S., 1989–1991. Illustrated moths of Taiwan. Taipei: Taiwan Mus. 5 v. Pages (in Chinese).

Curtis, J., 1829. A Guide to an arrangement of British Insects; being a Catalogue of all the named species hitherto discovered in Great Britain and Ireland. 256 pp., London.

Dantart, J., 2000. Llista sistemàtica dels geomètrids de Catalunya (Lepidoptera: Geometridae). *Treb. Soc. Cat. Lep.* **15**: 121–182 (in Catalan with English summary).

Dietze, K., 1904. Beiträge zur Kenntnis der Eupithecien. *Dt. ent. Z., Iris* [1903] **16** (2): 331–387, pls. III-V.

——, 1908. Beiträge zur Kenntnis der Eupithecien. *Dt. ent. Z., Iris* **21** (2–3): 153–201, pls. II, III.

——, 1910. Biologie der Eupithecien. I. Berlin, Tafn. 1–82.

——, 1913. Biologie der Eupithecien. II. Berlin, 173 S., 86 Taf.

Djakonov, A.M., 1908. K faune Geometridae Semirechja i Semipalatinskoj oblasti [Data to the fauna of Geometridae of Semirechje and Semipalatinsk region]. *Russ. entomol. obozr.* **1908** (1): 25–31 (in Russian).

——, 1926. Zur Kentniss der Geometriden Fauna des Minussinsk. *Jahrb. Martjanov. Staatsmus. Minussinsk* **4** (1): 1–78 (in Russian).

——, 1927. Berichtigungen und Zusätze zur Geometriden Fauna des Minussinsk. *Jahrb. Martjanov. Staatsmus. Minussinsk* 5 (1): 53–58 (in Russian).
——, 1929. Entomologische Ergebnisse der schwedischen Kamtchatka-Expedition 1920–1922. *Arkiv för Zoologi* **21A** (1): 1–24, Taf. 1–2.
——, 1931. Fauna pjadenits (Lepidoptera, Geometridae) Kamtchatki [Die Geometridenfauna von Kamtshatka]. *Ann. Mus. Zool. Acad. Sci. USSR* **32**: 385–410.
——, 1936. Schwedisch-chinesische wissenschaftliche Expedition nach den nordwestlichen Provinzen Chinas. 57. Lepidoptera. 5. Geometridae. *Arkiv för Zoologi* **27A** (39): 1–67.
Doubleday, H., 1856. List of the British Eupitheciae, with Notes on some of the Species. *The Zoologist*, **14**: 5139–5144.
Duponchel, P.A.J., 1830. Histoire naturelle des Lépidoptères ou papillons de France. Tome 8 (1). Nocturnes 5: 1–598, pls. CLXXI-CCX.
Funakoshi, S., 2008. Geometrid species feeding on *Quercus glauca* Thunb. (Fagaceae). *Yugato* **191**: 23–29 (in Japanese).
Galsworthy, A.C., 1999. New and revised eupitheciine species (Geometridae, Larentiinae) from Hong Kong and South East Asia. *Trans. lepid. Soc. Japan* **50** (3): 223–234.
Galsworthy, A. & V. Mironov, 2005. *Eupithecia atrisignis* Butler, 1889 (Lepidoptera, Geometridae), its relatives, and related problems. *Trans. lepid. Soc. Japan* **56** (3): 223–236.
Guenée, M.A., [1858] 1857. *In*: Boisduval, M.M. & M.A. Guenée. Histoire Naturelle des Insectes. Species général des Lépidoptères. Uranides et Phalénites, T. 10 (2). Paris, 584 pp.
Gumppenberg, C.F. v., 1888–1890. Systema Geometrarum zonae temperatioris septentrionalis. Systematische Bearbeitung der Spanner der nördlichen gemässigten Zone, II, III. *Nova Acta Ksl. Leop.-Carol. Dt. Akad. Naturforsch.* **54** (4): 125–255, 269–431.
Hampson, G.F., 1895. The Fauna of British India, including Ceylon and Burma (Moths) **3**. London, xxviii+546 pp.
——, 1896. The Fauna of British India, including Ceylon and Burma (Moths) **4**. London, xxviii+594 pp.
——, 1907. The Moths of India. Supplementary paper to the volumes in "The Fauna of British India", Series III, part IX. *J. Bombay Nat. Hist. Soc.* **18** (1): 27–53.
Hashimoto, S., 1981. Notes on five species of the genus *Eupithecia* (Geometridae). *The Jap. Heterocerists' J.* **110**: 147–149.
Hedemann, W., 1881. Beitrag zur Lepidopteren-Fauna des Amur-Landes. *Horae Soc. ent. Ross.* **16**: 257–272, pl. XIII.
Heinemann, v. H. 1859. Die Schmetterlinge Deutschlands und der Schweiz. Bd. 1 (Grossschmetterlinge): i-xxiii, 1–850, 1–118, Braunschweig.
Heppner, J.B. & H. Inoue, 1992. Lepidoptera of Taiwan. Vol. 1 (2): Checklist, Sci. Publ., 276 pp.
Herbulot, C., 1984. Nouveaux Larentiinae du Nepal (Lep. Geometridae). *Miscellanea ent.* **50** (2): 41–45.
Herrich-Schäffer, G.A.W., 1848 [1847]. Systematische Bearbeitung der Schmetterlinge von Europa, zugleich als Text, Revision und Supplement. Die Spanner **3**: 1–184, 91 tabn. Regensburg, 1847.
——, 1861. Die Arten der Spannergattung Eupithecia. *Correspondenzblatt für Sammler von Insecten, insbesondere von Schmetterlingen* **2** (16): 121–125; (17): 129–133.
Herting, B., 1976. A Catalogue of Parasites and Predators of terrestrial arthropods. Sect. A, Host or prey/enemy. Vol. 7 Lepidoptera, Part 2 (Macrolepidoptera). Commonwealth agricultural Bureaux. (Commonwealth institute of Biol. Control), G. Britain, 221 pp.
Hirashima, Y., 1989. Check List Japanese Insects II: 100–1001.
Holloway, J.D., 1976. Moths of Borneo with special reference to Mount Kinabalu. Kuala Lumpur: Malayan Nat. Soc. pp. 1–264.
——, 1997. The moths of Borneo: family Geometridae, subfamilies Sterrhinae and Larentiinae. *Malay. Nat. J.* **51**: 1–242, pls 1–12, 608 figs. [The Moths of Borneo, part 10].

Inoue, H., 1941. On some Geometridae of Corea. *Mushi* 14 (1): 11–28.
——, 1946. A Catalogue of the Geometridae of Corea. *Bull. Lepid. Soc. Japan.* 1 (2): 19–59.
——, 1954a. New Geometridae from Japan. 1. *Kontyû* 22 (1/2): 29–35.
——, 1954b. New Geometridae from Japan. 2. *Kontyû* 22 (3/4): 71–79, pls. vi-vii.
——, 1955. Descriptions and records of some Japanese Geometridae. *Tinea* 2 (1/2): (73)-(89), pls. vi-vii.
——, 1956. Synonymic notes on some moths from Japan and adjacent countries. *Bull. Lab. zool. Gen. e Agr. "F. Silvestri" Portici* 33: 657–672.
——, 1956b. Check List of Lepidoptera of Japan. Part 3: 219–364. Rikusuisha, Tokyo.
——, 1958. Descriptions and records of some Japanese Geometridae (II). *Tinea* 4 (2): (241)-(256), pl. xxxiv.
——, 1963. Descriptions and records of some Japanese Geometridae (III). *Tinea* 6 (1/2): (29)-(39), pl. vii.
——, 1965. Descriptions and records of some Japanese Geometridae (IV). *Tinea* 7 (1): (102)-(111), pl. xviii.
——, 1970. Some new species and subspecies of the Geometridae from Taiwan (Lepidoptera). *Bull. Fac. domestic Sci., Otsuma Wom. Univ.* 6: 1–5, pls. i–iii.
——, 1972. Two new species of the Geometridae from Japan (Lepidoptera). *Tinea* 9 (1): 231–234, pls 72–73.
——, 1977. Catalogue of the Geometridae of Japan (Lepidoptera). *Bull. Fac. domestic Sci., Otsuma Wom. Univ.* 13: 227–346.
——, 1978. New and unrecorded species of the Geometridae from Taiwan with some synonymic notes (Lepidoptera). *Bull. Fac. domestic Sci., Otsuma Wom. Univ.* 14: 203–254.
——, 1979. Revision of the genus *Eupithecia* of Japan, Part 1 (Lepidoptera, Geometridae). *Bull. Fac. domestic Sci., Otsuma Wom. Univ.* 15: 157–224.
——, 1980. Revision of the genus *Eupithecia* of Japan, Part 2 (Lepidoptera, Geometridae). *Bull. Fac. domestic Sci., Otsuma Wom. Univ.* 16: 153–213.
——, 1982. Geometridae of eastern Nepal based on the collection of the lepidopterological research expedition to Nepal Himalaya by the Lepidopterological Society of Japan in 1963. Part II. *Bull. Fac. domestic Sci., Otsuma Wom. Univ.* 18: 129–190.
——, 1986. Descriptions and records of some Japanese Geometridae (VI). *Tinea* 12 (7): 45–71.
——, 1987. Geometridae of eastern Nepal based on the collection of the Lepidopterological Research expedition to Nepal Himalaya by the Lepidopterological Society of Japan in 1963. Part III. *Bull. Fac. domestic Sci., Otsuma Wom. Univ.* 23: 215–270.
——, 1988. The genus *Eupithecia* Curtis from Taiwan (Lepidoptera: Geometridae). *Bull. Fac. domestic Sci., Otsuma Wom. Univ.* 24: 323–375.
——, 1992. *In*: Heppner, J.B. & H. Inoue. Lepidoptera of Taiwan. Vol. 1, Part 2: Checklist. 276 pp. Gainesville, Florida: Association for Tropical Lepidoptera and Scientific Publishers.
——, 1994. The Geometridae of the Ogasawara Islands (Lepidoptera). [Moths of Ogasawara (Bonin) Islands, Part II]. *Bull. Fac. domestic Sci., Otsuma Wom. Univ.* 30: 195–223.
——, 1996. Four new species of the genus *Eupithecia* (Geometridae, Larentiinae) from Kashmir and Pakistan. *Trans. lepid. Soc. Japan* 47 (4): 237–242.
——, 1998. Four unrecorded species of *Eupithecia* (Geometridae, Larentiinae) from Japan. *Jap. Heterocerists' J.* 200: 433–437.
——, 2000a. *Eupithecia* Curtis (Geometridae, Larentiinae) from Nepal. *In*: Haruta, T. (ed.), Moths of Nepal, part 6. *Tinea* 16 (Supplement 1): 27–44.
——, 2000b. Descriptions of males of *Eupithecia caliginea* Butler and *E. fujisana* Inoue with a list of species-group taxa of *Eupithecia* changed and newly joined after 1982 to Japan (Geometridae). *Yugato* 160: 49–53 (in Japanese).
——, 2002. Notes on tribe Eupitheciini (Geometridae, Larentiinae) from Japan, Taiwan and Southeast Asia, with descriptions of ten new species. *Tinea* 17 (2): 92–112.

Inoue, H., Sugi, S., Kuroko, H., Moriuti, S. & A. Kawabe, 1982. *Moths of Japan*. Tokyo: Kodansha. Vol. **1**: 1–968; Vol. **2**: 1–556, pls. 1–392. [in Japanese; English summary].

International Code of Zoological Nomenclature, 1999. Fourth edition, adopted by the International Union of Biological Sciences. The International Trust for Zoological Nomenclature, c/o The Natural History Museum, London, XXX + 306 pp.

Joannis, J. de. 1929. Lépidoptères Hétérocères du Tonkin (2e partie). *Ann. Soc. ent. France* **XCVIII**: 361–552, pls. 3–6.

Kaila, L., Viidalepp, J., Mikkola, K. & V. Mironov, 1996. Geometridae (Lepidoptera) from the Tian-Shan Mountains in Kazakhstan and Kyrgyzstan, with descriptions of three new species and one new subspecies. *Acta Zool. Fennica* **200**: 57–82.

Kameda, M., 1995. Four species of the Larentiinae (Geometridae) from the sou[h]thern part of Hokkaido. *Yugato* **141**: 139–140 (in Japanese).

——, 2002. Records of *Eupithecia* (Geometridae, Larentiinae) from the southern part of Hokkaido. *Yugato* **170**: 137–142 (in Japanese).

——, 2006. Records of the genus *Eupithecia* (Geometridae, Larentiinae) from Hokkaido. *Yugato* **184**: 48–50 (in Japanese).

——, 2007. *Eupithecia extensaria* (Freyer) (Geometridae) captured at Southern Hokkaido. *Yugato* **189**: 99 (in Japanese).

——, 2008. Records of *Eupithecia* (Geometridae, Larentiinae) collected by Mr H. Kogi from Hokkaido. *Yugato* **192**: 37–42 (in Japanese).

Kameda, M. & H. Kogi, 2008. *Eupithecia caliginea* Butler (Geometridae) newly recorded from Hokkaido. *Yugato* **191**: 18 (in Japanese).

Kaneko, T., 2006. Moths of Sado Island, Niigata Prefecture. *Yugato*, Suppl. **10**: 1–56 (in Japanese).

Korshunov, Yu.P. & J.R. Viidalepp, 1982. The Geometrid moths (Lepidoptera, Geometridae) of Khakasia. Part 2. *In*: Poleznye i vrednye nasekomye Sibiri. Novosibirsk, Publ. "Nauka", pp. 101–107 (in Russian).

Kostjuk, I.Yu., 1992. On the fauna of Geometrid moth (Lepidoptera, Geometridae) of S.-E. Transbaikalia. *In*: Insects of Dahuria and adjacent territories. Moscow, pp. 52–64 (in Russian).

Kostjuk, I.Yu., Budashkin, Yu.I. & M.I. Golovushkin, 1994. The Lepidoptera of the Dahursky Nature Reserve (An annotated Checklist). Shmalhausen Inst. Zool., Acad. Sci. Ukraine, Kiev, 36 pp. (in Russian).

——, 1994. *Eupithecia addictata*, an Asian pug moth confused in Europe with *E. thalictrata* (Lepidoptera: Geometridae). *Eur. J. Entomol.* **91** (3): 335–348, figs 1–40.

Kurentzov, A.I. & L.A. Ivliev, 1960. O vrediteljah kedrovogo stlanika na Kamchatke [About pests of mountain pine in Kamchatka]. *Izvestia Sib. otdel. AN USSR*, **11**: 97–103 (in Russian).

Laever, E. De, 1956. Études sur le Genre *Eupithecia* Curtis. *Bonn. zool. Beitr.* 7 (1–3): 237–247, pls. i-iv.

Lederer, J., 1960. Études sur le Genre *Eupithecia* Curtis. *Bonn. zool. Beitr.* **11** (1): 114–123, pls. v–vii.

——, 1853. Die Spanner. *Verh. Zool.-bot. Ver. Wien* **3**: 165–270, 2 tafn.

——, 1853. Lepidopterologisches aus Sibirien. *Verh. Zool.-bot. Ver. Wien* **3**: 351–386, pls. 1–7.

Leech, J.H., 1897. On Lepidoptera Heterocera from China, Japan, and Korea. Part II. Family Geometridae; Subfamilies Oenochrominae, Orthostixinae, Larentiinae, Acidaliinae, and Geometrinae. *Annals Mag. Nat Hist., incl. Zoology, Botany and Geology (6th ser.)* **20** (115): 65–110.

——, 1931. Die Insektenfauna Islands und ihre Probleme. *Zool. Bidrag. Uppsala* **13**: 1–599.

Matsumura, S., 1905. Catalogus Insectorum Japonicum (Lepidoptera). Tokyo, 307 pp.

Mayr, E., 1969. Principles of Systematic Zoology. McGraw-Hill Book Company. 428 pp.

——, 1977. Evolution and the Diversity of Life. The Belknap Press of Harvard Univ. Press, Cambridge, Massachusetts and London, England. Second printing, 721 pp.

McDunnough, J., 1929. Synonymic notes on Canadian *Eupithecia* (Geomet., Lepid.). *Can. Entomol.* **61** (3): 59–69.
——, 1932. Note on two Canadian species of *Eupithecia* (Geom., Lepid.). *Can. Entomol.* **64** (1): 2.
——, 1936. Further notes on the synonymy of North American *Eupithecia* (Lepid., Geom.). *Can. Entomol.* **68** (11): 257–259.
——, 1949. Revision of the North American species of the genus *Eupithecia* (Lepidoptera, Geometridae). *Bull. Amer. Mus. Nat. Hist.* **93** (8): 533–734, pl. 26–32.
McGuffin, W.C., 1958. Larvae of the Nearctic Larentiinae (Lepidoptera: Geometridae). *Can. Entomol.* **90** (Suppl. 8): 1–104.
Meyrick, E., 1892. On the classification of the Geometrina of the European Fauna. Part 1. *Trans. ent. Soc. Lond.* **1892**: (53)-(140), pl. III.
Mikkola, K., 1985. The Geometroidea and Noctuoidea described by Carl Clerck (Lepidoptera). *Ent. Scand.* **16**: 121–129.
——, 1994. Inferences about the function of genitalia in the genus *Eupithecia*, with description of a new organ (Lepidoptera, Geometridae). *Nota lepid.*, Suppl. **5**: 73–78.
Millière, P., 1869 (1869–1874). Iconographie et description de Chenilles et Lépidoptères inédits. *Annls Soc. Linn. Lyon* **3** (23–35): 1–488, 101–154 pls.
Mironov, V.G., 1986. New and little known species of the Geometrid moths (Lepidoptera, Geometridae) from Inner Tien-Shan Mountains. *Trudy VEO* **67**: 94–100 (in Russian).
——, 1989a. *Eupithecia* (Lepidoptera, Geometridae) from Southern Siberia, with description of a new species. *Ann. ent. Fennici* **55**: 145–152.
——, 1989b. New and little known species of the Geometrid moths of the genus *Eupithecia* Curt. (Lepidoptera, Geometridae) from Middle Asia. *Ent. obozr.* **68** (2): 369–375 (in Russian).
——, 1989c. New *Eupithecia* species (Lepidoptera, Geometridae) from the Middle Asia and Mongolia. *Vestnik zool.* **1989** (3): 12–19 (in Russian).
——, 1989d. A review of Geometrid moths of the group '*gueneata*' of genus *Eupithecia* Curt. (Lepidoptera, Geometridae). *Proc. Zool. Inst., Leningrad* **200**: 110–135 (in Russian).
——, 1990. A Systematic Catalogue of tribe *Eupitheciini* (Lepidoptera, Geometridae) of the USSR, I. *Ent. obozr.* **69** (3): 656–670 (in Russian).
——, 1991. A Systematic Catalogue of tribe Eupitheciini (Lepidoptera, Geometridae) of the USSR, II. *Ent. obozr.* **70** (1): 157–167 (in Russian).
——, 2001. Problems in the study of the tribe Eupitheciini (Lepidoptera: Geometridae, Larentiinae). *In*: Hausmann, A. & R. Trusch (ed.). Proceedings of the Forum Herbulot 2001 Neotropical Geometridae: Approaches to a Modern System on Genus and Tribe Level (8.3.-9.3.2001). *Spixiana* **24** (3): 199–200.
——, 2003. Larentiinae II (Perizomini and Eupitheciini). – *In*: Hausmann, A. (ed.): The Geometrid Moths of Europe **4**: 1–463.
——, 2005. Family Geometridae – Pjadenitzy, 4. Subfam. Larentiinae, tribe Eupitheciini. In: Ler P.A. (ed.). Key to the insects of Russian Far East. Vol. V. Trichoptera and Lepidoptera. Pt 5. – Vladivostok. Dal'nauka, pp. 470–504 (in Russian).
Mironov, V. G., Beljaev, E.A. & S.V. Vasilenko, 2007. Family Geometridae. *In*: Sinev, S. Yu. (ed.). Catalogue of the Lepidoptera of Russia. St. Petersburg – Moscow, KMK Scientific Press Ltd. pp. 190–226 (in Russian).
Mironov, V.G. & A.C. Galsworthy, 2007. The genus *Eupithecia* (Lepidoptera, Geometridae) in Taiwan: an updated survey. *Trans. lepid. Soc. Japan* **58** (3): 341–363.
——, 2009a. A survey of the genus *Eupithecia* (Lepidoptera, Geometridae) in mainland South East Asia: Part I. *Trans. lepid. Soc. Japan* **60** (2): 93–116.
——, 2009b. A survey of the genus *Eupithecia* (Lepidoptera, Geometridae) in mainland South East Asia: Part II. *Trans. lepid. Soc. Japan* **60** (3): 167–188.
——, 2010a. Further notes on *Eupithecia* (Lepidoptera, Geometridae) from Nepal and the Indian subcontinent. *Trans. lepid. Soc. Japan* **61** (2): 137–172.

——, 2010b. Some taxonomic changes to the Japanese *Eupithecia* fauna (Lepidoptera, Geometridae). *Tinea* 27(2): 74–81.
——, 2012. A generic level review of *Eupithecia* Curtis and some closely related genera based on the Palaearctic, Nearctic and Oriental fauna (Lepidoptera, Geometridae, Larentiinae). *Zootaxa* 3587: 46–64.
Mironov, V.G., Galsworthy, A.C. & U. Ratzel, 2008a. A survey of the *Eupithecia* fauna (Lepidoptera, Geometridae) of the Western Himalayas: Part I. *Trans. lepid. Soc. Japan* 59 (1): 55–77.
——, 2008b. A survey of the *Eupithecia* fauna (Lepidoptera, Geometridae) of the Western Himalayas: Part II. *Trans. lepid. Soc. Japan* 59 (2): 117–143.
——, 2008c. A survey of the *Eupithecia* fauna (Lepidoptera, Geometridae) of the Western Himalayas: Part III. *Trans. lepid. Soc. Japan* 59 (3): 201–224.
Mironov, V., Galsworthy, A.C. & D. Xue, 2004a. New species of *Eupithecia* (Lepidoptera, Geometridae) from China, part I. *Trans. lepid. Soc. Japan* 55 (1): 39–57.
——, 2004b. New species of *Eupithecia* (Lepidoptera, Geometridae) from China, part II. *Trans. lepid. Soc. Japan* 55 (2): 117–132.
——, 2004c. New species of *Eupithecia* (Lepidoptera, Geometridae) from China, part III. *Trans. lepid. Soc. Japan* 55 (3): 225–242.
——, 2004d. New species of *Eupithecia* (Lepidoptera, Geometridae) from China, part IV. *Trans. lepid. Soc. Japan* 55 (4): 285–300.
——, 2006. New species of *Eupithecia* (Lepidoptera, Geometridae) from China, part V. *Trans. lepid. Soc. Japan* 57 (4): 335–353.
——, 2011. New species of *Eupithecia* (Lepidoptera, Geometridae) from China, part VI. *Lep. Science* **62**: 12–32.
Mizutani, Y., 2006. Additional records of *Eupithecia extensaria leuca* Dietze (Geometridae) from Hokkaido. *Yugato* **185**: 126 (in Japanese).
Moore, F., 1888. Descriptions of new Indian Lepidopterous Insects from the collection of the late Mr. W.S. Atkinson, M.A., F.L.S., & c. Heterocera (Pyralidae, Crambidae, Geometridae, Tortricidae, Tineidae). Calcutta / London, pp. 199–299, 8 pls.
Murase, M., 1996. Larvae of two Larentiinae (Geometridae) found on the flowers of *Farfugium japonicum* (Compositae) in Wakayama Prefecture. *Yugato* **145**: 61–62 (in Japanese).
——, 1997. On the larvae of two geometrid species in Wakayama Prefecture. *Yugato* **149**: 129–130 (in Japanese).
Murase, M., 2000. Larvae of Geometridae found on the flowers in Wakayama Prefecture II. *Yugato* **161**: 97–98 (in Japanese).
Oh, S.H., 1992. Taxonomic Revision of the Genus *Eupithecia* (Lepidoptera: Gometridae: Larentiinae) from Korea (I). *The Korean J. Appl. Ent.* **31** (4): 440–449.
——, 1993. Taxonomic Review of the Genus *Eupithecia* (Lepidoptera: Geometridae: Larentiinae) from Korea (II). *The Korean J. Appl. Ent.* **32** (4): 461–474.
Packard, A.S., 1876. A monograph of the Geometrid Moths or Phalaenidae of the United States. *Report US Geol. Survey of the territories* **10**: i–iv, 607, i-xiii pls.
Petersen, W., 1909. Ein Beitrag zur Kenntnis der Gattung *Eupithecia* Curt. *Dt. ent. Z., Iris* **22**: 203–314, 28 pls.
Pierce, F.N., 1914. The Genitalia of the Group Geometridae of the Lepidoptera of the British Islands. Liverpool, XXIX & 88 pp.+ 48 Pl.
Prout, L.B., 1907. Notes on the genus Eupithecia. *The Entomologist* **40** (531): 169–175.
——, 1915 (1912–1916). Larentiinae (part.). In Seitz, A.: The Macrolepidoptera of the World **4**. Stuttgart, 479 pp. + 25 pls.
——, 1926a. New *Geometridae. Novit. zool.* **33** (1): 1–32.
——, 1926b. On a collection of moths of the family Geometridae from Upper Burma made by Captain A.E. Swann. Parts I–II. *J. Bombay nat. Hist. Soc.* **31** (183): 129–146; 306–322, 2 pls.
——, 1926c. New Indian Geometridae. *Mem. Dep. Agric. India* (Ent.) **9** (8): 247–257.

——, 1929. A Catalogue of the Lepidoptera of Hainan. Geometridae. *Bull. Hill Mus.* (*A Magazine of Lepidopterology*), [1930] **4** (1): 125–144.
——, 1931. New Geometridae from the Indo-Australian Region. *Novit. zool.* 37 (1): 18–34.
——, 1934–1954. Brephinae, Oenochrominae, Geometrinae, Sterrhinae, Larentiinae. *In* Seitz, A.: The Macrolepidoptera of the World **4** (Suppl.). Stuttgart, 766 pp. + 18 pls.
——, 1958. New species of Indo-Australian Geometridae. *Bull. Brit. Mus.* (*Nat. Hist.*), *Entomology*, **6** (12): 367–463, figs. 1–72.
Püngeler, R., 1897. Kenntniss der Geometridenfauna Japans. *Dt. ent. Z., Iris* **10** (2): 361–371.
Riley, A.M., 1986. A review of the status of *Eupithecia goossensiata* Mab. (the Ling Pug) and *E. absinthiata* Cl. (the Wormwood Pug) (Lep.: Geometridae). *The Entomologist's Rec. J. Var.* **98** (5–6): 85–89.
Riley, A.M. & G. Prior, 2003. British and Irish Pug Moths (Lepidoptera: Geometridae, Larentiinae, Eupitheciini) – a guide to their identification and biology. Harley Books, England, 264 pp.
Rindge, F.H., 1952. Taxonomic and life history notes on North American *Eupithecia* (Lepidoptera, Geometridae). *Amer. Mus. Novit.* **1569**: 1–27.
Robinson, G.S., Ackery, P.R., Kitching, I.J., Beccaloni, G.W. & L.M. Hermández, 2001. Hostplants of the moth and butterfly caterpillars of the Oriental Region. Publ. by Natural History Museum/Southdene Sdn Bhd, London/Kuala Lumpur, 744 pp.
Sato, R., 1995. *Eupithecia jezonica* Matsumura, a distinct species to be separated from *E. sophia* Butler (Lepidoptera: Geometridae). *Yugato* **138**: 1–5 (in Japanese).
——, 2001a. Type specimens of the Geometridae (Lepidoptera) described by Matsumura in the Hokkaido University Insect Collection, Japan. *Insecta Matsumurana*, N.S. **58**: 115–138, 23 figs.
——, 2001b. Revision of some geometrid species from Sakhalin by Matsumura (1925), excluding his new taxa (Lepidoptera). *Insecta Matsumurana*, N.S. **58**: 139–143.
Sato, R., Kameda, M. & N. Doi, 1995. Some additional records of *Eupithecia maenamiella* Inoue (Geometridae, Larentiinae), with description of the male. *Yugato* **141**: 137–138 (in Japanese).
Sato, R., Yamada, Y. & T. Naito, 1997. Additional records of *Eupithecia gummaensis* Inoue (Geometridae, Larentiinae), with description of the male. *Yugato* **149**: 84–86 (in Japanese).
Schütze, E., 1956. *Eupithecia*-Studien VII und VIII. Die *semigraphata*-Gruppe (Lepidoptera, Geometridae). *Z. Wien. ent. Ges.* **41** (Bd **67**) (11): 306–320; (12): 328–339, Tafn. 33–35.
——, 1961a. Lepidoptera der Deutschen Nepal-Expedition 1955. Gattung *Eupithecia* Curtis. *Veröff. zool. StSamml. Münch.* **6**: 179–183.
——, 1961b. Weitere Eupithecien aus Iran und Arabien (Lep., Geom.). *Mitt. Münchn. Ent. Ges.* **51**: 58–71, Tafn. i-viii.
Scoble, M.J., Pitkin, L.M., Parsons, M.S., Honey, M.R. & B.R. Pitkin, 1999. Geometrid Moths of the World: A catalogue (Lepidoptera, Geometridae). Vol. 1 and 2. CSIRO Publishing and Apollo Books, Stenstrup. 1016 pp. (+ 129 pp. of Index).
Shin, Y.-H., 1996. Synonymic list and distribution of the Geometridae of Korea (Lepidoptera). Publ. by Center for Insect Systematics, 153 pp.
Shin, Y.-H., Park, K.-T. & S.-H. Nam, 1983. Illustrated Flora & Fauna of Korea. Vol. 27. Insecta (9): 1–1053, 48 pls.
Shin, Y.-H. & S.H. Kim, 1993. Notes on six unrecorded geometrid moths (Lepidoptera) from Korea. *J. Amat. Lepid. Soc. Korea* **6**: 9–12.
Staudinger, O., 1861. I. Macrolepidoptera. *In*: Staudinger, O. & M. Wocke: Catalog der Lepidopteren Europa's und der angrenzenden länder. Dresden, pp. i–xvi & 1–84.
——, 1882. Beitrag zur Lepidopteren-Fauna Central-Asiens. *Stett. Ent. Z.* **43** (1–3): 35–78.
——, 1885. Einige neue Lepidopteren des europäischen Faunengebietes. *Stett. Ent. Z.* **46** (7–9): 349–352.
——, 1892. Neue Arten und Varietäten von paläarktischen Geometriden. *Dt. ent. Z., Iris* **5** (1): 141–260.
——, 1897a. Ueber Lepidoptera von Uliassutai. *Dt. ent. Z., Iris* (1896) **9** (2): 240–283.

——, 1897b. Die Geometriden des Amurgebiets. *Dt. ent. Z., Iris* **10** (1): 1–122, pls. I–III.

——, 1900. Ueber Lepidopteren aus dem östlichsten Thian Schan-Gebiet. *Dt. ent. Z., Iris* [1899] **12** (2): 331–351.

Staudinger, O. & H. Rebel, 1901. Catalog der Lepidopteren des Palaearktischen Faunengebietes. Teil I. Berlin, XXXII & 411 pp.

Stephens, J.F., 1829a. A Systematic Catalogue of British Insects: being an attempt to arrange all the hitherto discovered indigenous insects in accordance with their natural affinities. Insecta Haustellata. Ordo I: (8). Lepidoptera. London, pp. 1–231.

——, 1829b. Illustrations of British Entomology or, a Synopsis of indigenous Insects: containing their generic and specific distinctions. Haustellata 3: 1–333, pls. 25–32. London.

Swinhoe, C., 1892. New species of Heterocera from Khasia Hills. Part II. *Trans. ent. Soc. Lond.* **1892** (1): 1–20, pl. I.

——, 1895. New Species of Indian Epiplemidae, Geometridae, Thyrididae, and Pyralidae. *Ann. Mag. nat. Hist.* (6) **16** (no. 94): 293–304.

Thompson, W.R., 1945. A Catalogue of the Parasites and Predators of Insect Pests. Sect. **1** (Parasite Host Catalogue), Part **6** (Parasites of the Lepidoptera). The Imperial Parasite Service. Belleville, Ontario, Canada, 131–258 pp.

Tomisawa, A., 2004. Moths collected in the subalpine zone of Mt Hakusan. *Yugato* 177: 91–98 (in Japanese).

Treitschke, F., 1828 [1827]. Die Schmetterlinge von Europa. Geometridae. Bd. 6 (1–2): i–viii, 1–319.

Viidalepp, J., 1973. Zwei neue Eupithecien-arten (Lepidoptera, Geometridae) aus der Mongolischen Volksrepublik. *Ent. obozr.* **52** (2): 397–399 (in Russian).

——, 1974. Die *Eupithecia*-Arten aus Sibirien in der Sammlung der Staats-Universität zu Tartu (Lepidoptera, Geometridae). *TRÜ Toimetised* 327. Zooloogia-alaseid töid **8**: 86–93 (in Russian).

——, 1975. On the fauna of Geometrid moths (Lepidoptera, Geometridae) of the Mongolia People's Republic. *Insects of Mongolia* **3**: 438–490 (in Russian).

——, 1976. New genera and species of Geometrid moths (Lepidoptera, Geometridae) from Southern Siberia and Mongolia. *Insects of Mongolia* **4**: 381–402 (in Russian).

——, 1978. A list of Geometridae (Lepidoptera) of the USSR, III. *Ent. obozr.* **57** (4): 752–761 (in Russian).

——, 1979. On the fauna of Lepidoptera of Tuva ASSR, IV. Geometridae. *TRÜ Toimetised* 483. Zooloogia-alaseid töid **12**: 79–133 (in Russian).

——, 1988. The fauna of Geometrid Moths of the mountains of Central Asia. Moscow, Publ. "Nauka", 240 pp. (in Russian).

——, 1996. Checklist of the Geometridae (Lepidoptera) of the former U.S.S.R. Apollo Books Aps., Stenstrup, Denmark. 111 pp.

Viidalepp, J. & V.P. Soljanikov, 1977. On the geometrid moths of the northern part of the Mongolian People's Republic. *Insects of Mongolia* **5**: 620–641 (in Russian).

Viidalepp, J. & V. Mironov, 1988a. Pugs of the Soviet Far East (Lepidoptera, Geometridae: *Eupithecia*). I. *Proc. Acad. Sci. Estonian SSR, Biology,* **37** (3): 200–214 (in Russian).

——, 1988b. Pugs of the Soviet Far East (Lepidoptera, Geometridae: *Eupithecia*). II. *Proc. Acad. Sci. Estonian SSR, Biology,* **37** (4): 281–293 (in Russian).

——, 1990. Pugs of the Soviet Far East (Lepidoptera, Geometridae: *Eupithecia*). III. *Proc. Acad. Sci. Estonian SSR, Biology,* **39** (2): 112–120 (in Russian).

Vojnits, A.M., 1972. New Taxa in the *Eupithecia haworthiata*-Group (Lepidoptera: Geometridae) (Studies on Palaearctic *Eupithecia* Species I.). *Ann. hist.-nat. Mus. natn hung.* **64**: 299–302.

——, 1973. New Species in the *Eupithecia haworthiata*-group (Studies on Palaearctic *Eupithecia* Species, Lepidoptera, Geometridae, II.). *Ann. hist.-nat. Mus. natn hung.* **65**: 249–256.

——, 1974a. New *Eupithecia* species (Lepidoptera: Geometridae) from China. Studies on Palaearctic *Eupithecia* species. III. *Acta zool. Acad. Sci. hung.* **20** (1–2): 219–226.

——, 1974b. Geometridae: *Eupithecini* I. (Lepidoptera). *Acta zool. Acad. Sci. hung.* **20** (3–4): 443–448.
——, 1975. Geometridae: *Eupithecini* II. (Lepidoptera). *Acta zool. Acad. Sci. hung.* **21** (3–4): 447–453.
——, 1976a. New species of the *Eupithecia bohatschi* group from China (Lepidoptera: Geometridae). [Studies on Palaearctic *Eupithecia* species V]. *Acta zool. Acad. Sci. hung.* **22** (1–2): 197–211.
——, 1976b. A revision of the species group *Eupithecia sinuosaria* Ev. (Lepidoptera: Geometridae). [Studies on Palaearctic *Eupithecia* species VI]. *Acta zool. Acad. Sci. hung.* **22** (3–4): 415–423.
——, 1977a. New *Eupithecia* species and subspecies from Asia and North Africa (Lepidoptera: Geometridae). [Studies on Palaearctic *Eupithecia* species VII]. *Acta zool. Acad. Sci. hung.* **23** (1–2): 227–236.
——, 1977b. Geometridae *Eupithecini* III. (Lepidoptera). *Acta zool. Acad. Sci. hung.* **23** (3–4): 461–485.
——, 1978. Investigations in the "*Eupithecia scalptata-syriacata*" group (Lepidoptera: Geometridae). *Acta zool. Acad. Sci. hung.* **24** (3–4): 431–438.
——, 1979a. New and rare *Eupithecia* species from China (Lepidoptera: Geometridae). *Acta zool. Acad. Sci. hung.* **25** (1–2): 193–211.
——, 1979b. New and rare *Eupithecia* species from China (Lepidoptera: Geometridae). II. *Acta zool. Acad. Sci. hung.* **25** (3–4): 425–439.
——, 1980a. Data to the *Eupithecia* fauna of China (Lepidoptera: Geometridae). *Acta zool. Acad. Sci. hung.* **26** (4): 433–449.
——, 1980b. On some Chinese *Eupithecia* material collected by H. Höne (Lepidoptera, Geometridae). *Acta zool. Acad. Sci. hung.* **27**(3–4): 407–430.
——, 1981a. Data to the *Eupithecia* fauna of Nepal. I (Lepidoptera: Geometridae). [Studies on Palaearctic *Eupithecia* species XIII]. *Acta zool. Acad. Sci. hung.* **27** (1–2): 217–238.
——, 1981b. On some chinese *Eupithecia* material collected by H. Höne (Lepidoptera: Geometridae). *Acta zool. Acad. Sci. hung.* **27** (3–4): 407–430.
——, 1981c. Data to the *Eupithecia* fauna of Asia (Lepidoptera, Geometridae). [Studies on Palaearctic *Eupithecia* species XV]. *Ann. hist.-nat. Mus. natn hung.* **73**: 221–237.
——, 1982. A revision of the "*Eupithecia innotata* group", I. (Lepidoptera, Geometridae). [Studies on Palaearctic *Eupithecia* species XV]. *Ann. hist.-nat. Mus. natn hung.* **74**: 217–239.
——, 1983. Data to the *Eupithecia* fauna of Nepal. II (Lepidoptera: Geometridae). [Studies on Palaearctic *Eupithecia* species XVI]. *Acta zool. Acad. Sci. hung.* **29** (1–3): 261–282.
——, 1984a. New data to the *Eupithecini* fauna of China based on Höne's collections. [Studies on Palaearctic *Eupithecia* species XX]. *Acta zool. hung.* **30** (1–2): 213–240.
——, 1984b. Investigations in the "*Eupithecia inepta-sacrosancta*" and the "*E. lasciva*" groups (Lepidoptera: Geometridae). [Studies on Palaearctic *Eupithecia* species XX]. *Acta zool. hung.* **30** (3–4): 523–544.
——, 1985. Revision der Paläarktischen Eupithecini-Arten I. [Studies on Palaearctic *Eupithecia* species XXIII]. *Acta zool. hung.* **31** (1–3): 267–273.
——, 1988a. Data to the *Eupithecia* fauna of Nepal. III (Lepidoptera: Geometridae). [Studies on Palaearctic *Eupithecia* species]. *Acta zool. hung.* **34** (1): 37–53.
——, 1988b. New *Eupithecia* species from Soviet Central Asia (Lepidoptera: Geometridae). *Ann. hist.-nat. Mus. natn hung.* **80**: 79–90.
Vojnits, A., 1994. New Data on the *Eupithecia*-Fauna of Chile (Lepidoptera, Geometridae). *Entomofauna* **15** (32): 361–376.
Vojnits, A. & E. De Laever, 1973. Revision of the "*Eupithecia suboxydata-subbrunneata*" group (Lepidoptera: Geometridae). *Acta zool. Acad. Sci. hung.* **19** (3–4): 427–444.
——, 1974. *Eupithecia szelenyica* sp. nov. from China (Lepidoptera: Geometridae). *Folia ent. hung.* (ser. nova) **27** (1): 229–231.
——, 1978. Eupitheciini from Korea and China (Lepidoptera). [Studies on Palaearctic *Eupithecia* species VIII]. *Acta zool. Acad. Sci. hung.* **24** (1–2): 225–252.

Walker, F., 1862a. Catalogue of Lepidoptera Heterocera. Fifth series. *List of the specimens of Lepidopterous Insects in the collection of the British Museum.* **24** (Geometrites): 1020–1280, London.

——, 1862b. Catalogue of Lepidoptera Heterocera. Fifth series. *List of the specimens of Lepidopterous Insects in the collection of the British Museum.* **26** (Geometrites): 1478–1796, London.

——, 1866. Catalogue of Lepidoptera Heterocera. Seventh series. *List of the specimens of Lepidopterous Insects in the collection of the British Museum.* **35** (suppl. 5) (Geometrites): 1534–2040, London.

Wang, H.Y., 1997. Geometer Moths of Taiwan and It's Allied Species from the Neighboring Countries, Vol. 1, Taiwan Museum, Taipei, 405 pp.

Warren, W., 1888. On Lepidoptera collected by Major Yerbury in Western India in 1886 and 1887. *Proc. zool. Soc. Lond.* **1888**: 292–339.

——, 1896. New Indian Epiplemidae and Geometridae. *Novit. zool.* **3** (3): 307–321.

——, 1897. New genera and species of Drepanidae, Thyrididae, Epiplemidae, Uraniidae, and Geometridae in the Tring Museum. *Novit. zool.* **4** (2): 195–262.

——, 1901. Drepanulidae, Uraniidae, and Geometridae from the Palaearctic and Indo-Australian Regions. *Novit. zool.* **8** (3): 190–201.

——, 1903. New Drepanulidae, Thyrididae, Uraniidae, and Geometridae from the Oriental Region. *Novit. zool.* **10** (2): 255–270.

——, 1904. New American Thyrididae, Uraniidae and Geometridae. *Novit. zool.* **11**: 493–582.

Wehrli, E., 1929. Beitrag zur Geometriden-Fauna von Minussinsk (Sibirien, Gouv. Jenissej). *Jahrb. Martjanov. Staatsmus. Minussinsk* **6** (1): 8–30.

Wileman, A.E., 1911. New and unrecorded species of Lepidoptera Heterocera from Japan. *Trans. ent. Soc. London*: **1911** (part II): 189–406, pls. xxx–xxxi.

Wileman, A.E. & R. South, 1917. New species of Noctuidae and Geometridae from Formosa. *The Entomologist* **50** (646): 53–57.

Wiltshire, E.P., 1952. Early stages of Palearctic Lepidoptera, X. *Bull. Soc. Fouad Ier Entom.* **36**: 175–185, with Plate.

Wood, W., 1854. Index Entomologicus; or, A Complete Illustrated Catalogue, consisting of upwards of two thousand accurately coloured figures of the Lepidopterous Insects of Great Britain. London, 298 pp.

Yoon, I.B., Nam, S.H. & D.S. Jung, 1982. A study of Moth Fauna in Tonghae City Area, Kangwondo. *J. Sci. Tech. Korea Univ.* **23**: 20–24.

INDEX TO SCIENTIFIC NAMES OF TAXA TREATED

The valid senior names of species fully treated in this work are in bold type. Species numbers assigned to them in the text are given first in bold type (and also correspond to the numbers on the illustrations), following which are page numbers for the main entry for fully treated species and species groups in italics, and other references in normal type.

PLATES

PLATE 1 (2× LIFE SIZE)

1. *Eupithecia subtacincta* Hampson, China, Taiwan
2. *Eupithecia cichisa* Prout, China, Sichuan
3. *Eupithecia aenigma* Mironov & Galsworthy, China, Yunnan, holotype
4. *Eupithecia zibellinata* Christoph, Russia, Amur
5. *Eupithecia olgae* Mironov, Kirghizstan (reared)
6. *Eupithecia illaborata* Dietze, China, Xinjiang
7. *Eupithecia rubellata* Dietze, China, Qinghai
8. *Eupithecia bohatschi* Staudinger, Russia, Primorje
9. *Eupithecia boneta* Mironov & Galsworthy, China, Yunnan, holotype
10. *Eupithecia jezonica* Matsumura, Russia, S. Primorye (reared)
11. *Eupithecia sophia* Butler, Japan
12. *Eupithecia lineisdistincta* Vojnits, Nepal
13. *Eupithecia albigutta* Prout, Nepal
14. *Eupithecia rubridorsata* Hampson, West Bhutan
15. *Eupithecia ustata* Moore, China, Yunnan, holotype *E. konradi*
16. *Eupithecia delaeveri* Vojnits, Nepal
17. *Eupithecia likiangi likiangi* Vojnits, China, Yunnan, holotype

17a. *Eupithecia likiangi pedisequa* Vojnits, China, Taiwan, holotype

18. *Eupithecia sinicaria* Leech, China, Sichuan, holotype
19. *Eupithecia ruficorpus* Warren, India, Cherrapunji
20. *Eupithecia szelenyica* Vojnits, China, Sichuan
21. *Eupithecia liberata* Inoue, Nepal, paratype
22. *Eupithecia yoshimotoi* Inoue, China, Taiwan
23. *Eupithecia buysseata* Mironov & Galsworthy, China, Yunnan, holotype
24. *Eupithecia kuroshio* Inoue, Japan
25. *Eupithecia chingana* Wehrli, China, Inner Mongolia, holotype
26. *Eupithecia brevifasciaria* Leech, China, Sichuan

27a. *Eupithecia eurytera* Prout, China, Sichuan, allotype

27b. *Eupithecia eurytera* Prout, China, Beijing

28. *Eupithecia rufescens* Butler, China, Zhejiang
29. *Eupithecia leamariae* Ratzel, West Bhutan
30. *Eupithecia abietaria* Goeze, Japan

1
2
3
4
5
6
7
8
9
10
11
12
13
14
15
19
17
17a
18
16
20
21
22
23
24
25
27a
26
27b
28
29
30

PLATE 2 (2× LIFE SIZE)

31. *Eupithecia gigantea* Staudinger, Russia, Ussuri
32. *Eupithecia interrubrescens* Hampson, China, Tibet, holotype
33. *Eupithecia costalis* Walker, Thailand, Chiangmai
34. *Eupithecia jermyi* Vojnits, Vietnam
35. *Eupithecia lini* Mironov & Galsworthy, China, Taiwan
36. *Eupithecia lucigera* Butler, Nepal
37. *Eupithecia missionerata* Mironov & Galsworthy, China, Tibet
38. *Eupithecia proterva* Butler, Japan
39. *Eupithecia subbreviata* Staudinger, Russia, Askold, syntype
40. *Eupithecia megaproterva* Inoue, China, Taiwan
41. *Eupithecia verprota* Mironov & Galsworthy, China, Shaanxi, holotype
42. *Eupithecia clavifera* Inoue, Japan, holotype
43. *Eupithecia wangi* Mironov & Galsworthy, China, Guangxi, paratype
44. *Eupithecia signigera* Butler, Japan
45. *Eupithecia russula* Mironov & Galsworthy, China, Tibet
46. *Eupithecia flavoapicaria* Inoue, China, Taiwan
47. *Eupithecia blenna* Mironov & Galsworthy, China, Shaanxi, holotype
48. *Eupithecia convexa* Inoue, China, Taiwan
49. *Eupithecia opicata* Mironov & Galsworthy, China, Gansu, holotype
50. *Eupithecia costiconvexa* Inoue, Japan, holotype
51. *Eupithecia concava* Mironov & Galsworthy, China, Taiwan, holotype
52. *Eupithecia aritai* Inoue, Japan

31
32
33
34
35
37
36
39
40
38
42
41
43
45
44
48
46
47
49
50
51
52

PLATE 3 (2× LIFE SIZE)

53. *Eupithecia admiranda* Mironov & Galsworthy, China, Shaanxi
54. *Eupithecia hashimotoi* Inoue, China, Taiwan
55. *Eupithecia flavimacula* Mironov & Galsworthy, China, Taiwan
56. *Eupithecia amplificata* Inoue, Nepal
57. *Eupithecia asempiterna* Inoue, Nepal (57a and b colour variants)
58. *Eupithecia wardi* Prout, China, Tibet
59. *Eupithecia minusculata* Alphéraky, European Russia
60. *Eupithecia opistographata* Dietze, China, Xinjiang, holotype
61. *Eupithecia viata* Mironov & Galsworthy, China, Gansu, holotype
62. *Eupithecia mustangata* Schütze, Nepal
63. *Eupithecia leucostaxis* Prout, Nepal
64. *Eupithecia ochracea* Warren, Nepal
65. *Eupithecia homogrammata* Dietze, Far Eastern Russia, syntype
66. *Eupithecia plumbeolata* Haworth, China, Xinjiang
67. *Eupithecia severa* Mironov & Galsworthy, China, Shaanxi
68. *Eupithecia spissata* Mironov & Galsworthy, China, Sichuan
69. *Eupithecia cordata* Mironov & Galsworthy, China, Yunnan
70. *Eupithecia turpis* Vojnits, China, Yunnan (holotype)
71. *Eupithecia peguensis* Prout, Nepal (paratype)
72. *Eupithecia apta* Mironov & Galsworthy, China, Yunnan (holotype)
73. *Eupithecia burselongata* Mironov & Galsworthy, China, Yunnan
74. *Eupithecia minutula* Mironov & Galsworthy, China, Yunnan
75. *Eupithecia impavida* Vojnits, China, Sichuan
76. *Eupithecia fervida* Vojnits, China, Yunnan (holotype)

53
55
54
56
57a
57b
58
59
61
60
63
66
62
67
68
64
65
69
71
72
70
74
75
73
76

PLATE 4 (2× LIFE SIZE)

77. *Eupithecia inexhausta* Vojnits, China, Sichuan
78. *Eupithecia hongxiangae* Mironov & Galsworthy, China, Gansu (holotype)
79. *Eupithecia alishana* Inoue, China, Taiwan
80. *Eupithecia antiqua* Mironov & Galsworthy, China, Yunnan
81. *Eupithecia assulata* Bastelberger, Vietnam
82. *Eupithecia adoranda* Vojnits, China, Henan
83. *Eupithecia granata* Vojnits, China, Yunnan (holotype)
84. *Eupithecia amicula* Mironov & Galsworthy, China, Sichuan
85. *Eupithecia hoenehermanni* Mironov & Galsworthy, China, Yunnan
86. *Eupithecia extinctata* Dietze, China, Tibet (syntype)
87. *Eupithecia consortaria* Leech, China, Sichuan (holotype)
88. *Eupithecia vana* Vojnits, China, Shaanxi (holotype *E. abrepta*)
89. *Eupithecia nirvana* Mironov & Galsworthy, China, Sichuan (paratype)
90. *Eupithecia hilariata* Dietze, Kyrgyzstan
91. *Eupithecia hesperina* Mironov & Galsworthy, China, Yunnan
92. *Eupithecia dissertata* Püngeler, Europe (syntype)
93. *Eupithecia carpophilata* Staudinger, Russia, Ussuri (syntype)
94. *Eupithecia albispumata* Warren, China, Sichuan
95. *Eupithecia astricta* Inoue, China, Taiwan (holotype)

96a. *Eupithecia abbreviata* Stephens, Europe

96b. *Eupithecia abbreviata* Stephens, China, Sichuan

97. *Eupithecia pusillata* Denis & Schiffermüller, Europe
98. *Eupithecia kozlovi* Viidalepp, China, Qinghai
99. *Eupithecia exrubicunda* Inoue, China, Taiwan (holotype)

100a. *Eupithecia lanceata* Hübner, China, Gansu

100b. *Eupithecia lanceata* Hübner, European Russia

77
78
79
80
82
81
83
84
86
85
87
89
88
90
92
91
93
95
94
96a
97
99
96b
98
100a
100b

PLATE 5 (2× LIFE SIZE)

101. *Eupithecia kudoi* Inoue, China, Taiwan
102. *Eupithecia daemionata* Dietze, Japan (syntype)
103. *Eupithecia habermani* Viidalepp & Mironov, Mongolia (paratype)
104. *Eupithecia emendata* Vojnits, Nepal (holotype)
105. *Eupithecia dolichoptera*, Galsworthy & Mironov, China, Taiwan
106. *Eupithecia phantastica* Mironov & Galsworthy, China, Taiwan (paratype)
107. *Eupithecia atuni* Mironov & Galsworthy, China, Yunnan (paratype)
108. *Eupithecia funerea* Inoue, China, Taiwan (holotype)
109. *Eupithecia fulvidorsata* Mironov & Galsworthy, China, Shaanxi (holotype)
110. *Eupithecia okadai* Inoue, Japan
111. *Eupithecia delicata* Mironov & Galsworthy, China, Sichuan (paratype)
112. *Eupithecia camilla* Mironov & Galsworthy, China, Shaanxi (holotype)
113. *Eupithecia julia* Mironov & Galsworthy, China, Sichuan (holotype)
114. *Eupithecia fortis* Mironov & Galsworthy, China, Yunnan (holotype)
115. *Eupithecia lariciata* Freyer, Japan
116. *Eupithecia emanata* Dietze, Russia, S. Primorje
117. *Eupithecia honesta* Mironov & Galsworthy, China, Tibet (paratype)
118. *Eupithecia pediba* Vojnits, China, Yunnan (holotype *E. epileptica*)
119. *Eupithecia argentea* Mironov & Galsworthy, China, Sichuan (holotype)
120. *Eupithecia primitiva* Mironov & Galsworthy, China, Gansu (holotype)
121. *Eupithecia sacrivicae* Mironov & Galsworthy, China, Hunan (paratype)
122. *Eupithecia horrida* Vojnits, China, Yunnan (holotype)
123. *Eupithecia stomachosa* Vojnits, China, Shaanxi (holotype)
124. *Eupithecia sola* Vojnits, Nepal (holotype)
125. *Eupithecia marmorea* Vojnits, Nepal
126. *Eupithecia albimedia* Mironov & Galsworthy, China, Yunnan (paratype)
127. *Eupithecia hebes*, Mironov & Galsworthy, China, Shanxi (holotype)

101
102
103
104
105
106
107
108
109
110
111
112
113
114
115
116
117
118
119
120
121
122
123
124
126
127
125

PLATE 6 (2× LIFE SIZE)

128. *Eupithecia stataria* Inoue, China, Taiwan
129. *Eupithecia tripunctaria* Herrich-Schäffer, China, Shanxi
130. *Eupithecia barteli* Dietze, Pakistan
131. *Eupithecia tenuisquama* Warren, Nepal
132a. *Eupithecia amandae* Mironov & Galsworthy ♂, China, Yunnan (holotype)
132b. *Eupithecia amandae* Mironov & Galsworthy ♀, China, Yunnan (paratype)
133. *Eupithecia blandula* Mironov & Galsworthy, China, Taiwan (paratype)
134. *Eupithecia puella* Mironov & Galsworthy, China, Yunnan (holotype)
135. *Eupithecia werneri* Vojnits, China, Yunnan(paratype)
136. *Eupithecia batangi* Vojnits, China, Tibet (paratype)
137. *Eupithecia sacrosancta* Vojnits, China, Shaanxi
138. *Eupithecia placida* Vojnits, China, Yunnan (paratype)
139. *Eupithecia laudabilis* Vojnits, China, Yunnan
140. *Eupithecia manifesta* Mironov & Galsworthy, China, Yunnan (holotype)
141. *Eupithecia intolerabilis* Vojnits, China, Yunnan (holotype)
142. *Eupithecia mentita* Vojnits, China, Yunnan (paratype *E. depressa disiuncta*)
143. *Eupithecia fenita* Mironov & Galsworthy, China, Yunnan
144. *Eupithecia refertissima* Vojnits, China, Tibet
145. *Eupithecia avara* Vojnits, China, Yunnan
146. *Eupithecia caduca* Vojnits, China, Yunnan (holotype)
147. *Eupithecia propagata* Prout, India (holotype)
148. *Eupithecia balintszolti* Vojnits, Laos
149. *Eupithecia hainanensis* Mironov & Galsworthy, Thailand
150. *Eupithecia groenblomi* Urbahn, European Russia (reared)
151. *Eupithecia selinata* Herrich-Schäffer, Far Eastern Russia
152. *Eupithecia tamara* Mironov & Pekarsky, China, Sichuan (holotype)
153. *Eupithecia arenosa* Vojnits, China, Yunnan (holotype)
154. *Eupithecia levata* Mironov & Galsworthy, China, Yunnan (holotype)
155. *Eupithecia molestissima* Vojnits, China, Yunnan

128
129
130
131
132a
132b
133
134
135
136
137
138
139
140
141
142
143
144
145
146
147
148
149
150
151
152
153
154
155

PLATE 7 (2× LIFE SIZE)

156. *Eupithecia actaeata* Walderdorff, Russia, S. Primorje
157. *Eupithecia actrix* Mironov & Galsworthy, China, Shaanxi
158. *Eupithecia uliata* Staudinger, China, Inner Mongolia
159. *Eupithecia detritata* Staudinger, Russia, S. Primorje
160. *Eupithecia jinboi* Inoue, China, Japan (holotype)
161. *Eupithecia pellicata* Mironov & Galsworthy, China, Taiwan (holotype)
162. *Eupithecia irreperta* Vojnits, China, Yunnan (holotype)
163. *Eupithecia vivida* Vojnits, China, Yunnan (holotype)
164. *Eupithecia quadripunctata* Warren, Nepal
165. *Eupithecia eximia* Vojnits, China, Yunnan (paratype)
166. *Eupithecia nepalata* Schütze, Nepal
167. *Eupithecia fatigata* Vojnits, China, Yunnan (paratype)
168. *Eupithecia repentina* Vojnits & de Laever, China, Hunan
169. *Eupithecia virgaureata* Doubleday, Russia, S. Primorje
170. *Eupithecia taiwana* Wileman & South, China, Taiwan
171. *Eupithecia matrona* Mironov & Galsworthy, China, Gansu (paratype)
172. *Eupithecia rhadine* Mironov & Galsworthy, China, Taiwan (holotype)
173. *Eupithecia cotidiana* Vojnits, China, Shaanxi (holotype)
174. *Eupithecia krampli* Vojnits, China, Yunnan (holotype)
175. *Eupithecia inopinata* Vojnits, China, Yunnan (paratype)
176. *Eupithecia dealbata* Inoue, China, Taiwan
177. *Eupithecia lata* Vojnits, Nepal (holotype)
178. *Eupithecia karapinensis* Wileman & South, China, Taiwan
179. *Eupithecia kobayashii* Inoue, Russia, S. Primorje
180. *Eupithecia incorrupta* Vojnits, China, Yunnan (holotype)

156
157
158
160
159
161
163
162
164
166
165
167
169
168
170
172
171
174
173
175
176
177
178
179
180

PLATE 8 (2× LIFE SIZE)

181. *Eupithecia leucenthesis* Prout, China, Tibet
182. *Eupithecia pimpinellata* Hübner, Russia, S. Siberia
183. *Eupithecia denotata* Hübner, Kyrgyzstan
184. *Eupithecia chesiata* Dietze, Kyrgyzstan
185. *Eupithecia mundiscripta* Warren, China, Hong Kong (holotype *E. sekkongensis*)
186. *Eupithecia robiginascens robiginascens* Prout, China, Sichuan
186a. *Eupithecia robiginascens flexicornuta* Inoue, China, Taiwan (holotype)
187. *Eupithecia luctuosa* Mironov & Galsworthy, China, Fujian (holotype)
188. *Eupithecia qinlingata* Mironov & Galsworthy, China, Shaanxi (holotype)
189. *Eupithecia svetlanae* Mironov & Pekarsky, China, Sichuan (holotype)
190. *Eupithecia contraria* Vojnits, Nepal
191. *Eupithecia rajata* Guenée, India, Khasi hills
192. *Eupithecia ebriosa* Vojnits, China, Jiangxi (holotype)
193. *Eupithecia vasta* Vojnits, China, Yunnan
194. *Eupithecia lobbichlerata* Schütze, China, Yunnan
195. *Eupithecia coniurata* Vojnits, China, Yunnan
196. *Eupithecia cuneata* Vojnits, Thailand
197. *Eupithecia particeps* Vojnits, Nepal (holotype)
198. *Eupithecia sempiterna* Vojnits, China, Yunnan
199. *Eupithecia serpentigena* Mironov & Galsworthy, China, Sichuan (holotype)
200. *Eupithecia raniata* Prout, India, Assam
201. *Eupithecia atrisignis* Butler, Myanmar
202. *Eupithecia pyricoetes* Prout, India, Sikkim (holotype)
203. *Eupithecia suspiciosa* Vojnits, Nepal (holotype)
204. *Eupithecia nuceistrigata* Bastelberger, China, Taiwan
205. *Eupithecia tricrossa* Prout, India, Assam
206. *Eupithecia yazakii* Inoue, Japan (holotype)
207. *Eupithecia suboxydata* Staudinger, Russia, Primorye
208. *Eupithecia formosa* Vojnits & de Laever, China, Yunnan
209. *Eupithecia subbrunneata* Dietze, Russia, Primorye

182
181
183
185
184
186a
187
190
186b
188
189
191
192
193
194
195
196
197
198
199
200
201
202
203
204
205
206
207
209
208

PLATE 9 (2× LIFE SIZE)

210. *Eupithecia innotata* Hufnagel, Russia, Tarbagatay Mts.
211. *Eupithecia parallelaria* Bohatsch, Turkmenistan
212. *Eupithecia corroborata* Dietze, Tadjikistan
213. *Eupithecia praesignata* Bohatsch, Kyrgyzstan
214. *Eupithecia paupera* Dietze, China, Xinjiang (lectotype)
215. *Eupithecia mitigata* Dietze, Kazakhstan
216. *Eupithecia vicariata* Dietze, Mongolia
217. *Eupithecia magica* Mironov & Galsworthy, China, Tibet (holotype)
218. *Eupithecia relaxata* Dietze, Kyrgyzstan
219. *Eupithecia centaureata* Denis & Schiffermüller, Kyrgyzstan
220. *Eupithecia exquisita* Vojnits, China, Shanxi (paratype)
221. *Eupithecia brunneilutea* Mironov & Galsworthy, China, Yunnan (holotype)
222. *Eupithecia despectaria* Lederer, Uzbekistan
223. *Eupithecia molybdaena* Mironov & Galsworthy, China, Shaanxi (holotype)
224. *Eupithecia gratiosata* Herrich-Schäffer, Russia, N. Caucasus
225. *Eupithecia subpulchrata* Alphéraky, China, Xinjiang
226. *Eupithecia perendina* Vojnits, China, Yunnan
227. *Eupithecia lamata* Mironov & Galsworthy, China, Qinghai (holotype)
228. *Eupithecia ancillata* Mironov & Galsworthy, China, Shaanxi (holotype)
229. *Eupithecia simpliciata* Haworth, Kirgizstan
230. *Eupithecia sinuosaria* Eversmann, China, Yunnan
231. *Eupithecia pallescens* Dietze, China, Xinjiang (holotype)
232. *Eupithecia conterminata* Lienig & Zeller, Japan
233. *Eupithecia insignioides* Wehrli, China, Shaanxi
234. *Eupithecia costimacularia* Leech, China, Guangxi
235. *Eupithecia assimilata* Doubleday, China, Xinjiang
236. *Eupithecia vulgata* Haworth, China, Xinjiang

211
210
212
214
215
213
217
218
216
220
221
219
224
222
223
225
228
226
227
229
234
230
232
235
233
231
236

PLATE 10 (2× LIFE SIZE)

237. *Eupithecia andrasi* Mironov & Galsworthy, China, Shaanxi
238. *Eupithecia nigristriata* Mironov & Galsworthy, China, Qinghai
239. *Eupithecia sacrimontis* Prout, China, Sichuan
240. *Eupithecia citraria* Mironov & Galsworthy, China, Yunnan (holotype)
241. *Eupithecia nobilitata* Staudinger, China, Tibet
242. *Eupithecia sutiliata* Christoph, Iran
243. *Eupithecia biornata* Christoph, China, Xinjiang
244. *Eupithecia kuldschaensis* Staudinger, China, Xinjiang (syntype)
245. *Eupithecia rubeni* Viidalepp, Russia, Primorje
246. *Eupithecia rebeli* Bohatsch, Uzbekistan
247. *Eupithecia thalictrata* Püngeler, Russia, S. Primorje
248. *Eupithecia dura* Vojnits, China, Yunnan (holotype)
249. *Eupithecia addictata* Dietze, Russia, S. Primorje
250. *Eupithecia exiguata* Hübner, Russia, S. Siberia
251. *Eupithecia hannemanni* Vojnits, China, Shanxi
252. *Eupithecia bella* Staudinger, Russia, Primorje
253. *Eupithecia insolita* Vojnits, China, Yunnan
254. *Eupithecia duplex* Sterneck, China, Yunnan (paratype *E. pollens*)
255. *Eupithecia tectaria* Mironov & Galsworthy, China, Fujian (holotype)
256. *Eupithecia mandschurica* Staudinger, Russia, Primorje (lectotype)
257. *Eupithecia intricata* Zetterstedt, Kazakhstan
258. *Eupithecia lupa* Mironov & Galsworthy, China, Taiwan (holotype)

259a. *Eupithecia satyrata* Hübner form *juldusi* Dietze, China, Xinjiang

259b. *Eupithecia satyrata* Hübner form *serenata* Staudinger, China, Qinghai

240
237
241
242
238
243
239
244
245
247
249
246
248
250
251
252
253
254
255
256
259a
257
259b
258

PLATE 11 (2× LIFE SIZE)

260. *Eupithecia pernotata* Guenée, Japan (holotype *E. antaggregata*)
261. *Eupithecia amplexata* Christoph, Russia, Vladivostock (syntype)
262a. *Eupithecia extensaria* Freyer, Kyrgyzstan
262b. *Eupithecia extensaria* Freyer form *leuca* Dietze, China, Inner Mongolia
263. *Eupithecia absinthiata* Clerck, Russia, Priamurje
264. *Eupithecia interpunctaria* Inoue, China, Shaanxi
265. *Eupithecia korbi* Dietze, Russia, Primorje
266. *Eupithecia kurilensis* Bryk, China, Jilin
267. *Eupithecia ensifera* Mironov & Galsworthy, China, Qinghai (holotype)
268. *Eupithecia ultrix* Mironov & Galsworthy, China, Yunnan (holotype)
269. *Eupithecia cervina* Mironov & Galsworthy, China, Tibet
270. *Eupithecia impolita* Vojnits, China, Shanxi (holotype)
271. *Eupithecia neosatyrata* Inoue, China, Shaanxi
272. *Eupithecia indissolubilis* Vojnits, China, Shaanxi (holotype *E. subita*)
273. *Eupithecia nephelata* Staudinger, China, Xinjiang (syntype)
274. *Eupithecia exactata* Staudinger, China, Xinjiang
275. *Eupithecia omissa* Mironov & Galsworthy, China, Shaanxi
276. *Eupithecia succenturiata* Linnaeus, China, Xinjiang
277. *Eupithecia marginata* Staudinger, China, Xinjiang (syntype)
278. *Eupithecia masculina* Inoue, China, Taiwan
279. *Eupithecia subumbrata* Denis & Schiffermüller, Kyrgyzstan
280. *Eupithecia subfuscata* Haworth, China, Sichuan
281. *Eupithecia costipicta* Warren, Pakistan
282. *Eupithecia mediocincta* Mironov & Galsworthy, China, Shaanxi

260
261
262a
263
264
262b
265
266
267
268
269
270
271
272
273
274
275
276
277
278
279
280
281
282

PLATE 12 (2× LIFE SIZE)

283. *Eupithecia chui* Inoue, China, Taiwan
284. *Eupithecia tibetana* Mironov & Galsworthy, China, Tibet (holotype)
285. *Eupithecia tepida* Mironov & Galsworthy, China, Shaanxi
286. *Eupithecia nishizawai* Inoue, China, Taiwan
287. *Eupithecia praecipitata* Vojnits, China, Shaanxi
288. *Eupithecia lasciva* Vojnits, China, Shaanxi
289. *Eupithecia egena* Vojnits, China, Yunnan (holotype *E. sublasciva*)
290. *Eupithecia lunatica* Vojnits, China, Yunnan (holotype)
291. *Eupithecia otiosa* Vojnits, China, Tibet (holotype)
292. *Eupithecia acutipapillata* Inoue, China, Taiwan (holotype)
293. *Eupithecia supercastigata* Inoue, Japan (holotype)
294. *Eupithecia concremata* Dietze, China, Qinghai (holotype)
295. *Eupithecia coribalteata* Mironov & Galsworthy, China, Qinghai (paratype)
296. *Eupithecia lilliputata* Mironov & Galsworthy, China, Yunnan (holotype)
297. *Eupithecia salubris* Mironov & Galsworthy, China, Shaanxi (holotype)
298. *Eupithecia villica* Mironov & Galsworthy, China, Yunnan
299. *Pareupithecia spadix* Inoue, Japan
300. *Eva flexa* Vojnits, China, Yunnan
301. *Girida rigida* Swinhoe, India, Khasis
302. *Mesoptila melanolopha* Swinhoe, India, Khasis

283
284
285
286
287
288
289
290
291
292
293
294
295
296
297
298
299
300
301
302

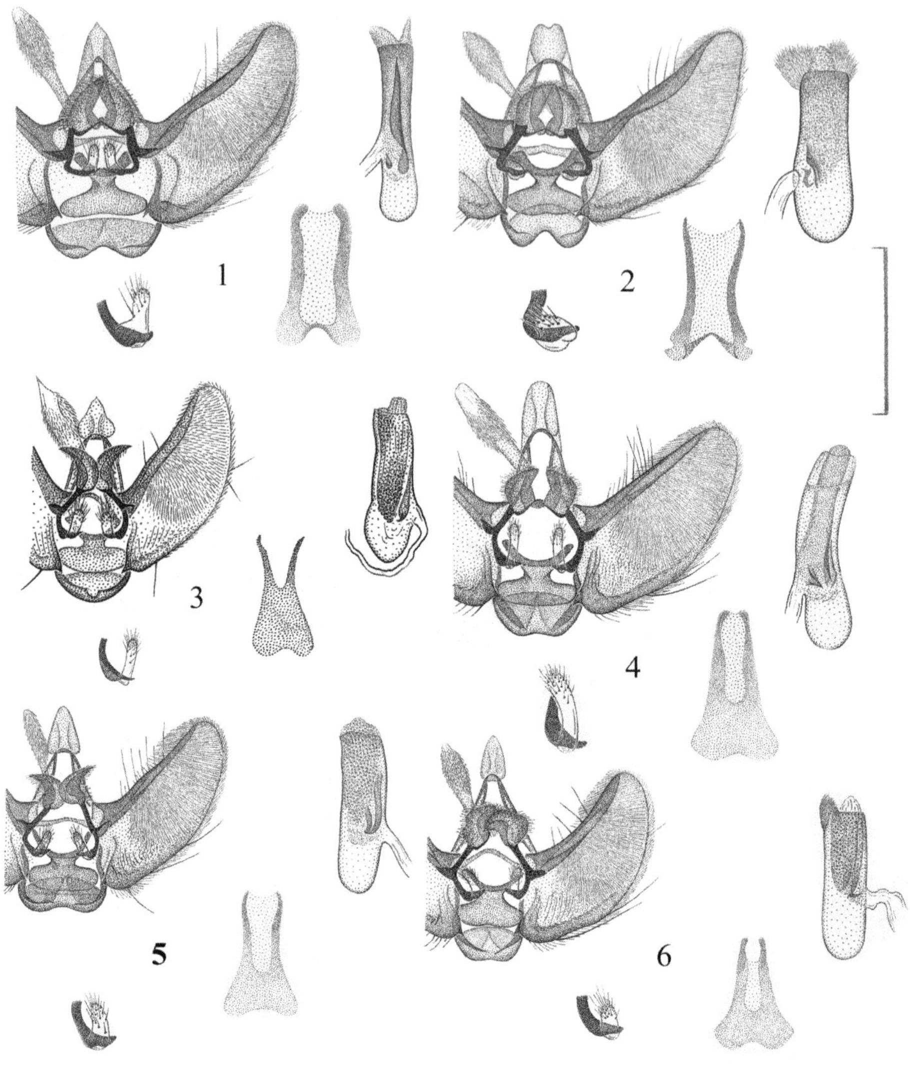

MALE GENITALIA (scale bar = 1mm; with sternum A8; papillae enlarged)

1. *Eupithecia subtacincta* Hmps., Russia, S. Primorje
2. *Eupithecia cichisa* Prout, China, Shaanxi
3. *Eupithecia aenigma* Mironov & Galsworthy, China, Yunnan
4. *Eupithecia zibellinata* Christoph, Russia, S. Primorje
5. *Eupithecia olgae* Mironov, Tajikistan, Pamir Mts.
6. *Eupithecia illaborata* Dietze, Kyrgyzstan, Naryn

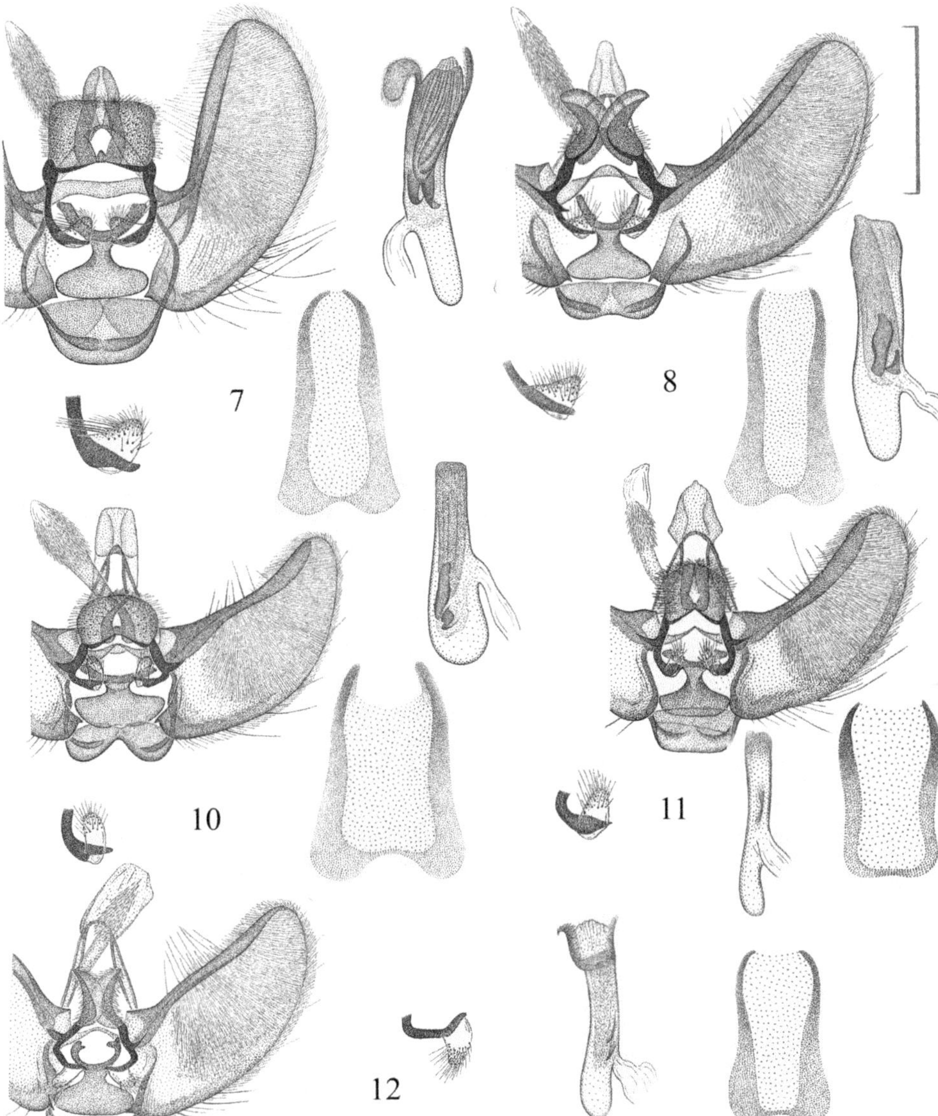

MALE GENITALIA (scale bar = 1mm; with sternum A8; papillae enlarged)

7. *Eupithecia rubellata* Dietze, Kyrgyzstan, Sary-Jaz
8. *Eupithecia bohatschi* Staudinger, Russia, S. Priamurje
10. *Eupithecia jezonica* Matsumura, Russia, S. Primorje
11. *Eupithecia sophia* Butler, Russia, S. Primorje
12. *Eupithecia lineisdistincta* Vojnits, Nepal, Solu Khumbu Himal

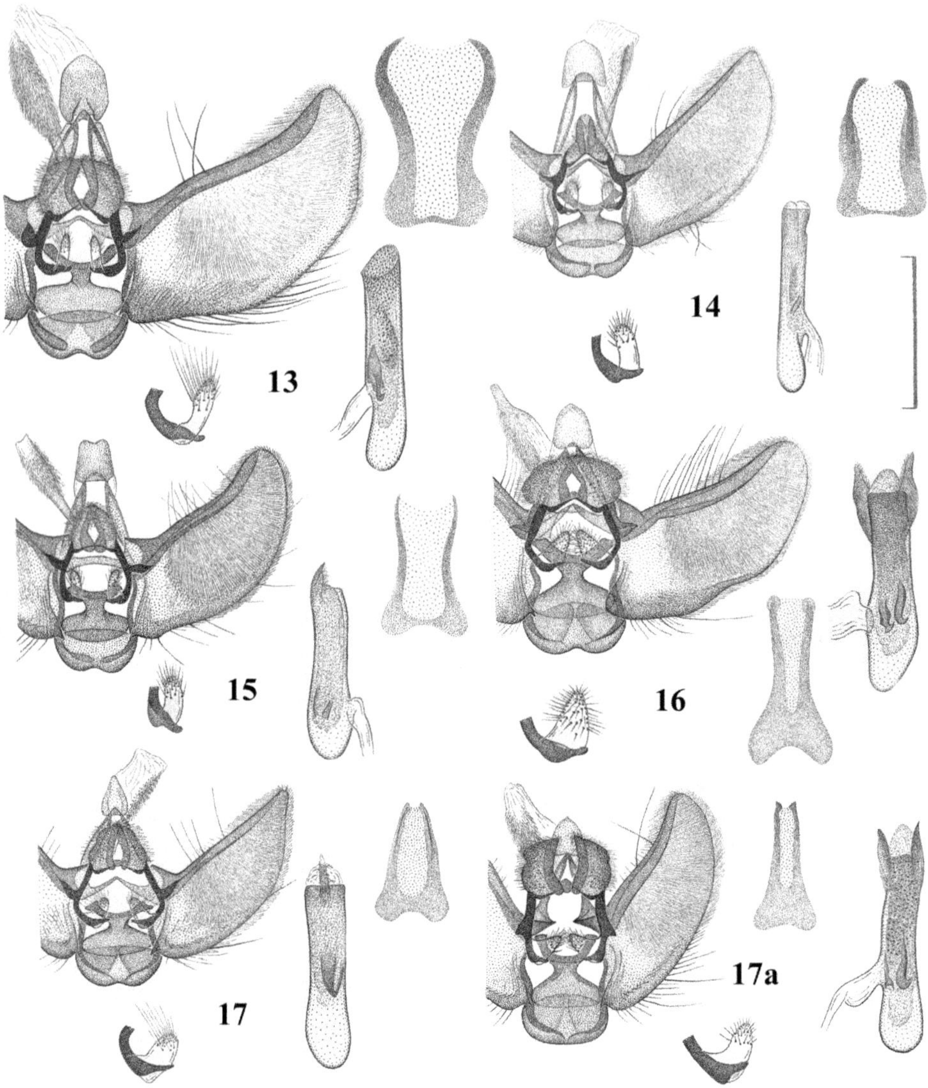

MALE GENITALIA (scale bar = 1mm; with sternum A8; papillae enlarged)

13. *Eupithecia albigutta* Prout, China, Yunnan
14. *Eupithecia rubridorsata* Hmps., China, Yunnan
15. *Eupithecia ustata* Moore, China, Tibet
16. *Eupithecia delaeveri* Vojnits, Nepal, Gandaki
17. *Eupithecia likiangi likiangi* Vojnits, China, Yunnan
17a. *Eupithecia likiangi pedisequa* ssp. n., China, Taiwan

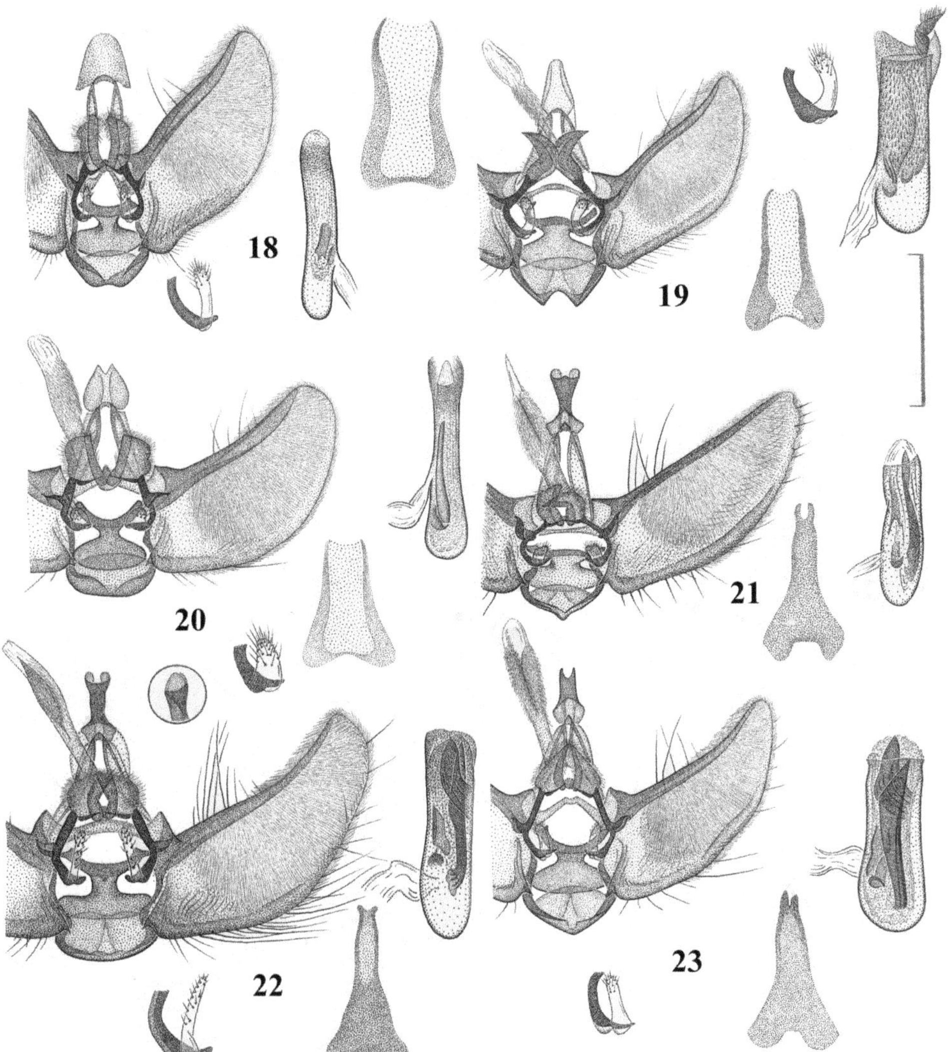

MALE GENITALIA (scale bar = 1mm; with sternum A8; papillae, lateral view of uncus enlarged)

18. *Eupithecia sinicaria* Leech, China, Sichuan
19. *Eupithecia ruficorpus* Warren, China, Tibet
20. *Eupithecia szelenyica* Vojnits, China, S. Shaanxi
21. *Eupithecia liberata* Inoue, China, Yunnan
22. *Eupithecia yoshimotoi* Inoue, China, Taiwan
23. *Eupithecia buysseata* Mironov & Galsworthy, China, Yunnan

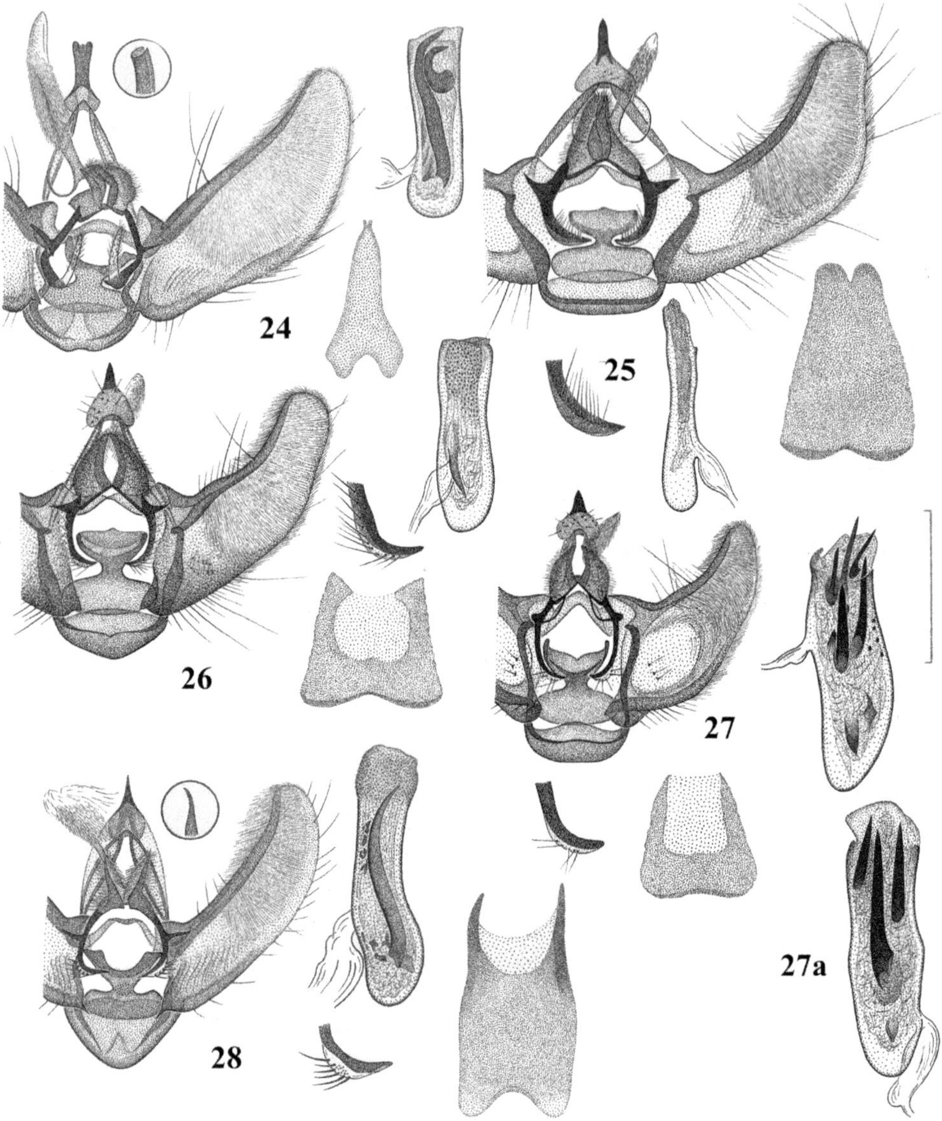

MALE GENITALIA (scale bar = 1mm; with sternum A8; papillae, lateral view of uncus enlarged)

24. *Eupithecia kuroshio* Inoue, Japan, Shikoku
25. *Eupithecia chingana* Wehrli, China, Beijing
26. *Eupithecia brevifasciaria* Leech, China, Sichuan
27. and 27a *Eupithecia eurytera* Prout, China, Beijing
28. *Eupithecia rufescens* Butler, China, Zhejiang

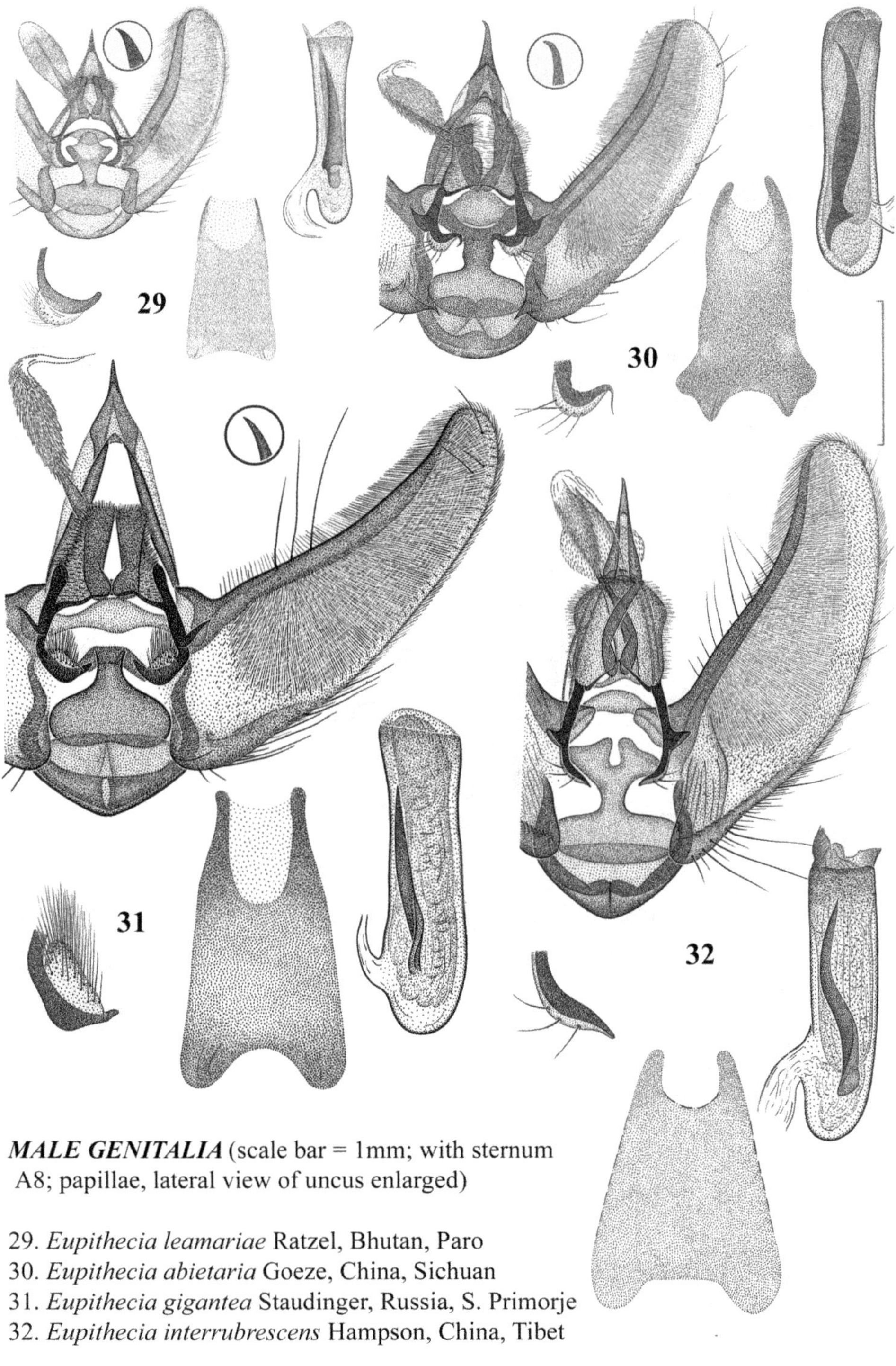

MALE GENITALIA (scale bar = 1mm; with sternum A8; papillae, lateral view of uncus enlarged)

29. *Eupithecia leamariae* Ratzel, Bhutan, Paro
30. *Eupithecia abietaria* Goeze, China, Sichuan
31. *Eupithecia gigantea* Staudinger, Russia, S. Primorje
32. *Eupithecia interrubrescens* Hampson, China, Tibet

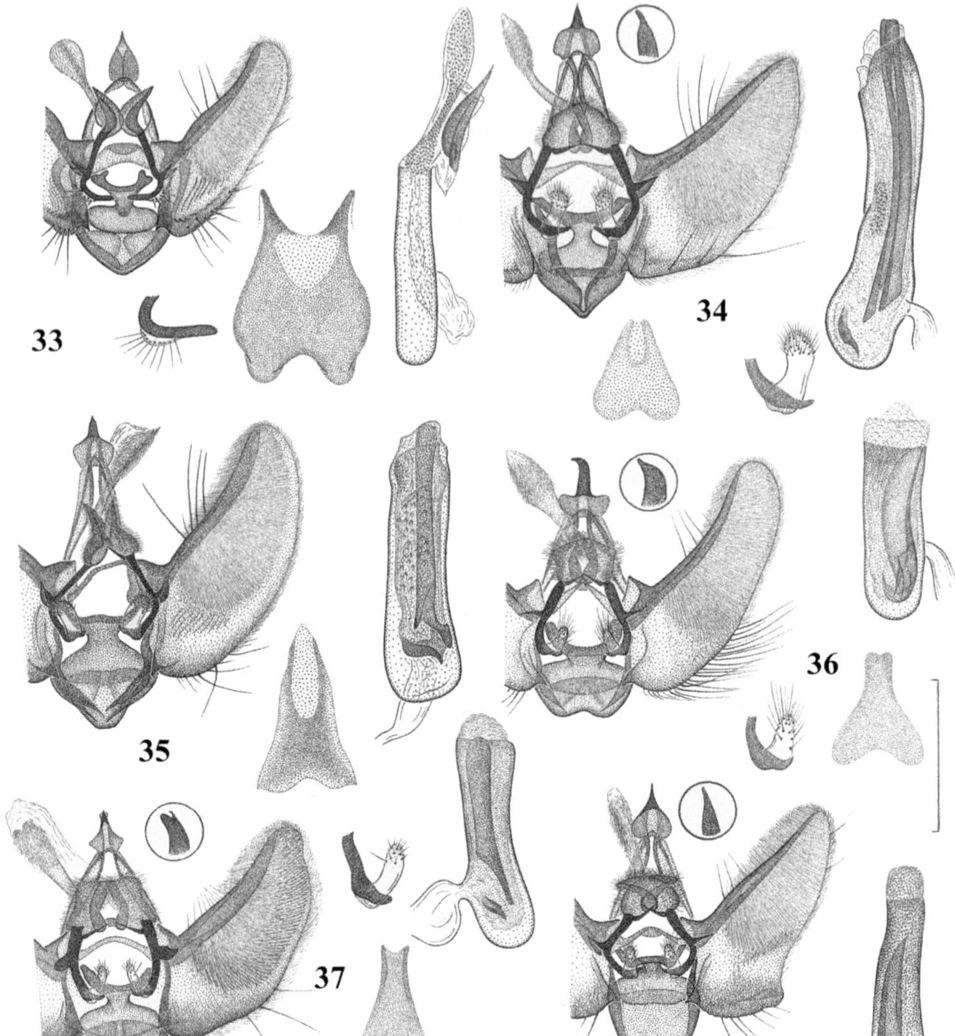

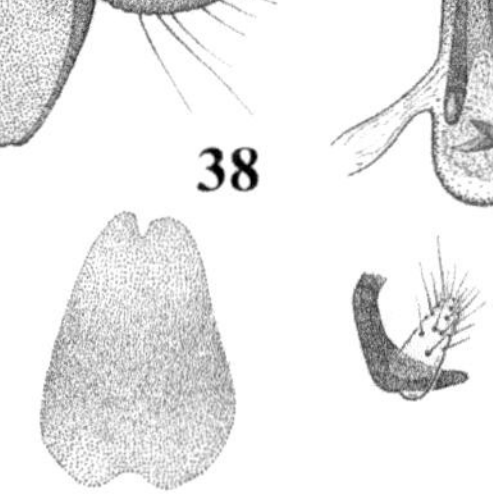

MALE GENITALIA (scale bar = 1mm; with sternum A8; papillae, lateral view of uncus enlarged)

33. *Eupithecia costalis* Walker, China, Taiwan
34. *Eupithecia jermyi* Vojnits, Vietnam, Lao Cai Prov.
35. *Eupithecia lini* Mironov & Galsworthy, China, Taiwan
36. *Eupithecia lucigera* Butler, Nepal, Bagmati
37, *Eupithecia missionerata* Mironov & Galsworthy, China, Tibet
38. *Eupithecia proterva* Butler, Russia, S. Primorje

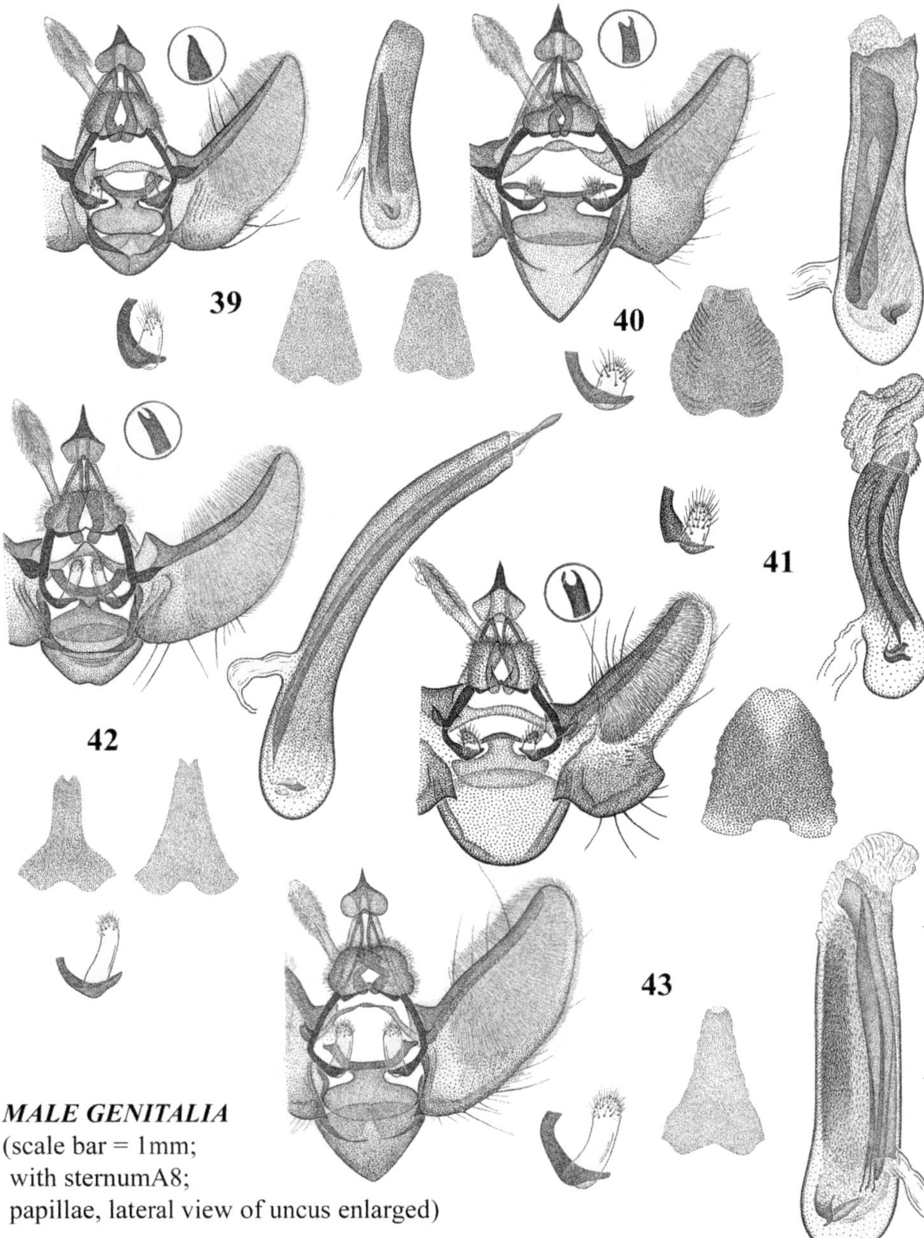

MALE GENITALIA
(scale bar = 1mm;
with sternumA8;
papillae, lateral view of uncus enlarged)

39. *Eupithecia subbreviata* Staudinger, Russia, Primorje (variation sternite A8)
40. *Eupithecia megaproterva* Inoue, China, Taiwan
41. *Eupithecia verprota* Mironov & Galsworthy, China, Shaanxi
42. *Eupithecia clavifera* Inoue, China, Shaanxi (variation sternite A8)
43. *Eupithecia wangi* Mironov & Galsworthy, China, Guangxi

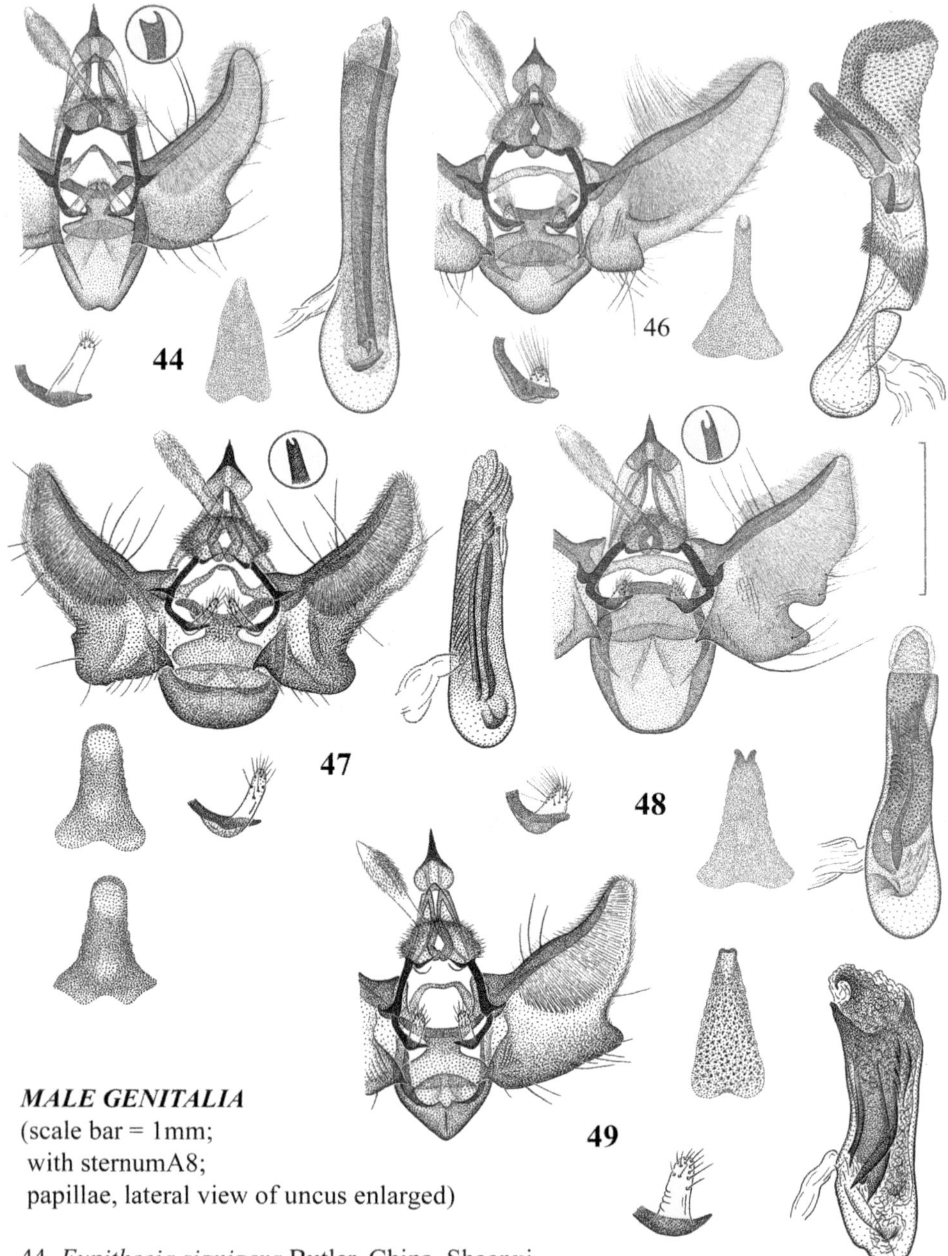

MALE GENITALIA
(scale bar = 1mm;
with sternumA8;
papillae, lateral view of uncus enlarged)

44. *Eupithecia signigera* Butler, China, Shaanxi
46. *Eupithecia flavoapicaria* Inoue, China, Taiwan
47. *Eupithecia blenna* Mironov & Galsworthy, China, Sichuan (variation sternite A8)
48. *Eupithecia convexa* Inoue, China, Taiwan
49. *Eupithecia opicata* Mironov & Galsworthy, China, Gansu

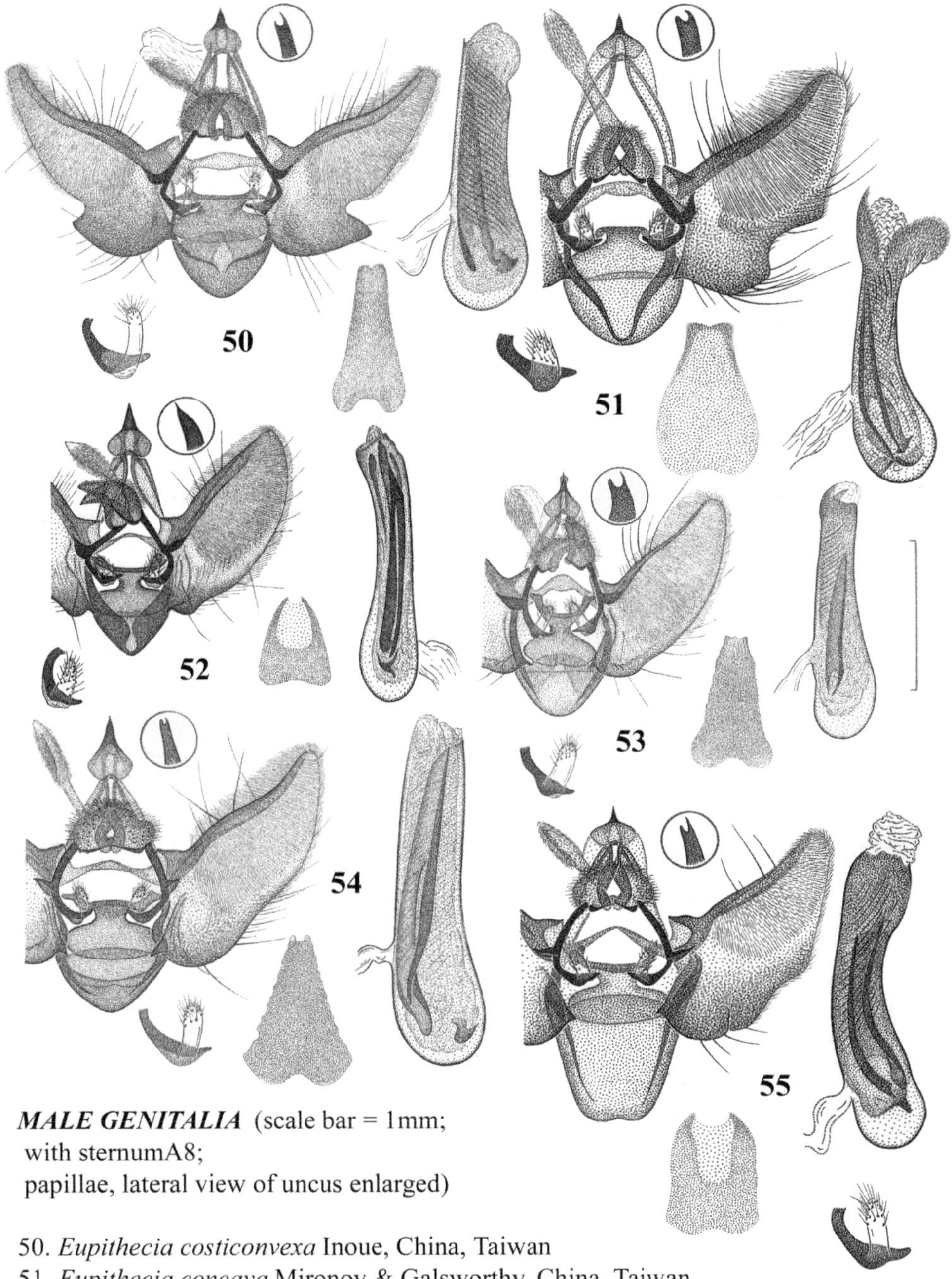

MALE GENITALIA (scale bar = 1mm;
with sternumA8;
papillae, lateral view of uncus enlarged)

50. *Eupithecia costiconvexa* Inoue, China, Taiwan
51. *Eupithecia concava* Mironov & Galsworthy, China, Taiwan
52. *Eupithecia aritai* Inoue, Japan
53. *Eupithecia admiranda* Mironov & Galsworthy, China, Shaanxi
54. *Eupithecia hashimotoi* Inoue, China, Taiwan
55. *Eupithecia flavimacula* Mironov & Galsworthy, China, Taiwan

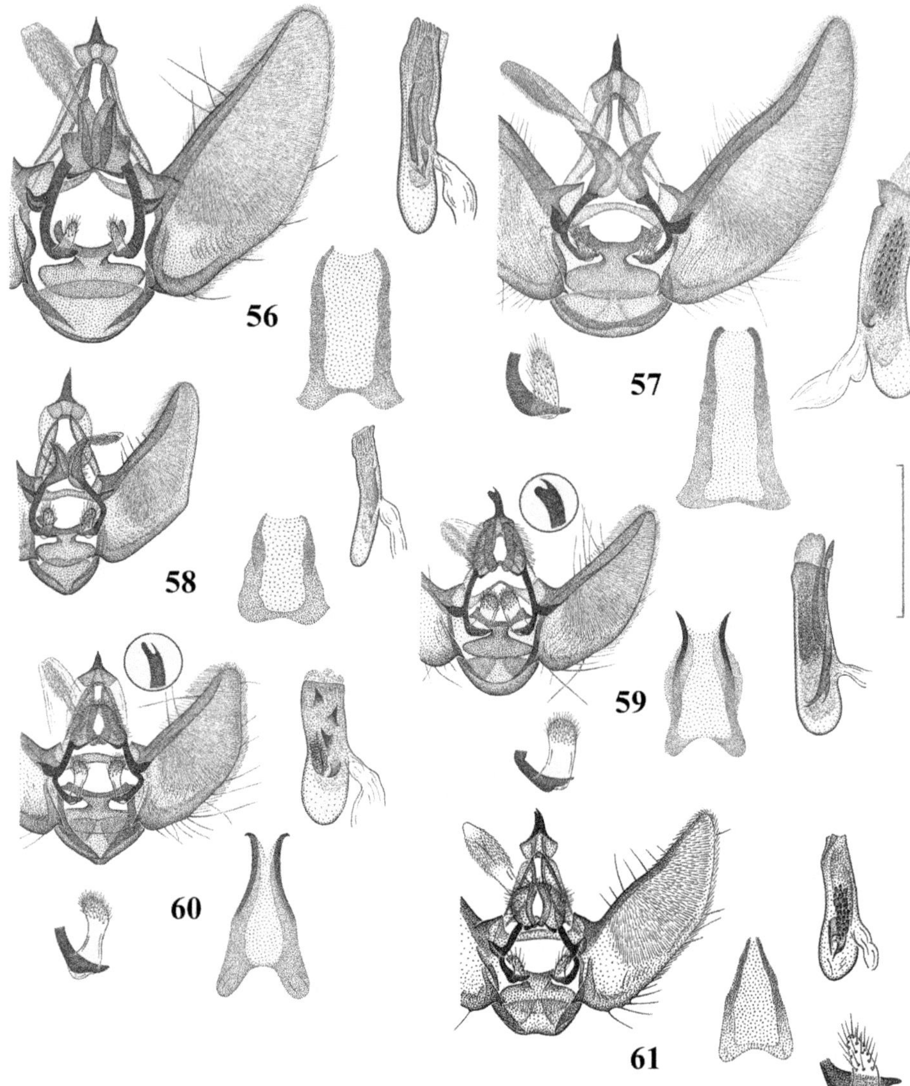

MALE GENITALIA (scale bar = 1mm; with sternumA8; papillae, lateral view of uncus enlarged)

56. *Eupithecia amplificata* Inoue, Nepal, Bagmati, Langtang Himal
57. *Eupithecia asempiterna* Inoue, Nepal, Annapurna Himal
58. *Eupithecia wardi* Prout, China, Tibet
59. *Eupithecia minusculata* Alphéraky, Turkmenistan, Dashoguz
60. *Eupithecia opistographata* Dietze, Turkmenistan, Dashoguz
61. *Eupithecia viata* Mironov & Galsworthy, China, Gansu

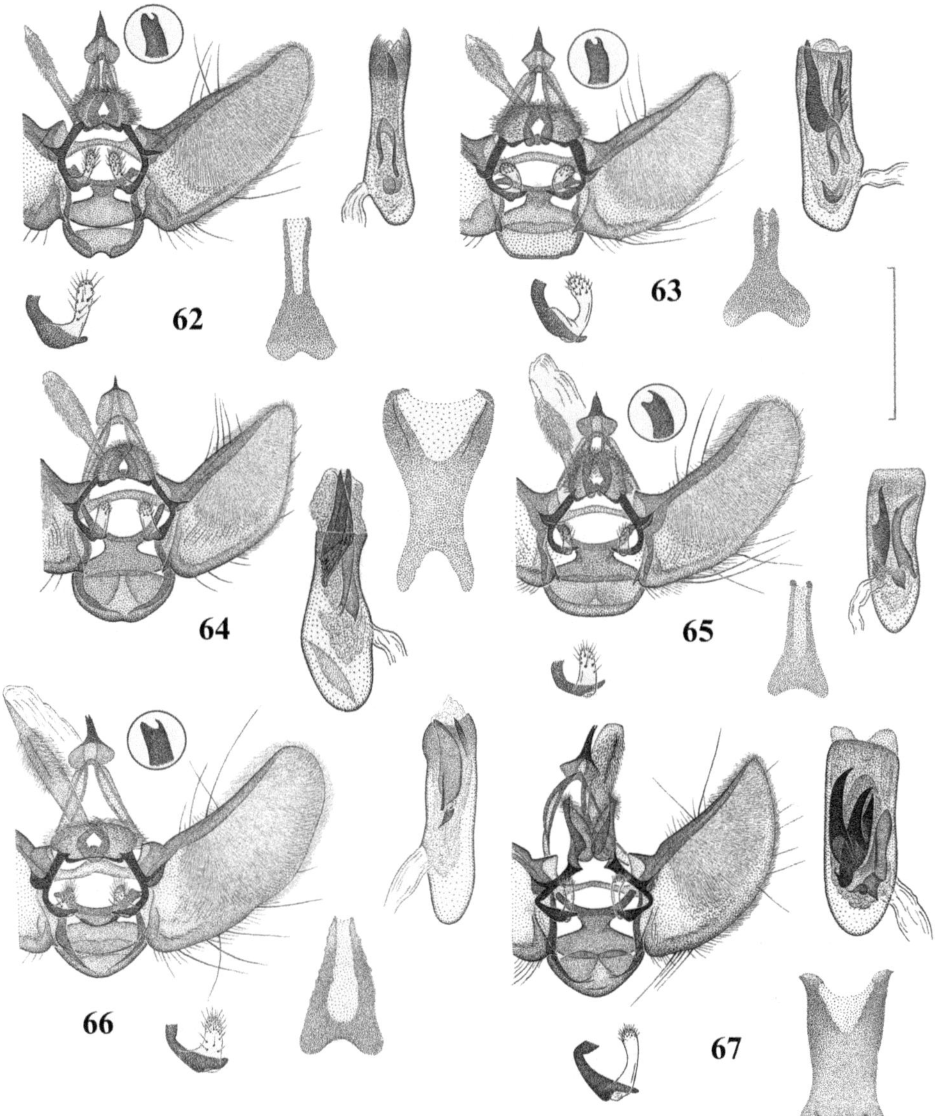

MALE GENITALIA (scale bar = 1mm; with sternumA8; papillae, lateral view of uncus enlarged)

62. *Eupithecia mustangata* Schütze, Pakistan, Gilgit, Deosai Pass
63, *Eupithecia leucostaxis* Prout, Nepal, Annapurna Himal
64. *Eupithecia ochracea* Warren, N.E.India, Khasis
65. *Eupithecia homogrammata* Dietze, Russia, S. Primorje
66. *Eupithecia plumbeolata* Haworth, Kyrgyzstan, Naryn
67. *Eupithecia severa* Mironov & Galsworthy, China, Shaanxi

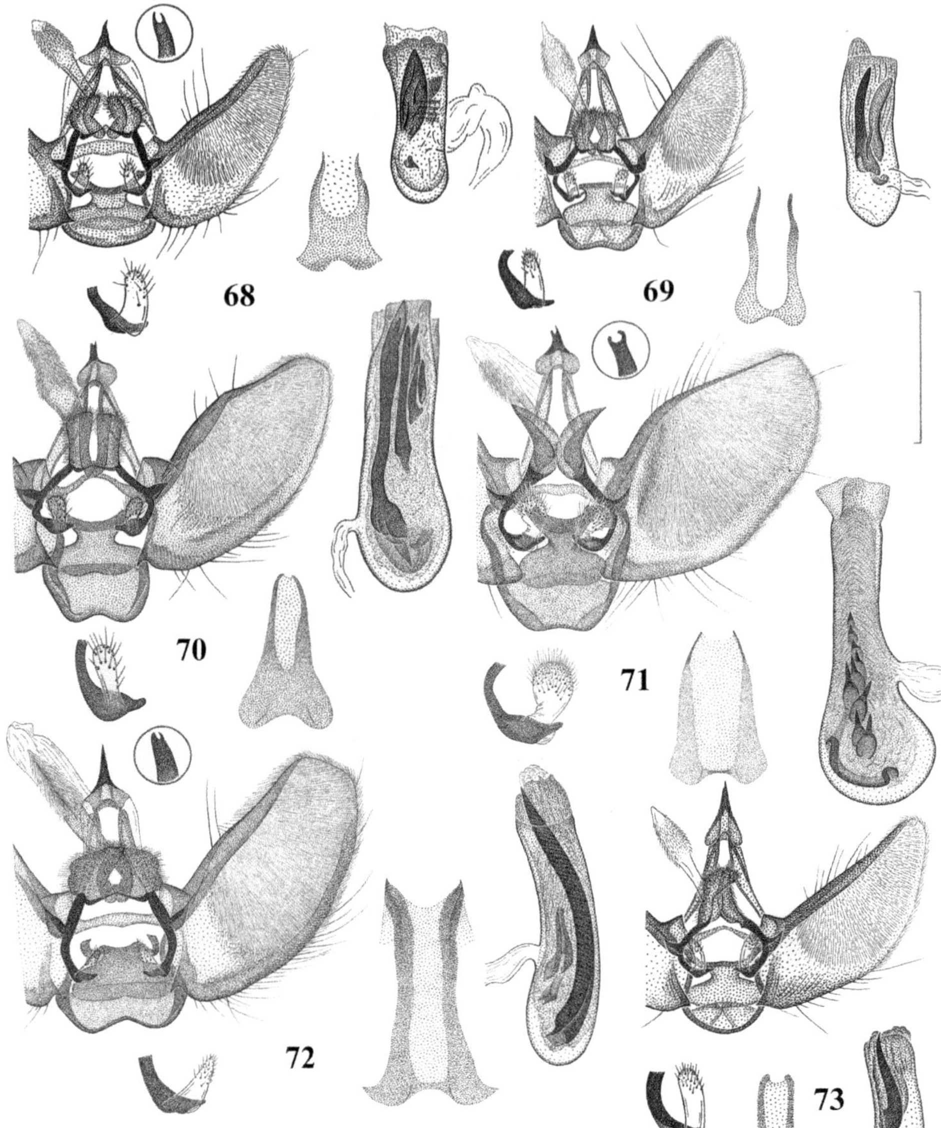

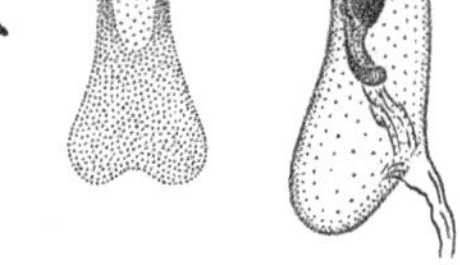

MALE GENITALIA (scale bar = 1mm; with sternumA8; papillae, lateral view of uncus enlarged)

68. *Eupithecia spissata* Mironov & Galsworthy, China, Sichuan
69. *Eupithecia cordata* Mironov & Galsworthy, China, Tibet
70. *Eupithecia turpis* Vojnits, China, Yunnan
71. *Eupithecia peguensis* Prout, Nepal, Gandaki, Ganesh Himal
72. *Eupithecia apta* Mironov & Galsworthy, China, Yunnan
73. *Eupithecia burselongata* Mironov & Galsworthy, China, Yunnan

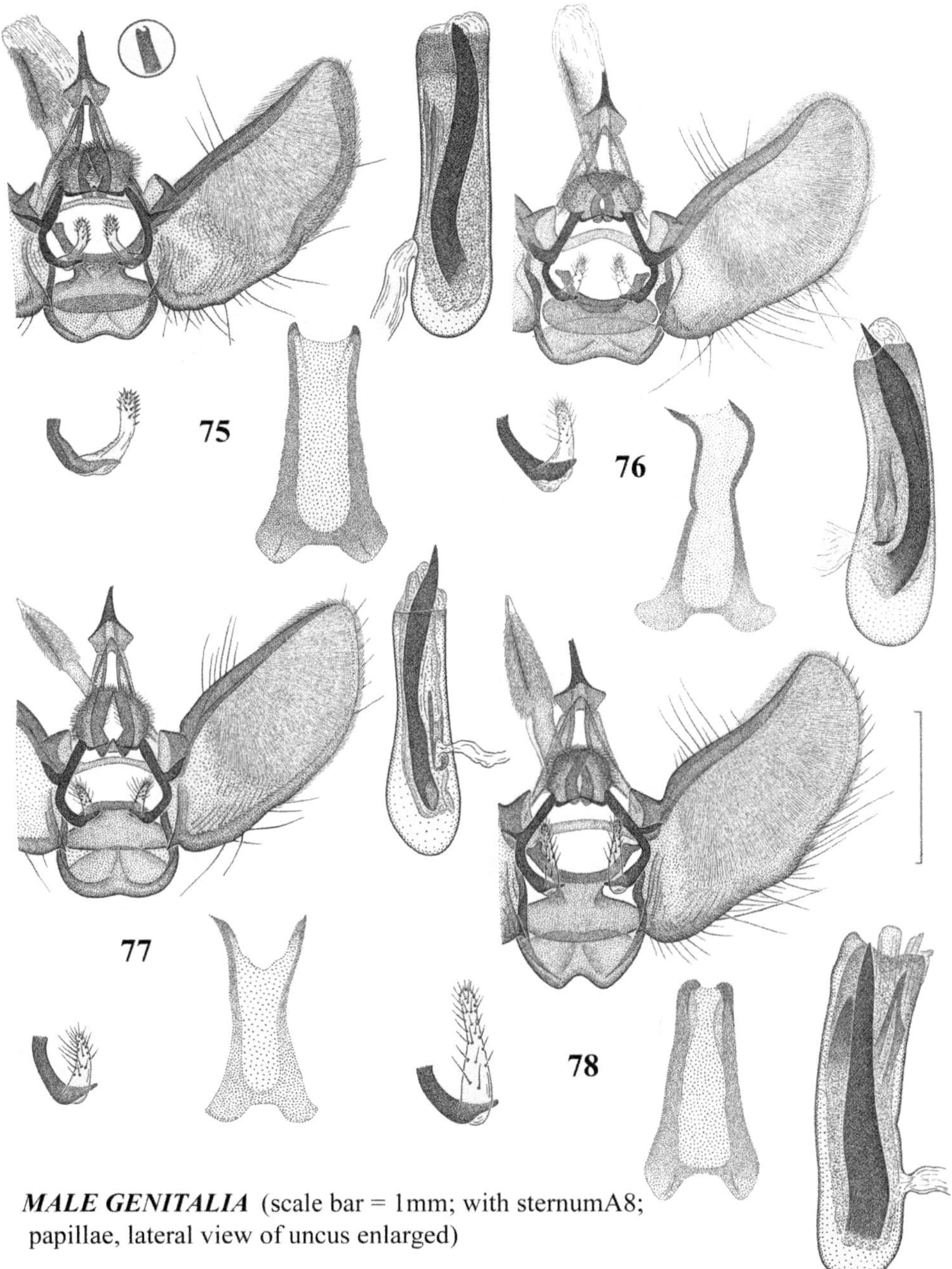

MALE GENITALIA (scale bar = 1mm; with sternumA8; papillae, lateral view of uncus enlarged)

75. *Eupithecia impavida* Vojnits, Pakistan, Kaghan Valley
76. *Eupithecia fervida* Vojnits, China, Yunnan
77. *Eupithecia inexhausta* Vojnits, China, Yunnan
78. *Eupithecia hongxiangae* Mironov & Galsworthy, China, Henan

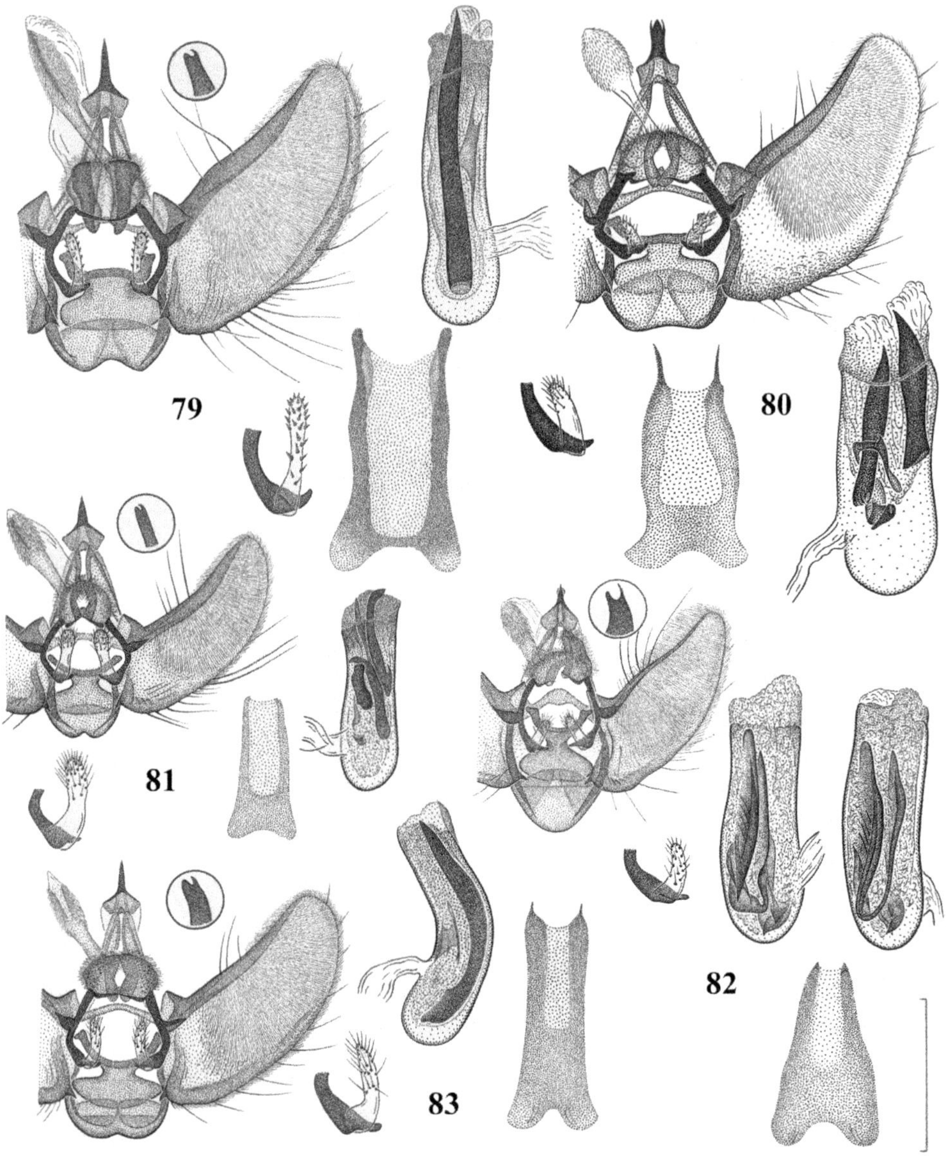

MALE GENITALIA (scale bar = 1mm; with sternumA8; papillae, lateral view of uncus enlarged)

79. *Eupithecia alishana* Inoue, China, Taiwan
80. *Eupithecia antiqua* Mironov & Galsworthy, China, Yunnan
81. *Eupithecia assulata* Bastelberger, China, Taiwan
82. *Eupithecia adoranda* Vojnits, China, Shanxi (aedeagus on right, Henan)
83. *Eupithecia granata* Vojnits, China, Yunnan

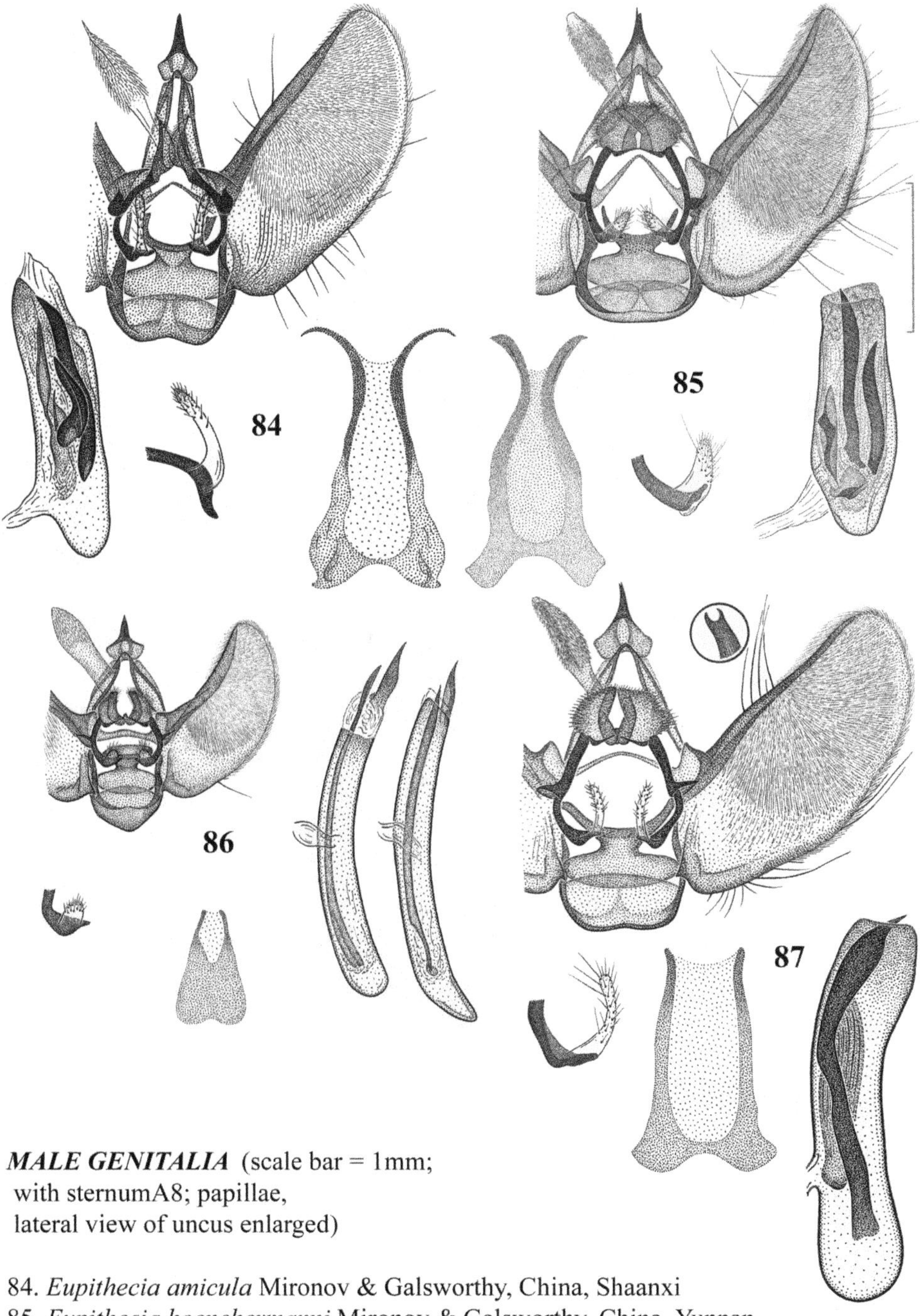

MALE GENITALIA (scale bar = 1mm;
with sternumA8; papillae,
lateral view of uncus enlarged)

84. *Eupithecia amicula* Mironov & Galsworthy, China, Shaanxi
85. *Eupithecia hoenehermanni* Mironov & Galsworthy, China, Yunnan
86. *Eupithecia extinctata* Dietze, China, Tibet (aedeagus 2 exx.)
87. *Eupithecia consortaria* Leech, Russia, Primorye

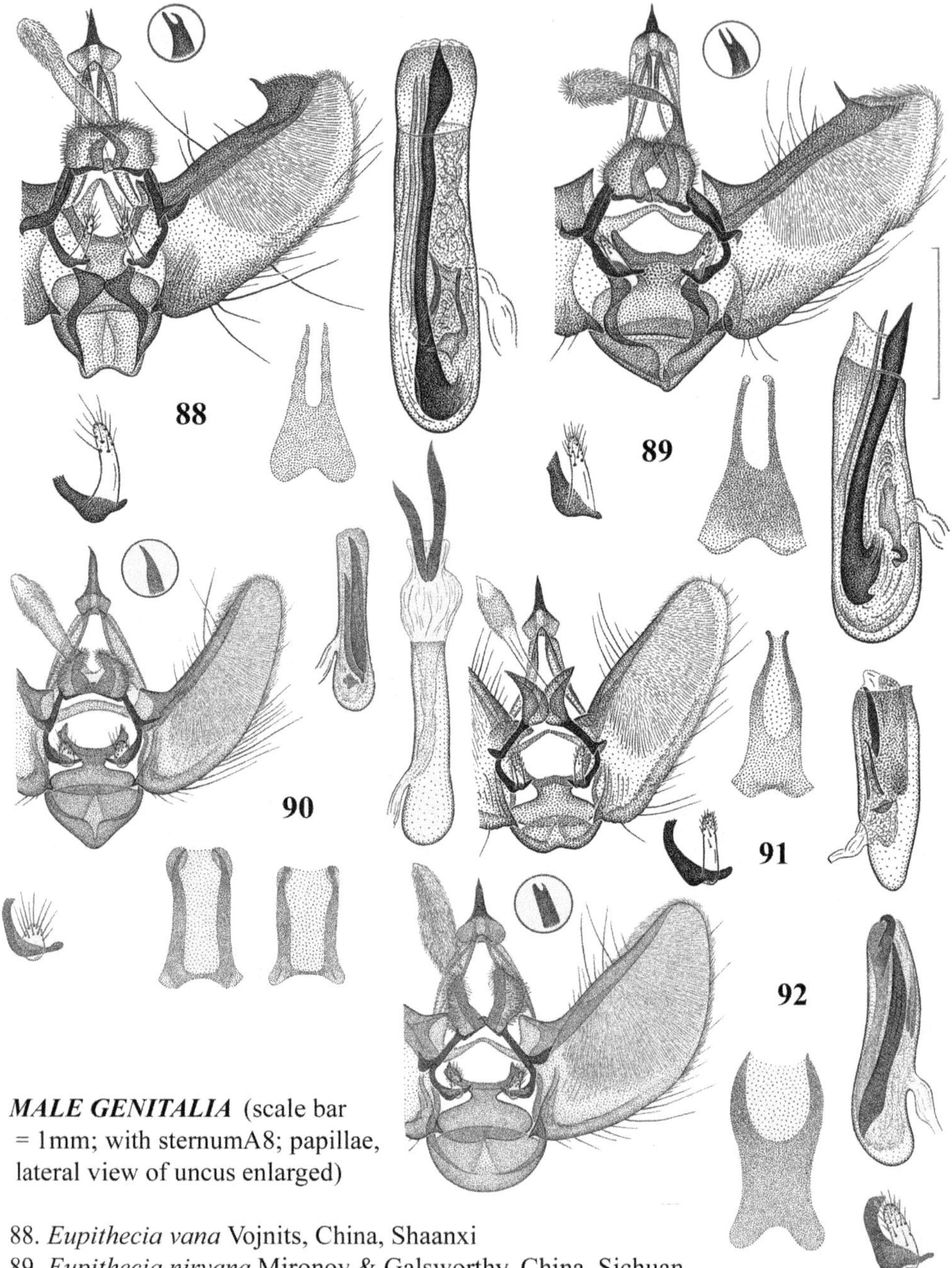

MALE GENITALIA (scale bar = 1mm; with sternumA8; papillae, lateral view of uncus enlarged)

88. *Eupithecia vana* Vojnits, China, Shaanxi
89. *Eupithecia nirvana* Mironov & Galsworthy, China, Sichuan
90. *Eupithecia hilariata* Dietze, Uzbekistan, Tchimgan mountains (aedeagus everted and uneverted, variation st8)
91. *Eupithecia hesperina* Mironov & Galsworthy, China, Yunnan
92. *Eupithecia dissertata* Püngeler, Mongolia, Dornot aimag

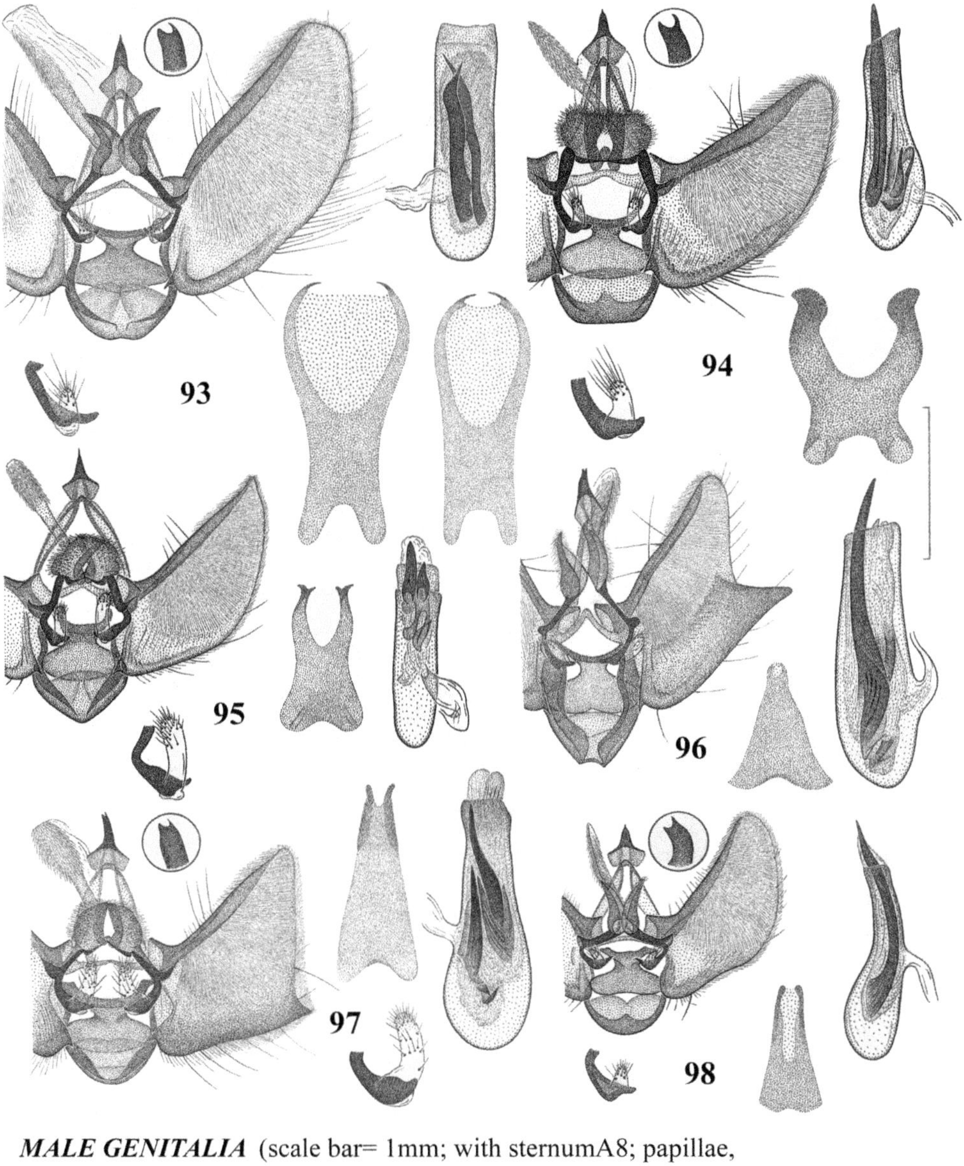

MALE GENITALIA (scale bar= 1mm; with sternumA8; papillae, lateral view of uncus enlarged)

93. *Eupithecia carpophilata* Staudinger, Mongolia, Urga (variation sternite A8)
94. *Eupithecia albispumata* Warren, China, Sichuan
95. *Eupithecia astricta* Inoue, China, Taiwan
96. *Eupithecia abbreviata* Stephens, China, Sichuan
97. *Eupithecia pusillata* Denis & Schiffermüller, Russia, S. Urals
98. *Eupithecia kozlovi* Viidalepp, China, Qinghai

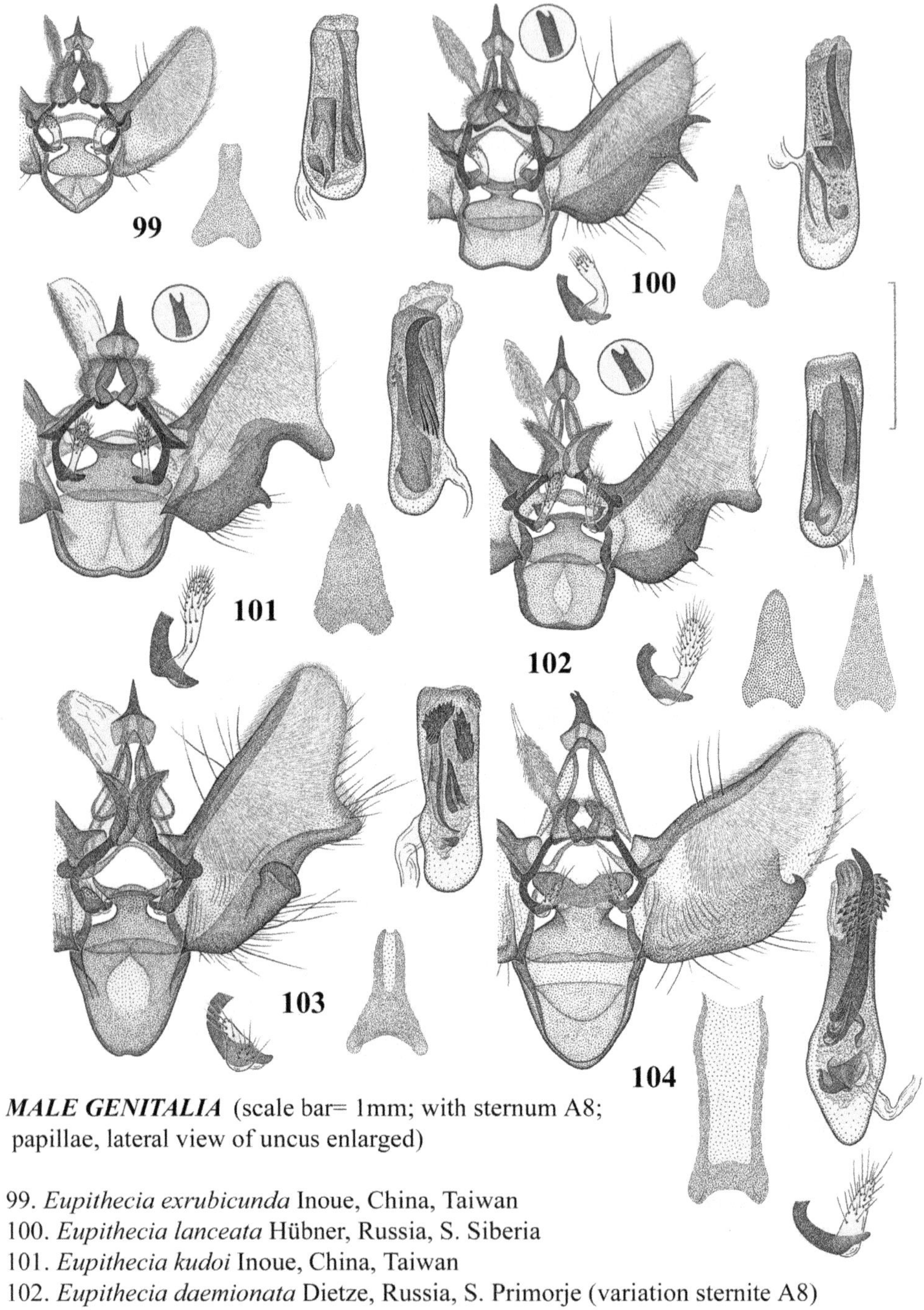

MALE GENITALIA (scale bar= 1mm; with sternum A8; papillae, lateral view of uncus enlarged)

99. *Eupithecia exrubicunda* Inoue, China, Taiwan
100. *Eupithecia lanceata* Hübner, Russia, S. Siberia
101. *Eupithecia kudoi* Inoue, China, Taiwan
102. *Eupithecia daemionata* Dietze, Russia, S. Primorje (variation sternite A8)
103. *Eupithecia habermani* Viidalepp & Mironov, China, Gansu
104. *Eupithecia emendata* Vojnits, Nepal , Sagarmatha (holotype)

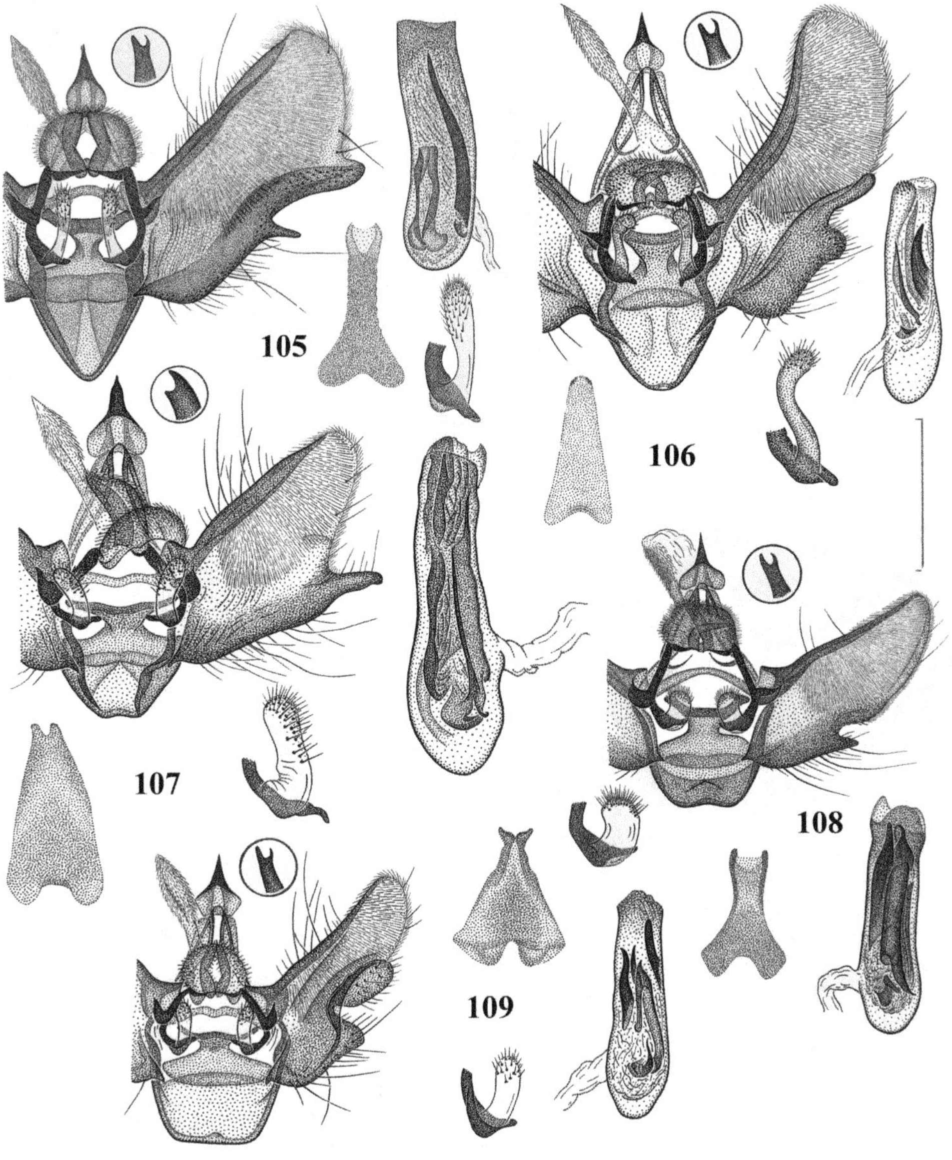

MALE GENITALIA (scale bar= 1mm; with sternum A8; papillae, lateral view of uncus enlarged)

105. *Eupithecia dolichoptera*, Galsworthy & Mironov, China, Taiwan
106. *Eupithecia phantastica* Mironov & Galsworthy, China, Zhejiang
107. *Eupithecia atuni* Mironov & Galsworthy, China, Yunnan
108. *Eupithecia funerea* Inoue, China, Taiwan
109. *Eupithecia fulvidorsata* Mironov & Galsworthy, China, Shaanxi

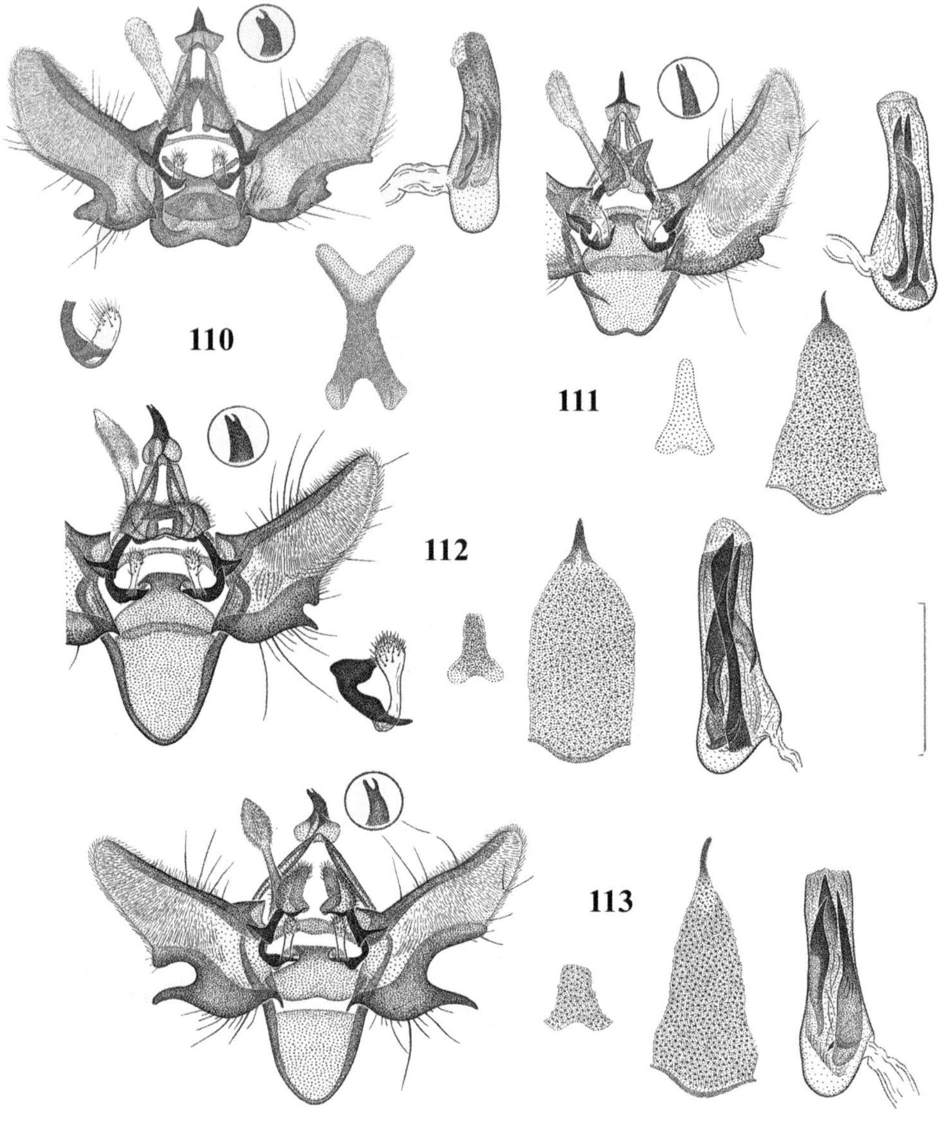

MALE GENITALIA (scale bar= 1mm; with sternum A8, and in some cases tergite A8 on right; papillae, lateral view of uncus enlarged)

110. *Eupithecia okadai* Inoue, China, Shaanxi
111. *Eupithecia delicata* Mironov & Galsworthy, China, Sichuan
112. *Eupithecia camilla* Mironov & Galsworthy, China, Shaanxi
113. *Eupithecia julia* Mironov & Galsworthy, China, Sichuan

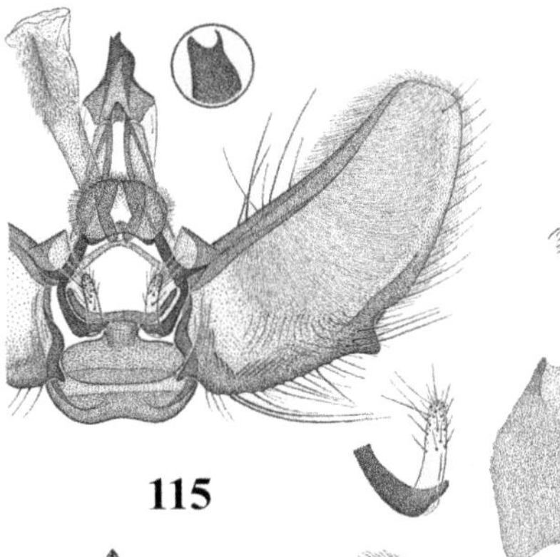

115

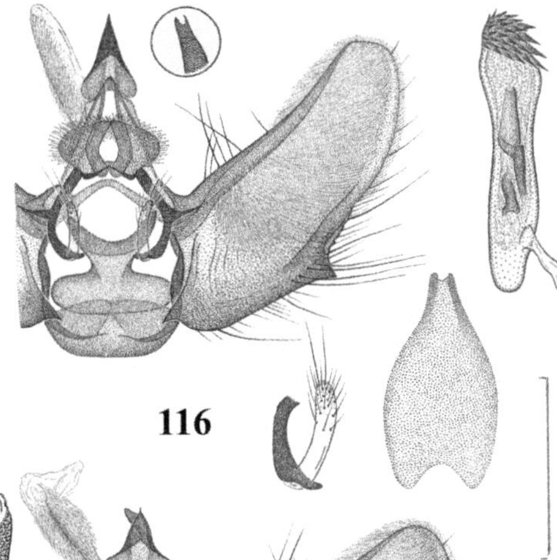

116

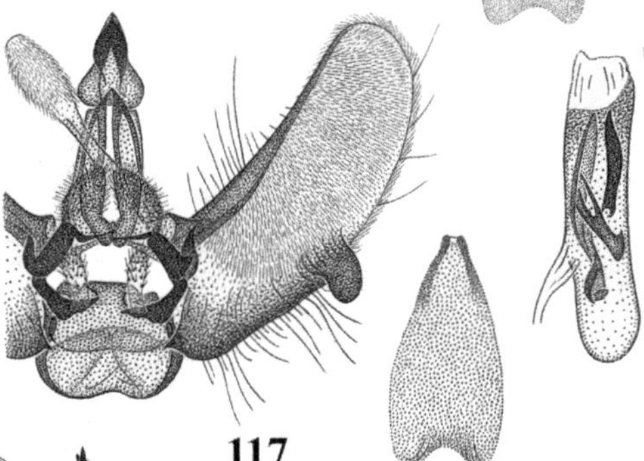

117

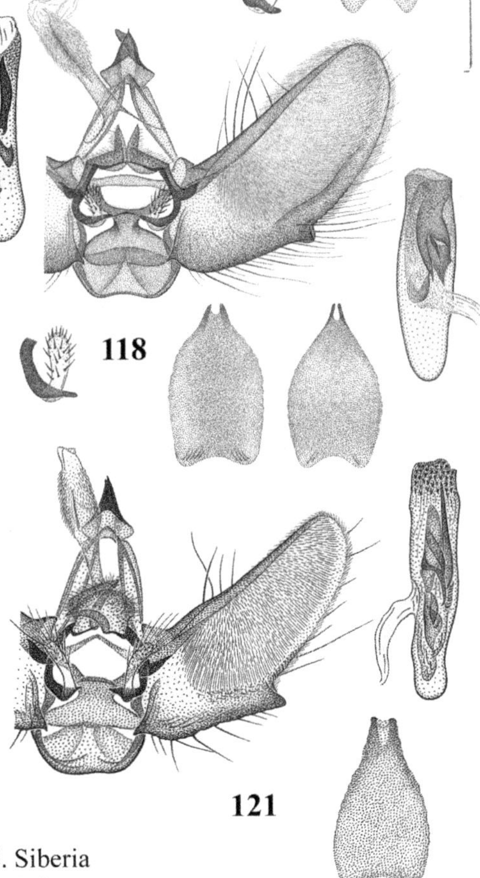

118

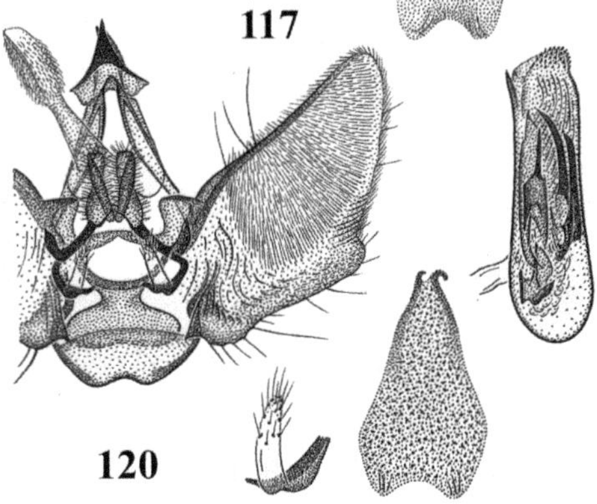

120

121

MALE GENITALIA (scale bar= 1mm; with sternum A8; papillae, lateral view of uncus enlarged)

115. *Eupithecia lariciata* Freyer, Russia, S. Siberia
116. *Eupithecia emanata* Dietze, Russia, S. Primorye
117. *Eupithecia honesta* Mironov & Galsworthy, China, Yunnan
118. *Eupithecia pediba* Vojnits, China, Yunnan (variation, sternite A8)
120. *Eupithecia primitiva* Mironov & Galsworthy, China, Gansu
121. *Eupithecia sacrivicae* Mironov & Galsworthy, China, Hubei

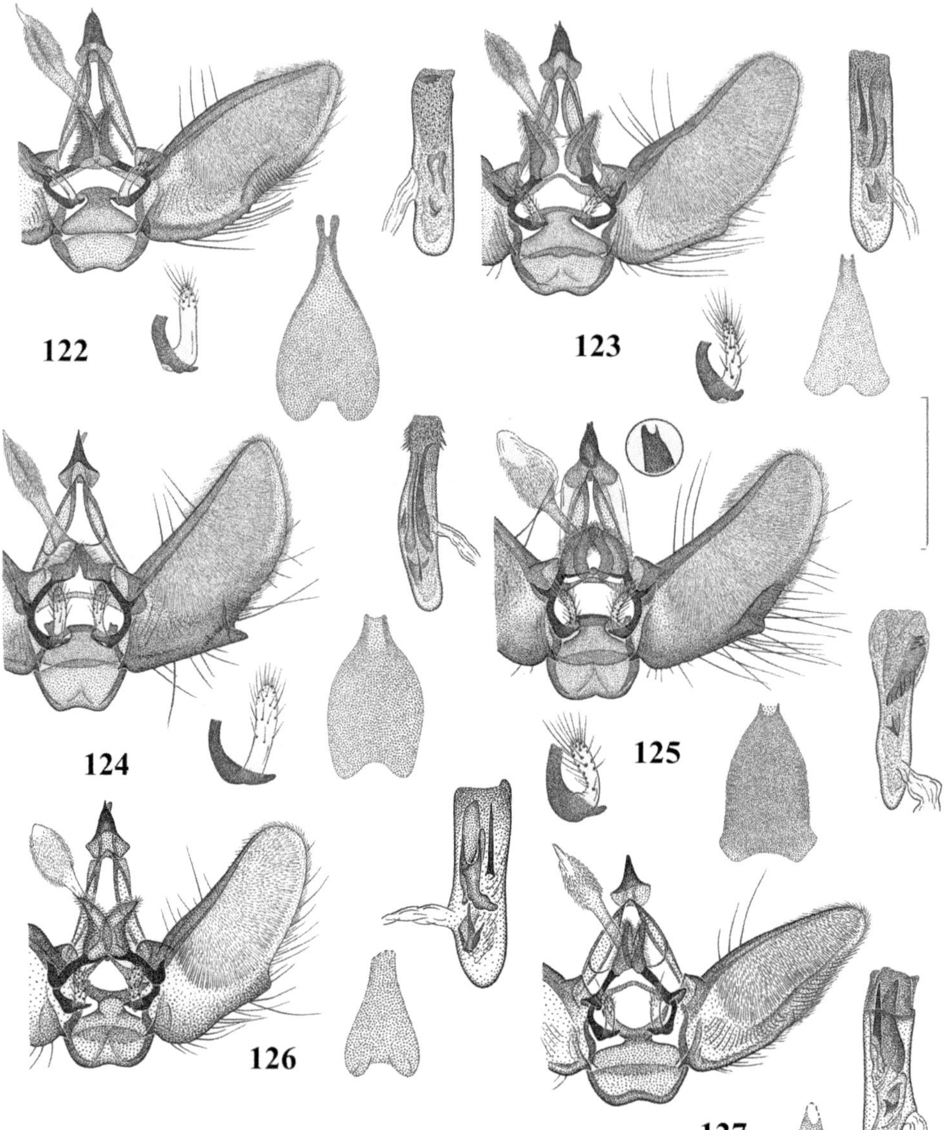

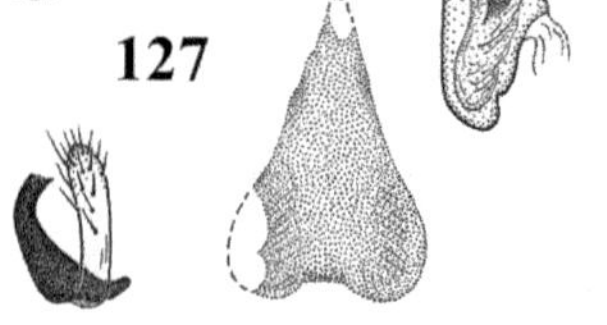

MALE GENITALIA (scale bar= 1mm; with sternum A8; papillae, lateral view of uncus enlarged)

122. *Eupithecia horrida* Vojnits, China, Yunnan
123. *Eupithecia stomachosa* Vojnits, China, Shanxi
124. *Eupithecia sola* Vojnits, Nepal, Sagarmatha (holotype)
125. *Eupithecia marmorea* Vojnits, Nepal, Sagarmatha
126. *Eupithecia albimedia* Mironov & Galsworthy, China, Yunnan
127. *Eupithecia hebes* Mironov & Galsworthy, China, Shanxi

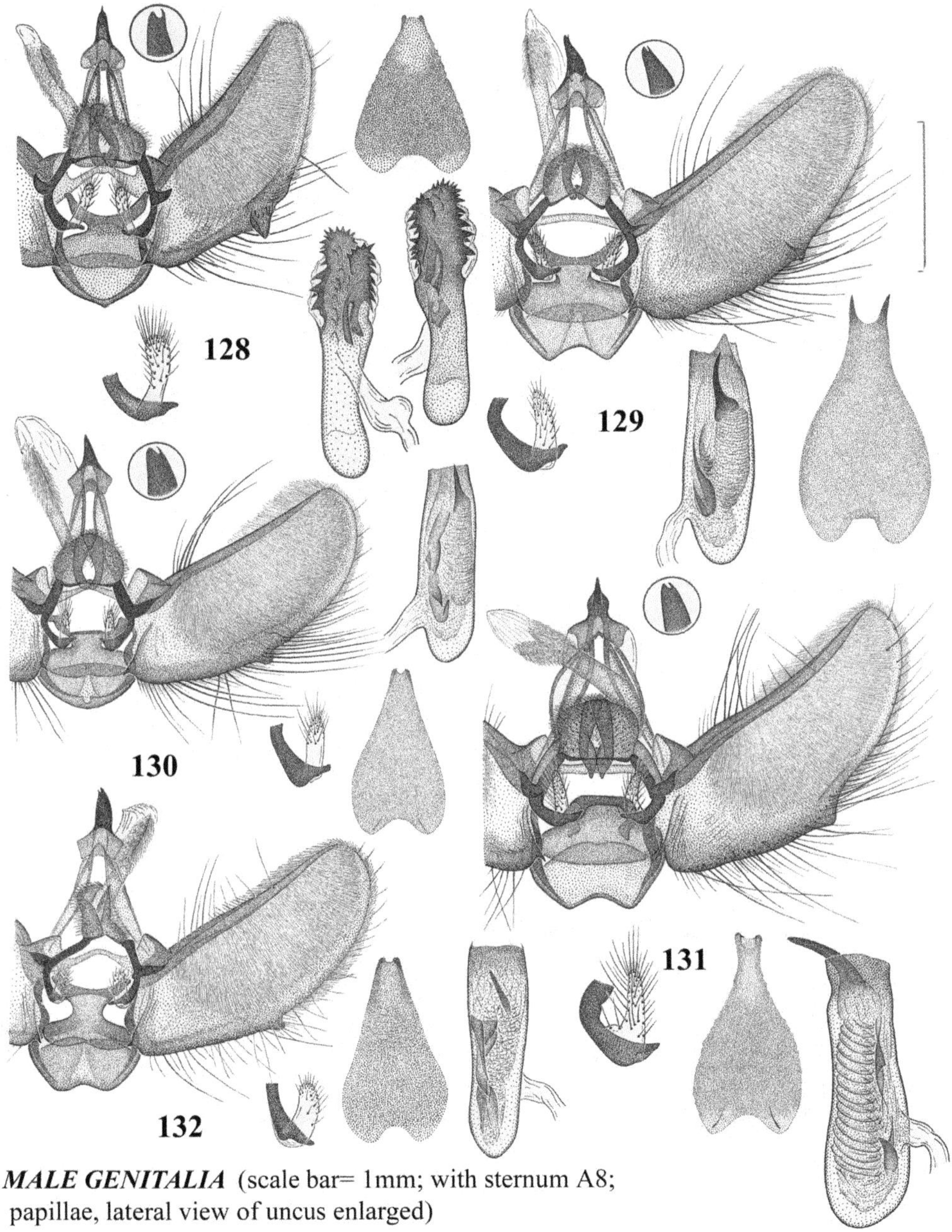

MALE GENITALIA (scale bar= 1mm; with sternum A8; papillae, lateral view of uncus enlarged)

128. *Eupithecia stataria* Inoue, China, Taiwan (dorsal and ventral views)
129. *Eupithecia tripunctaria* Herrich-Schäffer, Russia, S. Primorje
130. *Eupithecia barteli* Dietze, Kyrgyzstan, Naryn
131. *Eupithecia tenuisquama* Warren, Japan, Honshu
132. *Eupithecia amandae* Galsworthy & Mironov, China, Yunnan

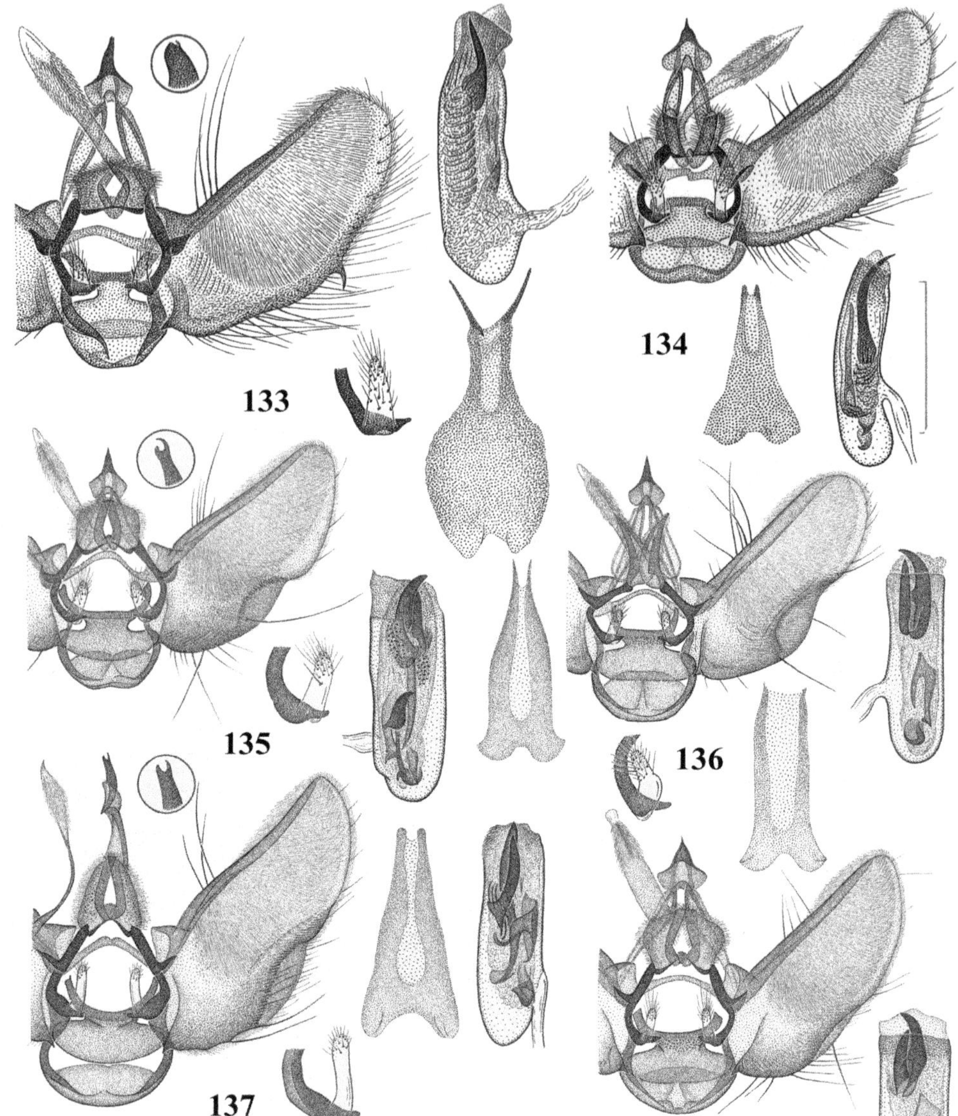

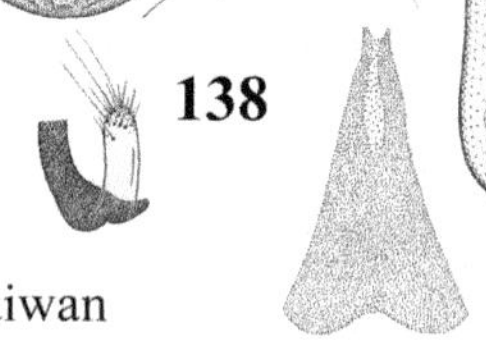

MALE GENITALIA (scale bar= 1mm; with sternum A8; papillae, lateral view of uncus enlarged)

133. *Eupithecia blandula* Mironov & Galsworthy, China, Taiwan
134. *Eupithecia puella* Mironov & Galsworthy, China, Yunnan
135. *Eupithecia werneri* Vojnits, China, Yunnan
136. *Eupithecia batangi* Vojnits, China, Tibet
137. *Eupithecia sacrosancta* Vojnits, China, Shaanxi
138. *Eupithecia placida* Vojnits, China, Yunnan

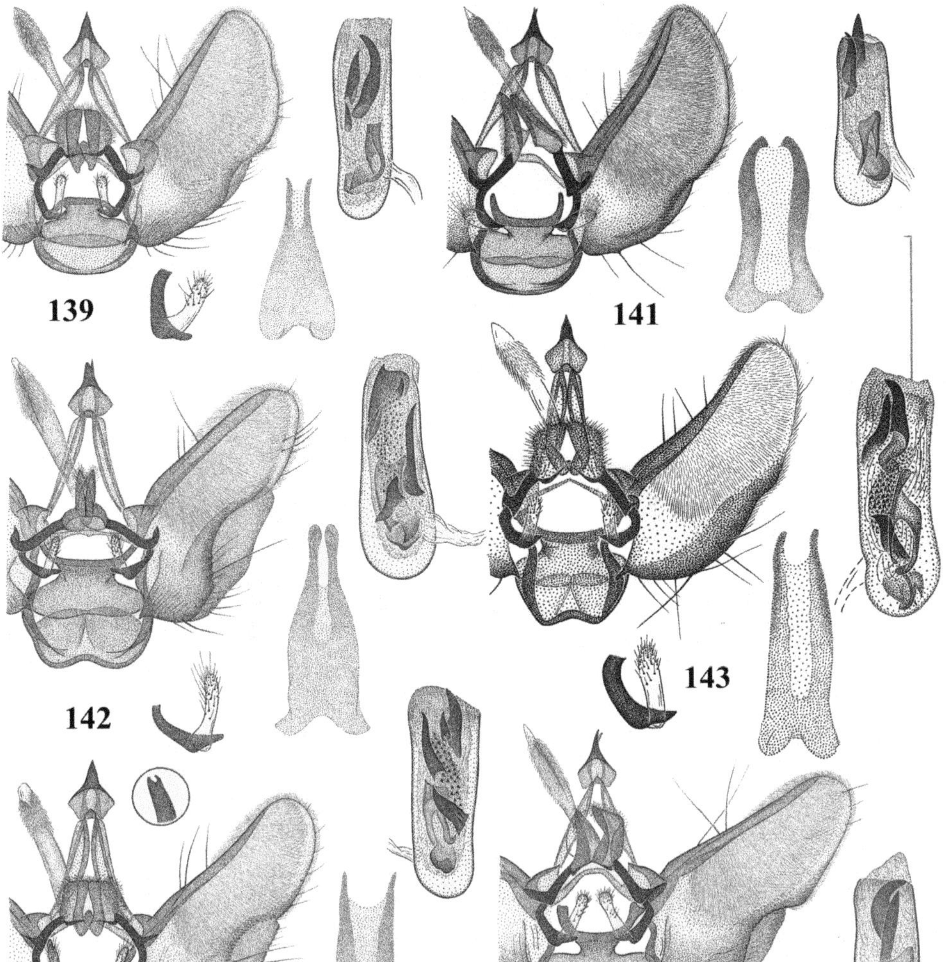

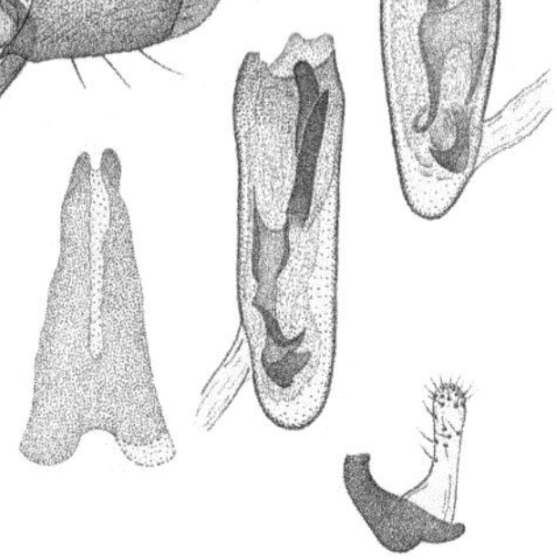

MALE GENITALIA (scale bar= 1mm; with sternum A8; papillae, lateral view of uncus enlarged)

139. *Eupithecia laudabilis* Vojnits, China, Yunnan
141. *Eupithecia intolerabilis* Vojnits, China, Yunnan
142. *Eupithecia mentita* Vojnits, China, Yunnan
143. *Eupithecia fenita* Mironov & Galsworthy, China, Yunnan
144. *Eupithecia refertissima* Vojnits, China, Yunnan
145. *Eupithecia avara* Vojnits, China, Yunnan (aedeagus from 2 angles)

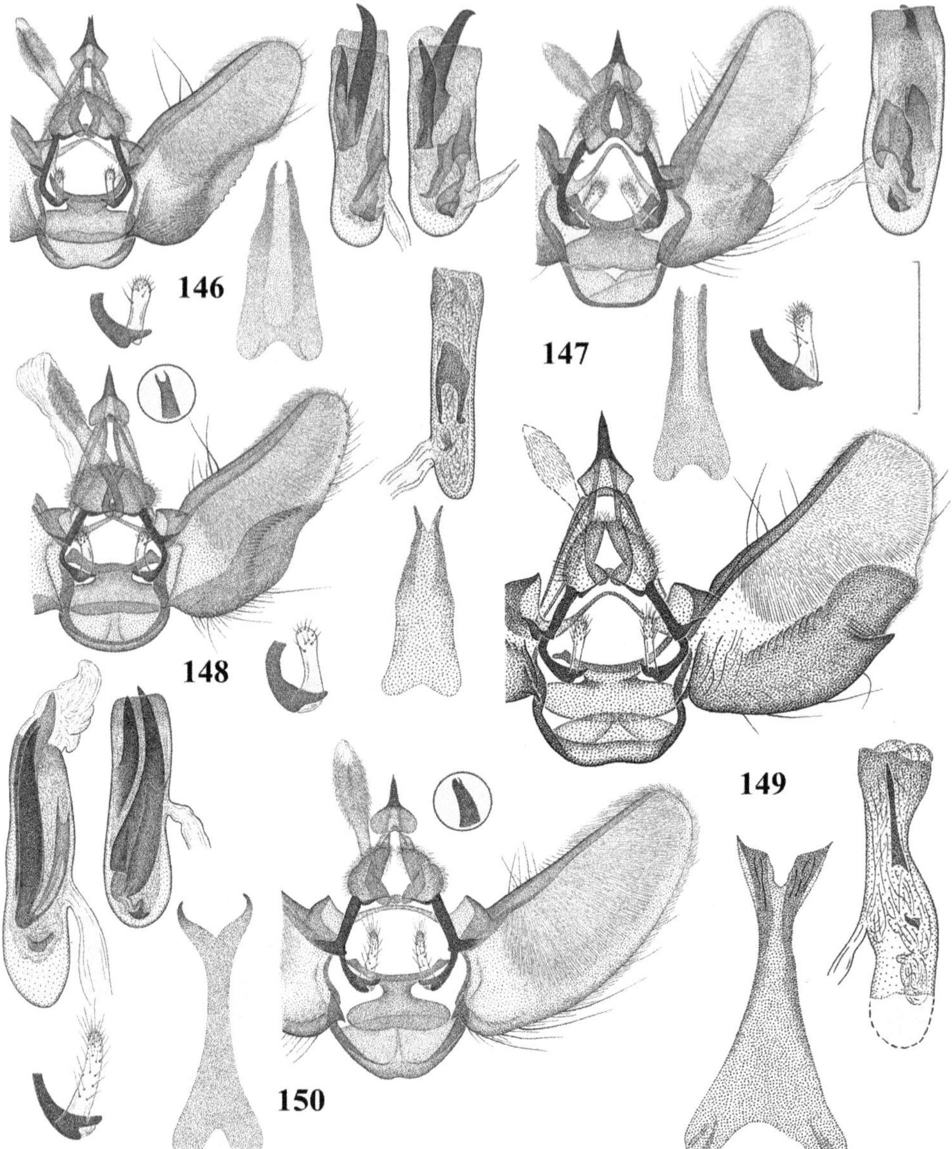

MALE GENITALIA (scale bar= 1mm; with sternum A8; papillae, lateral view of uncus enlarged)

146. *Eupithecia caduca* Vojnits, China, Yunnan (aedeagus 2 variants)
147. *Eupithecia propagata* Prout, N. India, Uttar Pradesh
148. *Eupithecia balintszolti* Vojnits, N. Thailand, Chiangmai
149. *Eupithecia hainanensis* Mironov & Galsworthy, China, Hainan
150. *Eupithecia groenblomi* Urbahn, Russia, Primorje (aedeagus 2 variants)

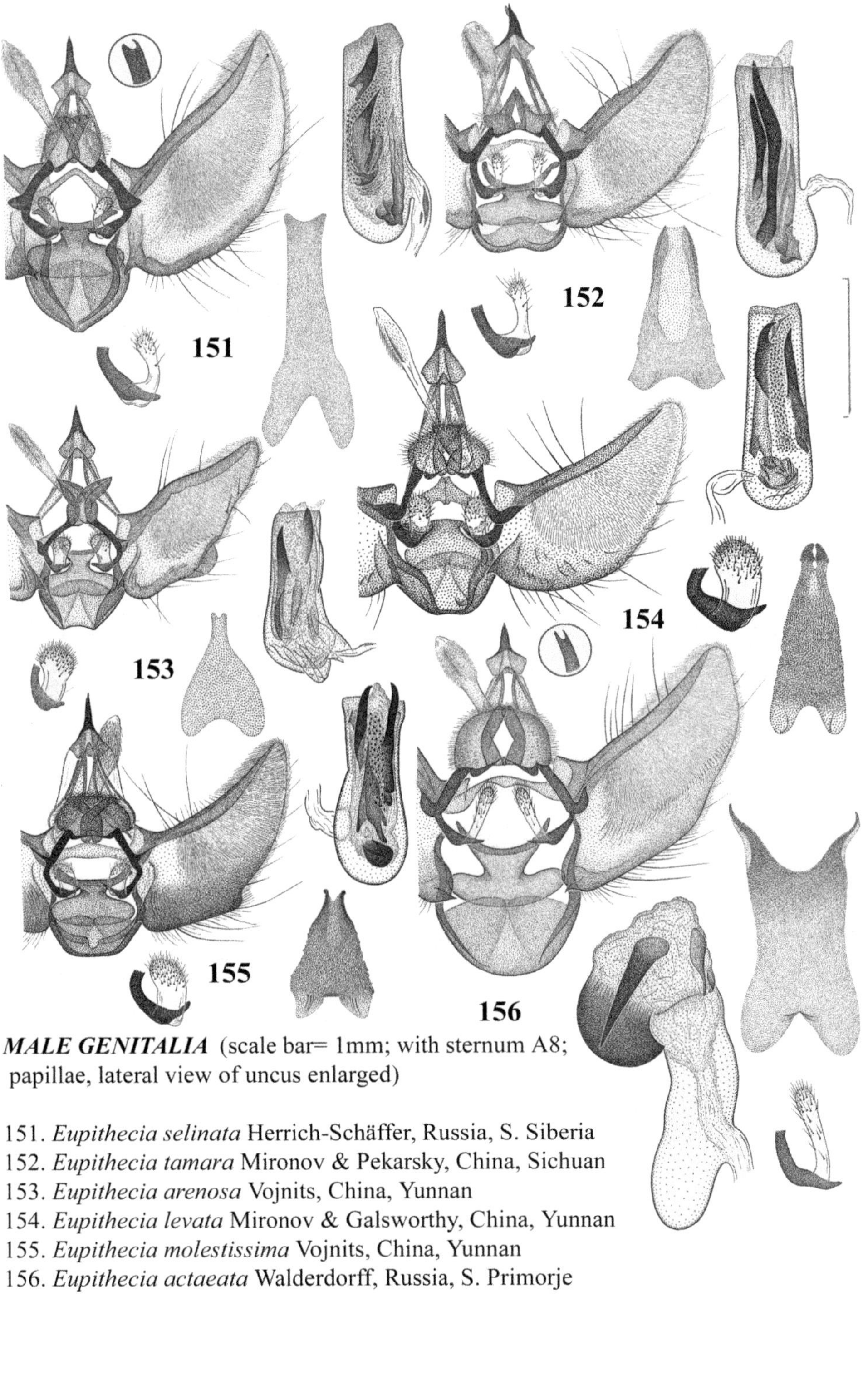

MALE GENITALIA (scale bar= 1mm; with sternum A8; papillae, lateral view of uncus enlarged)

151. *Eupithecia selinata* Herrich-Schäffer, Russia, S. Siberia
152. *Eupithecia tamara* Mironov & Pekarsky, China, Sichuan
153. *Eupithecia arenosa* Vojnits, China, Yunnan
154. *Eupithecia levata* Mironov & Galsworthy, China, Yunnan
155. *Eupithecia molestissima* Vojnits, China, Yunnan
156. *Eupithecia actaeata* Walderdorff, Russia, S. Primorje

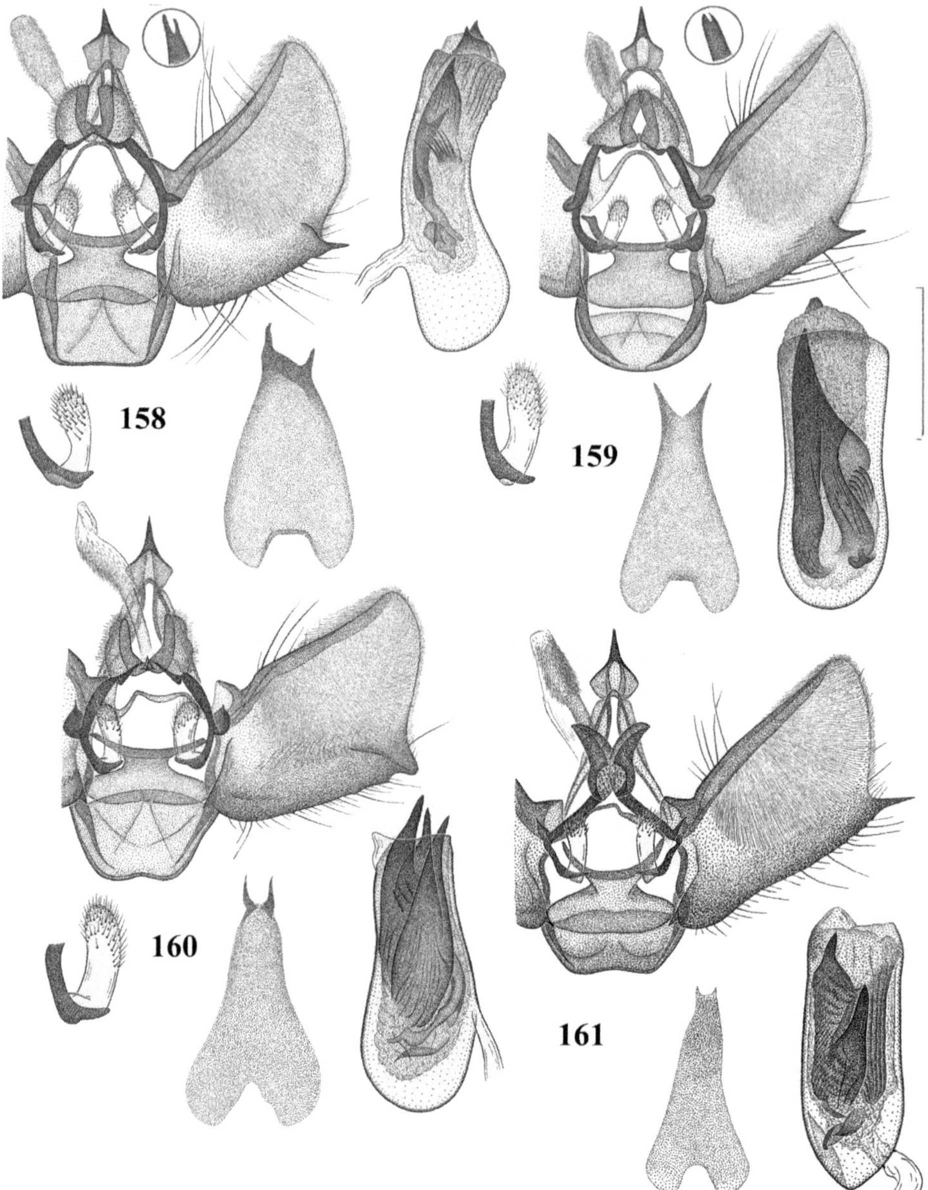

MALE GENITALIA (scale bar= 1mm; with sternum A8; papillae, lateral view of uncus enlarged)

158. *Eupithecia uliata* Staudinger, China, Qinghai
159. *Eupithecia detritata* Staudinger, Russia, S. Primorje
160. *Eupithecia jinboi* Inoue, Japan, Honshu
161. *Eupithecia pellicata* Mironov & Galsworthy, China, Taiwan

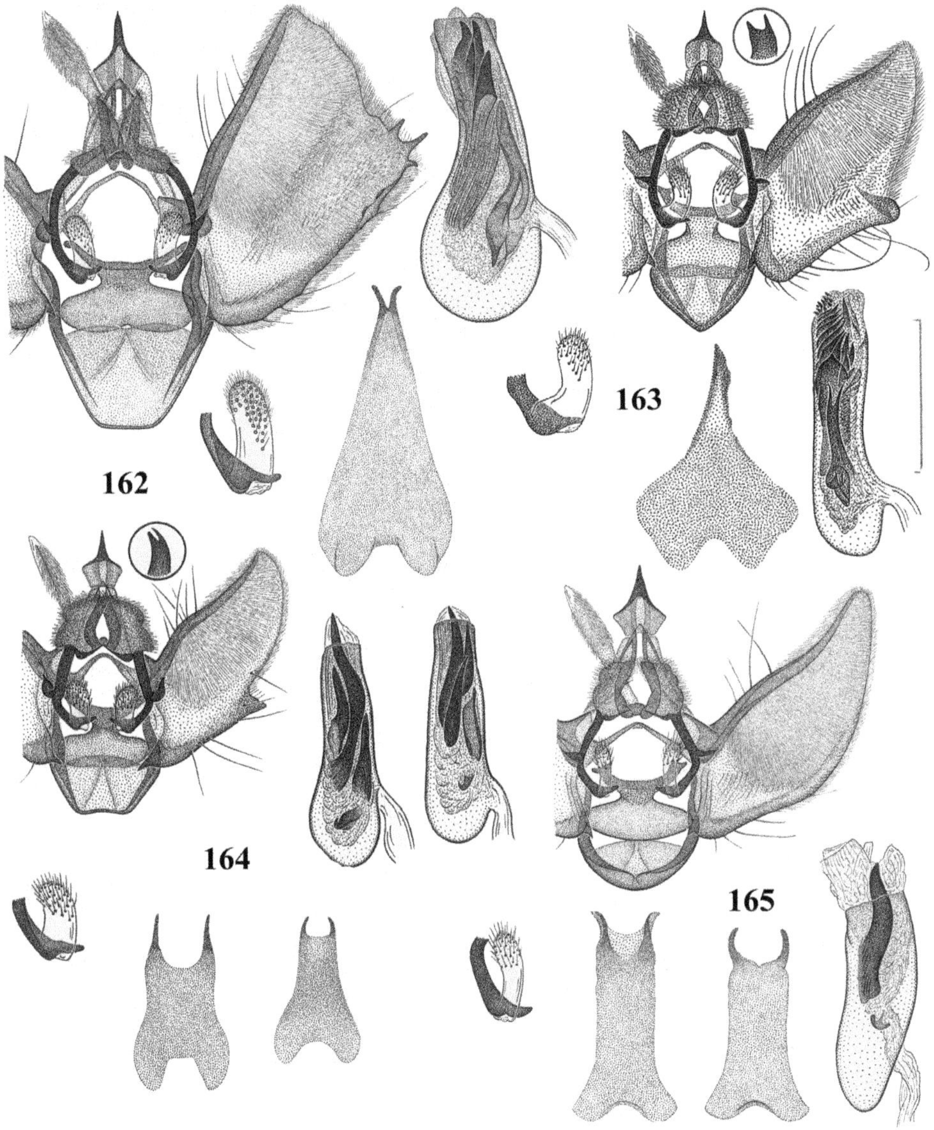

MALE GENITALIA (scale bar= 1mm; with sternum A8; papillae, lateral view of uncus enlarged)

162 *Eupithecia irreperta* Vojnits, China, Yunnan
163. *Eupithecia vivida* Vojnits, Nepal, Kathmandu
164. *Eupithecia quadripunctata* Warren, Kashmir (aedeagus and sterniteA8 on right from far east Russia)
165. *Eupithecia eximia* Vojnits, China, Yunnan (variation sternite A8)

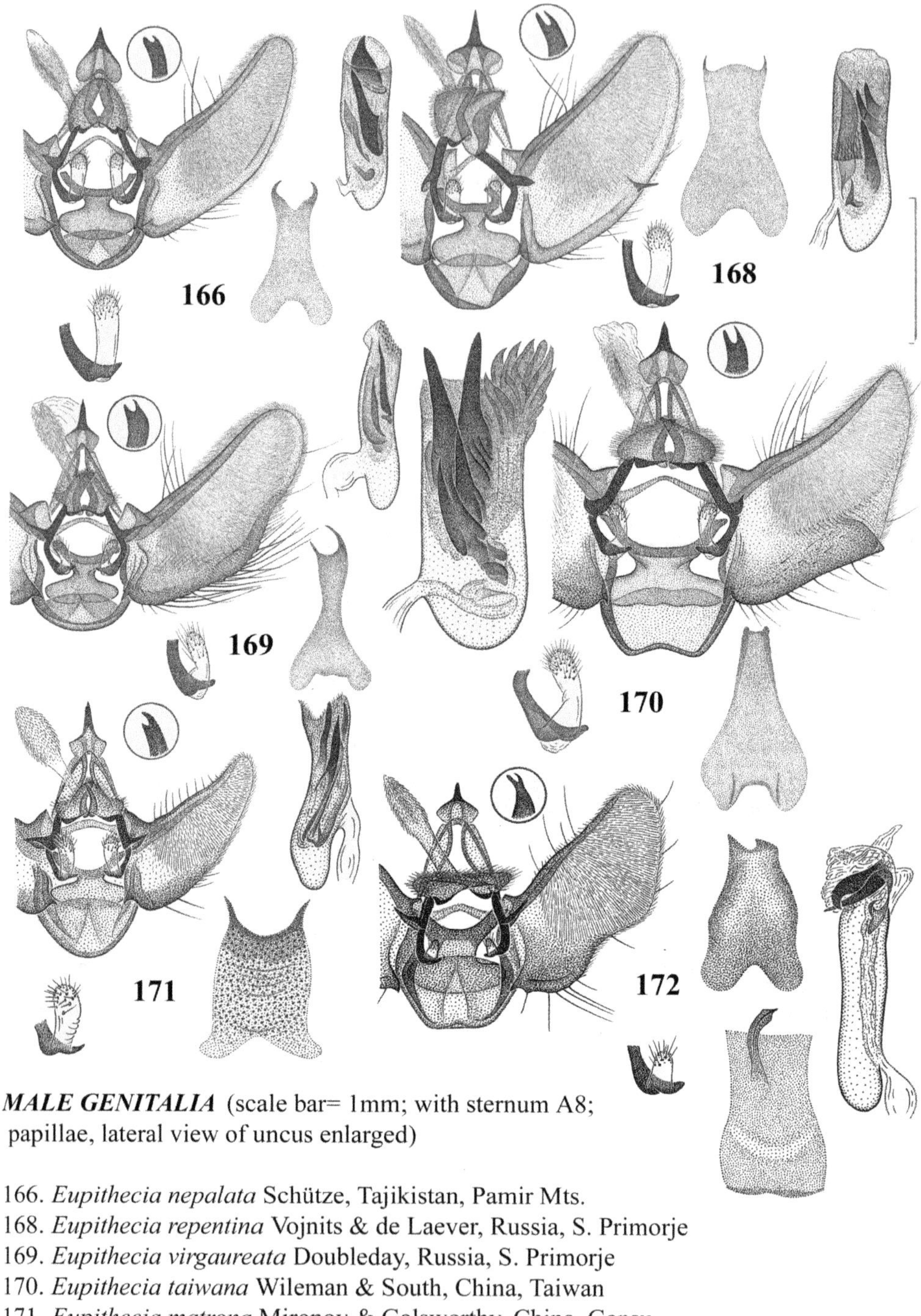

MALE GENITALIA (scale bar= 1mm; with sternum A8; papillae, lateral view of uncus enlarged)

166. *Eupithecia nepalata* Schütze, Tajikistan, Pamir Mts.
168. *Eupithecia repentina* Vojnits & de Laever, Russia, S. Primorje
169. *Eupithecia virgaureata* Doubleday, Russia, S. Primorje
170. *Eupithecia taiwana* Wileman & South, China, Taiwan
171. *Eupithecia matrona* Mironov & Galsworthy, China, Gansu
172. *Eupithecia rhadine* Mironov & Galsworthy, China, Taiwan (tergite A8 below)

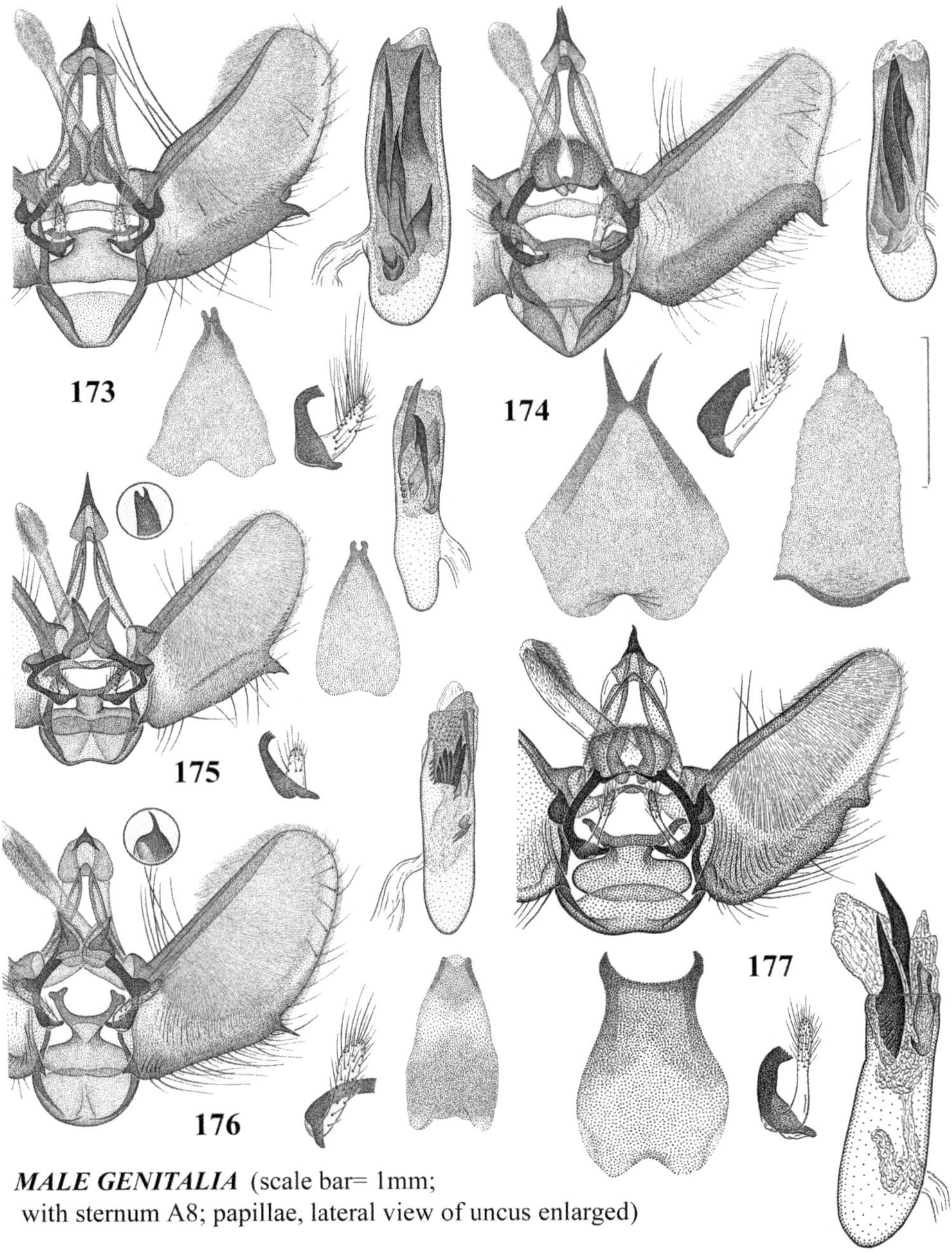

MALE GENITALIA (scale bar= 1mm;
with sternum A8; papillae, lateral view of uncus enlarged)

173. *Eupithecia cotidiana* Vojnits, China, Gansu
174. *Eupithecia krampli* Vojnits, China, Sichuan (tergite A8 on right)
175. *Eupithecia inopinata* Vojnits, China, Sichuan
176. *Eupithecia dealbata* Inoue, China, Taiwan
177. *Eupithecia lata* Vojnits, China, Tibet

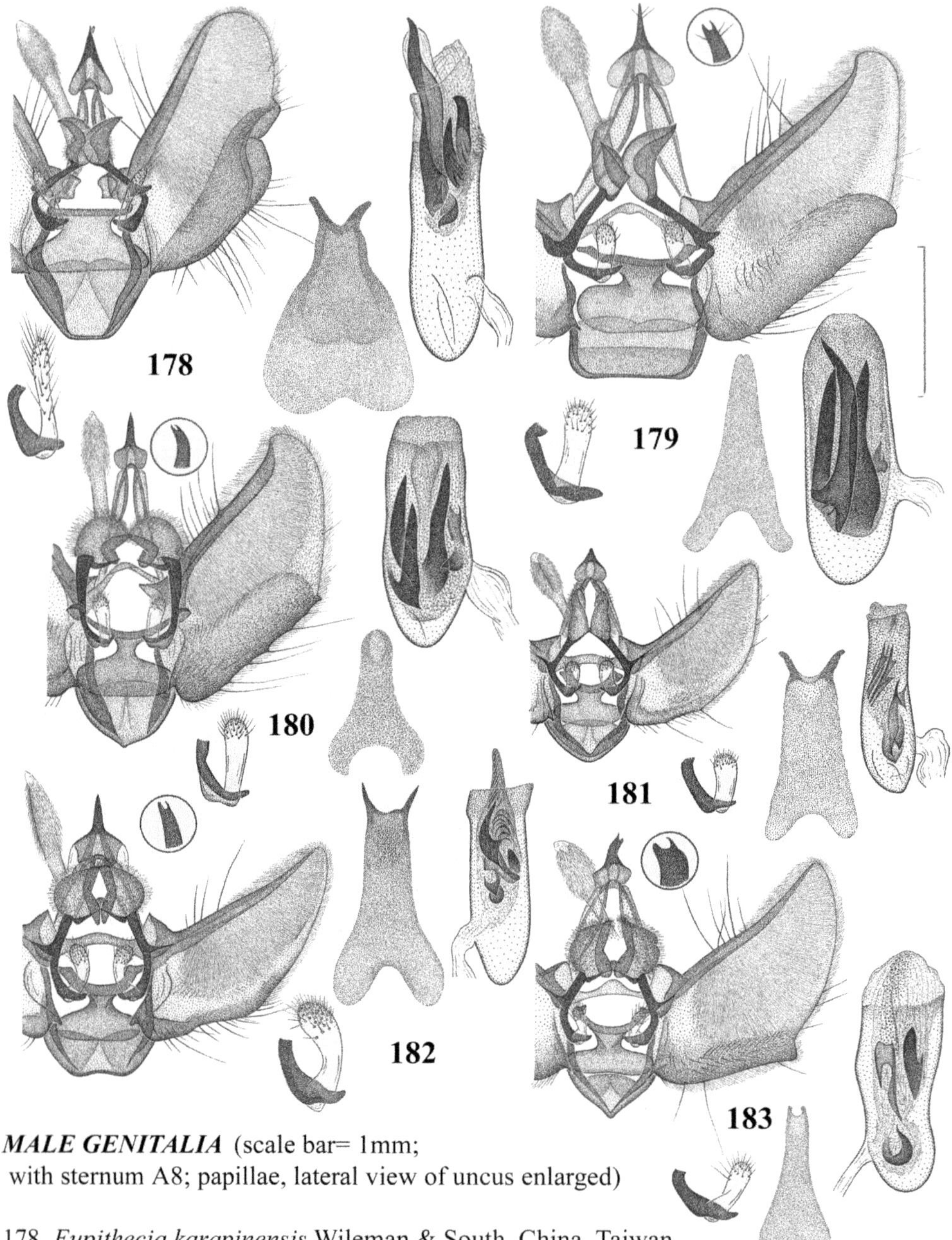

MALE GENITALIA (scale bar= 1mm;
with sternum A8; papillae, lateral view of uncus enlarged)

178. *Eupithecia karapinensis* Wileman & South, China, Taiwan
179. *Eupithecia kobayashii* Inoue, Russia, S. Primorje
180. *Eupithecia incorrupta* Vojnits, China, Yunnan
181. *Eupithecia leucenthesis* Prout, China, Sichuan
182. *Eupithecia pimpinellata* Hübner, Kazakhstan, Zailijsky Alatau Mts.
183. *Eupithecia denotata* Hübner, Kazakhstan, Zailijsky Alatau Mts.

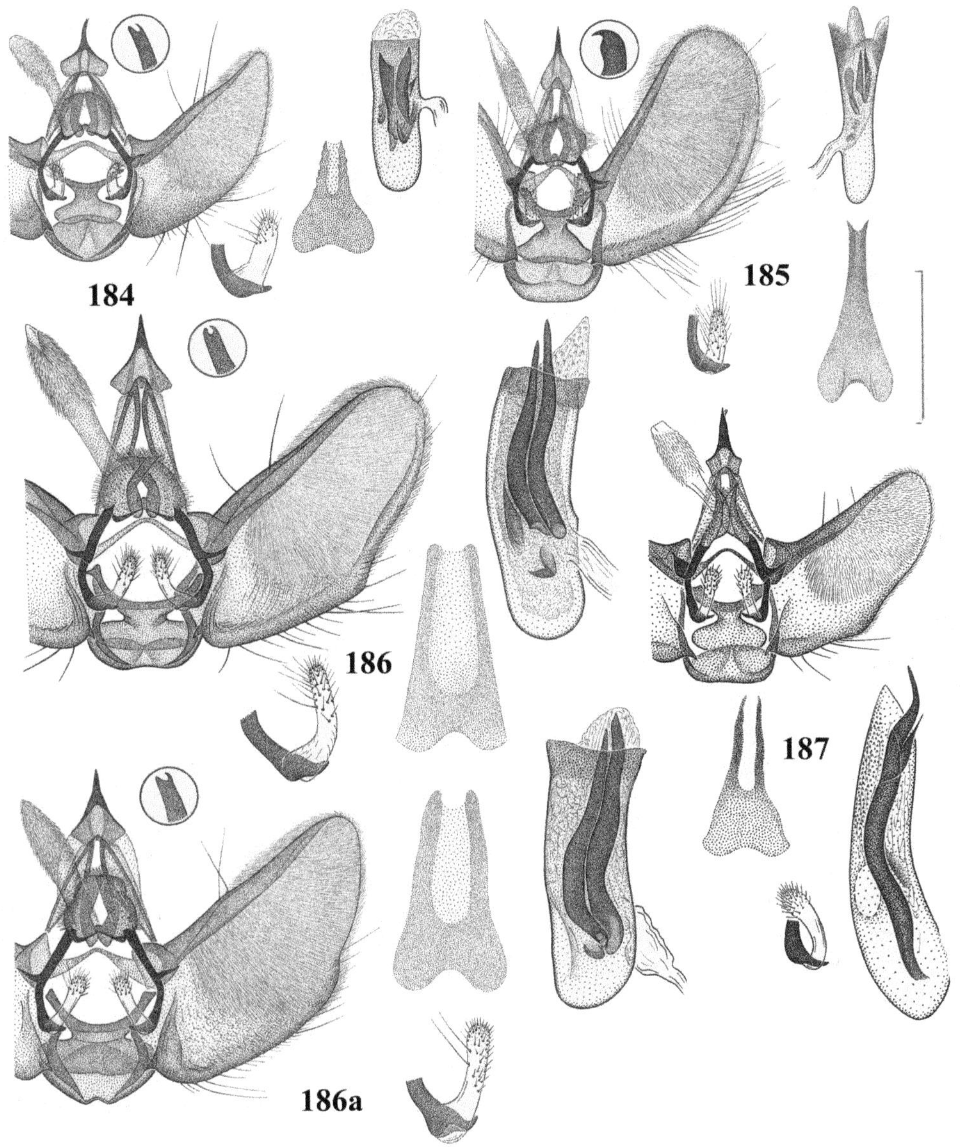

MALE GENITALIA (scale bar= 1mm;
with sternum A8; papillae, lateral view of uncus enlarged)

184. *Eupithecia chesiata* Dietze, Kyrgyzstan, Tian Shan Mts.
185. *Eupithecia mundiscripta* Warren, China, Taiwan
186. *Eupithecia robiginascens robiginascens* Prout, China, Sichuan
186a *Eupithecia robiginascens flexicornuta* Inoue, China, Taiwan
187. *Eupithecia luctuosa* Mironov & Galsworthy, China, Fujian

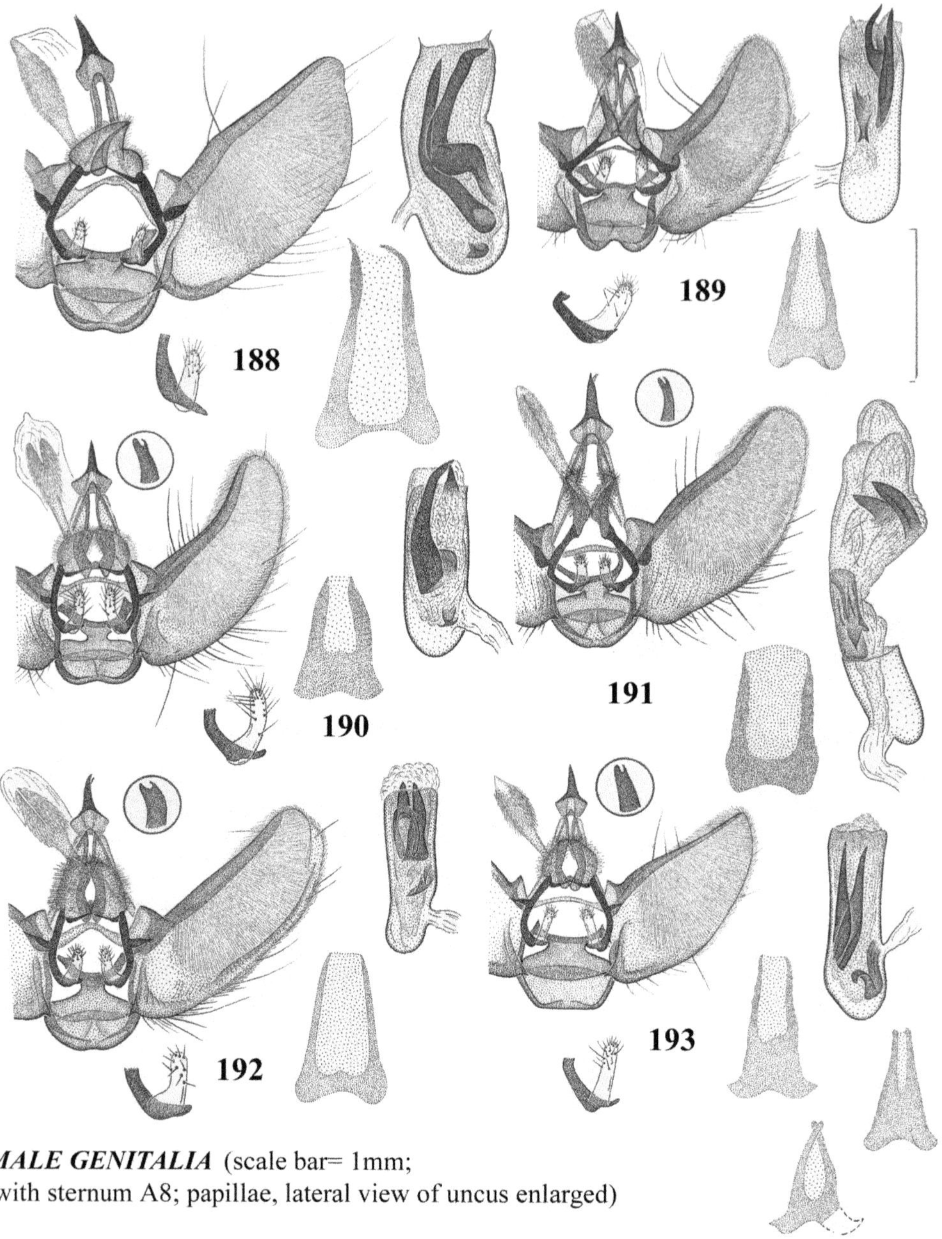

MALE GENITALIA (scale bar= 1mm;
with sternum A8; papillae, lateral view of uncus enlarged)

188. *Eupithecia qinlingata* Mironov & Galsworthy, China, Shaanxi
189. *Eupithecia svetlanae* Mironov & Pekarsky, China, Sichuan
190. *Eupithecia contraria* Vojnits, Nepal, Sagarmatha
191. *Eupithecia rajata* Guenée, NE India, Khasis
192. *Eupithecia ebriosa* Vojnits, China, Sichuan
193. *Eupithecia vasta* Vojnits, China, Yunnan (holotype; variation sternite A8)

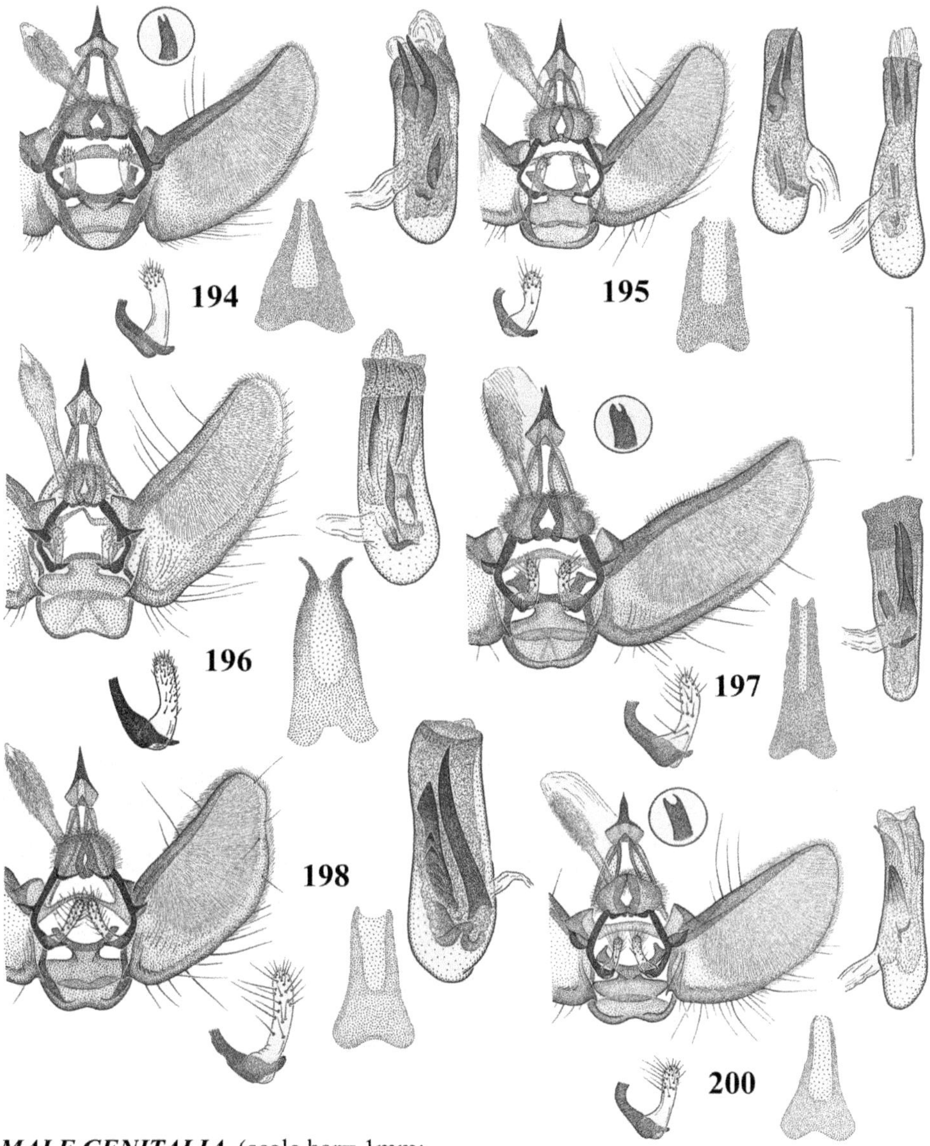

MALE GENITALIA (scale bar= 1mm;
with sternum A8; papillae, lateral view of uncus enlarged)

194. *Eupithecia lobblicherata* Schütze, China, Sichuan
195. *Eupithecia coniurata* Vojnits, China, Tibet (aedeagus variation)
196. *Eupithecia cuneata* Vojnits, China, Yunnan
197. *Eupithecia particeps* Vojnits, Pakistan, Khagan Valley
198. *Eupithecia sempiterna* Vojnits, China, Yunnan
200. *Eupithecia raniata* Prout, Myanmar , Chin State

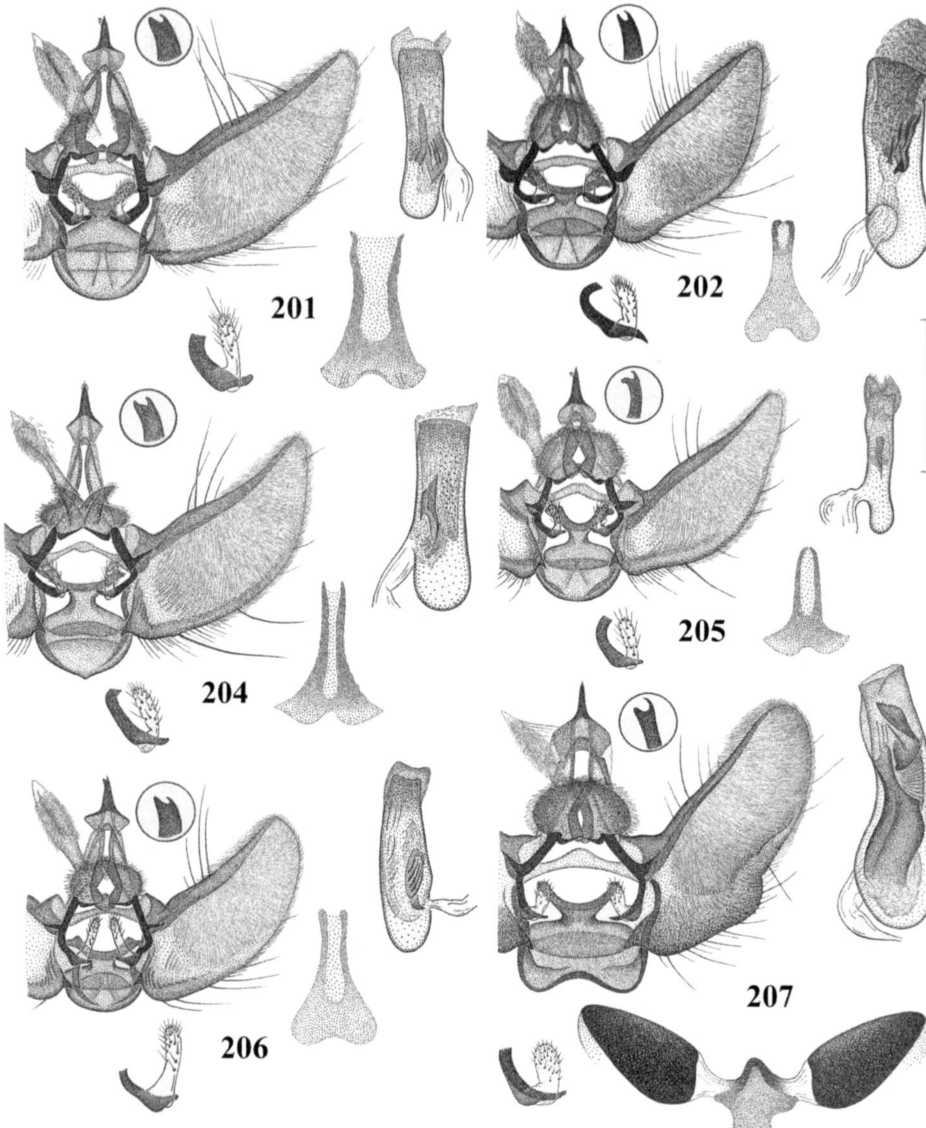

MALE GENITALIA (scale bar= 1mm; with sternum A8; papillae, lateral view of uncus enlarged)

201. *Eupithecia atrisignis* Butler, Nepal, Sagarmatha
202. *Eupithecia pyricoetes* Prout, India, W. Bengal
204. *Eupithecia nuceistrigata* Bastelberger, China, Taiwan
205. *Eupithecia tricrossa* Prout, Nepal, Sagarmatha
206. *Eupithecia yazakii* Inoue, China, Taiwan
207. *Eupithecia suboxydata* Staudinger, Russia, S. Primorje

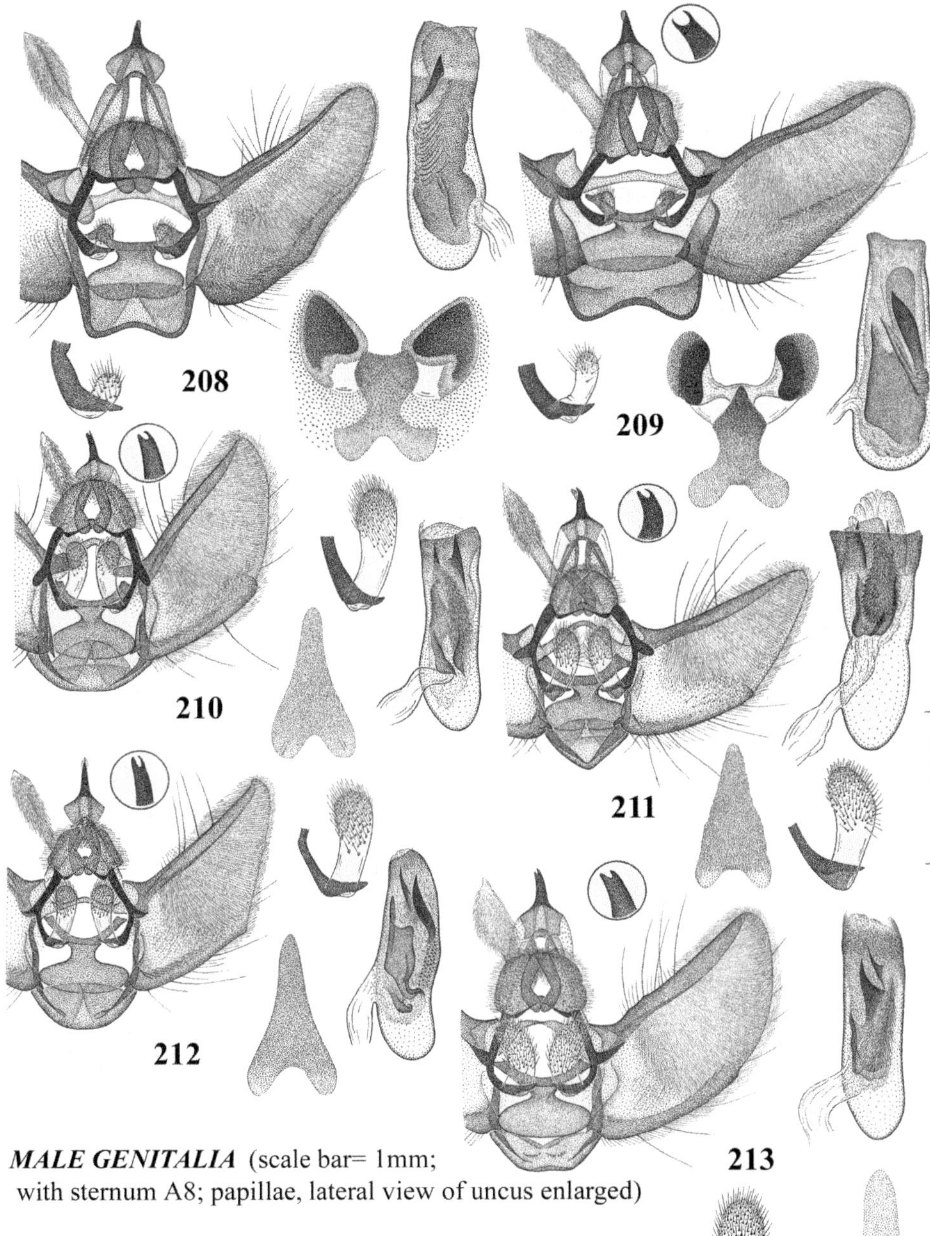

MALE GENITALIA (scale bar= 1mm; with sternum A8; papillae, lateral view of uncus enlarged)

208. *Eupithecia formosa* Vojnits, China, Yunnan
209. *Eupithecia subbrunneata* Dietze, Russia, S. Primorje
210. *Eupithecia innotata* Hufnagel, Kazakhstan, Tarbagatay Mts.
211. *Eupithecia parallelaria* Bohatsch, Pakistan/India, Kashmir
212. *Eupithecia corroborata* Dietze, Kyrgyzstan, Sary-dzhaz valley
213. *Eupithecia praesignata* Bohatsch, Kyrgyzstan, Naryn

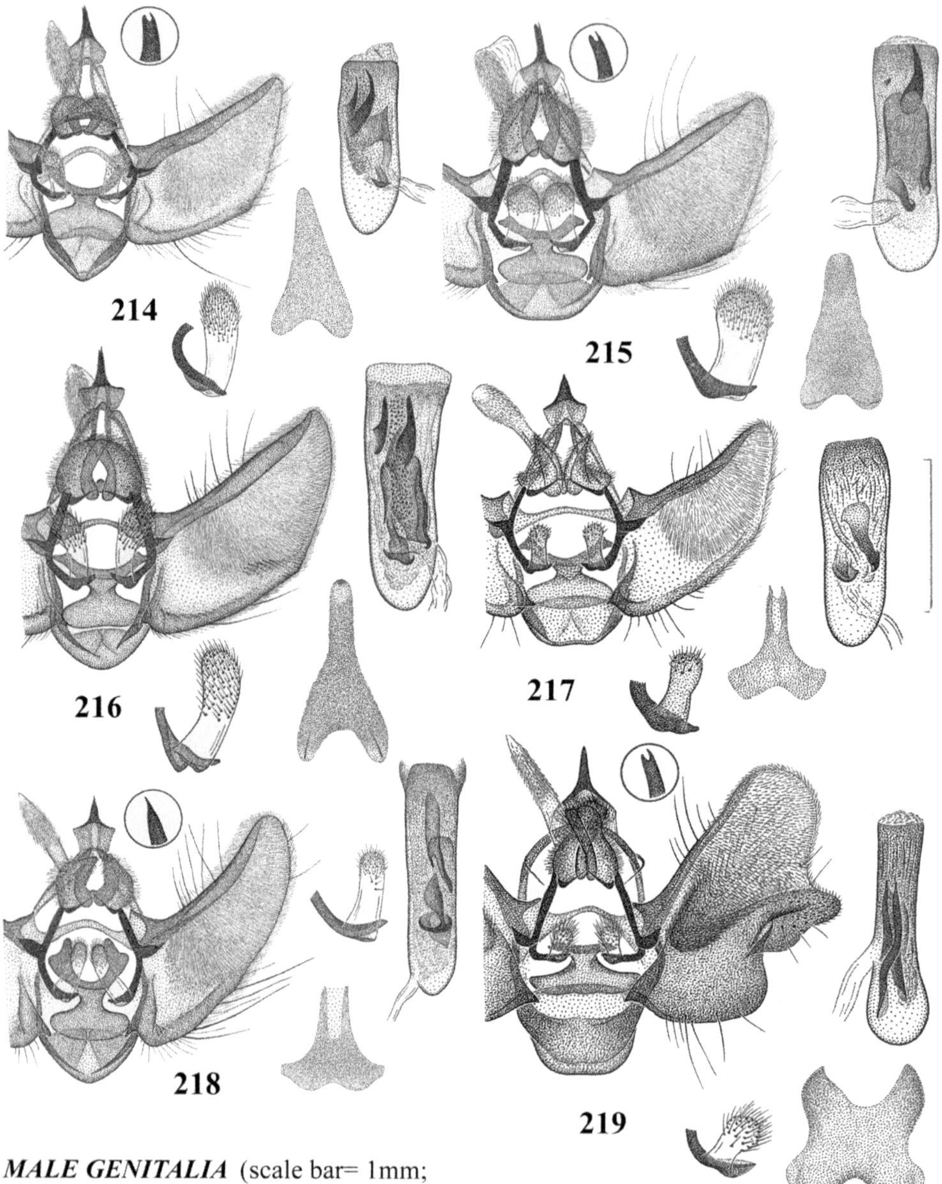

MALE GENITALIA (scale bar= 1mm;
with sternum A8; papillae, lateral view of uncus enlarged)

214. *Eupithecia paupera* Dietze, China, Xinjiang
215. *Eupithecia mitigata* Dietze, India, Ladakh
216. *Eupithecia vicariata* Dietze, Uzbekistan, Zarafshan
217. *Eupithecia magica* Mironov & Galsworthy, China, Tibet
218. *Eupithecia relaxata* Dietze, Tajikistan, Pamir Mts.
219. *Eupithecia centaureata* [Denis & Schiffermüller], China, Taiwan

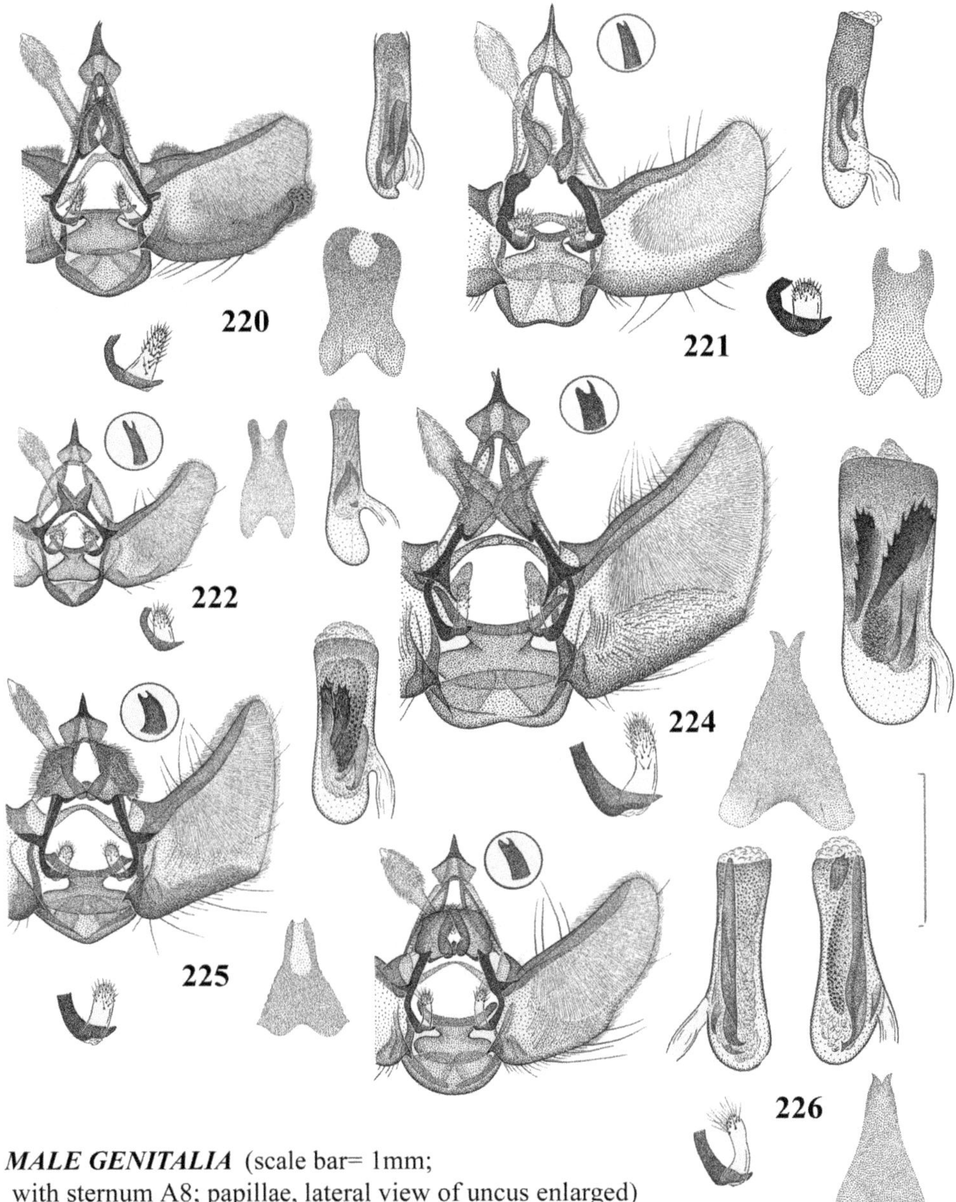

MALE GENITALIA (scale bar= 1mm; with sternum A8; papillae, lateral view of uncus enlarged)

220. *Eupithecia exquisita* Vojnits, China, Shanxi
221. *Eupithecia brunneilutea* Mironov & Galsworthy, China, Yunnan
222. *Eupithecia despectaria* Lederer, Tajikistan, Ghissar Mts.
224. *Eupithecia gratiosata* Herrich-Schäffer, Turkmenistan, Sjunt Mts.
225. *Eupithecia subpulchrata* Alphéraky, Kazakhstan, Betpak Dala Desert
226. *Eupithecia perendina* Vojnits, China, Yunnan (aedeagus from both sides)

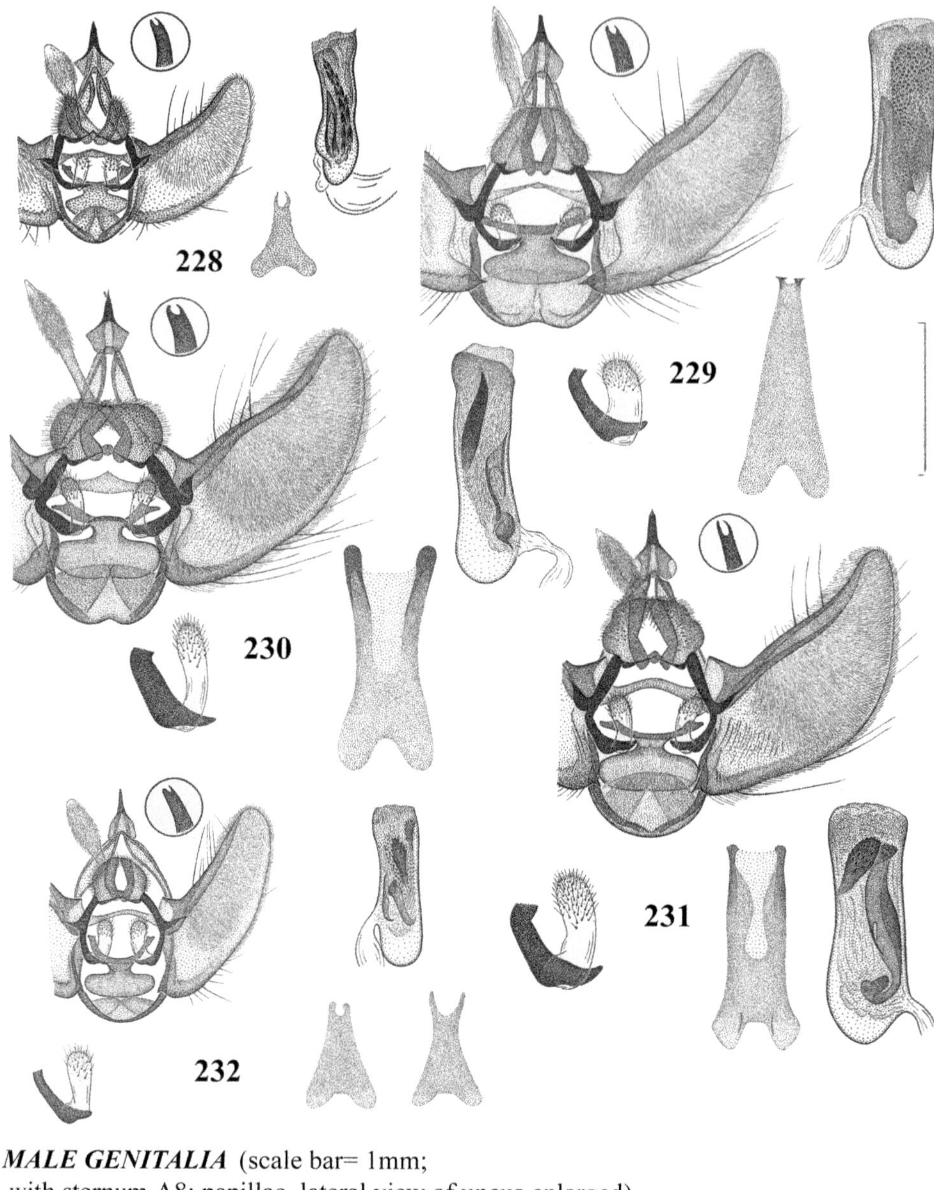

MALE GENITALIA (scale bar= 1mm;
with sternum A8; papillae, lateral view of uncus enlarged)

228. *Eupithecia ancillata* Mironov & Galsworthy, China, Shaanxi
229. *Eupithecia simpliciata* Haworth, Kyrgyzstan, Naryn
230. *Eupithecia sinuosaria* Eversmann, Russia, S. Siberia
231. *Eupithecia pallescens* Dietze, Kyrgyzstan, Naryn
232. *Eupithecia conterminata* Lienig & Zeller, Russia, S. Primorje (variation sternite A8)

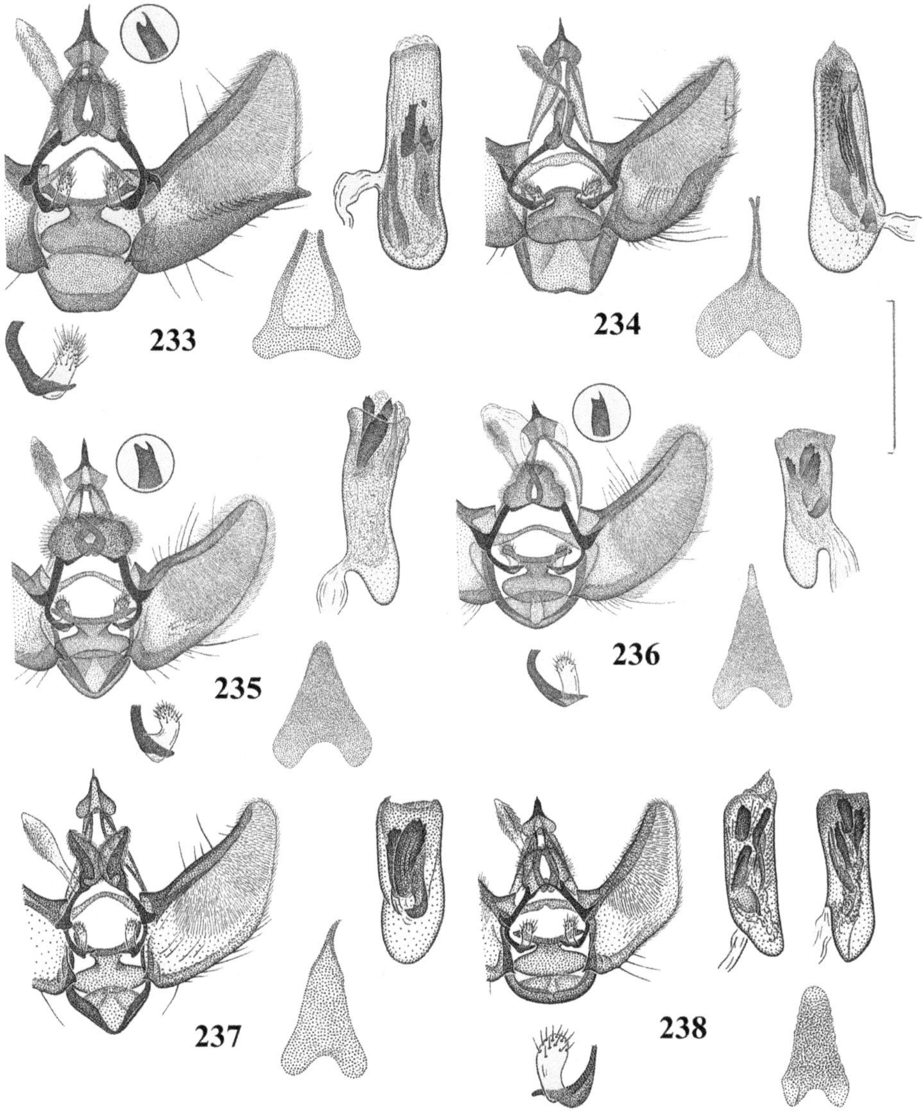

MALE GENITALIA (scale bar= 1mm;
with sternum A8; papillae, lateral view of uncus enlarged)

233. *Eupithecia insignioides* Wehrli, Russia, S. Primorje
234. *Eupithecia costimacularia* Leech, Japan, Kyushyu
235. *Eupithecia assimilata* Doubleday, Kyrgyzstan, Suusamyr valley
236. *Eupithecia vulgata* Haworth, Kyrgyzstan, Suusamyr valley
237. *Eupithecia andrasi* Mironov & Galsworthy, China, Shaanxi
238. *Eupithecia nigristriata* Mironov & Galsworthy, China, Qinghai (aedeagus from dorsal and ventral sides)

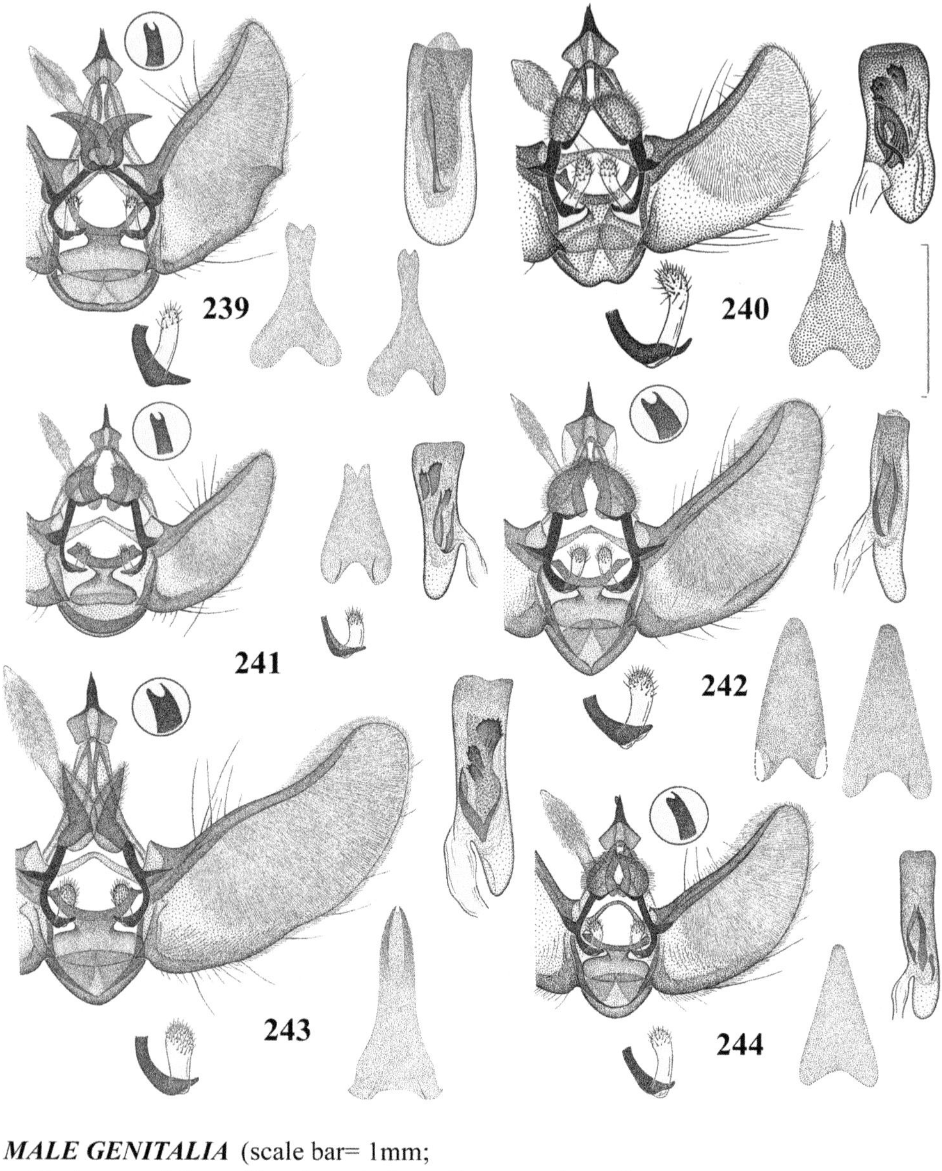

MALE GENITALIA (scale bar= 1mm;
with sternum A8; papillae, lateral view of uncus enlarged)

239. *Eupithecia sacrimontis* Prout, China, Shaanxi (variation sternite A8)
240. *Eupithecia citraria* Mironov & Galsworthy, China, Yunnan
241. *Eupithecia nobilitata* Staudinger, Kazakhstan, Djair Mts.
242. *Eupithecia sutiliata* Christoph, Kazakhstan, Dzhulek
243. *Eupithecia biornata* Christoph, S. Russia, Taganrog
244. *Eupithecia kuldschaensis* Staudinger, Mongolia, Govi-Altai aimag

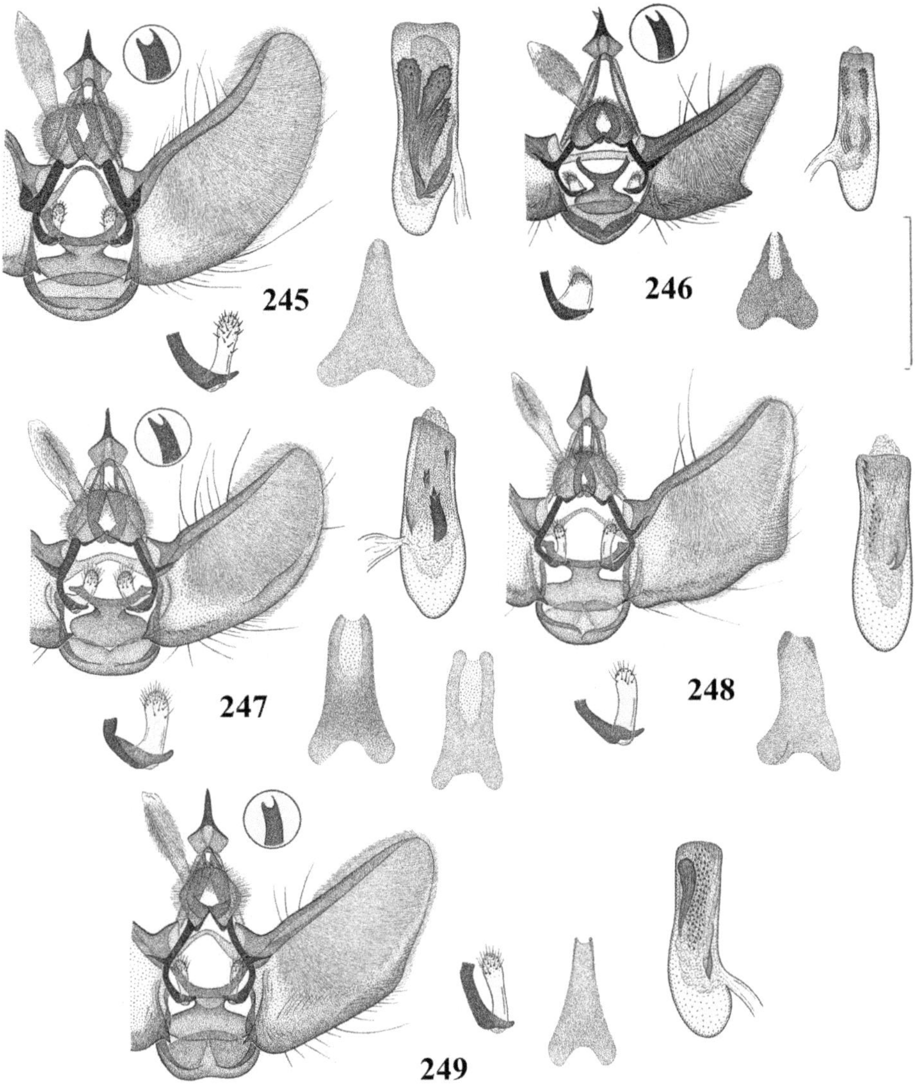

MALE GENITALIA (scale bar= 1mm;
with sternum A8; papillae, lateral view of uncus enlarged)

245. *Eupithecia rubeni* Viidalepp, Russia, S. Siberia
246. *Eupithecia rebeli* Bohatsch, Tajikistan, Alai Mts.
247. *Eupithecia thalictrata* Püngeler, Russia, Altai
248. *Eupithecia dura* Vojnits, China, Yunnan
249. *Eupithecia addictata* Dietze, Russia, S. Primorje

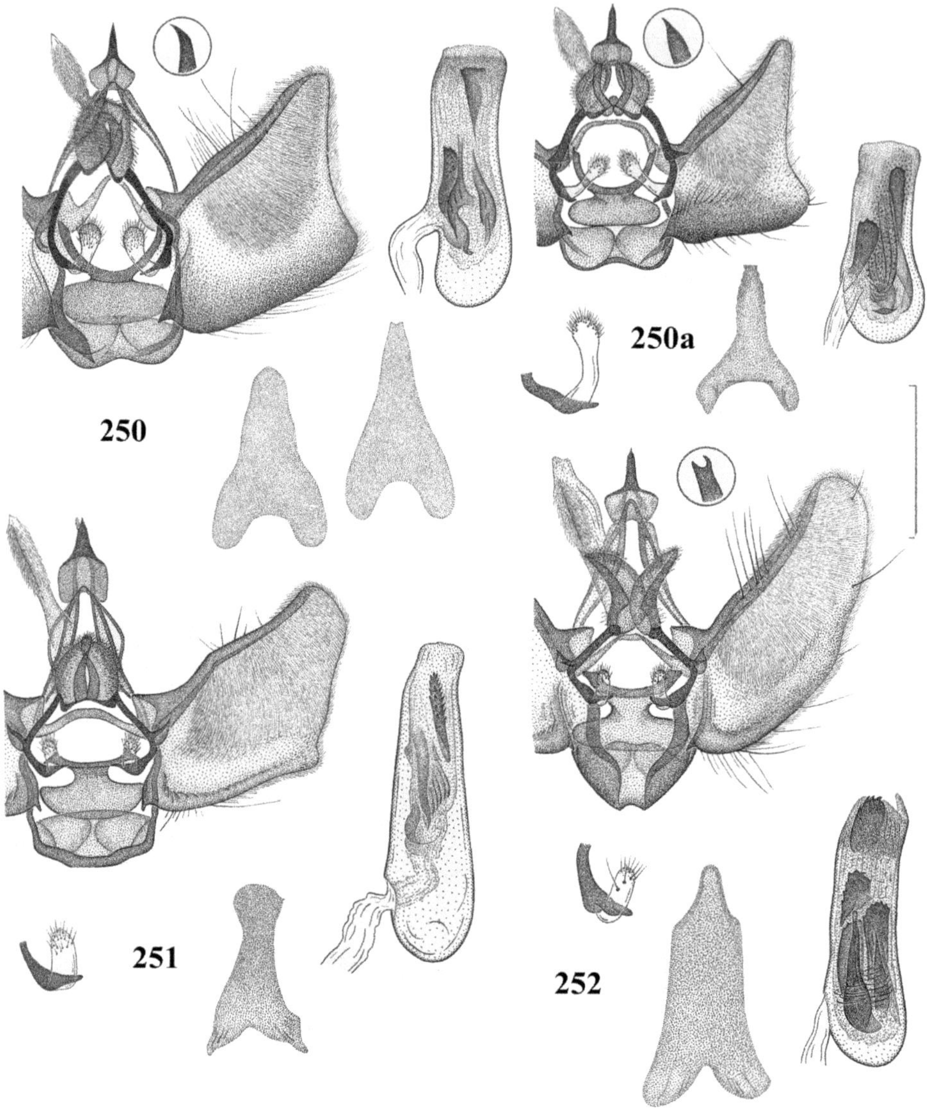

MALE GENITALIA (scale bar= 1mm;
with sternum A8; papillae, lateral view of uncus enlarged)

250. *Eupithecia exiguata* Hübner, Russia, S. Primorje (variation sternite A8)
250a. *Eupithecia exiguata* Hübner, Russia, Altai Mts
251. *Eupithecia hannemanni* Vojnits, China, Shaanxi
252. *Eupithecia bella* Staudinger, Russia, S. Primorje

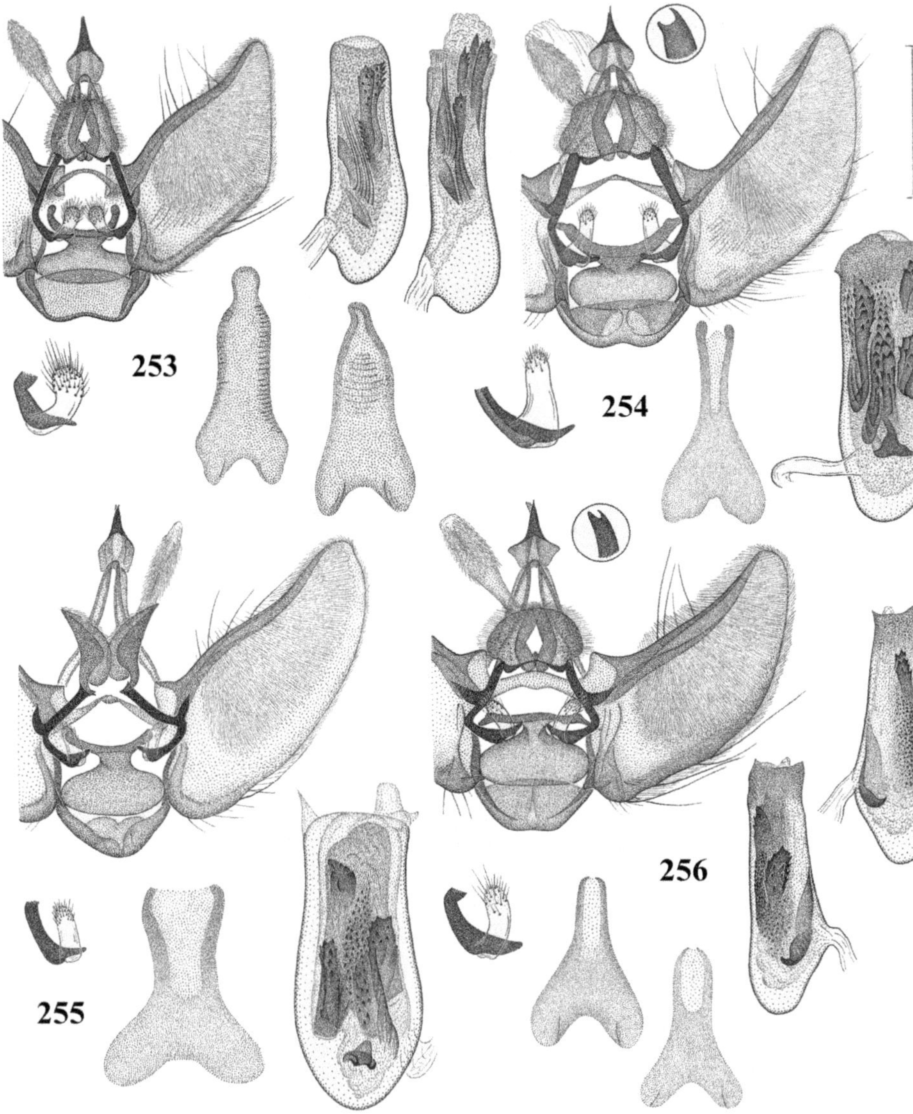

MALE GENITALIA (scale bar= 1mm;
with sternum A8; papillae, lateral view of uncus enlarged)

253. *Eupithecia insolita* Vojnits, China, Sichuan (aedeagus and sternite A8 on right, from Tibet)
254. *Eupithecia duplex* Sterneck, China, Sichuan
255. *Eupithecia tectaria* Mironov & Galsworthy, China, Fujian
256. *Eupithecia mandschurica* Staudinger, Russia, S. Primorje (aedeagus from 2 angles, variation sternite A8)

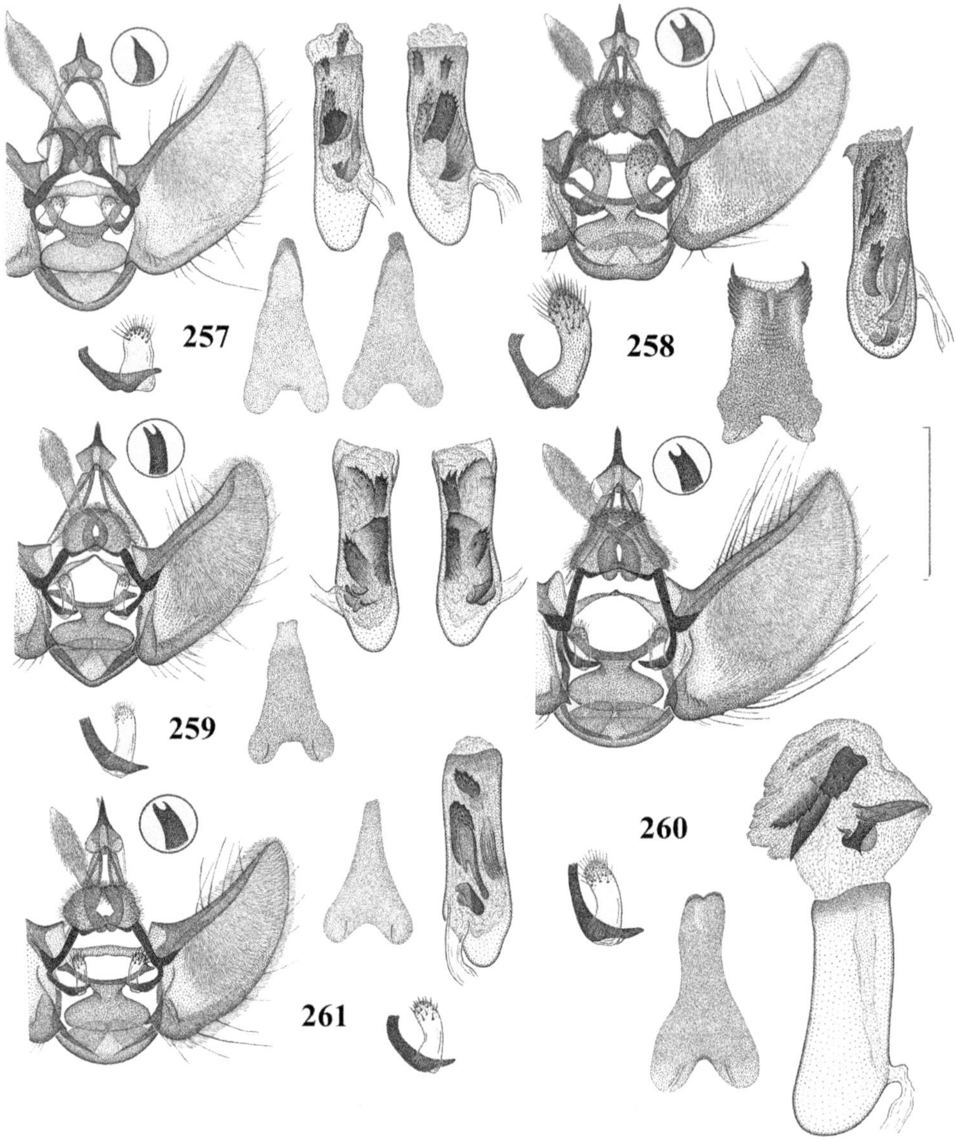

MALE GENITALIA (scale bar= 1mm;
with sternum A8; papillae, lateral view of uncus enlarged)

257. *Eupithecia intricata* Zetterstedt, Kazakhstan, Zailijsky Alatau Mts. (variation aedeagus and sternite A8)
258. *Eupithecia lupa* Mironov & Galsworthy, China, Taiwan
259. *Eupithecia satyrata* Hübner, Kazakhstan, Tarbagatay Mts. (aedeagus from 2 angles)
260. *Eupithecia pernotata* Guenée, Russia, S. Primorje
261. *Eupithecia amplexata* Christoph, Russia, S. Primorje

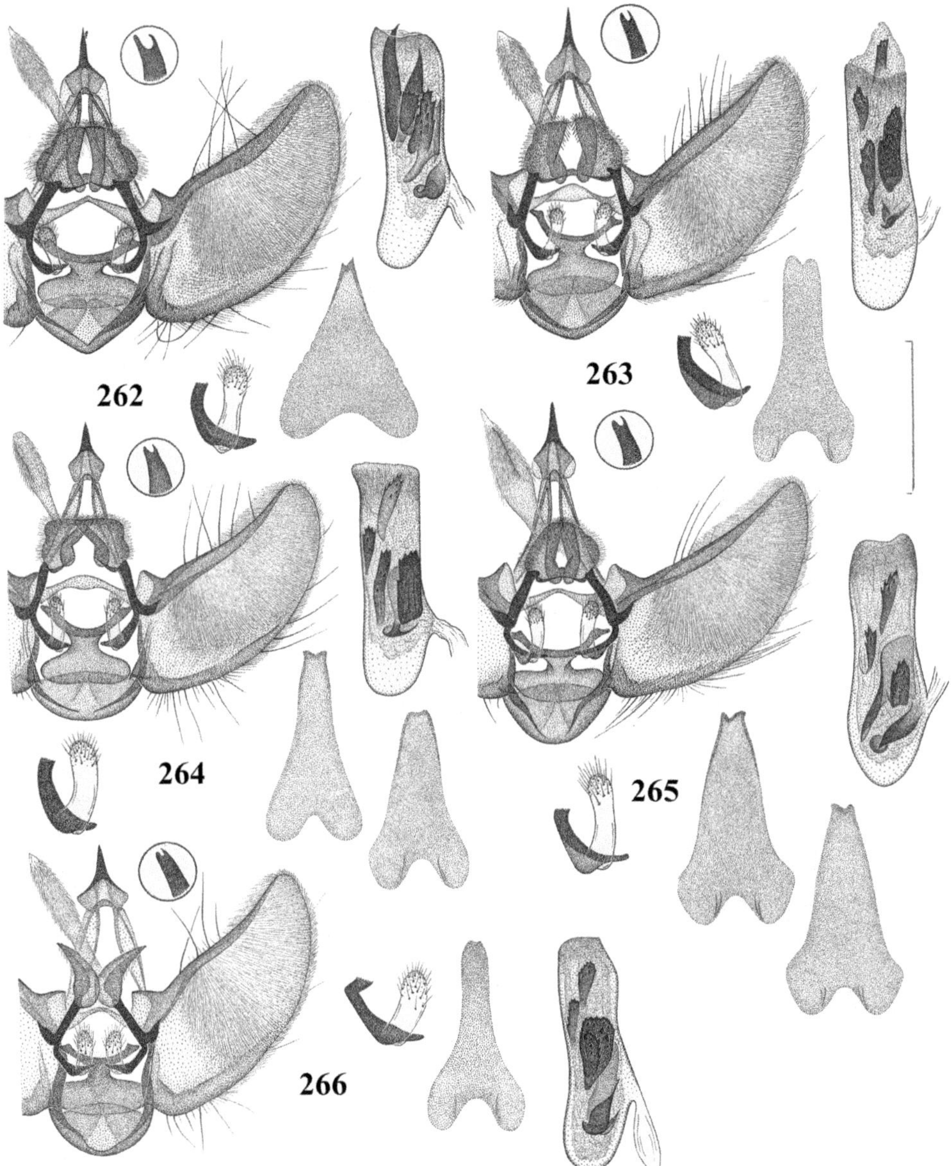

MALE GENITALIA (scale bar= 1mm; with sternum A8; papillae, lateral view of uncus enlarged)

262. *Eupithecia extensaria* Freyer, North Korea
263. *Eupithecia absinthiata* Clerck, Russia, Priamurje
264. *Eupithecia interpunctaria* Inoue, Russia, S. Primorje (variation sternite A8)
265. *Eupithecia korbi* Dietze, Russia, S. Primorje (variation sternite A8)
266. *Eupithecia kurilensis* Bryk, Russia, Sakhalin

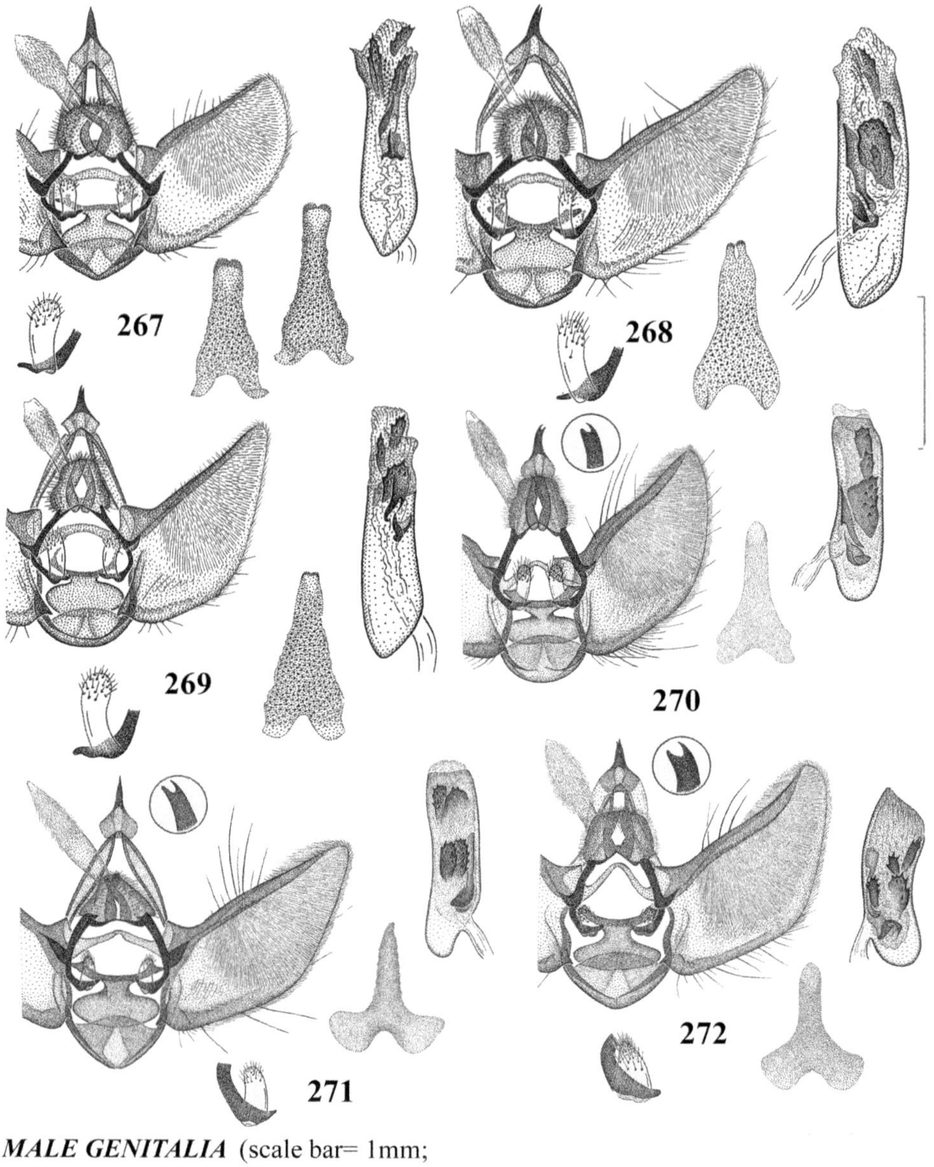

MALE GENITALIA (scale bar= 1mm;
with sternum A8; papillae, lateral view of uncus enlarged)

267. *Eupithecia ensifera* Mironov & Galsworthy, China, Qinghai (variation sternite A8)
268. *Eupithecia ultrix* Mironov & Galsworthy, China, Yunnan
269. *Eupithecia cervina* Mironov & Galsworthy, China, Tibet
270. *Eupithecia impolita* Vojnits, China, Shanxi
271. *Eupithecia neosatyrata* Inoue, Russia, S. Siberia
272. *Eupithecia indissolubilis* Vojnits, Russia, Altai mts.

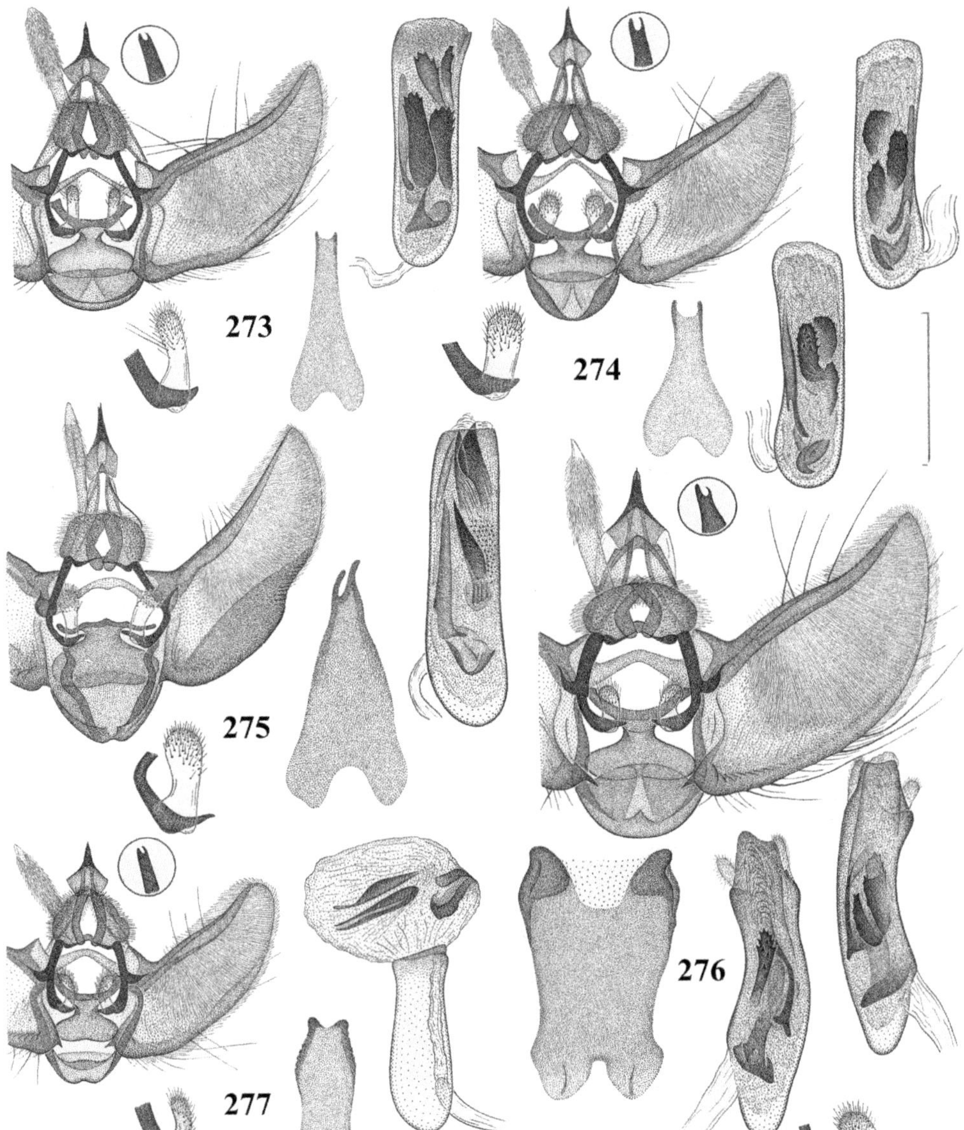

MALE GENITALIA (scale bar= 1mm; with sternum A8; papillae, lateral view of uncus enlarged)

273. *Eupithecia nephelata* Staudinger, Mongolia, N. Gobi desert
274. *Eupithecia exactata* Staudinger, Tajikistan, Pamir Mts.
275. *Eupithecia omissa* Mironov & Galsworthy, China, Shaanxi
276. *Eupithecia succenturiata* Linnaeus, Russia, S. Siberia
277. *Eupithecia marginata* Staudinger, Tajikistan, Darwaz Mts.

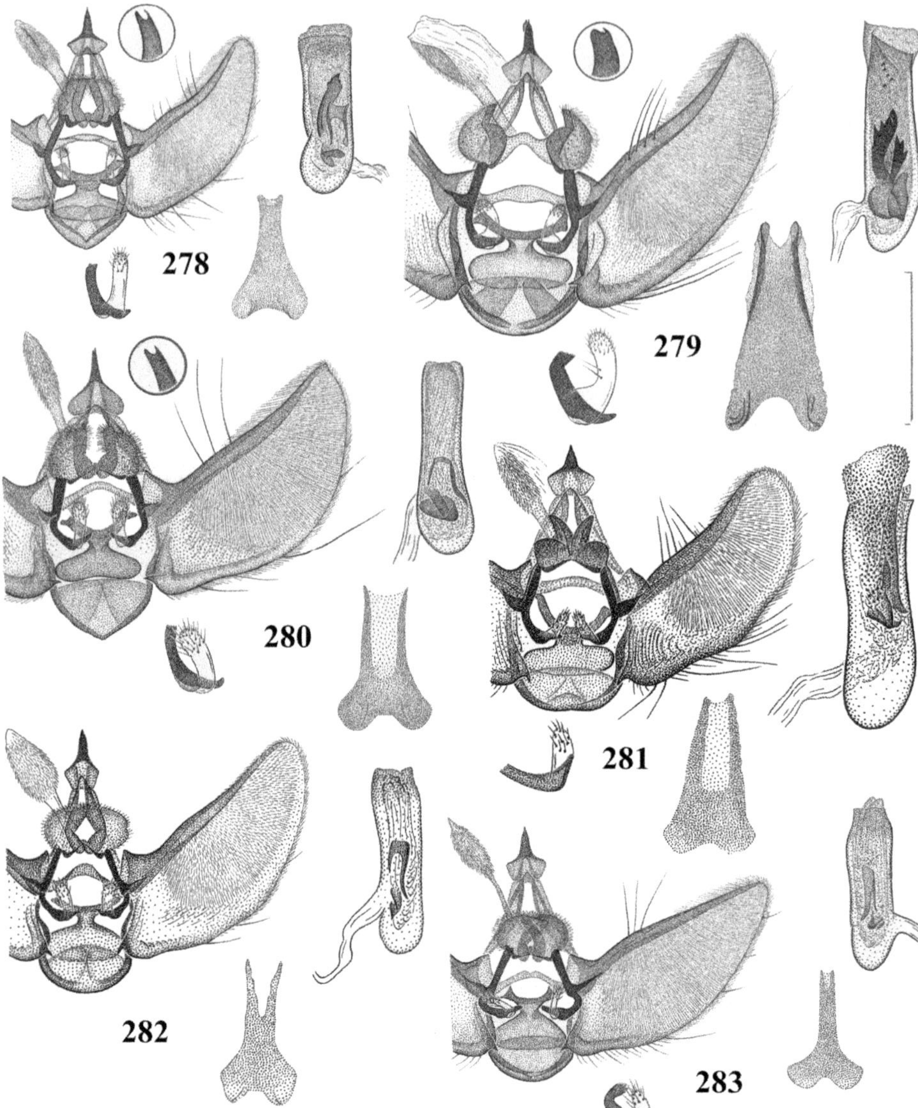

MALE GENITALIA (scale bar= 1mm;
with sternum A8; papillae, lateral view of uncus enlarged)

278. *Eupithecia masculina* Inoue, China, Taiwan
279. *Eupithecia subumbrata* Denis & Schiffermüller, Kyrgyzstan, Naryn
280. *Eupithecia subfuscata* Haworth, Russia, S. Primorje
281. *Eupithecia costipicta* Warren, Pakistan/India, Kashmir
282. *Eupithecia mediocincta* Mironov & Galsworthy, China, Gansu
283. *Eupithecia chui* Inoue, China, Taiwan

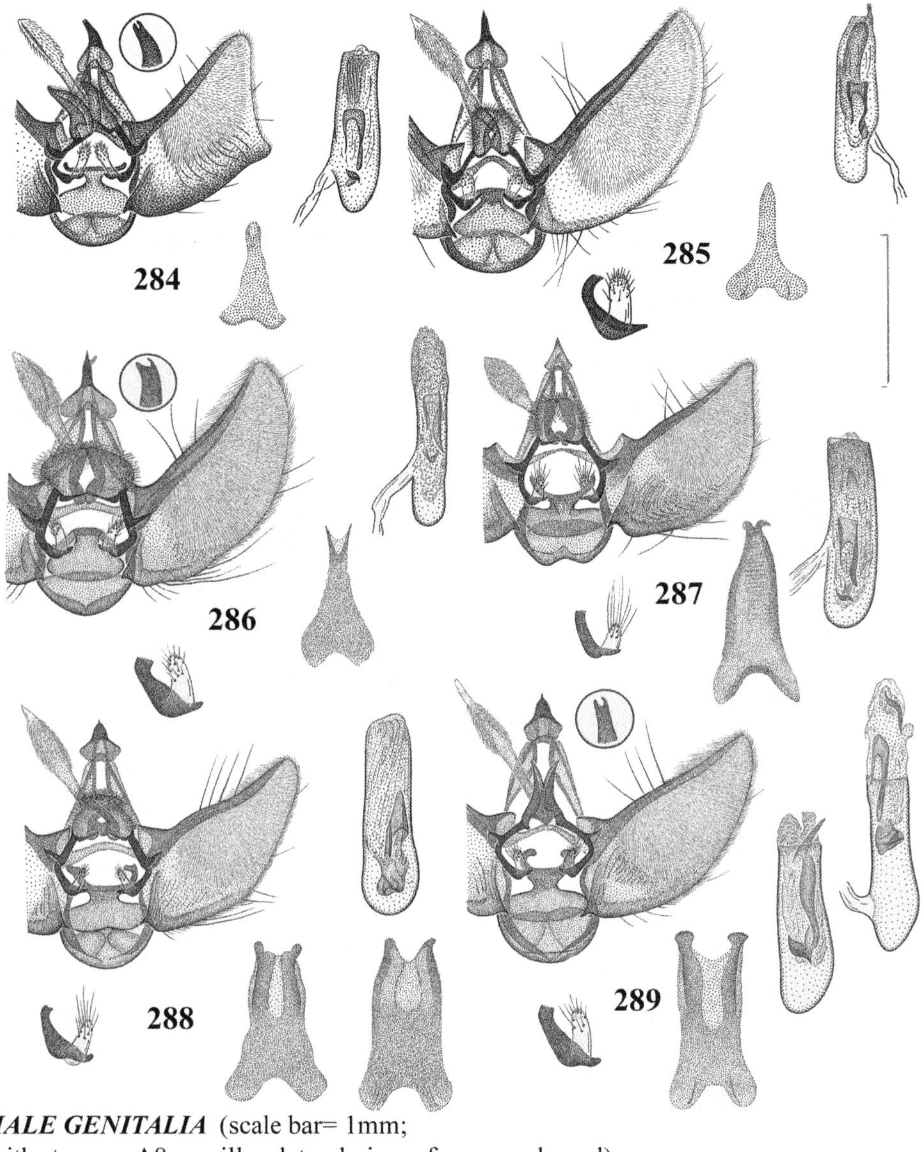

MALE GENITALIA (scale bar= 1mm;
with sternum A8; papillae, lateral view of uncus enlarged)

284. *Eupithecia tibetana* Mironov & Galsworthy, China, Tibet
285. *Eupithecia tepida* Mironov & Galsworthy, China, Shaanxi
286. *Eupithecia nishizawai* Inoue, China, Taiwan
287. *Eupithecia praecipitata* Vojnits, China, Tibet
288. *Eupithecia lasciva* Vojnits, China, Shaanxi
289. *Eupithecia egena* Vojnits, China, Yunnan (aedeagus uneverted and semi-everted)

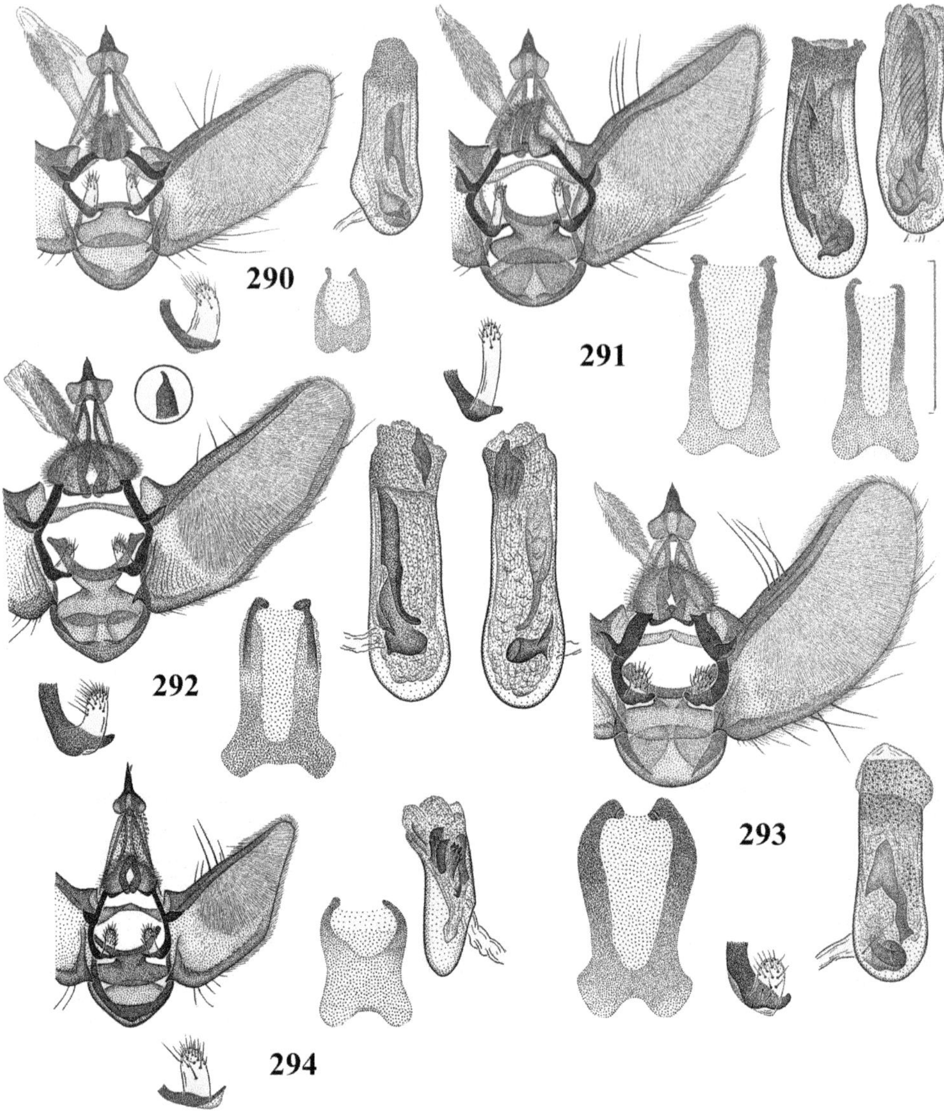

MALE GENITALIA (scale bar= 1mm;
with sternum A8; papillae, lateral view of uncus enlarged)

290. *Eupithecia lunatica* Vojnits, China, Yunnan
291. *Eupithecia otiosa* Vojnits, China, Yunnan (aedeagus from both sides, variation sternite A8
292. *Eupithecia acutipapillata* Inoue, China, Taiwan (aedeagus from both sides)
293. *Eupithecia supercastigata* Inoue, China, Zhejiang
294. *Eupithecia concremata* Dietze, China, Sichuan

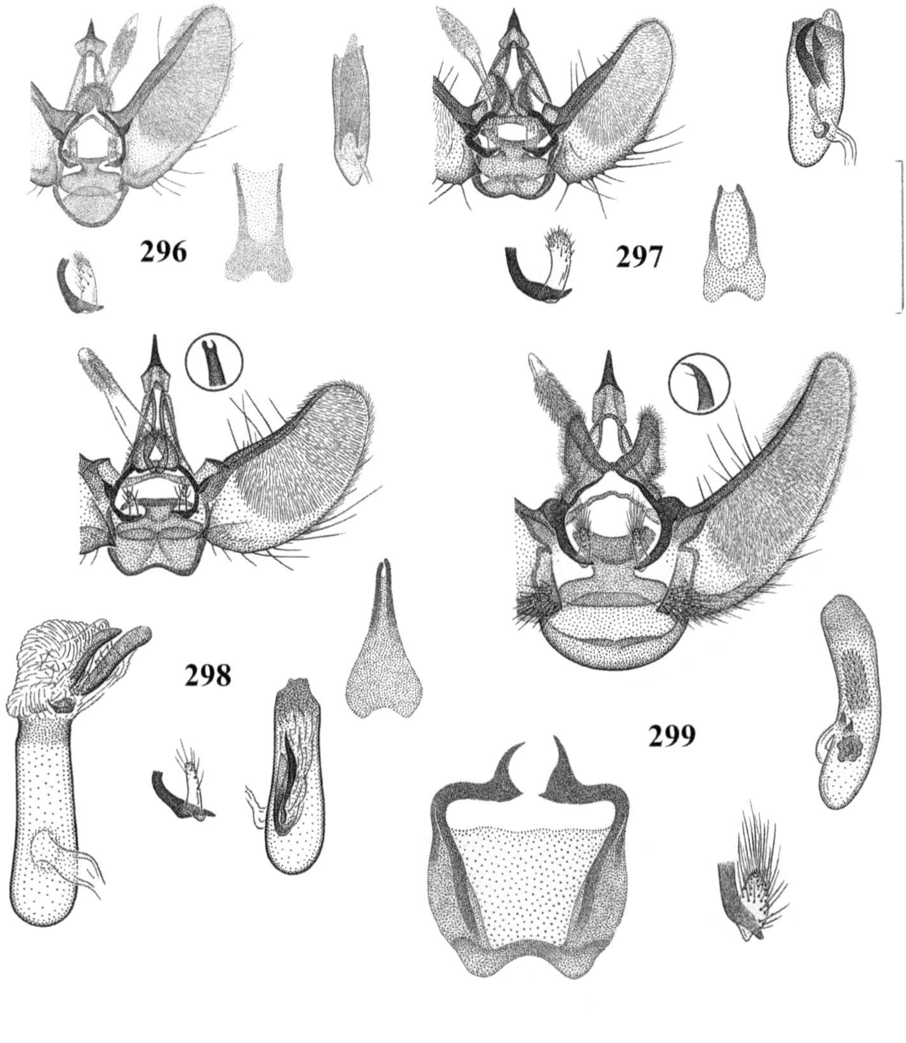

MALE GENITALIA (scale bar= 1mm;
with sternum A8; papillae, lateral view of uncus enlarged)

296. *Eupithecia lilliputata* Mironov & Galsworthy, China, Yunnan
297. *Eupithecia salubris* Mironov & Galsworthy, China, Shaanxi
298. *Eupithecia villica* Mironov & Galsworthy, China, Yunnan (vesica everted and uneverted)
299. *Pareupithecia spadix* Inoue, Russia, S. Primorje

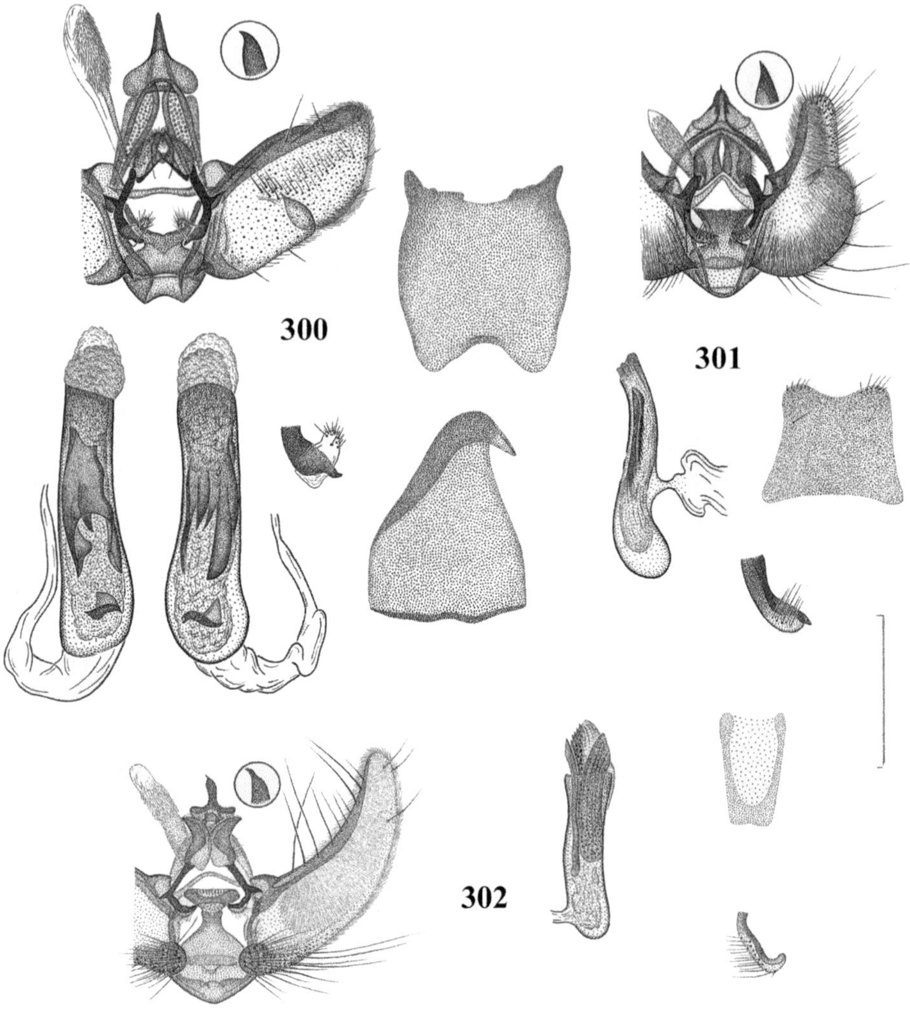

MALE GENITALIA (scale bar= 1mm;
with sternum A8; papillae, lateral view of uncus enlarged)

300. *Eva flexa* Vojnits, China, Yunnan (aedeagus from dorsal and ventral aspects, 8th sternite above, 8th tergite below)
301. *Girida rigida* Swinhoe India, Nilgiri Hills
302. *Mesoptila melanolopha* Swinhoe, China, Taiwan

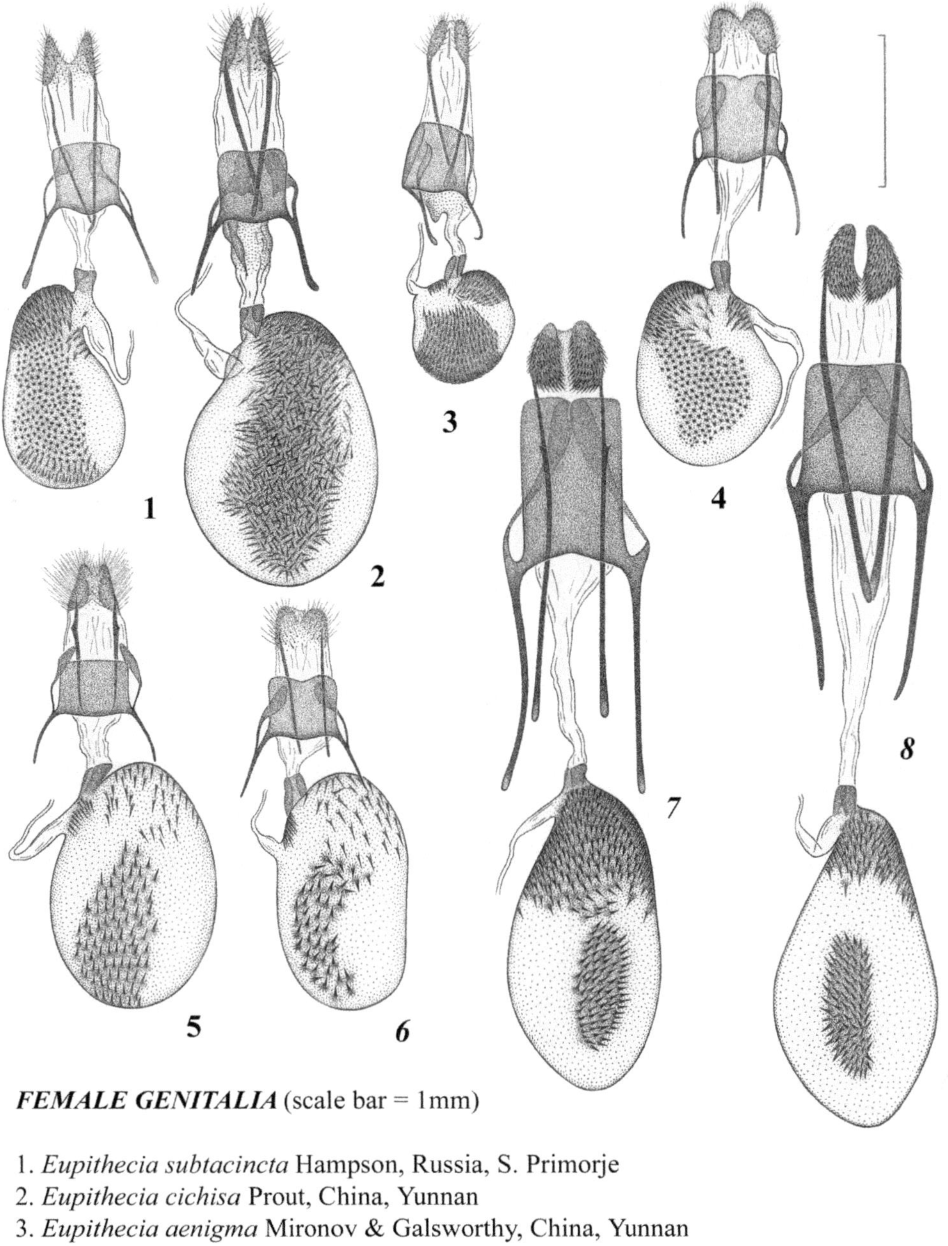

FEMALE GENITALIA (scale bar = 1mm)

1. *Eupithecia subtacincta* Hampson, Russia, S. Primorje
2. *Eupithecia cichisa* Prout, China, Yunnan
3. *Eupithecia aenigma* Mironov & Galsworthy, China, Yunnan
4. *Eupithecia zibellinata* Christoph, Russia, S. Primorje
5. *Eupithecia olgae* Mironov, Kyrgyzstan, Naryn
6. *Eupithecia illaborata* Dietze, Kyrgyzstan, Naryn
7 *Eupithecia rubellata* Dietze, Kyrgyzstan, Issyk-kul
8 *Eupithecia bohatschi* Staudinger, Russia, Priamurje

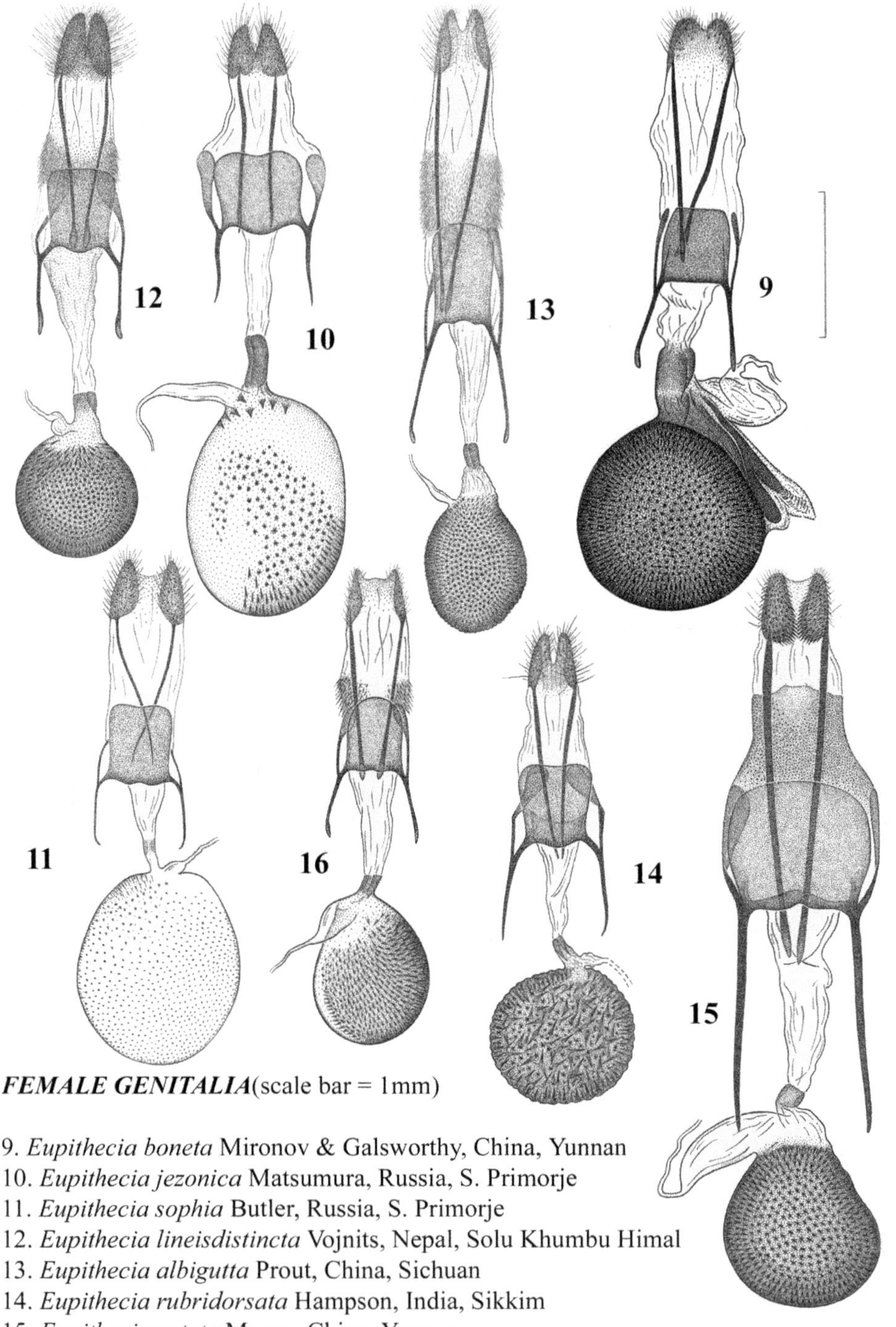

FEMALE GENITALIA(scale bar = 1mm)

9. *Eupithecia boneta* Mironov & Galsworthy, China, Yunnan
10. *Eupithecia jezonica* Matsumura, Russia, S. Primorje
11. *Eupithecia sophia* Butler, Russia, S. Primorje
12. *Eupithecia lineisdistincta* Vojnits, Nepal, Solu Khumbu Himal
13. *Eupithecia albigutta* Prout, China, Sichuan
14. *Eupithecia rubridorsata* Hampson, India, Sikkim
15. *Eupithecia ustata* Moore, China, Yunnan
16. *Eupithecia delaeveri* Vojnits, Nepal, Gandaki

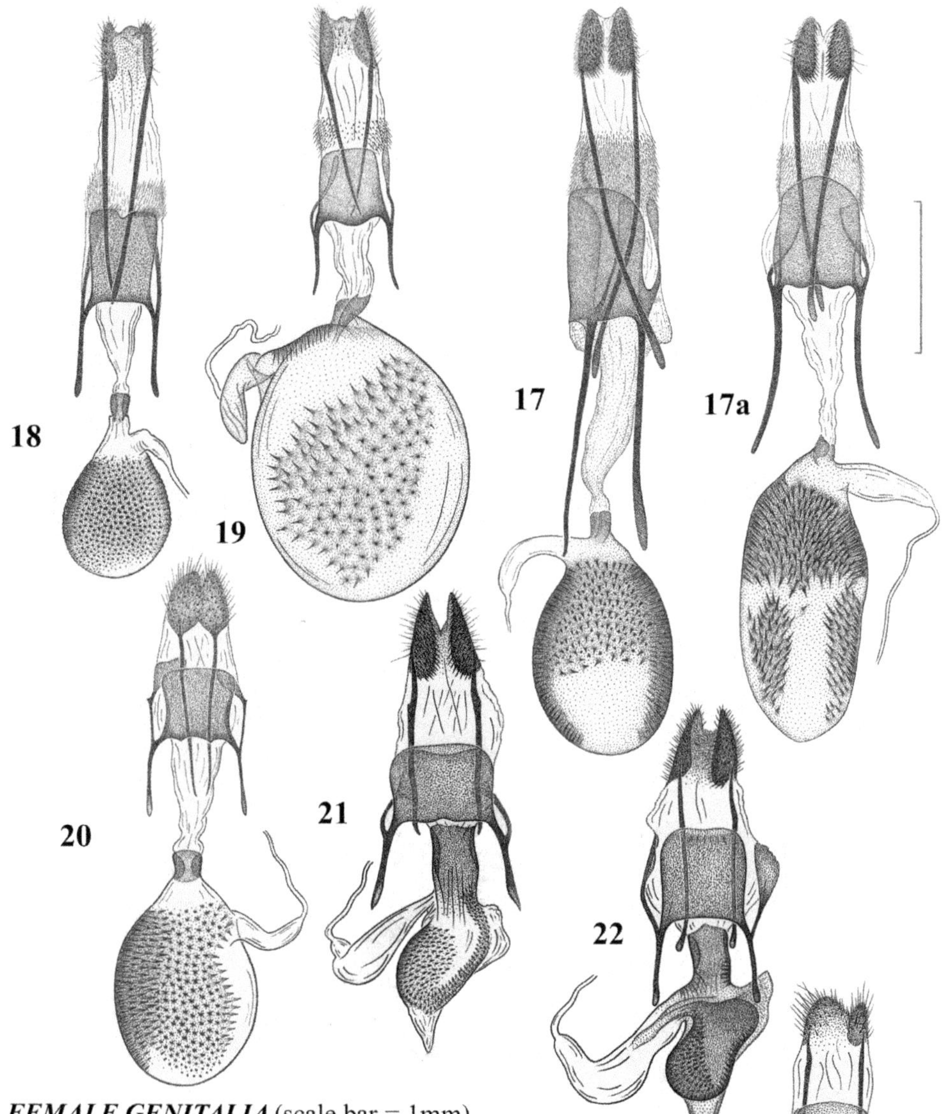

FEMALE GENITALIA (scale bar = 1mm)

17. *Eupithecia likiangi likiangi* Vojnits, China, Yunnan
17a.*Eupithecia likiangi pedisequa* ssp.n., China, Taiwan (paratype)
18.*Eupithecia sinicaria* Leech, China, Tibet
19.*Eupithecia ruficorpus* Warren, China, Tibet
20.*Eupithecia szelenyica* Vojnits, China, Shaanxi
21.*Eupithecia liberata* Inoue, Nepal, Kathmandu
22.*Eupithecia yoshimotoi* Inoue, China, Taiwan
24.*Eupithecia kuroshio* Inoue, China, Taiwan

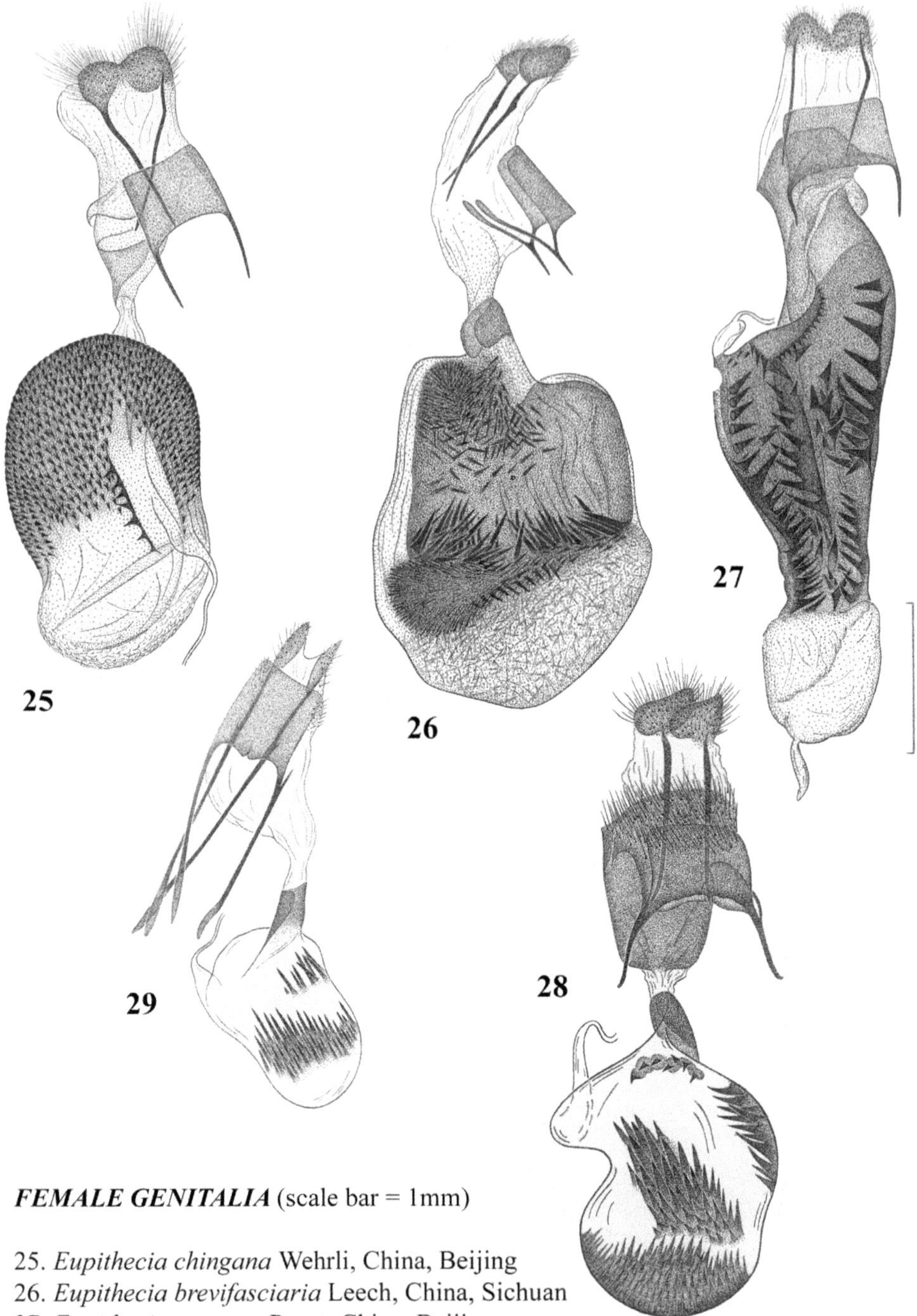

FEMALE GENITALIA (scale bar = 1mm)

25. *Eupithecia chingana* Wehrli, China, Beijing
26. *Eupithecia brevifasciaria* Leech, China, Sichuan
27. *Eupithecia eurytera* Prout, China, Beijing
28. *Eupithecia rufescens* Butler, China, Zhejiang
29. *Eupithecia leamariae* Ratzel, Bhutan, Paro (holotype)

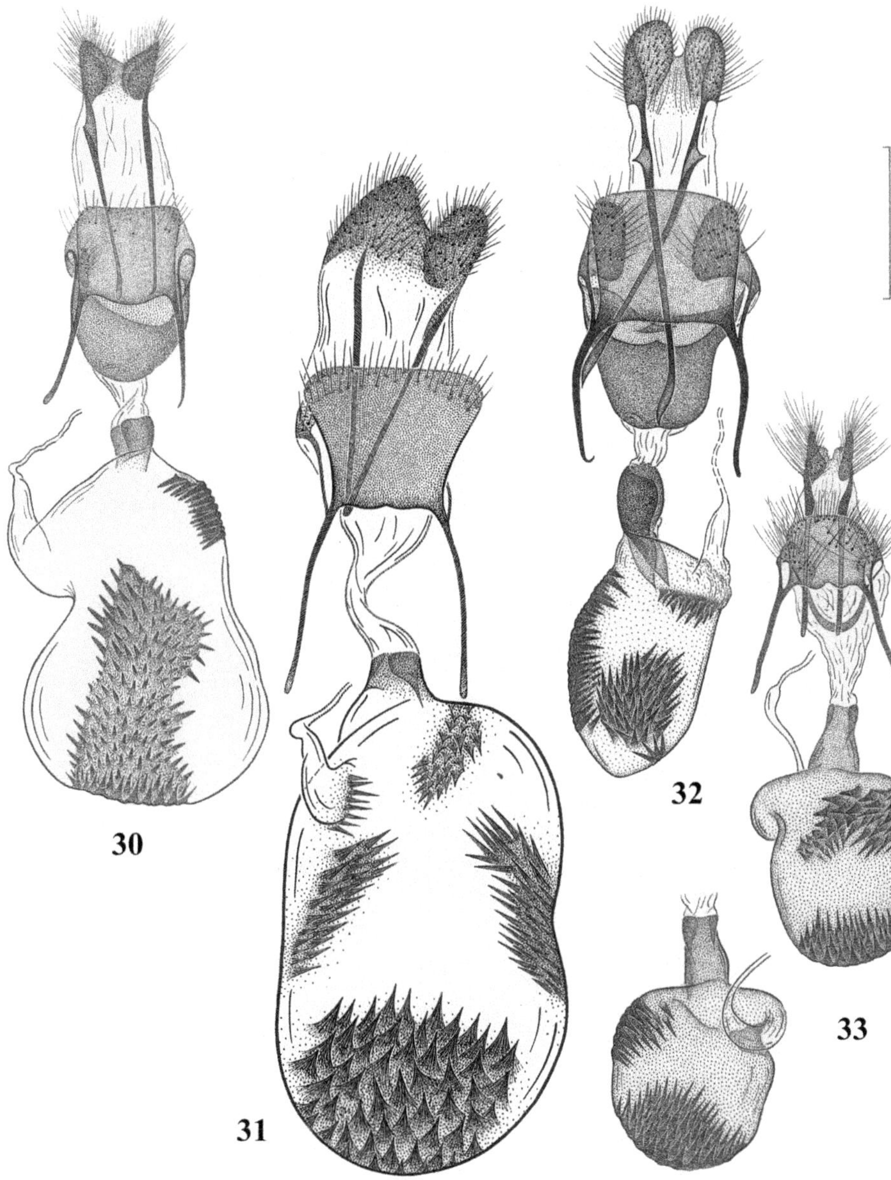

FEMALE GENITALIA (scale bar = 1mm)

30. *Eupithecia abietaria* Goeze, Russia, S. Primorje
31. *Eupithecia gigantea* Staudinger, Russia, S. Primorje
32. *Eupithecia interrubrescens* Hampson, China, Tibet
33. *Eupithecia costalis* Walker, China, Taiwan (figure below left ventral aspect)

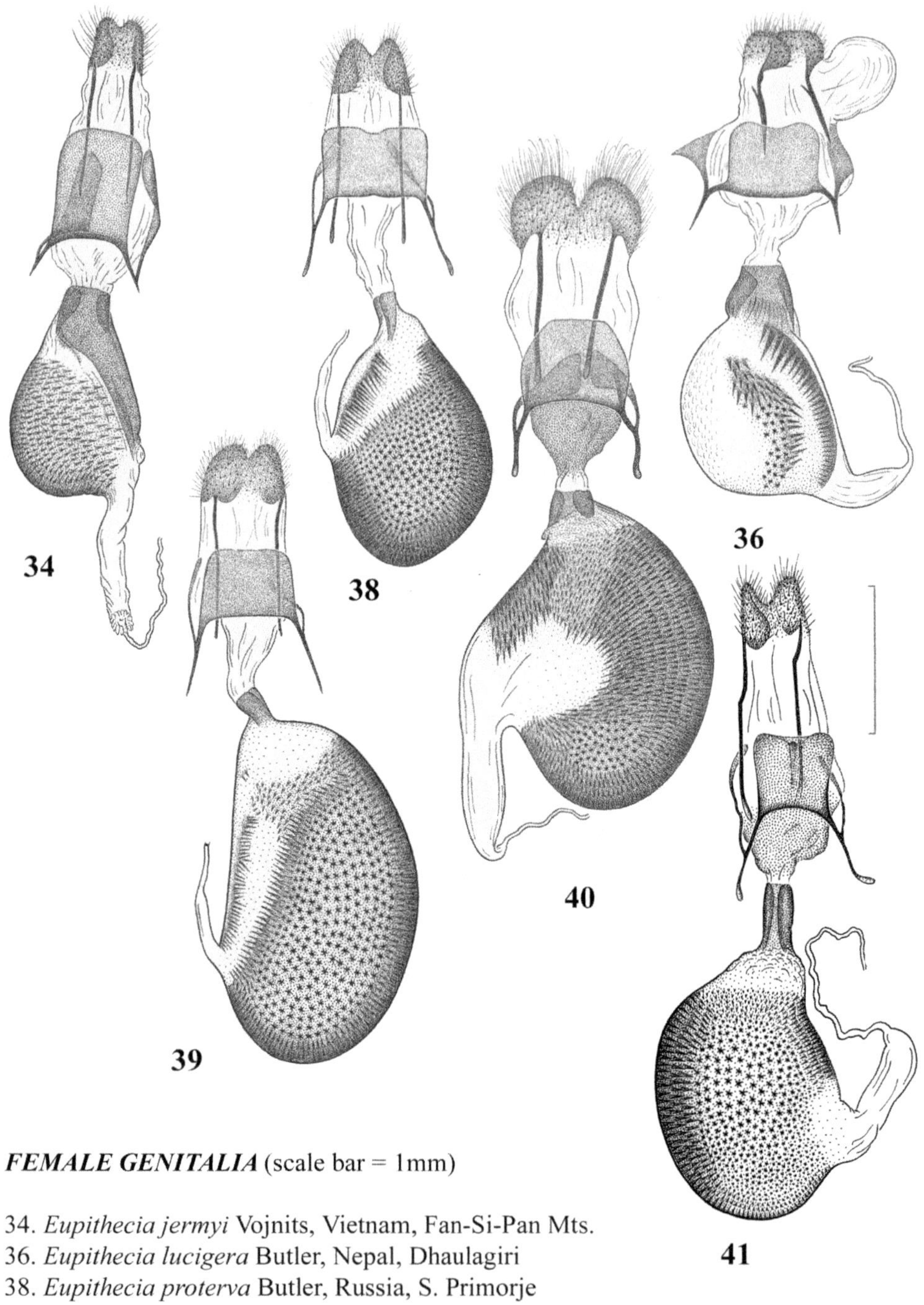

FEMALE GENITALIA (scale bar = 1mm)

34. *Eupithecia jermyi* Vojnits, Vietnam, Fan-Si-Pan Mts.
36. *Eupithecia lucigera* Butler, Nepal, Dhaulagiri
38. *Eupithecia proterva* Butler, Russia, S. Primorje
39. *Eupithecia subbreviata* Staudinger, Russia, S. Primorje
40. *Eupithecia megaproterva* Inoue, China, Taiwan
41. *Eupithecia verprota* Mironov & Galsworthy, China, Shaanxi (paratype)

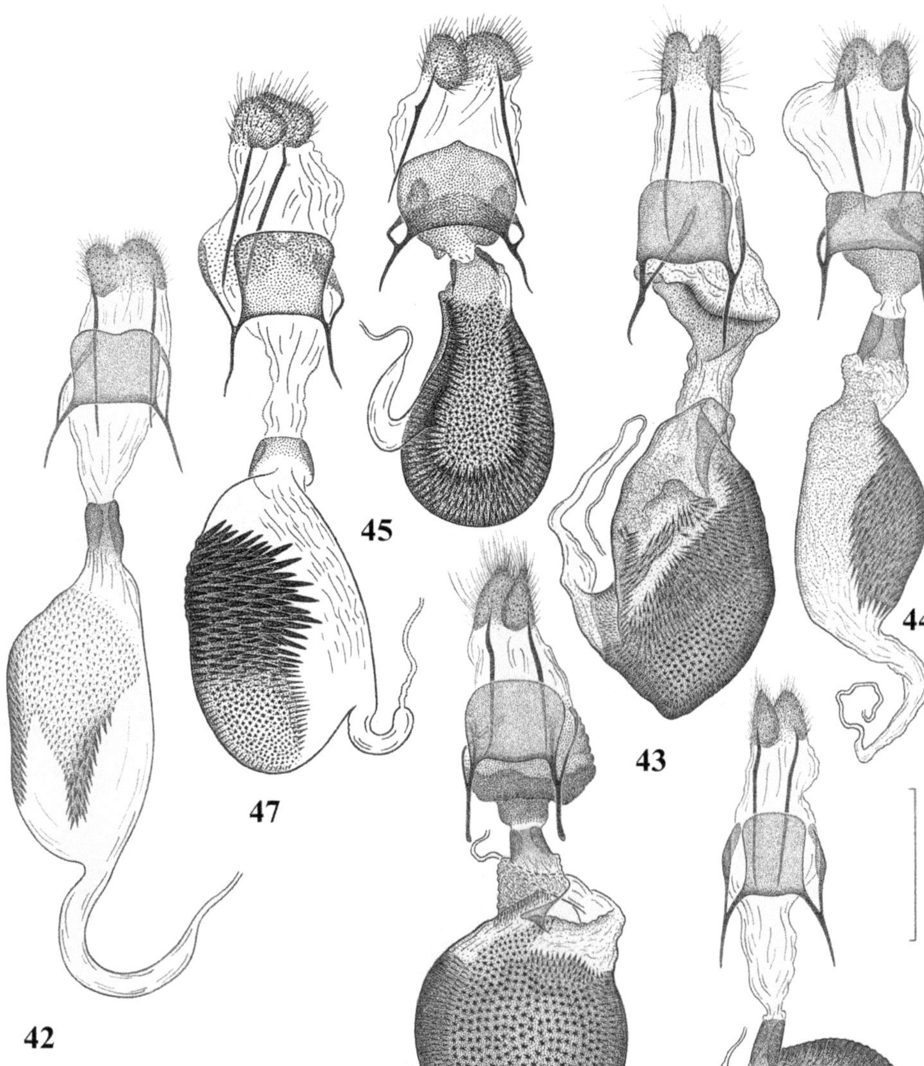

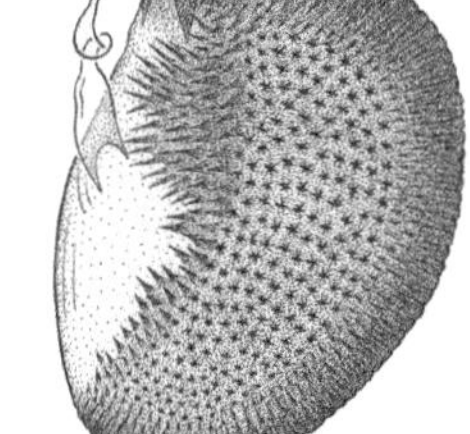

FEMALE GENITALIA (scale bar = 1mm)

42. *Eupithecia clavifera* Inoue, Russia, S. Primorje
43. *Eupithecia wangi* Mironov & Galsworthy, China, Guangxi
44. *Eupithecia signigera* Butler, China, Shaanxi
45. *Eupithecia russula* Mironov & Galsworthy, China, Tibet (holotype)
46. *Eupithecia flavoapicaria* Inoue, China, Taiwan
47. *Eupithecia blenna* Mironov & Galsworthy, China, Shaanxi (holotype)
48. *Eupithecia convexa* Inoue, China, Taiwan

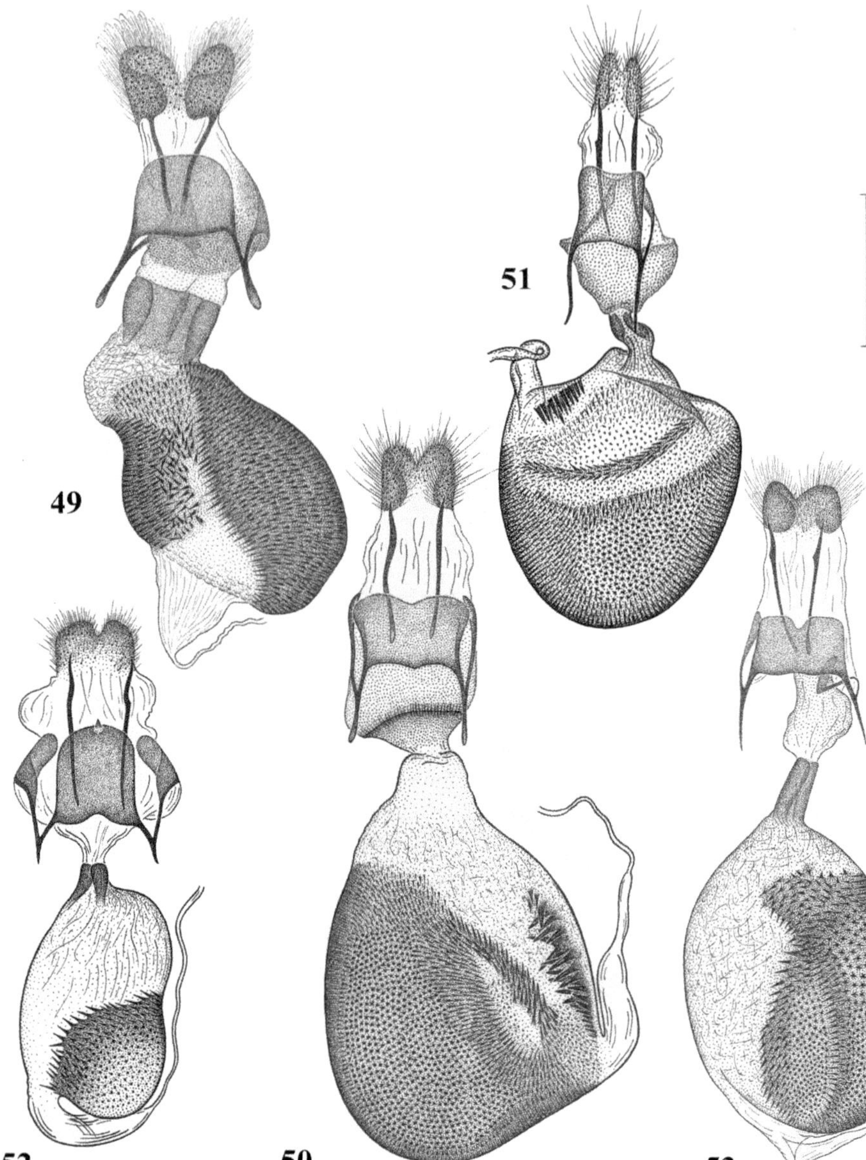

FEMALE GENITALIA (scale bar = 1mm)

49. *Eupithecia opicata* Mironov & Galsworthy, China, Shaanxi
50. *Eupithecia costiconvexa* Inoue, N. Thailand, Prov. Chiang Mai
51. *Eupithecia concava* Mironov & Galsworthy, China, Taiwan
52. *Eupithecia aritai* Inoue, China, Taiwan
53. *Eupithecia admiranda* Mironov & Galsworthy, China, Shaanxi

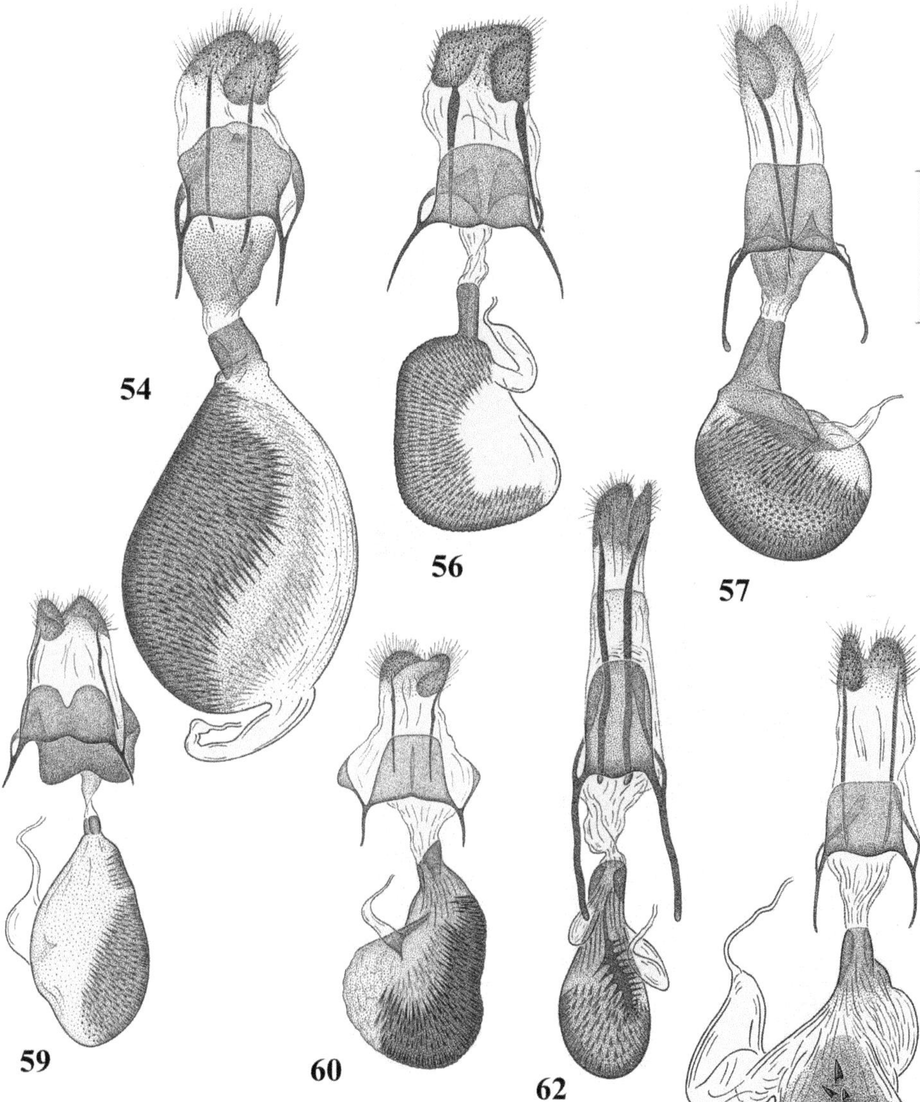

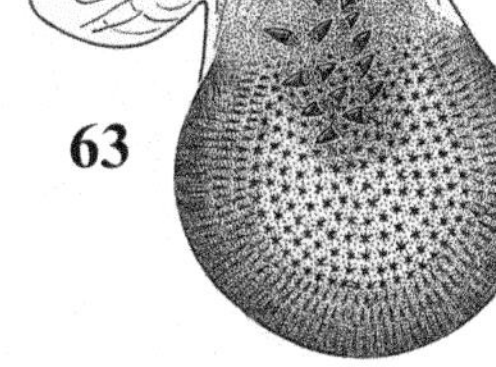

FEMALE GENITALIA (scale bar = 1mm)

54. *Eupithecia hashimotoi* Inoue, China, Taiwan
56. *Eupithecia amplificata* Inoue, China, Yunnan
57. *Eupithecia asempiterna* Inoue, Nepal, Gandaki
59. *Eupithecia minusculata* Alphéraky, China, Xinjiang (holotype)
60. *Eupithecia opistographata* Dietze, Turkmenistan, Dashoguz
62. *Eupithecia mustangata* Schütze, Pakistan, Gilgit-Baltistan
63. *Eupithecia leucostaxis* Prout, China, Yunnan

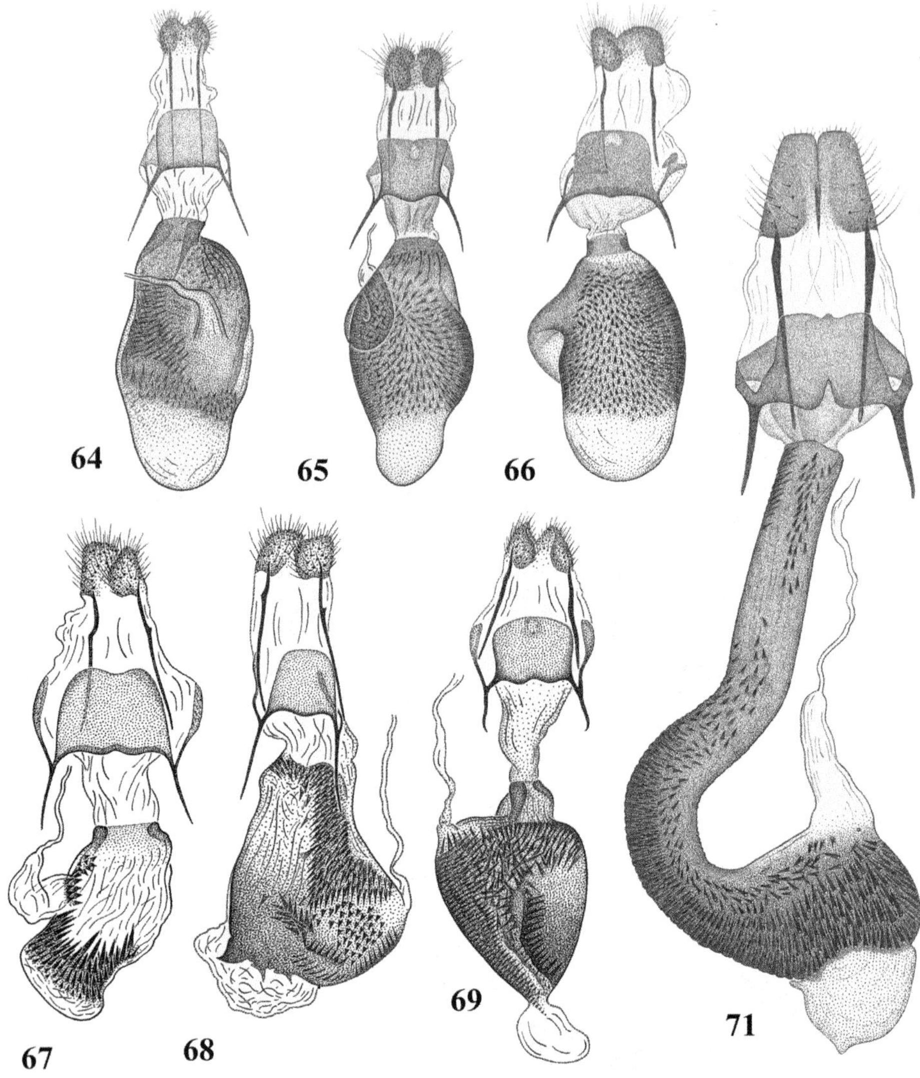

FEMALE GENITALIA (scale bar = 1mm)

64. *Eupithecia ochracea* Warren, Pakistan, North West Frontier Province
65. *Eupithecia homogrammata* Dietze, Russia, S. Primorje
66. *Eupithecia plumbeolata* Haworth, Kyrgyzstan, Naryn
67. *Eupithecia severa* Mironov & Galsworthy, China, Shaanxi (holotype)
68. *Eupithecia spissata* Mironov & Galsworthy, China, Sichuan (holotype)
69. *Eupithecia cordata* Mironov & Galsworthy, China, Yunnan (holotype)
71. *Eupithecia peguensis* Prout, Nepal, Gandaki

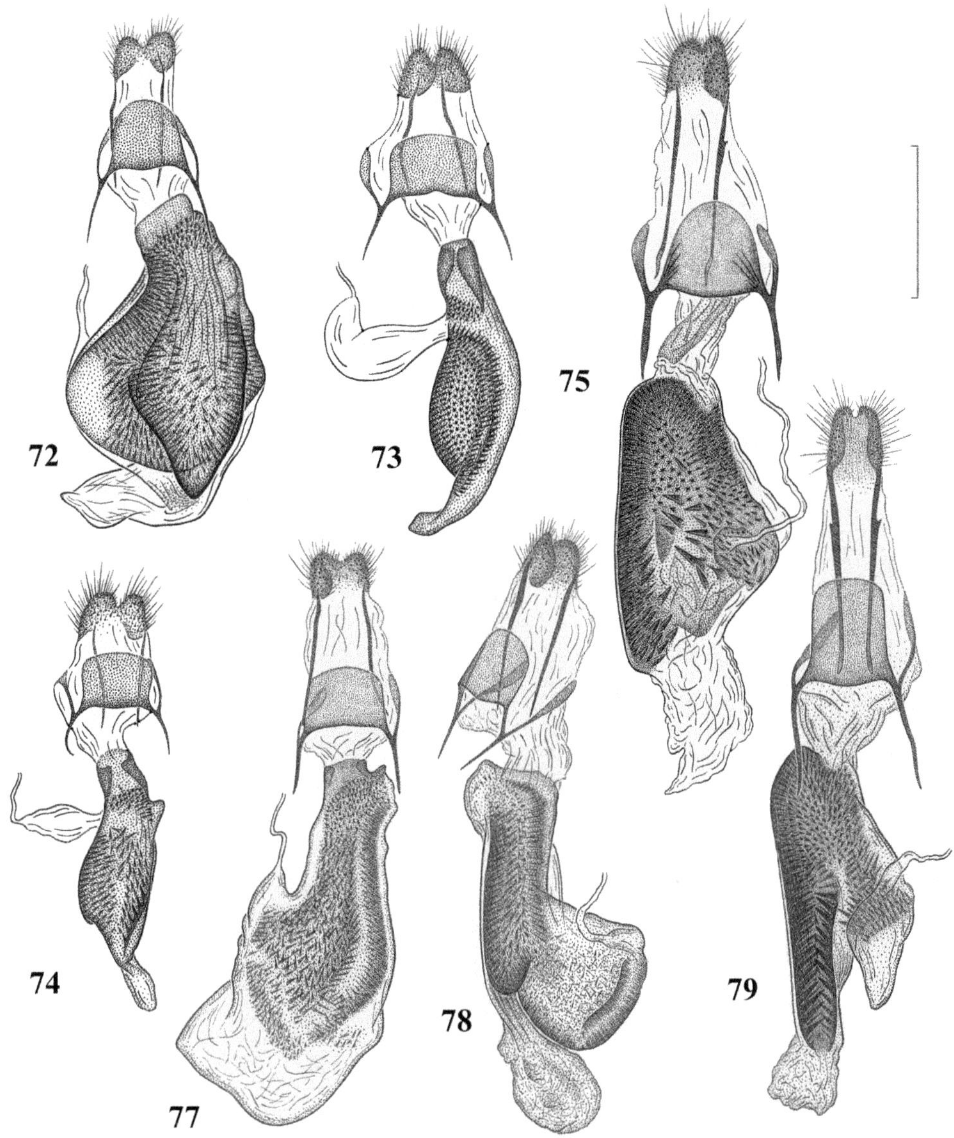

FEMALE GENITALIA (scale bar = 1mm)

72. *Eupithecia apta* Mironov & Galsworthy, China, Yunnan
73. *Eupithecia burselongata* Mironov & Galsworthy, China, Yunnan
74. *Eupithecia minutula* Mironov & Galsworthy, China, Yunnan
75. *Eupithecia impavida* Vojnits, Pakistan
77. *Eupithecia inexhausta* Vojnits, China, Sichuan
78. *Eupithecia hongxiangae* Mironov & Galsworthy, China, Gansu
79. *Eupithecia alishana* Inoue, China, Taiwan

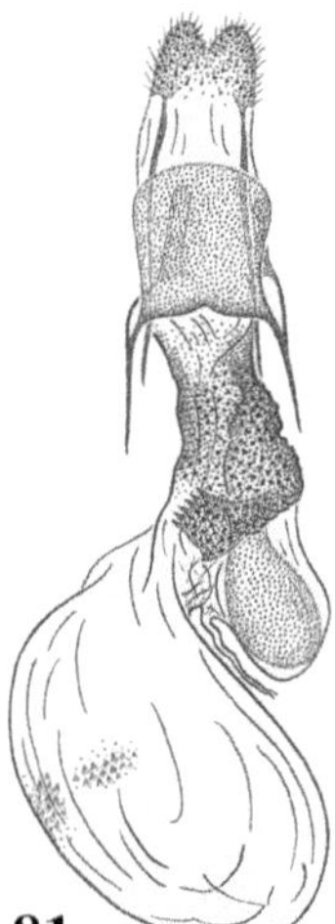

81

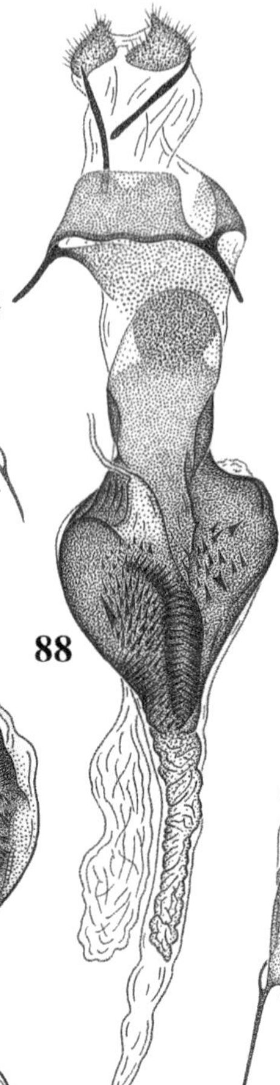

88

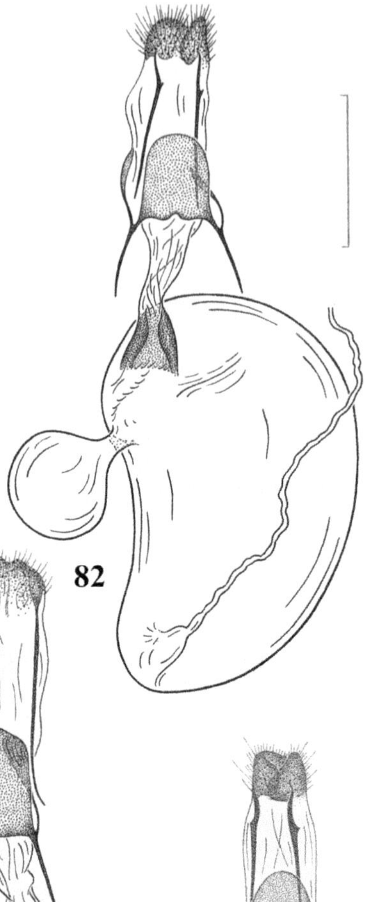

82

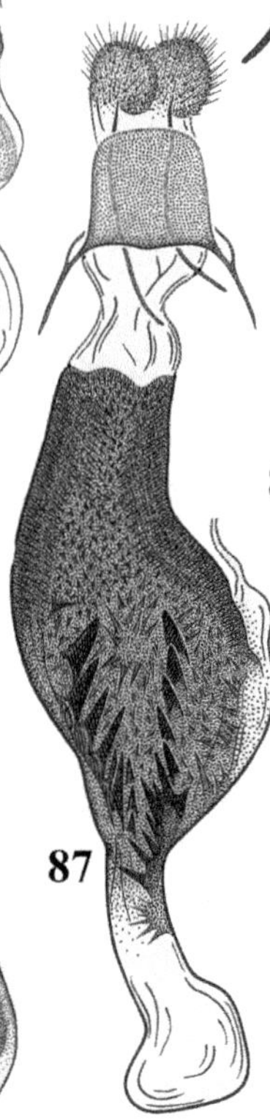

87

83

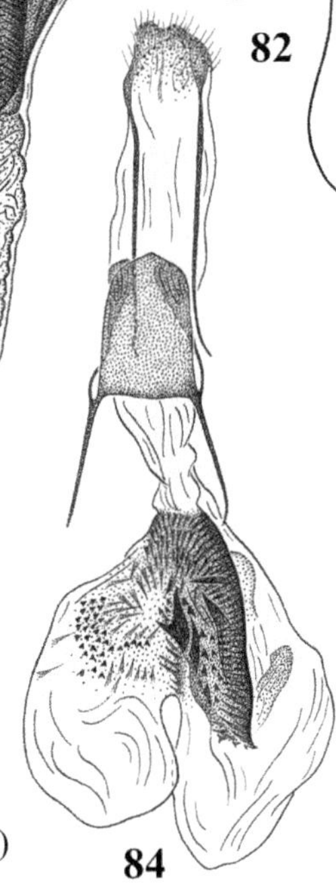

84

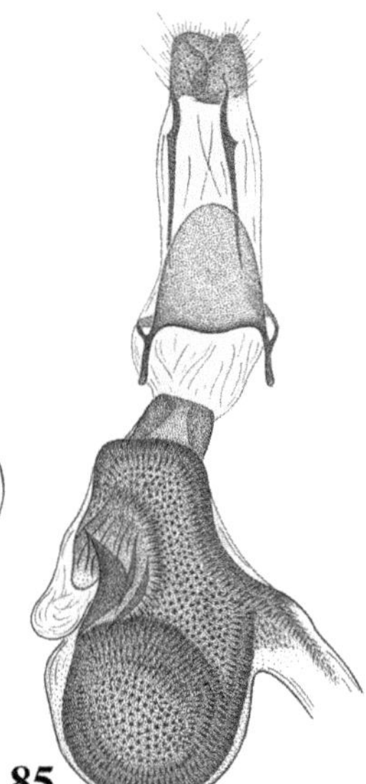

85

FEMALE GENITALIA (scale bar = 1mm)

81. *Eupithecia assulata* Bastelberger, China, Taiwan
82. *Eupithecia adoranda* Vojnits, China, Shaanxi
83. *Eupithecia granata* Vojnits, China, Yunnan
84. *Eupithecia amicula* Mironov & Galsworthy, China, Shaanxi
85. *Eupithecia hoenehermanni* Mironov & Galsworthy, China, Yunnan
87. *Eupithecia consortaria* Leech, Russia, Primorje
88. *Eupithecia vana* Vojnits, China, Shaanxi

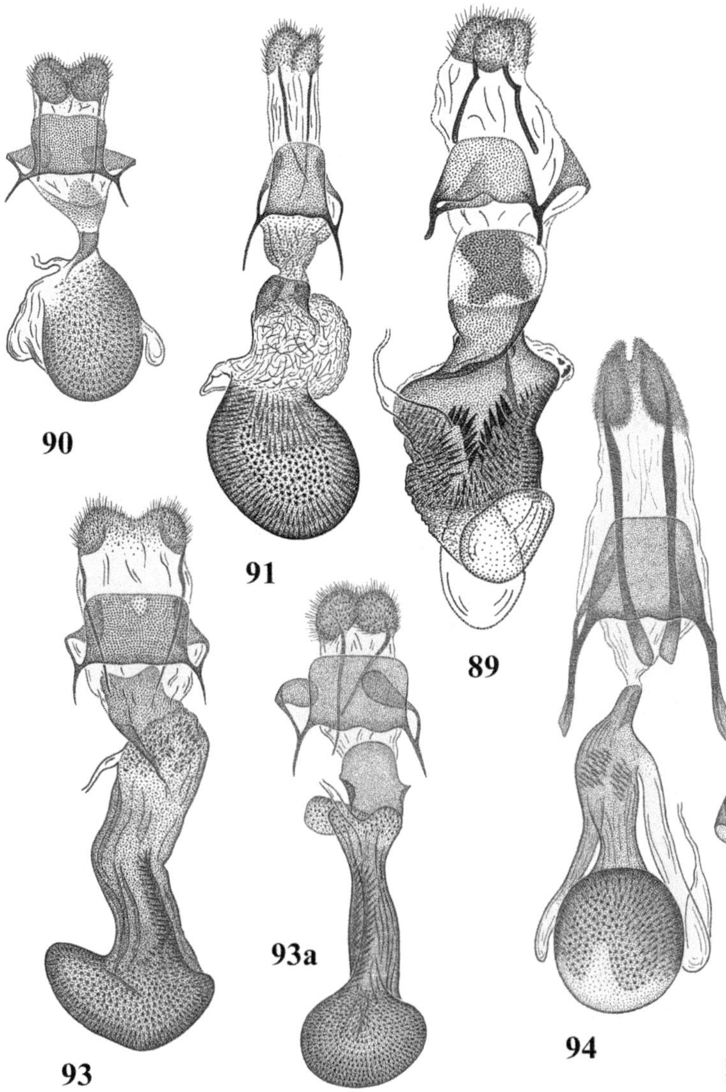

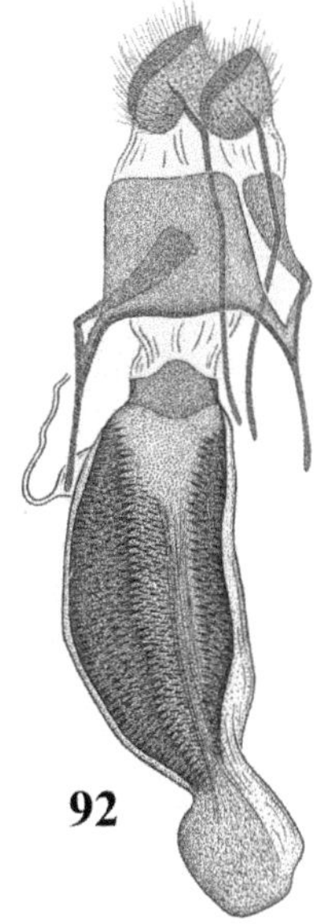

FEMALE GENITALIA (scale bar = 1mm)

89. *Eupithecia nirvana* Mironov & Galsworthy, China, Sichuan
90. *Eupithecia hilariata* Dietze, Tajikistan, Pamir Mts.
91. *Eupithecia hesperina* Mironov & Galsworthy, China, Yunnan
92. *Eupithecia dissertata* Püngeler, Russia, Dahuria
93. *Eupithecia carpophilata* Staudinger, Russia, S. Siberia, Baikal
93a. *Eupithecia carpophilata* Staudinger, Russia, S. Priamurje
94. *Eupithecia albispumata* Warren, China, Sichuan
96. *Eupithecia abbreviata* Stephens, Russia, W. Caucasus

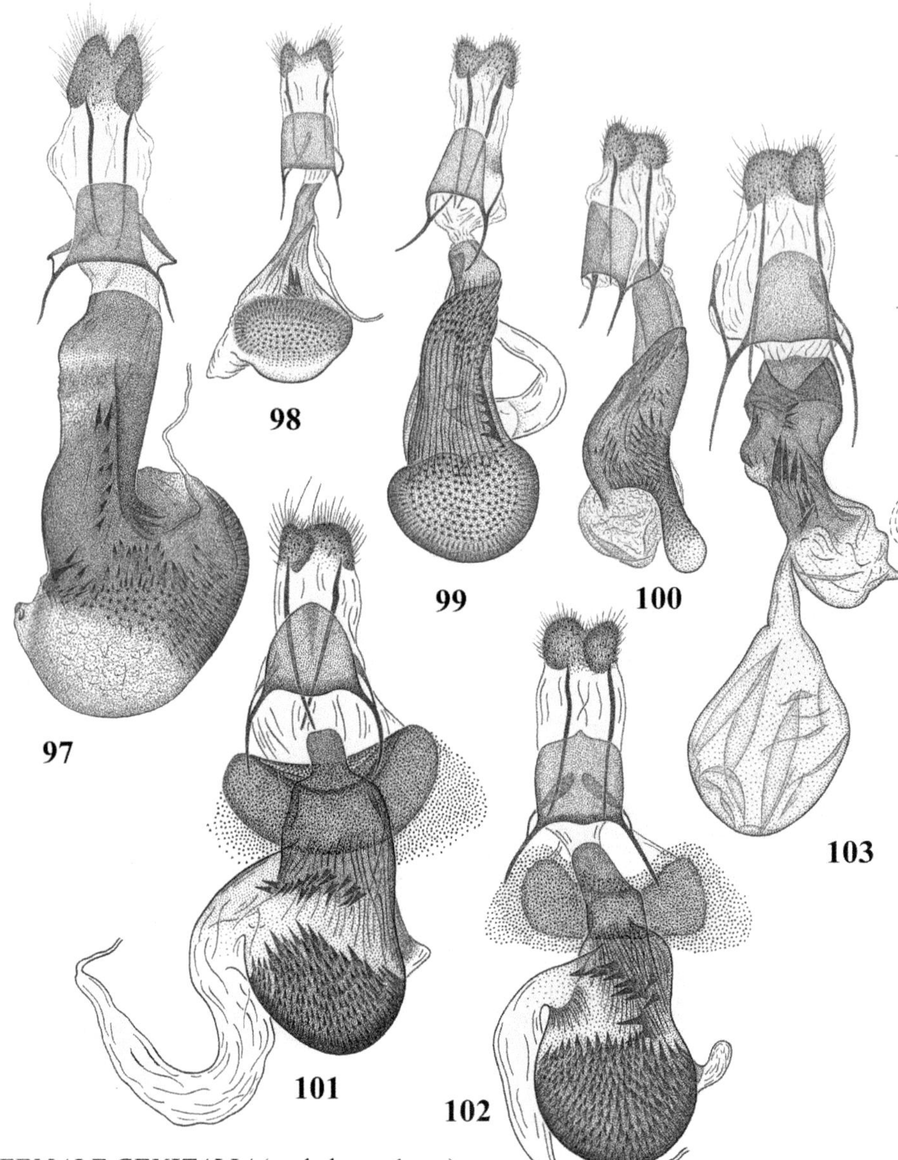

FEMALE GENITALIA (scale bar = 1mm)

97. *Eupithecia pusillata* Denis & Schiffermüller, Russia, Khabarovsky kraj
98. *Eupithecia kozlovi* Viidalepp, China, Qinghai
99. *Eupithecia exrubicunda* Inoue, China, Taiwan
100. *Eupithecia lanceata* Hübner, China, Gansu
101. *Eupithecia kudoi* Inoue, China, Taiwan
102. *Eupithecia daemionata* Dietze, Russia, Vladivostok
103. *Eupithecia habermani* Viidalepp & Mironov, China, Gansu

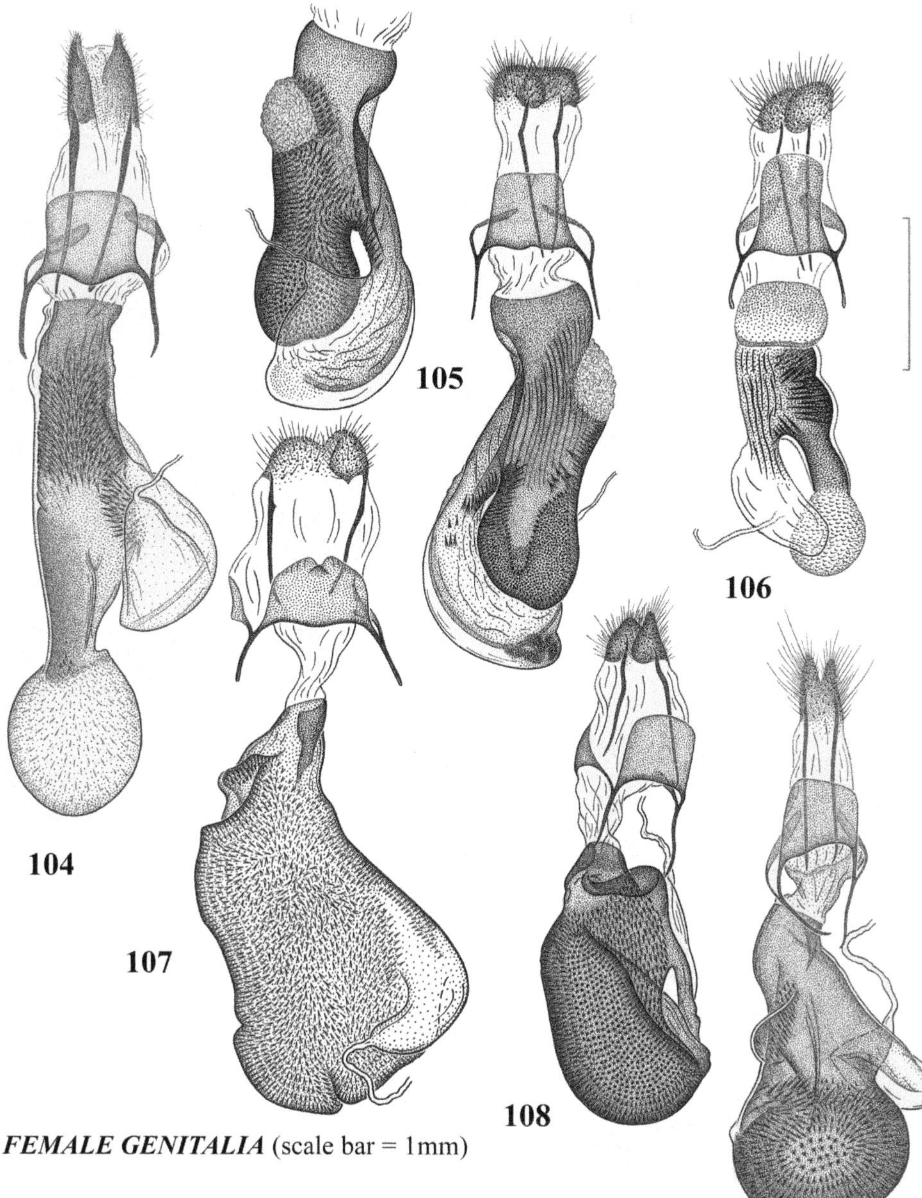

FEMALE GENITALIA (scale bar = 1mm)

104. *Eupithecia emendata* Vojnits, China, Tibet
105. *Eupithecia dolichoptera*, Galsworthy & Mironov , China, Taiwan (ventral view upper left)
106. *Eupithecia phantastica* Mironov & Galsworthy, China, Taiwan
107. *Eupithecia atuni* Mironov & Galsworthy, China, Yunnan
108. *Eupithecia funerea* Inoue, China, Taiwan
109. *Eupithecia fulvidorsata* Mironov & Galsworthy, China, Shaanxi

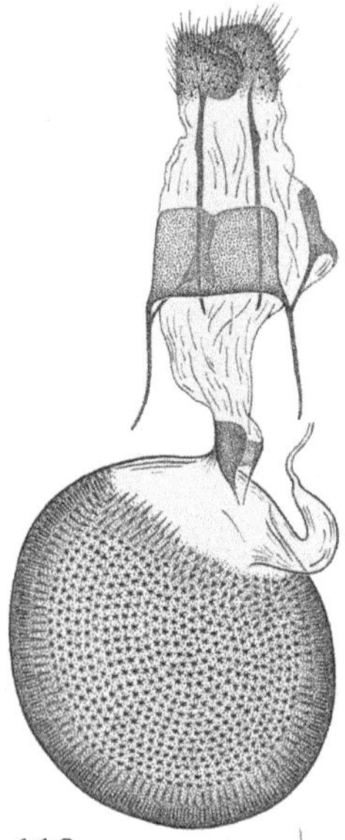

110

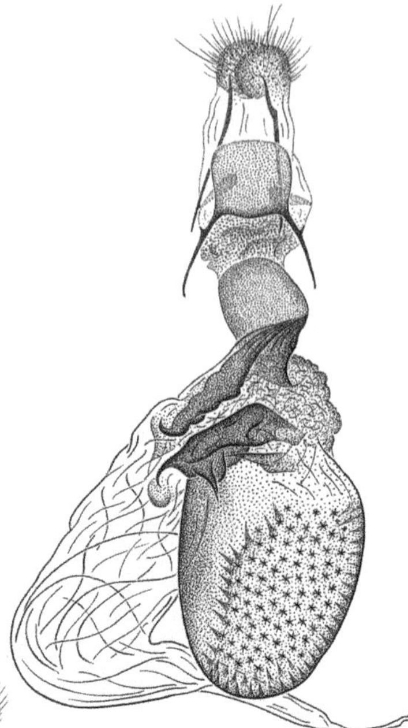

111

112

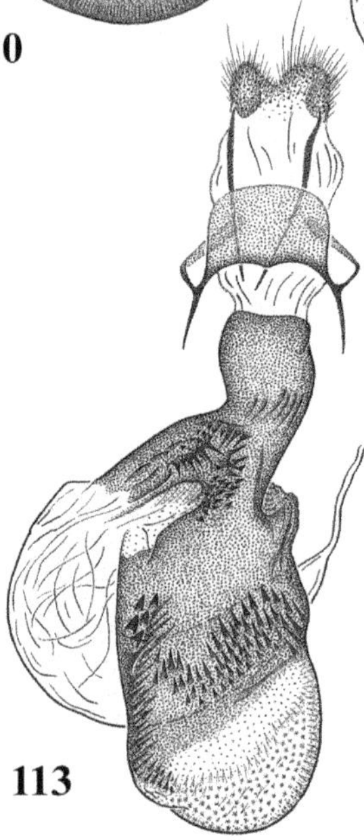

113

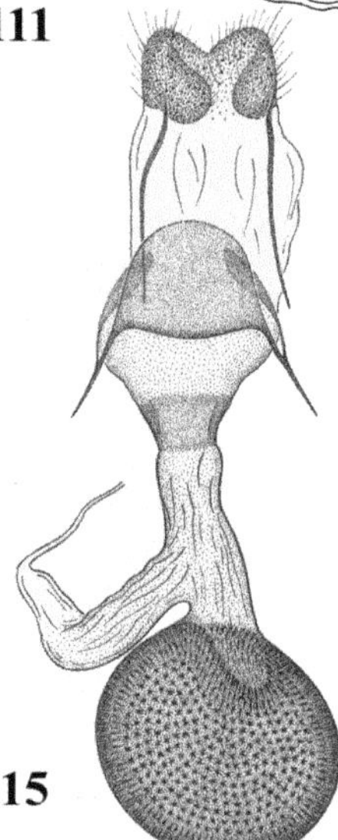

115

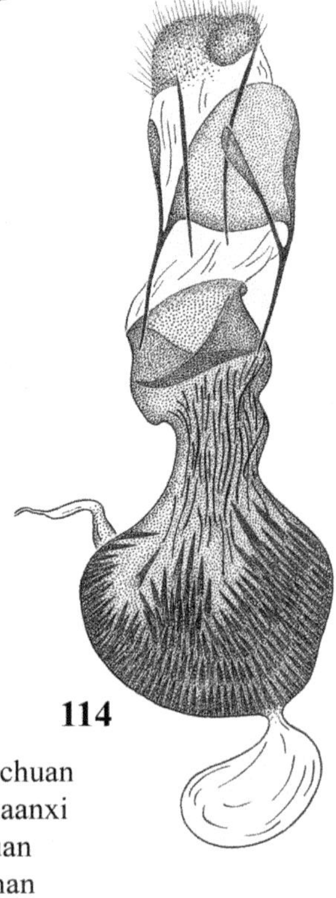

114

FEMALE GENITALIA (scale bar = 1mm)

110. *Eupithecia okadai* Inoue, Japan
111. *Eupithecia delicata* Mironov & Galsworthy, China, Sichuan
112. *Eupithecia camilla* Mironov & Galsworthy, China, Shaanxi
113. *Eupithecia julia* Mironov & Galsworthy, China, Sichuan
114. *Eupithecia fortis* Mironov & Galsworthy, China, Yunnan
115. *Eupithecia lariciata* Freyer, Russia, S. Siberia

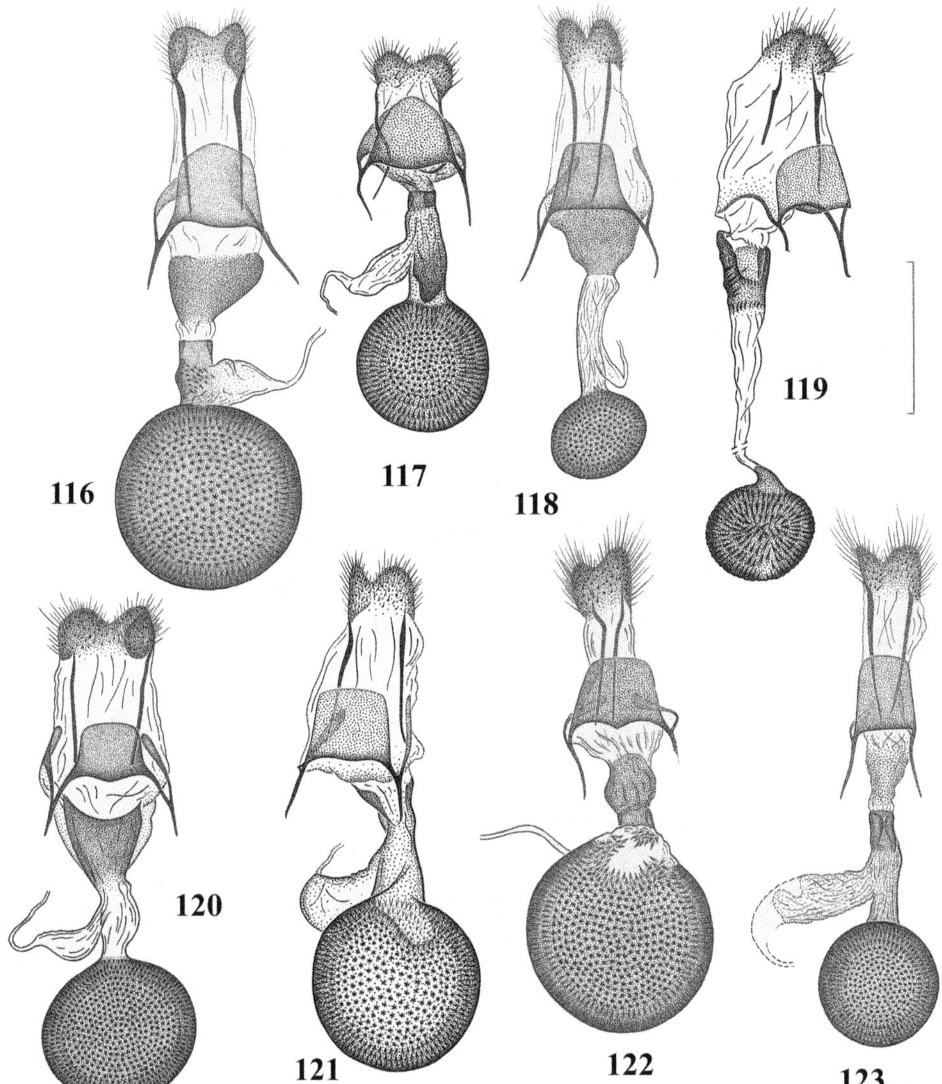

FEMALE GENITALIA (scale bar = 1mm)

116. *Eupithecia emanata* Dietze, Russia, S. Primorje
117. *Eupithecia honesta* Mironov & Galsworthy, China, Yunnan
118. *Eupithecia pediba* Vojnits, China, Tibet
119. *Eupithecia argentea* Mironov & Galsworthy, China, Sichuan
120. *Eupithecia primitiva* Mironov & Galsworthy, India, Kashmir
121. *Eupithecia sacrivicae* Mironov & Galsworthy, China, Shaanxi
122. *Eupithecia horrida* Vojnits, China, Yunnan
123. *Eupithecia stomachosa* Vojnits, China, Shaanxi

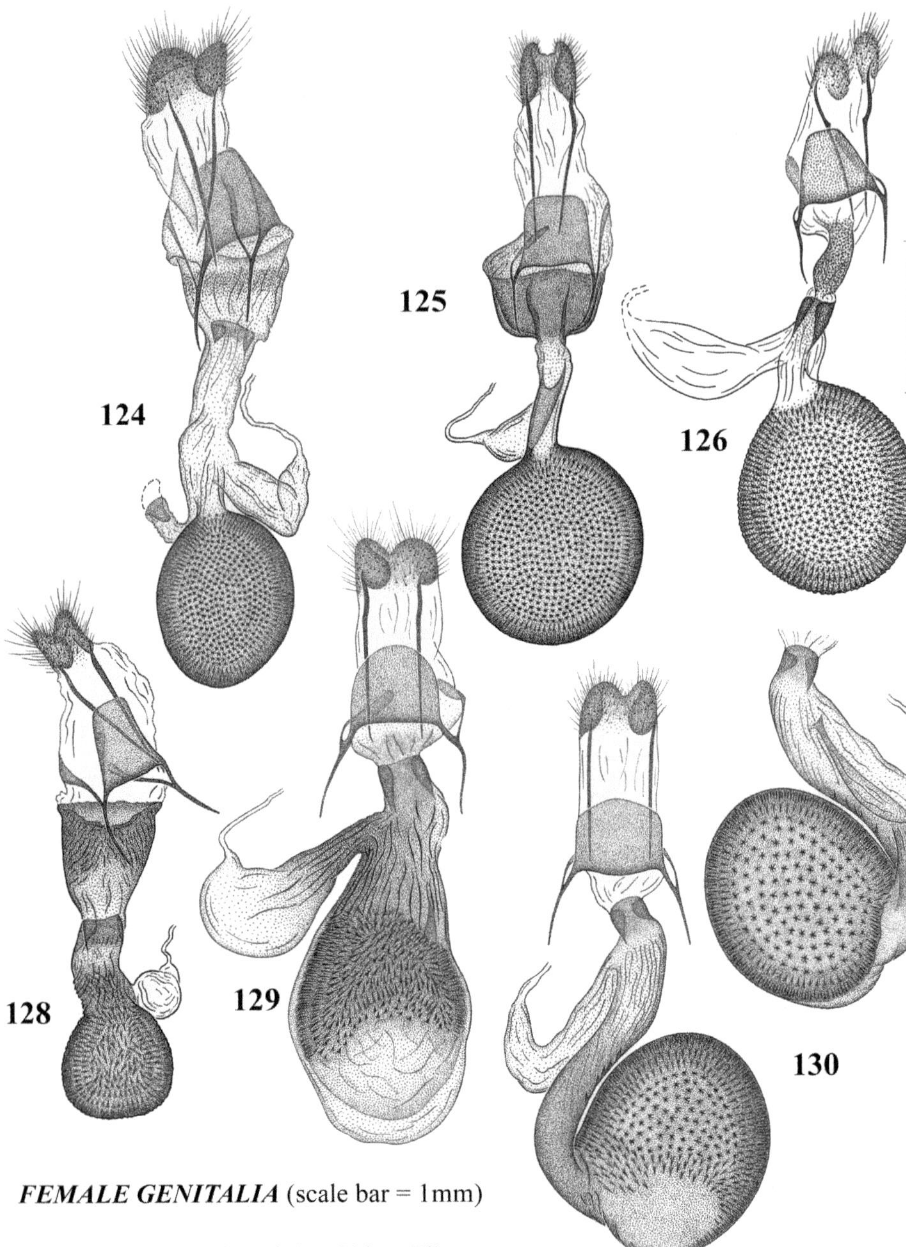

FEMALE GENITALIA (scale bar = 1mm)

124. *Eupithecia sola* Vojnits, China, Tibet
125. *Eupithecia marmorea* Vojnits, China, Tibet
126. *Eupithecia albimedia* Mironov & Galsworthy, China, Yunnan
128. *Eupithecia statari*a Inoue, China, Taiwan
129. *Eupithecia tripunctaria* Herrich-Schäffer, Russia, S. Primorje
130. *Eupithecia barteli* Dietze, Kazakhstan, Uralsk (ventral view upper right)

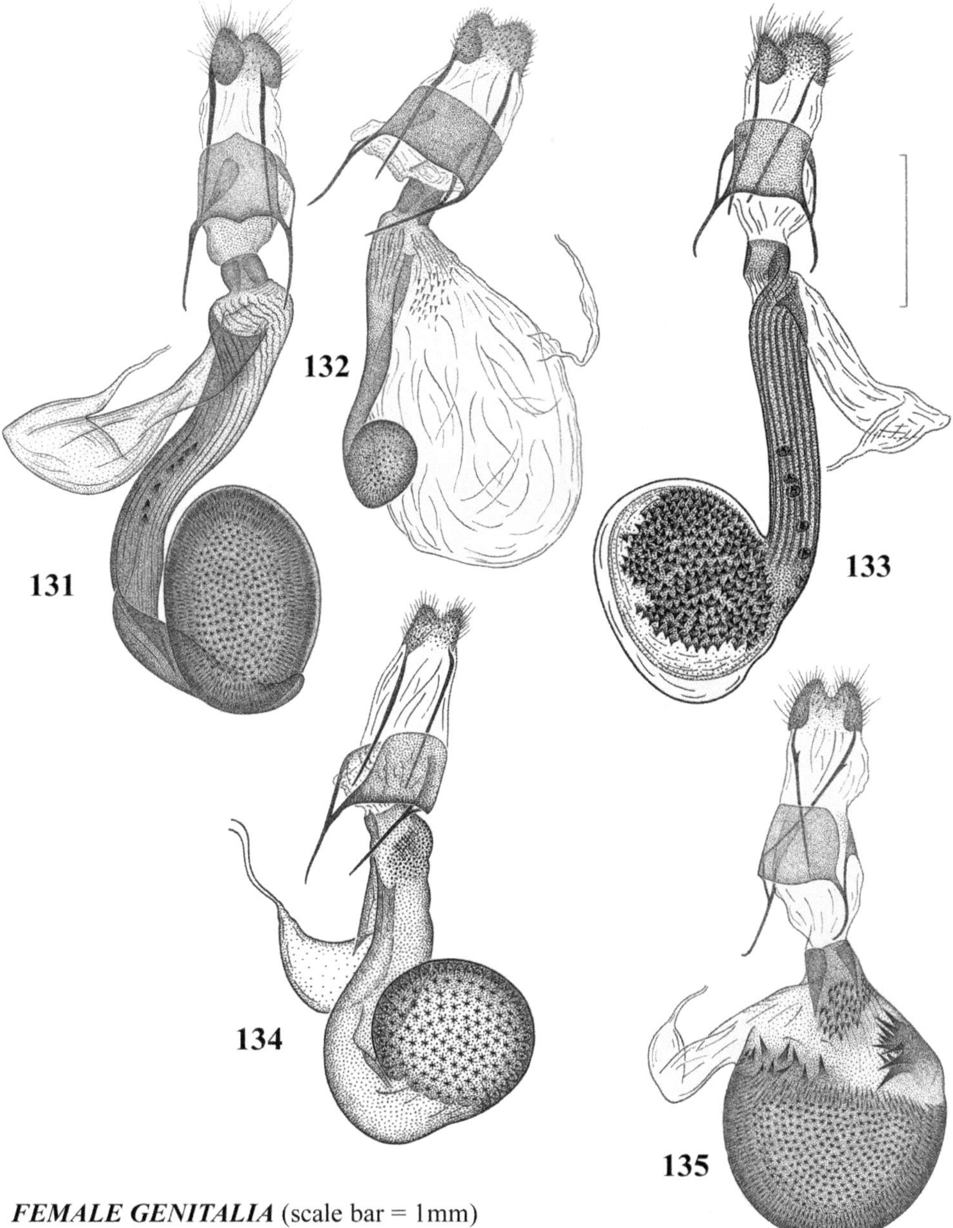

FEMALE GENITALIA (scale bar = 1mm)

131. *Eupithecia tenuisquama* Warren, Nepal, Ganesh Himal
132. *Eupithecia amandae* Galsworthy & Mironov, China, Yunnan
133. *Eupithecia blandula* Mironov & Galsworthy, China, Taiwan
134. *Eupithecia puella* Mironov & Galsworthy, China, Yunnan
135. *Eupithecia werneri* Vojnits, China, Yunnan

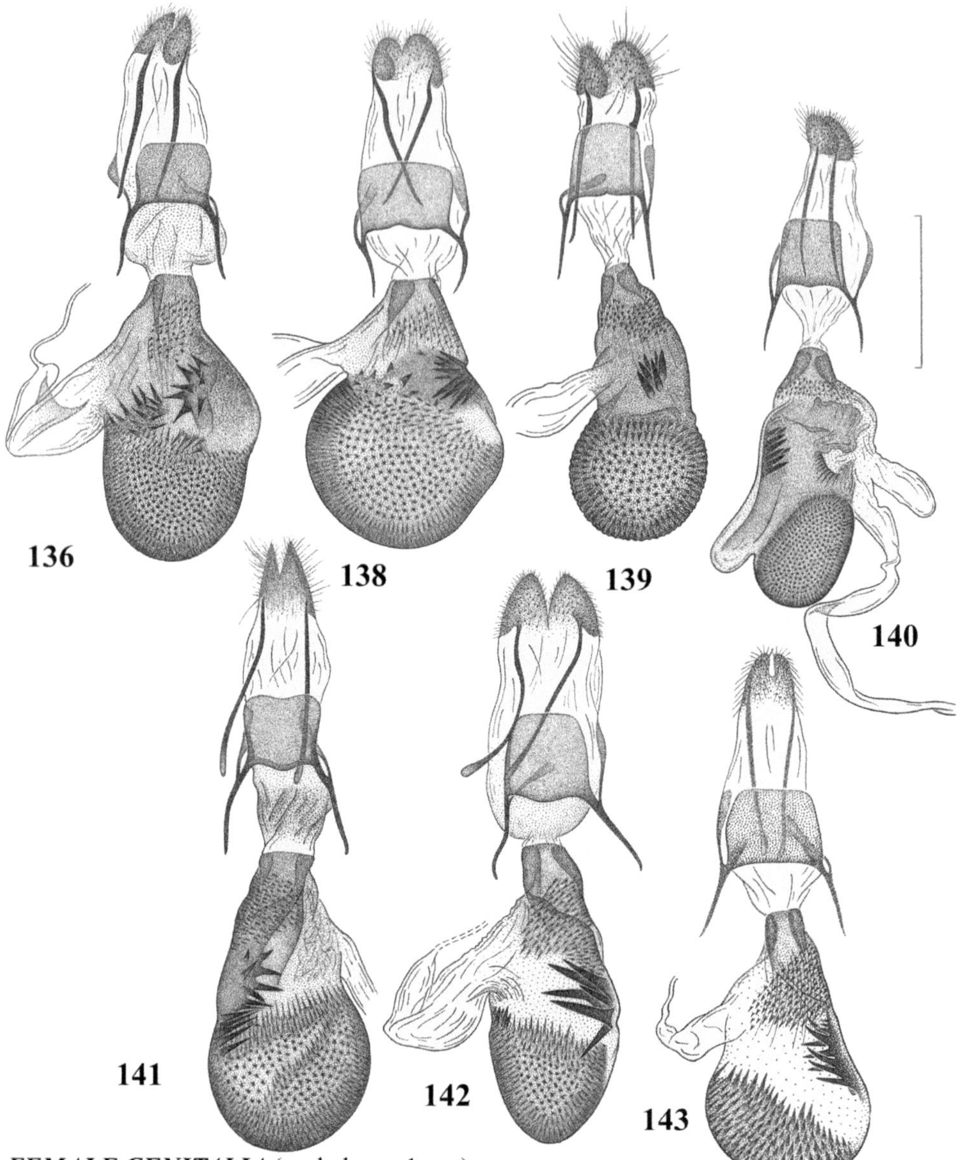

FEMALE GENITALIA (scale bar = 1mm)

136. *Eupithecia batangi* Vojnits, China, Tibet
138. *Eupithecia placida* Vojnits, China, Yunnan
139. *Eupithecia laudabilis* Vojnits, China, Yunnan
140. *Eupithecia manifesta* Mironov & Galsworthy, China, Yunnan (paratype)
141. *Eupithecia intolerabilis* Vojnits, China, Yunnan (holotype)
142. *Eupithecia mentita* Vojnits, China, Yunnan
143. *Eupithecia fenita* Mironov & Galsworthy, China, Yunnan (paratype)

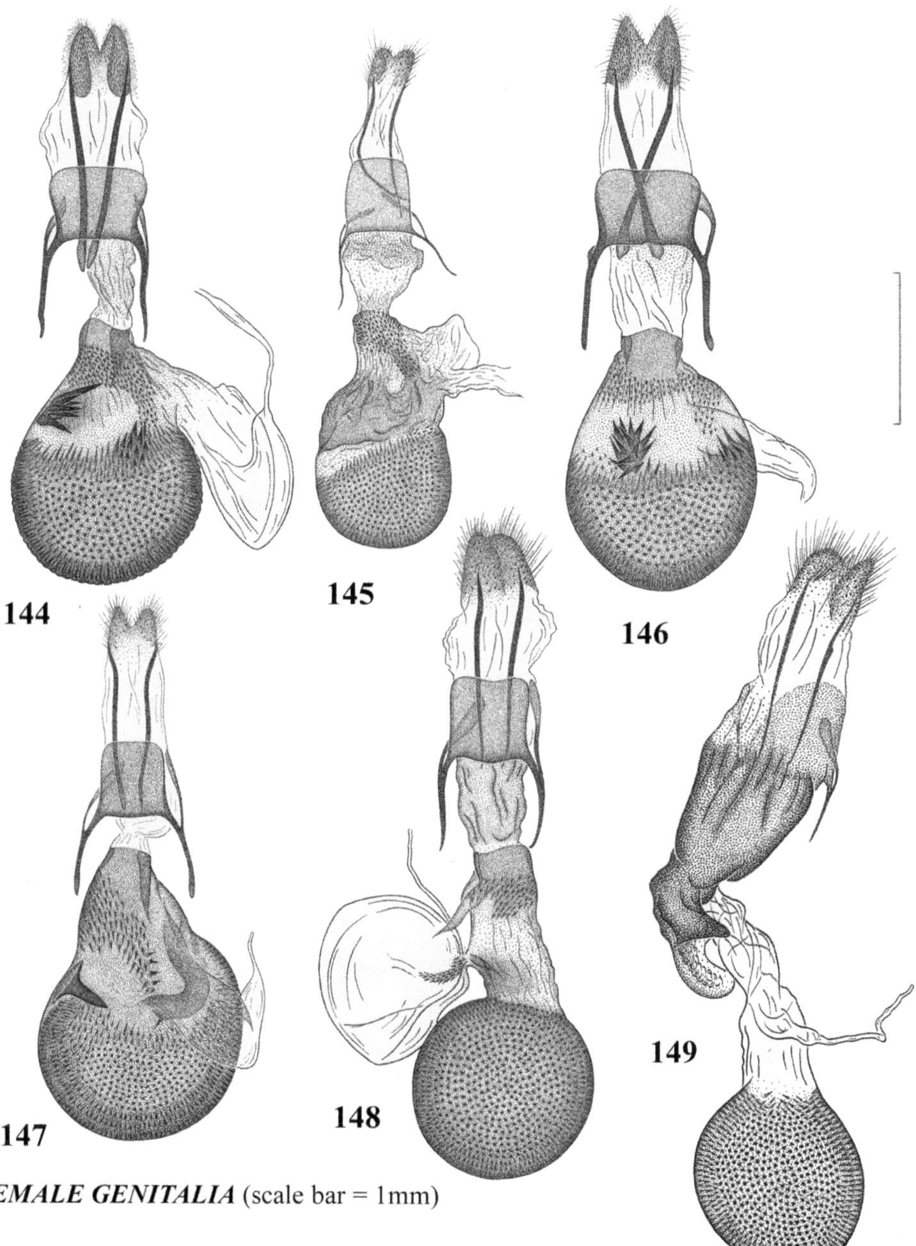

144 145 146 147 148 149

FEMALE GENITALIA (scale bar = 1mm)

144. *Eupithecia refertissima* Vojnits, China, Tibet
145. *Eupithecia avara* Vojnits, China, Yunnan
146. *Eupithecia caduca* Vojnits, China, Yunnan
147. *Eupithecia propagata* Prout, Nepal, Gandaki
148. *Eupithecia balintszolti* Vojnits, Thailand, Chiangmai
149. *Eupithecia hainanensis* Mironov & Galsworthy, China, Hainan (paratype)

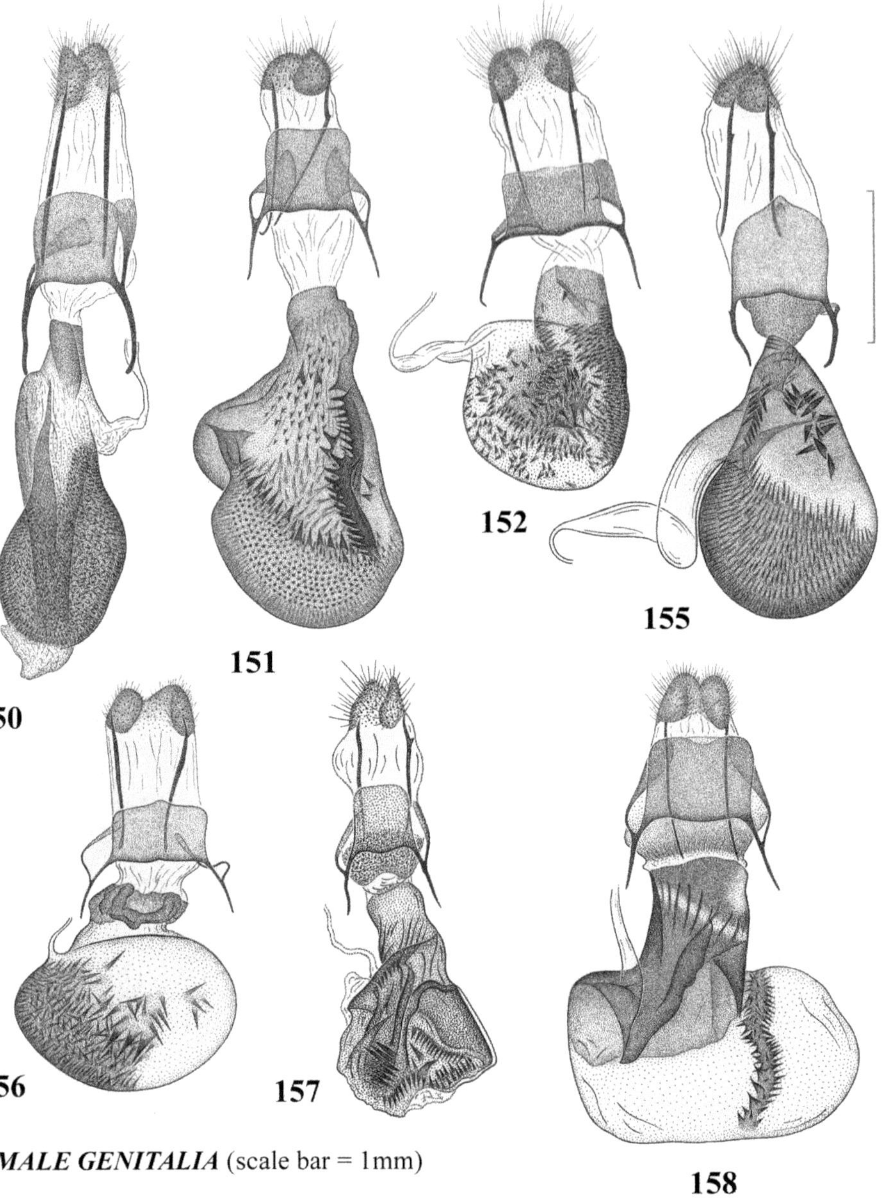

FEMALE GENITALIA (scale bar = 1mm)

150. *Eupithecia groenblomi* Urbahn, China, Qinghai
151. *Eupithecia selinata* Herrich-Schäffer, Russia, S. Siberia
152. *Eupithecia tamara* Mironov & Pekarsky, China, Sichuan (holotype)
155. *Eupithecia molestissima* Vojnits, China, Yunnan
156. *Eupithecia actaeata* Walderdorff, Russia, S. Primorje
157. *Eupithecia actrix* Mironov & Galsworthy, China, Shaanxi (holotype)
158. *Eupithecia uliata* Staudinger, Russia, S. Primorje

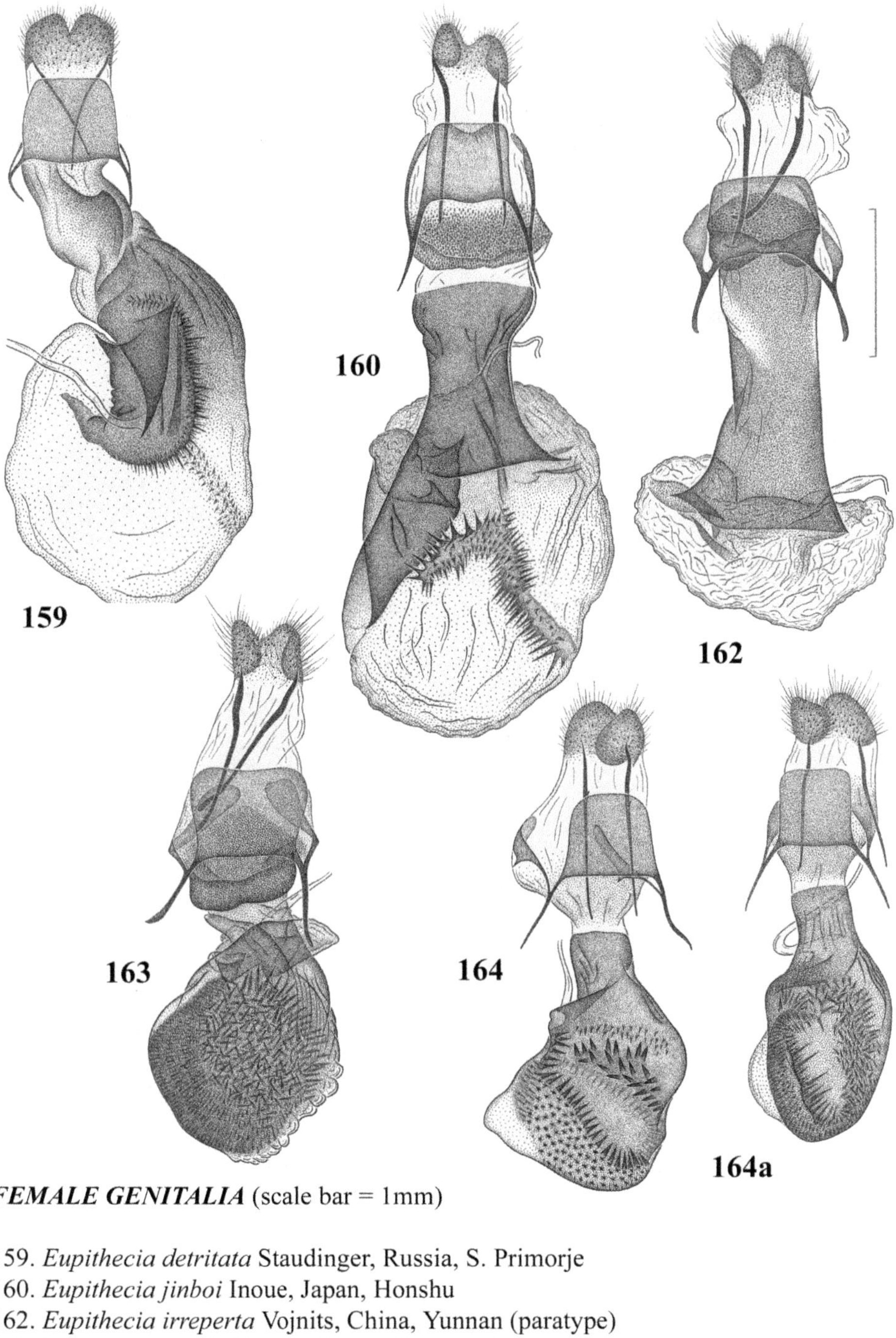

FEMALE GENITALIA (scale bar = 1mm)

159. *Eupithecia detritata* Staudinger, Russia, S. Primorje
160. *Eupithecia jinboi* Inoue, Japan, Honshu
162. *Eupithecia irreperta* Vojnits, China, Yunnan (paratype)
163. *Eupithecia vivida* Vojnits, China, Yunnan (holotype)
164, 164a. *Eupithecia quadripunctata* Warren, Russia, S. Primorje (variation)

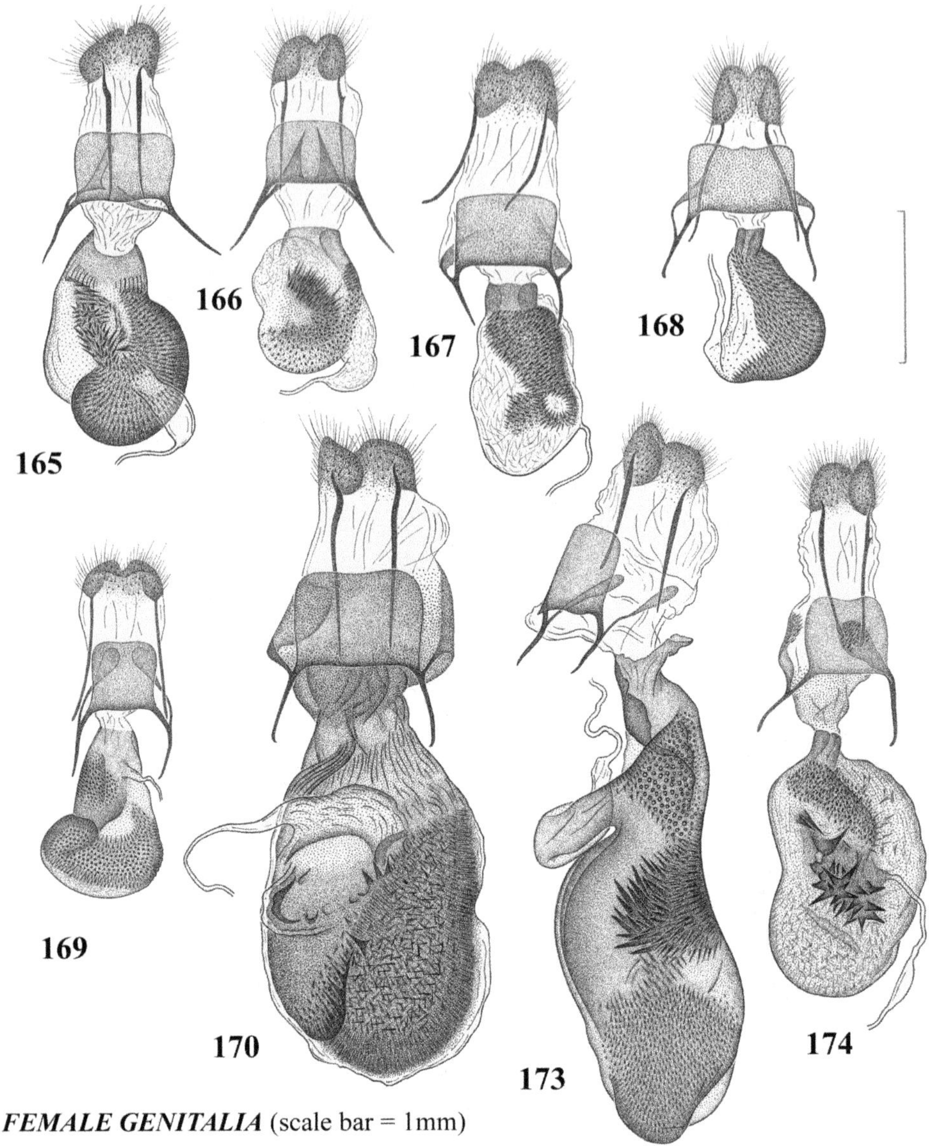

FEMALE GENITALIA (scale bar = 1mm)

165. *Eupithecia eximia* Vojnits, China, Sichuan
166. *Eupithecia nepalata* Schütze, Pakistan, Deosai Pass
167. *Eupithecia fatigata* Vojnits & de Laever, China, Yunnan (holotype)
168. *Eupithecia repentina* Vojnits & de Laever, Russia, S. Primorje
169. *Eupithecia virgaureata* Doubleday, Russia, S. Primorje
170. *Eupithecia taiwana* Wileman & South, China, Taiwan
173. *Eupithecia cotidiana* Vojnits, China, Gansu
174. *Eupithecia krampli* Vojnits, China, Sichuan

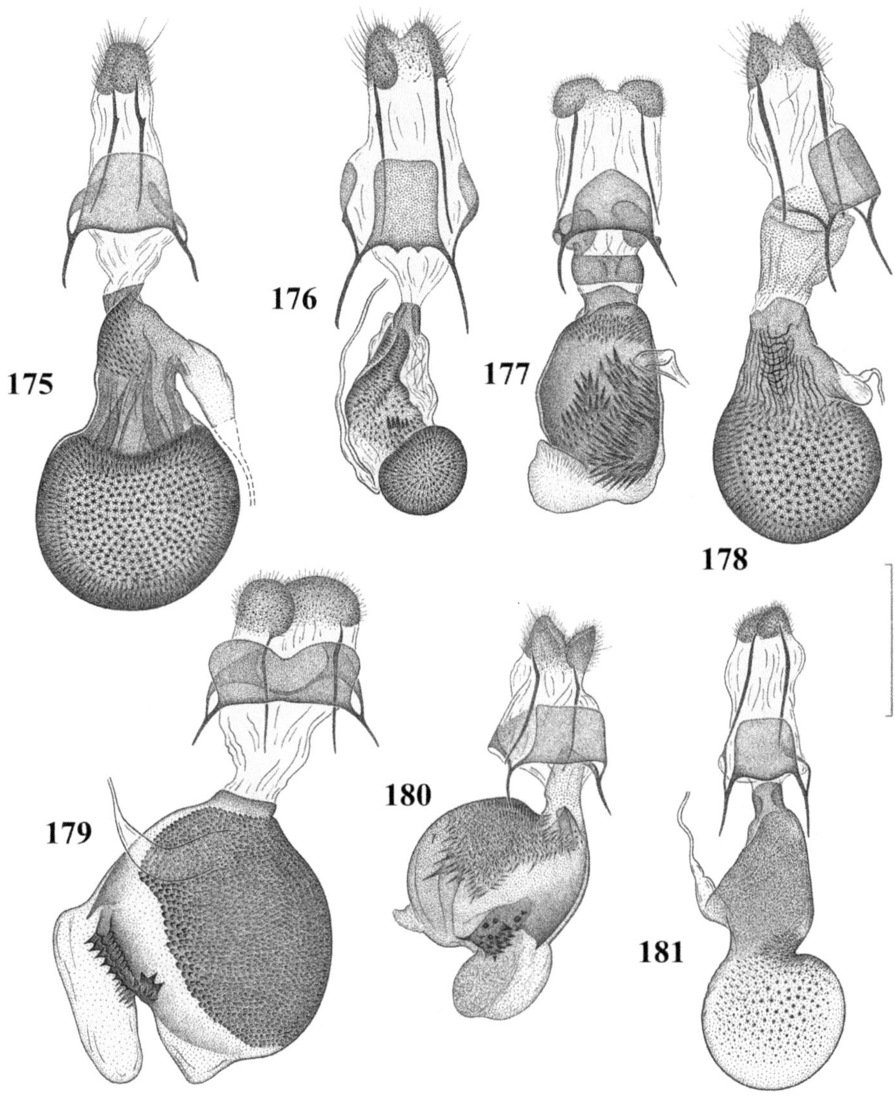

FEMALE GENITALIA (scale bar = 1mm)

175. *Eupithecia inopinata* Vojnits, China, Yunnan
176. *Eupithecia dealbata* Inoue, China, Taiwan
177. *Eupithecia lata* Vojnits, Nepal, Sagarmatha
178. *Eupithecia karapinensis* Wileman & South, China, Taiwan
179. *Eupithecia kobayashii* Inoue, Russia, S. Primorje
180. *Eupithecia incorrupta* Vojnits, China, Sichuan
181. *Eupithecia leucenthesis* Prout, Nepal, Sagarmatha

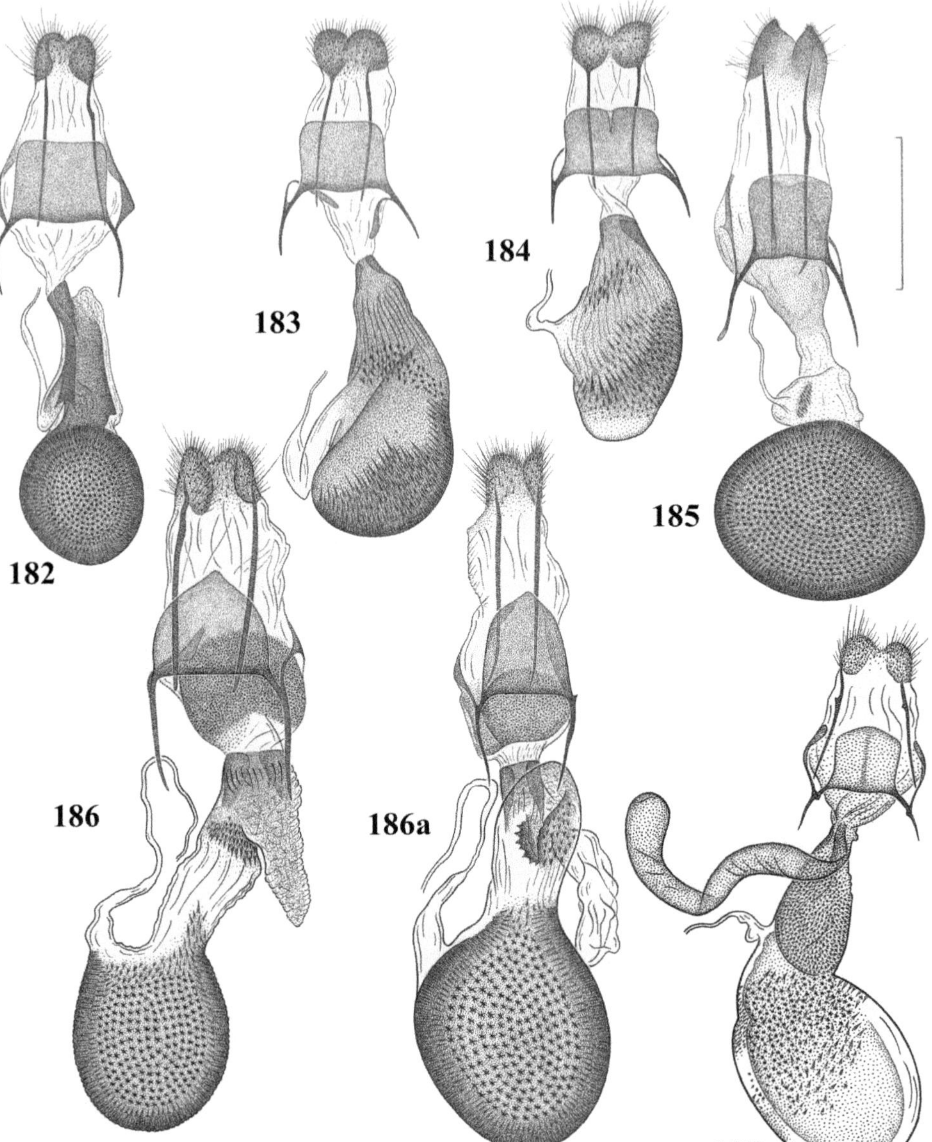

FEMALE GENITALIA (scale bar = 1mm)

182. *Eupithecia pimpinellata* Hübner, Mongolia, Central Aimag
183. *Eupithecia denotata* Hübner, Kazakhstan, Zailijsky Alatau Mts.
184. *Eupithecia chesiata* Dietze, Tajikistan, Pamir Mts.
185. *Eupithecia mundiscripta* Warren, China, Taiwan
186. *Eupithecia robiginascens robiginascens* Prout, China, Sichuan
186a. *Eupithecia robiginascens flexicornuta* Inoue, China, Taiwan
187. *Eupithecia luctuosa* Mironov & Galsworthy, China, Fujian (holotype)

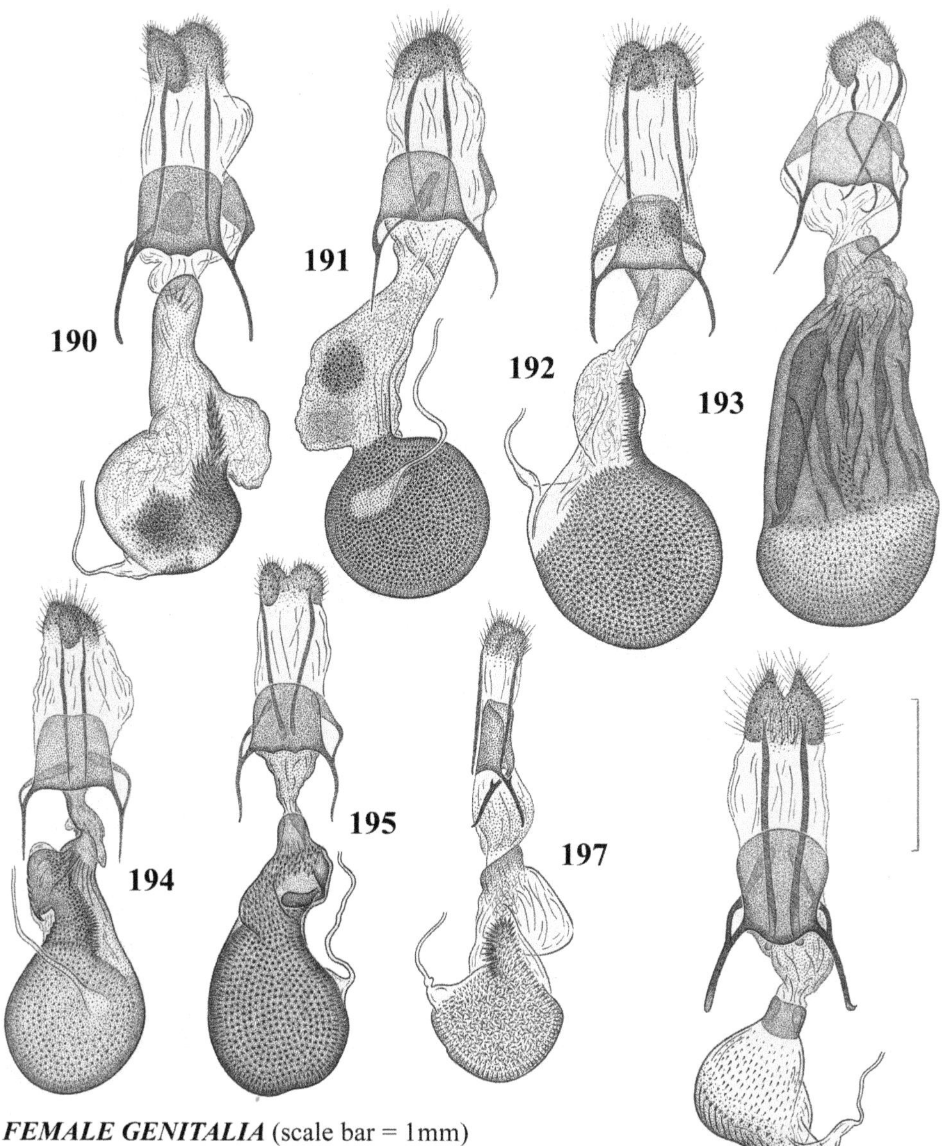

FEMALE GENITALIA (scale bar = 1mm)

190. *Eupithecia contraria* Vojnits, Nepal, Dhaulagiri
191. *Eupithecia rajata* Guenée, India, Sikkim
192. *Eupithecia ebriosa* Vojnits, China, Sichuan
193. *Eupithecia vasta* Vojnits, China, Yunnan
194. *Eupithecia lobbichlerata* Schütze, China, Sichuan
195. *Eupithecia coniurata* Vojnits, China, Sichuan
196. *Eupithecia cuneata* Vojnits, Thailand, Chiangmai
197. *Eupithecia particeps* Vojnits, China, Tibet

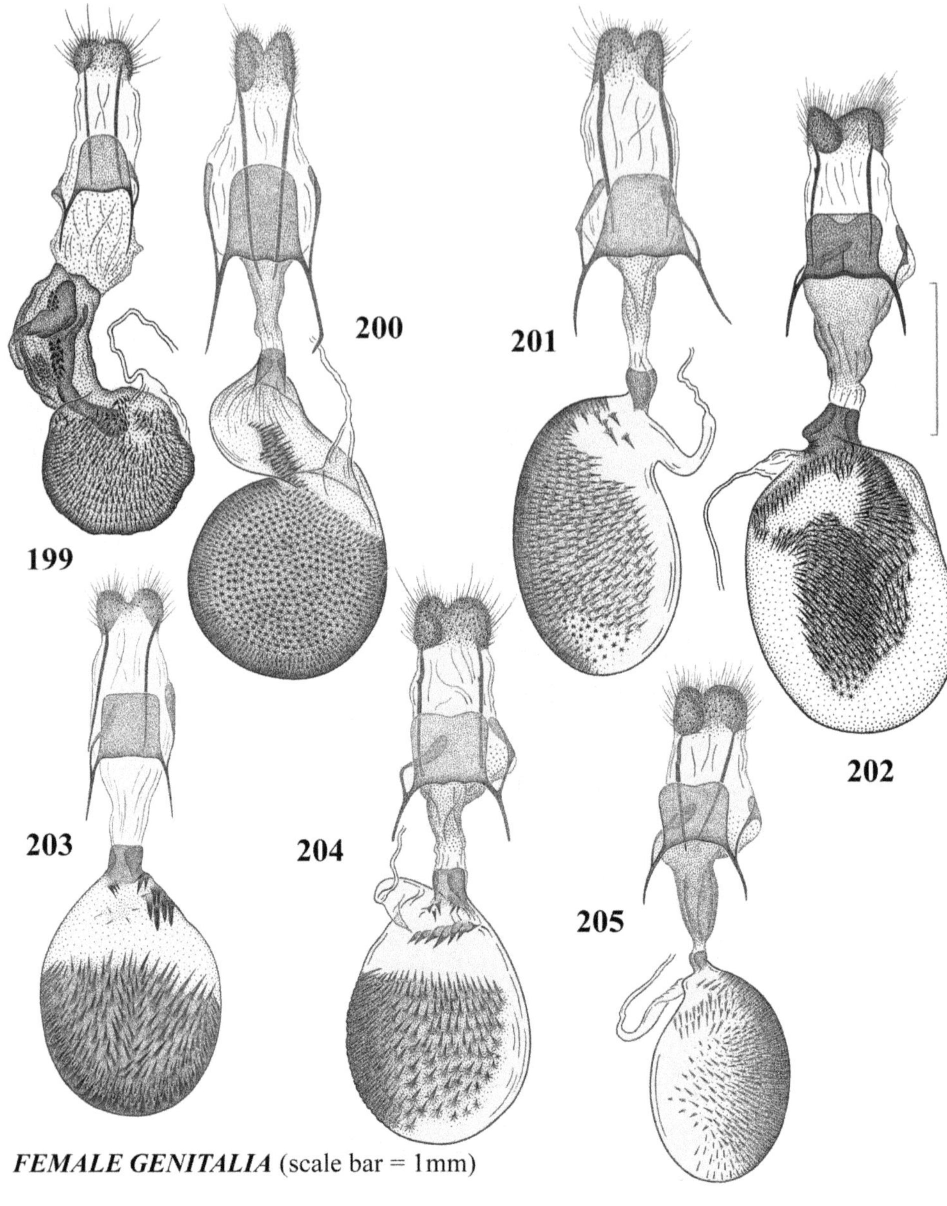

FEMALE GENITALIA (scale bar = 1mm)

199. *Eupithecia serpentigena* Mironov & Galsworthy, China, Sichuan (holotype)
200. *Eupithecia raniata* Prout, Nepal, Kathmandu
201. *Eupithecia atrisignis* Butler, India, Sikkim
202. *Eupithecia pyricoetes* Prout, India, Sikkim
203. *Eupithecia suspiciosa* Vojnits, Nepal, Janakpur (holotype)
204. *Eupithecia nuceistrigata* Bastelberger, China, Taiwan
205. *Eupithecia tricrossa* Prout, India, Sikkim

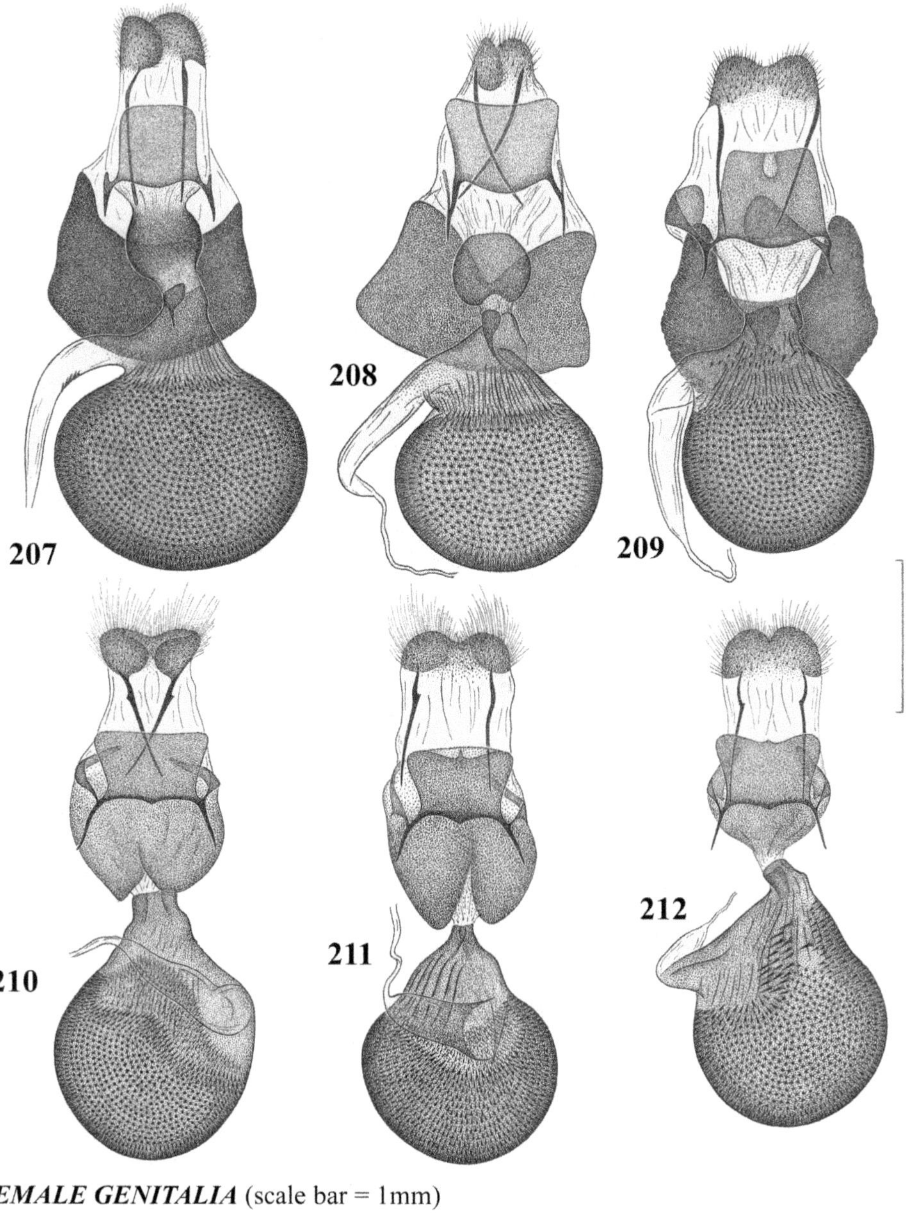

FEMALE GENITALIA (scale bar = 1mm)

207. *Eupithecia suboxydata* Staudinger, Russia, S. Primorje
208. *Eupithecia formosa* Vojnits, China, Yunnan (paratype)
209. *Eupithecia subbrunneata* Dietze, Russia, S. Primorje
210. *Eupithecia innotata* Hufnagel, Kazakhstan, Tarbagatay Mts.
211. *Eupithecia parallelaria* Bohatsch, Pakistan/India, valley of Indus
212. *Eupithecia corroborata* Dietze, Kyrgyzstan, Sary-dzhaz valley

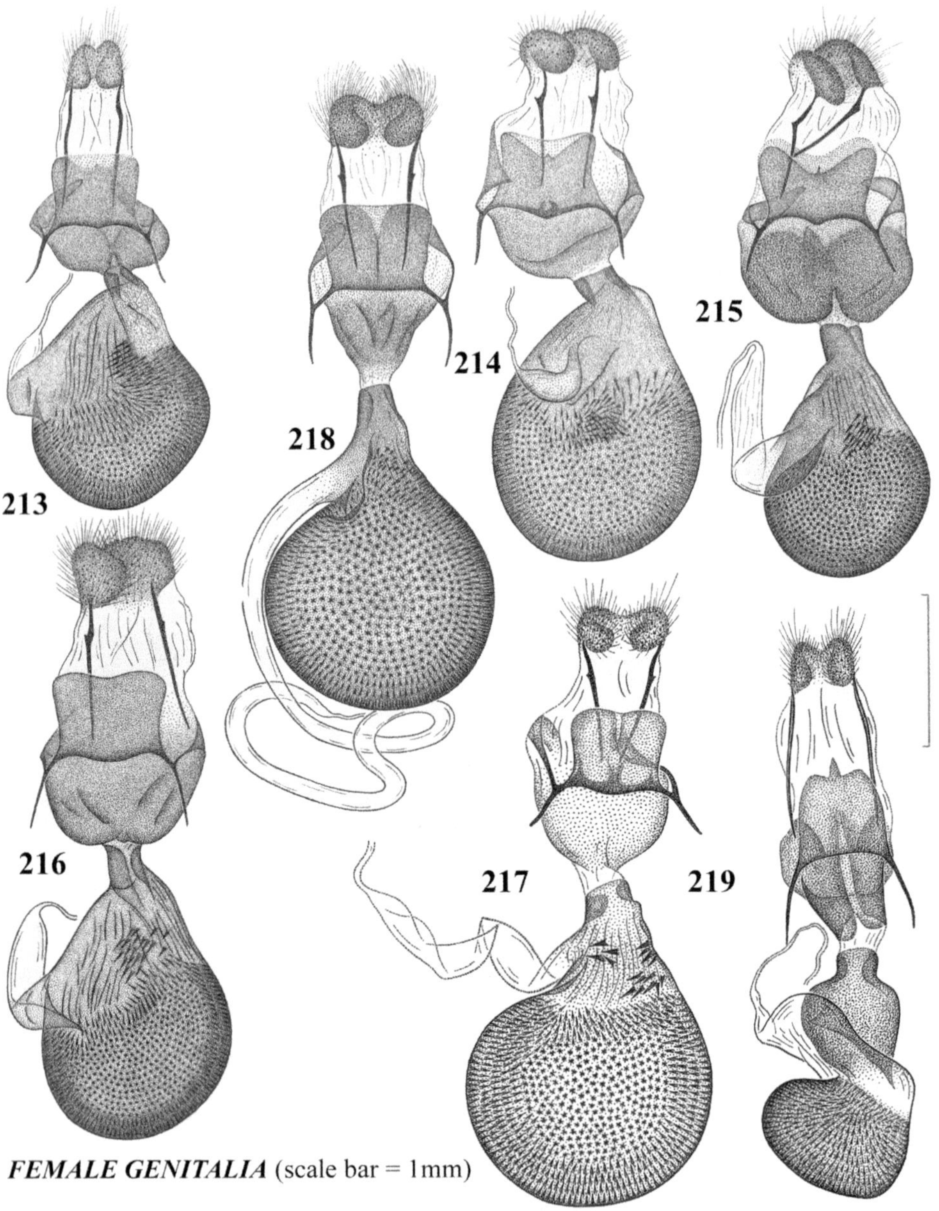

FEMALE GENITALIA (scale bar = 1mm)

213. *Eupithecia praesignata* Bohatsch, Kyrgyzstan, Alai Mts.
214. *Eupithecia paupera* Dietze, Kazakhstan, Zailijsky Alatau Mts.
215. *Eupithecia mitigata* Dietze, N. India, Ladakh Mts.
216. *Eupithecia vicariata* Dietze, Kazakhstan, Yarkent
217. *Eupithecia magica* Mironov & Galsworthy, China, Tibet (paratype)
218. *Eupithecia relaxata* Dietze, Kyrgyzstan, Naryn
219. *Eupithecia centaureata* Denis & Schiffermüller, China, Guangdong

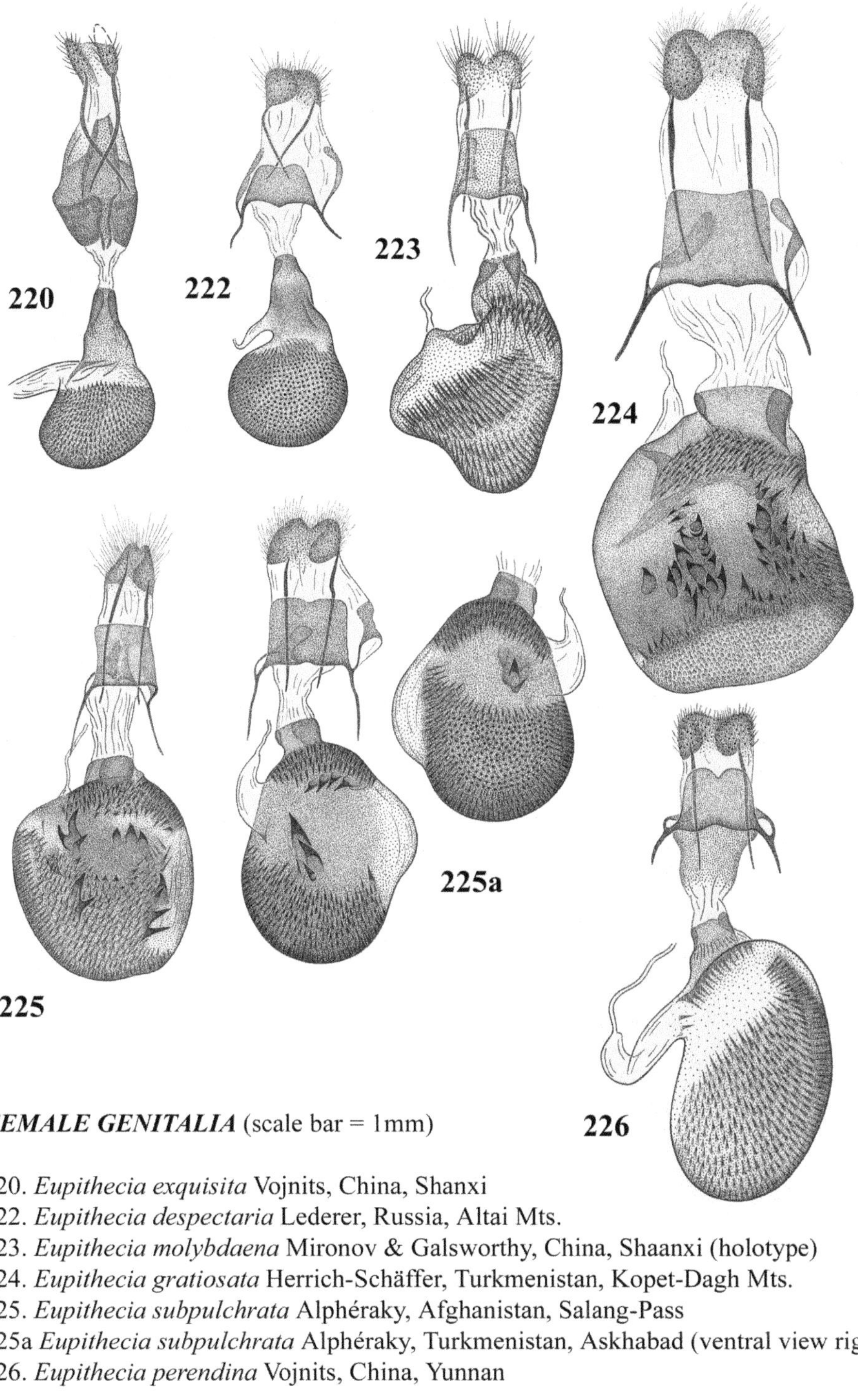

FEMALE GENITALIA (scale bar = 1mm)

220. *Eupithecia exquisita* Vojnits, China, Shanxi
222. *Eupithecia despectaria* Lederer, Russia, Altai Mts.
223. *Eupithecia molybdaena* Mironov & Galsworthy, China, Shaanxi (holotype)
224. *Eupithecia gratiosata* Herrich-Schäffer, Turkmenistan, Kopet-Dagh Mts.
225. *Eupithecia subpulchrata* Alphéraky, Afghanistan, Salang-Pass
225a *Eupithecia subpulchrata* Alphéraky, Turkmenistan, Askhabad (ventral view right)
226. *Eupithecia perendina* Vojnits, China, Yunnan

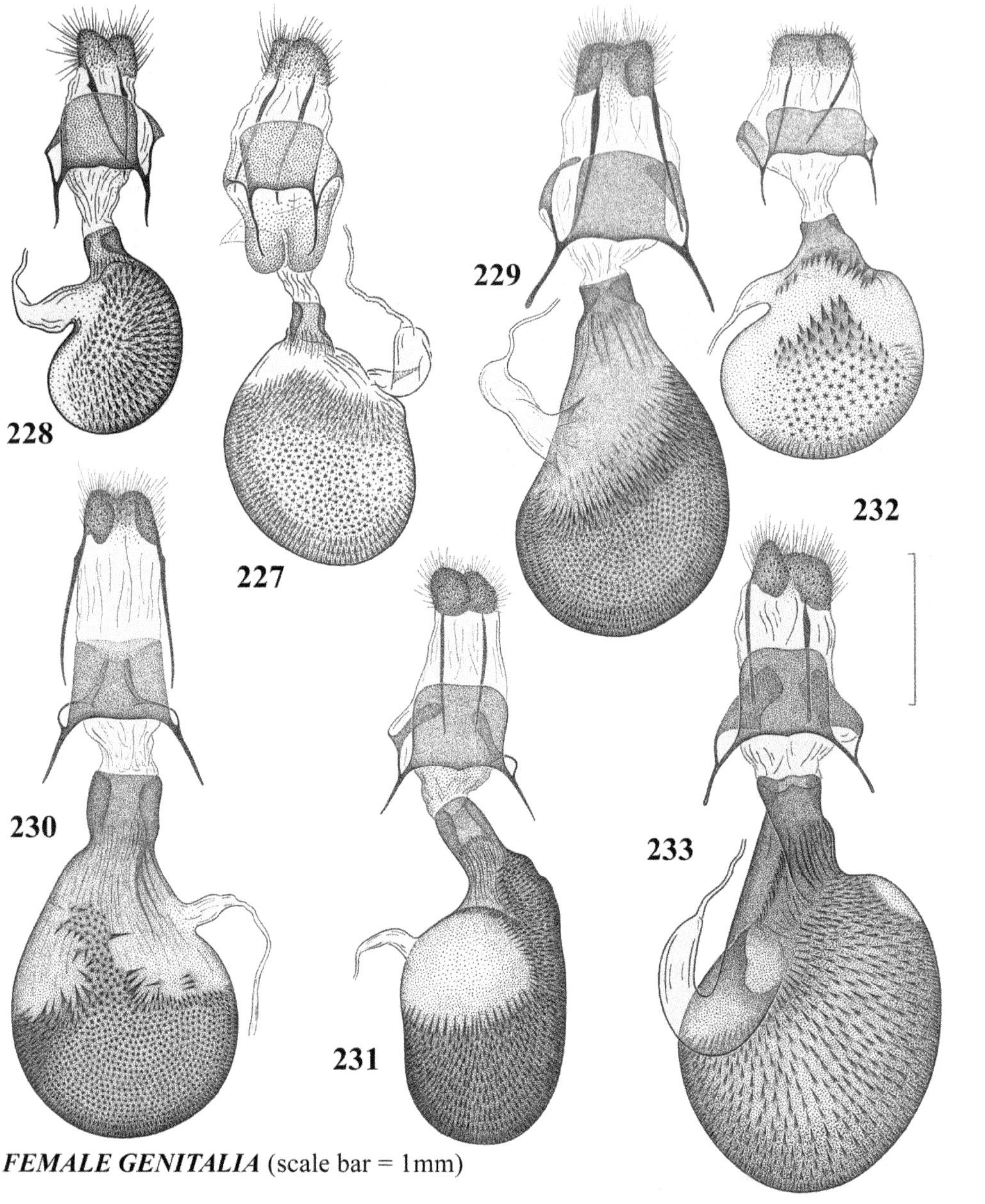

FEMALE GENITALIA (scale bar = 1mm)

227. *Eupithecia lamata* Mironov & Galsworthy, China, Qinghai (holotype)
228. *Eupithecia ancillata* Mironov & Galsworthy, China, Shaanxi (paratype)
229. *Eupithecia simpliciata* Haworth, Kyrgyzstan, Naryn
230. *Eupithecia sinuosaria* Eversmann, Russia, S. Siberia
231. *Eupithecia pallescens* Dietze, Kyrgyzstan, Naryn
232. *Eupithecia conterminata* Lienig & Zeller, Russia, S. Primorje
233. *Eupithecia insignioides* Wehrli, Russia, S. Primorje

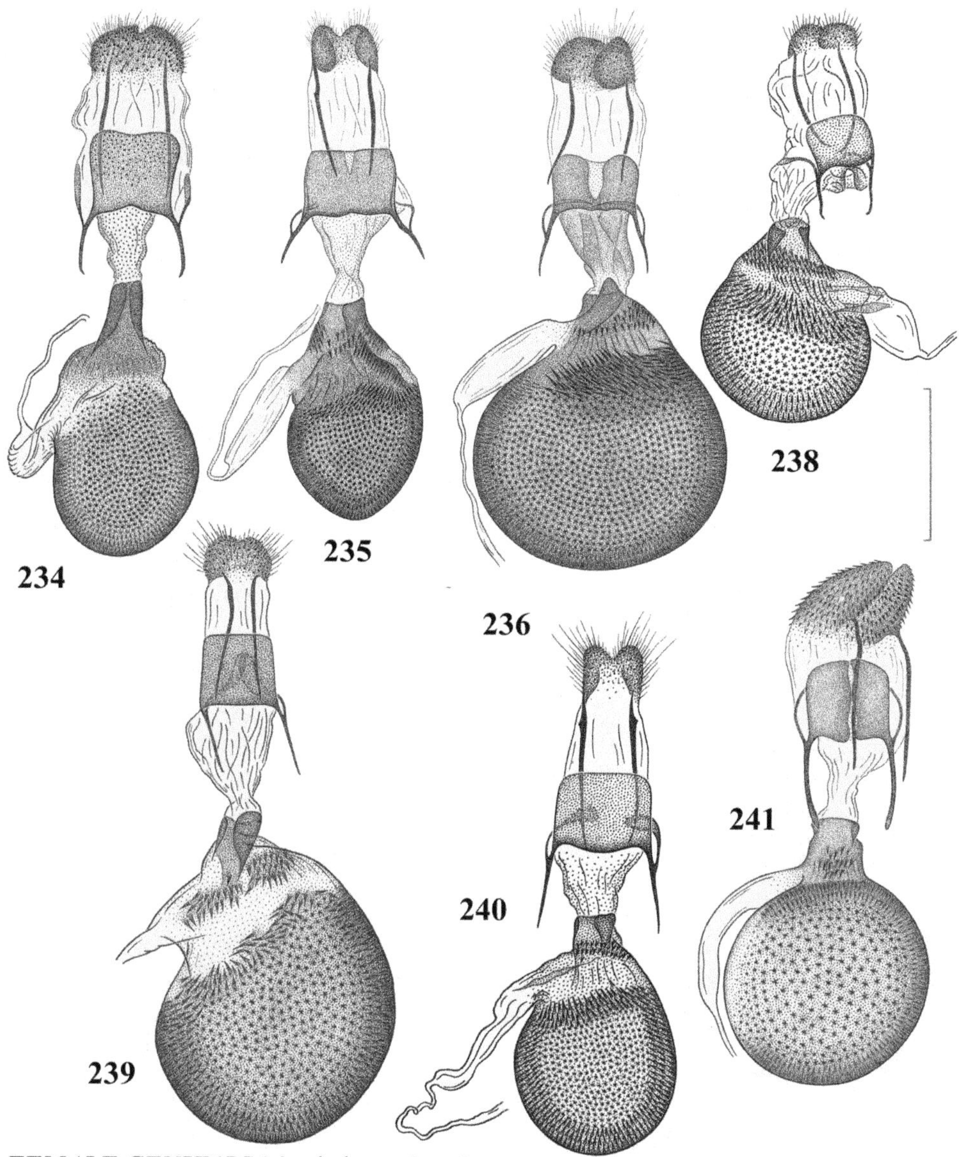

FEMALE GENITALIA (scale bar = 1mm)

234. *Eupithecia costimacularia* Leech, China, Taiwan
235. *Eupithecia assimilata* Doubleday, Russia, Altai Mts.
236. *Eupithecia vulgata* Haworth, Kyrgyzstan, Tien-Shan Mts.
238. *Eupithecia nigristriata* Mironov & Galsworthy, China, Qinghai (paratype)
239. *Eupithecia sacrimontis* Prout, China, Shaanxi
240. *Eupithecia citraria* Mironov & Galsworthy, China, Yunnan (paratype)
241. *Eupithecia nobilitata* Staudinger, Kazakhstan, Djair Mts.

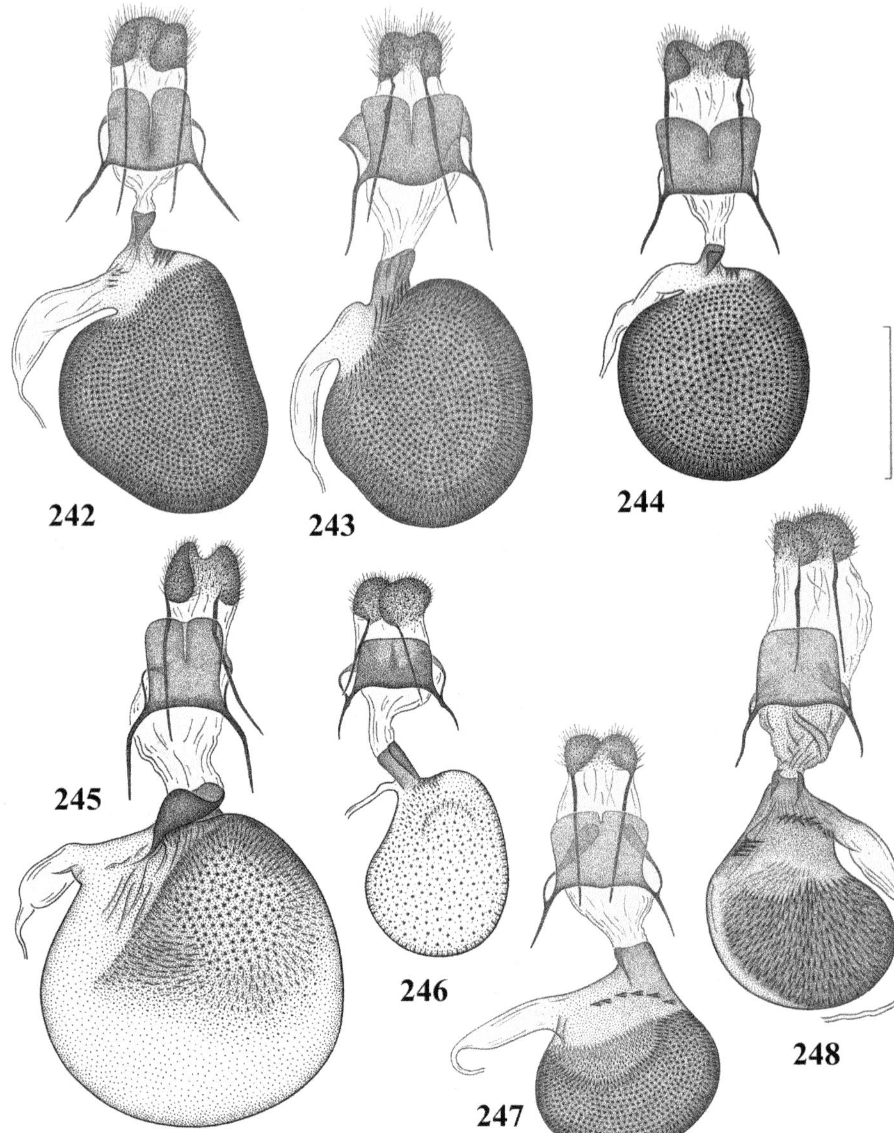

FEMALE GENITALIA (scale bar = 1mm)

242. *Eupithecia sutiliata* Christoph, Kazakhstan, Dzhulek
243. *Eupithecia biornata* Christoph, European Russia, Taganrog
244. *Eupithecia kuldschaensis* Staudinger, China, Xinjiang, Kuldscha (syntype)
245. *Eupithecia rubeni* Viidalepp, S. Siberia, Tuva (paratype)
246. *Eupithecia rebeli* Bohatsch, Tajikistan, Alai Mts.
247. *Eupithecia thalictrata* Püngeler, Russia, Priamurje
248. *Eupithecia dura* Vojnits, China, Yunnan

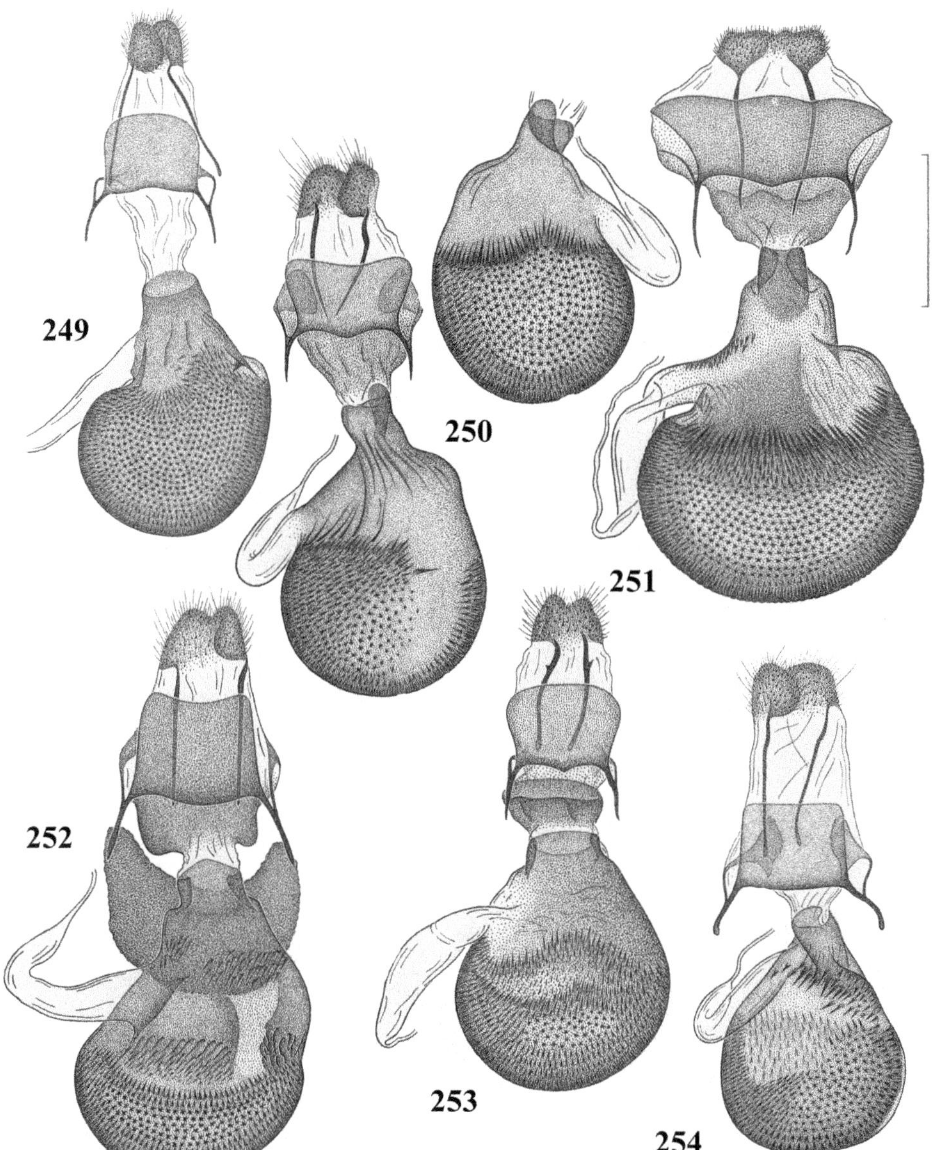

FEMALE GENITALIA (scale bar = 1mm)

249. *Eupithecia addictata* Dietze, Russia, Priamurje
250. *Eupithecia exiguata* Hübner, Russia, Altai Mts. (ventral view above right)
251. *Eupithecia hannemanni* Vojnits, China, Shanxi
252. *Eupithecia bella* Staudinger, Russia, S. Primorje
253. *Eupithecia insolita* Vojnits, China, Sichuan (paratype)
254. *Eupithecia duplex* Sterneck, China, Sichuan

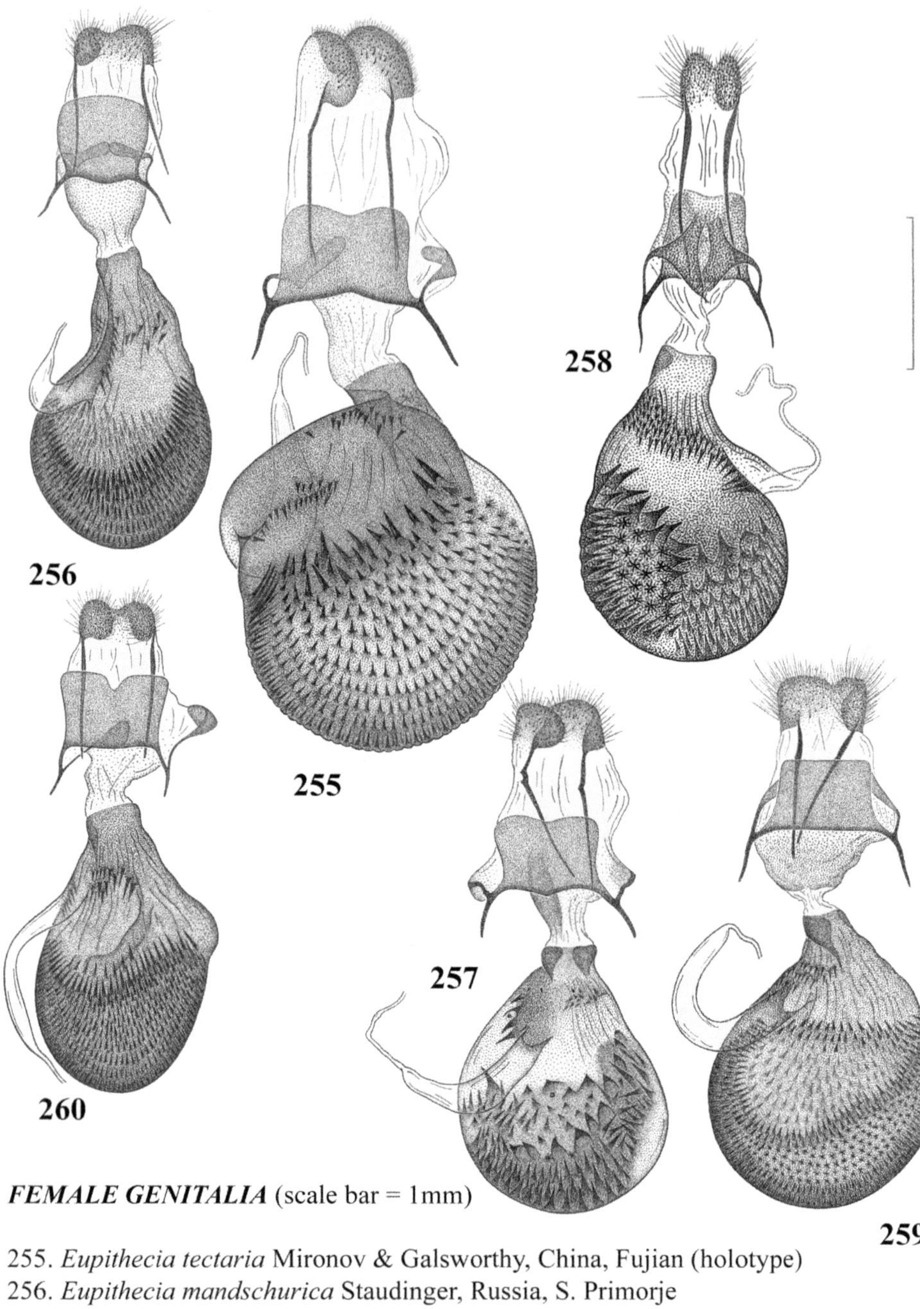

FEMALE GENITALIA (scale bar = 1mm)

255. *Eupithecia tectaria* Mironov & Galsworthy, China, Fujian (holotype)
256. *Eupithecia mandschurica* Staudinger, Russia, S. Primorje
257. *Eupithecia intricata* Zetterstedt, Kazakhstan, prov. Almaty
258. *Eupithecia lupa* Mironov & Galsworthy, China, Taiwan
259. *Eupithecia satyrata* Hübner, Kazakhstan, Tarbagatay Mts.
260. *Eupithecia pernotata* Guenée, Russia, S. Primorje

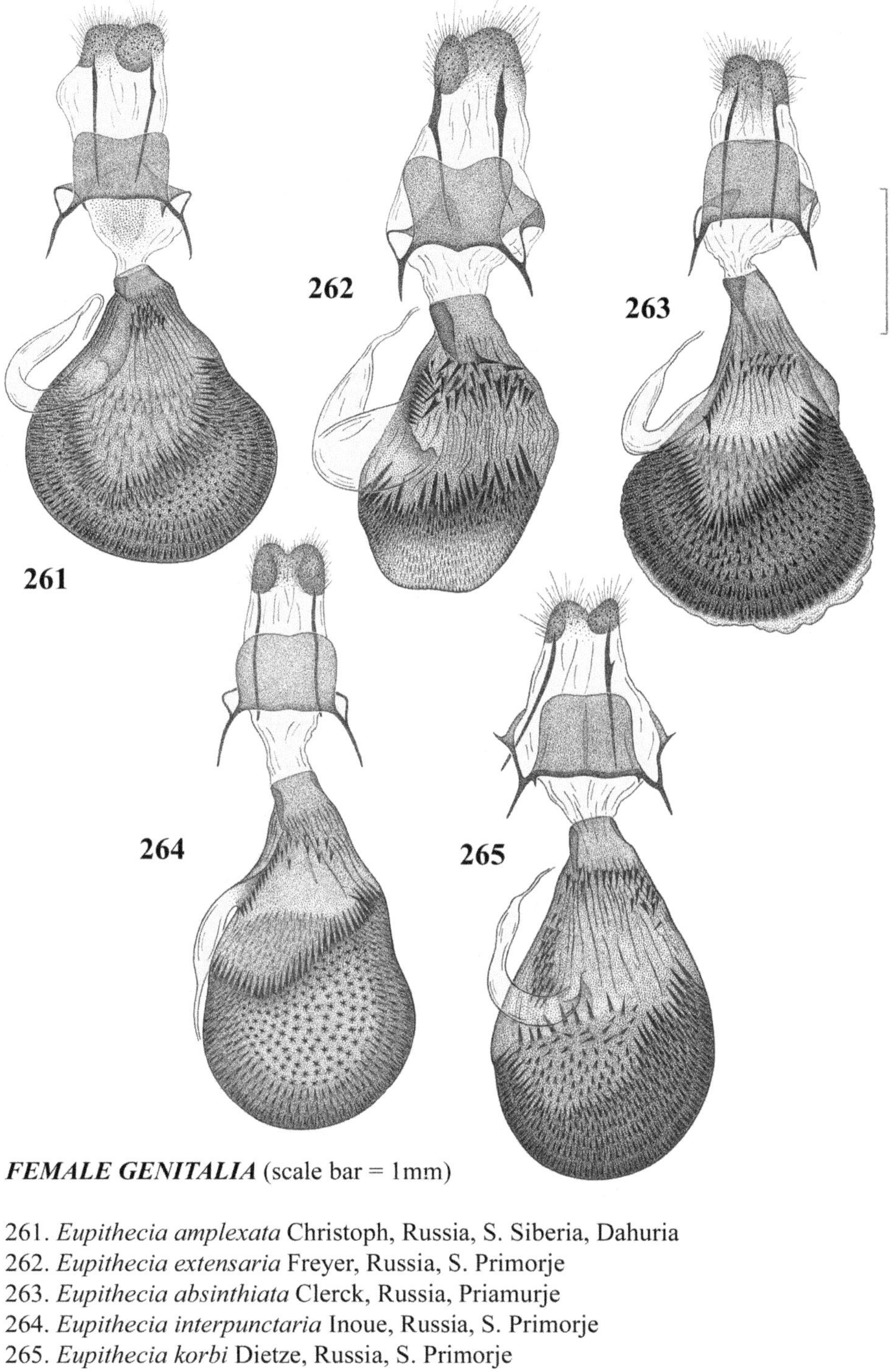

FEMALE GENITALIA (scale bar = 1mm)

261. *Eupithecia amplexata* Christoph, Russia, S. Siberia, Dahuria
262. *Eupithecia extensaria* Freyer, Russia, S. Primorje
263. *Eupithecia absinthiata* Clerck, Russia, Priamurje
264. *Eupithecia interpunctaria* Inoue, Russia, S. Primorje
265. *Eupithecia korbi* Dietze, Russia, S. Primorje

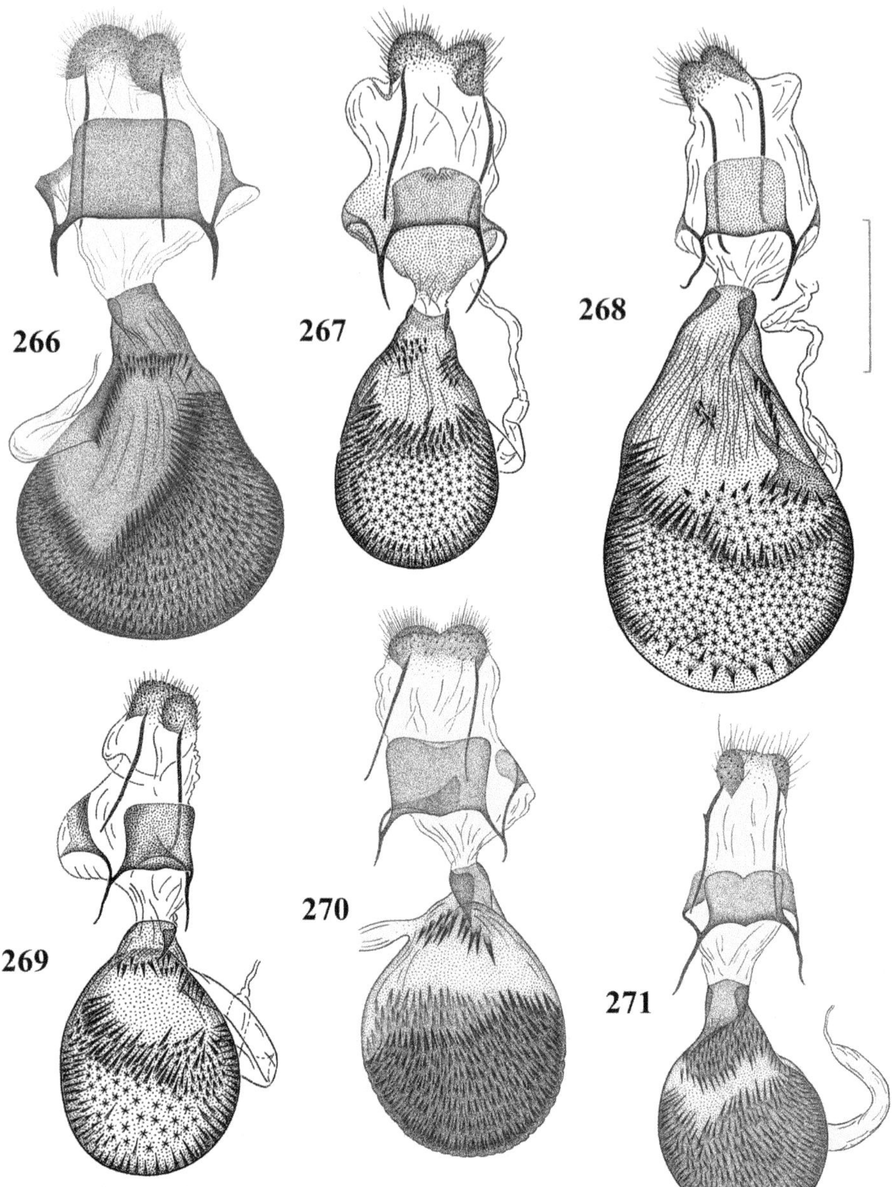

FEMALE GENITALIA (scale bar = 1mm)

266. *Eupithecia kurilensis* Bryk, Russia, Sakhalin
267. *Eupithecia ensifera* Mironov & Galsworthy, China, Qinghai (holotype)
268. *Eupithecia ultrix* Mironov & Galsworthy, China, Sichuan (paratype)
269. *Eupithecia cervina* Mironov & Galsworthy, China, Tibet (paratype)
270. *Eupithecia impolita* Vojnits, China, Shanxi (paratype)
271. *Eupithecia neosatyrata* Inoue, Japan, Shikoku (paratype)

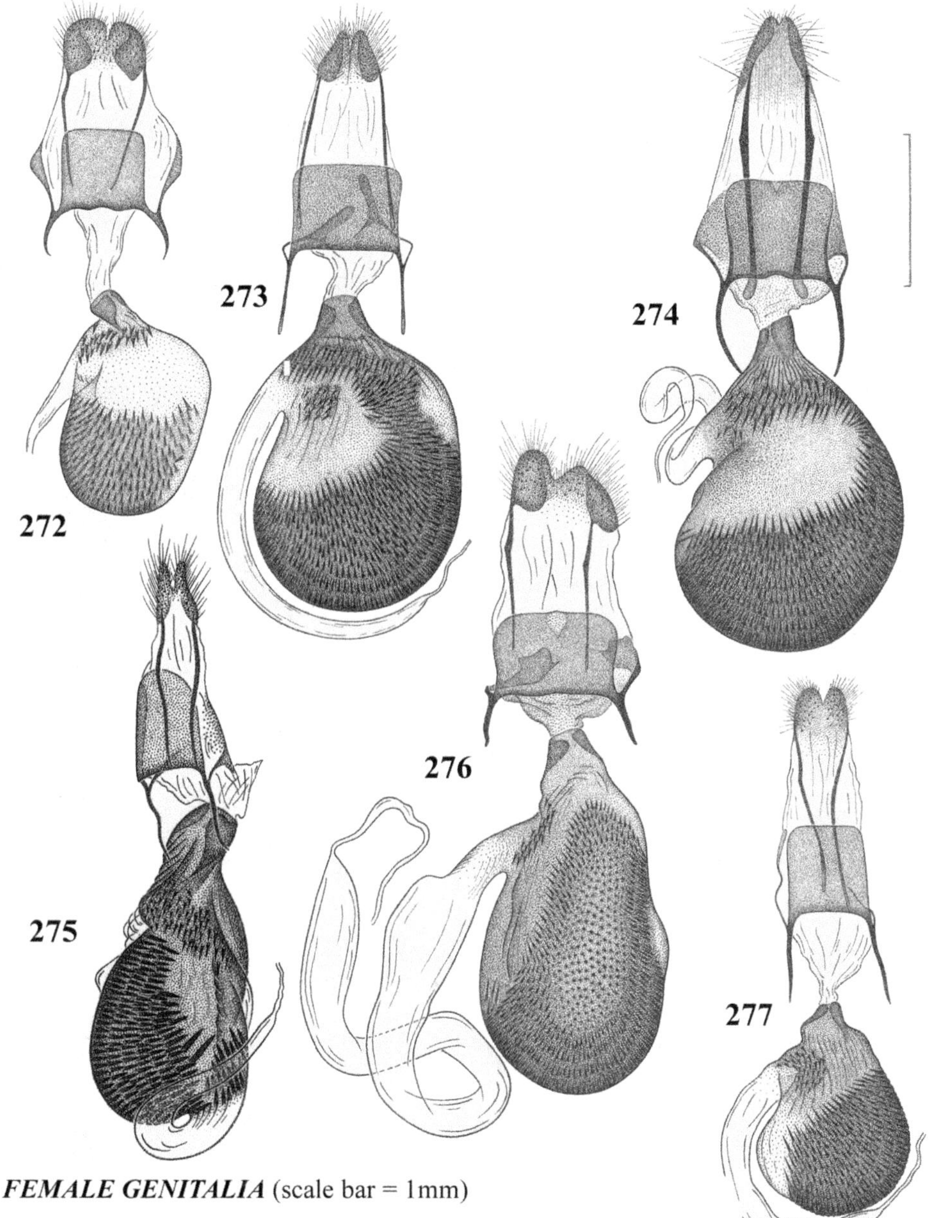

FEMALE GENITALIA (scale bar = 1mm)

272. *Eupithecia indissolubilis* Vojnits, Russia, Altai Mts.
273. *Eupithecia nephelata* Staudinger, Mongolia, N. Gobi Desert (Kholt)
274. *Eupithecia exactata* Staudinger, Kyrgyzstan, Naryn
275. *Eupithecia omissa* Mironov & Galsworthy, China, Shaanxi
276. *Eupithecia succenturiata* Linnaeus, Russia, Altai Mts.
277. *Eupithecia marginata* Staudinger, Tajikistan, Pamir Mts.

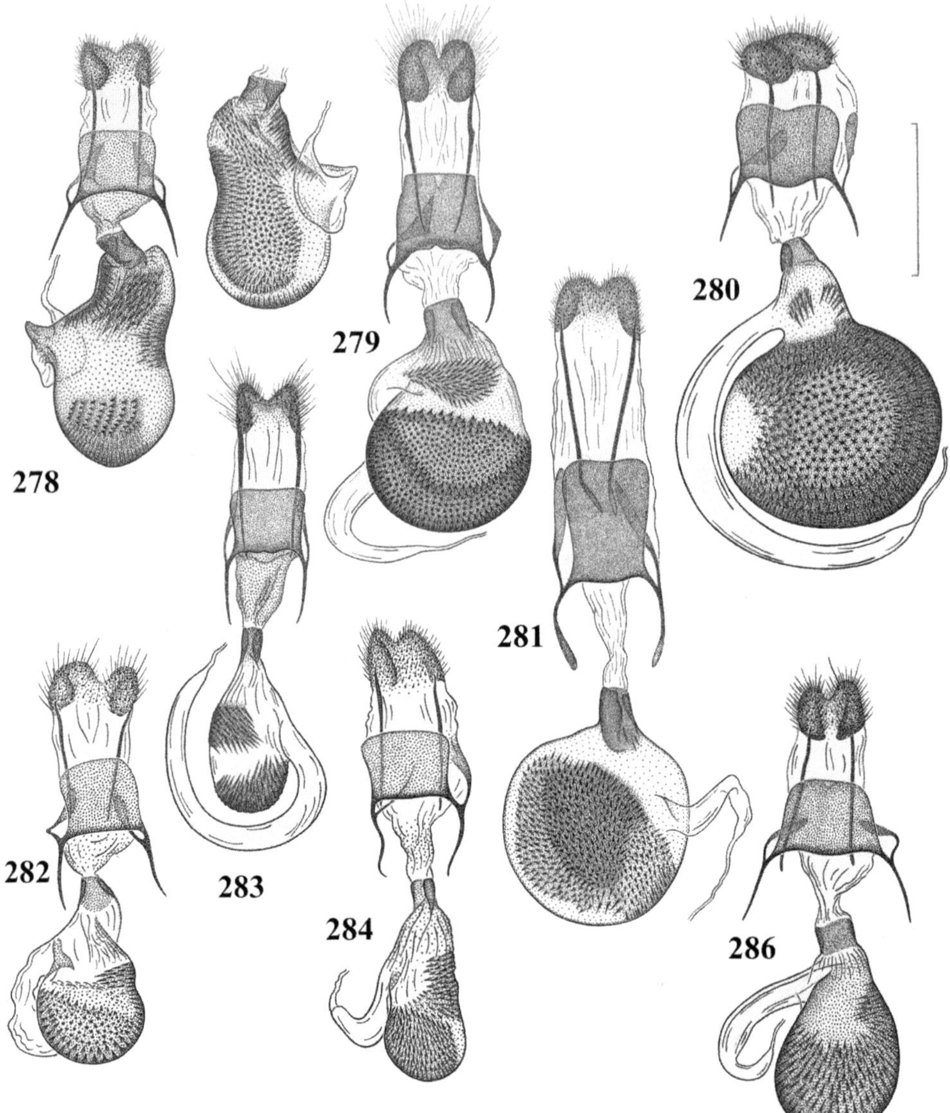

FEMALE GENITALIA (scale bar = 1mm)

278. *Eupithecia masculina* Inoue, China, Taiwan (ventral view above right)
279. *Eupithecia subumbrata* Denis & Schiffermüller, Kyrgyzstan, Naryn
280. *Eupithecia subfuscata* Haworth, Russia, S. Primorje
281. *Eupithecia costipicta* Warren, Pakistan, Gilgit-Baltistan, Deosai Mts.
282. *Eupithecia mediocincta* Mironov & Galsworthy, China, Gansu (paratype)
283. *Eupithecia chui* Inoue, China, Taiwan
284. *Eupithecia tibetana* Mironov & Galsworthy, China, Tibet (paratype)
286. *Eupithecia nishizawai* Inoue, China, Taiwan

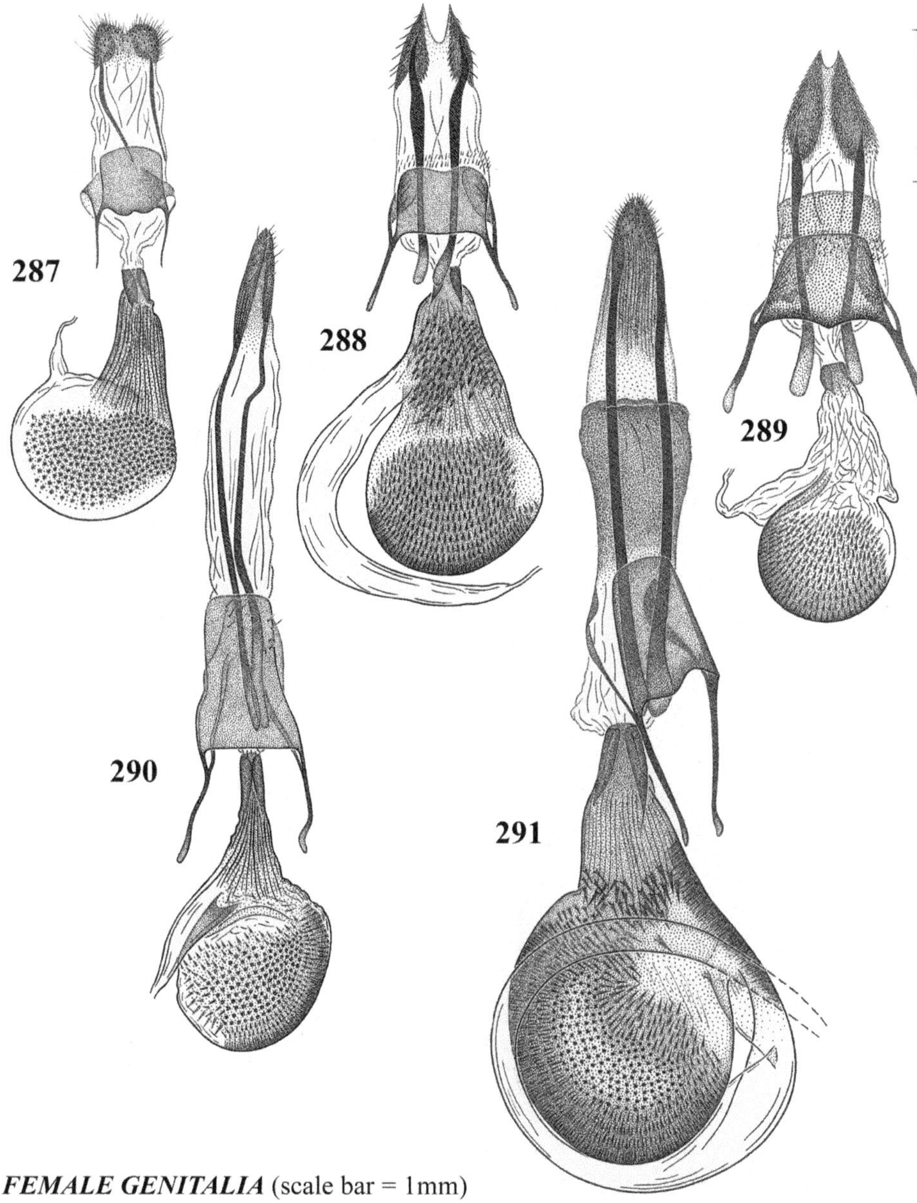

FEMALE GENITALIA (scale bar = 1mm)

287. *Eupithecia praecipitata* Vojnits, China, Shaanxi (holotype)
288. *Eupithecia lasciva* Vojnits, China, Shaanxi (paratype)
289. *Eupithecia egena* Vojnits, China, Yunnan
290. *Eupithecia lunatica* Vojnits, China, Yunnan (paratype)
291. *Eupithecia otiosa* Vojnits, China, Sichuan

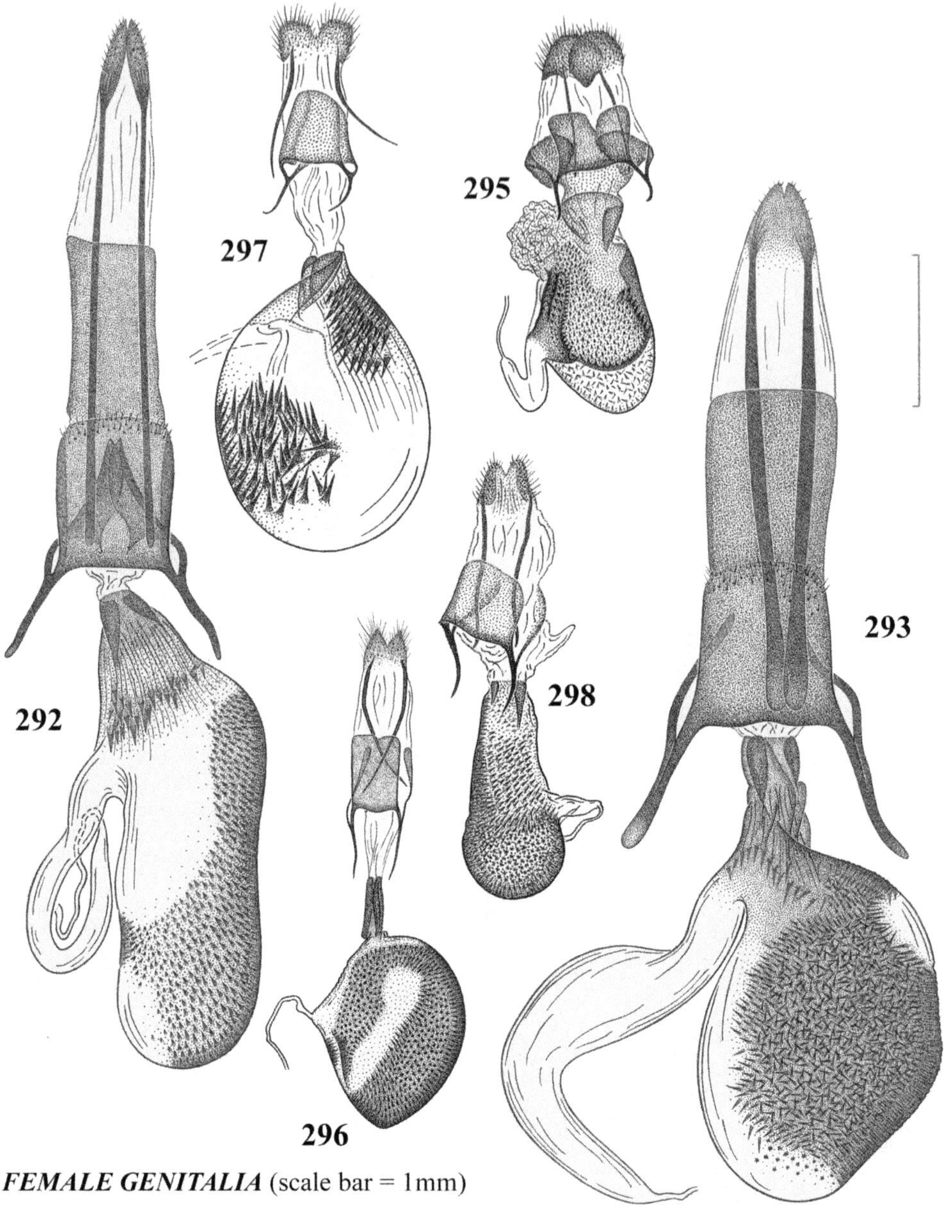

FEMALE GENITALIA (scale bar = 1mm)

292. *Eupithecia acutipapillata* Inoue, China, Taiwan
293. *Eupithecia supercastigata* Inoue, Korea
295. *Eupithecia coribalteata* Mironov & Galsworthy, China, Gansu (holotype)
296. *Eupithecia lilliputata* Mironov & Galsworthy, China, Yunnan (holotype)
297. *Eupithecia salubris* Mironov & Galsworthy, China, Shaanxi (paratype)
298. *Eupithecia villica* Mironov & Galsworthy, China, Tibet (holotype)

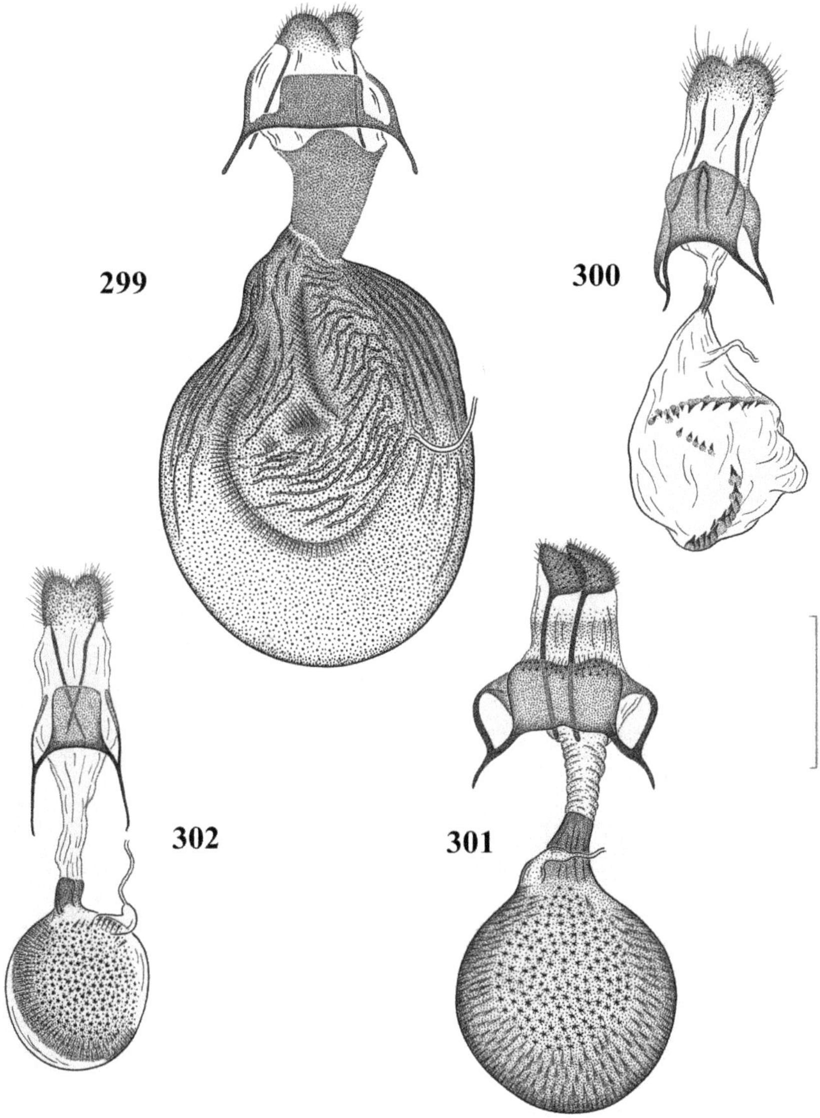

FEMALE GENITALIA (scale bar = 1mm)

299. *Pareupithecia spadix*, Inoue, Russia, S. Primorje
300. *Eva flexa* Vojnits, China, Yunnan
301. *Girida rigida* Swinhoe, India, Nilgiri Hills
302. *Mesoptila melanolopha* Swinhoe, China, Taiwan

www.ingramcontent.com/pod-product-compliance
Ingram Content Group UK Ltd.
Pitfield, Milton Keynes, MK11 3LW, UK
UKHW022208051026
15624UKWH00001B/58

9 789004 254510